de Gruyter Studies in Mathematics 17

Editors: Heinz Bauer · Jerry L. Kazdan · Eduard Zehnder

de Gruyter Studies in Mathematics

Francesco Altomare · Michele Campiti

Korovkin-type Approximation Theory and its Applications

Walter de Gruyter
Berlin · New York 1994

Authors

Francesco Altomare
Michele Campiti
Dipartimento di Matematica
Università degli Studi di Bari
I-70125 Bari, Italy

Series Editors

Heinz Bauer	Jerry L. Kazdan	Eduard Zehnder
Mathematisches Institut	Department of Mathematics	ETH-Zentrum/Mathematik
der Universität	University of Pennsylvania	Rämistraße 101
Bismarckstraße 1½	209 South 33rd Street	CH-8092 Zürich
D-91054 Erlangen, Germany	Philadelphia, PA 19104-6395, USA	Switzerland

1991 Mathematics Subject Classification: 41-02; 41A10, 41A25, 41A35, 41A36, 41A63, 41A65, 42A10, 46A55, 46E15, 46E27, 47A58, 47B38, 47B65; 46J10, 46L89, 47D06, 47B48, 47C10, 60F, 60J25, 60J35, 60J60

Key words: Positive Radon measures, Choquet boundaries, Korovkin-type theorems, positive linear projections, finitely defined operators, lattice homomorphisms, approximation by positive linear operators, rate of convergence, probabilistic methods in approximation theory, positive approximation processes, Feller semigroups, degenerate diffusion equations.

♾ Printed on acid-free paper which falls within the guidelines of the ANSI to ensure permanence and durability.

Library of Congress Cataloging-in-Publication Data

Korovkin-type approximation theory and its applications /
  Francesco Altomare, Michele Campiti.
      p.   cm. − (De Gruyter studies in mathematics : 17)
  Includes bibliographical references and index.
  ISBN 3-11-014178-7
  1. Approximation theory.   I. Altomare, Francesco, 1951−
  II. Campiti, Michele, 1959−   .   III. Series.
  QA221.K68   1994
  511'.42−dc20   94-16853
                                                                        CIP

Die Deutsche Bibliothek − Cataloging-in-Publication Data

Altomare, Francesco: Korovkin type approximation theory and
its applications / Francesco Altomare ; Michele Campiti. − Berlin ; New York: de Gruyter, 1994
  (De Gruyter studies in mathematics ; 17)
  ISBN 3-11-014178-7
NE: Campiti, Michele:; GT

Printed in Germany.
Typesetting: ASCO Trade Ltd. Hong Kong. Printing: Gerike GmbH, Berlin.
Binding: Lüderitz & Bauer GmbH, Berlin. Cover design: Rudolf Hübler, Berlin.

Affectionately and gratefully dedicated to
Raffaella, Bianca Maria and Gianluigi
and to Giusy and my parents

Preface

Since their discovery the simplicity and, at the same time, the power of the classical theorems of Korovkin impressed several mathematicians.

During the last thirty years a considerable amount of research extended these theorems to the setting of different function spaces or more general abstract spaces such as Banach lattices, Banach algebras, Banach spaces and so on.

This work, in fact, delineated a new theory that we may now call Korovkin-type approximation theory.

At the same time, strong and fruitful connections of this theory have been revealed not only with classical approximation theory, but also with other fields such as functional analysis, harmonic analysis, measure theory, probability theory and partial differential equations.

This has been accomplished by a large number of mathematicians ranging from specialists in approximation theory to functional analysts.

A selected part of the theory is already documented in the monographs of Donner [1982] and Keimel and Roth [1992].

With this book we hope to contribute further to the subject by presenting a modern and comprehensive exposition of the main aspects of the theory in spaces of continuous functions (vanishing at infinity, respectively) defined on a compact space (a locally compact space, respectively), together with its main applications.

We have chosen to treat these function spaces since they play a central role in the whole theory and are the most useful for the applications.

Besides surveying both classical and recent results in the field, the book also contains a certain amount of new material. In any case, the majority of the results appears in a book for the first time.

We are happy to acknowledge our indebtedness to several friends and colleagues.

First, we would like to thank Hubert Berens, Heinz H. Gonska, Silvia Romanelli and Yurji A. Shashkin for reading a large part of the manuscript, for their fruitful suggestions and for their help in correcting mistakes.

We are also grateful to Ferdinand Beckhoff and Michael Pannenberg, to whom we asked to write Appendices B and A, respectively, for their collaboration in outlining the development of the theory in the setting of Banach algebras.

We are particularly indebted to Ferdinand Beckhoff, George Maltese, Rainer Nagel, Ioan Rasa and Rouslan K. Vasil'ev who read and checked the entire manuscript, gave valuable advice and criticisms and kindly corrected several mistakes and inaccuracies as well as our poor English. To them we extend our particular warm thanks.

We want to express our deep gratitude to Heinz Bauer not only for his interest in this work, for reading several chapters and for making several remarks, but also for inviting us to publish the book in the prestigious series De Gruyter Studies in Mathematics of which he is co-editor.

We thank him and the other editors of the series for accepting the monograph and Walter De Gruyter & Co. for producing it according to their usual high standard quality.

Finally we express our great affection and gratitude to Raffaella, Bianca Maria and Gianluigi and to Giusy, for their patience and understanding as well as for their constant encouragement over all these years without which this monograph would have never been completed.

We dedicate the book to them.

Bari, October 1993   *Francesco Altomare*
                                                                Michele Campiti

Contents

Introduction

Positive approximation processes play a fundamental role in approximation theory and appear in a very natural way in many problems dealing with the approximation of continuous functions, especially when one requires further qualitative properties, such as monotonicity, convexity, shape preservation and so on.

In 1953, P.P. Korovkin discovered the, perhaps, most powerful and, at the same time, simplest criterion in order to decide whether a given sequence $(L_n)_{n \in \mathbb{N}}$ of positive linear operators on the space $\mathscr{C}([0,1])$ is an approximation process, i.e., $L_n(f) \to f$ uniformly on $[0,1]$ for every $f \in \mathscr{C}([0,1])$.

In fact, it is sufficient to verify that $L_n(f) \to f$ uniformly on $[0,1]$ only for $f \in \{1, x, x^2\}$.

Starting with this result, during the last thirty years a considerable number of mathematicians have extended Korovkin's theorem to other function spaces or, more generally, to abstract spaces, such as Banach lattices, Banach algebras, Banach spaces and so on.

This work, in fact, delineated a new theory that we may now call Korovkin-type approximation theory (in short, KAT).

At the same time, strong and fruitful connections of this theory have also been revealed not only with classical approximation theory, but also with other fields such as functional analysis (abstract Choquet boundaries and convexity theory, uniqueness of extensions of positive linear forms, convergence of sequences of positive linear operators in Banach lattices, structure theory of Banach lattices, convergence of sequences of linear operators in Banach algebras and in C^*-algebras, structure theory of Banach algebras, approximation problems in function algebras), harmonic analysis (convergence of sequences of convolution operators on function spaces and function algebras on (locally) compact topological groups, structure theory of topological groups), measure theory and probability theory (weak convergence of sequences of positive Radon measures and positive approximation processes constructed by probabilistic methods), and partial differential equations (approximation of solutions of Dirichlet problems and of diffusion equations).

After the pioneer work of P.P. Korovkin and his students E.N. Morozov and V.I. Volkov, that came to light in the fifties, a decisive step toward the modern development of Korovkin-type approximation theory was carried out by Yu.A. Shashkin when, in the sixties, he characterized the finite Korovkin sets in the space $\mathscr{C}(X)$, X compact metric space, in many respects and, in particular, in terms of geometric properties of state spaces.

The development of KAT in $\mathscr{C}(X)$-spaces was also pursued and amplified by Wulbert [1968], Berens and Lorentz [1973], [1975], and Bauer [1973]. In particular, Bauer expanded the investigation of Korovkin subspaces by using, in a systematic way, suitable enveloping functions, previously considered in connection with abstract Dirichlet problems.

As a matter of fact, these methods allowed Bauer [1973], [1974] to characterize Korovkin subspaces also in the framework of adapted spaces.

This was the first systematic study of Korovkin subspaces carried out in spaces of continuous functions on locally compact Hausdorff spaces.

This line of investigation led Bauer and Donner [1978], [1986] to the development of a satisfactory parallel theory in the space $\mathscr{C}_0(X)$ of all real-valued continuous functions vanishing at infinity on a locally compact space X.

On the other hand, in the seventies and eighties, Korovkin-type approximation theory rapidly grew along many other directions including other classical function Banach spaces, such as $L^p(X, \mu)$-spaces, and more abstract spaces such as locally convex ordered spaces and Banach lattices, Banach algebras, Banach spaces and so on.

In the specific setting of $L^p(X, \mu)$-spaces nowadays we have very satisfactory results. Noteworthy achievements were obtained by several mathematicians and culminated in the (in many respects) conclusive results of Donner [1980], [1981], [1982].

KAT has been well developed also in the framework of Banach lattices and locally convex vector lattices as is documented, for instance, in Donner's book [1982]; there, theorems on the extensions of positive linear operators are fruitfully used as a main tool.

In this context, fundamental contributions have been carried out by the Russian school (notably, M.A. Krasnosel'skii, E.A. Lifshits, S.S. Kutateladze, A.M. Rubinov, R.K. Vasil'ev) and by the German school (especially, K. Donner, H.O. Flösser, E. Scheffold and M. Wolff).

As far as we know, the development of the theory is still incomplete in the context of Banach algebras (especially in the non commutative case) and, even more so, in Banach spaces, although some attempts have been made to frame the different problems in a more systematic way, for instance, by Altomare [1982a], [1982c], [1984], [1986], [1987a], Pannenberg [1985], [1992], Limaye and Namboodiri [1979], [1986], Labsker [1971], [1972], [1982], [1985], [1989a].

Very recently, Keimel and Roth [1992] presented a unified approach to Korovkin-type approximation theory in the framework of so-called locally convex cones.

The reader will find a quite complete picture of what has been achieved in these fields in Appendix D, where we present a subject classification of KAT, which reflects the main lines of its development.

All references in the final bibliography concerned with KAT, are classified according to this subject classification; the classification numbers are indicated by the prefix SC.

Furthermore, in the same Appendix D we also include a subject index with a list of all references pertaining to every subsection of the subject classification.

However, in spite of our efforts, we are sure that the list of references is not complete. We apologize for possible errors and omissions due to lack of accurate information.

The main purpose of this book is to present a modern and comprehensive exposition of the main aspects of Korovkin-type approximation theory in the spaces $\mathscr{C}_0(X)$ (X locally compact non-compact space) and $\mathscr{C}(X)$ (X compact space), together with its main applications.

The function spaces we have chosen to treat play a central role in the whole theory and are the most useful for the applications in the various univariate, multivariate and infinite dimensional settings.

However we occasionally give some results concerning $L^p(X, \mu)$-spaces too.

The book is mainly intended as a reference text for research workers in the field; a large part of it can also serve as a textbook for a graduate level course.

The organization of the material does not follow the historical development of the subject and allows us to present the most important part of the theory in a concise way.

As a prerequisite, we require a basic knowledge of the theory of Radon measures on locally compact spaces as well as some standard topics from functional analysis such as various Hahn-Banach extension and separation theorems, the Krein-Milman theorem and Milman's converse theorem.

For the reader's convenience and to make the exposition self-contained, we collect all these prerequisites in Chapter 1.

However in some few sections, such as Sections 4.3, 5.2, 6.1 and 6.2, in order to present some significant applications of Korovkin approximation theory, we have also required a solid background on measures on topological spaces and the Riesz representation theorem, on some basic principles of probability theory, on Choquet's integral representation theory and on C_0-semigroups of bounded linear operators.

The definitions and the results pertaining to these topics are briefly reviewed also in Chapter 1 in some starred sections.

Thus a starred section or subsection in principle is not essential for the whole of the book but it serves only for a particular (notable) application that will be indicated in the same section.

Chapters 2, 3 and 4 are devoted to the main aspects of Korovkin-type approximation theory in $\mathscr{C}_0(X)$ and $\mathscr{C}(X)$-spaces.

The fundamental problem consists in studying, for a given positive linear operator $T: \mathscr{C}_0(X) \to \mathscr{C}_0(Y)$, those subspaces H of $\mathscr{C}_0(X)$ (if any) which have the remarkable property that every arbitrary equicontinuous net of positive linear operators (or positive contractions) from $\mathscr{C}_0(X)$ into $\mathscr{C}_0(Y)$ converges strongly to T whenever it converges to T on H.

Such subspaces are called Korovkin subspaces for T.

Historically, this problem (and related ones) was first considered when T is the identity operator; this classical case is developed in many respects in Chapter 4.

In the same chapter we also point out the strong interplay between KAT and Choquet's integral representation theory, as well as Stone-Weierstrass-type theorems.

Furthermore we present a detailed analysis of the existence of finite dimensional Korovkin subspaces and we give some estimates of the minimal dimension of such subspaces in terms of the small inductive dimension of the underlying space as well as of other topological parameters.

In Chapter 3 we characterize Korovkin subspaces for an arbitrary positive linear operator by emphasizing, among other things, additional properties, such as universal Korovkin-type properties with respect to positive linear operators, monotone operators and linear contractions.

We also consider other important classes of operators, such as positive projections, finitely defined operators and lattice homomorphisms.

The results concerning positive projections lead to some applications to Bauer simplices and to potential theory.

Finitely defined operators are important in this context because they are the only positive linear operators which can admit finite dimensional Korovkin subspaces.

Several characterizations are provided for this case and the interplay between Korovkin subspaces for finitely defined operators and Chebyshev systems is stressed.

Our main approach in developing the theory uses the basic idea, whose quintessence goes back to Korovkin, of studying approximation problems for equicontinuous nets of positive linear forms (Radon measures). This study, in fact, is carried out in Chapter 2. We deal with both the general case when the limit functionals are arbitrary bounded positive Radon measures, and the case when they are discrete or Dirac measures. The latter leads directly to the study of Choquet boundaries.

Chapters 5 and 6 are mainly concerned with applications to:
– Approximation of continuous functions by means of positive linear operators.
– Approximation and representation of the solutions of particular partial differential equations of diffusion type, by means of powers of positive linear operators.

More precisely, in Chapter 5 we give the first and best-known applications of Korovkin-type approximation theory, namely to the approximation of continuous functions defined on real intervals (bounded or not).

Throughout the chapter we describe different kinds of positive approximation processes.

Particular care is devoted to probabilistic-type operators, discrete-type operators, convolution operators for periodic functions and summation methods.

In general our results concern the uniform convergence on the whole interval or on compact subsets of it.

However in some cases we also investigate the convergence in L^p-spaces or in suitable weighted function spaces.

For almost all the specific approximation processes we consider in Chapter 5, we give estimates of the rate of convergence in terms of the classical modulus of continuity and, in some cases, of the second modulus of smoothness.

These estimates are not the sharpest possible but, on the other hand, an adequate analysis of improving them or of using more suitable moduli of smoothness would have gone too far for the purpose of this book.

For more details concerning rates of convergence of the specific approximation processes considered in Chapter 5 or of other more general ones, we refer, for instance, other than to the pioneering book of Korovkin [1960], also to the excellent books of Butzer and Nessel [1971], De Vore [1972], Ditzian and Totik [1987], Lorentz [1986 a] and Sendov and Popov [1988], for the univariate case as well as to the articles of Censor [1971] and Nishishiraho [1977], [1982b], [1983], [1987] for the multivariate and the infinite dimensional cases, respectively (see also Keimel and Roth [1992]).

In the final Chapter 6 we present a detailed analysis of some further sequences of positive linear operators that have been studied recently. These operators seem to play a non negligible role in some fine aspects of approximation theory. They connect the theory of C_0-semigroups of operators, partial differential equations and Markov processes.

The main examples we consider are the Bernstein-Schnabl operators, the Stancu-Schnabl operators and the Lototsky-Schnabl operators.

All these operators are constructed by means of a positive projection acting on the space of continuous functions on a convex compact set.

This general framework has the advantage of unifying the presentation of various well-known approximation processes and, at the same time, of providing new ones both in univariate and multivariate settings and in the infinite dimensional case, e.g., Bauer simplices.

After a careful analysis of the approximation properties of these operators, both from a qualitative and a quantitative point of view, a discussion follows of their monotonicity properties as well as their preservation of some global smoothness properties of functions, e.g., Hölder continuity.

Subsequently we show how these operators are strongly connected with initial and (Wentcel-type) boundary value problems in the theory of partial differential equations.

In fact, we prove that there exists a uniquely determined Feller semigroup that can be represented in terms of powers of the operators with which we are dealing.

The infinitesimal generator of the semigroup is explicitly determined in a core of its domain and, in the finite dimensional case, it turns out to be an elliptic second-order differential operator which degenerates on the Choquet boundary of the range of the projection.

Consequently we derive a representation and some qualitative properties of the solutions of the Cauchy problems which correspond to these diffusion equations.

We also emphasize the probabilistic meaning of our results by describing the transition function and the asymptotic behavior of the Markov processes governed by the above mentioned diffusion equations.

In Appendices A and B, written by M. Pannenberg and F. Beckhoff respectively, some of the main developments of Korovkin-type approximation theory in the setting of Banach algebras (commutative or not) are outlined essentially without proofs.

There the reader will have the opportunity to realize once again how Korovkin-type approximation theory, besides having its own interest, may also be used for solving problems of other important fields, such as Banach algebras and particular function algebras on locally compact abelian groups.

In Appendix C we list several concrete examples of Korovkin sets and determining sets. This list could be useful for rapidly checking those Korovkin sets that are most appropriate for the applications.

Finally we close every section with historical notes, giving credit and detailed references to supplementary results, so that, except in a few cases, we do not give references in the text. However, any inaccuracy or omission for historical details or in assigning priorities is unintentional and we apologize for possible errors.

In a diagram we also indicate some of the main connections among the various sections of the book.

After looking closely at the above mentioned subject classification in Appendix D, the reader will clearly see that the topics we have selected are not exhaustive with respect to the complete theory.

We have not dealt with certain other aspects of the theory, some of which have been indicated at the beginning of this introduction.

We also have to mention, for their particular interest and value of further investigations, those results concerning spaces of differentiable functions and spaces of continuous affine functions on convex compact subsets. The latter subject has been recently studied by Dieckmann [1993]. There the reader will also find a rather complete survey on this topic.

Although the aim of the book is to survey both classical and recent results in the field, the reader will find a certain amount of new material. In any case, the majority of the results presented here appears in a book for the first time.

We hope this monograph may serve not only to illustrate how effectively Korovkin-type approximation theory acts as a contact point between approximation theory and other areas of researches, notably functional analysis, but it may also lead to further investigations and to new applications to the theory of approximation by positive linear operators.

Interdependence of sections

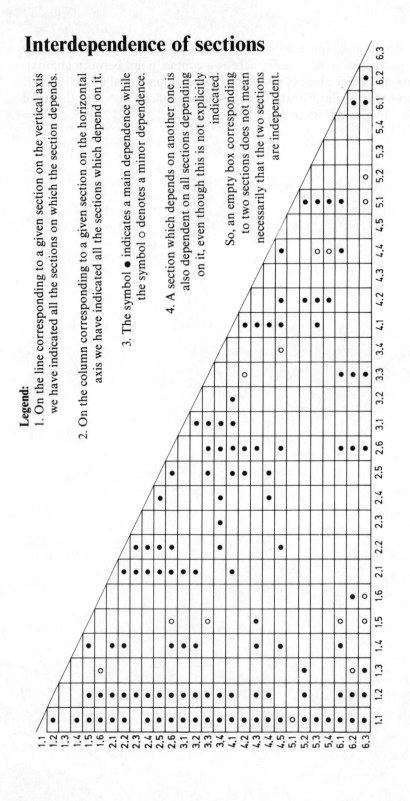

Legend:

1. On the line corresponding to a given section on the vertical axis we have indicated all the sections on which the section depends.

2. On the column corresponding to a given section on the horizontal axis we have indicated all the sections which depend on it.

3. The symbol ● indicates a main dependence while the symbol ○ denotes a minor dependence.

4. A section which depends on another one is also dependent on all sections depending on it, even though this is not explicitly indicated.

So, an empty box corresponding to two sections does not mean necessarily that the two sections are independent.

Legend: In each square we have considered the interdependence among the sections of a chapter and all the preceding ones.

A dotted line indicates a minor dependence.

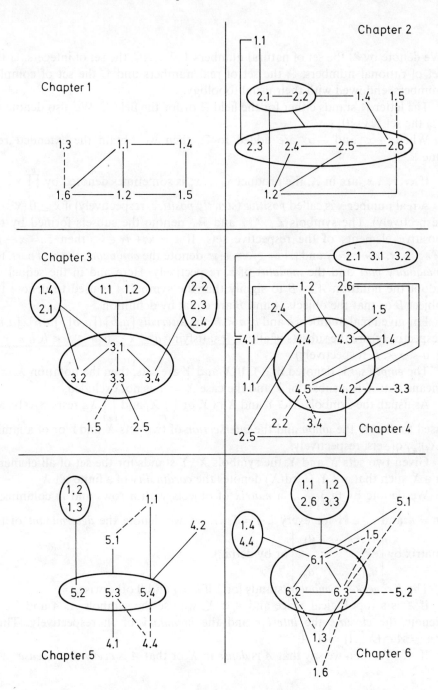

Notation

We denote by $\mathbb{N}$ the set of natural numbers $1, 2, \ldots$, $\mathbb{Z}$ the set of integers, $\mathbb{Q}$ the set of rational numbers, $\mathbb{R}$ the set of real numbers and $\mathbb{C}$ the set of complex numbers, endowed with their usual topology.

The letter $\mathbb{K}$ stands either for the field $\mathbb{R}$ or for the field $\mathbb{C}$. We also denote by $\mathbb{N}_0$ the set $\mathbb{N} \cup \{0\}$.

When $+\infty$ and $-\infty$ are added to $\mathbb{R}$, then we obtain the extended real line $\tilde{\mathbb{R}}$.

If $x_1, \ldots, x_p$ are in $\mathbb{K}$, the product $x_1 \ldots x_p$ is sometimes denoted by $\prod_{i=1}^{p} x_i$.

A real number x is called *positive* (*strictly positive*, respectively) if $x \geq 0$ ($x > 0$, respectively). The symbols $\mathbb{Z}_+$, $\mathbb{Q}_+$ and $\mathbb{R}_+$ denote the subsets formed by the positive elements of the respective sets. If $z = x + iy \in \mathbb{C}$, then $\bar{z} := x - iy$, $\mathscr{R}e\, z := x$, $\mathscr{I}m\, z := y$ and $|z| := \sqrt{x^2 + y^2}$ denote the *conjugate*, the *real part*, the *imaginary part* and the *modulus* of z, respectively. Here and in the sequel we adopt the notation $A := B$ to signify that the symbol A is used to denote the object B or that the objects A and B are equal by definition.

For given real numbers a and b, $a < b$, the *intervals* $[a, b]$ ($[a, b[$, $]a, b]$, $]a, b[$, respectively) are the subsets of all $x \in \mathbb{R}$ satisfying $a \leq x \leq b$ ($a \leq x < b$, $a < x \leq b$, $a < x < b$, respectively).

The *empty set* is denoted by $\varnothing$. If X and Y are sets, then the notation $X \subset Y$ means that X is a subset of Y and the case $X = Y$ is not excluded.

As usual, the symbols $X \cup Y$ and $X \cap Y$, or $\bigcup_{i \in I} X_i$ and $\bigcap_{i \in I} X_i$, respectively, are used to denote the *union* and the *intersection* of two sets X and Y or of a family $(X_i)_{i \in I}$ of sets, respectively.

Given two sets X and Y, the symbols $X \setminus Y$ stands for the set of all elements $x \in X$ such that $x \notin Y$. $\mathrm{Card}(X)$ denotes the *cardinality* of a finite set X.

We denote by $(a_{ij})_{\substack{1 \leq i \leq n \\ 1 \leq j \leq m}}$ a *matrix* of objects with n rows and m columns; if $m = n$ and $a_{ij} \in \mathbb{K}$ for every $i, j = 1, \ldots, n$, we denote the *determinant* of the

matrix by $\begin{vmatrix} a_{11} & \cdots & a_{1n} \\ \vdots & \vdots & \vdots \\ a_{n1} & \cdots & a_{nn} \end{vmatrix}$ or by $\det(a_{ij})$.

The *Kronecker symbol* δ_{ij} stands for 1 if $i = j$ and 0 otherwise.

If X is a topological space and $A \subset X$, we use the symbols $\bar{A}$, $\mathring{A}$ and ∂A to denote the *closure*, the *interior* and the *boundary* of A, respectively. Thus, $\partial A := \bar{A} \cap (\overline{X \setminus A}) = \bar{A} \setminus \mathring{A}$.

If $\bar{A} = X$, then we say that A is *dense* in X or that A is *everywhere dense*.

If X is a vector space over the field $\mathbb{K}$ and $A \subset X$, we denote by $\mathscr{L}(A)$ the *linear subspace generated by* A, i.e., the intersection of all linear subspaces containing A.

If $A \subset X$ and $a \in X$, the set $A + a := \{x + a \,|\, x \in A\}$ denotes a *translate* of A. Moreover $-A := \{-x \,|\, x \in A\}$ and, if $B \subset X$, $A + B := \{x + y \,|\, x \in A, y \in B\}$ and $A - B := \{x - y \,|\, x \in A, y \in B\}$.

If E is a function space and $A \subset E$, we also put $A + \mathbb{R}_+ := \{f + \alpha \,|\, f \in A, \alpha \in \mathbb{R}_+\}$.

The symbol $\dim(E)$ denotes the algebraic dimension of a vector space E.

The *cartesian product* $X \times Y$ of two sets X and Y is the set of all ordered pairs (x, y) with $x \in X$ and $y \in Y$.

More generally, if $(X_i)_{i \in I}$ is a family of sets, the *cartesian product* $\prod_{i \in I} X_i$ of the family consists of all families $(x_i)_{i \in I}$ where $x_i \in X_i$ for every $i \in I$.

If I is finite, say $I = \{1, \ldots, p\}$, then the cartesian product $\prod_{i=1}^{p} X_i$ is often identified with the set of all p-uples $(x_i, \ldots, x_p)$ of elements where $x_i \in X_i$ $(1 \leq i \leq p)$.

The cartesian product of p copies of the same set X is denoted by X^p.

The *unit circle* $\mathbb{T}$ and the *unit disk* $\mathbb{D}$ are, respectively, the subsets $\mathbb{T} := \{(x, y) \in \mathbb{R}^2 \,|\, x^2 + y^2 = 1\} = \{z \in \mathbb{C} \,|\, |z| = 1\}$ and $\mathbb{D} := \{(x, y) \in \mathbb{R}^2 \,|\, x^2 + y^2 \leq 1\} = \{z \in \mathbb{C} \,|\, |z| \leq 1\}$. The *unit sphere* in $\mathbb{R}^{p+1}$ is denoted by $\mathbb{S}_p$.

If $(X_i)_{i \in I}$ is a family of topological spaces, then the cartesian product $\prod_{i \in I} X_i$ endowed with the product topology is called the *product space* of the family $(X_i)_{i \in I}$.

As usual, the symbol $f \colon X \to Y$ denotes a *mapping* from a set X into a set Y. Sometimes we also use the symbol $x \mapsto f(x)$. If $f \colon X \to Y$ is a mapping, for every $A \subset X$ and $B \subset Y$ we set

$$f(A) := \{y \in Y \,|\, \text{there exists } x \in A \text{ such that } f(x) = y\}$$

and

$$f^{-1}(B) := \{x \in X \,|\, f(x) \in B\}.$$

The subsets $f(A)$ and $f^{-1}(B)$ are called the *image* of A and the *inverse image of* B under f, respectively.

More generally, the symbol $f \colon D_X(f) \to Y$ stands for a mapping from a subset $D_X(f)$ of X into Y. The subset $D_X(f)$ is called the *domain* of the mapping f and the image $f(D_X(f))$ of $D_X(f)$ is called the *range* of f. If $Y = X$, $D_X(f)$ will be simply denoted by $D(f)$.

Given a mapping $f \colon X \to Y$ and a subset A of X, the restriction of f to A is denoted by $f_{|A}$. Moreover, considering another mapping $g \colon Y \to Z$ from Y into a set Z, the *composition* of f and g is denoted by $g \circ f$. The p-th power $(p \geq 1)$ of a mapping $f: X \to X$ is defined as

$$f^p = \begin{cases} f, & \text{if } p = 1, \\ f^{p-1} \circ f, & \text{if } p \geq 2. \end{cases}$$

Generally, throughout the book the mappings are denoted by small letters, say $f, g, h, \ldots$. However, when we deal with random variables acting on probability spaces as well as with linear mappings (almost always called *linear operators*) acting on vector spaces, we use capital letters, say $Y, Z, \ldots$, and $L, S, T, \ldots$, respectively.

Sometimes for a given linear mapping $T: E \to F$ acting from a vector space E into a vector space F, the value of T at a point $f \in E$ is denoted by Tf instead of $T(f)$, if no confusion can arise. Furthermore the composition of T with another linear mapping $S: F \to G$ is also denoted by ST instead of $S \circ T$.

A linear mapping from a vector space (over $\mathbb{K}$) into $\mathbb{K}$ is also called a *linear form* or a *functional* on E.

A mapping $f: X \to Y$ is said to be *injective* if $f(x) = f(y)$ implies $x = y$ for every $x, y \in X$. If $f(X) = Y$, i.e., for every $y \in Y$ there exists $x \in X$ such that $f(x) = y$, we say that f is *surjective*.

If f is both injective and surjective, we say that f is *bijective*.

A mapping $f: X \to Y$ is bijective if and only if is *invertible*, i.e., there exists a (unique) mapping $g: Y \to X$ such that $g(f(x)) = x$ and $f(g(y)) = y$ for every $x \in X$ and $y \in Y$.

The mapping g is called the *inverse* of f and it is denoted by f^{-1}.

A mapping from a set X into $\mathbb{R}$ ($\tilde{\mathbb{R}}$, $\mathbb{C}$, respectively) is called a *real-valued* (*numerical*, *complex-valued*, respectively) *function on X*. When we simply speak of a *function on X* we always mean a real-valued function on X.

If A is contained in a set X, then the *characteristic function* of A is the function $\mathbf{1}_A: X \to \mathbb{R}$ defined by putting for every $x \in X$

$$\mathbf{1}_A(x) := \begin{cases} 1, & \text{if } x \in A, \\ 0, & \text{if } x \notin A. \end{cases}$$

The constant function on X of constant value 1 is denoted by $\mathbf{1}$.

Giving two functions $f: X \to \mathbb{R}$ and $g: X \to \mathbb{R}$, we use the symbols $\sup(f, g)$ and $\inf(f, g)$ to denote the functions on X defined by putting

$$\sup(f, g)(x) := \sup(f(x), g(x)),$$

$$\inf(f, g)(x) := \inf(f(x), g(x))$$

for every $x \in X$. These functions are also denoted by $f \vee g$ and $f \wedge g$, respectively.

More generally, if $f_1, \ldots, f_p$ are real functions on X we define the functions $\sup_{1 \leq i \leq p} f_i \colon X \to \mathbb{R}$ and $\inf_{1 \leq i \leq p} f_i \colon X \to \mathbb{R}$ by

$$\left(\sup_{1 \leq i \leq p} f_i \right)(x) := \sup_{1 \leq i \leq p} f_i(x), \quad \left(\inf_{1 \leq i \leq p} f_i \right)(x) := \inf_{1 \leq i \leq p} f_i(x) \quad (x \in X).$$

These functions are also denoted by $f_1 \vee \cdots \vee f_p$ and $f_1 \wedge \cdots \wedge f_p$, respectively.

If $f \colon X \to \mathbb{R}$, the *positive part* f^+ of f, the *negative* part f^- of f and the *absolute value* $|f|$ of f are defined as

$$f^+ := \sup(f, 0), \quad f^- := \sup(-f, 0), \quad |f| := \sup(f, -f).$$

Clearly we have $f = f^+ - f^-$ and $|f| = f^+ + f^-$.

If $f_1, \ldots, f_p$ are functions on X, sometimes we use the symbol $\prod_{i=1}^{p} f_i$ to denote the real-valued function on X defined by

$$\left(\prod_{i=1}^{p} f_i \right)(x) := \prod_{i=1}^{p} f_i(x) = f_1(x) \ldots f_p(x) \quad (x \in X).$$

A function $f \colon X \to \mathbb{R}$ is said to be *positive* if $f(x) \geq 0$ for every $x \in X$. Moreover we say that f is *strictly positive* if $f(x) > 0$ for every $x \in X$.

If $f \colon X \to \mathbb{R}$ and $g \colon X \to \mathbb{R}$ are functions on X, we write $f \leq g$ ($f < g$, respectively) if $f(x) \leq g(x)$ ($f(x) < g(x)$, respectively) for every $x \in X$.

A real function $f \colon I \to \mathbb{R}$ defined on a real interval I is called *increasing* (*decreasing, strictly increasing, strictly decreasing*, respectively) if $f(x) \leq f(y)$ ($f(y) \leq f(x)$, $f(x) < f(y)$, $f(y) < f(x)$, respectively) for every $x, y \in I$ satisfying $x < y$. A *monotone* (*strictly monotone*, respectively) function is a function which is indifferently increasing or decreasing (strictly increasing or strictly decreasing, respectively).

The symbols o and O are the usual Landau symbols. Thus, if $(x_n)_{n \in \mathbb{N}}$ and $(y_n)_{n \in \mathbb{N}}$ are sequences of real numbers, the symbol $x_n = o(y_n)$, $n \to \infty$, means that $x_n/y_n \to 0$ as $n \to \infty$, while the symbol $x_n = O(y_n)$, $n \to \infty$, means that there exists a constant $M > 0$ such that $|x_n/y_n| \leq M$ for every $n \in \mathbb{N}$.

A similar meaning must be attributed to the symbols $f(x) = o(g(x))$, $x \to x_0$, and $f(x) = O(g(x))$, $x \to x_0$, where f and g are functions defined on a subset X of a topological space and x_0 is a limit point for X.

The symbols $\Rightarrow, \Leftarrow, \Leftrightarrow$ stand for the usual logical implication symbols. Thus (a) $\Rightarrow$ (b), (a) $\Leftarrow$ (b) and (a) $\Leftrightarrow$ (b) mean that statement (a) implies statement (b), statement (b) implies statement (a) and statement (a) is equivalent to statement (b), respectively.

The section numbered, say, by a.b is the b-th section of Chapter a.

Definitions, lemmas, propositions and theorems are numbered by three digits, say a.b.c, where a denotes the number of the chapter, b that of the section and c is the progressive number within the section.

Formulas are numbered by an index of the form (a.b.c) where the digits a, b, c have the above specified meaning.

Sections, formulas, theorems, definitions, etc., are referred to by their corresponding numbers.

The end of a proof is indicated by the symbol □.

Chapter 1

Preliminaries

The main aim of this introductory chapter is to present the general notation and definitions we shall use throughout the book.

To make the exposition self-contained we also review those prerequisites which are necessary for a full understanding.

The topics are Radon measures, locally convex spaces and some basic aspects of general topology.

They have been selected primarily in view of our needs and are presented without pretence of completeness and without proofs.

Throughout the book we shall attempt to give various applications of Korovkin-type approximation theory. However some of them require a solid background also from other branches of analysis, e.g., measures on topological spaces, integration theory with respect to Radon measures, basic principles of probability theory, Choquet's integral representation theory and C_0-semigroups of bounded linear operators.

For the sake of completeness we also review the definitions and results pertaining to these topics and we include them in starred sections.

Thus a starred section or subsection is not essential for the whole book but will be only used for a particular (important) application of the Korovkin-type approximation theory.

1.1 Topology and analysis

In this section we present the definitions and the main properties of compact and locally compact spaces. We shall also introduce the main function spaces which we shall be concerned with in the sequel. For more details see Bourbaki [1965] and Engelking [1980] or the short and elegant Section 7.4 of Bauer [1981].

We begin by recalling the notions of net and filter.

A *filter* on a set I is a collection $\mathscr{F}$ of non-empty subsets of I, which is closed under the formation of finite intersections and such that, if $F \in \mathscr{F}$ and $F \subset G \subset I$, then $G \in \mathscr{F}$.

If $(I, \leq)$ is a *directed set*, i.e., $\leq$ is a partial ordering on I such that for every $i, j \in I$ there exists $\lambda \in I$ satisfying $i \leq \lambda$ and $j \leq \lambda$, then the set $\mathscr{F}_{\leq}$ of all subsets F of I for which there exists $i_0 \in I$ such that $\{i \in I | i_0 \leq i\} \subset F$, is a filter on I and it is called the *filter of sections on I*.

If $\mathscr{F}_1$ and $\mathscr{F}_2$ are filters on I, we shall say that $\mathscr{F}_2$ is *finer* than $\mathscr{F}_1$ if $\mathscr{F}_1 \subset \mathscr{F}_2$ or, equivalently, if for every $F_1 \in \mathscr{F}_1$ there exists $F_2 \in \mathscr{F}_2$ such that $F_2 \subset F_1$.

A *filtered family* $(x_i)_{i \in I}^{\mathscr{F}}$ of a set X is a family $(x_i)_{i \in I}$ of elements of X such that on the index set I there is fixed a filter $\mathscr{F}$.

A *net* (or *generalized sequence*) on X is a family $(x_i)_{i \in I}^{\leq}$ of elements of X such that on the set I there is a partial ordering $\leq$ with respect to which $(I, \leq)$ is a directed set.

Given a topological space X, we say that a filtered family $(x_i)_{i \in I}^{\mathscr{F}}$ *converges to a point* $x_0 \in X$ if for every neighborhood V of x_0 there exists $F \in \mathscr{F}$ such that $x_i \in V$ for each $i \in F$. In this case, x_0 is called a *limit* of the filtered family.

A filtered family is said to be *convergent* if it converges to some point.

If X is a *Hausdorff space*, i.e., for every pair of distinct points $x_1, x_2 \in X$ there exist neighborhoods V_1 and V_2 of x_1 and x_2, respectively, such that $V_1 \cap V_2 = \varnothing$, then every convergent filtered family $(x_i)_{i \in I}^{\mathscr{F}}$ converges to a unique limit $x_0 \in X$. In this case we shall write

$$\lim_{\substack{\mathscr{F} \\ i \in I}} x_i = x_0. \tag{1.1.1}$$

Sometimes, the notation $x_i \overset{\mathscr{F}}{\to} x_0$ will be also used, if no confusion can arise.

If $(x_i)_{i \in I}^{\leq}$ is a net, we say that $(x_i)_{i \in I}^{\leq}$ converges to a point $x_0 \in X$, if $(x_i)_{i \in I}^{\mathscr{F}_{\leq}}$ converges to x_0. Explicitly this means that for every neighborhood V of x_0 there exists $i_0 \in I$ such that $x_i \in V$ for every $i \in I, i \geq i_0$. The point x_0 is called a *limit* of the net $(x_i)_{i \in I}^{\mathscr{F}}$.

If X is Hausdorff, then x_0 is unique and we shall write

$$\lim_{\substack{\leq \\ i \in I}} x_i = x_0, \quad \text{or} \quad x_i \to x_0. \tag{1.1.2}$$

If X and Y are topological spaces, then a mapping $f: X \to Y$ is continuous at a point $x_0 \in X$ if and only if for every net $(x_i)_{i \in I}^{\leq}$ in X which converges to x_0, $(f(x_i))_{i \in I}^{\leq}$ converges to $f(x_0)$.

Given a topological space X, a numerical function $f: X \to \mathbb{R} \cup \{+\infty\}$ ($f: X \to \mathbb{R} \cup \{-\infty\}$, respectively) is said to be *lower semi-continuous* (*upper semi-continuous*, respectively) *at a point* $x_0 \in X$ if for every $\lambda \in \mathbb{R}$ satisfying $f(x_0) > \lambda$ ($f(x_0) < \lambda$, respectively) there exists a neighborhood V of x_0 such that $f(x) > \lambda$ ($f(x) < \lambda$, respectively) for every $x \in V$.

A function f is said to be *lower semi-continuous* (*upper semi-continuous*, respectively) if it is lower semi-continuous (upper semi-continuous, respectively) at every point of X.

In fact, f is lower semi-continuous (upper semi-continuous, respectively) if and only if the subset $\{x \in X \mid f(x) > \lambda\}$ is open for every $\lambda \in \mathbb{R}$ (the subset $\{x \in X \mid f(x) < \lambda\}$ is open for every $\lambda \in \mathbb{R}$, respectively) or, equivalently, the subset $\{x \in X \mid f(x) \le \lambda\}$ is closed for every $\lambda \in \mathbb{R}$ (the subset $\{x \in X \mid f(x) \ge \lambda\}$ is closed for every $\lambda \in \mathbb{R}$, respectively).

Moreover, a function $f \colon X \to \mathbb{R}$ is continuous if and only if it is both lower and upper semi-continuous.

A topological space X is called a *compact space* if every open cover of X has a finite subcover.

A subset of a topological space is said to be *compact* (*relatively compact*, respectively) if it is compact in the relative topology (if its closure is compact, respectively).

A useful characterization of compact spaces may be stated in terms of filtered families. More precisely, a topological space X is compact if and only if for every filtered family $(x_i)_{i \in I}^{\mathscr{F}}$ in X there exists a filter $\mathscr{F}'$ on I finer than $\mathscr{F}$ such that $(x_i)_{i \in I}^{\mathscr{F}'}$ is convergent in X.

If X is *metrizable*, i.e., the topology of X is induced by a metric on X, then X is compact if and only if every sequence of points of X admits a convergent subsequence.

Every compact metrizable space is *separable*, i.e., X contains a dense countable subset.

Let X be a compact Hausdorff space and denote by $\mathscr{T}_1$ its topology. If $\mathscr{T}_2$ is another Hausdorff topology on X such that $\mathscr{T}_2 \subset \mathscr{T}_1$, then necessarily $\mathscr{T}_2 = \mathscr{T}_1$.

Every compact Hausdorff space X is *normal*, i.e., each closed subset of X possesses a fundamental system of closed neighborhoods.

In this case, the *Tietze's extension theorem* holds (see Choquet [1969, Theorem 6.1]).

1.1.1 Theorem (*Tietze*). *If X is a normal space, then every continuous function from a closed subset of X is continuously extendable to X.*

Finally, we recall that, if X is compact, then every increasing (decreasing, respectively) sequence of lower semi-continuous (upper semi-continuous, respectively) functions on X converging pointwise to a continuous function, converges uniformly on X as well (*Dini's theorem*) (for a proof see Engelking [1989, Lemma 3.2.18]).

A topological space X is said to be *locally compact* if each of its point possesses a compact neighborhood.

In fact, if X is locally compact and Hausdorff, then each point of X has a fundamental system of compact neighborhoods.

The spaces $\mathbb{R}^p$, $p \ge 1$, as well as discrete spaces and compact spaces are examples of locally compact spaces.

Every locally compact Hausdorff space which is *countable at infinity* (i.e., it is the union of a sequence of compact subsets of X), is normal.

Furthermore, a locally compact Hausdorff space which has a *countable base* (i.e., there exists a countable family of open subsets such that every open subset is the union of some subfamily of this countable family) is necessarily metrizable, complete (and, hence, normal) and separable.

Conversely, a metrizable locally compact Hausdorff space which is countable at infinity has a countable base and, hence, is separable.

For technical reasons, it will be often useful to consider the (*Alexandrov*) *one-point-compactification X_ω of X*, which is defined as $X_\omega := X \cup \{\omega\}$, where ω is an object which does not belong to X (ω is often called the *point at infinity of X*).

The topology $\mathcal{T}_\omega$ on X_ω is defined as

$$\mathcal{T}_\omega := \mathcal{T} \cup \{X_\omega \backslash K \,|\, K \subset X, K \text{ compact}\}, \tag{1.1.3}$$

where $\mathcal{T}$ denotes the topology on X.

The topological space $(X_\omega, \mathcal{T}_\omega)$ is a Hausdorff compact space and X is open in X_ω. Furthermore, if X is non compact, then X is dense in X_ω.

For example, the one-point-compactification of $\mathbb{R}^p$, $p \geq 1$, is homeomorphic to the unit sphere of $\mathbb{R}^{p+1}$.

If a net $(x_i)_{i \in I}^{\leq}$ of elements of X converges to ω in X_ω, we say that the net $(x_i)_{i \in I}^{\leq}$ *converges to the point at infinity of X*. This means that for every compact subset K of X there exists $i_0 \in I$ such that $x_i \in X \backslash K$ for every $i \in I, i \geq i_0$.

Analogously, if $f \colon X \to \mathbb{R}$ and $\alpha \in \mathbb{R}$, we say that *f converges to α at the point at infinity*, if for every $\varepsilon > 0$ there exists a compact subset K of X such that $|f(x) - \alpha| \leq \varepsilon$ for every $x \in X \backslash K$. If this is the case, we shall write

$$\lim_{x \to \omega} f(x) = \alpha. \tag{1.1.4}$$

Given a set X, we shall denote by $\mathscr{B}(X)$ the Banach space of all real-valued bounded functions defined on X, endowed with the norm of the *uniform convergence* (briefly, the *sup-norm*) defined by

$$\|f\| := \sup_{x \in X} |f(x)| \quad \text{for every } f \in \mathscr{B}(X). \tag{1.1.5}$$

If X is a topological space, $\mathscr{C}(X)$ denotes the space of all real-valued continuous functions on X. Furthermore, we set

$$\mathscr{C}_b(X) := \mathscr{C}(X) \cap \mathscr{B}(X). \tag{1.1.6}$$

The space $\mathscr{C}_b(X)$, endowed with the sup-norm, is a Banach space.

Given a locally compact space X, we shall denote by $\mathscr{C}_0(X)$ the space of all functions $f \in \mathscr{C}(X)$ which *vanish at infinity*.

A function $f \in \mathscr{C}(X)$ vanishes at infinity if for every real number $\varepsilon > 0$ the set $\{x \in X \mid |f(x)| \geq \varepsilon\}$ is compact or, equivalently, if for every real number $\varepsilon > 0$ there exists a compact subset K of X such that $|f(x)| \leq \varepsilon$ for every $x \in X \setminus K$.

In other words, $f \in \mathscr{C}_0(X)$ if and only if $\lim\limits_{x \to \omega} f(x) = 0$ or, what is the same, the function $\tilde{f} \colon X_\omega \to \mathbb{R}$ defined by

$$\tilde{f}(x) := \begin{cases} f(x), & \text{if } x \in X, \\ 0, & \text{if } x = \omega, \end{cases} \tag{1.1.7}$$

is continuous in X_ω.

Sometimes, the above function $\tilde{f}$ is called the *canonical extension* of f to X_ω.

Note that if $f \in \mathscr{C}_0(X)$ and $(x_i)_{i \in I}^{\leq}$ is a net of elements of X converging to the point at infinity of X, then $(f(x_i))_{i \in I}^{\leq}$ converges to 0.

Clearly, if X is compact, then $\mathscr{C}_0(X) = \mathscr{C}(X) = \mathscr{C}_b(X)$.

In general, $\mathscr{C}_0(X)$ is a closed subspace of $\mathscr{C}_b(X)$ and, hence, endowed with the sup-norm, it is a Banach space. Unless otherwise stated, we shall always consider the space $\mathscr{C}_0(X)$ endowed with this norm.

On $\mathscr{C}_0(X)$ we shall also consider the natural ordering induced by the cone

$$\mathscr{C}_0^+(X) := \{ f \in \mathscr{C}_0(X) \mid f(x) \geq 0 \text{ for every } x \in X \}. \tag{1.1.8}$$

If $f \in \mathscr{C}_0(X)$, then $f^+, f^-, |f| \in \mathscr{C}_0(X)$ and

$$\|f\| = \||f|\| = \max\{\|f^+\|, \|f^-\|\}. \tag{1.1.9}$$

Thus $\mathscr{C}_0(X)$ is a Banach lattice. We recall, indeed, that a *normed lattice* E is a vector lattice (see Section 1.4) endowed with a *lattice norm* $\|\cdot\|$, i.e.,

$$|f| \leq |g| \quad \Rightarrow \quad \|f\| \leq \|g\| \quad \text{for every } f, g \in E. \tag{1.1.10}$$

If E is a Banach space for a lattice norm, then we say that E is a *Banach lattice*.

Other than $\mathscr{C}_0(X)$, standard examples of Banach lattices are the spaces $\mathscr{C}_b(X)$, $\mathscr{B}(X)$ and $L^p(X, \mu)$, $1 \leq p \leq +\infty$ (see Section 1.2).

An important property of Banach lattices concerns the automatic continuity of positive linear operators acting on them.

More precisely, if E is a Banach lattice and if F is another normed lattice, then every *positive* linear operator $T \colon E \to F$ (i.e., $T(f) \geq 0$ for every $f \in E$, $f \geq 0$) is continuous.

Furthermore, if $E = \mathscr{C}(X)$, X compact, then

$$\|T\| = \|T(\mathbf{1})\|, \tag{1.1.11}$$

where $\mathbf{1}$ denotes the constant function 1.

In particular every *positive linear form* on a Banach lattice E, i.e., every positive linear mapping from E into $\mathbb{R}$, is continuous.

Another function space playing a fundamental role in this book is the space $\mathscr{K}(X)$ of all real-valued continuous functions $f: X \to \mathbb{R}$ whose *support*

$$\operatorname{Supp}(f) := \overline{\{x \in X \,|\, f(x) \neq 0\}} \qquad (1.1.12)$$

is compact.

In other words, a continuous function $f: X \to \mathbb{R}$ belongs to $\mathscr{K}(X)$ if it vanishes on the complement of a suitable compact subset of X.

The space $\mathscr{K}(X)$ is dense in $\mathscr{C}_0(X)$ and, if X is compact, obviously coincides with $\mathscr{C}(X)$.

Besides the topology induced by the sup-norm, on $\mathscr{K}(X)$ there are other important locally convex topologies, such as the inductive topology and the projective topology. However, throughout the book we shall only make use of the first one. We refer to Choquet [1969, Section 16] for a detailed study of the properties of the other topologies.

If X is a locally compact Hausdorff space, then there are sufficiently many functions in $\mathscr{K}(X)$. This is a consequence of the following result, whose proof can be found for instance in Bauer [1981, Lemma 7.4.2].

1.1.2 Theorem. *For every compact subset K of X and for every open subset U containing K, there exists $g \in \mathscr{K}(X)$ such that $0 \leq g \leq 1$, $g = 1$ on K and $\operatorname{Supp}(g) \subset U$.*

1.2 Radon measures

Radon measures are a powerful tool which is fruitfully used in several branches of analysis such as probability theory, potential theory and integral representation theory. In this book they play a central role.

For more details the reader is referred to Bourbaki [1969] or Choquet [1969]. For a modern approach see also Bauer [1992] and Anger and Portenier [1992].

A *Radon measure* on a locally compact Hausdorff space X is a linear form $\mu: \mathscr{K}(X) \to \mathbb{R}$ satisfying the following property:

For any compact subset K of X there exists $M_K \geq 0$ such that $|\mu(f)| \leq M_K \|f\|$ for every $f \in \mathscr{K}(X)$ having its support contained in K.

The space of all Radon measures on X will be denoted by $\mathscr{M}(X)$. Thus $\mathscr{M}(X)$ is the dual space of the locally convex space $\mathscr{K}(X)$ endowed with the inductive topology.

As a matter of fact, in this book we shall restrict ourselves mainly to *bounded Radon measures*. They are those Radon measures $\mu \in \mathcal{M}(X)$ which are continuous with respect to the sup-norm. In this case the *norm* of μ is defined to be the number

$$\|\mu\| := \sup \{|\mu(f)|\,|\,f \in \mathcal{K}(X), \|f\| \leq 1\}. \tag{1.2.1}$$

A bounded Radon measure μ is said to be *contractive* if $\|\mu\| \leq 1$.

The space of all bounded Radon measures will be denoted by $\mathcal{M}_b(X)$.

Furthermore, we shall denote by $\mathcal{M}^+(X)$ the cone of all positive Radon measures. Thus, $\mu \in \mathcal{M}^+(X)$ if $\mu \in \mathcal{M}(X)$ and $\mu(f) \geq 0$ for every $f \in \mathcal{K}(X)$, $f \geq 0$. In fact, every positive linear form on $\mathcal{K}(X)$ is automatically in $\mathcal{M}^+(X)$.

Finally we set

$$\mathcal{M}_b^+(X) := \mathcal{M}^+(X) \cap \mathcal{M}_b(X), \tag{1.2.2}$$

and

$$\mathcal{M}_1^+(X) := \{\mu \in \mathcal{M}_b^+(X)|\,\|\mu\| = 1\}. \tag{1.2.3}$$

The elements of $\mathcal{M}_1^+(X)$ are also called *probability Radon measures*.

It is easy to see that every $\mu \in \mathcal{M}_b(X)$ ($\mu \in \mathcal{M}_b^+(X)$, respectively) can be extended to a (unique) continuous (positive, respectively) linear form on $\mathcal{C}_0(X)$ that we shall continue to denote by μ.

If X is compact, then $\mathcal{M}(X) = \mathcal{M}_b(X)$ (and, hence, $\mathcal{M}^+(X) = \mathcal{M}_b^+(X)$). Moreover, for every $\mu \in \mathcal{M}^+(X)$, $\|\mu\| = \mu(\mathbf{1})$. Conversely, if $\mu \in \mathcal{M}(X)$ and $\|\mu\| = \mu(\mathbf{1})$, then $\mu \in \mathcal{M}^+(X)$.

Another simple but useful property which is satisfied by every positive Radon measure $\mu \in \mathcal{M}^+(X)$ is the so-called *Cauchy-Schwarz inequality*, i.e.,

$$\mu(|fg|) \leq \sqrt{\mu(f^2)\mu(g^2)}, \tag{1.2.4}$$

which holds for every $f, g \in \mathcal{K}(X)$ (respectively, for every $f, g \in \mathcal{C}_0(X)$ provided $\mu \in \mathcal{M}_b^+(X)$).

As a matter of fact the same inequality holds by replacing $\mathcal{K}(X)$ (or $\mathcal{C}_0(X)$) with an arbitrary vector sublattice E of continuous functions on X and by considering a positive linear form $L: E \to \mathbb{R}$ on E (we recall that a linear subspace E of a vector lattice F is said to be a *sublattice* of F if for every $f, g \in E$ the supremum and the infimum of f and g in F lie in E). In this case one has

$$L(|fg|) \leq \sqrt{L(f^2)L(g^2)}, \tag{1.2.5}$$

for every $f, g \in E$ such that $fg, f^2, g^2 \in E$.

The simplest examples of Radon measures on X are the Dirac measures.

More precisely, given $x \in X$, the *Dirac measure* at x is the (bounded) Radon measure ε_x defined by

$$\varepsilon_x(f) := f(x) \qquad \text{for every } f \in \mathcal{K}(X) \text{ (or, } f \in \mathcal{C}_0(X)). \tag{1.2.6}$$

In fact, $\varepsilon_x \in \mathcal{M}_1^+(X)$.

A linear combination of Dirac measures is called a *discrete measure* on X. Thus, discrete measures are those bounded Radon measures on X of the form

$$\mu := \sum_{i=1}^{n} \lambda_i \varepsilon_{x_i}, \tag{1.2.7}$$

where $n \geq 1$, $x_1, \ldots, x_n \in X$ and $\lambda_1, \ldots, \lambda_n \in \mathbb{R}$. In this case μ is positive if and only if every λ_i is positive.

Furthermore

$$\|\mu\| = \sum_{i=1}^{n} |\lambda_i|. \tag{1.2.8}$$

Other important examples of Radon measures can be constructed as follows.

Let $\mu \in \mathcal{M}(X)$ and consider $g \in \mathcal{C}(X)$. Then, the linear form $v \colon \mathcal{K}(X) \to \mathbb{R}$ defined by

$$v(f) := \mu(f \cdot g) \quad \text{for every } f \in \mathcal{K}(X), \tag{1.2.9}$$

is a Radon measure on X. It is called the *measure with density g relative to μ* and it is denoted by $g \cdot \mu$.

If μ and g are bounded (respectively, positive), then $g \cdot \mu$ is bounded (respectively, positive) and

$$\|g \cdot \mu\| \leq \|g\| \, \|\mu\|. \tag{1.2.10}$$

By a simple method it is possible to extend every bounded positive Radon measure $\mu \in \mathcal{M}_b^+(X)$ to a positive Radon measure $\tilde{\mu}$ on the one-point-compactification X_ω of X. The measure $\tilde{\mu}$ is defined by

$$\tilde{\mu}(g) := \mu(g_{|X} - g(\omega)) + g(\omega) \|\mu\| \quad \text{for every } g \in \mathcal{C}(X_\omega). \tag{1.2.11}$$

Clearly we have $\|\tilde{\mu}\| = \|\mu\|$ and $\tilde{\mu}(\tilde{f}) = \mu(f)$ for every $f \in \mathcal{C}_0(X)$.

We say that a Radon measure $\mu \in \mathcal{M}(X)$ is *zero* on an open subset U of X if $\mu(f) = 0$ for every function $f \in \mathcal{K}(X)$ whose support is contained in U.

We shall denote by $\mathfrak{A}(\mu)$ the collection of all open subsets of X on which μ is zero.

The *support* of the measure μ is then defined to be the subset

$$\mathrm{Supp}(\mu) := X \setminus \bigcup_{U \in \mathfrak{A}(\mu)} U. \tag{1.2.12}$$

Thus $\mathrm{Supp}(\mu)$ is a closed subset of X and, in fact, is the complement of the largest open subset of X on which μ is zero.

Clearly a point $x_0 \in X$ belongs to $\mathrm{Supp}(\mu)$ if for every neighborhood V of x_0 there exists $f \in \mathcal{K}(X)$ such that $\mathrm{Supp}(f) \subset V$ and $\mu(f) \neq 0$.

On the other hand, $x_0 \notin \mathrm{Supp}(\mu)$ if there exists an open neighborhood V of x_0 on which μ is zero.

It is also clear that $\mathrm{Supp}(\mu) = \varnothing$ if and only if $\mu = 0$.

Here, we list some of the main properties of supports which we shall use later. For a proof see Bourbaki [1969, Chapter III, Section 2] and Choquet [1969, Section 11].

1.2.1 Theorem. *Let X be a locally compact Hausdorff space and let $\mu \in \mathcal{M}(X)$. Then*

(1) $\mathrm{Supp}(\mu + \upsilon) \subset \mathrm{Supp}(\mu) \cup \mathrm{Supp}(\upsilon)$ *for every* $\upsilon \in \mathcal{M}(X)$. *If μ and υ are positive we have equality in the above inclusion.*

(2) *If $f \in \mathcal{K}(X)$ and $f = 0$ on $\mathrm{Supp}(\mu)$, then $\mu(f) = 0$. If $\mu \in \mathcal{M}_b(X)$, then the same property holds for every $f \in \mathcal{C}_0(X)$.*

(3) *If $f, g \in \mathcal{K}(X)$ (or $f, g \in \mathcal{C}_0(X)$ provided $\mu \in \mathcal{M}_b(X)$) and $f = g$ on $\mathrm{Supp}(\mu)$, then $\mu(f) = \mu(g)$.*

(4) *If $\mu \in \mathcal{M}^+(X)$ and $f \in \mathcal{K}(X)$ (or $f \in \mathcal{C}_0(X)$ provided $\mu \in \mathcal{M}_b^+(X)$) then $\mu(f) \geq 0$ if $f \geq 0$ on $\mathrm{Supp}(\mu)$.*

(5) *If $\mu \in \mathcal{M}^+(X)$ and $f \in \mathcal{K}(X)$, $f \geq 0$ (or $f \in \mathcal{C}_0^+(X)$ provided $\mu \in \mathcal{M}_b^+(X)$) and if $\mu(f) = 0$, then $f = 0$ on $\mathrm{Supp}(\mu)$.*

(6) *For every $g \in \mathcal{C}(X)$, $\mathrm{Supp}(g \cdot \mu) = \overline{\{x \in \mathrm{Supp}(\mu) \mid g(x) \neq 0\}} \subset \mathrm{Supp}(g) \cap \mathrm{Supp}(\mu)$.*

(7) *If $x_1, \ldots, x_n$ are distinct points of X, $n \geq 1$, then $\mathrm{Supp}(\mu) = \{x_1, \ldots, x_n\}$ if and only if $\mu = \sum_{i=1}^{n} \lambda_i \varepsilon_{x_i}$ for some $\lambda_1, \ldots, \lambda_n \in \mathbb{R} \setminus \{0\}$.*

(8) *If $\mathrm{Supp}(\mu)$ is compact, then μ is bounded.*

On the space $\mathcal{M}(X)$ we shall consider the *vague topology* which is, by definition, the coarsest topology on $\mathcal{M}(X)$ for which all the mappings $\varphi_f (f \in \mathcal{K}(X))$ are continuous, where

$$\varphi_f(\mu) := \mu(f) \quad \text{for every } f \in \mathcal{K}(X) \text{ and } \mu \in \mathcal{M}(X). \tag{1.2.13}$$

It is a locally convex topology and, in fact, is the weak*-topology of the dual space of the locally convex space $\mathcal{K}(X)$ (see Section 1.4).

A net $(\mu_i)_{i \in I}^{\leq}$ in $\mathcal{M}(X)$ converges to a Radon measure $\mu \in \mathcal{M}(X)$ with respect to the vague topology if $\lim_{\substack{\leq \\ i \in I}} \mu_i(f) = \mu(f)$ for every $f \in \mathcal{K}(X)$. In this case we also say that $(\mu_i)_{i \in I}^{\leq}$ *converges vaguely* to μ.

Endowed with the vague topology, $\mathcal{M}(X)$ is a locally convex Hausdorff space. Furthermore in general $\mathcal{M}(X)$ is not metrizable. If X has a countable base, then $\mathcal{M}^+(X)$ is metrizable and separable.

A useful characterization of *vaguely compact subsets of* $\mathcal{M}(X)$ (i.e., those subsets which are compact with respect to the vague topology) is indicated below.

We also notice that a subset $\mathfrak{A}$ of $\mathcal{M}(X)$ is *vaguely bounded* if the set $\{\mu(f) | \mu \in \mathfrak{A}\}$ is bounded for every $f \in \mathcal{K}(X)$.

Moreover, $\mathfrak{A}$ is *strongly bounded* if for every compact subset K of X there exists $M_K \geq 0$ such that $|\mu(f)| \leq M_K \|f\|$ for every $\mu \in \mathfrak{A}$ and for every $f \in \mathcal{K}(X)$ satisfying $\mathrm{Supp}(f) \subset K$.

For a proof of the next result see Choquet [1969, Theorem 12.6].

1.2.2 Theorem. *A subset $\mathfrak{A}$ of $\mathcal{M}(X)$ is relatively vaguely compact in $\mathcal{M}(X)$ if and only if it is vaguely bounded or, equivalently, if and only if it is strongly bounded.*

From this criterion it follows that for every $r > 0$ the set $\{\mu \in \mathcal{M}_b(X) | \|\mu\| \leq r\}$ is vaguely compact in $\mathcal{M}(X)$. Moreover, if X is compact, then $\{\mu \in \mathcal{M}^+(X) | \|\mu\| = r\}$ is vaguely compact too.

Via the vague topology every Radon measure can be approximated by discrete measures. More precisely we have the following result, whose proof can be found again in Choquet [1969, Theorem 12.11].

1.2.3 Theorem (*Approximation theorem*). *Let X be a locally compact Hausdorff space. Then the following assertions hold:*
(1) *For every $\mu \in \mathcal{M}(X)$ there exists a net $(\mu_i)_{i \in I}^{\leq}$ of discrete Radon measures which converges vaguely to μ. Moreover, if μ is positive, every μ_i can be chosen positive too.*
(2) *If $\mu \in \mathcal{M}(X)$ and $\mathrm{Supp}(\mu)$ is compact, then there exists a net $(\mu_i)_{i \in I}^{\leq}$ of discrete Radon measures vaguely convergent to μ such that $\|\mu_i\| = \|\mu\|$ and $\mathrm{Supp}(\mu_i) \subset \mathrm{Supp}(\mu)$ for every $i \in I$.*
Furthermore, if μ is positive, the measures μ_i can also be chosen positive.

*Measures on topological spaces. The Riesz representation theorem

In this subsection we briefly discuss some properties of Borel and Baire measures on a topological space together with Riesz's representation theorem.

This last theorem will be used only in Section 5.2 to give a probabilistic interpretation of Korovkin's theorem.

In the sequel it is assumed that the reader has a thorough familiarity with the essential notions of measure theory and integration theory. As a reference the reader could consult Bauer [1981, part I], [1992].

If X is a topological space, we shall denote by $\mathfrak{B}(X)$ the σ-algebra of all *Borel sets* in X, i.e., the σ-algebra generated by the open subsets of X.

We shall also denote by $\mathfrak{B}_0(X)$ the σ-algebra of *Baire sets* of X, which is defined as the smallest σ-algebra in X with respect to which all continuous functions on X are measurable.

In general, $\mathfrak{B}_0(X) \subsetneqq \mathfrak{B}(X)$. If X is metrizable, then the two σ-algebras coincide.

A measure v on $\mathfrak{B}(X)$ (on $\mathfrak{B}_0(X)$, respectively) is called *regular* if for every $B \in \mathfrak{B}(X)$

$$v(B) = \inf\{v(U)|B \subset U, U \text{ open}\} = \sup\{v(K)|K \subset B, K \text{ compact}\}, \quad (1.2.14)$$

(respectively, for every $B \in \mathfrak{B}_0(X)$

$$v(B) = \inf\{v(U)|B \subset U, U \text{ open}, U \in \mathfrak{B}_0(X)\}$$

$$= \sup\{v(K)|K \subset B, K \text{ compact}, K \in \mathfrak{B}_0(X)\}). \quad (1.2.15)$$

Furthermore we say that a measure v on $\mathfrak{B}(X)$ (on $\mathfrak{B}_0(X)$, respectively) is a *Borel measure* on X (a *Baire measure* on X, respectively) if

$$v(K) < +\infty \quad \text{for every compact subset } K \text{ of } X \quad (1.2.16)$$

(respectively,

$$v(K) < +\infty \quad \text{for every compact subset } K \in \mathfrak{B}_0(X)). \quad (1.2.17)$$

If X is a *Polish space*, i.e., it has a countable base and its topology is defined by a metric with respect to which it is complete, then every finite Borel measure on X is regular (note that locally compact Hausdorff spaces with a countable base are Polish spaces).

If X is a locally compact Hausdorff space which is countable at infinity, then every Baire measure on X is regular and σ-finite.

We now state the Riesz's representation theorem which establishes a one-to-one correspondence between Radon measures and Baire (Borel) measures.

For a proof see Bourbaki [1969] and Bauer [1981, Theorem 7.5.4].

1.2.4 Theorem (*Riesz's representation theorem*). *Let X be a locally compact Hausdorff space. Given $\mu \in \mathscr{M}^+(X)$, then*
(1) *If μ is bounded, there exists a unique regular finite Borel measure v on X such that $\mathscr{K}(X) \subset \mathscr{L}^1(X, \mathfrak{B}(X), v)$ and $\mu(f) = \int f\, dv$ for every $f \in \mathscr{K}(X)$.*

(2) *If X is countable at infinity, there exists a unique (regular and σ-finite) Baire measure v on X such that $\mathscr{K}(X) \subset \mathscr{L}^1(X, \mathscr{B}_0(X), v)$ and $\mu(f) = \int f \, dv$ for every $f \in \mathscr{K}(X)$.*

If X is countable at infinity and metrizable, then an explicit construction of the measure v is indicated in the next subsection.

From now on we shall assume that X is a locally compact Hausdorff space which is countable at infinity.

On the basis of the above theorem, by a common abuse of notation, we shall continue to denote by $\mathscr{M}^+(X)$ the cone of all Baire measures on X. Furthermore, the subset of all $\mu \in \mathscr{M}^+(X)$ satisfying $\mu(X) < +\infty$ ($\mu(X) = 1$, respectively) will be denoted by $\mathscr{M}_b^+(X)$ ($\mathscr{M}_1^+(X)$, respectively).

A net $(\mu_i)_{i \in I}^{\leq}$ in $\mathscr{M}_b^+(X)$ is said to be *weakly convergent* to a measure $\mu \in \mathscr{M}_b^+(X)$ if for every $f \in \mathscr{C}_b(X)$

$$\lim_{i \in I}{}_{\leq} \int f \, d\mu_i = \int f \, d\mu. \tag{1.2.18}$$

If X is a Polish space and if $(\mu_i)_{i \in I}^{\leq}$ converges weakly to μ, then we obtain

$$\lim_{i \in I}{}_{\leq} \int f \, d\mu_i = \int f \, d\mu \tag{1.2.19}$$

for every Borel-measurable, bounded, μ-a.e. continuous function $f \colon X \to \mathbb{R}$.

Finally note that, in general, the weak convergence in $\mathscr{M}_b^+(X)$ can be derived from a topology on $\mathscr{M}_b^+(X)$ that is called the *weak topology* on $\mathscr{M}_b^+(X)$.

As a simple criterion to decide if a sequence of measures with densities converges weakly, we mention the following result which is often referred to as *Scheffé's theorem*:

Let $\mu \in \mathscr{M}^+(X)$ and consider a sequence $(g_n)_{n \in \mathbb{N}}$ of positive μ-integrable numerical functions on X converging pointwise to a positive μ-integrable function g on X. If $\int g_n \, d\mu \to \int g \, d\mu$, then the sequence of measures with densities $(g_n \cdot \mu)_{n \in \mathbb{N}}$ converges weakly to $g \cdot \mu$.

In fact, for every $f \in \mathscr{C}_b(X)$

$$\left| \int f \, d(g_n \cdot \mu) - \int f \, d(g \cdot \mu) \right| = \left| \int (fg_n - fg) \, d\mu \right| \leq \|f\| \int |g_n - g| \, d\mu$$

$$= \|f\| \int (g_n + g - 2(g_n \wedge g)) \, d\mu \to 0,$$

since $\int g_n \wedge g \, d\mu \to \int g \, d\mu$ by Lebesgue's dominated convergence theorem.

As a matter of fact the same result holds even if $g_n \to g$ μ-*stochastically*, i.e., for every $\varepsilon > 0$ and for every Baire subset A of finite measure $\mu(A)$ one has

$$\lim_{n \to \infty} \mu(\{|g_n - g| \geq \varepsilon\} \cap A) = 0$$

(see Bauer [1992, Lemma 21.6]).

*Integration with respect to Radon measures. $L^p(X, \mu)$-spaces

As we have seen in the above subsection, there is a one-to-one correspondence between positive Radon measures and regular Baire (Borel) measures.

Without using this correspondence, it is nevertheless possible to develop an integration theory with respect to Radon measures that leads, in particular, to the construction of the same $L^p(X, \mu)$-spaces that one obtains via the classical procedure by starting from the corresponding regular Baire (Borel) measures.

Below we shall briefly indicate the most salient points of this integration theory. For complete details see Bourbaki [1969, Chapter IV] or Choquet [1969, Section 11].

However the reader who is not interested in Korovkin-type theorems in $L^p(X, \mu)$-spaces can directly proceed to the other sections.

In the sequel we shall fix a locally compact Hausdorff space X and a positive Radon measure $\mu \in \mathcal{M}^+(X)$.

We set

$$\mathcal{K}^+(X) := \{f \in \mathcal{K}(X) | f \geq 0\} \tag{1.2.20}$$

and

$$\mathcal{I}^+(X) := \{f : X \to \tilde{\mathbb{R}} | f \text{ is lower semi-continuous and positive}\}. \tag{1.2.21}$$

For every $f \in \mathcal{I}^+(X)$ we put

$$\int^* f \, d\mu := \sup\{\mu(g) | g \in \mathcal{K}^+(X), g \leq f\} \in \mathbb{R}_+ \cup \{+\infty\}. \tag{1.2.22}$$

The number $\int^* f \, d\mu$ is called the *upper integral of f with respect to μ*. Of course,

$$\int^* g \, d\mu = \mu(g) \quad \text{if } g \in \mathcal{K}^+(X). \tag{1.2.23}$$

To assign an upper integral to every positive function we proceed as follows. If $f: X \to \mathbb{R}_+ \cup \{+\infty\}$ is an arbitrary function we set

$$\int^* f \, d\mu := \inf \left\{ \int^* g \, d\mu \,\middle|\, g \in \mathscr{I}^+(X), f \leq g \right\} \in \mathbb{R}_+ \cup \{+\infty\}. \quad (1.2.24)$$

Again, $\int^* f \, d\mu$ is called the *upper integral of f with respect to* μ and coincides with the one defined by (1.2.22) provided $f \in \mathscr{I}^+(X)$.

Now, fix $p \in \mathbb{R}$, $1 \leq p$, and set for every $f: X \to \mathbb{R}$

$$N_p(f) := \left(\int^* |f|^p d\mu \right)^{1/p}. \quad (1.2.25)$$

Furthermore let

$$\mathscr{F}^p(X, \mu) := \{ f: X \to \mathbb{R} \,|\, N_p(f) < +\infty \}. \quad (1.2.26)$$

Then, N_p is a seminorm on $\mathscr{F}^p(X, \mu)$ and $\mathscr{K}(X) \subset \mathscr{F}^p(X, \mu)$, because of (1.2.23) and the equality $\mathscr{K}(X) = \mathscr{K}^+(X) - \mathscr{K}^+(X)$.

We denote by $\mathscr{L}^p(X, \mu)$ the closure of $\mathscr{K}(X)$ in $\mathscr{F}^p(X, \mu)$ with respect to the seminorm N_p. Finally, we set

$$L^p(X, \mu) := \mathscr{L}^p(X, \mu)/\mathscr{N}_p, \quad (1.2.27)$$

where $\mathscr{N}_p$ is the equivalence relation on $\mathscr{L}^p(X, \mu)$ defined by

$$f \mathscr{N}_p g \quad \Leftrightarrow \quad N_p(f - g) = 0 \quad (f, g \in \mathscr{L}^p(X, \mu).) \quad (1.2.28)$$

The space $L^p(X, \mu)$ endowed with the norm $\| \cdot \|_p$ inherited from N_p is a Banach space.

By embedding $\mathscr{L}^p(X, \mu)$ in $L^p(X, \mu)$ we have that $\mathscr{K}(X)$ is dense in $L^p(X, \mu)$.

The functions in $\mathscr{L}^1(X, \mu)$ (or, in $L^1(X, \mu)$) are also called *μ-integrable functions*.

The value attained by the unique extension of μ to $\mathscr{L}^1(X, \mu)$ in $f \in \mathscr{L}^1(X, \mu)$ is called the *integral of f with respect to* μ and is denoted by one of the following symbols

$$\int f \, d\mu, \quad \int f(x) \, d\mu(x), \quad \int f\mu. \quad (1.2.29)$$

If f, $g \in \mathscr{L}^1(X, \mu)$ and $f \mathscr{N}_1 g$, then $\int f \, d\mu = \int g \, d\mu$. So, the integral can be defined for every $f \in L^1(X, \mu)$.

For some classical criteria of integrability as well as for a complete treatment of the properties of the upper integral, the integral and $L^p(X, \mu)$-spaces see Bourbaki [1969, Chapter IV].

For example, if $f: X \to \mathbb{R}_+ \cup \{+\infty\}$ is lower semi-continuous and positive, then f is μ-integrable if and only if $\int^* f \, d\mu < +\infty$. If this is the case, then

$$\int f \, d\mu = \int^* f \, d\mu. \tag{1.2.30}$$

Because of (1.2.23) and (1.2.30), for every $f \in \mathcal{K}(X)$ we shall often denote the value $\mu(f)$ by $\int f \, d\mu$.

Given a subset A of X we shall set

$$\mu^*(A) := \int^* 1_A \, d\mu \in \mathbb{R}_+ \cup \{+\infty\}, \tag{1.2.31}$$

where 1_A denotes the characteristic function of A.

The value $\mu^*(A)$ is called the *outer measure* of A with respect to μ.

A subset A of X such that $\mu^*(A) = 0$ is called *negligible* (or *of measure zero*).

A property P of points of X is said to hold μ-*almost everywhere* (shortly, μ-a.e.) if the subsets of all points $x \in X$ for which $P(x)$ is false, is contained in a set of measure zero.

The *inner measure* of a subset A of X is defined as

$$\mu_*(A) := \sup\{\mu^*(K) | K \subset A, K \text{ compact}\}. \tag{1.2.32}$$

A subset A of X is called μ-*integrable* if 1_A is μ-integrable. In this case, we have

$$\mu_*(A) = \mu^*(A) = \int^* 1_A \, d\mu. \tag{1.2.33}$$

The space X is μ-integrable, i.e., the constant functions are μ-integrable, if and only if μ is bounded. Then,

$$\|\mu\| = \mu^*(X) = \int 1 \, d\mu. \tag{1.2.34}$$

Furthermore $\mathcal{C}_0(X)$ is a dense subspace of $\mathcal{L}^p(X, \mu)$ $(1 \le p < +\infty)$. Moreover for every $f \in \mathcal{C}_0(X)$

$$\left| \int f \, d\mu \right| \le \|f\| \, \|\mu\|. \tag{1.2.35}$$

A function $f: X \to \mathbb{R}_+ \cup \{+\infty\}$ is called μ-*measurable* if for every $\varepsilon > 0$ and for every compact subset Y of X there exists a compact subset K of X such that f is continuous on K and $\mu^*(Y \backslash K) < \varepsilon$.

The subspace of all μ-measurable functions $f: X \to \mathbb{R}$ such that

$$N_\infty(f) := \inf\{\alpha \in \mathbb{R}_+ \,|\, |f| \leq \alpha \ \mu\text{-a.e.}\} < +\infty \tag{1.2.36}$$

is denoted by $\mathscr{L}^\infty(X, \mu)$.

The quotient space $\mathscr{L}^\infty(x, \mu)/\mathcal{N}_\infty$, where $\mathcal{N}_\infty$ is the equivalence relation on $\mathscr{L}^\infty(X, \mu)$ defined by

$$f \mathcal{N}_\infty g \quad \Leftrightarrow \quad N_\infty(f - g) = 0 \quad \Leftrightarrow \quad f = g \ \mu\text{-a.e.}, \tag{1.2.37}$$

will be denoted by $L^\infty(X, \mu)$.

The space $L^\infty(X, \mu)$ endowed with the norm

$$\|f\|_\infty := N_\infty(f) \tag{1.2.38}$$

(where every $f \in L^\infty(X, \mu)$ is identified with an arbitrary representing function), is a Banach space.

Moreover, all $L^p(X, \mu)$-spaces $(1 \leq p \leq +\infty)$ endowed with the ordering

$$f \leq g \quad \text{if} \quad f \leq g \quad \mu\text{-a.e.} \tag{1.2.39}$$

are Banach lattices.

We say that a subset A of X is μ-*measurable* if $\mathbf{1}_A$ is μ-measurable. The set of all μ-measurable subsets of X will be denoted by $\mathfrak{B}^*(X)$. In fact, $\mathfrak{B}^*(X)$ is a σ-algebra and μ^* is a measure on $\mathfrak{B}^*(X)$. Moreover, $\mathfrak{B}^*(X)$ contains the σ-algebra $\mathfrak{B}(X)$ of Borel sets in X.

If X is countable at infinity and metrizable, then the unique Baire (or, equivalently, Borel) measure v on X determined from Riesz's representation theorem is, in fact, μ^*. Furthermore, $\mathscr{L}^1(X, \mathfrak{B}(X), \mu^*) = \mathscr{L}^1(X, \mu)$ and for every $f \in \mathscr{L}^1(X, \mu)$

$$\int f \, d\mu = \int f \, d\mu^* \tag{1.2.40}$$

(see Choquet [1969, Theorem 11.18]).

Restrictions and extensions of Radon measures

Again we fix a locally compact Hausdorff space X and a measure $\mu \in \mathcal{M}^+(X)$. Given a locally compact subset Y of X, there is a classical procedure to define a new Radon measure on Y, which is called the *restriction of μ to Y* and is denoted by $\mu_{|Y}$.

This restriction is defined by

$$\mu_{|Y}(f) := \int f^* \, d\mu \quad \text{for every } f \in \mathcal{K}(Y), \tag{1.2.41}$$

where

$$f^*(x) := \begin{cases} f(x), & \text{if } x \in Y, \\ 0, & \text{if } x \notin Y. \end{cases} \tag{1.2.42}$$

(Note that $(f^+)^*$ and $(f^-)^*$ are upper semi-continuous and, hence, μ-measurable; so, $f^* = (f^+)^* - (f^-)^*$ is μ-measurable, bounded and with compact support, so that it is μ-integrable).

When X is metrizable or countable at infinity, we may employ another procedure (which does not make use of integration theory) to define the restriction of μ to Y in the particular case when Y is compact and $\text{Supp}(\mu) \subset Y$.

In fact, since X is normal, given $f \in \mathcal{K}(Y) = \mathcal{C}(Y)$, by Tietze's theorem there exists a continuous extension $f_1 \colon X \to \mathbb{R}$ of f. After choosing $f_2 \in \mathcal{K}(X)$ such that $f_2 = 1$ on Y, clearly the function $g := f_1 f_2$ is another extension of f which belongs to $\mathcal{K}(X)$.

Note that, if g_1 and g_2 are two extensions of f belonging to $\mathcal{K}(X)$, then $\mu(g_1) = \mu(g_2)$ since $\text{Supp}(\mu) \subset Y$. So, we may define a Radon measure υ on Y by

$$\upsilon(f) := \mu(g) \quad \text{for every } f \in \mathcal{C}(Y), \tag{1.2.43}$$

where $g \in \mathcal{K}(X)$ is an arbitrary extension of f to X.

As it is easy to see, the measure υ coincides, in fact, with the measure $\mu_{|Y}$ defined by (1.2.41).

We shall use this approach mainly in the particular case when X is compact.

Finally let us make some remarks about a simple procedure to extend Radon measures.

Let Y be a closed subset of X and fix $\mu \in \mathcal{M}_b^+(Y)$. Then we may define a new bounded Radon measure $\bar{\mu}$ on X by

$$\bar{\mu}(f) := \mu(f_{|Y}) \quad \text{for every } f \in \mathcal{K}(X). \tag{1.2.44}$$

The measure $\bar{\mu}$ is called the *canonical extension of μ over X* and, by a common abuse of language, it is still denoted by μ.

Image Radon measures

Let X and Y be locally compact Hausdorff spaces. A continuous mapping $\varphi: X \to Y$ is called *proper* if for every compact subset K of Y, $\varphi^{-1}(K)$ is compact in X.

If X is compact, then every continuous mapping from X into Y is proper.

Furthermore, if $f: Y \to \mathbb{R}$, then $\mathrm{Supp}(f \circ \varphi) \subset \varphi^{-1}(\mathrm{Supp}(f))$ so that, if f is proper, we have

$$f \circ \varphi \in \mathcal{K}(X) \quad \text{for every } f \in \mathcal{K}(Y), \tag{1.2.45}$$

as well as

$$f \circ \varphi \in \mathcal{C}_0(X) \quad \text{for every } f \in \mathcal{C}_0(Y). \tag{1.2.46}$$

Given a Radon measure $\mu \in \mathcal{M}(X)$ and a proper mapping $\varphi: X \to Y$, we may consider the Radon measure υ on Y defined by

$$\upsilon(f) := \mu(f \circ \varphi) \quad \text{for every } f \in \mathcal{K}(Y). \tag{1.2.47}$$

The measure υ is called the *image of μ under the mapping φ* and it is denoted by $\varphi(\mu)$.

If μ is positive, $\varphi(\mu)$ is positive. If μ is bounded, then $\varphi(\mu)$ is bounded as well and $\|\varphi(\mu)\| \leq \|\mu\|$. Moreover, if, in addition, μ is positive, then $\|\varphi(\mu)\| = \|\mu\|$.

In general, $\mathrm{Supp}(\varphi(\mu)) \subset \varphi(\mathrm{Supp}(\mu))$ and we have equality if μ is positive.

Furthermore the mapping $\mu \mapsto \varphi(\mu)$ from $\mathcal{M}(X)$ into $\mathcal{M}(Y)$ is vaguely continuous.

Note that, if $\varphi: X \to Y$ is an arbitrary continuous mapping (or, more generally, measurable with respect to the σ-algebras $\mathfrak{B}(X)$ and $\mathfrak{B}(Y)$) and $\mu \in \mathcal{M}_b^+(X)$, we could define a Radon measure $\varphi^*(\mu)$ on $\mathcal{M}(Y)$ by

$$\varphi^*(\mu)(f) := \int f \circ \varphi \, d\mu \quad \text{for every } f \in \mathcal{K}(Y). \tag{1.2.48}$$

But, in this case, the mapping $\mu \mapsto \varphi^*(\mu)$ fails to be continuous.

For more details on proper mappings and image measures see Bourbaki [1965, § 10, n.1], [1969] and Choquet [1969, Section 13].

Tensor products of Radon measures and of positive operators

Let $(X_i)_{1 \leq i \leq p}$ be a finite family of locally compact Hausdorff spaces and consider the product space $\prod_{i=1}^{p} X_i$ endowed with the product topology.

The product space is a locally compact Hausdorff space and is compact if each X_i is compact.

For each $j = 1, \ldots, p$ we shall denote by $pr_j: \prod_{i=1}^{p} X_i \to X_j$ the *j-th projection* which is defined by

$$pr_j(x) := x_j \quad \text{for every } x = (x_i)_{1 \le i \le p} \in \prod_{i=1}^{p} X_i. \tag{1.2.49}$$

By a common abuse of notation, if $X \subset \prod_{i=1}^{p} X_i$, the restriction of each pr_j to X will be denoted by pr_j as well.

If finitely many functions $f_i: X_i \to \mathbb{R}$, $1 \le i \le p$, are given, we shall denote by $\bigotimes_{i=1}^{p} f_i: \prod_{i=1}^{p} X_i \to \mathbb{R}$ the new function defined by

$$\left(\bigotimes_{i=1}^{p} f_i \right)(x) := \prod_{i=1}^{p} f_i(x_i) \quad \text{for every } x = (x_i)_{1 \le i \le p} \in \prod_{i=1}^{p} X_i. \tag{1.2.50}$$

Thus we have

$$\bigotimes_{i=1}^{p} f_i = \prod_{i=1}^{p} f_i \circ pr_i. \tag{1.2.51}$$

Furthermore, if $j = 1, \ldots, p$ and $f_j: X_j \to \mathbb{R}$, then

$$f_j \circ pr_j = \bigotimes_{i=1}^{p} f_{i,j}, \tag{1.2.52}$$

where $f_{i,j} := \mathbf{1}$ if $i \ne j$, and $f_{i,j} := f_j$ if $i = j$.

Clearly, if for each $i = 1, \ldots, p$, $f_i \in \mathscr{K}(X_i)$, then $\bigotimes_{i=1}^{p} f_i \in \mathscr{K}\left(\prod_{i=1}^{p} X_i \right)$, because $\text{Supp}\left(\bigotimes_{i=1}^{p} f_i \right) = \prod_{i=1}^{p} \text{Supp}(f_i)$.

We shall denote by $\bigotimes_{i=1}^{p} \mathscr{K}(X_i)$ the linear subspace generated by $\left\{ \bigotimes_{i=1}^{p} f_i \mid f_i \in \mathscr{K}(X_i), i = 1, \ldots, p \right\}$. In fact, we have that $\bigotimes_{i=1}^{p} \mathscr{K}(X_i)$ is dense in $\mathscr{K}\left(\prod_{i=1}^{p} X_i \right)$ with respect to the sup-norm (see Choquet [1969, Lemma 13.8]).

Now, for every $i = 1, \ldots, p$ fix $\mu_i \in \mathscr{M}(X_i)$. Then there exists a uniquely determined Radon measure υ on $\prod_{i=1}^{p} X_i$ such that for every $(f_i)_{1 \le i \le p} \in \prod_{i=1}^{p} \mathscr{K}(X_i)$

$$\upsilon\left(\bigotimes_{i=1}^{p} f_i \right) = \prod_{i=1}^{p} \mu_i(f_i). \tag{1.2.53}$$

Such a measure is called the *tensor product* of the family $(\mu_i)_{1 \leq i \leq p}$ and is denoted by $\bigotimes_{i=1}^{p} \mu_i$ or $\mu_1 \otimes \cdots \otimes \mu_p$. Thus, if $f_i \in \mathcal{K}(X_i)$, $1 \leq i \leq p$, then

$$\left(\bigotimes_{i=1}^{p} \mu_i \right) \left(\bigotimes_{i=1}^{p} f_i \right) = \prod_{i=1}^{p} \mu_i(f_i). \tag{1.2.54}$$

Note that, if $f \in \mathcal{K}\left(\prod_{i=1}^{p} X_i \right)$, then for every $j = 1, \ldots, p - 1$ and $(x_1, \ldots, x_j) \in \prod_{i=1}^{j} X_i$ the function

$$x_j \mapsto \int \cdots \left(\int f(x_1, \ldots, x_j, x_{j+1}, \ldots, x_p) \, d\mu_p(x_p) \right) \ldots d\mu_{j+1}(x_{j+1})$$

from X_j into $\mathbb{R}$ is continuous and has compact support. As a matter of fact it is possible to show that

$$\left(\bigotimes_{i=1}^{p} \mu_i \right)(f) = \int \left(\cdots \left(\int f(x_1, \ldots, x_p) \, d\mu_p(x_p) \right) \cdots \right) d\mu_1(x_1). \tag{1.2.55}$$

We shall also denote the right-hand side of (1.2.55) by

$$\int \cdots \int f(x_1, \ldots, x_p) \, d\mu_1(x_1) \ldots d\mu_p(x_p). \tag{1.2.56}$$

In fact, from the above formula (1.2.55) one can also deduce the *Fubini's theorem* for functions $f \colon \prod_{i=1}^{p} X_i \to \mathbb{R}$ which are $\left(\bigotimes_{i=1}^{p} \mu_i \right)$-integrable, namely

$$\int f \, d\left(\bigotimes_{i=1}^{p} \mu_i \right) = \int \cdots \int f(x_1, \ldots, x_p) \, d\mu_1(x_1) \ldots d\mu_p(x_p). \tag{1.2.57}$$

Note also that, if every μ_i is positive, then $\bigotimes_{i=1}^{p} \mu_i$ is positive. Furthermore, if each X_i is compact and $\mu_i(\mathbf{1}) = 1$, then, because of (1.2.51) and (1.2.54), for each $j = 1, \ldots, p$ we obtain

$$pr_j \left(\bigotimes_{i=1}^{p} \mu_i \right) = \mu_j. \tag{1.2.58}$$

Some of the main properties of tensor products of measures are listed below (for a proof see Bourbaki [1969, Chapter III, Section 4] or Choquet [1969, Section 13]).

1.2.5 Proposition. *Let $(X_i)_{1 \leq i \leq p}$ be a finite family of locally compact Hausdorff spaces and for every $i = 1, \ldots, p$ fix $\mu_i \in \mathcal{M}(X_i)$.*
Then the following statement hold:

(1) $\mathrm{Supp}\left(\bigotimes\limits_{i=1}^{p} \mu_i \right) = \prod\limits_{i=1}^{p} \mathrm{Supp}(\mu_i).$

(2) *If each μ_i is bounded then $\bigotimes\limits_{i=1}^{p} \mu_i$ is bounded and* $\left\| \bigotimes\limits_{i=1}^{p} \mu_i \right\| = \prod\limits_{i=1}^{p} \| \mu_i \|.$

(3) *(Commutativity property) If $\sigma: \{1, \ldots, p\} \to \{1, \ldots, p\}$ is a permutation, then*
$$\bigotimes\limits_{i=1}^{p} \mu_{\sigma(i)} = \bigotimes\limits_{i=1}^{p} \mu_i.$$

(4) *(Associativity property) If $(I_k)_{1 \leq k \leq n}$ is a partition of $\{1, \ldots, p\}$, then*
$$\bigotimes\limits_{k=1}^{n} \left(\bigotimes\limits_{i \in I_k} \mu_i \right) = \bigotimes\limits_{i=1}^{p} \mu_i.$$

Note that, in general, the mapping $(\mu_i)_{1 \leq i \leq p} \mapsto \bigotimes\limits_{i=1}^{p} \mu_i$ from $\prod\limits_{i=1}^{p} \mathcal{M}(X_i)$ into $\mathcal{M}\left(\prod\limits_{i=1}^{p} X_i \right)$ is *not continuous* with respect to the product topology (of the vague topologies) and the vague topology on $\mathcal{M}\left(\prod\limits_{i=1}^{p} X_i \right)$.

However, we have the following useful result (see Choquet [1969, Proposition 13.12]).

1.2.6 Proposition. *Given a finite family $(X_i)_{1 \leq i \leq p}$ of locally compact Hausdorff spaces, then the mapping $(\mu_i)_{1 \leq i \leq p} \mapsto \bigotimes\limits_{i=1}^{p} \mu_i$ from $\prod\limits_{i=1}^{p} \mathcal{M}^+(X_i)$ into $\mathcal{M}^+\left(\prod\limits_{i=1}^{p} X_i \right)$ is continuous with respect to the product topology (of the vague topologies) and the vague topology on $\mathcal{M}^+\left(\prod\limits_{i=1}^{p} X_i \right)$.*

By using tensor products of measures we can also construct positive linear operators on spaces of continuous functions on the product space.

First, note that to every positive linear operator defined on a suitable function space, it is possible to associate a family of positive Radon measures.

More precisely, given two locally compact Hausdorff spaces X and Y, consider two function spaces E and F on X and Y, respectively (i.e., E and F are vector subspaces of continuous functions on X and Y, respectively). Furthermore suppose that $\mathcal{K}(X) \subset E$.

Given a positive linear operator $T: E \to F$, for every $y \in Y$ we consider the linear form $\mu_y^T: \mathcal{K}(X) \to \mathbb{R}$ defined by

$$\mu_y^T(f) := Tf(y) \quad \text{for every } f \in \mathcal{K}(X). \tag{1.2.59}$$

Then μ_y^T is positive and, hence, it is a Radon measure on X.

More generally, if $\mu: F \to \mathbb{R}$ is a positive linear form, we shall denote by $T(\mu)$ the positive Radon measure on X defined by

$$T(\mu)(f) := \mu(T(f)) \quad \text{for every } f \in \mathcal{K}(X). \tag{1.2.60}$$

Thus $\mu_y^T = T(\varepsilon_{y|F})$.

Now, let $(X_i)_{1 \leq i \leq p}$ and $(Y_i)_{1 \leq i \leq p}$ be two finite families of locally compact Hausdorff spaces. For every $i = 1, \ldots, p$ let us consider a positive linear operator $T_i: \mathcal{K}(X_i) \to \mathcal{C}(Y_i)$.

Then we define a linear operator $T: \mathcal{K}\left(\prod_{i=1}^{p} X_i\right) \to \mathcal{C}\left(\prod_{i=1}^{p} Y_i\right)$ by

$$Tf(y) := \left(\bigotimes_{i=1}^{p} \mu_{y_i}^{T_i}\right)(f) = \int \cdots \int f(x_1, \ldots, x_p) \, d\mu_{y_1}^{T_1}(x_1) \ldots d\mu_{y_p}^{T_p}(x_p) \tag{1.2.61}$$

for every $f \in \mathcal{K}\left(\prod_{i=1}^{p} X_i\right)$ and $y = (y_i)_{1 \leq i \leq p} \in \prod_{i=1}^{p} Y_i$, where the measures $\mu_{y_i}^{T_i}$ are defined as in (1.2.59). (Note that $T(f)$ is continuous by virtue of Proposition 1.2.6).

The operator T is positive and is denoted by $\bigotimes_{i=1}^{p} T_i$. It is also called the *tensor product of the family* $(T_i)_{1 \leq i \leq p}$.

Thus $\bigotimes_{i=1}^{p} T_i: \mathcal{K}\left(\prod_{i=1}^{p} X_i\right) \to \mathcal{C}\left(\prod_{i=1}^{p} Y_i\right)$ and for every $f \in \mathcal{K}\left(\prod_{i=1}^{p} X_i\right)$ and $y = (y_i)_{1 \leq i \leq p} \in \prod_{i=1}^{p} Y_i$,

$$\left(\bigotimes_{i=1}^{p} T_i\right)(f)(y) = \left(\bigotimes_{i=1}^{p} \mu_{y_i}^{T_i}\right)(f) = \int \cdots \int f(x_1, \ldots, x_p) \, d\mu_{y_1}^{T_1}(x_1) \ldots d\mu_{y_p}^{T_p}(x_p). \tag{1.2.62}$$

In particular, taking (1.2.59) and (1.2.54) into account, for every $(f_i)_{1 \leq i \leq p} \in \prod_{i=1}^{p} \mathcal{K}(X_i)$ we have

$$\left(\bigotimes_{i=1}^{p} T_i\right)\left(\bigotimes_{i=1}^{p} f_i\right) = \bigotimes_{i=1}^{p} T_i(f_i). \tag{1.2.63}$$

*1.3 Some basic principles of probability theory

Here we survey some classical material on probability theory that will be used mainly in Section 5.2. For more details and proofs see for instance Bauer [1981, part II] and Feller [1957], [1966].

Random variables

Consider a *probability space* $(\Omega, \mathscr{F}, P)$, i.e., $\mathscr{F}$ is a σ-algebra in the set Ω and P is a measure on $\mathscr{F}$ such that $P(\Omega) = 1$, and let $(\Omega', \mathscr{F}')$ be a measurable space. A *random variable* from Ω into Ω' is a mapping $Z: \Omega \to \Omega'$ which is *measurable* with respect to $\mathscr{F}$ and $\mathscr{F}'$ (i.e., $Z^{-1}(B) \in \mathscr{F}$ for every $B \in \mathscr{F}'$).

When $\Omega' = \mathbb{R}$ and $\mathscr{F}' = \mathfrak{B}(\mathbb{R})$ we shall speak of real random variables on Ω. The set of all real random variables will be denoted by $M(\Omega)$.

If $Z: \Omega \to \Omega'$ is a random variable from Ω into Ω', the image measure $Z(P)$ is called the *distribution* of Z (with respect to P) or the *probability law* of Z and is denoted by P_Z.

Thus P_Z is a probability measure on $\mathscr{F}'$ and for every $B \in \mathscr{F}'$

$$P_Z(B) := P\{Z^{-1}(B)\}. \tag{1.3.1}$$

The subset $Z^{-1}(B)$ and the number $P\{Z^{-1}(B)\}$ are often denoted by $\{Z \in B\}$ and $P\{Z \in B\}$, respectively.

If μ is a probability measure on $\mathscr{F}'$ and $P_Z = \mu$, we also say that Z is *distributed according to* μ.

If $Z \in M(\Omega)$, then $P_Z \in \mathscr{M}_1^+(\mathbb{R})$. More generally, if Ω' is a locally compact Hausdorff space which is countable at infinity and $\mathscr{F}' = \mathfrak{B}_0(\Omega')$, then $P_Z \in \mathscr{M}_1^+(\Omega')$ for every random variable $Z: \Omega \to \Omega'$.

In this case, if Ω' has a countable base, then $\mathrm{Supp}(P_Z) \subset \overline{X(\Omega)}$ (we recall that if $\mu \in \mathscr{M}^+(\Omega')$, the *support* $\mathrm{Supp}(\mu)$ of μ is the complement of the largest open subset of Ω' of measure zero with respect to μ).

If $Z: \Omega \to \Omega'$ is a random variable and if $f: \Omega' \to \tilde{\mathbb{R}}$ is positive and $\mathscr{F}'$-measurable, then

$$\int_{\Omega'} f \, dP_Z = \int_{\Omega} f \circ Z \, dP. \tag{1.3.2}$$

Furthermore a $\mathscr{F}'$-measurable function $f: \Omega' \to \tilde{\mathbb{R}}$ is P_Z-integrable if and only if $f \circ Z$ is P-integrable. In this case (1.3.2) holds as well.

If a real random variable $Z: \Omega \to \mathbb{R}$ is positive or P-integrable, then we set

$$E(Z) := \int_{\Omega} Z \, dP = \int_{-\infty}^{+\infty} x \, dP_Z(x) \tag{1.3.3}$$

and we call $E(Z)$ the *expected value* of Z.

If the random variable $Z: \Omega \to \mathbb{R}$ is P-integrable, we call

$$\mathrm{Var}(Z) := E((Z - E(Z))^2) \in \tilde{\mathbb{R}}_+, \tag{1.3.4}$$

and

$$\sigma(Z) := \sqrt{\mathrm{Var}(Z)}, \tag{1.3.5}$$

the *variance* and the *standard deviation* (or *dispersion*) of Z, respectively.

A real random variable Z is P-integrable and has finite variance if and only if it is square-integrable. If this is the case, then

$$\mathrm{Var}(Z) = E(Z^2) - E(Z)^2 = \int_{-\infty}^{+\infty} x^2 \, dP_Z(x) - \left(\int_{-\infty}^{+\infty} x \, dP_Z(x) \right)^2. \tag{1.3.6}$$

If $Z: \Omega \to \mathbb{R}^p$ is a random variable with components $Z_1, \ldots, Z_p$ (i.e., $Z(\omega) = (Z_1(\omega), \ldots, Z_p(\omega))$ for every $\omega \in \Omega$), we set

$$E(Z) := (E(Z_1), \ldots, E(Z_p)) \in \mathbb{R}^p, \tag{1.3.7}$$

provided every Z_i is integrable. Furthermore, if every Z_i is square integrable, we also set

$$\mathrm{Var}(Z) := \sum_{i=1}^{p} \mathrm{Var}(Z_i) = E(\|Z\|^2) - \|E(Z)\|^2. \tag{1.3.8}$$

Let $Z: \Omega \to \mathbb{R}$ be a real random variable such that $P\{Z \in \mathbb{N}_0\} = 1$. For every $n \in \mathbb{N}_0$ we set $\alpha_n := P\{Z = n\} \left(\text{hence } \sum_{n=0}^{\infty} \alpha_n = 1 \right)$.

The *probability generating function* of Z is the function $g_Z: [-1, 1] \to \mathbb{R}$ defined by

$$g_Z(t) := \sum_{n=0}^{\infty} \alpha_n t^n = E(t^Z) = \int_{-\infty}^{+\infty} t^x \, dP_Z(x) \quad \text{for every } t \in [-1, 1]. \tag{1.3.9}$$

In this case we have

$$E(Z) = \sum_{n=1}^{\infty} n\alpha_n = g_Z'(1) \qquad (1.3.10)$$

and

$$\text{Var}(Z) = g_Z''(1) + g_Z'(1) - g_Z'(1)^2. \qquad (1.3.11)$$

If $\text{Var}(Z) < +\infty$, then

$$\text{Var}(Z) = \sum_{n=1}^{\infty} n^2\alpha_n - \left(\sum_{n=1}^{\infty} n\alpha_n\right)^2. \qquad (1.3.12)$$

More generally, if for every $p \in \mathbb{N}$, we denote by $m_p(Z)$ the *p-th factorial moment* of Z, i.e.,

$$m_p(Z) := E(Z(Z-1)\ldots(Z-p+1)) = \int_{-\infty}^{+\infty} x(x-1)\ldots(x-p+1)\,dP_Z(x),$$
$$(1.3.13)$$

then

$$m_p(Z) = g_Z^{(p)}(1). \qquad (1.3.14)$$

A random variable $Z: \Omega \to \mathbb{R}^p$ is said to be *discretely distributed* if there exists a sequence $(a_n)_{n \in \mathbb{N}_0}$ in $\mathbb{R}^p$ and a sequence $(\alpha_n)_{n \in \mathbb{N}_0}$ in $\mathbb{R}_+$ satisfying $\sum_{n=0}^{\infty} \alpha_n = 1$, such that

$$P_Z = \sum_{n=0}^{\infty} \alpha_n \varepsilon_{a_n}, \qquad (1.3.15)$$

where, for every $n \in \mathbb{N}_0$, $\varepsilon_{a_n} \in \mathcal{M}_1^+(\mathbb{R}^p)$ denotes the *unit mass* at a_n, i.e., for every $B \in \mathfrak{B}(\mathbb{R}^p)$

$$\varepsilon_{a_n}(B) := \begin{cases} 1, & \text{if } a_n \in B, \\ 0, & \text{if } a_n \notin B. \end{cases} \qquad (1.3.16)$$

In particular we have that $P\{Z = a_n\} = \alpha_n$ for every $n \in \mathbb{N}_0$. Furthermore a measurable function $f: \mathbb{R}^p \to \tilde{\mathbb{R}}$ is P_Z-integrable if and only if $\sum_{n=0}^{\infty} \alpha_n |f(a_n)| < +\infty$. In this case

$$\int_{\mathbb{R}^p} f\,dP_Z = \sum_{n=0}^{\infty} \alpha_n f(a_n). \qquad (1.3.17)$$

The same formula holds if f is positive (not necessarily integrable).

In the case $p = 1$ we have that Z is integrable (square-integrable, respectively) if and only if $\sum_{n=0}^{\infty} \alpha_n |a_n| < +\infty \left(\sum_{n=0}^{\infty} \alpha_n a_n^2 < +\infty, \text{ respectively} \right)$.

Furthermore

$$E(Z) = \sum_{n=0}^{\infty} \alpha_n a_n \qquad (1.3.18)$$

and

$$\text{Var}(Z) = \sum_{n=0}^{\infty} \alpha_n a_n^2 - \left(\sum_{n=0}^{\infty} \alpha_n a_n \right)^2. \qquad (1.3.19)$$

Important examples of discretely distributed real random variables are the *binomial* or *Bernoulli* random variables. A real random variable Z is said to be a Bernoulli random variable with parameters n and p ($n \geq 1$, $0 \leq p \leq 1$) if it is distributed according to the *binomial* or *Bernoulli* distribution

$$\beta_{n,p} := \sum_{k=0}^{n} \binom{n}{k} p^k (1-p)^{n-k} \varepsilon_k. \qquad (1.3.20)$$

In this case

$$E(Z) = np \quad \text{and} \quad \text{Var}(Z) = np(1-p). \qquad (1.3.21)$$

Furthermore for every $t \in [-1, 1]$

$$g_Z(t) = (pt + (1-p))^n. \qquad (1.3.22)$$

Another class of important discretely distributed real random variables are the *Poisson random variables*. A Poisson random variable Z with parameter $\alpha > 0$ is a real random variable which is distributed according to the *Poisson distribution* with parameter α

$$\pi_\alpha := \sum_{k=0}^{\infty} \exp(-\alpha) \frac{\alpha^k}{k!} \varepsilon_k. \qquad (1.3.23)$$

We have

$$E(Z) = \text{Var}(Z) = \alpha \qquad (1.3.24)$$

and

$$g_Z(t) = \exp(\alpha(t-1)) \quad \text{for every } t \in [-1, 1].$$ (1.3.25)

As an example of discretely distributed random variables on $\mathbb{R}^p$ with $p \geq 2$, consider finitely many positive real numbers $r_1, \ldots, r_p$ satisfying $\sum_{i=1}^{p} r_i \leq 1$ and fix $n \in \mathbb{N}$.

Every random variable Z on $\mathbb{R}^p$ distributed according to

$$\sum_{\substack{h_1, \ldots, h_p \geq 0 \\ h_1 + \cdots + h_p \leq n}} \frac{n!}{h_1! \ldots h_p! (n - h_1 - \cdots - h_p)!}$$

$$\times r_1^{h_1} \ldots r_p^{h_p} (1 - r_1 - \cdots - r_p)^{n - h_1 - \cdots - h_p} \varepsilon_{(h_1, \ldots, h_p)}$$ (1.3.26)

is called a *multinomial random variable* of order $p + 1$ with parameters $n, r_1, \ldots, r_p$.

In this case $E(Z) = (nr_1, \ldots, nr_p)$ and $\mathrm{Var}(Z) = n \sum_{i=1}^{p} r_i(1 - r_i)$.

A random variable $Z: \Omega \to \mathbb{R}^p$ is said to be *Lebesgue-continuous* if P_Z is λ_p-continuous, i.e., $P\{Z \in B\} = 0$ for every $B \in \mathfrak{B}(\mathbb{R}^p)$ such that $\lambda_p(B) = 0$. Here λ_p denotes the *Lebesgue-Borel measure* in $\mathbb{R}^p$.

By the Radon-Nikodym theorem (see, e.g., Bauer [1981, Theorem 2.9.10]), if Z is Lebesgue continuous, then there exists a Borel-measurable positive function $g: \mathbb{R}^p \to \tilde{\mathbb{R}}$ satisfying $\int_{\mathbb{R}^p} g(x)\,dx = 1$ such that $P_Z = g \cdot \lambda_p$, i.e., for every $B \in \mathfrak{B}(\mathbb{R}^p)$

$$P\{Z \in B\} = \int_B g(x)\,dx.$$ (1.3.27)

The function g is also called the *probability density* of Z.

A measurable function $f: \mathbb{R}^p \to \tilde{\mathbb{R}}$ is P_Z-integrable if and only if fg is λ_p-integrable. In this case

$$\int_{\mathbb{R}^p} f\,dP_Z = \int_{\mathbb{R}^p} f(x)g(x)\,dx.$$ (1.3.28)

This formula also holds provided f is positive and measurable.

In particular, when $p = 1$, we obtain that Z is integrable (square-integrable, respectively) if and only if $\int_{-\infty}^{+\infty} |x|g(x)\,dx < +\infty$ ($\int_{-\infty}^{+\infty} x^2 g(x)\,dx < +\infty$, respectively). Then

$$E(Z) = \int_{-\infty}^{+\infty} xg(x)\,dx$$ (1.3.29)

and

$$\mathrm{Var}(Z) = \int_{-\infty}^{+\infty} x^2 g(x)\, dx - \left(\int_{-\infty}^{+\infty} x g(x)\, dx \right)^2. \qquad (1.3.30)$$

The most important Lebesgue-continuous random variables are the normal ones. We recall that a *normal* or *Gaussian random variable* Z with parameters α and σ^2 ($\alpha \in \mathbb{R}$, $\sigma > 0$) is a real Lebesgue-continuous random variable having as probability density the function

$$g_{\alpha,\sigma^2}(t) := (2\pi\sigma^2)^{-1/2} \exp\left(-\frac{(t-\alpha)^2}{2\sigma^2} \right) \quad (t \in \mathbb{R}). \qquad (1.3.31)$$

In this case

$$E(Z) = \alpha \quad \text{and} \quad \mathrm{Var}(Z) = \sigma^2. \qquad (1.3.32)$$

Some properties of real random variables can be also described in terms of their distribution functions. Actually, given a real random variable Z, the *distribution function* of Z is the function $F_Z \colon \mathbb{R} \to \mathbb{R}$ defined by

$$F_Z(x) := P\{Z < x\} \quad \text{for every } x \in \mathbb{R}. \qquad (1.3.33)$$

The function F_Z is increasing and *left-continuous* $\Big($ i.e., $\lim\limits_{x \to a^-} F_Z(x) = F_Z(a)$ for every $a \in \mathbb{R}\Big)$ and satisfies the conditions $\lim\limits_{x \to -\infty} F_Z(x) = 0$ and $\lim\limits_{x \to +\infty} F_Z(x) = 1$.

Furthermore $P\{a \le Z < b\} = F_Z(b) - F_Z(a)$ provided $a < b$, and F_Z is continuous in a point $a \in \mathbb{R}$ if and only if $P\{Z = a\} = 0$.

We also recall that, if $F \colon \mathbb{R} \to \mathbb{R}$ is an increasing left-continuous function converging to 0 as $x \to -\infty$ and to 1 as $x \to +\infty$, then there exists a (in general, non-unique) random variable Z such that $F_Z = F$.

A crucial notion in probability theory is that of independence of random variables.

Consider an arbitrary family $(Z_i)_{i \in I}$ of real random variables defined on the same probability space $(\Omega, \mathscr{F}, P)$.

The family $(Z_i)_{i \in I}$ is said to be *independent* if for every finite subset $J \subset I$ and for every finite family $(B_i)_{i \in I}$ in $\mathfrak{B}(\mathbb{R})$ we have

$$P\left(\bigcap_{i \in J} \{Z_i \in B_i\} \right) = \prod_{i \in J} P\{Z_i \in B_i\}. \qquad (1.3.34)$$

If this is the case, then for every finite subset $J \subset I$ the following statements hold:

(1) If we consider the random variable $\bigotimes_{i \in J} Z_i \colon \Omega \to \mathbb{R}^J$ defined as $\left(\bigotimes_{i \in J} Z_i\right)(\omega) = (Z_i(\omega))_{i \in J}$ for every $\omega \in \Omega$ and if $\bigotimes_{i \in J} P_{Z_i}$ denotes the tensorial product of the family $(P_{Z_i})_{i \in J}$, then

$$P_{\bigotimes_{i \in J} Z_i} = \bigotimes_{i \in J} P_{Z_i}. \tag{1.3.35}$$

(2) If either all Z_i are positive or all Z_i are integrable, then

$$E\left(\prod_{i \in J} Z_i\right) = \prod_{i \in J} E(Z_i). \tag{1.3.36}$$

(3) If all Z_i are integrable, then

$$\mathrm{Var}\left(\sum_{i \in J} Z_i\right) = \sum_{i \in J} \mathrm{Var}(Z_i). \tag{1.3.37}$$

(4) If $\mathop{\bigstar}\limits_{i \in J} P_{Z_i}$ denotes the convolution product of the family $(P_{Z_i})_{i \in J}$, then

$$P_{\sum_{i \in J} Z_i} = \mathop{\bigstar}\limits_{i \in J} P_{Z_i}. \tag{1.3.38}$$

(5) If we consider the probability generating function $g_{\sum_{i \in J} Z_i} \colon [-1, 1] \to \mathbb{R}$ of $\sum_{i \in J} Z_i$, then

$$g_{\sum_{i \in J} Z_i} = \prod_{i \in J} g_{Z_i}. \tag{1.3.39}$$

From (1.3.38) it easily follows that if Z_1 and Z_2 are two independent binomial (Poisson, normal, respectively) random variables with parameters n, p and m, p (α and β, α, σ^2 and β, τ^2, respectively), then $Z_1 + Z_2$ is a binomial (Poisson, normal, respectively) random variable with parameters $n + m$, p ($\alpha + \beta$, $\alpha + \beta$ and $\sigma^2 + \tau^2$, respectively).

Similarly, one can prove that, if $Z_1, \ldots, Z_n$ are independent normal random variables with the same parameters 0 and σ^2, then the random variable $\chi^2_{n,\sigma^2} := \sum_{i=1}^{n} Z_i^2$ is Lebesgue-continuous and its probability density is the function

$$g_{n,\sigma^2}(t) := \begin{cases} \dfrac{t^{n/2-1} \exp(-t/2\sigma^2)}{(2\sigma^2)^{n/2} \Gamma(n/2)}, & \text{if } t > 0, \\[2mm] 0, & \text{if } t \le 0, \end{cases} \tag{1.3.40}$$

where $\Gamma(t) := \int_0^{+\infty} x^{t-1} \exp(-x)\, dx$ $(t > 0)$ denotes the *gamma function*.

The random variable χ^2_{n,σ^2} is also called a *chi-squared random variable with n degrees of freedom and parameter σ^2*.

It is well known that

$$E(\chi^2_{n,\sigma^2}) = n\sigma^2 \quad \text{and} \quad \text{Var}(\chi^2_{n,\sigma^2}) = 2n\sigma^4. \tag{1.3.41}$$

Finally, by using the Kolmogorov's theorem about the existence of the infinite tensorial product of probability spaces, it is possible to show that, if $((\Omega_i, \mathscr{F}_i, P_i))_{i \in I}$ is an arbitrary family of probability spaces, then there exist a probability space $(\Omega, \mathscr{F}, P)$ and an independent family $(Z_i)_{i \in I}$ of random variables, $Z_i \colon \Omega \to \Omega_i$ ($i \in I$), such that $P_{Z_i} = P_i$ for every $i \in I$ (for a proof see Bauer [1981, Corollary 5.4.5]).

Convergence of random variables

Consider again a probability space $(\Omega, \mathscr{F}, P)$. On the space $M(\Omega)$ of all real random variables on Ω we shall consider three different concepts of convergence that can be derived from suitable topologies.

A sequence $(Z_n)_{n \in \mathbb{N}}$ of real random variables is said to be *P-almost surely convergent* to a random variable Z if there exists a *P*-negligible set $N \subset \Omega$ (i.e., $N \in \mathscr{F}$ and $P(N) = 0$) such that

$$\lim_{n \to \infty} Z_n(\omega) = Z(\omega) \quad \text{for every } \omega \in \Omega \setminus N. \tag{1.3.42}$$

We say that $(Z_n)_{n \in \mathbb{N}}$ converges *P-stochastically* (or *stochastically*) to Z if

$$\lim_{n \to \infty} P\{|Z_n - Z| \geq \varepsilon\} = 0 \quad \text{for every } \varepsilon > 0. \tag{1.3.43}$$

Finally we say that $(Z_n)_{n \in \mathbb{N}}$ *converges in distribution* to Z if

$$\lim_{n \to \infty} P_{Z_n} = P_Z \quad \text{weakly.} \tag{1.3.44}$$

The logical relations between these concepts of convergence are as follows:

$$(P\text{-almost surely convergence}) \Rightarrow (P\text{-stochastic convergence}) \Rightarrow$$
$$\Rightarrow (\text{convergence in distribution}).$$

In addition, if Z is *P*-almost surely constant and if $(Z_n)_{n \in \mathbb{N}}$ converges in distribution to Z, then $(Z_n)_{n \in \mathbb{N}}$ converges *P*-stochastically to Z.

Note also that $(Z_n)_{n \in \mathbb{N}}$ converges in distribution to Z if and only if $\lim_{n \to \infty} F_{Z_n}(x) = F_Z(x)$ for every point $x \in \mathbb{R}$ at which F_Z is continuous, or if the sequence of the Fourier transforms $(\hat{P}_{Z_n})_{n \in \mathbb{N}}$ converges pointwise to $\hat{P}_Z$ (*continuity theorem of* P. Lévy).

By Scheffé's theorem (see Section 1.2), if every Z_n and Z are Lebesgue-continuous with probability densities g_n and g respectively and if $g_n \to g$ P-almost everywhere, then $Z_n \to Z$ in distribution.

For the same reasons, if every Z_n and Z are discretely distributed and have a common support $\{a_k | k \in \mathbb{N}_0\} \subset \mathbb{R}$, i.e.,

$$P_{Z_n} = \sum_{k=0}^{\infty} \alpha_{n,k} \varepsilon_{a_k} \ (n \in \mathbb{N}) \quad \text{and} \quad P_Z = \sum_{k=0}^{\infty} \alpha_k \varepsilon_{a_k},$$

then $Z_n \to Z$ in distribution provided $\lim_{n \to \infty} \alpha_{n,k} = \alpha_k$ for each $k \in \mathbb{N}_0$.

A sequence $(Z_n)_{n \in \mathbb{N}}$ of integrable real random variables is said to obey the *strong law of large numbers* (the *weak law of large numbers*, respectively) if

$$\lim_{n \to \infty} \frac{1}{n} \sum_{i=1}^{n} (Z_i - E(Z_i)) = 0 \quad P\text{-almost surely} \tag{1.3.45}$$

(respectively,

$$\lim_{n \to \infty} \frac{1}{n} \sum_{i=1}^{n} (Z_i - E(Z_i)) = 0 \quad \text{stochastically}). \tag{1.3.46}$$

Two celebrated *theorems of Kolmogorov* show that an independent sequence $(Z_n)_{n \in \mathbb{N}}$ of real integrable random variables obeys the strong law of large numbers if

$$\sum_{n=1}^{\infty} \frac{\text{Var}(Z_n)}{n^2} < +\infty, \tag{1.3.47}$$

or alternatively, if the Z_n's are *identically distributed* (i.e., $P_{Z_n} = P_{Z_m}$ for every n, $m \in \mathbb{N}$).

On the other hand, a sequence $(Z_n)_{n \in \mathbb{N}}$ of real integrable *pairwise uncorrelated* random variables (i.e., $E(Z_n Z_m) = E(Z_n)E(Z_m)$ for every $n \neq m$) obeys the weak law of large numbers if

$$\lim_{n \to \infty} \frac{1}{n^2} \sum_{i=1}^{n} \text{Var}(Z_i) = 0, \tag{1.3.48}$$

(*theorem of Markov*) or alternatively, if the Z_n's are identically distributed (*theorem of Khinchin*).

For more details see Bauer [1981, Chapter 6 and Sections 7.7 and 8.2].

1.4 Selected topics on locally convex spaces

In this section we survey some classical results on locally convex vector spaces such as various Hahn-Banach extension and separation theorems, the Krein-Milman theorem and Milman's converse theorem.

As a reference for these topics see for instance Choquet [1969] and Horváth [1966].

Let E be a topological vector space over the field $\mathbb{K}$ of real or complex numbers. A point $x \in E$ is said to be a *convex combination* of n given points $x_1, \ldots,$ $x_n \in E$ if $x = \sum_{i=1}^{n} \lambda_i x_i$ for some $\lambda_1, \ldots, \lambda_n \geq 0$ such that $\sum_{i=1}^{n} \lambda_i = 1$.

A subset X of E is said to be *convex* if $\lambda x + (1 - \lambda)y \in X$ for every $x, y \in X$ and $\lambda \in [0,1]$ or, equivalently, if $\sum_{i=1}^{n} \lambda_i x_i \in X$ for every finite family $(x_i)_{1 \leq i \leq n}$ of elements of X and for every $\lambda_1, \ldots, \lambda_n \geq 0$ such that $\sum_{i=1}^{n} \lambda_i = 1$.

The *convex hull* of a subset X of E is, by definition, the smallest convex subset of E containing X and is denoted by $\mathrm{co}(X)$. In fact, we have

$$\mathrm{co}(X) = \left\{ \sum_{i=1}^{n} \lambda_i x_i \,\middle|\, x_i \in X, \lambda_i \geq 0, i = 1, \ldots, n \text{ and } \sum_{i=1}^{n} \lambda_i = 1 \right\}. \qquad (1.4.1)$$

In general, if X is compact, $\mathrm{co}(X)$ is not compact. If E is Hausdorff and complete, then the closure $\overline{\mathrm{co}}(X)$ of $\mathrm{co}(X)$ is compact.

A topological vector space is said to be *locally convex* if the origin possesses a fundamental system of convex neighborhoods.

In fact, in a locally convex vector space the origin possesses a fundamental system of balanced, closed and convex neighborhoods (we recall that a subset X of E is said to be *balanced* if $\lambda x \in X$ for each $x \in X$ and $\lambda \in \mathbb{K}, |\lambda| \leq 1$).

Given a topological vector space E, we shall denote by E' the space of all continuous linear forms on E and call it the *dual space* of E.

On E' we shall often consider the *weak*-topology* $\sigma(E', E)$ which is, by definition, the coarsest topology on E' for which all the linear forms $\psi_x (x \in E)$ are continuous, where

$$\psi_x(\varphi) := \varphi(x) \quad \text{for every } x \in E \text{ and } \varphi \in E'. \qquad (1.4.2)$$

Thus a net $(\varphi_i)_{i \in I}^{\leq}$ in E' converges to an element φ in $(E', \sigma(E', E))$ if and only if $\lim_{\substack{\leq \\ i \in I}} \varphi_i(x) = \varphi(x)$ for every $x \in E$.

The dual space E', endowed with the weak*-topology, is a locally convex Hausdorff space. Furthermore, if for every $x \in E$, $x \neq 0$, there exists $\varphi \in E'$ such that $\varphi(x) \neq 0$, then the dual space of $(E', \sigma(E', E))$ can be identified with E itself, that is, if a linear form $\psi: E' \to \mathbb{K}$ is continuous for the topology $\sigma(E', E)$, then there exists a unique $x \in E$ such that $\psi = \psi_x$.

A useful criterion to decide if a subset $\mathfrak{A}$ of E' is $\sigma(E', E)$-compact is furnished by the Alaoglu-Bourbaki theorem.

We recall that a subset $\mathfrak{A}$ of E' is *equicontinuous* if for every $\varepsilon > 0$ there exists a neighborhood V of the origin of E such that $|\varphi(x)| \leq \varepsilon$ for every $x \in V$ and $\varphi \in \mathfrak{A}$.

More generally, a subset $\mathfrak{A}$ of linear mappings from E into another topological vector space F is said to be *equicontinuous* if for every neighborhood W of the origin of F there exists a neighborhood V of the origin of E such that $u(x) \in W$ for every $x \in V$ and $u \in \mathfrak{A}$.

If E and F are both normed spaces, then $\mathfrak{A}$ is equicontinuous if and only if each $u \in \mathfrak{A}$ is continuous and $\sup\{\|u\| \,|\, u \in \mathfrak{A}\} < +\infty$.

If E is a Banach space and $\mathfrak{A}$ is a subset of continuous linear mappings such that $\sup\{\|u(x)\| \,|\, u \in \mathfrak{A}\} < +\infty$ for every $x \in E$, then $\mathfrak{A}$ is equicontinuous (*uniform boundedness principle*) (for a proof see Choquet [1969, Theorem 7.4]).

1.4.1 Theorem (*Alaoglu-Bourbaki*). *Given a topological vector space E, then every equicontinuous subset of E' is relatively compact for the weak*-topology.*

For a proof of the above result see Hórvath [1966, Chapter 3, Section 4, Theorem 1].

In the sequel we shall present several extension and separation theorems.

The first one concerns the extension of positive linear forms.

To this end we recall that an *ordered vector space* is a real vector space E endowed with a partial ordering $\leq$ satisfying the following properties:

$$x + z \leq y + z \quad \text{for every } x, y, z \in E, x \leq y, \tag{1.4.3}$$

and

$$\lambda x \leq \lambda y \quad \text{for every } x, y \in E, x \leq y \text{ and } \lambda \geq 0. \tag{1.4.4}$$

A *vector lattice* is an ordered vector space E such that for every $x, y \in E$ there exists $\sup\{x, y\}$ in E.

In this case we set $x \vee y := \sup\{x, y\}$, $x \wedge y := -\sup\{-x, -y\} = \inf\{x, y\}$, $|x| := \sup\{-x, x\}$, $x^+ := \sup\{x, 0\}$ and $x^- := \sup\{-x, 0\}$.

A linear form $\varphi: E \to \mathbb{R}$ is said to be positive if $\varphi(x) \geq 0$ for every $x \in E, x \geq 0$.

The proof of the next result can be found in Choquet [1969, Theorem 34.2].

1.4.2 Theorem. *Let E be an ordered vector space and F a subspace of E such that for every $x \in E$ there exists $y \in F$ satisfying $x \leq y$.*

Given a positive linear form $\varphi: F \to \mathbb{R}$, *then for every* $x \in E$ *and* $\alpha \in \mathbb{R}$ *satisfying*

$$\sup_{\substack{y \in F \\ y \leq x}} \varphi(y) \leq \alpha \leq \inf_{\substack{z \in F \\ x \leq z}} \varphi(z),$$

there exists a positive linear form $\tilde{\varphi}$ *on* E *satisfying* $\tilde{\varphi}(x) = \alpha$ *and extending* φ *over* E.

From this result one can derive the classical Hahn-Banach theorem. To state it we recall that, given a vector space E, a mapping $p: E \to \mathbb{R}$ is said to be *sublinear* if

$$p(\lambda x) = \lambda p(x) \quad \text{for every } x \in E \text{ and } \lambda \geq 0, \tag{1.4.5}$$

and

$$p(x + y) \leq p(x) + p(y) \quad \text{for every } x, y \in E. \tag{1.4.6}$$

A *seminorm* $p: E \to \mathbb{R}$ is a sublinear mapping such that $p(-x) = p(x)$ for each $x \in E$.

1.4.3 Theorem (*Hahn-Banach*). *Let* E *be a real vector space and* $p: E \to \mathbb{R}$ *a sublinear mapping. If* F *is a subspace of* E *and* $\varphi: F \to \mathbb{R}$ *is a linear form satisfying* $\varphi \leq p_F$, *then there exists a linear form* $\tilde{\varphi}: E \to \mathbb{R}$ *satisfying* $\tilde{\varphi} \leq p$ *and extending* φ *over* E.

The following corollaries are direct consequences of the above result. For a proof we refer to Choquet [1969, Section 21] or Hórvath [1966, Chapter 3, Section 1].

1.4.4 Corollary. *Given a real locally convex space* E *and a continuous seminorm* $p: E \to \mathbb{R}$, *then for every* $x_0 \in E$ *there exists* $\varphi \in E'$ *such that* $\varphi(x_0) = p(x_0)$.
   Then, if E *is Hausdorff, for every* $x_0 \in E$ *there exists* $\varphi \in E'$ *such that* $\varphi(x_0) \neq 0$.

1.4.5 Corollary. *Let* E *be a real normed space,* F *a subspace of* E *and* $\varphi \in F'$. *Then there exists* $\tilde{\varphi} \in E'$ *such that* $\tilde{\varphi}_F = \varphi$ *and* $\|\tilde{\varphi}\| = \|\varphi\|$.

From Corollary 1.4.5 (that holds also for complex normed spaces) it follows in particular that, if E is a normed space and $x_0 \in E$, $x_0 \neq 0$, then there always exists $\varphi \in E'$ such that $\varphi(x_0) = \|x_0\|$ and $\|\varphi\| = 1$.
   The Hahn-Banach theorem has a wide range of applications.

However, in some cases the domination is required only on a cone and the sublinear mapping may attain the value $+\infty$.

In these settings several extension theorems are available. Here we shall quote a useful one due to Anger and Lembcke [1974].

Let E be a real locally convex Hausdorff space and $P \subset E$ a *convex cone* (i.e., $x + y \in P$ and $\lambda x \in P$ for every $x, y \in P$ and $\lambda > 0$) such that $0 \in P$.

A mapping $p: P \to \mathbb{R} \cup \{+\infty\}$ which satisfies (1.4.5) and (1.4.6) for every x, $y \in P$ and $\lambda \geq 0$ is said to be a *hypolinear mapping* (in (1.4.5) the convention $0 \cdot (+\infty) = 0$ must be observed).

If $p(P) \subset \mathbb{R}$ and $p(x + y) = p(x) + p(y)$ for every $x, y \in P$, p is called *linear*.

Consider two convex cones P and C of E such that $0 \in P \cap C$. Consider a hypolinear mapping $p: P \to \mathbb{R} \cup \{+\infty\}$ and a linear mapping $\varphi: C \to \mathbb{R}$.

For every $x \in E$ we set

$$p_\varphi(x) := \inf\{p(x_1) + \varphi(y_1) - \varphi(y_2)|x_1 \in P, y_1, y_2 \in C, x_1 + y_1 - y_2 = x\} \tag{1.4.7}$$

(with the convention $\inf \varnothing = +\infty$) and

$$\hat{p}_\varphi(x) := \liminf_{y \to x} p_\varphi(y). \tag{1.4.8}$$

Then $\hat{p}_\varphi$ is the largest lower semi-continuous minorant of p_φ.

1.4.6 Theorem (*Anger-Lembcke*). *If* $\hat{p}_\varphi(0) > -\infty$, *then for every* $x \in E$ *and for every* $\alpha \in \;]-\hat{p}_\varphi(-x), \hat{p}_\varphi(x)[$ *there exists* $\tilde{\varphi} \in E'$ *such that* $\tilde{\varphi}_{|C} = \varphi$, $\tilde{\varphi} \leq p$ *on* P *and* $\varphi(x) = \alpha$.

Note that, if $C = \{0\}$, $\varphi = 0$ and $P = E$, then $p_\varphi = p$ so that $\hat{p}_\varphi = p$ provided p is lower semi-continuous on E. Accordingly we have the following result.

1.4.7 Corollary. *Let E be a real locally convex Hausdorff space and $p: E \to \mathbb{R} \cup \{+\infty\}$ a hypolinear lower semi-continuous mapping. Then for every $x \in E$ and $\alpha \in \;]-p(-x), p(x)[$ there exists $\varphi \in E'$ such that $\varphi \leq p$ and $\varphi(x) = \alpha$.*

Other important consequences of the Hahn-Banach theorem concern the separation of convex sets.

Let E be a real vector space. An *affine subspace* G of E is a translate of a subspace of E, i.e., there exist a subspace F of E and $a \in E$ such that $G = F + a := \{x + a|x \in F\}$.

A *hyperplane* in E is a subspace of E of codimension 1. A subspace G of E is a hyperplane if and only if there exists a linear form $\varphi: E \to \mathbb{R}$ such that $G = \{x \in E|\varphi(x) = 0\}$.

Note that each hyperplane is either closed or dense in E.

An *affine hyperplane* is a translate of a hyperplane. Thus an affine hyperplane is necessarily of the form $\{x \in E | \varphi(x) = \lambda\}$ where φ is a linear form on E and $\lambda \in \mathbb{R}$.

Given two subsets U and V of E, we say that a hyperplane $G = \{x \in E | \varphi(x) = \lambda\}$ *separates* U and V (*separates strictly* U and V, respectively) if

$$U \subset \{x \in E | \varphi(x) \geq \lambda\} \quad \text{and} \quad V \subset \{x \in E | \varphi(x) \leq \lambda\} \qquad (1.4.9)$$

(respectively,

$$U \subset \{x \in E | \varphi(x) > \lambda\} \quad \text{and} \quad V \subset \{x \in E | \varphi(x) < \lambda\}). \qquad (1.4.10)$$

In the next result we collect some important separation theorems (see Choquet [1969, Section 21]).

1.4.8 Theorem. *Let E be a real topological vector space. Then the following statements hold*:
(1) *If U is a non-empty open convex subset of E and G an affine subspace of E disjoint from U, then there exists a closed affine hyperplane disjoint from U which contains G.*
(2) *If U and V are non-empty disjoint convex subsets of E with U open, then there exists a closed affine hyperplane which separates U and V.*
    If V is also open, the separation is strict.
(3) *If E is locally convex and Hausdorff and if U and V are disjoint closed convex subsets of E with U compact, then there exists a closed affine hyperplane which separates strictly U and V.*

The above separation theorems can be fruitfully used in the approximation of lower semi-continuous convex functions on convex compact sets.

Let E be a locally convex Hausdorff space and K a convex compact subset of E.

A function $u: K \to \mathbb{R}$ is said to be *affine* (*convex, concave*, respectively) if for every $x, y \in K$ and $\lambda \in [0, 1]$

$$u(\lambda x + (1 - \lambda)y) = \lambda u(x) + (1 - \lambda)u(y), \qquad (1.4.11)$$

$(u(\lambda x + (1 - \lambda)y) \leq \lambda u(x) + (1 - \lambda)u(y), u(\lambda x + (1 - \lambda)y) \geq \lambda u(x) + (1 - \lambda)u(y)$, respectively).

In this case, for every $x_1, \ldots, x_n \in K$, $n \geq 2$, and $\lambda_1, \ldots, \lambda_n \geq 0$ satisfying $\sum_{i=1}^{n} \lambda_i = 1$, one has

$$u\left(\sum_{i=1}^{n} \lambda_i x_i\right) = \sum_{i=1}^{n} \lambda_i u(x_i) \tag{1.4.12}$$

$$\left(u\left(\sum_{i=1}^{n} \lambda_i x_i\right) \leq \sum_{i=1}^{n} \lambda_i u(x_i), u\left(\sum_{i=1}^{n} \lambda_i x_i\right) \geq \sum_{i=1}^{n} \lambda_i u(x_i), \text{ respectively}\right).$$

We shall denote by $A(K)$ the closed subspace of $\mathscr{C}(K)$ consisting of all continuous affine functions on K.

Moreover we shall denote by $A(K, E)$ the subspace of all restrictions to K of continuous affine functions on E. Note that

$$A(K, E) = \{\varphi_{|K} + \lambda \,|\, \varphi \in E', \lambda \in \mathbb{R}\}. \tag{1.4.13}$$

1.4.9 Theorem. *Let E be a real locally convex Hausdorff space and K a convex compact subset of E. Then the following statements hold:*
(1) *If $u: K \to \mathbb{R}$ is a convex lower semi-continuous function, then for every $x \in K$*

$$u(x) = \sup\{a(x) \,|\, a \in A(K, E), a < u\}.$$

(2) *(Mokobodzki) If $u: K \to \mathbb{R}$ is an upper semi-continuous function, then for every $x \in K$*

$$u(x) = \inf\{v(x) \,|\, v \text{ convex continuous function}, u < v\}.$$

Moreover, if u is also convex, then there exists a net $(u_i)_{i \in I}^{\leq}$ of convex continuous functions on K which converges pointwise to u.
(3) *If $u \in A(K)$, then there exists an increasing sequence in $A(K, E)$ which converges uniformly to u. Thus, $A(K, E)$ is dense in $A(K)$.*

For a proof of the above theorem see Alfsen [1971, Proposition I.1.2, Corollary I.1.5 and Proposition I.5.1].

Now we list some properties of convex compact sets including the classical Krein-Milman theorem.

Let K be a convex compact subset of a locally convex Hausdorff space. A point $x_0 \in K$ is said to be an *extreme point* of K if $K \backslash \{x_0\}$ is convex or, equivalently, if for every $x_1, x_2 \in K$ and $\lambda \in \,]0, 1[$ satisfying $x_0 = \lambda x_1 + (1 - \lambda)x_2$, it necessarily follows that $x_0 = x_1 = x_2$.

The set of all extreme points of K will be denoted by $\partial_e K$.

For example, if X is a compact Hausdorff space and $K = \mathscr{M}_1^+(X)$ endowed with the vague topology, then

$$\partial_e K = \{\varepsilon_x \,|\, x \in X\}, \tag{1.4.14}$$

while, if $K = \{\mu \in \mathcal{M}(X) | \, \|\mu\| \leq 1\}$, then

$$\partial_e K = \{\varepsilon_x | x \in X\} \cup \{-\varepsilon_x | x \in X\}. \tag{1.4.15}$$

An important topological property of $\partial_e K$ which was proved by Choquet (see Choquet [1969, Theorem 27.9]) is that $\partial_e K$ is a Baire space in the relative topology (i.e., the intersection of every sequence of dense open subsets of $\partial_e K$ in dense in $\partial_e K$).

Moreover, if K is metrizable, $\partial_e K$ is a countable intersection of open subsets of K and hence it is a complete metric space.

We also recall that a *ray* of a locally convex Hausdorff space E is a subset δ of E of the form

$$\delta = \delta_{x_0} := \{\lambda x_0 | \lambda > 0\}, \tag{1.4.16}$$

where $x_0 \in E$, $x_0 \neq 0$.

Clearly every affine subspace which does not contain the origin and which has non-empty intersection with a ray δ intersects the ray in a unique point.

Given a convex cone C of E, a ray $\delta \subset C$ is said to be an *extreme ray* of C if $C \backslash \delta$ is convex.

This also means that every affine subspace G of E not containing the origin and having non-empty intersection with δ intersects δ in an extreme point of $G \cap C$.

If C is *proper*, i.e., $C \cap (-C) = \{0\}$, then a ray δ of C is an extreme ray if and only if for every $a \in \delta$ and $b \in C$ such that $a - b \in C$, there exists $\lambda \geq 0$ such that $b = \lambda a$ (and hence $b \in \delta$).

If the convex cone C has a compact *base* K (i.e., K is the intersection of C with an affine hyperplane and for every $y \in C$ there exist $x \in K$ and $\lambda > 0$ such that $y = \lambda x$), then a ray δ is an extreme ray of C if and only if $\delta = \delta_{x_0}$ for some $x_0 \in \partial_e K$ (Choquet [1953]).

Note that in general a convex cone may have no extreme rays, while every convex compact subset has extreme points. This will follow from the Bauer's maximum principle for convex compact sets (see Choquet [1969, Theorem 25.9]).

1.4.10 Theorem (*Bauer's maximum principle*). *Let K be a convex compact subset of a real locally convex Hausdorff space and $u: K \to \mathbb{R}$ a convex upper semi-continuous function. Then there exists $x_0 \in \partial_e K$ such that $u(x_0) = \max\{u(x) | x \in K\}$.*

In particular, $\partial_e K$ is non-empty.

In addition to the above result, the next comparison principle is useful in comparing semi-continuous convex and concave functions. For a proof see Bauer [1963].

1.4.11 Theorem (*Bauer*). *Let K be a convex compact subset of a real locally convex Hausdorff space and $u: K \to \mathbb{R}$ and $v: K \to \mathbb{R}$ respectively a convex and a concave semi-continuous function (that is, either lower or upper semi-continuous). If $u \le v$ on $\partial_e K$, then $u \le v$ on K.*

By using Theorem 1.4.10 and Theorem 1.4.8, (3), one can easily prove the famous Krein-Milman theorem and its converse (see Choquet [1969, Theorem 25.12, Corollary 25.14]).

1.4.12 Theorem. *Let K be a convex compact subset of a real locally convex Hausdorff space.*
   Then the following statements hold:
(1) *(Krein-Milman)* $K = \overline{\mathrm{co}}(\partial_e K)$.
(2) *(Milman's Converse of the Krein-Milman theorem) If Q is a closed subset of K such that $K = \overline{\mathrm{co}}(Q)$, then $\partial_e K \subset Q$.*

As a matter of fact, if K is a convex compact subset of $\mathbb{R}^p$ we have

$$K = \mathrm{co}(\partial_e K). \tag{1.4.17}$$

(*Minkowski's theorem*).
   However in the finite dimensional case we have sharper results (see, e.g., Cheney [1982, Chapter 1, Section 5] and Bonnesen and Fenchel [1934, p. 9]).

1.4.13 Theorem. *If K is a convex compact subset of $\mathbb{R}^p$ then the following statements hold:*
(1) *(Carathéodory) For every $x \in K$ there exist at most $p + 1$ extreme points $x_1, \ldots, x_{p+1}$ of K and $\lambda_1, \ldots, \lambda_{p+1} \ge 0$ with $\sum_{i=1}^{p+1} \lambda_i = 1$, such that $x = \sum_{i=1}^{p+1} \lambda_i x_i$.*
(2) *(Fenchel) If $p \ge 2$ and $K = \mathrm{co}(Q)$ with Q compact and connected, then for every $x \in K$ there exist at most p extreme points $x_1, \ldots, x_p$ of K and $\lambda_1, \ldots, \lambda_p \ge 0$ with $\sum_{i=1}^{p} \lambda_i = 1$, such that $x = \sum_{i=1}^{p} \lambda_i x_i$.*

We conclude this section with some remarks concerning the state space of a function space (see Choquet [1969, Section 29]).
   Let X be a compact Hausdorff space and let H be a subspace of $\mathscr{C}(X)$ which contains the constants and *separates the points* of X, i.e., for every $x, y \in X$, $x \ne y$, there exists $h \in H$ such that $h(x) \ne h(y)$.
   Let us denote by H' the dual of H endowed with the weak*-topology $\sigma(H', H)$. Moreover we shall denote by H'_+ the cone of all positive linear forms on H.
   The *state space* of H is, by definition, the subset of H'_+

$$K := \{\mu \in H'_+ \,|\, \|\mu\| = \mu(1) = 1\}. \tag{1.4.18}$$

The subset K is convex. Furthermore, by the Alaoglu-Bourbaki theorem 1.4.1, it is also compact (for $\sigma(H', H)$).

Consider the continuous mapping $\Phi\colon X \to H'$ defined by

$$\Phi(x) := \varepsilon_{x|H} \quad \text{for every } x \in X. \tag{1.4.19}$$

Since H separates the points of X, Φ is injective and $\Phi(X) \subset K$. The mapping Φ is often called the *canonical embedding* of X into the state space of H.

Some important properties of Φ are listed below. First we have

$$K = \overline{\text{co}}(\Phi(X)). \tag{1.4.20}$$

Hence, by Milman's converse theorem 1.4.12, $\partial_e K \subset \Phi(X)$. In Section 2.6 we shall see that $\Phi^{-1}(\partial_e K)$ coincides with the so-called Choquet boundary of H.

Finally

$$H = \{\varphi \circ \Phi | \varphi \in (H')'\}, \tag{1.4.21}$$

and hence

$$\overline{H} = \{a \circ \Phi | a \in A(K)\}. \tag{1.4.22}$$

*1.5 Integral representation theory for convex compact sets

In this section we sketch the essential aspects of Choquet's beautiful theory of integral representation for convex compact sets. This theory will be mainly applied in Section 4.3.

For more details, developments and applications we refer the reader to Choquet [1969], Alfsen [1971], Phelps [1966], Asimow and Ellis [1980].

Choquet-Bishop-de Leeuw integral representation theory

Let E be a real locally convex Hausdorff space and consider a convex compact subset K of E.

By using the approximation theorem 1.2.3, it is possible to show that for every $\mu \in \mathcal{M}^+(K)$ there exists a unique point $r(\mu) \in E$ such that

$$\varphi(r(\mu)) = \mu(\varphi_{|K}) \quad \text{for every } \varphi \in E'. \tag{1.5.1}$$

The point $r(\mu)$ is called the *resultant of* μ and belongs to K provided $\mu \in \mathcal{M}_1^+(K)$.

For every $\mu \in \mathcal{M}^+(K)$, $\mu \neq 0$, the resultant $r\left(\dfrac{\mu}{\|\mu\|}\right) \in K$ is also called the *barycenter of* μ.

If $\mu \in \mathcal{M}_1^+(K)$ and $x \in K$ and if $r(\mu) = x$, we also say that the point x is *represented* by μ.

Since the topology of K coincides with the coarsest topology on K for which all functions $\varphi_{|K}(\varphi \in E')$ are continuous, the mapping $\mu \mapsto r(\mu)$ is continuous from $\mathcal{M}_1^+(K)$ endowed with the vague topology, into K.

Furthermore, if $\mu = \sum_{i=1}^n \lambda_i \varepsilon_{x_i}$ is discrete, then $r(\mu) = \sum_{i=1}^n \lambda_i x_i$ and hence $r(\mu) \in \text{co}(\text{Supp}(\mu))$.

From this and from the approximation theorem 1.2.3, (2), it follows that

$$r(\mu) \in \overline{\text{co}}(\text{Supp}(\mu)) \quad \text{for every } \mu \in \mathcal{M}_1^+(K). \tag{1.5.2}$$

It is of interest also to note the next result (see Choquet [1969, Lemma 26.14]).

1.5.1 Proposition. *If* $\mu \in \mathcal{M}_1^+(K)$, *then there exists a net* $(\mu_i)_{i \in I}^{\leq}$ *of discrete measures in* $\mathcal{M}_1^+(K)$ *such that* $\mu_i \to \mu$ *vaguely and* $r(\mu_i) = r(\mu)$ *for every* $i \in I$.

From (1.5.1), (1.4.13) and Theorem 1.4.9, it also follows that if $\mu \in \mathcal{M}_1^+(K)$, then

$$a(r(\mu)) = \mu(a) \quad \text{for every } a \in A(K) \tag{1.5.3}$$

as well as

$$u(r(\mu)) \leq \int u \, d\mu \tag{1.5.4}$$

for every convex lower semi-continuous function $u: K \to \mathbb{R}$.

Formula (1.5.3) is referred to as the *barycenter formula*. As a matter of fact, if $\overline{\text{co}}(\text{Supp}(\mu)) = K$, then the continuous affine functions on K are the only concave continuous functions which satisfy (1.5.3). This follows from the following result of Rasa [1986].

1.5.2 Theorem (*Rasa*). *If* $\mu \in \mathcal{M}_1^+(K)$ *and if* $f \in \mathscr{C}(K)$ *is a concave function such that* $\mu(f) = f(r(\mu))$, *then* f *is affine on* $\overline{\text{co}}(\text{Supp}(\mu))$.

Trivially every point $x \in K$ is the barycenter of some $\mu \in \mathcal{M}_1^+(K)$, e.g., $\mu = \varepsilon_x$. The problem becomes more interesting and important if we require that the representing measure μ is supported by $\partial_e K$ (see the definition before Theorem 1.5.5).

This problem was solved by Choquet when K is metrizable and by Bishop and de Leeuw in the general case.

Note, however, that because of the Krein-Milman theorem, for every $x \in K$ there always exists $\mu \in \mathcal{M}_1^+(K)$ supported by $\overline{\partial_e K}$ such that $r(\mu) = x$.

In order to explain the results of Choquet and Bishop and de Leeuw we need some preliminaries.

We shall denote by $P(K)$ the cone of all convex continuous real-valued functions on K and by $Q(K)$ the cone of all convex lower semi-continuous functions with values in $\mathbb{R} \cup \{+\infty\}$.

The cone $P(K)$ induces an ordering $<$ on $\mathcal{M}^+(K)$ defined by

$$\mu < \nu \quad \text{if} \quad \mu(u) \le \nu(u) \quad \text{for every } u \in P(K) \tag{1.5.5}$$

$(\mu, \nu \in \mathcal{M}^+(K))$.

The ordering $<$ is sometimes called the *Choquet-Meyer ordering*.

If $\mu < \nu$ we also say that ν is *more diffuse* than μ. Clearly, if $\mu < \nu$, then $\|\mu\| = \|\nu\|$ and $r(\mu) = r(\nu)$.

Furthermore, by using Theorem 1.4.9, (1), one can also show that, if $\mu < \nu$, then

$$\mu(u) \le \nu(u) \quad \text{for every } u \in Q(K). \tag{1.5.6}$$

The ordering $<$ is inductive so that by Zorn's lemma there exist maximal elements. A Radon measure $\mu \in \mathcal{M}^+(K)$ which is maximal with respect to $<$ will be simply called a *maximal measure* or a *boundary measure*.

To characterize boundary measures let us introduce the following enveloping functions.

For every bounded function $f: K \to \mathbb{R}$ and for every $x \in K$ we set

$$\check{f}(x) := \sup\{u(x) | u \in P(K), u \le f\} \tag{1.5.7}$$

and

$$\hat{f}(x) := -(-f)\check{}(x) = \inf\{v(x) | -v \in P(K), f \le v\}. \tag{1.5.8}$$

Clearly, $\check{f}, -\hat{f} \in Q(K)$. Furthermore $f = \check{f}$ (respectively, $f = \hat{f}$) if and only if $f \in Q(K)$ (respectively, $-f \in Q(K)$). Consequently

$$\hat{\hat{f}} = \hat{f} \quad \text{and} \quad \check{\check{f}} = \check{f} \quad \text{for every } f \in \mathscr{C}(K). \tag{1.5.9}$$

By using Theorem 1.4.9, (1), it is possible to show that for every $x \in K$

$$\check{f}(x) = \sup\{a(x) | a \in A(K), a \le f\} \tag{1.5.10}$$

and

$$\hat{f}(x) = \inf\{b(x)|b \in A(K), f \le b\}. \qquad (1.5.11)$$

1.5.3 Proposition. *If* μ, $v \in \mathcal{M}^+(K)$, *then* $\mu < v$ *if and only if* $\int \check{f}\,d\mu \le v(f)$ *or, equivalently,* $\mu(f) \le \int \hat{f}\,dv$ *for every* $f \in \mathscr{C}(K)$.

A first characterization of maximal measures is indicated below (see Choquet [1969, Theorem 26.16]).

1.5.4 Theorem (*Mokobodzki*). *A Radon measure* $\mu \in \mathcal{M}_1^+(K)$ *is maximal if and only if* $\mu(f) = \int \hat{f}\,d\mu$ *for every* $f \in \mathscr{C}(K)$ *or, equivalently,* $\mu(f) = \int \hat{f}\,d\mu$ *for every* $f \in P(K)$.

Thus, if $\mu \in \mathcal{M}^+(K)$ is maximal, we also have $\mu(f) = \int \check{f}\,d\mu$ for every $f \in \mathscr{C}(K)$.

Another characterization of boundary measures can be expressed in terms of the so-called bordering sets.

Given $u \in P(K)$, the *bordering set* of u is, by definition, the subset of K

$$K_u := \{x \in K | \hat{u}(x) = u(x)\}. \qquad (1.5.12)$$

Note that K_u is the intersection of a sequence of open subsets of K. In fact, it is possible to show that

$$\partial_e K = \bigcap_{u \in P(K)} K_u. \qquad (1.5.13)$$

Furthermore, if $u \in P(K)$ is *strictly convex*, i.e.,

$$u(\lambda x + (1 - \lambda)y) < \lambda u(x) + (1 - \lambda)u(y)$$

for every $x, y \in K$, $x \ne y$ and $\lambda \in \,]0, 1[$,

then

$$\partial_e K = K_u. \qquad (1.5.14)$$

On the other hand, by *Hervé's theorem*, a convex compact subset admits a strictly convex continuous function if and only if it is metrizable (see Alfsen [1971, Theorem I.4.3]).

Accordingly, if K is metrizable, $\partial_e K$ is the intersection of a sequence of open subsets of K.

If $\mu \in \mathscr{M}^+(K)$ and $X \subset K$, we say that μ is *supported* by X if $\mu^*(K \setminus X) = 0$. If X is closed, this also means that $\mathrm{Supp}(\mu) \subset X$.

Here we have another characterization of maximal measures (see Choquet [1969, Theorem 27.4 and Corollary 27.5]).

1.5.5 Theorem. *Let $\mu \in \mathscr{M}^+(K)$. Then the following statements hold:*
(1) *The measure μ is maximal if and only if μ is supported by K_u for every $u \in P(K)$. If this is the case, then $\mathrm{Supp}(\mu) \subset \overline{\partial_e K}$.*
(2) *If μ is supported by $\partial_e K$, then μ is maximal.*
(3) *If K is metrizable, then μ is maximal if and only if μ is supported by $\partial_e K$.*

Note that, since $(\mathscr{M}^+(K), <)$ is inductive, then by Zorn's lemma for every $\mu \in \mathscr{M}^+(K)$ there exists a maximal measure $v \in \mathscr{M}^+(K)$ such that $\mu < v$.

In particular for every $x \in K$ there exists $\mu \in \mathscr{M}^+(K)$, μ maximal, such that $\varepsilon_x < \mu$ (and hence $\|\mu\| = 1$ and $r(\mu) = r(\varepsilon_x) = x$).

Combining these remarks with the previous results we arrive at the famous integral representation theorem due to Choquet [1956] for metrizable convex compact sets and to Bishop and de Leeuw [1959] in the general case. For a proof we refer to Choquet [1969, Theorem 27.6].

1.5.6 Theorem (*Choquet-Bishop-de Leeuw*). *Every point of a convex compact set K is represented by a maximal probability Radon measure on K.*

In other words, for every $x \in K$ there exists a Radon measure $\mu \in \mathscr{M}_1^+(K)$ supported by every bordering set K_u ($u \in P(K)$) (supported by $\partial_e K$, provided K is metrizable, respectively) such that $\mu(a) = a(x)$ for every $a \in A(K)$.

Choquet simplices and uniqueness of representing measures

The question of uniqueness of representing measures in Theorem 1.5.6 was solved by Choquet.

Let K be a convex compact subset of a real locally convex Hausdorff space E. We set

$$\tilde{K} := \{(\lambda x, \lambda) \in E \times \mathbb{R} \mid x \in K, \lambda > 0\}. \tag{1.5.15}$$

In fact, $\tilde{K}$ is a cone of $E \times \mathbb{R}$ and $\tilde{K} - \tilde{K}$ is a subspace of $E \times \mathbb{R}$.

On $\tilde{K} - \tilde{K}$ we may consider the partial ordering $\leq$ induced by $\tilde{K}$. Thus for every $z_1, z_2 \in \tilde{K} - \tilde{K}$ we have $z_1 \leq z_2$ if $z_2 - z_1 \in \tilde{K}$.

We say that K is a *Choquet simplex* if $\tilde{K} - \tilde{K}$ is a lattice with respect to the above defined partial ordering.

It is possible to show that, if we put $\mathscr{G}(K) := \{\lambda K + a | \lambda \geq 0, a \in E\}$, then K is a Choquet simplex if and only if for every $U, V \in \mathscr{G}(K)$ the intersection $U \cap V$ is either empty or belongs to $\mathscr{G}(K)$.

As a matter of fact, this last property was assumed by Choquet as the original definition of a simplex (see Choquet [1956]) and the proof of the equivalence can be found in Choquet [1956] and Kendall [1962].

Thus in $\mathbb{R}^p$ the Choquet simplices are the convex hulls of k affinely independent points, $k \leq p + 1$. (We recall that k points $x_1, \ldots, x_k \in \mathbb{R}^p$ are *affinely independent* if for every $\lambda_1, \ldots, \lambda_k \in \mathbb{R}$ satisfying $\sum_{i=1}^{k} \lambda_i x_i = 0$ and $\sum_{i=1}^{k} \lambda_i = 0$ one necessarily has $\lambda_1 = \cdots = \lambda_k = 0$).

As in the finite dimensional case the Choquet simplices are exactly those convex compact sets for which the integral representation in terms of extreme points is unique.

1.5.7 Theorem (*Choquet-Meyer*). *Given a convex compact subset K of a real locally convex Hausdorff space, the following statements are equivalent:*

(i) *K is a Choquet simplex.*

(ii) *For every $x \in K$ there exists a unique maximal measure $\mu \in \mathscr{M}_1^+(K)$ such that $r(\mu) = x$.*

(iii) *$\hat{u}$ is affine on K for every $u \in P(K)$.*

(iv) *$(u + v)\hat{} = \hat{u} + \hat{v}$ for every $u, v \in P(K)$.*

(v) *If $\mu \in \mathscr{M}_1^+(K)$ is maximal, then $\mu(u) = \hat{u}(r(\mu))$ for every $u \in P(K)$.*

For a proof we refer to Choquet [1969, Theorem 28..4].

If K is a Choquet simplex, for every $x \in K$ we shall denote by μ_x the unique maximal probability Radon measure on K which represents x.

By the above statement (v) we have

$$\mu_x(u) = \hat{u}(x) \quad \text{for every } u \in P(K). \qquad (1.5.16)$$

A Choquet simplex K such that $\partial_e K$ is closed will be called a *Bauer simplex*.

Bauer simplices occur, e.g., when one wants to solve the Dirichlet problem on $\partial_e K$ for every continuous function $f: \partial_e K \to \mathbb{R}$ (see below).

A proof of the next result can be found in Alfsen [1971, Theorems II.4.1 and II.4.2] (see also Corollary 4.3.3).

1.5.8 Theorem (*Bauer*). *Given a convex compact subset K of a real locally convex Hausdorff space, the following statements are equivalent:*

(i) *K is a Bauer simplex.*

(ii) *For every $x \in K$ there exists a unique $\mu \in \mathscr{M}_1^+(K)$ such that $\mathrm{Supp}(\mu) \subset \overline{\partial_e K}$ and $r(\mu) = x$.*

(iii) *K is a Choquet simplex and the mapping $x \to \mu_x$ is continuous from K into $\mathcal{M}_1^+(K)$.*

(iv) *$\hat{u} \in A(K)$ for every $u \in P(K)$.*

 (v) *The space $A(K)$ endowed with the natural order is a lattice.*

(vi) *Every continuous function $f: \partial_e K \to \mathbb{R}$ can be extended to a (unique) function $\tilde{f} \in A(K)$.*

Furthermore, if one of the statements (i)–(vi) *holds, then for every continuous function $f: \partial_e K \to \mathbb{R}$ and for every $x \in K$*

$$\tilde{f}(x) = \mu_x(\bar{f}), \tag{1.5.17}$$

where $\bar{f}$ is an arbitrary continuous extension of f to K.

The problem of the extension of a continuous real-valued function on $\partial_e K$ to a function in $A(K)$ is referred to as the *Dirichlet problem on the extreme points of K.*

By the above theorem we see that this problem is solvable for every continuous function on $\partial_e K$ if and only if K is a Bauer simplex.

For some conditions ensuring that the Dirichlet problem on $\partial_e K$ is solvable for a given continuous function $f: \partial_e K \to \mathbb{R}$ see Alfsen [1971] (see also Corollary 4.3.2).

Note that, if K is a Bauer simplex, then the operator $T: \mathscr{C}(K) \to \mathscr{C}(K)$ defined by

$$Tf(x) := \mu_x(f) \quad \text{for every } f \in \mathscr{C}(K) \text{ and } x \in K \tag{1.5.18}$$

is linear, positive and $T(\mathscr{C}(K)) \subset A(K)$. Furthermore $T(a) = a$ for every $a \in A(K)$ and T is a *projection*, i.e., $T = T \circ T$. Moreover T is the unique positive projection on $\mathscr{C}(K)$ having $A(K)$ as its range because of Proposition 3.3.1 and Bauer's maximum principle. In fact, such a projection can be constructed only for Bauer simplices according to the following result due to Rogalski [1968] (see also Altomare [1977]).

1.5.9 Corollary. *Given a convex compact subset K of a locally convex Hausdorff space, the following statements are equivalent:*

 (i) *K is a Bauer simplex.*

(ii) *There exists a (unique) positive linear projection $T: \mathscr{C}(K) \to \mathscr{C}(K)$ such that $T(\mathscr{C}(K)) = A(K)$.*

The projection T is also called the *canonical positive projection* associated with the Bauer simplex K and it is explicitly defined by (1.5.18).

*1.6 C_0-semigroups of operators and abstract Cauchy problems

The theory of C_0-semigroups of bounded linear operators is a beautiful example of a rich and powerful mathematical theory that, using abstract and general principles of functional analysis and operator theory, succeeds in clarifying and in solving deep problems from probability, potential theory, partial differential equations, ergodic theory, mathematical physic and population dynamics.

The reader wishing to be acquainted with the basic theory of C_0-semigroups and with some of its applications as well should consult, for instance, Belleni-Morante [1979], Bratteli and Jørgensen [1984], Butzer and Berens [1967], van Casteren [1985], Clément et al. [1987], Davies [1980], J. A. Goldstein [1985], Hille and Phillips [1957], R. Nagel (Ed.) [1986], Pazy [1983].

Here we shall content ourselves with simply reviewing those results that will be used in Sections 6.2 and 6.3.

C_0-semigroups and their generators

Let E be a Banach space over the field $\mathbb{K}$.

We recall that a linear operator S from E into another Banach space F is said to be *bounded* if $\sup\{\|S(f)\| \,|\, \|f\| \le 1\} < +\infty$. In fact S is bounded if and only if S is continuous.

A linear operator $S: E \to F$ is said to be a *linear contraction* if $\|S(f)\| \le 1$ for every $f \in E$, $\|f\| \le 1$.

We shall denote by $L(E)$ the space of all bounded linear operators from E into E.

The space $L(E)$ endowed with the norm

$$\|S\| := \sup\{\|S(f)\| \,|\, \|f\| \le 1\} \quad (S \in L(E)), \tag{1.6.1}$$

is a Banach space.

On $L(E)$ we shall also consider the *strong topology*. A net $(S_i)_{i \in I}^{\le}$ of elements of $L(E)$ *converges strongly* to $S \in L(E)$ if $S_i(f) \to S(f)$ for every $f \in E$.

A *semigroup* (or a *one-parameter semigroup*) of bounded linear operators on E is a family $(T(t))_{t \ge 0}$ of elements of $L(E)$ such that

$$T(0) = I_E, \tag{1.6.2}$$

$$T(s + t) = T(s)T(t) \quad \text{for every } s, t \ge 0, \tag{1.6.3}$$

where I_E denotes the *identity operator* on E, i.e., $I_E(f) := f$ for every $f \in E$.

A semigroup $(T(t))_{t \geq 0}$ on E is said to be *strongly continuous* if for every $t_0 \geq 0$ and $f \in E$

$$\lim_{t \to t_0} \|T(t)f - T(t_0)f\| = 0. \tag{1.6.4}$$

Actually, because of (1.6.3), property (1.6.4) holds for every $t_0 \geq 0$ and $f \in E$ if and only if $\lim_{t \to 0^+} \|T(t)f - f\| = 0$ for every $f \in E$.

A strongly continuous semigroup will be called simply a C_0-*semigroup*.

A semigroup $(T(t))_{t \geq 0}$ on E is said to be *uniformly continuous* if for every $t_0 \geq 0$

$$\lim_{t \to t_0} \|T(t) - T(t_0)\| = 0, \tag{1.6.5}$$

or, equivalently, if $\lim_{t \to 0^+} \|T(t) - I_E\| = 0$.

If $(T(t))_{t \geq 0}$ is a C_0-semigroup, then there always exist $\omega \in \mathbb{R}$ and $M \geq 1$ such that

$$\|T(t)\| \leq M \exp(\omega t) \quad \text{for every } t \geq 0. \tag{1.6.6}$$

The *growth bound* of the semigroup is then defined by

$$\omega_0 := \inf\{\omega \in \mathbb{R} \,|\, \text{There exists } M \geq 0 \text{ such that } \|T(t)\| \leq M \exp(\omega t)$$

$$\text{for every } t \geq 0\}. \tag{1.6.7}$$

Thus, $\omega_0 \in \mathbb{R} \cup \{-\infty\}$.

A *contraction C_0-semigroup* is a C_0-semigroup $(T(t))_{t \geq 0}$ of linear contractions (i.e., $\|T(t)\| \leq 1$ for every $t \geq 0$).

If E is a Banach lattice and every $T(t)$ is a positive linear operator, then $(T(t))_{t \geq 0}$ is called a *positive C_0-semigroup*.

Given a C_0-semigroup $(T(t))_{t \geq 0}$, we may consider the following linear operator $A : D(A) \to E$ defined on the linear subspace

$$D(A) := \left\{ f \in E \,|\, \text{There exists } \lim_{t \to 0^+} \frac{T(t)f - f}{t} \in E \right\}, \tag{1.6.8}$$

by

$$Af := \lim_{t \to 0^+} \frac{T(t)f - f}{t} \quad \text{for every } f \in D(A). \tag{1.6.9}$$

The operator A is called the *generator* of the semigroup $(T(t))_{t \geq 0}$.

Note that if the semigroup $(T(t))_{t \geq 0}$ is uniformly continuous then $D(A) = E$ and A is bounded. Conversely, if $A \in L(E)$, putting

$$T(t) := \exp(tA) := \sum_{n=0}^{\infty} \frac{t^n}{n!} A^n \quad (t \geq 0), \qquad (1.6.10)$$

we have that $(T(t))_{t \geq 0}$ is a semigroup whose generator is A.

To state some properties of $(A, D(A))$ let us recall that a linear operator $A: D(A) \to E$ defined on a subspace $D(A)$ of E is said to be *closed* if $D(A)$ endowed with the *graph norm*

$$\| f \|_A := \| f \| + \| Af \|, \quad (f \in D(A)), \qquad (1.6.11)$$

becomes a Banach space. More explicitly this means that for every sequence $(f_n)_{n \in \mathbb{N}}$ of elements of $D(A)$ such that $f_n \to f \in E$ and $Af_n \to g \in E$ it follows that $f \in D(A)$ and $Af = g$.

In other words, the graph $\{(f, Af)| f \in D(A)\}$ is closed in $E \times E$.

Given a linear operator $A: D(A) \to E$, we say that a linear operator $B: D(B) \to E$ defined on a subspace $D(B)$ of E is *an extension* of A if $D(A) \subset D(B)$ and $Af = Bf$ for every $f \in D(A)$.

We say that a linear operator $A: D(A) \to E$ is *closable* if there exists a closed extension of A.

This also means that if $(f_n)_{n \in \mathbb{N}}$ is a sequence in $D(A)$ such that $f_n \to 0$ and $(Af_n)_{n \in \mathbb{N}}$ is convergent, then $Af_n \to 0$.

If $A: D(A) \to E$ is closable, the smallest closed extension $\bar{A}: D(\bar{A}) \to E$ of A is called the *closure* of A.

We have

$$D(\bar{A}) = \{ f \in E | \text{There exists } (f_n)_{n \in \mathbb{N}} \text{ in } D(A) \text{ such that } f_n \to f$$

$$\text{and } (Af_n)_{n \in \mathbb{N}} \text{ is convergent}\}, \qquad (1.6.12)$$

and

$$\bar{A}f := \lim_{n \to \infty} Af_n, \qquad (1.6.13)$$

for every $f \in D(\bar{A})$, $(f_n)_{n \in \mathbb{N}}$ being an arbitrary sequence in $D(A)$ such that $f_n \to f$ and $(Af_n)_{n \in \mathbb{N}}$ is convergent.

A *core* for a linear operator $A: D(A) \to E$ is a linear subspace D_0 of $D(A)$ which is dense in $D(A)$ with respect to the graph norm.

If A is closed and $\lambda I_{D(A)} - A$ is invertible for some $\lambda \in \mathbb{K}$, then the inverse $(\lambda I_{D(A)} - A)^{-1}$ of $\lambda I_{D(A)} - A$ is continuous from E into $(D(A), \|\cdot\|_A)$ by virtue of the *open-mapping theorem*[1] hence a fortiori $(\lambda I_{D(A)} - A)^{-1} \in L(E)$.

In this case a subspace D_0 of $D(A)$ is a core for A if and only if $(\lambda I_{D(A)} - A)(D_0)$ is dense in E.

After these preliminaries we state some basic properties of generators of C_0-semigroups (see, e.g., R. Nagel (Ed.) [1986, Section A-I-1] and Pazy [1983, Sect. 1.2]).

1.6.1 Theorem. *Let* $A: D(A) \to E$ *be the generator of a C_0-semigroup* $(T(t))_{t \geq 0}$ *on* E. *Then the following statements hold:*

(1) $D(A)$ *is dense in E and A is closed.*

(2) *If a subspace D_0 of $D(A)$ is dense in E and $T(t)(D_0) \subset D_0$ for every $t \geq 0$, then D_0 is a core for A.*

(3) $T(t)(D(A)) \subset D(A)$ *for every $t \geq 0$.*

(4) *Given $f \in E$, the mapping $\xi_f: \mathbb{R}_+ \to E$ defined by $\xi_f(t) := T(t)f$ $(t \geq 0)$ is differentiable in $\mathbb{R}_+$ if and only if $f \in D(A)$.*

   If this is the case, then $\xi_f'(t_0) = AT(t_0)f = T(t_0)Af$ for every $t_0 \geq 0$.

(5) *For every $t \geq 0$ and $f \in E$, $\int_0^t T(s)f\,ds \in D(A)$ and $A \int_0^t T(s)f\,ds = T(t)f - f$.*

(6) *For every $t \geq 0$ and $f \in D(A)$, $\int_0^t T(s)Af\,ds = T(t)f - f$.*

Given a closed operator $A: D(A) \to E$ we shall denote by $\rho(A)$ the *resolvent set* of A, i.e.,

$$\rho(A) := \{\lambda \in \mathbb{K} \mid \lambda I_{D(A)} - A \text{ is invertible}\}. \tag{1.6.14}$$

The *spectrum* $\sigma(A)$ is defined as

$$\sigma(A) := \mathbb{K} \setminus \rho(A). \tag{1.6.15}$$

If $\mathbb{K} = \mathbb{R}$, $\sigma(A)$ may be empty.

If $\lambda \in \rho(A)$ we shall denote by $R(\lambda, A)$ the inverse of $\lambda I_{D(A)} - A$.

As explained above, by the open-mapping theorem, $R(\lambda, A) \in L(E)$ for every $\lambda \in \rho(A)$.

We also point out the so-called *resolvent identity*:

$$R(\lambda, A) - R(\mu, A) = (\mu - \lambda)R(\lambda, A)R(\mu, A) \quad \text{for every } \lambda, \mu \in \rho(A). \tag{1.6.16}$$

---

[1] In the form needed by us this theorem says that, if E and F are Banach spaces and if $T: E \to F$ is a continuous bijective linear mapping, then the inverse mapping T^{-1} is continuous. For a proof see Hórvath [1966, Chapter 1, Section 9, Theorem 1].

If A is the generator of a C_0-semigroup $(T(t))_{t \geq 0}$, then $\rho(A) \neq \varnothing$. Furthermore the *spectral bound* $s(A)$ of A defined by

$$s(A) := \sup\{\mathscr{R}e\, \lambda \,|\, \lambda \in \sigma(A)\} \tag{1.6.17}$$

is finite.

Moreover, if ω_0 denotes the growth bound of $(T(t))_{t \geq 0}$, we have

$$s(A) \leq \omega_0, \tag{1.6.18}$$

and for every $\lambda \in \mathbb{K}$, $\mathscr{R}e\, \lambda > \omega_0$ and $f \in E$

$$R(\lambda, A)f = \int_0^{+\infty} \exp(-\lambda t)\, T(t)f\, dt, \tag{1.6.19}$$

as well as

$$R(\lambda, A)\, Af = A\, R(\lambda, A)f = \int_0^{+\infty} \exp(-\lambda t)\, T(t)Af\, dt, \tag{1.6.20}$$

provided $f \in D(A)$.

By means of the resolvent operators $R(\lambda, A)$ it is possible to represent the semigroup $(T(t))_{t \geq 0}$ in several manners. Here we mention the exponential formula:

$$T(t)f = \lim_{n \to \infty} \left(\frac{n}{t} R\left(\frac{n}{t}, A\right)\right)^n f \quad \text{for every } f \in E \text{ and } t > 0, \tag{1.6.21}$$

and this limit is uniform on every bounded interval of $]0, +\infty[$.

One of the most important application of the theory of semigroups of bounded linear operators concerns linear differential equations in Banach spaces.

Given a closed linear operator $A: D(A) \to E$, consider the *abstract Cauchy problem*

$$\begin{cases} \dot{u}(t) = Au(t), & t \geq 0 \\ u(0) = u_0, & u_0 \in D(A). \end{cases} \tag{1.6.22}$$

A solution of problem (1.6.22) is a continuous differentiable function $u: [0, +\infty[\to E$ such that $u(0) = u_0$, $u(t) \in D(A)$ for every $t \geq 0$ and $\dot{u}(t) = Au(t)$ for every $t \geq 0$, where $\dot{u}(t)$ denotes the derivative of u at t.

Of course, when E is a concrete function space, then (1.6.22) may come from a concrete Cauchy problem associated with a partial differential equation.

The close connection between the solvability of the problem (1.6.22) and the theory of C_0-semigroups is indicated in the following result, whose proof can be found, e.g., in R. Nagel (Ed.) [1986, Chapter A-II, Corollary 1.2].

1.6.2 Theorem. *Let $A: D(A) \to E$ be a closed linear operator. Then the following statements are equivalent:*
(i) *$\rho(A) \neq \varnothing$ and for every $u_0 \in D(A)$ there exists a unique solution $u(\cdot)$ of the abstract Cauchy problem (1.6.22).*
(ii) *A is the generator of a C_0-semigroup $(T(t))_{t \geq 0}$ on E.*
   Moreover, if one of the above statements is true, then for every $t \geq 0$, $u(t) = T(t)u_0$.

Generation theorems

On account of Theorem 1.6.2 it is essential to have criteria to decide when a given linear operator A is the generator of a C_0-semigroup.

Here we present some of the most important ones (for a proof see, e.g., R. Nagel (Ed.) [1986, Chapter A-II, Theorems 1.7 and 2.11] or Butzer and Berens [1967, Theorem 1.3.6] and Pazy [1983, Chapter 1, Theorem 4.3]).

1.6.3 Theorem (*Hille-Yosida*). *A linear operator $A: D(A) \to E$ is the generator of a C_0-semigroup $(T(t))_{t \geq 0}$ on E satisfying $\|T(t)\| \leq \exp(\omega t)$ for every $t \geq 0$ and for some $\omega \in \mathbb{R}$ if and only if the following properties hold:*
(1) *A is closed and $D(A)$ is dense in E.*
(2) *$]\omega, +\infty[\subset \rho(A)$ and $\|R(\lambda, A)\| \leq \dfrac{1}{\lambda - \omega}$ for every $\lambda > \omega$.*

In fact, a more general result can be stated.

1.6.4 Theorem (*Feller-Miyadera-Phillips*). *A linear operator $A: D(A) \to E$ is the generator of a C_0-semigroup $(T(t))_{t \geq 0}$ on E satisfying $\|T(t)\| \leq M \exp(\omega t)$ for every $t \geq 0$ and for some $M > 0$ and $\omega \in \mathbb{R}$ if and only if the following properties hold:*
(1) *A is closed and $D(A)$ is dense in E.*
(2) *$]\omega, +\infty[\subset \rho(A)$ and $\|R(\lambda, A)^n\| \leq \dfrac{M}{(\lambda - \omega)^n}$ for every $\lambda > \omega$ and $n \geq 1$.*

From Theorem 1.6.3 one easily obtains the original characterization of generators of contraction C_0-semigroups (i.e., $\omega = 0$), due to Hille and Yosida. Next we shall see another characterization of such generators.

First let us recall the definition of dissipative operator. For every $f \in E$ we set

$$\mathfrak{A}(f) := \{\varphi \in E' | \varphi(f) = \|f\|^2 = \|\varphi\|^2\}. \tag{1.6.23}$$

By the Hahn-Banach theorem $\mathfrak{A}(f)$ is non-empty.

A linear operator $A: D(A) \to E$ is called *dissipative* if for every $f \in D(A)$ there exists $\varphi \in \mathfrak{A}(f)$ such that $\mathscr{R}e\,\varphi(Af) \leq 0$.

Equivalently this also means that

$$\|(\lambda I_{D(A)} - A)f\| \geq \lambda \|f\| \quad \text{for every } f \in D(A) \text{ and } \lambda > 0. \tag{1.6.24}$$

1.6.5 Theorem (*Lumer-Phillips*). *Let $A: D(A) \to E$ be a linear operator such that $D(A)$ is dense in E. Then A is the generator of a contraction C_0-semigroup if and only if A is dissipative and the range $(\lambda_0 I_{D(A)} - A)(D(A))$ is equal to E for some $\lambda_0 > 0$.*

In this case $(\lambda I_{D(A)} - A)(D(A)) = E$ for every $\lambda > 0$ and $\mathscr{R}e\,\varphi(Af) \leq 0$ for every $f \in D(A)$ and $\varphi \in \mathfrak{A}(f)$.

1.6.6 Corollary. *Let $A: D(A) \to E$ be a linear operator such that $D(A)$ is dense in E. Then the following statements are equivalent:*

(i) *A is dissipative and the range $(\lambda I_{D(A)} - A)(D(A))$ is dense in E for some $\lambda > 0$.*

(ii) *A is closable and the closure of A is the generator of a contraction C_0-semigroup.*

Finally we mention two generation (and approximation) theorems for discrete semigroups that are very important in the study of powers of positive approximation processes.

The first one is due to Trotter [1958] (see also Pazy [1983, Chapter 3, Theorem 6.7]).

1.6.7 Theorem (*Trotter*). *Let $(L_n)_{n \in \mathbb{N}}$ be a sequence of bounded linear operators on E and let $(\rho_n)_{n \in \mathbb{N}}$ be a decreasing sequence of positive real numbers tending to 0. Suppose that there exist $M \geq 0$ and $\omega \in \mathbb{R}$ such that*

$$\|L_n^k\| \leq M \exp(\omega \rho_n k) \quad \text{for every } k, n \in \mathbb{N}. \tag{1.6.25}$$

Let $A: D(A) \to E$ be the linear operator defined as

$$Af := \lim_{n \to \infty} \frac{L_n f - f}{\rho_n}, \tag{1.6.26}$$

for every $f \in D(A) := \left\{ g \in E \Big| \text{There exists } \lim_{n \to \infty} \frac{L_n g - g}{\rho_n} \in E \right\}.$

If $D(A)$ is dense in E and if the range $(\lambda I_{D(A)} - A)(D(A))$ is dense in E for some $\lambda > \omega$, then A is closable and the closure of A is the generator of a C_0-semigroup $(T(t))_{t \geq 0}$ such that for every $t \geq 0$ and for every sequence of positive integers $(k(n))_{n \in \mathbb{N}}$ satisfying $\lim_{n \to \infty} k(n)\rho_n = t$, one has

$$T(t) = \lim_{n \to \infty} L_n^{k(n)} \quad strongly. \tag{1.6.27}$$

In particular $\|T(t)\| \leq M \exp(\omega t)$ for every $t \geq 0$ and, if we take $k(n) := [t/\rho_n]$, where $[t/\rho_n]$ denotes the integer part of t/ρ_n,

$$T(t) = \lim_{n \to \infty} L_n^{[t/\rho_n]} \quad strongly, \tag{1.6.28}$$

and this last convergence is uniform on bounded intervals of $\mathbb{R}_+$.

Another result in the same direction was obtained by Schnabl [1972] (see also Nishishiraho [1978]).

1.6.8 Theorem (*Schnabl*). *Let $(L_n)_{n \in \mathbb{N}}$ be a sequence of linear contractions on E and let $(\rho_n)_{n \in \mathbb{N}}$ be a decreasing sequence of positive real numbers tending to 0.*

Consider the linear operator $A: D(A) \to E$ defined by (1.6.26) and suppose that there exists a family $(F_i)_{i \in I}$ of finite dimensional subspaces of $D(A)$ which are invariant under each L_n (i.e., $L_n(F_i) \subset F_i$ for every $i \in I$ and $n \geq 1$) and whose union is dense in E.

Then A is closable and the closure of A is the generator of a contraction C_0-semigroup which satisfies (1.6.27) and (1.6.28) for every $t \geq 0$ and for every sequence $(k(n))_{n \in \mathbb{N}}$ of positive integers satisfying $\lim_{n \to \infty} k(n)\rho_n = t$.

For further extensions of these results (even to the setting of locally convex spaces) see Becker and Nessel [1975], T. G. Kurtz [1970] and Seidman [1970].

Positive semigroups on $\mathscr{C}(X)$ and Markov processes

Here we briefly list some noteworthy properties of positive C_0-semigroups on $\mathscr{C}(X)$, X being a compact Hausdorff space.

For more details on positive semigroups on $\mathscr{C}(X)$ or on more general Banach lattices see R. Nagel (Ed.) [1986].

In the sequel we fix a compact Hausdorff space X.

Let $(T(t))_{t \geq 0}$ be a C_0-semigroup on $\mathscr{C}(X)$ and let $A: D(A) \to \mathscr{C}(X)$ be its generator. Then we have the following result which is due to Derndinger (see, e.g., R. Nagel (Ed.) [1986, Chapter B-IV, Theorem 1.1]).

1.6.9 Theorem (*Derndinger*). *If every operator* $T(t)$ *is positive, then* $s(A) = \omega_0 \in \sigma(A)$ *(see (1.6.7) and (1.6.17)). In particular, if* $s(A) < 0$, *then* $\lim_{t \to +\infty} \exp(\varepsilon t) \|T(t)\| = 0$ *for some* $\varepsilon > 0$.

Another interesting result was obtained by Feller [1952] (see also R. Nagel (Ed.) [1986, Chapter B-II, Theorem 1.6] for further extensions).

1.6.10 Theorem (*Feller*). *The following statements are equivalent:*
(i) *For every* $t \geq 0$, $T(t)$ *is positive and* $T(t)\mathbf{1} = \mathbf{1}$.
(ii) $A\mathbf{1} = 0$ *and* A *satisfies the positive maximum principle, i.e., for every* $f \in D(A)$ *and* $x_0 \in X$ *such that* $\sup_{x \in X} f(x) = f(x_0) \geq 0$, *one has* $Af(x_0) \leq 0$.

In the study of Lototsky-Schnabl operators (see Chapter 6) the following result is very useful. It is due to Arendt (see R. Nagel (Ed.) [1986, Chapter B-II, Theorem 1.20]) for the positive case and to Dorroh [1966] for the contractive one.

1.6.11 Theorem (*Arendt-Dorroh*). *Suppose that every* $T(t)$ *is positive (respectively, contractive). If* $\alpha \in \mathscr{C}(X)$ *and* $\alpha(x) > 0$ *for every* $x \in X$, *then the operator* αA *defined on the domain* $D(\alpha A) := D(A)$ *is the generator of a positive* C_0-*semigroup (respectively, contraction* C_0-*semigroup).*

Finally we conclude this section by reviewing some basic results on Markov processes and their connections with Feller semigroups. For more details the interested reader is referred, e.g., to Bauer [1981, Chapter 12] and Taira [1988].

From now on we shall suppose that X is a compact subset of a metric space (X_0, d) that we assume to be non-compact.

Let $(\Omega, \mathscr{U})$ be a measurable space and suppose that for every $x \in X$ a probability measure P^x is assigned on $\mathscr{U}$. Furthermore consider a family $(Z_t)_{t \geq 0}$ of $(\mathscr{U}, \mathfrak{B}(X))$-random variables from Ω into X such that

$$\text{the function } x \mapsto P^x(A) \text{ is Borel measurable for every } A \in \mathscr{U}, \quad (1.6.29)$$

and

$$P^x\{Z_{s+t} \in B | \mathscr{U}_s\} = P^{Z_s}\{Z_t \in B\} \quad P^x\text{-a.e.,} \quad (1.6.30)$$

for every $s, t \geq 0$, $x \in X$ and $B \in \mathfrak{B}(X)$, where $\mathscr{U}_s$ denotes the σ-algebra of the events up to time s, i.e., the σ-algebra generated by $(Z_u)_{0 \leq u \leq s}$, $P^{Z_s}\{Z_t \in B\}$ denotes the random variable $\omega \mapsto P^{Z_s(\omega)}\{Z_t \in B\}$ and $P^x\{Z_{s+t} \in B | \mathscr{U}_s\}$ is the *conditional probability* of $\{Z_{s+t} \in B\}$ *given* $\mathscr{U}_s$, i.e., the (almost surely uniquely determined) $\mathscr{U}_s$-measurable and positive functions on Ω such that for every

$A \in \mathscr{U}_s$

$$\int_A P^x \{ Z_{s+t} \in B | \mathscr{U}_s \} \, dP^x = P^x (A \cap \{ Z_{s+t} \in B \}).$$

Under these assumptions we shall call the quadruple

$$(\Omega, \mathscr{U}, (P^x)_{z \in X}, (Z_t)_{t \geq 0}), \tag{1.6.31}$$

a *Markov process* with state space X.

The Markov process is called *normal* if

$$P^x \{ Z_0 = x \} = 1 \quad \text{for every } x \in X. \tag{1.6.32}$$

Intuitively we may think of a particle which moves in X after a random experiment $\omega \in \Omega$. Then for every experiment $\omega \in \Omega$, $Z_t(\omega)$ expresses the position of the particle at time $t \geq 0$.

If $A \in \mathscr{U}$, then $P^x(A)$ may be interpreted as the probability of the event A if the particle starts at position $x \in X$.

In particular $P^x \{ Z_t \in B \}$ is the probability that a particle starting at position x will be found in the set B at time t.

The intuitive meaning of condition (1.6.30) is that the future behavior of a particle, knowing its history up to the time t is the same as the behavior of a particle starting at $X_t(\omega)$.

Given a Markov process (1.6.31), for every $\omega \in \Omega$ the mapping $t \mapsto Z_t(\omega)$ is called a *path* of the process.

Furthermore the random variable $\zeta \colon \Omega \to \mathbb{R}$ defined by

$$\zeta(\omega) := \inf \{ t \geq 0 | Z_t(\omega) \in \partial X \} \quad \text{for each } \omega \in \Omega \tag{1.6.33}$$

is called the *lifetime* of the process.

A Markov process is called *right-continuous* if for every $x \in X$

$$P^x \{ \omega \in \Omega | \text{the path } t \mapsto Z_t(\omega) \text{ is right-continuous on } [0, +\infty[\} = 1. \tag{1.6.34}$$

The Markov process is called *continuous* if for every $x \in X$

$$P^x \{ \omega \in \Omega | \text{the path } t \mapsto Z_t(\omega) \text{ is continuous on } [0, \zeta(\omega)[\} = 1. \tag{1.6.35}$$

Now we proceed to discuss the concept of kernel which is strongly related to Markov processes.

A *Markov kernel* on X is a function $Q \colon X \times \mathscr{B}(X) \to \mathbb{R}$ such that

the function $x \mapsto Q(x, B)$ is Borel measurable for each $B \in \mathscr{B}(X)$, (1.6.36)

and

the function $B \mapsto Q(x, B)$ is a Borel measure on X satisfying $Q(x, X) = 1$,

$$(1.6.37)$$

for each $x \in X$.

A *Markov transition function* on X is a family $(P_t)_{t \geq 0}$ of Markov kernels such that

$$P_{s+t}(x, B) = \int P_t(x, dy) P_s(y, B) \left(:= \int P_s(\cdot, B) \, dP_t(x, \cdot) \right) \qquad (1.6.38)$$

for every $s, t \geq 0$, $x \in X$ and $B \in \mathfrak{B}(X)$.

Formula (1.6.38) is called the *Chapman-Kolmogorov equation*. From it we also obtain $P_s(x, X) \leq P_t(x, X)$ if $s \geq t$. The Markov transition function $(P_t)_{t \geq 0}$ is called *normal* if

$$P_0(x, \{x\}) = 1 \quad \text{for every } x \in X. \qquad (1.6.39)$$

The strong link between Markov processes and Markov transition functions is expressed by the following result.

1.6.12 Theorem. *If a normal Markov process* (1.6.31) *with state space X is given, then by putting*

$$P_t(x, B) = P^x\{Z_t \in B\} \quad \text{for every } t \geq 0, x \in X \text{ and } B \in \mathfrak{B}(X), \quad (1.6.40)$$

one obtains a normal Markov transition function on X (note that $P_t(x, \cdot)$ is the distribution $P^x_{Z_t}$ of Z_t with respect to P^x (see (1.3.1))).

Conversely, if $(P_t)_{t \geq 0}$ is a normal Markov transition function on X, there exists a normal Markov process with state space X which satisfies (1.6.40).

For a proof see Bauer [1981, Theorems 12.4.5 and 12.4.6].

From the viewpoint of functional analysis the transition functions are more convenient than Markov processes. In fact, it is possible to associate to them a semigroup of bounded linear operators on the space

$$\mathscr{B}_0(X) := \{ f \colon X \to \mathbb{R} \,|\, f \text{ is Borel measurable and bounded} \} \qquad (1.6.41)$$

endowed with the sup-norm.

Indeed, if $(P_t)_{t \geq 0}$ is a Markov transition function on X, for every $t \geq 0$ consider the positive linear operator $T(t) \colon \mathscr{B}_0(X) \to \mathscr{B}_0(X)$ defined by

$$T(t)(f)(x) := \int P_t(x, dy) f(y) \left(:= \int f \, dP_t(x, \cdot) \right) \quad \text{for every } f \in \mathscr{B}_0(X) \text{ and } x \in X.$$

$$(1.6.42)$$

By using (1.6.38) and (1.6.39) one easily shows that $(T(t))_{t\geq 0}$ is a semigroup of contractions on $\mathscr{B}_0(X)$.

This semigroup is called the *transition semigroup* associated with $(P_t)_{t\geq 0}$ or with the Markov process (1.6.31), if $(P_t)_{t\geq 0}$ is given by (1.6.40).

In this case, according to (1.3.2), we also have

$$T(t)(f)(x) = \int f \circ Z_t \, dP^x. \tag{1.6.43}$$

The Markov transition function $(P_t)_{t\geq 0}$ will be called a *Feller transition function* if $T(t)(\mathscr{C}(X)) \subset \mathscr{C}(X)$ for every $t \geq 0$.

In this case we shall denote again by $T(t)$ the restriction of $T(t)$ to $\mathscr{C}(X)$.

Clearly $(P_t)_{t\geq 0}$ is a Feller transition function if and only if for every $t \geq 0$ the mapping $x \mapsto P_t(x, \cdot)$ is continuous from X into $\mathscr{M}^+(X)$.

However note that, in general, the semigroup $(T(t))_{t\geq 0}$ is not strongly continuous on $\mathscr{C}(X)$.

A Markov transition function $(P_t)_{t\geq 0}$ is said to be *uniformly stochastically continuous* on X if for every $\varepsilon > 0$

$$\lim_{t\to 0^+} \sup_{x\in X} 1 - P_t(x, B(x,\varepsilon)) = 0, \tag{1.6.44}$$

where $B(x,\varepsilon) := \{ y \in X \,|\, d(x,y) < \varepsilon \}$.

For these kinds of transition functions Theorem 1.6.12 can be improved considerably.

1.6.13 Theorem. *If $(P_t)_{t\geq 0}$ is a uniformly stochastically continuous normal Markov transition function, then there exists a right-continuous normal Markov process with state space X which satisfies (1.6.40) and whose paths have left-hand limits on $[0, \zeta[$ almost surely.*

Furthermore, if for every $\varepsilon > 0$

$$\lim_{t\to 0^+} \frac{1}{t} \sup_{x\in X} P_t(x, X \backslash B(x,\varepsilon)) = 0, \tag{1.6.45}$$

then the Markov process can be chosen so that its paths are almost surely continuous on $[0, \zeta[$.

For a proof see Dynkin [1961].

Our last result shows the connection between uniformly stochastically continuous transition functions and Feller semigroups.

We recall that a positive C_0-semigroup $(T(t))_{t\geq 0}$ on $\mathscr{C}(X)$ is called a *Feller semigroup* if $T(t)\mathbf{1} = \mathbf{1}$ for every $t \geq 0$.

The next result can be derived from Taira [1988, Theorem 9.2.6] (see also van Casteren [1985, Theorem 2.1]).

1.6.14 Theorem. *If $(P_t)_{t \geq 0}$ is a uniformly stochastically continuous normal Feller transition function, then the semigroup $(T(t))_{t \geq 0}$ defined by (1.6.42) is a Feller semigroup on $\mathscr{C}(X)$.*

Conversely, every Feller semigroup $(T(t))_{t \geq 0}$ on $\mathscr{C}(X)$ is the transition semigroup of a uniformly stochastically continuous normal Feller transition function $(P_t)_{t \geq 0}$ on X or, equivalently, of a right-continuous normal Markov process with state space X, whose paths have left-hand limits on $[0, \zeta[$ almost surely.

Furthermore, for every $t \geq 0$ and $x \in X$, the Borel measure $P_t(x, \cdot)$ is exactly the measure which corresponds to the Radon measure

$$\mu_{x,t}(f) := T(t)(f)(x) \quad (f \in \mathscr{C}(X)) \tag{1.6.46}$$

via the Riesz representation theorem.

On account of this bi-unique correspondence between Feller semigroups $(T(t))_{t \geq 0}$ and Feller transition functions $(P_t)_{t \geq 0}$, we see that, if $\mu \in \mathscr{M}^+(X)$ and if for every $t \geq 0$ we consider the Radon measure $T(t)(\mu)$ defined by

$$T(t)(\mu)(f) := \mu(T(t)(f)) \quad \text{for every } f \in \mathscr{C}(X), \tag{1.6.47}$$

(see (1.2.60)), then the unique Borel measure which corresponds to $T(t)(\mu)$ via the Riesz representation theorem is the measure $P_t(\tilde{\mu})$ defined by

$$P_t(\tilde{\mu})(B) := \int P_t(x, B) \, d\tilde{\mu}(x) \quad \text{for each } B \in \mathscr{B}(X), \tag{1.6.48}$$

$\tilde{\mu}$ being the Borel measure which corresponds to μ.

Intuitively, if μ gives the distribution of the initial position of the particle, then $P_t(\mu)$ (or $T(t)(\mu)$) gives the distribution of the position at time t.

Chapter 2

Korovkin-type theorems for bounded positive Radon measures

In this chapter we shall present some basic Korovkin-type theorems for bounded positive Radon measures on a locally compact Hausdorff space. From these results we shall derive later on the main Korovkin-type theorems for positive operators.

In Section 2.1 we investigate under which circumstances the vague convergence of an arbitrary equicontinuous net of bounded positive Radon measures is a consequence of its convergence on some special subspaces.

Similar problems are also treated by considering nets of positive contractive Radon measures.

In the remaining sections particular care will be devoted to the case when the limit measure is discrete or a Dirac measure at some point. The latter will directly lead to the study of Choquet boundaries.

2.1 Determining subspaces for bounded positive Radon measures

Let X be a locally compact Hausdorff space and fix a bounded positive Radon measure $\mu \in \mathcal{M}_b^+(X)$. In this section we are interested in studying those subsets H of $\mathcal{C}_0(X)$ which can serve as a test set to check the vague convergence to μ of nets of equicontinuous positive Radon measures.

2.1.1 Definition. *A subset H of $\mathcal{C}_0(X)$ is called a determining subset for μ with respect to bounded positive Radon measures or, more simply, a D_+-subset for μ, if it satisfies the following property:*

> *if $(\mu_i)_{i \in I}^{\leq}$ is an arbitrary net in $\mathcal{M}_b^+(X)$ satisfying $\sup_{i \in I} \|\mu_i\| < +\infty$*
> *and if $\lim_{i \in I}^{\leq} \mu_i(h) = \mu(h)$ for all $h \in H$, then $\lim_{i \in I}^{\leq} \mu_i(f) = \mu(f)$*
> *for every $f \in \mathcal{C}_0(X)$, i.e., $(\mu_i)_{i \in I}^{\leq}$ converges vaguely to μ.* $\qquad$ (2.1.1)

Obviously, H is a D_+-subset for μ if and only if the linear subspace $\mathcal{L}(H)$ generated by H is a D_+-subset for μ. For this reason, we restrict ourself to the case where H is a subspace of $\mathscr{C}_0(X)$ and, in this case, we shall say that H is a D_+-*subspace* for μ.

Of course, by virtue of (1.1.11) and by replacing, if necessary, the net $(\mu_i)_{i \in I}^{\leq}$ with a suitable subfamily $(\mu_i)_{i \geq i_0}^{\leq}$ $(i_0 \in I)$, the condition $\sup_{i \in I} \|\mu_i\| < +\infty$ in (2.1.1) is automatically satisfied if X is compact, $1 \in H$ and $\lim_{i \in I} \mu_i(h) = \mu(h)$ for every $h \in H$.

To establish whether a fixed subspace H of $\mathscr{C}_0(X)$ is a D_+-subspace for μ it is often useful to study and to determine the following *convergence subspace associated with μ and H*:

$$D_+(H,\mu) := \left\{ f \in \mathscr{C}_0(X) \middle| \lim_{i \in I} \mu_i(f) = \mu(f) \text{ for every net } (\mu_i)_{i \in I}^{\leq} \text{ in } \mathscr{M}_b^+(X) \right.$$

$$\left. \text{satisfying } \sup_{i \in I} \|\mu_i\| < +\infty \text{ and } \lim_{i \in I} \mu_i(h) = \mu(h) \text{ for every } h \in H \right\}.$$
$$(2.1.2)$$

Consequently, one tries to find conditions which ensure the equality $D_+(H,\mu) = \mathscr{C}_0(X)$, i.e., H is a D_+-subspace for μ.

Note that the subspace $D_+(H,\mu)$ can be also defined even if H is only a subset of functions and, of course, $D_+(H,\mu) = D_+(\mathscr{L}(H),\mu)$.

However, the study of the subspace $D_+(H,\mu)$ seems to have its own interest since it is the maximal subspace on which every equicontinuous net of bounded positive Radon measures converges to μ whenever it converges to μ on H.

We also introduce the *uniqueness subspace associated with μ and H* defined as:

$$U_+(H,\mu) := \{ f \in \mathscr{C}_0(X) \mid v(f) = \mu(f) \text{ for every } v \in \mathscr{M}_b^+(X) \text{ such that } v = \mu \text{ on } H \}.$$
$$(2.1.3)$$

Thus, $U_+(H,\mu)$ is the maximal subspace of $\mathscr{C}_0(X)$ containing H on which each bounded positive Radon measure coincides with μ whenever it is equal to μ on H. Consequently, $U_+(H,\mu)$ is equal to $\mathscr{C}_0(X)$ if and only if there exists a unique bounded positive Radon measure on X which extends the restriction $\mu_{|H}$ to $\mathscr{C}_0(X)$ (namely, μ itself).

We have the following preliminary result.

2.1.2 Proposition. *Let H be a subspace of $\mathscr{C}_0(X)$. Then*

$$D_+(H,\mu) = U_+(H,\mu).$$
$$(2.1.4)$$

Proof. The inclusion $D_+(H,\mu) \subset U_+(H,\mu)$ being obvious, we have only to prove the converse one. So, let $f \in U_+(H,\mu)$ and consider an equicontinuous net $(\mu_i)_{i \in I}^{\leq}$ in $\mathscr{M}_b^+(X)$ which converges pointwise to μ on H. If the net $(\mu_i(f))_{i \in I}^{\leq}$ does not converge to $\mu(f)$, then there exists $\varepsilon_0 > 0$ and for every $i \in I$ there exists $\alpha(i) \in I$, $\alpha(i) \geq i$, such that

$$|\mu_{\alpha(i)}(f) - \mu(f)| \geq \varepsilon_0. \tag{1}$$

The set $\{\mu_{\alpha(i)} | i \in I\}$ is strongly bounded in $\mathcal{M}(X)$ and so it is relatively compact. So, there exist a filter $\mathfrak{F}$ on I finer than the filter of sections on I and $\upsilon \in \mathcal{M}_b(X)$ such that $\lim_{\mathfrak{F} \atop i \in I} \mu_{\alpha(i)} = \upsilon$ vaguely. Consequently, υ is positive and for every $h \in H$, $\upsilon(h) = \lim_{\mathfrak{F} \atop i \in I} \mu_{\alpha(i)}(h) = \lim_{\leq \atop i \in I} \mu_i(h) = \mu(h)$. Since $f \in U_+(H, \mu)$, we have $\mu(f) = \upsilon(f) = \lim_{\mathfrak{F} \atop i \in I} \mu_{\alpha(i)}(f)$ in contradiction with (1). $\qquad \square$

Thus, because of Proposition 2.1.2, the study of the subspace $D_+(H, \mu)$ can be transferred to that of the subspace $U_+(H, \mu)$ which is easier to handle.

Furthermore, we point out that, if in the definition of $D_+(H, \mu)$ we drop the hypothesis of equicontinuity for the net of measures, then equality (2.1.4) is no longer true.

To show this consider the following example due to Amir and Ziegler [1978].

Example. Let $X := \,]0, 1]$ and denote by H the subspace of $\mathscr{C}_0(X)$ generated by the functions x^2, x^3 and x^4. If $x_0 \in X$ and if $\upsilon \in \mathcal{M}_b^+(X)$ satisfies $\upsilon = \varepsilon_{x_0}$ on H, then $\upsilon(h) = 0$ for $h(x) = x^2(x - x_0)^2$ $(x \in X)$ and therefore its support must be contained in $\{x_0\}$. Consequently there exists $\lambda \in \mathbb{R}$ such that $\upsilon = \lambda \varepsilon_{x_0}$ and by evaluating υ and $\lambda \varepsilon_{x_0}$ on the function x^2, we obtain $\lambda = 1$ and $\upsilon = \varepsilon_{x_0}$. Hence $U_+(H, \varepsilon_{x_0}) = \mathscr{C}_0(X)$. On the other hand, for every $n \in \mathbb{N}$, set $\mu_n := \varepsilon_{x_0} + n\varepsilon_{1/n}$. The sequence of the norms of $(\mu_n)_{n \in \mathbb{N}}$ is not bounded and $\lim_{n \to \infty} \mu_n(h) = \varepsilon_{x_0}(h)$ for every $h \in H$, but it does not converge vaguely to ε_{x_0}.

We begin our investigations on the subspaces $D_+(H, \mu)$ by considering first the case where X is compact.

We recall that a subspace H of $\mathscr{C}(X)$ is said to be *cofinal* in $\mathscr{C}(X)$ if for every $f \in \mathscr{C}(X)$ there exists $h \in H$ such that $f \leq h$, or, equivalently, if there exists $h_0 \in H$ such that $h_0(x) > 0$ for every $x \in X$.

For a cofinal subspace we may use the classical Hahn-Banach theorem to characterize the subspace $U_+(H, \mu)$.

2.1.3 Theorem. *Suppose X compact and consider a cofinal subspace H of $\mathscr{C}(X)$. For every $f \in \mathscr{C}(X)$ the following statements are equivalent:*

(i) $f \in U_+(H, \mu)$.

(ii) $\displaystyle\sup_{h \in H \atop h \leq f} \mu(h) = \mu(f) = \inf_{k \in H \atop f \leq k} \mu(k)$.

(iii) *For every $\varepsilon > 0$ there exist $h, k \in H$ such that $h \leq f \leq k$ and $\mu(k) - \mu(h) < \varepsilon$.*

Proof. For the sake of completeness we present the sketch of the proof of this classical result.

(i) $\Rightarrow$ (ii): By the Hahn-Banach theorem applied to the sublinear mapping $p\colon \mathscr{C}(X) \to \mathbb{R}$ defined by $p(g) := \inf_{\substack{k \in H \\ g \le k}} \mu(k)$ for every $g \in \mathscr{C}(X)$, there exists a linear form v on $\mathscr{C}(X)$ such that $v \le p$ and $v(f) = p(f)$. We note that, if $g \in \mathscr{C}(X)$ and $g \le 0$, then $v(g) \le p(g) \le 0$. So v is positive and hence $v \in \mathscr{M}^+(X)$. Moreover $v = \mu$ on H since $p = \mu$ on H. So $\mu(f) = v(f) = p(f)$, because of (i).

On the other hand, since $-f \in U_+(H,\mu)$, we also have $\mu(-f) = p(-f)$, that is
$$\mu(f) = -p(-f) = \sup_{\substack{h \in H \\ h \le f}} \mu(h).$$

(ii) $\Rightarrow$ (i): If $v \in \mathscr{M}^+(X)$ and $v = \mu$ on H, then from (ii) and from the inequalities $\sup_{\substack{h \in H \\ h \le f}} v(h) \le v(f) \le \inf_{\substack{k \in H \\ f \le k}} v(k)$ it follows that $v(f) = \mu(f)$.

The equivalence (ii) $\Leftrightarrow$ (iii) is obvious. $\qquad\square$

Theorem 2.1.3 and Proposition 2.1.2 allow us to characterize those cofinal subspaces of $\mathscr{C}(X)$ which are determining subspaces. In fact, the next result shows that, if X is compact, then every determining subspace is necessarily cofinal.

2.1.4 Theorem. *Suppose X compact and consider a subspace H of $\mathscr{C}(X)$. Then, the following statements are equivalent:*
(i) *H is a D_+-subspace for μ.*
(ii) *H is cofinal in $\mathscr{C}(X)$ and for every $f \in \mathscr{C}(X)$, $\sup_{\substack{h \in H \\ h \le f}} \mu(h) = \mu(f) = \inf_{\substack{k \in H \\ f \le k}} \mu(k)$.*
(iii) *H is cofinal in $\mathscr{C}(X)$ and for every $f \in \mathscr{C}(X)$ and $\varepsilon > 0$ there exist $h, k \in H$ such that $h \le f \le k$ and $\mu(k) - \mu(h) < \varepsilon$.*

Proof. By Theorem 2.1.3 and Proposition 2.1.2 we only have to show that, if H is a D_+-subspace for μ, then H is cofinal in $\mathscr{C}(X)$. Indeed, consider the canonical embedding $\Phi\colon X \to H'$ defined by $\Phi(x) := \varepsilon_{x|H}$ for every $x \in X$, where H' denotes the dual of H endowed with the weak*-topology, and set $K := \overline{\mathrm{co}}(\Phi(X))$. Thus K is convex and compact.

Suppose that H is a D_+-subspace for μ. We claim that the zero functional $0 \in H'$ does not belong to K. Otherwise, there exists a net $(\mu_\alpha)_{\alpha \in A}^{\le}$ in $\mathrm{co}(\Phi(X))$ which weak*-converges to 0. Then, for every $\alpha \in A$ there exist finitely many points $x_{1,\alpha}, \ldots, x_{n(\alpha),\alpha}$ in X and $\lambda_{1,\alpha}, \ldots, \lambda_{n(\alpha),\alpha}$ in $\mathbb{R}_+$ with $\sum_{i=1}^{n(\alpha)} \lambda_{i,\alpha} = 1$, such that
$$\mu_\alpha = \sum_{i=1}^{n(\alpha)} \lambda_{i,\alpha} \varepsilon_{x_{i,\alpha|H}}.$$

Put $\eta_\alpha := \sum_{i=1}^{n(\alpha)} \lambda_{i,\alpha} \varepsilon_{x_{i,\alpha}} \in \mathcal{M}^+(X)$. Then $\|\eta_\alpha\| = \eta_\alpha(\mathbf{1}) = 1$ and there exist $\eta \in \mathcal{M}^+(X)$ and a filter $\mathfrak{F}$ on A, finer than the filter of sections, such that $\lim_{\mathfrak{F}} \eta_\alpha = \eta$ vaguely. Clearly $\eta = 0$ on H and so, if we put $\upsilon := \mu + \eta$, we have
$\upsilon = \mu$ on H. Hence $\upsilon = \mu$ and so $\eta = 0$ on $\mathcal{C}(X)$, while $\eta(\mathbf{1}) = \lim_{\substack{\mathfrak{F} \\ \alpha \in A}} \eta_\alpha(\mathbf{1}) = 1$.

Having proved that $0 \notin K$, by using the separation theorem 1.4.8, (3), and the representation of the dual of H', we can choose a strictly positive number λ and an element h in H such that $\upsilon(h) > \lambda$ for every $\upsilon \in K$. In particular $h(x) > \lambda > 0$ for every $x \in X$ and this completes the proof. $\square$

Remark. By applying Theorem 2.1.4 to the particular case $\mu = 0$, we deduce that a subspace H of $\mathcal{C}(X)$ is a D_+-subspace for the null functional if and only if it is cofinal in $\mathcal{C}(X)$. Afterwards we shall characterize a similar property for locally compact spaces (see Corollary 2.1.8).

The next example shows that, if X is locally compact and non compact, a result similar to Theorem 2.1.4 cannot be expected.

Example. Let $X :=]0, 1]$ and let H be the linear subspace of $\mathcal{C}_0(X)$ generated by the two functions e_1 and e_2, where $e_k(x) := x^k$ for every $x \in X$, $k = 1, 2$. Then H is not cofinal in $\mathcal{C}_0(X)$ (for example, the function $\sqrt{x}$ cannot be majorized by any element of H). Nevertheless, H is a D_+-subspace for the Dirac measure ε_1.

To see this, take $\upsilon \in \mathcal{M}_b^+(X)$ such that $\upsilon = \varepsilon_1$ on H. Then $\upsilon(e_1 - e_2) = 0$. Hence the support $\operatorname{Supp}(\upsilon)$ of υ is contained in $\{x \in X | (e_1 - e_2)(x) = 0\} = \{1\}$. So, there exists $\lambda \geq 0$ such that $\upsilon = \lambda \varepsilon_1$; but, since $\upsilon(e_1) = 1$, we have $\lambda = 1$ and $\upsilon = \varepsilon_1$.

Our next goal is to present in the setting of locally compact spaces a general characterization of determining subspaces (not necessarily cofinal) similar to the one presented in Theorem 2.1.4.

We begin with some preliminary considerations.

From now on we fix a subspace H of $\mathcal{C}_0(X)$. Set

$$H_u^* := \{f \in \mathcal{C}_0(X) | \text{For every } \varepsilon > 0 \text{ there exists } k \in H \text{ such that } f \leq k + \varepsilon\},$$
$$\tag{2.1.5}$$

$$H_l^* := \{f \in \mathcal{C}_0(X) | \text{For every } \varepsilon > 0 \text{ there exists } h \in H \text{ such that } h - \varepsilon \leq f\}$$
$$\tag{2.1.6}$$

and

$$H^* := H_u^* \cap H_l^*.   \tag{2.1.7}$$

It is easy to verify that H_u^*, H_l^* and H^* are closed; moreover, if $v \in \mathcal{M}_b^+(X)$ and $f \in H_u^*$ ($f \in H_l^*$, respectively), then

$$v(f) - \varepsilon \|v\| \leq v(k) \quad \textit{for every } \varepsilon > 0 \textit{ and } k \in H \textit{ such that } f \leq k + \varepsilon \quad (2.1.8)$$

$$(v(h) \leq v(f) + \varepsilon \|v\| \quad \textit{for every } \varepsilon > 0 \textit{ and } h \in H \textit{ such that } h - \varepsilon \leq f, \quad (2.1.9)$$

respectively).

The following result is a technical preliminary to the main result.

2.1.5 Lemma. *Let f and k be functions in $\mathcal{C}_0(X)$ and let $\varepsilon > 0$. Then the following statements are equivalent*:

(i) $f \leq k + \varepsilon$.

(ii) $(f - k)^+ \leq \varepsilon$.

(iii) *There exists $u \in \mathcal{C}_0^+(X)$ such that $\|u\| \leq \varepsilon$ and $f \leq k + u$.*

(iv) *For every $\mu \in \mathcal{M}_b^+(X)$ with $\|\mu\| \leq 1$ one has $\mu(f) \leq \mu(k) + \varepsilon$.*

Proof. We only have to show the implication (iv) $\Rightarrow$ (ii), because the others are evident. In fact, suppose that (iv) holds and consider the continuous sublinear mapping $p: \mathcal{C}_0(X) \to \mathbb{R}$ defined by $p(g) := \|g^+\|$ for every $g \in \mathcal{C}_0(X)$.

By the Hahn-Banach theorem there exists $\mu \in \mathcal{M}_b(X)$ such that $\mu \leq p$ and $\mu(f - k) = p(f - k)$. We note that, if $g \in \mathcal{C}_0(X)$ and $g \leq 0$, then $\mu(g) \leq \|g^+\| = 0$ and hence μ is positive. Moreover, if $g \in \mathcal{C}_0(X)$ and $\|g\| \leq 1$, then $g \leq 1$ and, consequently, $\mu(g) \leq \|g^+\| \leq 1$ because $g^+ \leq 1$. So, since $\|\mu\| \leq 1$, $\mu(f - k) \leq \varepsilon$ and hence $(f - k)^+ \leq \|(f - k)^+\| \leq \varepsilon$. $\qquad\square$

We now introduce the enveloping functions in terms of which we shall characterize the subspace $U_+(H, \mu)$. These functions appear in some problems of integral representations and of extension of positive operators (see Donner [1982]).

Consider the functions $\hat{\mu}_H: \mathcal{C}_0(X) \to \mathbb{R} \cup \{+\infty\}$ and $\check{\mu}_H: \mathcal{C}_0(X) \to \mathbb{R} \cup \{-\infty\}$ defined by

$$\hat{\mu}_H(f) := \sup_{\varepsilon > 0} \left(\inf_{\substack{k \in H \\ f \leq k + \varepsilon}} \mu(k) \right), \qquad (2.1.10)$$

$$\check{\mu}_H(f) := -\hat{\mu}_H(-f) = \inf_{\varepsilon > 0} \left(\sup_{\substack{h \in H \\ h - \varepsilon \leq f}} \mu(h) \right) \qquad (2.1.11)$$

for every $f \in \mathcal{C}_0(X)$, where the convention inf $\varnothing = +\infty$ and sup $\varnothing = -\infty$ must be observed.

We list some elementary properties of $\hat{\mu}_H$ and $\check{\mu}_H$, whose proofs are straightforward.

$$\mu(f) - \varepsilon\|\mu\| \leq \hat{\mu}_H(f) \text{ and } \check{\mu}_H(f) \leq \mu(f) + \varepsilon\|\mu\|, \quad (f \in \mathscr{C}_0(X), \varepsilon > 0) \quad (2.1.12)$$

$$\check{\mu}_H(f) \leq \mu(f) \leq \hat{\mu}_H(f), \qquad\qquad (f \in \mathscr{C}_0(X)) \quad (2.1.13)$$

$$\check{\mu}_H(h) = \mu(h) = \hat{\mu}_H(h), \qquad\qquad (h \in H) \quad (2.1.14)$$

$$\hat{\mu}_H(f + g) \leq \hat{\mu}_H(f) + \hat{\mu}_H(g), \qquad\qquad (f, g \in \mathscr{C}_0(X)) \quad (2.1.15)$$

$$\hat{\mu}_H(tf) = t\,\hat{\mu}_H(f), \qquad\qquad (f \in \mathscr{C}_0(X), t \geq 0) \quad (2.1.16)$$

with the convention $0 \cdot (+\infty) = 0$,

$$\hat{\mu}_H(f) \leq \hat{\mu}_H(g) \qquad\qquad (f, g \in \mathscr{C}_0(X), f \leq g). \quad (2.1.17)$$

An important property of $\hat{\mu}_H$ is indicated below:

$$\text{The numerical function } \hat{\mu}_H \text{ is lower semi-continuous on } \mathscr{C}_0(X). \quad (2.1.18)$$

In fact, let us consider $f \in \mathscr{C}_0(X)$ and $\alpha \in \mathbb{R}$ such that $\alpha < \hat{\mu}_H(f)$. If $f \notin H_u^*$, then the subset $U := \mathscr{C}_0(X) \backslash H_u^*$ is an open neighborhood of f and for every $g \in U$ we have $\hat{\mu}_H(g) = +\infty > \alpha$.

If $f \in H_u^*$, there exists $\varepsilon > 0$ such that $\alpha < \inf\limits_{\substack{k \in H \\ f \leq k+\varepsilon}} \mu(k)$. Then, for all $g \in \mathscr{C}_0(X)$ with $\|g - f\| \leq \dfrac{\varepsilon}{2}$ and for all $h \in H$ such that $g \leq h + \dfrac{\varepsilon}{2}$, we have $f \leq \|g - f\| + g \leq h + \varepsilon$. So,

$$\alpha < \inf\limits_{\substack{k \in H \\ f \leq k+\varepsilon}} \mu(k) \leq \inf\limits_{\substack{h \in H \\ g \leq h+\varepsilon/2}} \mu(h) \leq \hat{\mu}_H(g),$$

and this completes the proof. $\qquad\qquad\qquad\qquad\qquad\qquad\qquad\qquad\qquad\square$

We are now in a position to show the following fundamental result.

2.1.6 Theorem. *For every $f \in \mathscr{C}_0(X)$ the following statements are equivalent:*
  (i) *$f \in U_+(H, \mu)$.*
  (ii) *$\check{\mu}_H(f) = \mu(f) = \hat{\mu}_H(f)$ (in particular $f \in H^*$).*
  (iii) *For every $\varepsilon > 0$ there exist $h, k \in H$ such that $h - \varepsilon \leq f \leq k + \varepsilon$ and $|\mu(k - h)| \leq \varepsilon$.*
  (iv) *For every $\varepsilon > 0$ there exist $h, k \in H$ such that $\|(f - k)^+\| \leq \varepsilon$, $\|(h - f)^+\| \leq \varepsilon$ and $|\mu(k - h)| \leq \varepsilon$.*
  (v) *For every $\varepsilon > 0$ there exist $h, k \in H$ and $u, v \in \mathscr{C}_0^+(X)$ such that $\|u\| \leq \varepsilon$, $\|v\| \leq \varepsilon$, $h - v \leq f \leq k + u$ and $|\mu(k - h)| \leq \varepsilon$.*

Proof. Statements (iii), (iv) and (v) are equivalent by virtue of Lemma 2.1.5.

(i) $\Rightarrow$ (ii): Let $f \in U_+(H, \mu)$. By (2.1.13) we know that $\breve{\mu}_H(f) \leq \mu(f) \leq \hat{\mu}_H(f)$. Suppose that $\breve{\mu}_H(f) < \hat{\mu}_H(f)$. From (2.1.15), (2.1.16) and (2.1.18) it follows that $\hat{\mu}_H$ is a hypolinear lower semi-continuous mapping on $\mathscr{C}_0(X)$ and so, as a consequence of the Anger-Lembcke theorem (see Corollary 1.4.7), for every fixed $\alpha \in \mathbb{R}$ with $\breve{\mu}_H(f) < \alpha < \hat{\mu}_H(f)$ there exists $\upsilon \in \mathscr{M}_b(X)$ such that $\upsilon(f) = \alpha$ and $\upsilon \leq \hat{\mu}_H$ on $\mathscr{C}_0(X)$. The measure υ is necessarily positive since for all $g \in \mathscr{C}_0(X)$, $g \leq 0$, we have $\upsilon(g) \leq \hat{\mu}_H(g) \leq \hat{\mu}_H(0) = 0$ (see (2.1.17) and (2.1.14)). Moreover, if $h \in H$, then by (2.1.14) it follows that $\upsilon(h) \leq \hat{\mu}_H(h) = \mu(h)$ and hence $\upsilon = \mu$ on H. Since $f \in U_+(H, \mu)$ we have $\mu(f) = \upsilon(f) = \alpha$ but this is not possible because α was chosen arbitrarily.

(ii) $\Rightarrow$ (iii): Let $\varepsilon > 0$ and set $\varepsilon' := \dfrac{\varepsilon}{1 + \|\mu\|}$. Since $\displaystyle\inf_{\substack{k \in H \\ f \leq k + \varepsilon'/2}} \mu(k) \leq \hat{\mu}_H(f) = \mu(f) < \mu(f) + \varepsilon/2$, there exists $k \in H$ such that

$$f \leq k + \varepsilon'/2 \quad \text{and} \quad \mu(k) < \mu(f) + \varepsilon/2.$$

Analogously one proves that there exists $h \in H$ such that

$$h - \varepsilon'/2 \leq f \quad \text{and} \quad \mu(f) - \varepsilon/2 < \mu(h).$$

From (2.1.8) and (2.1.9) we also have

$$\mu(f) - \varepsilon'(\|\mu\| + 1)/2 \leq \mu(k) \quad \text{and} \quad \mu(h) \leq \mu(f) + \varepsilon'(\|\mu\| + 1)/2$$

and hence $|\mu(k) - \mu(h)| \leq \varepsilon$.

(iii) $\Rightarrow$ (i): Let us consider $\upsilon \in \mathscr{M}_b^+(X)$ such that $\upsilon = \mu$ on H. For a given $\varepsilon > 0$ there exist $h, k \in H$ such that $h - \varepsilon \leq f \leq k + \varepsilon$ and $|\mu(k) - \mu(h)| \leq \varepsilon$. By applying (2.1.8) and (2.1.9) to the Radon measures υ and μ, we have

$$\upsilon(h) - \varepsilon\|\upsilon\| \leq \upsilon(f) \leq \varepsilon\|\upsilon\| + \upsilon(k)$$

and

$$\mu(h) - \varepsilon\|\mu\| \leq \mu(f) \leq \varepsilon\|\mu\| + \mu(k).$$

Since $\mu(h) = \upsilon(h)$ and $\mu(k) = \upsilon(k)$, we have

$$|\mu(f) - \upsilon(f)| \leq \mu(k) - \mu(h) + \varepsilon(\|\mu\| + \|\upsilon\|) \leq \varepsilon(1 + \|\mu\| + \|\upsilon\|).$$

Thus $\mu(f) = \upsilon(f)$ because ε was chosen arbitrarily. $\qquad\square$

The following corollary is a consequence of Proposition 2.1.2 and Theorem 2.1.6.

2.1.7 Corollary. *A subspace H of $\mathscr{C}_0(X)$ is a D_+-subspace for μ if and only if one of the statements* (i), (ii), (iii), (iv) *and* (v) *of Theorem 2.1.6 is satisfied for every $f \in \mathscr{C}_0(X)$.*

In particular, for $\mu = 0$, Theorem 2.1.6 yields $U_+(H,0) = H^*$. In addition, we can state this further result.

2.1.8 Corollary. *If H is a subspace of $\mathscr{C}_0(X)$, then the following statements are equivalent*:
 (i) *H is a D_+-subspace for the null functional.*
 (ii) $\inf\limits_{k \in H} \|(f - k)^+\| = 0 = \inf\limits_{h \in H} \|(h - f)^+\|$ *for every $f \in \mathscr{C}_0(X)$.*
(iii) $H^* = \mathscr{C}_0(X)$.
(iv) *For every $\varepsilon > 0$ and for every compact subset K of X there exist $h \in H$ and $u \in \mathscr{C}_0^+(X)$ such that $0 \le h + u$ on X, $1 \le h + u$ on K and $\|u\| \le \varepsilon$.*
 (v) *For every $\varepsilon > 0$ and for every compact subset K of X there exists $h \in H$ such that $0 \le h + \varepsilon$ on X and $1 \le h$ on K.*

Proof. The equivalences (i) $\Leftrightarrow$ (ii) $\Leftrightarrow$ (iii) follow from Corollary 2.1.7 and Theorem 2.1.6.
(iii) $\Rightarrow$ (v): Given a compact subset K of X, choose a function $f \in \mathscr{K}(X) \subset \mathscr{C}_0(X)$ such that $0 \le f \le 2$ and $f = 2$ on K. Since $f \in H^*$, for a fixed $0 < \varepsilon < 1$, there exists $h \in H$ such that $f \le h + \varepsilon$. Hence $h + \varepsilon \ge 0$ and $h \ge 1$ on K.
(v) $\Rightarrow$ (iv): If K is a compact subset of X and $\varepsilon > 0$, there exists $h \in H$ such that $h + \varepsilon \ge 0$ and $h \ge 1$ on K. Then, it suffices to set $u := h^-$.
(iv) $\Rightarrow$ (i): Let $\mu \in \mathscr{M}_b^+(X)$ and suppose that $\mu = 0$ on H. Given a positive function $f \in \mathscr{K}(X)$, $f \ne 0$, and denoting by K the compact support of f, for every $\varepsilon > 0$ there exist $h \in H$ and $u \in \mathscr{C}_0^+(X)$ such that $0 \le h + u$, $1 \le h + u$ on K and $\|u\| \le \varepsilon$. Then $f/\|f\| \le h + u$ and hence $0 \le \mu(f/\|f\|) \le \mu(h + u) = \mu(h) + \mu(u) \le \|\mu\|\|u\| \le \varepsilon\|\mu\|$. Since ε was arbitrary, we deduce $\mu(f) = 0$. But $\mathscr{K}(X) = \mathscr{K}^+(X) - \mathscr{K}^+(X)$, and hence $\mu = 0$ on $\mathscr{K}(X)$ and consequently on $\mathscr{C}_0(X)$. $\quad\square$

Remark. If X is discrete, then conditions (iv) and (v) of Corollary 2.1.8 are equivalent, respectively, to the statements
(iv)′ *For every $\varepsilon > 0$ and $x \in X$ there exist $h \in H$ and $u \in \mathscr{C}_0^+(X)$ such that $0 \le h + u$ on X, $1 \le h(x) + u(x)$ and $\|u\| \le \varepsilon$.*
 (v)′ *For every $\varepsilon > 0$ and $x \in X$ there exists $h \in H$ such that $0 \le h + \varepsilon$ on X and $1 \le h(x)$.*
The proof is straightforward and is based on the well-known fact that the compact subsets of a discrete space are the finite subsets.
An important example of discrete space is $X = \mathbb{N}$; in this case $\mathscr{C}_0(\mathbb{N}) = c_0$, the space of all sequences of real numbers converging to zero.

We conclude this section by briefly considering another property which is similar to (2.1.1). It concerns the convergence of nets of positive linear contractions toward a fixed positive contractive Radon measure.

2.1.9 Definition. *Let $\mu \in \mathcal{M}_b^+(X)$ and suppose that μ is contractive, i.e., $\|\mu\| \leq 1$.*
A subset (or, a subspace) H of $\mathcal{C}_0(X)$ is called a determining subset (or, a determining subspace) for μ with respect to positive contractive Radon measures, or, more shortly, a D_+^1-subset (or, a D_+^1-subspace) for μ if it satisfies (2.1.1) for nets of positive contractive Radon measures.

Explicitly this means that if $(\mu_i)_{i \in I}^{\leq}$ is an arbitrary net in $\mathcal{M}_b^+(X)$ such that $\sup_{i \in I} \|\mu_i\| \leq 1$ and $\lim_{i \in I}^{\leq} \mu_i(h) = \mu(h)$ for every $h \in H$, then $\lim_{i \in I}^{\leq} \mu_i(f) = \mu(f)$ for every $f \in \mathcal{C}_0(X)$.

Given a subset H of $\mathcal{C}_0(X)$, we shall consider the following subspaces:

$$D_+^1(H, \mu) := \left\{ f \in \mathcal{C}_0(X) \Big| \lim_{i \in I}^{\leq} \mu_i(f) = \mu(f) \text{ for every net } (\mu_i)_{i \in I}^{\leq} \text{ in } \mathcal{M}_b^+(X) \right.$$

$$\left. \text{satisfying } \sup_{i \in I} \|\mu_i\| \leq 1 \text{ and } \lim_{i \in I}^{\leq} \mu_i(h) = \mu(h) \text{ for every } h \in H \right\}$$

(2.1.19)

and

$$U_+^1(H, \mu) := \{ f \in \mathcal{C}_0(X) | v(f) = \mu(f) \text{ for every } v \in \mathcal{M}_b^+(X), \|v\| \leq 1$$

$$\text{such that } v = \mu \text{ on } H \}.$$

(2.1.20)

Thus $U_+^1(H, \mu)$ is the maximal closed subspace of $\mathcal{C}_0(X)$ on which μ has a unique positive contractive extension.

By arguing as in the proof of Proposition 2.1.2 it is easy to show that

$$D_+^1(H, \mu) = U_+^1(H, \mu).$$

(2.1.21)

From this last equality it follows that in general $D_+(H, \mu) \subsetneqq D_+^1(H, \mu)$.

Example. Consider the space $\mathcal{C}_0(]0, 1])$ and the subspace H of $\mathcal{C}_0(]0, 1])$ generated by the two functions e_1 and e_2 defined by $e_k(x) := x^k$ for every $x \in]0, 1]$ and $k = 1, 2$. Set $\mu := \varepsilon_{x_0}$ where $x_0 := 2/3$. Then $U_+^1(H, \mu) = \mathcal{C}_0(]0, 1])$ because, if $v \in \mathcal{M}_b^+(]0, 1])$, $\|v\| \leq 1$ and $v(h) = h(x_0)$ for every $h \in H$, then, denoting by $\tilde{v} \in \mathcal{M}^+([0, 1])$ the canonical extension of v defined by (1.2.11) and considering the function $h_0 := (e_1 - x_0)^2 = e_2 - 2x_0 e_1 + x_0^2$, we have $0 \leq \tilde{v}(h_0) \leq e_2(x_0) -$

$2x_0 e_1(x_0) + x_0^2 \|\tilde{v}\| \leq x_0^2 - 2x_0^2 + x_0^2 = 0$. Hence $\text{Supp}(\tilde{v}) \subset \{x_0\}$ and so $\tilde{v} = \lambda \varepsilon_{x_0}$ where $\lambda \in [0,1]$. But $x_0 = \mu(e_1) = v(e_1) = \tilde{v}(e_1) = \lambda x_0$ and so $\lambda = 1$, i.e., $\tilde{v} = \varepsilon_{x_0}$ and hence $v = \varepsilon_{x_0}$.

On the other hand $U_+(H,\mu) \neq \mathscr{C}_0(]0,1])$ (in fact $U_+(H,\mu) = H$ (see Theorem 4.1.2 and Proposition 4.1.16)), because for the positive bounded Radon measure $v \in \mathscr{M}_b^+(]0,1])$ defined by $v(f) := \dfrac{4}{3} \displaystyle\int_0^1 f(t)\,dt$ for every $f \in \mathscr{C}_0(]0,1])$, we have $v = \mu$ on H but $v(e_3) \neq \mu(e_3)$ where $e_3(x) := x^3$ for every $x \in]0,1]$.

However, we also remark that, *if X is compact, $1 \in H$ and $\mu(1) \leq 1$, then*, on account of (1.1.11), (2.1.4) and (2.1.21), we have

$$D_+^1(H,\mu) = D_+(H,\mu). \tag{2.1.22}$$

Our next goal is again to characterize the subspace $U_+^1(H,\mu)$ in terms of suitable enveloping functions.

Let us define the functions $\hat{\mu}_H^1 \colon \mathscr{C}_0(X) \to \mathbb{R}$ and $\check{\mu}_H^1 \colon \mathscr{C}_0(X) \to \mathbb{R}$ by

$$\hat{\mu}_H^1(f) := \inf\{\mu(k) + \alpha \,|\, k \in H, \alpha \geq 0, f \leq k + \alpha\}, \tag{2.1.23}$$

$$\check{\mu}_H^1(f) := -\hat{\mu}_H^1(-f) = \sup\{\mu(h) - \alpha \,|\, h \in H, \alpha \geq 0, h - \alpha \leq f\} \tag{2.1.24}$$

for every $f \in \mathscr{C}_0(X)$.

By using Lemma 2.1.5 it is easy to show that for every $f \in \mathscr{C}_0(X)$

$$\hat{\mu}_H^1(f) = \inf\{\mu(k) + \|(f - k)^+\| \,|\, k \in H\}$$

$$= \inf\{\mu(k) + \|u\| \,|\, k \in H, u \in \mathscr{C}_0^+(X), f \leq k + u\} \tag{2.1.25}$$

as well as

$$\check{\mu}_H^1(f) = \sup\{\mu(h) - \|(h - f)^+\| \,|\, h \in H\}$$

$$= \sup\{\mu(h) - \|v\| \,|\, h \in H, v \in \mathscr{C}_0^+(X), h - v \leq f\}. \tag{2.1.26}$$

Some other properties of the function $\hat{\mu}_H^1$ are listed below. We omit the straightforward proofs for the sake of brevity.

$$\check{\mu}_H^1(f) \leq \mu(f) \leq \hat{\mu}_H^1(f), \quad (f \in \mathscr{C}_0(X)) \tag{2.1.27}$$

$$\check{\mu}_H^1(h) = \mu(h) = \hat{\mu}_H^1(h), \quad (h \in H) \tag{2.1.28}$$

$$|\hat{\mu}_H^1(f) - \hat{\mu}_H^1(g)| \leq \|f - g\|, \quad (f,g \in \mathscr{C}_0(X)) \tag{2.1.29}$$

the function $\hat{\mu}_H^1$ is increasing and sublinear on $\mathscr{C}_0(X)$. \hfill (2.1.30)

2.1.10 Theorem. *For every $f \in \mathscr{C}_0(X)$ the following statements are equivalent:*

(i) $f \in U^1_+(H, \mu)$.

(ii) $\check{\mu}^1_H(f) = \mu(f) = \hat{\mu}^1_H(f)$.

(iii) *For every $\varepsilon > 0$ there exist $h_1, h_2 \in H$ and $\alpha_1, \alpha_2 \in \mathbb{R}_+$ such that $h_1 - \alpha_1 \leq f \leq h_2 + \alpha_2$ and $|\mu(h_2 - h_1) + \alpha_1 + \alpha_2| \leq \varepsilon$.*

(iv) *For every $\varepsilon > 0$ there exist $h_1, h_2 \in H$ such that $|\mu(h_2 - h_1) + \|(f - h_2)^+\| + \|(h_1 - f)^+\|| \leq \varepsilon$.*

(v) *For every $\varepsilon > 0$ there exist $h_1, h_2 \in H$ and $u_1, u_2 \in \mathscr{C}_0^+(X)$ such that $h_1 - u_1 \leq f \leq h_2 + u_2$ and $|\mu(h_2 - h_1) + \|u_1\| + \|u_2\|| \leq \varepsilon$.*

Consequently, H is a D^1_+-subspace for μ if and only if one of the statements (i), (ii), (iii), (iv) *and* (v) *is satisfied for every $f \in \mathscr{C}_0(X)$.*

Proof. The equivalences (ii) ⇔ (iii), (ii) ⇔ (iv) and (ii) ⇔ (v) follow from (2.1.23), (2.1.24), (2.1.25) and (2.1.26). To prove the implication (i) ⇒ (ii) it is sufficient to follow the lines of the proof of the implication (i) ⇒ (ii) of Theorem 2.1.6 by using the Hahn-Banach theorem instead of the Anger-Lembcke theorem. Finally, we prove the implication (iii) ⇒ (i). Fix $v \in \mathscr{M}^+_b(X)$, $\|v\| \leq 1$, such that $v = \mu$ on H. Given $\varepsilon > 0$, there exist $h_1, h_2 \in H$ and $\alpha_1, \alpha_2 \in \mathbb{R}_+$ such that $h_1 - \alpha_1 \leq f \leq h_2 + \alpha_2$ and $|\mu(h_2 - h_1) + \alpha_1 + \alpha_2| \leq \varepsilon$. From (2.1.8) and (2.1.9) it follows

$$\mu(h_1) - \alpha_1 \leq \mu(h_1) - \alpha_1\|\mu\| \leq \mu(f) \leq \mu(h_2) + \alpha_2\|\mu\| \leq \mu(h_2) + \alpha_2,$$

and, analogously,

$$v(h_1) - \alpha_1 \leq v(h_1) - \alpha_1\|v\| \leq v(f) \leq v(h_2) + \alpha_2\|v\| \leq v(h_2) + \alpha_2.$$

Since $v(h_1) = \mu(h_1)$ and $v(h_2) = \mu(h_2)$, we infer

$$|\mu(f) - v(f)| \leq \mu(h_2) - \mu(h_1) + \alpha_2 + \alpha_1 \leq \varepsilon$$

and hence $\mu(f) = v(f)$ since ε was chosen arbitrarily.   □

Remark. From Theorem 2.1.6 it follows, in particular, that $D_+(H, \mu) \subset H^*$. Hence $H^* = \mathscr{C}_0(X)$ if H is a D_+-subspace for some $\mu \in \mathscr{M}^+_b(X)$. On the contrary, the inclusion $D^1_+(H, \mu) \subset H^*$ does not hold in general, even if X is compact.

Take, in fact, $X = \mathbb{N}$ and consider the positive Radon measure $\mu_0 \in \mathscr{M}^+_b(\mathbb{N})$ defined as $\mu_0 := \sum_{n=1}^{\infty} 2^{-n}\varepsilon_n$ and the subspace $H := \{f \in \mathscr{C}_0(\mathbb{N}) | \mu_0(f) = 0\}$ of $\mathscr{C}_0(\mathbb{N})$.

Fix $n \in \mathbb{N}$ and let $\mu \in \mathscr{M}^+_b(\mathbb{N})$ such that $\|\mu\| \leq 1$ and $\mu(h) = h(n)(= \varepsilon_n(h))$ for each $h \in H$. We show that $\mu(f) = f(n)$ for each $f \in \mathscr{C}_0(\mathbb{N})$.

In fact we observe first that for each positive function $f \in \mathscr{C}_0(\mathbb{N})$ we can consider $h \in H$ such that $h \leq f$ and $h(n) = f(n)$ (for instance, define h to be equal to

f on $\mathbb{N}\setminus\{n+1\}$ and put $h(n+1):=-2^{n+1}\sum_{\substack{j=1\\j\neq n+1}}^{\infty}2^{-j}f(j))$. This yields $f(n)=$ $h(n)=\mu(h)\leq\mu(f)$. Now, assume that $f\in\mathscr{C}_0(\mathbb{N})$ is a positive function such that $f(n)<\mu(f)$ and consider the positive function $g\in\mathscr{C}_0(\mathbb{N})$ such that $g(n):=\|f\|-f(n)$ and $g:=0$ elsewhere. Then, $\mu(g)=\mu(g+f)-\mu(f)\leq$ $\|g+f\|-\mu(f)<\|f\|-f(n)=g(n)$ and this contradicts the previous inequality $g(n)\leq\mu(g)$.

We can now conclude that $\mu(f)=f(n)$ for each $f\in\mathscr{C}_0(\mathbb{N})$ by applying the above arguments to the positive and negative part of f.

Since $n\in\mathbb{N}$ is arbitrary, we obtain that H is a D_+^1-subspace for ε_n for every $n\in\mathbb{N}$. Moreover, we observe that $H^*=H$ since the positive Radon measure $\mu_0\in\mathscr{M}_b^+(\mathbb{N})$ defined above satisfies $\mu_0(f)=0$ if and only if $f\in H$.

Adapting the same reasoning, one can show that, if $X=\mathbb{N}\cup\{\omega\}$ is the one-point compactification of $\mathbb{N}$ and if we put $\mu_0:=\frac{1}{2}\varepsilon_\omega+\frac{1}{2}\sum_{n=1}^{\infty}\frac{1}{2^n}\varepsilon_n$ and $H:=$ $\{f\in\mathscr{C}(X)|\mu_0(f)=0\}$, then $H^*=H$ while H is a D_+^1-subspace for ε_x for every $x\in X$.

Notes and references

The idea of introducing the enveloping functions (2.1.10) and (2.1.25) is essentially due to Bauer [1978b] and to Bauer and Donner [1978], [1986] (see also Donner [1982]) who used them only relatively to Dirac measures. For a positive Radon measure (or, more generally, for a positive linear form defined on an ordered locally convex space) they were introduced and studied by Altomare [1989] to whom Theorems 2.1.6 and 2.1.10 are due. Other enveloping functions related to the same circle of ideas or in connection with other problems of Korovkin-type approximation theory were considered by Flösser [1978], Flösser, Irmisch and Roth [1981], Watanabe [1977] and Keimel and Roth [1988], [1992]. Proposition 2.1.2 and the proof of Theorem 2.1.4 are taken from Ferguson and Rusk [1976]. Corollary 2.1.8 and the examples indicated in the Remark to Theorem 2.1.10 are due to Bauer and Donner [1986].

2.2 Determining subspaces for discrete Radon measures

In this section we shall characterize the determining subspaces for discrete Radon measures by means of a "quasi peak point" criterion.

In the subsequent sections we shall also indicate constructive methods to obtain them and we shall point out their connections with Chebyshev subspaces.

In the sequel X will denote a locally compact Hausdorff space. We preliminarily note that, if we want that a Radon measure $\mu \in \mathcal{M}_b^+(X)$ admits a finite dimensional determining subspace, then necessarily we have to suppose that μ is discrete.

2.2.1 Proposition. *Let $\mu \in \mathcal{M}_b^+(X)$, $\mu \neq 0$, and consider an n-dimensional subspace H of $\mathscr{C}_0(X)$. Suppose that H separates the points of X and, if X is not compact, that for every $x \in X$ there exists $h \in H$ such that $h(x) \neq 0$.*

Then there exist finitely many points $x_1, \ldots, x_p \in X$, $p \leq n + 1$, and positive numbers $\lambda_1, \ldots, \lambda_p$ such that $\sum_{i=1}^{p} \lambda_i \leq \|\mu\|$ and $\mu = \sum_{i=1}^{p} \lambda_i \varepsilon_x$ on H.

Consequently, if H is a D_+-subspace for μ, or a D_+^1-subspace for μ, provided $\|\mu\| \leq 1$, then $\mu = \sum_{i=1}^{p} \lambda_i \varepsilon_{x_i}$.

Proof. Without loss of generality we may suppose that $\|\mu\| = 1$. As a first step suppose that X is compact. Fix an algebraic base $h_1, \ldots, h_n$ of H and consider the continuous injective mapping $\Phi^*: X \to \mathbb{R}^n$ defined by

$$\Phi^*(x) := (h_1(x), \ldots, h_n(x)) \quad \text{for every } x \in X.$$

Then $\Phi^*(X)$ and its convex hull $K := \mathrm{co}(\Phi(X))$ are compact subsets of $\mathbb{R}^n$. Furthermore, by Milman's converse of the Krein-Milman theorem (see Theorem 1.4.12) we have $\partial_e K \subset \Phi^*(X)$.

Consider now the probability Radon measure $\mu^* \in \mathcal{M}_1^+(K)$ defined by $\mu^*(f) := \mu(f \circ \Phi^*)$ for every $f \in \mathscr{C}(K)$.

Denoting by $r(\mu^*) \in K$ the resultant of μ^* (see (1.5.1)), by Carathéodory's theorem 1.4.13, (1), there exist at most $n + 1$ points $x_1, \ldots, x_{n+1} \in X$ and $\lambda_1, \ldots,$ $\lambda_{n+1} \in \mathbb{R}_+$ satisfying $\sum_{i=1}^{n+1} \lambda_i = 1$, such that $r(\mu^*) = \sum_{i=1}^{n+1} \lambda_i \Phi^*(x_i)$.

Accordingly for every $j = 1, \ldots, n$ we obtain

$$\mu(h_j) = \mu^*(\mathrm{pr}_j) = \mathrm{pr}_j(r(\mu^*)) = \sum_{i=1}^{n+1} \lambda_i \, \mathrm{pr}_j(\Phi^*(x_i)) = \sum_{i=1}^{n+1} \lambda_i h_j(x_i),$$

where $\mathrm{pr}_j: \mathbb{R}^n \to \mathbb{R}$ denotes the j-th projection (see (1.2.49)).

So $\mu = \sum\limits_{i=1}^{n+1} \lambda_i \varepsilon_{x_i}$ on H.

Suppose now that X is non compact and consider the one-point compactification $X_\omega := X \cup \{\omega\}$ of X. For every $f \in \mathscr{C}_0(X)$ we shall denote by $\tilde{f} \colon X_\omega \to \mathbb{R}$ the continuous function defined by (1.1.7).

Moreover consider the positive Radon measure $\tilde{\mu} \in \mathscr{M}^+(X_\omega)$ defined by (1.2.11). Then $\|\tilde{\mu}\| = \|\mu\| = 1$ and $\mu(f) = \tilde{\mu}(\tilde{f})$ for every $f \in \mathscr{C}_0(X)$.

Since $\tilde{H} := \{\tilde{h} \mid h \in H\}$ separates the points of X_ω, by the above reasoning there exist $x_1, \dots, x_{n+1} \in X_\omega$ and $\lambda_1, \dots, \lambda_{n+1} \in \mathbb{R}_+$ such that $\sum\limits_{i=1}^{n+1} \lambda_i = 1$ and $\tilde{\mu} = \sum\limits_{i=1}^{n+1} \lambda_i \varepsilon_{x_i}$ on $\tilde{H}$. Set $I := \{i = 1, \dots, n+1 \mid x_i \neq \omega\}$. Then $I \neq \varnothing$ otherwise $\mu = 0$, and $\mu = \sum\limits_{i \in I} \lambda_i \varepsilon_{x_i}$ on H. $\qquad\square$

Remarks. 1. As the proof of the above theorem shows, if X is compact, then $\sum\limits_{i=1}^{n+1} \lambda_i = \|\mu\|$.

2. If X is compact and connected (or if X is locally compact and X_ω is connected, respectively) then, by using Fenchel's theorem 1.4.13, (2), instead of Charathéodory's theorem, we see that in Proposition 2.2.1 it suffices to choose at most n points $x_1, \dots, x_n \in X$ and n positive numbers $\lambda_1, \dots, \lambda_n$.

From now on we fix a subspace H of $\mathscr{C}_0(X)$ and n distinct points $x_1, \dots, x_n$ of X.

We set

$$C_+(x_1, \dots, x_n) := \{\mu \in \mathscr{M}_b^+(X) \mid \mathrm{Supp}(\mu) \subset \{x_1, \dots, x_n\}\}$$

$$= \left\{\mu \in \mathscr{M}_b^+(X) \mid \text{there exist } \alpha_1, \dots, \alpha_n \in \mathbb{R}_+ \text{ such that } \mu = \sum_{i=1}^{n} \alpha_i \varepsilon_{x_i}\right\},$$

$$C_+^1(x_1, \dots, x_n) := \left\{\mu \in C_+(x_1, \dots, x_n) \mid \mu = \sum_{i=1}^{n} \alpha_i \varepsilon_{x_i}\right.$$

$$\left. \text{where } \alpha_1, \dots, \alpha_n \in \mathbb{R}_+ \ \text{ and } \sum_{i=1}^{n} \alpha_i \leq 1\right\},$$

and

$$\tilde{C}_+^1(x_1, \dots, x_n) := \left\{\mu \in C_+^1(x_1, \dots, x_n) \mid \mu = \sum_{i=1}^{n} \alpha_i \varepsilon_{x_i}\right.$$

$$\left. \text{where } \alpha_1, \dots, \alpha_n \in \mathbb{R}_+ \ \text{ and } \sum_{i=1}^{n} \alpha_i = 1\right\}.$$

We have the following result.

2.2.2 Theorem. *A subspace H of $\mathscr{C}_0(X)$ is a D_+-subspace for every $\mu \in C_+(x_1,\ldots,x_n)$ if and only if it satisfies the following two conditions:*
(1) *For every $\varepsilon > 0$ and for every compact subset K of X satisfying $K \cap \{x_1,\ldots,x_n\} = \varnothing$, there exist $k \in H$ and $u \in \mathscr{C}_0^+(X)$ such that $\|u\| < \varepsilon, 0 \le k + u, 1 \le k + u$ on K and $k(x_i) + u(x_i) < \varepsilon$ for every $i = 1,\ldots,n$.*
(2) *There exist $h_1,\ldots,h_n \in H$ such that $\det(h_i(x_j)) \neq 0$.*

Proof. Suppose that H is a D_+-subspace for every $\mu \in C_+(x_1,\ldots,x_n)$. Fix $\mu = \sum_{i=1}^{n} \alpha_i \varepsilon_{x_i} \in C_+(x_1,\ldots,x_n), \mu \neq 0$, and put $\alpha := \min_{\substack{1 \le i \le n \\ \alpha_i \neq 0}} \alpha_i$ and $\beta := \sum_{i=1}^{n} \alpha_i$.

Given $\varepsilon > 0$ and a compact subset K of X satisfying $K \cap \{x_1,\ldots,x_n\} = \varnothing$, consider $f \in \mathscr{K}(X)$ such that $0 \le f \le 1$, $f(x_i) = 0$ for every $i = 1,\ldots,n$ and $f = 1$ on K. Since H is a D_+-subspace for the measure μ, from Corollary 2.1.7 it follows that $\mu(f) = \hat{\mu}_H(f)$, and hence (see (2.1.10)) there exists $k \in H$ such that $f \le k + \dfrac{\alpha\varepsilon}{2\beta}$ and $\mu(k) < \mu(f) + \dfrac{\alpha\varepsilon}{2}$.

Because of Lemma 2.1.5 there exists $u \in \mathscr{C}_0^+(X)$, $\|u\| \le \dfrac{\alpha\varepsilon}{2\beta}$, satisfying $f \le k + u$. Therefore, $0 \le k + u$, $1 \le k + u$ on K and $u(x_i) < \dfrac{\alpha\varepsilon}{2\beta}$ for every $i = 1,\ldots,n$.

Moreover, since $\mu(f) = 0$, we have $\mu(k) < \dfrac{\alpha\varepsilon}{2}$ and hence, for every $j = 1,\ldots,n$,

$$\alpha(k(x_j) + u(x_j)) \le \alpha_j(k(x_j) + u(x_j)) \le \sum_{i=1}^{n} \alpha_i k(x_i) + \alpha_i u(x_i)$$

$$= \mu(k) + \sum_{i=1}^{n} \alpha_i u(x_i) < \dfrac{\alpha\varepsilon}{2} + \dfrac{\alpha\varepsilon}{2} = \alpha\varepsilon.$$

So $k(x_j) + u(x_j) < \varepsilon$ and statement (1) is proved.

As regards statement (2), we note first that there exists $h_1 \in H$ such that $h_1(x_1) \neq 0$. Otherwise, if $0 = h(x_1) = \varepsilon_{x_1}(h)$ for every $h \in H$, then, since H is a D_+-subspace for ε_{x_1}, it would follow that $f(x_1) = 0$ for every $f \in \mathscr{C}_0(X)$ which is not possible.

Now, let us consider $m \in \mathbb{N}$, $m < n$ and suppose that there exist $h_1,\ldots,h_m \in H$ such that $\det(h_j(x_i))_{\substack{1 \le i \le m \\ 1 \le j \le m}} \neq 0$; we shall prove that there exists $h_{m+1} \in H$ such that $\det(h_j(x_i))_{\substack{1 \le i \le m+1 \\ 1 \le j \le m+1}} \neq 0$.

Otherwise, in fact, for every $h \in H$ we would have

$$
\begin{vmatrix}
h_1(x_1) & h_1(x_2) & \cdots & h_1(x_{m+1}) \\
\vdots & \vdots & \cdots & \vdots \\
h_m(x_1) & h_m(x_2) & \cdots & h_m(x_{m+1}) \\
h(x_1) & h(x_2) & \cdots & h(x_{m+1})
\end{vmatrix} = 0.
$$

By expanding this determinant with respect to the last row, we have $\sum_{i=1}^{m+1} \alpha_i h(x_i) = 0$, where the numbers α_i are independent of h. If we set $J := \{j \in \{1,\ldots,m+1\} \,|\, \alpha_j > 0\}$, we also have $\sum_{i \in J} \alpha_i \varepsilon_{x_i}(h) = -\sum_{\substack{i=1 \\ i \notin J}}^{m+1} \alpha_i \varepsilon_{x_i}(h)$ for every $h \in H$ (if $J = \varnothing$ or $J = \{1,\ldots,m+1\}$, the relevant sum must be assumed equal to zero).

Hence the same equality holds for every $f \in \mathscr{C}_0(X)$, i.e.,

$$
\begin{vmatrix}
h_1(x_1) & h_1(x_2) & \cdots & h_1(x_{m+1}) \\
\vdots & \vdots & \cdots & \vdots \\
h_m(x_1) & h_m(x_2) & \cdots & h_m(x_{m+1}) \\
f(x_1) & f(x_2) & \cdots & f(x_{m+1})
\end{vmatrix} = 0
$$

for every $f \in \mathscr{C}_0(X)$. But this is not possible since, if we choose $f \in \mathscr{K}(X)$ such that $f(x_{m+1}) = 1$ and $f(x_i) = 0$ for every $i = 1,\ldots,m$, then the determinant is different from zero.

By finite induction we easily deduce statement (2).

Conversely, suppose that statements (1) and (2) are satisfied.

Let us consider $\mu = \sum_{i=1}^n \alpha_i \varepsilon_{x_i} \in C_+(x_1,\ldots,x_n)$ and $v \in \mathscr{M}_b^+(X)$ and suppose that $\mu = v$ on H. To prove that $\mathrm{Supp}(v) \subset \{x_1,\ldots,x_n\}$, fix $y \in \mathrm{Supp}(v)$ and suppose that $y \notin \{x_1,\ldots,x_n\}$. Then we can choose a compact neighborhood K of y such that $K \cap \{x_1,\ldots,x_n\} = \varnothing$.

Given $\varepsilon > 0$, there exist $k \in H$ and $u \in \mathscr{C}_0^+(X)$ which verify the properties indicated in statement (1).

Moreover, since $y \in \mathrm{Supp}(v)$, we can choose $f \in \mathscr{K}(X)$, $0 \le f \le 1$, such that the support of f is contained in K and $v(f) > 0$. Consequently, $f \le k + u$ and hence

$$
0 < v(f) \le v(k + u) \le v(k) + \|v\|\varepsilon = \sum_{i=1}^n \alpha_i k(x_i) + \|v\|\varepsilon \le \left(\sum_{i=1}^n \alpha_i + \|v\| \right) \varepsilon.
$$

Since $\varepsilon > 0$ is arbitrary, we have a contradiction.

Thus, we have proved the inclusion $\text{Supp}(v) \subset \{x_1, \ldots, x_n\}$. Consequently, there exist $\beta_1, \ldots, \beta_n \in \mathbb{R}_+$ such that $v = \sum_{i=1}^{n} \beta_i \varepsilon_{x_i}$. By using statement (2) we deduce that $\alpha_i = \beta_i$ for every $i = 1, \ldots, n$ and hence $v = \mu$. $\square$

Remark. If X is discrete, then it is easy to show that statement (1) in Theorem 2.2.2 can be reformulated as

(1)′ *For every $\varepsilon > 0$ and $x \in X$, $x \notin \{x_1, \ldots, x_n\}$, there exist $k \in H$ and $u \in \mathscr{C}_0^+(X)$ such that $\|u\| < \varepsilon$, $0 \leq k + u$, $1 \leq k(x) + u(x)$ and $k(x_i) + u(x_i) < \varepsilon$ for every $i = 1, \ldots, n$.*

(see Remark to Corollary 2.1.8).

In the compact case, by using Theorem 2.1.4 instead of Corollary 2.1.7, we have a simpler characterization of determining subspaces for discrete measures. The details of the proof are left to the reader.

2.2.3 Theorem. *Suppose X compact. Then the following statements are equivalent:*

(i) *The subspace H is a D_+-subspace for every $\mu \in C_+(x_1, \ldots, x_n)$.*

(ii) (1) *For every $\varepsilon > 0$ and for every compact subset K of X satisfying $K \cap \{x_1, \ldots, x_n\} = \varnothing$ there exists $k \in H$ such that $0 \leq k$, $1 \leq k$ on K and $k(x_i) < \varepsilon$ for every $i = 1, \ldots, n$.*

 (2) *There exist $h_1, \ldots, h_n \in H$ such that $\det(h_i(x_j)) \neq 0$.*

(iii) (1) *For every $\varepsilon > 0$ and for every finite family of open neighborhoods $U_1, U_2, \ldots, U_n$ of $x_1, x_2, \ldots, x_n$, respectively, there exists $k \in H$ such that*

$$0 \leq k, \ 1 \leq k \text{ on } X \setminus \bigcup_{i=1}^{n} U_i \text{ and } k(x_i) < \varepsilon \text{ for every } i = 1, \ldots, n.$$

 (2) *There exist $h_1, \ldots, h_n \in H$ such that $\det(h_i(x_j)) \neq 0$.*

Remarks. 1. An inspection of the proof of Theorem 2.2.2 shows that, if we suppose that there exist $h_1, \ldots, h_n \in H$ such that $\det(h_i(x_j)) \neq 0$, then, given a Radon measure $\mu \in C_+(x_1, \ldots, x_n)$, $\mu \neq 0$, the subspace H is a D_+-subspace for μ *if and only if statement* (1) *of Theorem 2.2.2 (part* (1) *of* (ii) *of Theorem 2.2.3, in the compact case, respectively) is satisfied.*

Thus, if this is the case, then H is a D_+-subspace for every $v \in C_+(x_1, \ldots, x_n)$.

2. Another geometric characterization of statement (i) of Theorem 2.2.3 can be found in Proposition 2.6.6.

As regards the characterization of determining subspaces with respect to positive contractive Radon measures for discrete measures only partial results are available.

2.2.4 Theorem. *Suppose that there exist* $h_1, \ldots, h_n \in H$ *such that* $\det(h_i(x_j)) \neq 0$.
Then, given $\mu = \sum_{i=1}^{n} \alpha_i \varepsilon_{x_i} \in C_+^1(x_1, \ldots, x_n)$, $\mu \neq 0$, *the following statements are equivalent*:

(i) *The subspace* H *is a* D_+^1-*subspace for* μ.

(ii) *For every* $\varepsilon > 0$ *and for every compact subset* K *of* X *satisfying* $K \cap \{x_1, \ldots, x_n\} = \emptyset$, *there exist* $k \in H$ *and* $u \in \mathscr{C}_0^+(X)$ *such that* $\|u\| < \varepsilon + \sum_{i=1}^{n} \alpha_i u(x_i)$, $0 \leq k + u$, $1 \leq k + u$ *on* K *and* $k(x_i) + u(x_i) < \varepsilon$ *for every* $i = 1, \ldots, n$.

Proof. (i) $\Rightarrow$ (ii): Given $\varepsilon > 0$ and a compact subset K of X such that $K \cap \{x_1, \ldots, x_n\} = \emptyset$, consider $f \in \mathscr{K}(X)$ satisfying $0 \leq f \leq 1$, $f = 0$ on $\{x_1, \ldots, x_n\}$ and $f = 1$ on K. Set $\alpha := \min_{\substack{1 \leq i \leq n \\ \alpha_i > 0}} \alpha_i > 0$. Then, since $\mu(f) = \hat{\mu}_H^1(f)$ (see Theorem 2.1.10), there exist $k \in H$ and $u \in \mathscr{C}_0^+(X)$ such that $f \leq k + u$ and $\mu(k) + \|u\| < \mu(f) + \alpha\varepsilon = \alpha\varepsilon$.

Therefore $0 \leq k + u$ on X and $1 \leq k + u$ on K. Moreover, since $0 \leq \mu(k) + \mu(u)$, we have $\|u\| < \alpha\varepsilon - \mu(k) \leq \alpha\varepsilon + \mu(u) \leq \varepsilon + \sum_{i=1}^{n} \alpha_i u(x_i)$. Finally, for every $i = 1, \ldots, n$

$$\alpha(k(x_j) + u(x_j)) \leq \sum_{i=1}^{n} \alpha_i(k(x_i) + u(x_i)) = \mu(k) + \mu(u) \leq \mu(k) + \|u\| < \varepsilon\alpha$$

and hence $k(x_j) + u(x_j) < \varepsilon$.

(ii) $\Rightarrow$ (i): Consider a measure $v \in \mathscr{M}_b^+(X)$, $\|v\| \leq 1$, satisfying $\mu = v$ on H. Fix $y \in \mathrm{Supp}(v)$ and suppose that $y \notin \{x_1, \ldots, x_n\}$. Then there exists a compact neighborhood K of y such that $K \cap \{x_1, \ldots, x_n\} = \emptyset$. Given $\varepsilon > 0$, let us consider $k \in H$ and $u \in \mathscr{C}_0^+(X)$ satisfying the properties indicated in statement (ii). Consequently, after choosing $f \in \mathscr{K}(X)$, $0 \leq f \leq 1$, such that the support of f is contained in K and $v(f) > 0$, we have $f \leq k + u$ and hence

$$0 < v(f) \leq v(k) + v(u) \leq \mu(k) + \|u\| < \sum_{i=1}^{n} \alpha_i k(x_i) + \varepsilon + \sum_{i=1}^{n} \alpha_i u(x_i)$$

$$= \sum_{i=1}^{n} \alpha_i(k(x_i) + u(x_i)) + \varepsilon \leq 2\varepsilon.$$

But this is not possible, since ε was chosen arbitrarily.

Hence, this proves that $\mathrm{Supp}(v) \subset \{x_1, \ldots, x_n\}$ and so, by arguing as in the proof of Theorem 2.2.2, we conclude that $\mu = v$. $\qquad\square$

Remark. As noted in the Remark to Theorem 2.2.2, when the space X is discrete, statement (ii) of Theorem 2.2.4 is also equivalent to the following one:

(ii)′ *For every $\varepsilon > 0$ and $x \in X$, $x \notin \{x_1, \ldots, x_n\}$, there exist $k \in H$ and $u \in \mathscr{C}_0^+(X)$ such that*

$$\|u\| < \varepsilon + \sum_{i=1}^{n} \alpha_i u(x_i), \, 0 \le k + u, \, 1 \le k(x) + u(x) \quad and \quad k(x_i) + u(x_i) < \varepsilon$$

for every $i = 1, \ldots, n$.

In Theorems 2.2.2, 2.2.3 and 2.2.4 the hypothesis that there exist n functions $h_1, \ldots, h_n \in H$ such that $\det(h_i(x_j)) \neq 0$ is crucial.

In the following proposition we indicate an equivalent statement which will be useful in the sequel.

2.2.5 Proposition. *In order that there exist $h_1, \ldots, h_n \in H$ such that*

$$\det(h_i(x_j)) \neq 0 \tag{2.2.1}$$

it is necessary and sufficient that for every $m = 1, \ldots, n$ and $\varepsilon > 0$ there exist $k \in H$ and $u \in \mathscr{C}_0^+(X)$ such that

$$0 \le k + u,$$

$$u(x_i) < \varepsilon \quad for \; every \; i = 1, \ldots, n,$$

$$1 \le k(x_m) + u(x_m), \tag{2.2.2}$$

$$k(x_i) + u(x_i) < \varepsilon \quad for \; every \; i = 1, \ldots, n, \, i \neq m.$$

Proof. Suppose that condition (2.2.1) is satisfied and fix $m \in \{1, \ldots, n\}$. Then, by solving the linear system

$$\sum_{i=1}^{n} \lambda_i h_i(x_j) = \delta_{m,j} \quad j = 1, \ldots, n,$$

where $\delta_{m,j} := 0$ for $j \neq m$ and $\delta_{m,m} := 1$, we can choose a linear combination k of $h_1, \ldots, h_n$ such that $k(x_j) = \delta_{m,j}$ for every $j = 1, \ldots, n$.

Given $\varepsilon > 0$, let us consider $f \in \mathscr{K}(X)$ such that $f \ge 0$ and $f(x_i) = \dfrac{\varepsilon}{2}$ for every $i = 1, \ldots, n$. Then the functions $u := \sup(f, -k)$ and k satisfy condition (2.2.2).

Conversely, by applying (2.2.2) in the case $m = 1$ and $\varepsilon = \dfrac{1}{2}$, we can choose $h_1 \in H$ such that $h_1(x_1) \neq 0$.

Now, let us consider $m \in \{1, \ldots, n-1\}$ such that there exist $h_1, \ldots, h_m \in H$ satisfying the condition $\det(h_i(x_j))_{\substack{1 \le i \le m \\ 1 \le j \le m}} \neq 0$; we prove that there exists $h_{m+1} \in H$ such that

$$\det(h_i(x_j))_{\substack{1 \le i \le m+1 \\ 1 \le j \le m+1}} \neq 0.$$

Otherwise, there would exist $\alpha_1, \ldots, \alpha_{m+1} \in \mathbb{R}$ (with $\alpha_{m+1} := \det(h_i(x_j))_{\substack{1 \le i \le m \\ 1 \le j \le m}} \neq 0$) such that

$$\sum_{i=1}^{m+1} \alpha_i h(x_i) = 0 \quad \text{for every } h \in H.$$

Hence, for all $\varepsilon > 0$, by choosing $k \in H$ and $u \in \mathscr{C}_0^+(X)$ which satisfy condition (2.2.2) for $m+1$, we would have

$$|\alpha_{m+1}| \le |\alpha_{m+1}| \cdot (k(x_{m+1}) + u(x_{m+1}))$$

$$= \left| -\sum_{i=1}^{m} \alpha_i k(x_i) - \sum_{i=1}^{m} \alpha_i u(x_i) + \sum_{i=1}^{m} \alpha_i u(x_i) + \alpha_{m+1} u(x_{m+1}) \right|$$

$$\le \sum_{i=1}^{m} |\alpha_i|(k(x_i) + u(x_i)) + \sum_{i=1}^{m+1} |\alpha_i| \cdot u(x_i) < 2\varepsilon \sum_{i=1}^{m+1} |\alpha_i|.$$

Since $\varepsilon > 0$ was chosen arbitrarily, we conclude $\alpha_{m+1} = 0$ in contradiction to our assumption.

Condition (2.2.1) now follows by finite induction. $\qquad \square$

Some other conditions ensuring that a subspace is a determining subspace for discrete measures are indicated in the next results.

2.2.6 Theorem. *Suppose that a subspace H of $\mathscr{C}_0(X)$ satisfies the following two conditions:*
(1) *There exist $h_1, \ldots, h_n \in H$ such that $\det(h_i(x_j)) \neq 0$.*
(2) *For every $x \in X$, $x \notin \{x_1, \ldots, x_n\}$, there exists $h \in H$ (respectively $h \in H + \mathbb{R}_+$) such that $h \ge 0$, $h(x) > 0$ and $h(x_1) = \cdots = h(x_n) = 0$.*
   Then, H is a D_+-subspace for every $\mu \in C_+(x_1, \ldots, x_n)$ (it is a D_+^1-subspace for every $\mu \in \tilde{C}_+^1(x_1, \ldots, x_n)$, respectively).

Proof. We shall verify statement (1) of Theorem 2.2.2 and statement (ii) of Theorem 2.2.4, respectively. In fact, consider a compact subset K of X such that

$K \cap \{x_1, \ldots, x_n\} = \varnothing$. For every $x \in K$ there exists $h \in H$ ($h \in H$ and $\beta \in \mathbb{R}_+$, respectively) such that $h \geq 0$, $0 < h(x)$ and $h(x_1) = \cdots = h(x_n) = 0$ ($0 \leq h + \beta$, $0 < h(x) + \beta$, $h(x_1) + \beta = \cdots = h(x_n) + \beta = 0$, respectively). From the compactness of K it turns out that there exist $k_1, \ldots, k_p \in H$ ($k_1, \ldots, k_p \in H$ and $\beta_1, \ldots, \beta_p \in \mathbb{R}_+$, respectively) such that

$$0 \leq k_i, \quad 0 = k_i(x_j) \quad \text{for every } i = 1, \ldots, p \quad \text{and } j = 1, \ldots, n$$

and

$$\text{for every } x \in K \text{ there exists } i = 1, \ldots, p \text{ such that } k_i(x) > 0,$$

$$(0 \leq k_i + \beta_i, \quad 0 = k_i(x_j) + \beta_i \quad \text{for every } i = 1, \ldots, p \text{ and } j = 1, \ldots, n$$

and

$$\text{for every } x \in K \quad \text{there exists } i = 1, \ldots, p \quad \text{such that } k_i(x) + \beta_i > 0,$$

respectively).

If we set $k_0 := \sum\limits_{i=1}^{p} k_i \left(k_0 := \sum\limits_{i=1}^{p} k_i \text{ and } \beta := \sum\limits_{i=1}^{p} \beta_i \in \mathbb{R}_+, \text{ respectively} \right)$, we have

$$0 \leq k_0, \quad 0 < k_0(x) \text{ for every } x \in K, \quad k_0(x_i) = 0 \text{ for every } i = 1, \ldots, n$$

$(0 \leq k_0 + \beta, 0 < k_0(x) + \beta$ for every $x \in K, k_0(x_i) + \beta = 0$ for every $i = 1, \ldots, n$,

respectively).

So, if we put $k := \dfrac{1}{\alpha} \cdot k_0$, where $\alpha := \min\limits_{x \in K} k_0(x)$, we see that statement (1) of Theorem 2.2.2 is satisfied with $u := 0$ and for every $\varepsilon > 0$.

In the respective case put $\alpha := \min\limits_{x \in K} k_0(x) + \beta$ and set $k := \dfrac{1}{\alpha} \cdot k_0$. Moreover, choosing $f \in \mathcal{K}(X)$ satisfying

$$0 \leq f \leq 1, \quad f = 1 \text{ on } K, \quad f(x_i) = 0 \quad \text{for every } i = 1, \ldots, n,$$

set $u := \sup\left(-k, \dfrac{\beta}{\alpha} \cdot f \right) \in \mathscr{C}_0^+(X)$. Then we have $0 \leq k + u$ on X, $1 \leq k + u$ on K, $\|u\| \leq \dfrac{\beta}{\alpha}$ and $u(x_i) = \dfrac{\beta}{\alpha}$ for every $i = 1, \ldots, n$.

Hence, statement (ii) of Theorem 2.2.4 is satisfied for every $\varepsilon > 0$ and for every

$$\mu = \sum\limits_{i=1}^{n} \alpha_i \varepsilon_{x_i} \in \tilde{C}_+^1(x_1, \ldots, x_n). \qquad \square$$

Remark. For a geometric interpretation of condition (2) of Theorem 2.2.6 in the case X compact, see Remark 3 to Corollary 2.6.5.

2.2.7 Corollary. *Let $h_1, \ldots, h_n$ be functions in $\mathscr{C}_0(X)$ satisfying $\det(h_i(x_j)) \neq 0$ and consider a positive function $h_0 \in \mathscr{C}_0(X)$ (a positive function of the form $h_0 + \beta$, where $h_0 \in \mathscr{C}_0(X)$ and $\beta \in \mathbb{R}_+$, respectively) which vanishes exactly on $x_1, \ldots, x_n$.*

Then the subspace of $\mathscr{C}_0(X)$ generated by the functions $h_0, h_1, \ldots, h_n$ is a D_+-subspace for every $\mu \in C_+(x_1, \ldots, x_n)$ (a D_+^1-subspace for every $\mu \in \tilde{C}_+^1(x_1, \ldots, x_n)$, respectively).

Examples. 1. A simple way to apply Corollary 2.2.7 (for $n \geq 2$) consists in finding n positive functions $f_1, \ldots, f_n$ in $\mathscr{C}_0(X)$ such that for every $i = 1, \ldots, n$, f_i vanishes only on x_i. Then, it is sufficient to set

$$h_0 := \prod_{i=1}^{n} f_i \quad \text{and} \quad h_i := \prod_{\substack{j=1 \\ j \neq i}}^{n} f_j \quad \text{for every } i = 1, \ldots, n. \qquad (2.2.3)$$

For example, if $X := \mathbb{R}^p$, we can choose the functions

$$f_i(x) := \|x - x_i\|^2 \exp(-\|x\|) \quad \text{for } i = 1, \ldots, n, \quad x \in \mathbb{R}^p, \qquad (2.2.4)$$

or, if (X, d) is a compact metric space, we can set

$$f_i(x) := d^2(x, x_i) \quad \text{for } i = 1, \ldots, n, \quad x \in X. \qquad (2.2.5)$$

2. Consider $X := [0, 1]$ and set $h_0(x) := x^\alpha - x^\beta$, $h_1(x) := 1 - x^\alpha$ and $h_2(x) := x^\beta$ for every $x \in [0, 1]$, where α and β are fixed real numbers satisfying $0 < \alpha < \beta$. Then the subspace generated by $\{h_0, h_1, h_2\}$ is a D_+-subspace for every $\mu \in C_+(0, 1)$.

3. Consider the interval $[-1, 1]$ and the functions

$$h_0(x) := x^{2p}(1 - x^{2p}), \quad h_1(x) := x^{2p}(1 - x^{2p-1}), \quad h_2(x) := 1 - x^{2p},$$

$$h_3(x) := x^{2p}(1 + x^{2p-1}),$$

where $p \in \mathbb{N}$ is fixed.

Then the subspace generated by h_0, h_1, h_2, h_3 is a D_+-subspace for every $\mu \in C_+(-1, 0, 1)$.

4. Consider $X := \mathbb{N}$. In this case $\mathscr{C}_0(\mathbb{N}) = c_0$ (see Remark to Corollary 2.1.8). Fix n integers $p_1, \ldots, p_n$ satisfying $p_1 < \cdots < p_n$ and consider $h_0 = (\alpha_p^0)_{p \in \mathbb{N}}$, $h_1 = (\alpha_p^1)_{p \in \mathbb{N}}, \ldots, h_n = (\alpha_p^n)_{p \in \mathbb{N}}$ in c_0 satisfying the following conditions:

(1) $\alpha_p^0 \geq 0$ and $\alpha_p^0 = 0$ only when $p \in \{p_1, \ldots, p_n\}$;
(2) $\det(\alpha_{p_j}^i) \neq 0$.

Then, the subspace of c_0 generated by the elements $h_0, h_1, \ldots, h_n$ is a D_+-subspace for every positive linear form μ defined as

$$\mu(x) := \sum_{i=1}^{n} \lambda_i x_{p_i}, \quad (x = (x_p)_{p \in \mathbb{N}} \in c_0),$$

where $\lambda_i \geq 0$ for each $i = 1, \ldots, n$.

In the next result we explicitly indicate a typical consequence of Corollary 2.2.7 that can be useful for concrete applications.

2.2.8 Proposition. Let $h_1, \ldots, h_n$ be functions in $\mathscr{C}_0(X)$ satisfying $\Delta_0 := \det(h_i(x_j)) \neq 0$ and consider a positive function $h_0 \in \mathscr{C}_0(X)$ which vanishes only on $x_1, \ldots, x_n$.
Then, an equicontinuous net $(\mu_i)_{i \in I}^{\leq}$ in $\mathscr{M}_b^+(X)$ converges vaguely to some discrete Radon measure supported by $\{x_1, \ldots, x_n\}$ if and only if the following $n + 1$ conditions are satisfied:

$$\lim_{i \in I}{}_{\leq} \mu_i(h_0) = 0, \tag{2.2.6}$$

and

$$\text{there exists } \beta_i := \lim_{i \in I}{}_{\leq} \mu_i(h_j) \quad \text{for every } j = 1, \ldots, n. \tag{2.2.7}$$

In this case, moreover, for every $f \in \mathscr{C}_0(X)$ we have

$$\lim_{i \in I}{}_{\leq} \mu_i(f) = \frac{1}{\Delta_0} \sum_{i=1}^{n} \Delta_i f(x_i),$$

where for every $i = 1, \ldots, n$, Δ_i denotes the determinant obtained from the determinant Δ_0 by replacing the i-th column by the column $(\beta_j)_{1 \leq j \leq n}$.

Proof. From Corollary 2.2.7 we know that the subspace H generated by $h_0, h_1, \ldots, h_n$ is a D_+-subspace for every $\mu \in C_+(x_1, \ldots, x_n)$.
Consider an equicontinuous net $(\mu_i)_{i \in I}^{\leq}$ in $\mathscr{M}_b^+(X)$ satisfying conditions (2.2.6) and (2.2.7). Set $\alpha_j := \dfrac{\Delta_j}{\Delta_0}$ for every $j = 1, \ldots, n$ and $\mu := \sum_{i=1}^{n} \alpha_i \varepsilon_{x_i}$.

Since $\sum_{i=1}^{n} \alpha_i h_j(x_i) = \beta_j$ for every $j = 1, \ldots, n$ and $\sum_{i=1}^{n} \alpha_i h_0(x_i) = 0$, we have

$$\lim_{i \in I} {}_\leq \mu_i(h) = \mu(h) \quad \text{for every } h \in H.$$

Set $\mu_1 := \sum_{i \in J} \alpha_i \varepsilon_{x_i}$ and $\mu_2 := \sum_{\substack{i=1 \\ i \notin J}}^{n} \alpha_i \varepsilon_{x_i}$, where $J := \{i = 1, \ldots, n | \alpha_i \geq 0\}$ and the sum is assumed to be zero if $J = \varnothing$ or $\{i = 1, \ldots, n | i \notin J\} = \varnothing$.

Then $\mu = \mu_1 + \mu_2$ and, consequently, the equicontinuous net of bounded positive Radon measures $(\mu_i - \mu_2)_{i \in I}^{\leq}$ converges to $\mu_1 \in C_+(x_1, \ldots, x_n)$ on H.

Hence $(\mu_i - \mu_2)_{i \in I}^{\leq}$ converges vaguely to μ_1 and the result follows. $\qquad \square$

Notes and references

The main results of this section are taken from Altomare [1989] who established them in the more general context of function spaces on a locally compact space.

Proposition 2.2.1 is a particular case of a general result due to Flösser [1979] for locally convex M-spaces or dual atomic locally convex vector lattices.

An abstract version of Proposition 2.2.8 can be found in Labsker [1971].

Campiti [1988a] characterized determining subspaces for positive Radon measures whose support is contained in a fixed countable discrete closed subset of a locally compact space.

One of his most important results runs as follows. Let $P := \{x_n | n \in \mathbb{N}\}$ be a countable closed discrete subset of X. For a given subspace H of $\mathscr{C}_0(X)$ the following statements are equivalent:

(i) *H is a D_+-subspace for every* $\mu = \sum_{n=1}^{\infty} \alpha_n \varepsilon_{x_n}$ *with* $(\alpha_n)_{n \in \mathbb{N}} \in \ell_+^1$ $\Big($*here ℓ_+^1 denotes the cone of all sequences $(\alpha_n)_{n \in \mathbb{N}}$ of positive real numbers such that* $\sum_{n=1}^{\infty} \alpha_n < +\infty \Big)$;

(ii) *For every $(\alpha_n)_{n \in \mathbb{N}} \in \ell_+^1$, for every compact subset K of X and for every null sequence $(\varepsilon_n)_{n \in \mathbb{N}}$ in $\mathbb{R}_+$, there exist a sequence $(h_n)_{n \in \mathbb{N}}$ in H and a sequence $(u_n)_{n \in \mathbb{N}}$ in $\mathscr{C}_0^+(X)$ satisfying*

$$\|u_n\| \leq \varepsilon_n, \quad 0 \leq h_n + u_n, \quad 1 \leq h_n + u_n \text{ on } K, \quad h_n + u_n \leq \varepsilon_n \text{ on } \{x_1, \ldots, x_n\} \backslash K$$

for every $n \in \mathbb{N}$ and

$$\sum_{\substack{p=n+1 \\ x_p \notin K}}^{\infty} \alpha_p(h_n(x_p) + u_n(x_p)) \leq \varepsilon_n$$

(iii) (1) *For each $(\alpha_n)_{n \in \mathbb{N}} \in \ell_+^1$ and for each $x \in X \backslash P$ there exist $M \in \mathbb{R}_+$ and a neighborhood U of x such that $U \cap P = \emptyset$ and, for every $\varepsilon > 0$, there exist a sequence $(h_n)_{n \in \mathbb{N}}$ in H and a sequence $(u_n)_{n \in \mathbb{N}}$ in $\mathscr{C}_0^+(X)$ satisfying*

$$\|u_n\| \leq \varepsilon, \quad 0 \leq h_n + u_n, \quad 1 \leq h_n + u_n \text{ on } U, \quad h_n + u_n \leq \varepsilon \text{ on } \{x_1, \ldots, x_n\}$$

for every $n \in \mathbb{N}$ and

$$\liminf_{n \to \infty} \sum_{p=n+1}^{\infty} \alpha_p(h_n(x_p) + u_n(x_p)) \leq M\varepsilon.$$

(2) *For each $(\alpha_n)_{n \in \mathbb{N}} \in \ell_+^1$ and for each $x \in P$ there exist $M \in \mathbb{R}_+$ such that for every $\varepsilon > 0$ there exist a sequence $(h_n)_{n \in \mathbb{N}}$ in H and a sequence $(u_n)_{n \in \mathbb{N}}$ in $\mathscr{C}_0^+(X)$ satisfying*

$$u_n \leq \varepsilon \text{ on } P, \quad 0 \leq h_n + u_n \text{ on } P, \quad 1 \leq h_n(x) + u_n(x),$$

$$h_n + u_n \leq \varepsilon \text{ on } \{x_1, \ldots, x_n\} \backslash \{x\}$$

for every $n \in \mathbb{N}$ and

$$\liminf_{n \to \infty} \sum_{p=n+1}^{\infty} \alpha_p(h_n(x_p) + u_n(x_p)) \leq M\varepsilon.$$

2.3 Determining subspaces and Chebyshev systems

The main aim of this short section is to point out the strong relations between D_+-subspaces and Chebyshev subspaces. The results we shall obtain will be very useful to investigate the so-called Korovkin subspaces of order n on a real interval or on the unit circle (see Sections 3.4 and 4.5).

We begin by recalling the definition of Chebyshev system.

2.3.1 Definition. *Let X be a compact space with at least n points and let $h_0, h_1, \ldots, h_n$ be continuous real functions on X. We say that the set of functions $h_0, h_1, \ldots, h_n$ is a Chebyshev system of order $n + 1$ on X if each linear combination $\lambda_0 h_0 + \cdots + \lambda_n h_n$ with coefficients $\lambda_0, \ldots, \lambda_n \in \mathbb{R}$ not simultaneously null has at most n distinct zeros on X.*

A subspace H of $\mathscr{C}(X)$ generated by a Chebyshev system of order $n + 1$ on X is called a Chebyshev subspace of order $n + 1$ in $\mathscr{C}(X)$.

Obviously, if $h_0, h_1, \ldots, h_n$ is a Chebyshev system, then $h_0, h_1, \ldots, h_n$ are linearly independent.

2.3.2 Proposition. *Let X be a compact Hausdorff space and let $h_0, h_1, \ldots, h_n \in \mathscr{C}(X)$. Then $h_0, h_1, \ldots, h_n$ form a Chebyshev system of order $n + 1$ on X if and only if the following condition holds:*

$$\det h_i(x_j) \neq 0 \quad \text{for each distinct points } x_0, x_1, \ldots, x_n \in X. \tag{2.3.1}$$

Proof. Suppose that $h_0, h_1, \ldots, h_n$ is a Chebyshev system on X and consider $n + 1$ distinct points $x_0, x_1, \ldots, x_n$ of X. To show (2.3.1) we shall prove that the system

$$\begin{cases} \lambda_0 h_0(x_0) + \cdots + \lambda_n h_n(x_0) = 0 \\ \vdots \qquad \vdots \quad \vdots \qquad \vdots \\ \lambda_0 h_0(x_n) + \cdots + \lambda_n h_n(x_n) = 0 \end{cases} \tag{1}$$

admits only the trivial solution.

In fact, if $\lambda_0, \ldots, \lambda_n$ satisfies system (1), then the function $h := \sum_{i=0}^{n} \lambda_i h_i$ vanishes at $x_0, x_1, \ldots, x_n$. Thus $h = 0$ and hence $\lambda_i = 0$ for every $i = 0, \ldots, n$.

Conversely assume that $h_0, h_1, \ldots, h_n$ satisfy condition (2.3.1). Obviously $h_0, h_1, \ldots, h_n$ must be linearly independent; moreover, if $h = \sum_{i=0}^{n} \lambda_i h_i$ is a non zero element of $\mathscr{L}(\{h_0, h_1, \ldots, h_n\})$ with $n + 1$ distinct zeros $x_0, x_1, \ldots, x_n$, then the corresponding system (1) would have a non trivial solution $\lambda_0, \ldots, \lambda_n$ and this contradicts condition (2.3.1). $\qquad \square$

Thus, on account of the above proposition, $h_0, \ldots, h_n$ is a Chebyshev system of order $n + 1$ on X if and only if for a given set of distinct points $x_0, \ldots, x_n \in X$ and real numbers $\alpha_0, \ldots, \alpha_n$ there is a unique interpolating function P in $\mathscr{L}(\{h_0, \ldots, h_n\})$ with prescribed values α_i at the point x_i, $i = 0, \ldots, n$. The function P is given by $P = \sum_{i=0}^{n} \lambda_i h_i$, where $\lambda_0, \ldots, \lambda_n$ is the unique solution of the system $\sum_{i=0}^{n} \lambda_i h_i(x_j) = \alpha_j \, (j = 0, \ldots, n)$.

However note that the existence (but not the uniqueness) of such interpolating function is also guaranteed if the points $x_0, \ldots, x_n$ are not distinct.

Chebyshev systems have also an interesting characterization in the theory of best polynomial approximation. Indeed, a classical *theorem of Haar* states that for every $f \in \mathscr{C}(X)$ there exists a unique best uniform approximation from the subspace generated by $h_0, \ldots, h_n \in \mathscr{C}(X)$ if and only if $h_0, \ldots, h_n$ is a Chebyshev system of order $n + 1$ on X (see, e.g., Cheney [1982, p. 81]).

In spite of these nice properties of Chebyshev systems, in general it is not possible to find Chebyshev systems on arbitrary compact spaces. Indeed, the celebrated *Mairhuber-Curtis theorem* states that a compact space admits a Chebyshev system of order $n + 1$ if and only if it is homeomorphic to a subset of the unit circle $\mathbb{T} := \{(x, y) \in \mathbb{R}^2 \mid x^2 + y^2 = 1\}$ in $\mathbb{R}^2$. Moreover, X can be homeomorphic to the entire circle if and only if n is even (see Mairhuber [1956] and Curtis [1959]).

In the remaining discussion about Chebyshev systems, we shall assume that X is a compact real interval $[a, b]$ or the unit circle $\mathbb{T}$ of $\mathbb{R}^2$.

2.3.3 Proposition. *Let X be a compact real interval $[a, b]$ or the unit circle. If H is a Chebyshev subspace of order $n + 1$ and if a function $h \in H$ has exactly n distinct zeros $x_1, \ldots, x_n$, then it necessarily changes sign at each x_i, $i = 1, \ldots, n$ (at each internal point x_i whenever $X = [a, b]$).*

Proof. Indeed, suppose to the contrary that h is positive on both sides of x_j for some $j = 1, \ldots, n$. Let $h_1 \in H$ satisfy the n conditions $h_1(x_i) = \delta_{i1}$, $i = 1, \ldots, n$ and consider two points $y_1, y_2 \in X$ such that the arc (or the segment, respectively) from y_1 to y_2 contains only the zero x_j and such that h and h_1 are both positive on y_1 and y_2. By multiplying, if necessary, the function h_1 by a positive real number, we may assume $0 < h_1(y_1) < h(y_1)$ and $0 < h_1(y_2) < h(y_2)$. Then the function $h - h_1 \in H$ vanishes at the $n - 1$ points x_i, $i = 1, \ldots, n$, $i \neq j$ and satisfies $(h - h_1)(y_1) > 0$, $(h - h_1)(x_j) < 0$ and $(h - h_1)(y_2) > 0$; consequently, it must vanish on two other points between y_1 and y_2 and this contradicts the fact that $h - h_1$ can have at most n zeros. $\qquad\square$

The following theorem is due to Krein and describes the connection between condition (2.3.1) and Theorem 2.2.6.

We also recall that a *complete Chebyshev system of order* $n + 1$ *on* X is a Chebyshev system $h_0, h_1, \ldots, h_n$ of order $n + 1$ on X such that for every $k = 1, \ldots, n$, $h_0, h_1, \ldots, h_k$ is again a Chebyshev system of order $k + 1$ on X. Furthermore, a *complete Chebyshev subspace of order* $n + 1$ *on* X is a subspace of $\mathscr{C}(X)$ generated by a complete Chebyshev system of order $n + 1$ on X.

If X is a compact real interval and $x \in [a, b]$, we put

$$\omega(x) := \begin{cases} 2, & \text{if } x \in \,]a, b[, \\ 1, & \text{if } x = a \text{ or } x = b, \end{cases} \tag{2.3.2}$$

while, if $X = \mathbb{T}$, we set $\omega(x) := 2$ for every $x \in \mathbb{T}$.

2.3.4 Theorem. *Let X be a compact real interval $[a, b]$ or the unit circle $\mathbb{T}$. For a given $n \in \mathbb{N}$ let H be a Chebyshev subspace of order $n + 1$ in $\mathscr{C}(X)$ generated by the functions $h_0, \ldots, h_n$. If $k = 1, \ldots, n$ and if $x_1, \ldots, x_k \in X$ are different points satisfying $\sum\limits_{i=1}^{k} \omega(x_i) \leq n$, then there exists $h \in H$ such that*

$$h \geq 0, \, h(x_1) = \cdots = h(x_k) = 0 \quad \text{and } h > 0 \quad \text{on } X \backslash \{x_1, \ldots, x_k\}. \tag{2.3.3}$$

If X is a compact real interval $[a, b]$, n is even and one of the endpoints a or b is in $\{x_1, \ldots, x_k\}$, then the function h may vanish at both the endpoints a and b. This exception cannot occur if H is a complete Chebyshev subspace of order $n + 1$ in $\mathscr{C}([a, b])$.

Proof. For the sake of completeness, we restate the proof given in Karlin and Studden [1964, Theorem I.5.1].

First, assume that X is a compact real interval $[a, b]$.

We begin with the case where $n = 2m + 1$ is odd and the points $x_1, \ldots, x_k$ are internal to $[a, b]$. It is enough to show the result only for $k = m$. In fact, if (2.3.3) holds for m and if we have $x_1, \ldots, x_k \in X$ with $k < m$, then we may consider two disjoint sets of supplementary points $x_1', \ldots, x_{m-k}' \in X$ and $x_1'', \ldots, x_{m-k}'' \in X$ distinct from $x_1, \ldots, x_k, a, b$, and apply (2.3.3) to the points $x_1, \ldots, x_k, x_1', \ldots, x_{m-k}'$ and $x_1, \ldots, x_k, x_1'', \ldots, x_{m-k}''$. So, we obtain two positive functions h_1 and h_2 in H which vanish only at $x_1, \ldots, x_k, x_1', \ldots, x_{m-k}'$ and at $x_1, \ldots, x_k, x_1'', \ldots, x_{m-k}''$, respectively; thus, the sum $h := h_1 + h_2$ is positive and vanishes exactly at $x_1, \ldots, x_k$.

So let us suppose $k = m$ and take m points $x_1, \ldots, x_m$ such that $a < x_1 < \cdots < x_m < b$. Let $\varepsilon > 0$ be such that the n points $s_1 := a$, $s_2 := x_1$, $s_3 := x_1 + \varepsilon, \ldots,$ $s_{2m} := x_m, s_n := x_m + \varepsilon$ are in increasing order and $s_n < b$. Since $h_0, \ldots, h_n$ is a Chebyshev system, the function

$$h_\varepsilon(x) := \begin{vmatrix} h_0(s_1) & \dots & h_0(s_n) & h_0(x) \\ h_1(s_1) & \dots & h_1(s_n) & h_1(x) \\ \vdots & \vdots & \cdots & \vdots \\ h_n(s_1) & \dots & h_n(s_n) & h_n(x) \end{vmatrix} \quad (x \in [a,b])$$

belongs to H and vanishes exactly on $s_1, \dots, s_n$. Thus, h_ε has n distinct zeros and therefore it changes sign at each point s_i, $i = 1, \dots, n$. By multiplying h_ε with a suitable constant, if necessary, we may assume that $h_\varepsilon = \sum_{i=0}^{n} \alpha_i(\varepsilon) h_i$ with $\sum_{i=0}^{n} \alpha_i^2(\varepsilon) = 1$ and $h_\varepsilon > 0$ on $]s_{2i-1}, s_{2i}[$, $i = 1, \dots, m$, and on $]s_n, b[$.

Note that the coefficients $\alpha_i(\varepsilon)$ depend continuously on ε and we may always suppose that they converge to some α_i when $\varepsilon \to 0$. So, letting $\varepsilon \to 0$, we obtain a positive function $f_1 = \sum_{i=0}^{n} \alpha_i h_i \in H$ which is non trivial $\left(\text{since } \sum_{i=0}^{n} \alpha_i^2 = 1 \right)$ and vanishes exactly at the points $a, x_1, \dots, x_m$. Indeed, if f_1 vanishes at a point x different from $a, x_1, \dots, x_m$, then we may consider another function $g \in H$ such that $g(a) = 0$, $g(x) = 0$ and $g(x_i) = 1$ for every $i = 1, \dots, m$; then, for a sufficiently small $\lambda > 0$, the function $f_1 - \lambda g$ would have $2m + 2 = n + 1$ zeros and this is a contradiction.

In an analogous way, we can consider a positive function $f_2 \in H$ which vanishes only at the points $x_1, \dots, x_m, b$ and therefore the sum $h := f_1 + f_2$ is a positive element of H which vanishes exactly at the points $x_1, \dots, x_m$ and this completes the proof.

Now, suppose that $x_1 = a$ or $x_k = b$. Again we may assume $k = m$. Suppose, for example, that $a = x_1 < x_2 < \cdots < x_m < b$. Then, after choosing $\bar{x}_1$, $\bar{x}_2 \in [a, b]$ satisfying $x_m < \bar{x}_1 < \bar{x}_2 < b$, by virtue of the preceding case there exist positive functions f_1 and f_2 in H whose zeros are exactly $a, x_2, \dots, x_m, \bar{x}_1$ and $a, x_2, \dots, x_m, \bar{x}_2$, respectively. Then, the function $h := f_1 + f_2$ is positive and vanishes only in $a = x_1, x_2, \dots, x_m$.

Finally, suppose $a = x_1 < \cdots < x_k = b$. We may restrict ourselves to the maximal case $k = m + 1$. We consider $\varepsilon > 0$ such that the n points

$$s_1 := a = x_1, s_2 := x_1 + \varepsilon, \dots, s_{2m-1} := x_m, s_{2m} := x_m + \varepsilon, s_n := b = x_{m+1},$$

are in increasing order. As before, we consider a function h_ε in H which vanishes exactly at the n points $s_1, \dots, s_n$ and satisfies $h_\varepsilon = \sum_{i=0}^{n} \alpha_i(\varepsilon) h_i$ with $\sum_{i=0}^{n} \alpha_i(\varepsilon)^2 = 1$ and $h_\varepsilon > 0$ on $]s_{2i}, s_{2i+1}[$, $i = 1, \dots, m$. If ε tends to 0, we obtain a non trivial positive function $h \in H$ which vanishes exactly on $x_1, \dots, x_{m+1}$.

If $n = 2m$ is even, the proof is analogous except in the case $a = x_1 < \cdots < x_m < b$ or $a < x_1 < \cdots < x_m = b$, where we cannot ensure that the function h does not vanish at both the endpoints. This actually holds if H is a complete

Chebyshev subspace in $\mathscr{C}([a,b])$, since this hypothesis allows us to reduce the proof to the first part by considering the Chebyshev subspace in $\mathscr{C}([a,b])$ of order n generated by the functions $h_0,\ldots,h_{n-1}$.

Finally, assume that $X = \mathbb{T}$; we only have to consider the case where $n = 2m$ is even because of Mairhuber-Curtis theorem. In this case condition $\sum_{i=1}^{k} \omega(x_i) \leq n$ means that $k \leq m$. Moreover, we may always suppose that $k = m$. Fix an orientation on $\mathbb{T}$ and assume that the m points $x_1,\ldots,x_m$ satisfy $x_1 < \cdots < x_m$. Now, the proof is analogous to that of the preceding cases. For every sufficiently small $\varepsilon > 0$ consider n points $s_1 < \cdots < s_n$ of $\mathbb{T}$ such that for every $k = 1, \ldots, m$, $s_{2k-1} = x_k$ and $|s_{2k} - s_{2k-1}| \leq \varepsilon$.

Let $h_\varepsilon \in H$ be a positive function which vanishes exactly on $s_1, \ldots, s_n$. Thus h_ε changes sign at each point s_i, $i = 1,\ldots,n$, and, further, we can assume $h_\varepsilon = \sum_{i=0}^{n} \alpha_i(\varepsilon)h_i$ with $\sum_{i=0}^{n} \alpha_i(\varepsilon)^2 = 1$ and $h_\varepsilon > 0$ on $]s_{2i}, s_{2i+1}[$, $i = 1,\ldots,m$ (with the convention $s_{2m+1} = x_1$). If ε tends to 0, the function h_ε converges to a non trivial positive function $h \in H$ which vanishes exactly at the points $x_1,\ldots,x_m$. $\qquad\square$

We may now state the promised result which shows the close relation between Chebyshev subspaces and D_+-subspaces.

2.3.5 Theorem. *Let X be a compact real interval $[a,b]$ or the unit circle $\mathbb{T}$ and let H be a Chebyshev subspace of $\mathscr{C}(X)$ of order $n + 1$.*

Then, for every set of different points $x_1,\ldots,x_p \in X$ satisfying $\sum_{i=1}^{p} \omega(x_i) \leq n$ and for every $\mu \in C_+(x_1,\ldots,x_p)$ the space H is a D_+-subspace for μ.

Proof. Suppose first n to be odd. Consider p distinct points $x_1,\ldots,x_p \in X$ satisfying $\sum_{i=1}^{p} \omega(x_i) \leq n$ and fix $\mu = \sum_{i=1}^{p} \alpha_i \varepsilon_{x_i} \in C_+(x_1,\ldots,x_p)$. If $v \in \mathscr{M}^+(X)$ satisfies the relation $\mu = v$ on H, then, according to Theorem 2.3.4, after choosing $h \in H$ satisfying (2.3.3), we have $v(h) = 0$. So $\mathrm{Supp}(v)$ must be contained in $\{x_1,\ldots,x_p\}$ and hence $v = \sum_{i=1}^{p} \lambda_i \varepsilon_{x_i}$ for some $\lambda_1,\ldots,\lambda_p \in \mathbb{R}_+$. Consequently, for every $h \in H$ we obtain $\sum_{i=1}^{p} (\alpha_i - \lambda_i)h(x_i) = 0$.

Since $p \leq n + 1$, we may choose, if necessary, $(n + 1) - p$ distinct points $x_{p+1},\ldots,x_{n+1}$, so that we have $\sum_{i=1}^{n+1} \beta_i h(x_i) = 0$ for each $h \in H$, where

$$\beta_i := \begin{cases} \alpha_i - \lambda_i, & \text{if } 1 \leq i \leq p, \\ 0, & \text{if } p+1 \leq i \leq n+1. \end{cases}$$

Since H is a Chebyshev subspace of order $n + 1$, we must have $\beta_i = 0$ for every $i = 1, \ldots, n + 1$ and hence $\mu = v$. This shows that H is a D_+-subspace for μ.

Suppose now that n is even and $X = [a, b]$. Again, fix $\mu = \sum_{i=1}^{p} \alpha_i \varepsilon_{x_i} \in C_+(x_1, \ldots, x_p)$, where $a \leq x_1 < \cdots x_p \leq b$ and $\sum_{i=1}^{p} \omega(x_i) \leq n$, and consider $v \in \mathcal{M}^+([a, b])$ satisfying $\mu = v$ on H.

By virtue of Theorem 2.3.4 we may consider a function $h \in H$ satisfying (2.3.3). We cannot proceed as in above case when only one of the endpoints a or b is in $\{x_1, \ldots, x_p\}$, because in this case h may vanish at both a and b.

Thus suppose, for instance, that $a = x_1 < \cdots < x_p < b$. If $h(b) > 0$, then the conclusion follows as above.

If $h(b) = 0$, then consider the positive Radon measure $v_1 := v + \varepsilon_b$. Then, since $v_1(h) = v(h) + h(b) = 0$, again $\text{Supp}(v_1) \subset \{x \in [a, b] | h(x) = 0\} = \{x_1, \ldots, x_p, b\}$.

Consequently, there exist $\lambda_1, \ldots, \lambda_p, \lambda_{p+1} \in \mathbb{R}_+$ such that $v_1 = \sum_{i=1}^{p} \lambda_i \varepsilon_{x_i} + \lambda_{p+1} \varepsilon_b$. This implies that for every $u \in H$

$$\sum_{i=1}^{p} \alpha_i u(x_i) = \mu(u) = v(u) = v_1(u) - u(b) = \sum_{i=1}^{p} \lambda_i u(x_i) + \lambda_{p+1} u(b) - u(b),$$

i.e., $\sum_{i=1}^{p} \alpha_i u(x_i) + u(b) = \sum_{i=1}^{p} \lambda_i u(x_i) + \lambda_{p+1} u(b)$.

Note that, in this case, $2p - 1 = \sum_{i=1}^{p} \omega(x_i) \leq n$ and hence $p + 1 \leq n + 1$. Therefore, by arguing as in the preceding case, we again have $\alpha_i = \lambda_i$ for each $i = 1, \ldots, p$ and $\lambda_{p+1} = 1$. Thus $v = \sum_{i=1}^{p} \alpha_i \varepsilon_{x_i} = \mu$. $\qquad \square$

From the preceding theorem we may draw various examples of D_+-subspaces from the (long) list of Chebyshev systems. Here we recall some of them. In what follows the positive integer n is arbitrary and therefore the corresponding Chebyshev systems are all complete.

Examples. 1. (*Power functions*) The functions $1, x, \ldots, x^n$ constitute a Chebyshev system of order $n + 1$ on $[a, b]$. Indeed, for every $x_0, \ldots, x_n \in [a, b]$ such that $x_0 < \cdots < x_n$, $\det(x_j^i)$ is the Vandermonde determinant $\prod_{0 \leq i < j \leq n} (x_j - x_i)$ which is different from 0.

If $0 < a < b$ and if $\alpha_0, \ldots, \alpha_n$ are strictly positive real numbers such that $\alpha_0 < \cdots < \alpha_n$, then $x^{\alpha_0}, \ldots, x^{\alpha_n}$ ($x \in [a, b]$) is a Chebyshev system of order $n + 1$ on $[a, b]$.

2. (*Trigonometric polynomials*) The functions $1, \cos x, \ldots, \cos nx$ and the functions $\sin x, \ldots, \sin nx$ constitute Chebyshev systems on the interval $[0, \pi]$.

Moreover, using both Euler's formulas $\cos\theta = \dfrac{1}{2}(\exp(i\theta) + \exp(-i\theta))$, $\sin\theta$

$= \dfrac{1}{2i}(\exp(i\theta) + \exp(-i\theta))(\theta \in \mathbb{R})$, and De Moivre's formulas $z^n = \cos(n\theta) +$

$i\sin(n\theta)(z \in \mathbb{T})$, where θ denotes an argument in the polar coordinates of z, it follows that the functions $\mathbf{1}, \mathscr{R}e\,z, \mathscr{R}e\,z^2, \ldots, \mathscr{R}e\,z^n, \mathscr{I}m\,z, \mathscr{I}m\,z^2, \ldots, \mathscr{I}m\,z^n$ constitute a Chebyshev system of order $2n + 1$ on $\mathbb{T}$.

3. *(Cauchy kernels)* Let $0 < a < b$ and consider the kernel $K\colon [a,b] \times [a,b] \to \mathbb{R}$ defined by $K(t,x) := \dfrac{1}{t + x}$ for every $(t,x) \in [a,b] \times [a,b]$. If $a \le t_0 < t_1 < \cdots < t_n \le b$, then $K(t_0, \cdot), K(t_1, \cdot), \ldots, K(t_n, \cdot)$ is a Chebyshev system on $[a,b]$ of order $n + 1$.

In this case, for each $x_0, \ldots, x_n \in [a,b]$ such that $x_0 < x_1 < \cdots < x_n$ the determinant $\det(K(t_i, x_j))$ is given by

$$\frac{\displaystyle\prod_{\substack{i,j=0 \\ i<j}}^{n} (t_j - t_i)(x_j - x_i)}{\displaystyle\prod_{i,j=0}^{n} (t_i + x_j)}.$$

4. *(Gauss kernels)* Let $K\colon [a,b] \times [a,b] \to \mathbb{R}$ be defined by $K(t,x) := \exp(-(t-x)^2)$ for every $(t,x) \in [a,b] \times [a,b]$. In this case too, if $a \le t_0 < t_1 < \cdots < t_n \le b$, then $K(t_0, \cdot), K(t_1, \cdot), \ldots, K(t_n, \cdot)$ is a Chebyshev system on $[a,b]$ of order $n + 1$.

5. Other examples of Chebyshev systems may be found in Karlin and Studden [1964, I, Section 3].

Moreover, there are several methods to obtain Chebyshev systems from an assigned Chebyshev system $h_0, \ldots, h_n$ of order $n + 1$ on $[a,b]$.

For example, if $\varphi \in \mathscr{C}([a,b])$ is strictly positive, then $\varphi \cdot h_0, \ldots, \varphi \cdot h_n$ is again a Chebyshev system of order $n + 1$ on $[a,b]$.

In the same manner, if $\psi\colon [c,d] \to [a,b]$ is a strictly increasing continuous function then $h_0 \circ \psi, \ldots, h_n \circ \psi$ is a Chebyshev system of order $n + 1$ on $[c,d]$.

Notes and references

The material concerning Chebyshev subspaces is classical and can be found, for instance, in the monograph of Karlin and Studden [1964]. Theorem 2.3.4 is due to Krein and the proof is the same as given in Karlin and Studden.

Theorem 2.3.5 was obtained by Micchelli [1973a] for real intervals and by Rusk [1977] for the unit circle. They also show an interesting converse, in the sense that, if $\mu \in \mathcal{M}^+(X)$ and if a Chebyshev subspace of order $n + 1$ is a D_+-subspace for μ, then necessarily $\mu = \sum_{i=1}^{p} \alpha_i \varepsilon_{x_i}$ for some $\alpha_1, \ldots, \alpha_p \geq 0$ and $x_1, \ldots, x_p \in X$ satisfying $\sum_{i=1}^{p} \omega(x_i) \leq n$ (here X is a compact real interval or the unit circle).

2.4 Convergence subspaces associated with discrete Radon measures

In this section we study in more detail particular convergence subspaces associated with discrete measures.

As a consequence we indicate other possible ways to explicitly construct determining subspaces (or subsets) for discrete measures. We also point out that all the results of this section have a natural generalization in the context of commutative Banach algebras (see Romanelli [1989]). For other extensions see Altomare [1991] and Attalienti [1994].

Given a non-empty subset M of $\mathscr{C}_0(X)$ and $h \in \mathbb{N}$ we set

$$M^h := \{g^h | g \in M\}. \tag{2.4.1}$$

For every $n \geq 1$ and $f \in \mathscr{C}_0(X)$ we also put

$$Q_n(f, M) := \{f\} \cup \bigcup_{h=1}^{n} \{f \cdot f_1 \cdots f_h | f_j \in M \cup M^2, j = 1, \ldots, h\}, \tag{2.4.2}$$

$$P_n^*(M) := \bigcup_{h=1}^{n} \{f_1 \cdots f_h | f_j \in M \cup M^2, j = 1, \ldots, h\}. \tag{2.4.3}$$

If X is compact, we set

$$P_n(M) := Q_n(\{\mathbf{1}\}, M) = \{\mathbf{1}\} \cup \bigcup_{h=1}^{n} \{f_1 \cdots f_h | f_j \in M \cup M^2, j = 1, \ldots, h\}. \tag{2.4.4}$$

For a given $x \in X$ we put

$$M(x) := \{y \in X | h(y) = h(x) \quad \text{for every } h \in M\}. \tag{2.4.5}$$

Then $M(x)$ is a closed subset of X and $M(x) = \{x\}$ if and only if M *separates* x *in* X, i.e.,

for every $y \in X$, $y \neq x$, *there exists* $h \in M$ *such that* $h(x) \neq h(y)$. (2.4.6)

We also recall that a linear subspace A of $\mathscr{C}_0(X)$ is said to be a *subalgebra* of $\mathscr{C}_0(X)$ if $f \cdot g \in A$ for every $f, g \in A$.

The *closed subalgebra generated* by a subset M of $\mathscr{C}_0(X)$ is, by definition, the intersection of all closed subalgebras of $\mathscr{C}_0(X)$ containing M. We shall denote it by $\mathbf{A}(M)$.

We now present a characterization of the convergence subspace associated with $Q_n(f, M)$, $P_n^*(M)$ and $P_n(M)$ and with discrete measures supported by the set of points $x_1, \ldots, x_n$ of X that from now on we shall keep fixed.

2.4.1. Theorem. *Let M be a subset of $\mathscr{C}_0(X)$ and suppose that there exists $h_0 \in M$ such that*

$$h_0(x_i) \neq 0 \quad \text{for every } i = 1, \ldots, n \tag{2.4.7}$$

and, whenever $n > 1$,

$$h_0(x_i) \neq h_0(x_j) \quad \text{for every } i, j = 1, \ldots, n, \quad i \neq j. \tag{2.4.8}$$

Furthermore, let $f_0 \in \mathscr{C}_0(X)$ be a strictly positive function such that

$$f_0 \text{ is constant on each } M(x_i) \quad i = 1, \ldots, n. \tag{2.4.9}$$

Then

$$\bigcap_{\mu \in C_+(x_1,\ldots,x_n)} D_+(Q_n(f_0, M), \mu)$$

$$= \{ f \in \mathscr{C}_0(X) | f \text{ is constant on each } M(x_i), i = 1, \ldots, n \}. \tag{2.4.10}$$

Thus $\bigcap_{\mu \in C_+(x_1,\ldots,x_n)} D_+(Q_n(f_0, M), \mu)$ is a closed subalgebra of $\mathscr{C}_0(X)$ and so, denoting by $\mathbf{A}(M)$ the closed subalgebra generated by M,

$$\mathbf{A}(M) \subset D_+(Q_n(f_0, M), \mu) \quad \text{for every } \mu \in C_+(x_1, \ldots, x_n). \tag{2.4.11}$$

Finally, $Q_n(f_0, M)$ is a D_+-subset for every $\mu \in C_+(x_1, \ldots, x_n)$ if and only if

for every $i = 1, \ldots, n$ and for every $x \in X$, $x \neq x_i$,

there exists $h \in M$ such that $h(x) \neq h(x_i)$. $\tag{2.4.12}$

Proof. Clearly, if $f \in \bigcap_{\mu \in C_+(x_1,\ldots,x_n)} D_+(Q_n(f_0, M), \mu)$, then, given $i = 1, \ldots, n$ and $y \in M(x_i)$, on account of (2.4.9) we have $\varepsilon_y = \varepsilon_{x_i}$ on $M \cup \{f_0\}$, and hence on $Q_n(f_0, M)$; consequently $\varepsilon_y(f) = \varepsilon_{x_i}(f)$, i.e., $f(y) = f(x_i)$.

Conversely, suppose that a function $f \in \mathscr{C}_0(X)$ is constant on $M(x_i)$ for every $i = 1, \ldots, n$. Fix $\mu = \sum_{i=1}^{n} \alpha_i \varepsilon_{x_i} \in C_+(x_1, \ldots, x_n)$ and $v \in \mathscr{M}_b^+(X)$ and suppose that $\mu = v$ on $Q_n(f_0, M)$. We shall prove that

$$\text{Supp}(v) \subset \bigcup_{i=1}^{n} M(x_i). \tag{1}$$

In fact, we first note that, if $x \in X \setminus \bigcup_{i=1}^{n} M(x_i)$, then for every $i = 1, \ldots, n$ there exists $h_{i,x} \in M$ such that $h_{i,x}(x) \neq h_{i,x}(x_i)$. So, after putting

$$h_x := f_0 \cdot \prod_{i=1}^{n} (h_{i,x} - h_{i,x}(x_i))^2, \tag{2}$$

we have $h_x \in \mathcal{L}(Q_n(f_0, M))$ and

$$h_x \geq 0, \tag{3}$$

$$h_x(x) > 0, \tag{4}$$

$$h_x(x_i) = 0 \quad \text{for every } i = 1, \ldots, n. \tag{5}$$

Now consider $y \in \text{Supp}(v)$ and suppose that $y \notin \bigcup_{i=1}^{n} M(x_i)$; then there exists a compact neighborhood K of y such that $K \cap \bigcup_{i=1}^{n} M(x_i) = \varnothing$.

The above argument shows that for every $x \in K$ there exists $h_x \in \mathcal{L}(Q_n(f_0, M))$ satisfying (3), (4) and (5).

By the compactness of K there exist finitely many functions $h_1, \ldots, h_p$ in $\mathcal{L}(Q_n(f_0, M))$ such that

$$h_j \geq 0 \quad \text{for every } j = 1, \ldots, p, \tag{6}$$

$$h_j(x) > 0 \quad \text{for every } x \in K \text{ and for some } j = 1, \ldots, p, \tag{7}$$

$$h_j(x_i) = 0 \quad \text{for every } j = 1, \ldots, p \text{ and } i = 1, \ldots, n. \tag{8}$$

So, if we set $h := \sum_{j=1}^{p} h_j \in \mathcal{L}(Q_n(f_0, M))$, we deduce that

$$h \geq 0, \tag{9}$$

$$h(x) > 0 \quad \text{for every } x \in K, \tag{10}$$

$$h(x_i) = 0 \quad \text{for every } i = 1, \ldots, n. \tag{11}$$

We can also suppose that $h \geq 1$ on K. Since $y \in \text{Supp}(v)$, there exists a function $f \in \mathcal{K}(X)$, such that its support is contained in K, $0 \leq f \leq 1$ and $v(f) > 0$.

Hence $f \leq h$ and

$$0 < v(f) \leq v(h) = \mu(h) = \sum_{i=1}^{n} \alpha_i h(x_i) = 0,$$

and this leads to a contradiction. Thus, we have proved inclusion (1).

At this point, note that there exist $h_1, \ldots, h_n$ in the linear subspace generated by $Q_n(f_0, M)$ such that

$$\det(h_i(x_j)) \neq 0. \tag{12}$$

In fact, this is certainly true if $n = 1$ by virtue of (2.4.7). If $n > 1$, it is sufficient to take a function $h_0 \in M$ satisfying (2.4.8) and to put $h_p := f_0 h_0^p$ for every $p = 1, \ldots, n$, because

$$\det(h_p(x_q)) = f_0(x_1) \cdots f_0(x_n) h_0(x_1) \cdots h_0(x_n) \cdot \prod_{1 \leq p < q \leq n} (h_0(x_p) - h_0(x_q)) \neq 0.$$

Now, on account of (12), fix $\beta_1, \ldots, \beta_n \in \mathbb{R}$ satisfying $\sum_{i=1}^{n} \beta_i h_i(x_j) = f(x_j)$ $(j = 1, \ldots, n)$.

Consequently, if $x \in \mathrm{Supp}(v)$ and if $x \in M(x_j)$ for some $j = 1, \ldots, n$, we have

$$f(x) = f(x_j) = \sum_{i=1}^{n} \beta_i h_i(x_j) = \sum_{i=1}^{n} \beta_i h_i(x).$$

Thus $f = \sum_{i=1}^{n} \beta_i h_i$ on $\mathrm{Supp}(v)$ and hence

$$v(f) = \sum_{i=1}^{n} \beta_i v(h_i) = \sum_{i=1}^{n} \sum_{j=1}^{n} \beta_i \alpha_j h_i(x_j) = \sum_{j=1}^{n} \alpha_j \sum_{i=1}^{n} \beta_i h_i(x_j) = \sum_{j=1}^{n} \alpha_j f(x_j).$$

This reasoning shows that $f \in U_+(\mathscr{L}(Q_n(f_0, M)), \mu) = D_+(Q_n(f_0, M), \mu)$ (see Proposition 2.1.2). $\square$

Remark. If $f \in \mathscr{C}_0(X)$ is a strictly positive function such that (whenever $n > 1$) $f(x_i) \neq f(x_j)$ for every $i, j = 1, \ldots, n$, $i \neq j$, then we can apply Theorem 2.4.1 for $f_0 = f$ and $M = \{f\}$.

In this case $Q_n(f, M) = \{f, f^2, \ldots, f^{2n+1}\}$.

A result analogous to Theorem 2.4.1 for positive contractive Radon measures is indicated below.

2.4.2 Theorem. *Let M be a subset of $\mathscr{C}_0(X)$ and suppose that there exists $h_0 \in M$ satisfying (2.4.7) and (2.4.8). Then*

$$\bigcap_{\mu \in \tilde{C}^1_+(x_1,\ldots,x_n)} D^1_+(P_n^*(M), \mu) = \{f \in \mathscr{C}_0(X) | f \text{ is constant on each } M(x_i), i = 1,\ldots,n\}.$$

$$(2.4.13)$$

Hence

$$\mathbf{A}(M) \subset D^1_+(P_n^*(M), \mu) \quad \text{for every } \mu \in \tilde{C}^1_+(x_1,\ldots,x_n) \qquad (2.4.14)$$

and $P_n^*(M)$ is a D^1_+-subset for every $\mu \in \tilde{C}^1_+(x_1,\ldots,x_n)$ if and only if (2.4.12) holds.

Proof. The proof is analogous to the one of Theorem 2.4.1. We briefly sketch it. First, we show that for every $x \in X \backslash \bigcup_{i=1}^n M(x_i)$ there exists $h_x \in P_n^*(M) + \mathbb{R}_+$ satisfying (3), (4) and (5).

Take, in fact, $x \in X \backslash \bigcup_{i=1}^n M(x_i)$; then for every $i = 1,\ldots,n$ there exists $h_{i,x} \in M$ such that $h_{i,x}(x) \neq h_{i,x}(x_i)$; so it is sufficient to put $h_x := \beta_x + k_x$, where $\beta_x := \prod_{i=1}^n h_{i,x}^2(x_i)$ and $k_x := \left(\prod_{i=1}^n (h_{i,x} - h_{i,x}(x_i))^2 \right) - \beta_x$.

Now, if $\mu = \sum_{i=1}^n \alpha_i \varepsilon_{x_i} \in \tilde{C}^1_+(x_1,\ldots,x_n)$ and $\upsilon \in \mathscr{M}_b^+(X)$, $\|\upsilon\| \leq 1$, and if $\mu = \upsilon$ on $P_n^*(M)$, then again

$$\text{Supp}(\upsilon) \subset \bigcup_{i=1}^n M(x_i).$$

In fact, if K is a compact subset of X disjoint from $\bigcup_{i=1}^n M(x_i)$, then, by using the above arguments and the compactness of K, we infer the existence of $k \in P_n^*(M)$ and $\beta \in \mathbb{R}_+$ such that $k + \beta \geq 0$, $k + \beta \geq 1$ on K, $k(x_i) + \beta = 0$ for every $i = 1,\ldots,n$.

Then, for every function $f \in \mathscr{K}(X)$, $0 \leq f \leq 1$, whose support is contained in K, we have $f \leq k + \beta$ and so, by Lemma 2.1.5, there exists $u \in \mathscr{C}_0^+(X)$ such that $f \leq k + u$ and $\|u\| \leq \beta$.

Consequently

$$0 \leq \upsilon(f) \leq \upsilon(k + u) \leq \upsilon(k) + \|u\| \cdot \|\upsilon\| \leq \mu(k) + \beta = \sum_{i=1}^n \alpha_i(k(x_i) + \beta) = 0$$

and hence $\upsilon(f) = 0$.

The conclusion of the proof is now the same as the one of Theorem 2.4.1.

□

In the compact case, we can apply Theorem 2.4.1 to a subset M of $\mathscr{C}(X)$ satisfying (2.4.7) and (2.4.8) and to the function $f_0 = \mathbf{1}$.

Here we indicate how the same result holds under weaker hypotheses.

2.4.3 Theorem. *Let X be a compact space and let M be a subset of $\mathscr{C}(X)$. If $n > 1$, we further suppose that*

$$\text{for every } i = 1,\ldots,n \text{ there exists } h_i \in M \text{ such that } h_i(x_i) \neq h_i(x_j)$$

$$\text{for every } j = 1,\ldots,n, \quad j \neq i. \tag{2.4.15}$$

Then

$$\bigcap_{\mu \in C_+(x_1,\ldots,x_n)} D_+(P_n(M),\mu) = \{f \in \mathscr{C}(X) \mid f \text{ is constant on each } M(x_i), i = 1,\ldots,n\}.$$

$$\tag{2.4.16}$$

So

$$A(M) \subset D_+(P_n(M),\mu) \quad \text{for every } \mu \in C_+(x_1,\ldots,x_n), \tag{2.4.17}$$

and $P_n(M)$ is a D_+-subset for every $\mu \in C_+(x_1,\ldots,x_n)$ if and only if (2.4.12) holds.

Proof. The proof is the same as the one of Theorem 2.4.1 with $f_0 = \mathbf{1}$ provided we show that also in this case there exist $k_1,\ldots,k_n$ in the linear subspace generated by $P_n(M)$ such that $\det(k_i(x_j)) \neq 0$. In fact, if $n = 1$, take $k_1 = \mathbf{1}$. If $n > 1$, taking the functions $h_1,\ldots,h_n$ according to (2.4.15) and choosing for every $i = 1,\ldots,n$ a polynomial p_i of degree less than or equal to $n - 1$ such that $p_i(h_i)(x_j) = \delta_{ij}$, $i,j = 1,\ldots,n$, we have $\det(p_i(h_i)(x_j)) \neq 0$ and $p_i(h_i) \in \mathscr{L}(P_n(M))$ for each $i = 1,\ldots,n$.

□

Examples. 1. If X is a compact subset of $\mathbb{R}^p$ $(p \geq 2)$, one of the most simple subsets M to which one can apply Theorem 2.4.3 is the subset $M := \{\mathrm{pr}_1,\ldots,\mathrm{pr}_p\}$ of the projections on X (see (1.2.49)) provided, of course, (2.4.15) is satisfied.

2. If $f \in \mathscr{C}_0(X)$ is an injective function which never vanishes on X (this last condition is not necessary in the compact case) then Theorems 2.4.2 and 2.4.3 hold with $M = \{f\}$ for every $n \in \mathbb{N}$ and for every $x_1,\ldots,x_n \in X$.

Moreover, if f is strictly positive, then Theorem 2.4.1 holds with $f_0 = f$ and $M = \{f\}$ again for every $n \in \mathbb{N}$ and for every $x_1,\ldots,x_n \in X$.

We conclude this section by noting that, with the help of Theorem 2.2.6, it is possible to construct determining subsets smaller than those indicated in Theorems 2.4.1, 2.4.2 and 2.4.3, provided the subset M satisfies additional assumptions.

If $f \in \mathscr{C}_0(X)$ and $n \geq 1$ we set

$$f M^n := \{f \cdot g^n | g \in M\}. \tag{2.4.18}$$

2.4.4 Proposition. *Let M be a subset of $\mathscr{C}_0(X)$ satisfying the following property:*

for every $x \in X \setminus \{x_1, \ldots, x_n\}$ there exists $h \in M$ such that

$$h(x) \neq h(x_i) \quad i = 1, \ldots, n, \tag{2.4.19}$$

and let $f_0 \in \mathscr{C}_0(X)$ be a strictly positive function. Then the following properties hold:

(1) *If M satisfies (2.4.7) and (2.4.8), then $\{f_0\} \cup f_0 M \cup \cdots \cup f_0 M^{2n}$ is a D_+-subset for every $\mu \in C_+(x_1, \ldots, x_n)$, and $M \cup \cdots \cup M^{2n}$ is a D_+^1-subset for every $\mu \in \tilde{C}_+^1(x_1, \ldots, x_n)$.*

(2) *If X is compact and M satisfies (2.4.15), then $\{\mathbf{1}\} \cup M \cup \cdots \cup M^{2n}$ is a D_+-subset for every $\mu \in C_+(x_1, \ldots, x_n)$.*

Proof. It is enough to apply Theorem 2.2.6. As indicated in the proof of Theorems 2.4.1 and 2.4.3 conditions (2.4.7), (2.4.8) and (2.4.15) ensure the existence of n functions $h_1, \ldots, h_n$ belonging to $\mathscr{L}(f_0 M \cup \cdots \cup f_0 M^n)$ (or to $\mathscr{L}(\{\mathbf{1}\} \cup f_0 M \cup \cdots \cup f_0 M^n)$) in the compact case, respectively) such that $\det(h_i(x_j)) \neq 0$.

On the other hand, by (2.4.19) there exists a polynomial p of degree less than or equal to n such that $p(h(x)) = 1$ and $p(h(x_i)) = 0$, $i = 1, \ldots, n$.

So, $f_0 p^2(h)$ belongs to $\mathscr{L}(\{f_0\} \cup f_0 M \cup \cdots \cup f_0 M^{2n})$ ($p^2(h) \in \mathscr{L}(M \cup \cdots \cup M^{2n}) + \mathbb{R}_+$ or, if X is compact, $p^2(h) \in \mathscr{L}(\{\mathbf{1}\} \cup M \cup \cdots \cup M^{2n})$, respectively) and condition (2) of Theorem 2.2.6 is fulfilled. $\qquad \square$

2.5 Determining subspaces for Dirac measures

We continue our study of determining subspaces for discrete measures by considering a particular class of them, namely the Dirac measures.

The results of this section will be essential for the study of the convergence of nets of positive operators to the identity operator.

We begin by reformulating the main results of the previous sections in the case $n = 1$.

Here X will denote again a locally compact Hausdorff space, H is a fixed subspace of $\mathscr{C}_0(X)$ and x a given point of X.

On account of Proposition 2.1.2, Theorem 2.1.6 and Theorem 2.2.2 we have the following result.

2.5.1 Theorem. *Let H be a subspace of $\mathscr{C}_0(X)$. Then the following statements are equivalent*:

(i) *The subspace H is a D_+-subspace for the Dirac measure ε_x.*

(ii) *If $\mu \in \mathscr{M}_b^+(X)$ and $\mu = \varepsilon_x$ on H, then $\mu = \varepsilon_x$.*

(iii) $H^* = \mathscr{C}_0(X)$ *and for every $f \in \mathscr{C}_0(X)$,* $\displaystyle\inf_{\varepsilon > 0} \left(\sup_{\substack{h \in H \\ h - \varepsilon \leq f}} h(x) \right) = f(x) =$

$\displaystyle\sup_{\varepsilon > 0} \left(\inf_{\substack{k \in H \\ f \leq k + \varepsilon}} k(x) \right).$

(iv) *For every $f \in \mathscr{C}_0(X)$ and for every $\varepsilon > 0$ there exist $h, k \in H$ such that*

$$h - \varepsilon \leq f \leq k + \varepsilon \text{ and } |k(x) - h(x)| \leq \varepsilon.$$

(v) (1) *For every $\varepsilon > 0$ and for every compact subset K of X which does not contain x, there exist $k \in H$ and $u \in \mathscr{C}_0^+(X)$ such that*

$$\|u\| \leq \varepsilon, 0 \leq k + u, 1 \leq k + u \text{ on } K \text{ and } k(x) + u(x) < \varepsilon.$$

(2) *There exists $h \in H$ such that $h(x) \neq 0$.*

When X is compact we have a simpler result as indicated below.

2.5.2 Theorem. *Suppose X compact and let H be a subspace of $\mathscr{C}(X)$. Then the following statements are equivalent*:

(i) *The subspace H is a D_+-subspace for the Dirac measure ε_x.*

(ii) *If $\mu \in \mathscr{M}^+(X)$ and $\mu = \varepsilon_x$ on H, then $\mu = \varepsilon_x$.*

(iii) *The subspace H is cofinal in $\mathscr{C}(X)$ and $\displaystyle\sup_{\substack{h \in H \\ h \leq f}} h(x) = f(x) = \inf_{\substack{k \in H \\ f \leq k}} k(x)$ for every*
$f \in \mathscr{C}(X).$

(iv) *For every $f \in \mathscr{C}(X)$ and for every $\varepsilon > 0$ there exist $h, k \in H$ such that*

$$h \leq f \leq k \quad \text{and} \quad k(x) - h(x) \leq \varepsilon.$$

(v) (1) *For every $\varepsilon > 0$ and for every compact subset K of X which does not contain x, there exists $k \in H$ such that $0 \leq k, 1 \leq k$ on K and $k(x) < \varepsilon$.*

(2) *There exists $h \in H$ such that $h(x) \neq 0$.*

(vi) (1) *For every $\varepsilon > 0$ and for every open neighborhood U of x, there exists*
    $k \in H$ *such that* $0 \le k$, $1 \le k$ *on* $X \backslash U$ *and* $k(x) < \varepsilon$.
    (2) *There exists $h \in H$ such that $h(x) \ne 0$.*

From (2.1.21) and Theorems 2.1.10 and 2.2.4 we deduce the following further
result.

2.5.3 Theorem. *Let H be a subspace of $\mathscr{C}_0(X)$. Then the following statements are
equivalent:*
 (i) *The subspace H is a D_+^1-subspace for the Dirac measure ε_x.*
 (ii) *If $\mu \in \mathscr{M}_b^+(X)$, $\|\mu\| \le 1$ and $\mu = \varepsilon_x$ on H, then $\mu = \varepsilon_x$.*
 (iii) *For every $f \in \mathscr{C}_0(X)$,* $\displaystyle \sup_{\substack{h \in H, \alpha \ge 0 \\ h - \alpha \le f}} (h(x) - \alpha) = f(x) = \inf_{\substack{k \in H, \beta \ge 0 \\ f \le k + \beta}} (k(x) + \beta)$.
 (iv) *For every $f \in \mathscr{C}_0(X)$ and for every $\varepsilon > 0$ there exist $h, k \in H$ and $\alpha, \beta \ge 0$ such
    that*

$$h - \alpha \le f \le k + \beta \text{ and } k(x) - h(x) + \alpha + \beta \le \varepsilon.$$

 (v) (1) *For every $\varepsilon > 0$ and for every compact subset K of X which does not
    contain x, there exist $k \in H$ and $u \in \mathscr{C}_0^+(X)$ such that*

$$\|u\| \le \varepsilon + u(x), 0 \le k + u, 1 \le k + u \text{ on } K \quad \text{and} \quad k(x) + u(x) < \varepsilon.$$

 (2) *There exists $h \in H$ such that $h(x) \ne 0$.*

Another useful result is stated below and is a particular case of Theorem 2.2.6
for $n = 1$.

2.5.4 Theorem. *Suppose that the subspace H satisfies the following conditions:*
(1) *There exists $h \in H$ such that $h(x) \ne 0$.*
(2) *For every $y \in X$, $y \ne x$, there exists $k \in H$ ($k \in H + \mathbb{R}_+$, respectively) such that*

$$k \ge 0, k(y) > 0 \text{ and } k(x) = 0.$$

Then H is a D_+-subspace for ε_x (a D_+^1-subspace for ε_x, respectively).

Here we collect some examples together with some applications.

Examples. 1. The subspace generated by the two functions $h_1(x) := \exp(-\|x\|^2)$ and $h_2(x) := \|x\|^2 \exp(-\|x\|^2)$ $(x \in \mathbb{R}^p)$ is a D_+-subspace in $\mathscr{C}_0(\mathbb{R}^p)$ for ε_0.

As an application consider the sequence $(\mu_n)_{n \in \mathbb{N}}$ of bounded Radon measures on $\mathbb{R}^p$ defined by

$$\mu_n(f) := \left(\frac{n}{\pi}\right)^{p/2} \int_{\mathbb{R}^p} f(x) \exp(-n\|x\|^2) \, dx \quad \text{for every } f \in \mathscr{C}_0(\mathbb{R}^p) \text{ and } n \in \mathbb{N}.$$

(2.5.1)

Then $\lim\limits_{n \to \infty} \mu_n = \varepsilon_0$ vaguely.

In fact

$$\mu_n(h_1) = \left(\frac{n}{\pi}\right)^{p/2} \int_{\mathbb{R}^p} \exp(-(n+1)\|x\|^2) \, dx = \left(\frac{n}{n+1}\right)^{p/2} \to 1 = h_1(0)$$

and

$$\mu_n(h_2) = \left(\frac{n}{\pi}\right)^{p/2} \sum_{i=1}^{p} \int_{\mathbb{R}^p} x_i^2 \exp(-(n+1)\|x\|^2) \, dx$$

$$= \frac{p}{2(n+1)} \left(\frac{n}{n+1}\right)^{p/2} \to 0 = h_2(0);$$

hence the result follows.

2. From Theorem 2.5.4 it easily follows that the subspace generated by the functions $\mathbf{1}$ and $e_\alpha(x) := x^\alpha$ ($x \in [0,1], \alpha > 0$) is a D_+-subspace in $\mathscr{C}([0,1])$ both for ε_0 and ε_1.

3. Let (X, d) be a compact metric space. Given $x \in X$, apply Theorem 2.5.4 to the subspace H generated by $\mathbf{1}$ and h_x, where $h_x(y) := d^2(y, x)$ for every $y \in X$.

Hence H is a D_+-subspace for ε_x.

For $X = [0,1]$ the function h_x is defined by $h_x(t) := (t - x)^2$ for every $t \in [0,1]$.

By using this last result we determine the behavior of the powers of *Bernstein-Stancu* operators.

They are defined by

$$B_{n,\alpha,\beta}(f)(x) := \sum_{k=0}^{n} f\left(\frac{k+\alpha}{n+\beta}\right) \binom{n}{k} x^k (1-x)^{n-k}$$

(2.5.2)

for every $n \geq 1, f \in \mathscr{C}([0,1])$ and $x \in [0,1]$, where α and β are fixed real numbers such that $0 \leq \alpha \leq \beta$ and $\binom{n}{k} := \dfrac{n!}{k!(n-k)!}$ is the *binomial coefficient*.

When $\alpha = \beta = 0$ we obtain the classical *Bernstein polynomials*

$$B_n(f)(x) := \sum_{k=0}^{n} f\left(\frac{k}{n}\right)\binom{n}{k} x^k (1-x)^{n-k}. \tag{2.5.3}$$

The operators $B_{n,\alpha,\beta}$ are positive linear operators from $\mathscr{C}([0,1])$ into $\mathscr{C}([0,1])$. So for every $p \in \mathbb{N}$ we can consider the *p-th power* $B_{n,\alpha,\beta}^p$ of $B_{n,\alpha,\beta}$ which is defined as

$$B_{n,\alpha,\beta}^p := \begin{cases} B_{n,\alpha,\beta} & \text{if } p = 1, \\ B_{n,\alpha,\beta} \circ B_{n,\alpha,\beta}^{p-1} & \text{if } p \geq 2. \end{cases}$$

2.5.5 Proposition. *Let us consider the Bernstein-Stancu operators and the Bernstein operators defined by (2.5.2) and (2.5.3), respectively.*
The following statements hold:
(1) *For every $n \in \mathbb{N}$, $f \in \mathscr{C}([0,1])$ and $x \in [0,1]$,*

$$\lim_{p \to \infty} B_{n,\alpha,\beta}^p(f)(x) = f\left(\frac{\alpha}{\beta}\right), \tag{2.5.4}$$

provided $\beta > 0$, while, if $\beta = 0 (= \alpha)$,

$$\lim_{p \to \infty} B_n^p(f)(x) = (1-x)f(0) + xf(1). \tag{2.5.5}$$

(2) *If $(k(n))_{n \in \mathbb{N}}$ is a sequence of positive integers satisfying $\lim\limits_{n \to \infty} \dfrac{k(n)}{n} = +\infty$, then for every $f \in \mathscr{C}([0,1])$ and $x \in [0,1]$,*

$$\lim_{n \to \infty} B_{n,\alpha,\beta}^{k(n)}(f)(x) = f\left(\frac{\alpha}{\beta}\right), \tag{2.5.6}$$

provided $\beta > 0$, and, if $\beta = 0$,

$$\lim_{n \to \infty} B_n^{k(n)}(f)(x) = (1-x)f(0) + xf(1). \tag{2.5.7}$$

Proof. We shall use the following identities which follow easily from elementary properties of binomial coefficients:

$$\sum_{k=0}^{n} \binom{n}{k} x^k (1-x)^{n-k} = 1 \quad (x \in \mathbb{R}), \tag{2.5.8}$$

$$\sum_{k=0}^{n} \frac{k}{n} \binom{n}{k} x^k (1-x)^{n-k} = x \quad (x \in \mathbb{R}), \tag{2.5.9}$$

$$\sum_{k=0}^{n} \frac{k^2}{n^2} \binom{n}{k} x^k (1-x)^{n-k} = \frac{n-1}{n} x^2 + \frac{1}{n} x \quad (x \in \mathbb{R}). \tag{2.5.10}$$

Now, denoting by e_k ($k = 1, 2$) the functions defined by

$$e_k(t) := t^k \quad (t \in [0,1]), \tag{2.5.11}$$

we have

$$B_{n,\alpha,\beta}(1) = 1, \tag{2.5.12}$$

$$B_{n,\alpha,\beta}(e_1) = \frac{n}{n+\beta} e_1 + \frac{\alpha}{n+\beta} \mathbf{1}, \tag{2.5.13}$$

$$B_{n,\alpha,\beta}(e_2) = \frac{n(n-1)}{(n+\beta)^2} e_2 + \frac{(2\alpha+1)n}{(n+\beta)^2} e_1 + \frac{\alpha^2}{(n+\beta)^2} \mathbf{1}. \tag{2.5.14}$$

By induction, for every $p \geq 1$ we have

$$B_{n,\alpha,\beta}^p(1) = 1, \tag{2.5.15}$$

$$B_n^p(e_1) = e_1, \tag{2.5.16}$$

$$B_{n,\alpha,\beta}^p(e_1) = \left(\frac{n}{n+\beta}\right)^p e_1 + \frac{\alpha}{\beta}\left(1 - \left(\frac{n}{n+\beta}\right)^p\right) \mathbf{1}, \quad (\beta > 0), \tag{2.5.17}$$

$$B_n^p(e_2) = \left(\frac{n-1}{n}\right)^p e_2 + \left(1 - \left(\frac{n-1}{n}\right)^p\right) e_1, \tag{2.5.18}$$

$$B_{n,\alpha,\beta}^p(e_2) = \left(\frac{n(n-1)}{(n+\beta)^2}\right)^p e_2 + \frac{(2\alpha+1)}{\beta+1}\left(\frac{n}{n+\beta}\right)^p\left(1 - \left(\frac{n-1}{n+\beta}\right)^p\right) e_1$$

$$+ \frac{\alpha^2}{\beta^2}\left(1 - \frac{1+2\beta}{1+\beta}\left(\frac{n}{n+\beta}\right)^p + \frac{\beta}{1+\beta}\left(\frac{n(n-1)}{(n+\beta)^2}\right)^p\right) \mathbf{1}. \tag{2.5.19}$$

Hence (2.5.4) follows from (2.5.15), (2.5.17), (2.5.19) and from Example 3 to Theorem 2.5.4 by considering the sequence of positive Radon measures on $[0, 1]$ defined by

$$\mu_p(f) := B_{n,\alpha,\beta}^p(f)(x) \quad \text{for every } f \in \mathscr{C}([0, 1]) \text{ and } p \geq 1.$$

In this case, in fact, $\lim\limits_{p \to \infty} \mu_p(1) = 1$ and $\lim\limits_{p \to \infty} \mu_p(h_{\alpha/\beta}) = 0$ where $h_{\alpha/\beta}(t) :=$ $\left(t - \dfrac{\alpha}{\beta}\right)^2$ for $t \in [0, 1]$.

Analogously, (2.5.5) follows from (2.5.15), (2.5.16), (2.5.18) and Example 2 to Corollary 2.2.7.

Part (2) follows in a similar way because, if $\lim\limits_{n \to \infty} \dfrac{k(n)}{n} = +\infty$, then

$$\left(\frac{n}{n+1}\right)^{k(n)} = \exp\left(\frac{k(n)}{n} n \log\left(\frac{n}{n+1}\right)\right) \to 0 \quad \text{as } n \to \infty. \qquad \square$$

Remark. In the Examples to Theorem 4.2.7 we shall show that, in fact, formulas (2.5.5) and (2.5.7) hold uniformly on the interval $[0, 1]$.

Furthermore we shall see there that $(B_{n,\alpha,\beta})_{n \in \mathbb{N}}$ is a positive approximation process (see Example 2 to Theorem 4.2.7 and the introduction of Chapter 5).

To make more transparent some applications we prefer to state explicitly Theorems 2.4.1, 2.4.2 and 2.4.3 for the particular case $n = 1$.

We recall that if M is a subset of $\mathscr{C}_0(X)$ and $x \in X$, we have set

$$M(x) := \{y \in X \mid h(x) = h(y) \text{ for every } h \in M\}.$$

2.5.6 Theorem. *Let M be a subset of $\mathscr{C}_0(X)$, let $\mathrm{x} \in X$ and suppose that*

$$\text{there exists } h_0 \in M \text{ such that } h_0(x) \neq 0. \qquad (2.5.20)$$

Furthermore, let $f_0 \in \mathscr{C}_0(X)$ be a strictly positive function such that

$$f_0 \text{ is constant on } M(x). \qquad (2.5.21)$$

Then

$$D_+(\{f_0\} \cup f_0 M \cup f_0 M^2, \varepsilon_x) = \{f \in \mathscr{C}_0(X) \mid f \text{ is constant on } M(x)\}. \quad (2.5.22)$$

Hence $D_+(\{f_0\} \cup f_0 M \cup f_0 M^2, \varepsilon_x)$ is a closed subalgebra of $\mathscr{C}_0(X)$ and

$$\mathbf{A}(M) \subset D_+(\{f_0\} \cup f_0 M \cup f_0 M^2, \varepsilon_x). \qquad (2.5.23)$$

Moreover $\{f_0\} \cup f_0 M \cup f_0 M^2$ is a D_+-subset for ε_x if and only if M separates x in X (see (2.4.6)).

Similar results hold by replacing the subset $\{f_0\} \cup f_0 M \cup f_0 M^2$ with $M \cup M^2$ and $D_+(\{f_0\} \cup f_0 M \cup f_0 M^2, \varepsilon_x)$ with $D_+^1(M \cup M^2, \varepsilon_x)$ or, if X is compact, with $\{\mathbf{1}\} \cup M \cup M^2$ and $D_+(\{\mathbf{1}\} \cup M \cup M^2, \varepsilon_x)$ without assumption (2.5.20).

In some cases the number of elements of the determining subsets in Theorem 2.5.6 can be reduced further, namely when M is finite or countable or, provided X is compact, if $h(x) = 0$ for every $h \in M$.

2.5.7 Proposition. *Let* $M = \{h_n | n \in \mathbb{N}\}$ *be a finite or countable subset of* $\mathscr{C}_0(X)$ *such that the series*

$$u := \sum_{n=1}^{\infty} h_n^2 \tag{2.5.24}$$

converges uniformly on X *and consider* $x \in X$.
  Then the following statements are true:
(1) *If there exists* $n_0 \in \mathbb{N}$ *such that* $h_{n_0}(x) \neq 0$ *and if* $f_0 \in \mathscr{C}_0(X)$ *is a strictly positive function satisfying* (2.5.21), *then*

$$D_+(\{f_0\} \cup f_0 M \cup \{f_0 \cdot u\}, \varepsilon_x) = \{f \in \mathscr{C}_0(X) | f \text{ is constant on } M(x)\}, \tag{2.5.25}$$

$$D_+^1(M \cup \{u\}, \varepsilon_x) = \{f \in \mathscr{C}_0(X) | f \text{ is constant on } M(x)\} \tag{2.5.26}$$

and hence $\{f_0\} \cup f_0 M \cup \{f_0 \cdot u\}$ *is a* D_+-*subset for* ε_x *(respectively,* $M \cup \{u\}$ *is a* D_+^1-*subset for* ε_x) *if and only if* M *separates* x *in* X.
(2) *If* X *is compact, then*

$$D_+(\{\mathbf{1}\} \cup M \cup \{u\}, \varepsilon_x) = \{f \in \mathscr{C}(X) | f \text{ is constant on } M(x)\} \tag{2.5.27}$$

and $\{\mathbf{1}\} \cup M \cup \{u\}$ *is a* D_+-*subspace for* ε_x *if and only if* M *separates* x *in* X.

Proof. Clearly, by (2.5.22), we have

$$D_+(\{f_0\} \cup f_0 M \cup \{f_0 u\}, \varepsilon_x) \subset D_+(\{f_0\} \cup f_0 M \cup f_0 M^2, \varepsilon_x)$$

$$= \{f \in \mathscr{C}_0(X) | f \text{ is constant on } M(x)\}.$$

Conversely, if $f \in \mathscr{C}_0(X)$ is constant on $M(x)$, taking $\mu \in \mathscr{M}_b^+(X)$ such that $\mu(f_0) = f_0(x)$, $\mu(f_0 \cdot u) = f_0(x) \sum_{n=1}^{\infty} h_n^2(x)$ and $\mu(f_0 h_n) = f_0(x) h_n(x)$ for every $n \in \mathbb{N}$, then

$$0 \leq \sum_{n=1}^{\infty} \mu(f_0(h_n - h_n(x))^2)$$

$$= \mu\left(f_0 \sum_{n=1}^{\infty} h_n^2\right) - 2 \sum_{n=1}^{\infty} h_n(x)\mu(f_0 h_n) + \sum_{n=1}^{\infty} h_n^2(x)\mu(f_0) = 0.$$

Accordingly, $\mu(f_0(h_n - h_n(x))^2) = 0$ for every $n \in \mathbb{N}$. So, since f_0 is strictly positive,

$$\text{Supp}(\mu) \subset \bigcap_{n=1}^{\infty} \{y \in X | h_n(y) = h_n(x)\} = M(x).$$

Now, by (2.5.21), for every $y \in M(x)$ we obviously have $f(y) = \dfrac{f(x)}{f_0(x)} f_0(y)$, so $\mu(f) = \dfrac{f(x)}{f_0(x)} \mu(f_0) = f(x)$. This reasoning shows that $f \in D_+(\{f_0\} \cup f_0 M \cup \{f_0 u\}, \varepsilon_x)$.

In a similar way one may prove the other statements. □

Example. Consider the interval $[-\pi, \pi]$ and $x \in [-\pi, \pi]$. Then

$$D_+(\{1, \sin, \cos\}, \varepsilon_x)$$

$$= \begin{cases} \mathscr{C}([-\pi, \pi]), & \text{if } -\pi < x < \pi, \\ \{f \in \mathscr{C}([-\pi, \pi]) | f(-\pi) = f(\pi)\}, & \text{if } x = -\pi \text{ or } x = \pi. \end{cases}$$

2.5.8 Proposition. *Suppose X compact. Let M be a subset of $\mathscr{C}(X)$ and consider $x \in X$. Suppose that*

$$h(x) = 0 \quad \text{for every } h \in M. \tag{2.5.28}$$

Then

$$D_+(\{1\} \cup M^2, \varepsilon_x) = \{f \in \mathscr{C}(X) | f \text{ is constant on } M(x)\}. \tag{2.5.29}$$

Hence $\{1\} \cup M^2$ is a D_+-subset for ε_x if and only if M separates x in X.

Finally, if $M = \{h_n | n \in \mathbb{N}\}$ is finite or countable and $\sum\limits_{n=1}^{\infty} h_n^2$ converges uniformly on X, then similar statements hold by replacing M^2 with $\sum\limits_{n=1}^{\infty} h_n^2$.

Proof. Since $D_+(\{1\} \cup M^2, \varepsilon_x) \subset D_+(\{1\} \cup M \cup M^2, \varepsilon_x)$, by virtue of Theorem 2.5.6 one inclusion of the equality (2.5.29) is trivial.

On the other hand, take $f \in \mathscr{C}(X)$ and suppose that f is constant on $M(x) = \{y \in X | h(y) = 0$ for every $h \in M\}$; given $\mu \in \mathscr{M}^+(X)$ satisfying $\mu(1) = 1$ and $\mu(h^2) = h^2(x) = 0$ for every $h \in M$, we have

$$\operatorname{Supp}(\mu) \subset \bigcap_{h \in M} \{y \in X | h(y) = 0\} = M(x).$$

But, $f = f(x)\mathbf{1}$ on $M(x)$, so that $\mu(f) = f(x)\mu(\mathbf{1}) = f(x)$. This proves that $f \in D_+(\{1\} \cup M^2, \varepsilon_x)$. $\qquad\square$

Notes and references

The results of this section are a direct consequence of those of the preceding section.

However, some of them have been obtained by several authors by different methods.

Theorems 2.5.1 and 2.5.3 are essentially due to Bauer and Donner [1978], [1986].

Theorem 2.5.2 is due to Berens and Lorentz [1973], [1976] (see also Bauer [1973] and Baskakov [1961]). In the case when H is a closed subalgebra of $\mathscr{C}(X)$, the equivalence (ii) $\Leftrightarrow$ (vi) of Theorem 2.5.2 was obtained by Bishop and de Leeuw (see Phelps ([1966])). When X is compact, results similar to Theorem 2.5.3 were obtained by Berens and Lorentz [1976].

For a generalization of Theorem 2.5.3 to completely regular Hausdorff spaces we refer to Matsuda [1979].

Example 3 to Theorem 2.5.4, when X is a compact real interval, is due to Korovkin [1959], who used it to prove its celebrated theorem.

Bernstein-Stancu operators were introduced by Stancu in [1969a] and they form a positive approximation process in $\mathscr{C}([0,1])$, as we shall see in Example 2 to Theorem 4.2.7.

For a survey of the properties of these operators we also refer to Gonska and Meier-Gonska [1984].

Proposition 2.5.5 was proved by Marlewsky [1984] in the particular cases $\alpha = 0$, $\beta = 1$ and $\alpha = \beta = 1$.

2.6 Choquet boundaries

In the preceding sections we fixed a point $x \in X$ and we studied those subspaces of $\mathscr{C}_0(X)$ which are determining subspaces for ε_x.

Here we invert this approach. We fix a subspace H of $\mathscr{C}_0(X)$ and look for those points x of X such that H is a determining subspace for ε_x. These points constitute the so-called Choquet boundary of H.

This classical notion plays an important role not only in Korovkin-type approximation theory, but also in the theory of integral representations, in potential theory and in the study of function algebras.

In this section we deal with some of its fundamental properties.

2.6.1 Definition. *Given a linear subspace H of $\mathscr{C}_0(X)$, the Choquet boundary of H is the subset $\partial_H^+ X$ of all points $x \in X$ such that the subspace H is a D_+-subspace for ε_x.*

Moreover, we shall denote by $\partial_H^{1,+} X$ the subset of X consisting of all points $x \in X$ such that H is a D_+^1-subspace for ε_x.

In general $\partial_H^+ X \subsetneqq \partial_H^{1,+} X$ (see Example after Formula (2.1.21)).

But, if X is compact and $\mathbf{1} \in H$, then (see (2.1.22))

$$\partial_H^+ X = \partial_H^{1,+} X. \tag{2.6.1}$$

Theorem 2.5.1 (Theorem 2.5.2 in the compact case, respectively) gives some equivalent and useful statements characterizing the points of the Choquet boundary.

In the sequel we shall also see under which conditions $\partial_H^+ X$ is non-empty.

However, if we consider the one-point compactification $X_\omega := X \cup \{\omega\}$ of X, and if we set

$$\tilde{H} := \{\tilde{h} \in \mathscr{C}(X_\omega) \,|\, h \in H\} \tag{2.6.2}$$

(where $\tilde{h}$ is the canonical continuous extension of h to X_ω (see (1.1.7))) and

$$H_\omega := \mathscr{L}(\{\mathbf{1}\} \cup \tilde{H}), \tag{2.6.3}$$

then it is easy to show that

$$\partial_H^{1,+} X \subset \partial_{H_\omega}^+ X \cap X. \tag{2.6.4}$$

Moreover,

$$\omega \in \partial_{H_\omega}^+ X \Leftrightarrow H^* = \mathscr{C}_0(X). \tag{2.6.5}$$

On account of Theorem 2.5.4 we observe that the points $x \in X$ satisfying the following conditions (see (1) and (2) of Theorem 2.5.4):

(1) *There exists $h \in H$ such that $h(x) \neq 0$.*
(2) *For every $y \in X$, $y \neq x$, there exists $k \in H$ (respectively $k \in H + \mathbb{R}_+$) such that*

$$k \geq 0, k(y) > 0 \quad and \quad k(x) = 0.$$

are in $\partial_H^+ X$ (in $\partial_H^{1,+} X$, respectively). Such points are also called *peak-points for H* (*weak peak-points for H*, respectively).

In general, a point x_0 of $\partial_H^+ X$ is not a peak-point for H. For example, take

$$X := \{(x, y) \in \mathbb{R}^2 | -1 \leq x \leq 0, x^2 + y^2 \leq 1\}$$

$$\cup \{(x, y) \in \mathbb{R}^2 | 0 \leq x \leq 1, -1 \leq y \leq 1\}$$

and denote by H the subspace generated by $\mathbf{1}$ and the two projections pr_1 and pr_2 on X. Then $(0, 1)$ is an extreme point of X and hence $(0, 1) \in \partial_H^+ X$ by virtue of Proposition 2.6.3, while it is not a peak-point for H.

To emphasize these different notions, on account also of property (v) of Theorem 2.5.1 (or Theorems 2.5.2 and 2.5.3), the points in $\partial_H^+ X$ (in $\partial_H^{1,+} X$, respectively) are called *quasi peak-points for H* (*weak quasi peak-points for H*, respectively).

However, there are particular situations in which the set of peak-points for H coincides with $\partial_H^+ X$ (see Remark 3 to Corollary 2.6.5).

In general, if H contains the constants and X is metrizable and compact, then $\partial_H^+ X$ coincides with the closure of the set of all peak-points for H. This result is due to Bishop (see Phelps [1966, Corollary 8.5]). For other results related to these questions the reader is referred to Berens and Lorentz [1975], [1976].

We now present some important examples of Choquet boundaries. The first one is taken from potential theory. We refer to Helms [1969] for more details on the results and the notions quoted in the sequel.

Consider an open subset Ω of $\mathbb{R}^p$.

A numerical function $u: \Omega \to \mathbb{R} \cup \{+\infty\}$ is called *hyperharmonic* if it is lower semi-continuous and if, for every $x \in \Omega$ and $\varepsilon > 0$ satisfying $B(x, \varepsilon) \subset \Omega$,

$$\frac{1}{\sigma_p \varepsilon^{p-1}} \int_{\partial B(x, \varepsilon)} u \, d\sigma \leq u(x), \tag{2.6.6}$$

where $B(x, \varepsilon) := \{y \in \mathbb{R}^p | \|x - y\| < \varepsilon\}$. The symbols σ_p and σ denote the surface area of a ball of radius 1 and the surface area on $\partial B(x, \varepsilon)$, respectively.

A function $u: \Omega \to \mathbb{R} \cup \{+\infty\}$ is called *superharmonic* if it is hyperharmonic and it is finite on a relatively dense subset of Ω.

If Ω is connected, then an hyperharmonic function either is superharmonic or is identically $+\infty$.

It is possible to show that a function $u: \Omega \to \mathbb{R}$ having continuous second partial derivatives is superharmonic if and only if $\Delta u \le 0$ on Ω where Δ denotes the *Laplacian operator* acting on u, i.e.,

$$\Delta u := \sum_{i=1}^{p} \frac{\partial^2 u}{\partial x_i^2} \tag{2.6.7}$$

(see Helms [1969, Theorem 4.8].

A function $u: \Omega \to \mathbb{R} \cup \{-\infty\}$ is called *subharmonic* (*hypoharmonic*, respectively) if $-u$ is superharmonic (hyperharmonic, respectively).

Finally, a function $u: \Omega \to \mathbb{R}$ is called *harmonic* if it is both subharmonic and superharmonic.

Suppose now that Ω is bounded. Given a continuous function $f: \partial\Omega \to \mathbb{R}$, we define for every $x \in \Omega$

$$\overline{H}_f(x) := \inf \Big\{ u(x) | u: \Omega \to \mathbb{R} \cup \{+\infty\} \text{ hyperharmonic on } \Omega,$$

$$\text{bounded below on } \Omega \text{ and } f(x_0) \le \liminf_{y \to x_0} u(y) \quad \text{for every } x_0 \in \partial\Omega \Big\} \tag{2.6.8}$$

and

$$\underline{H}_f(x) := -\overline{H}_{-f}(x).$$

By a theorem of Wiener we know that $\underline{H}_f(x) = \overline{H}_f(x)$ for every $x \in \Omega$ and the function $H_f: \Omega \to \mathbb{R}$ defined by

$$H_f(x) := \underline{H}_f(x) = \overline{H}_f(x) \quad (x \in \Omega) \tag{2.6.9}$$

is harmonic on Ω (see Helms [1969, Theorem 8.11]). The function H_f is called the *generalized Perron-Wiener-Brelot solution of the Dirichlet problem for f*.

Moreover, we observe that the mapping $f \mapsto H_f$ is linear and positive.

A point $x_0 \in \partial\Omega$ is called *regular* if

$$\lim_{x \to x_0} H_f(x) = f(x_0) \quad \text{for every } f \in \mathscr{C}(\partial\Omega). \tag{2.6.10}$$

The subset of all regular points will be denoted by $r(\Omega)$.

The subset Ω will be called *regular* if $r(\Omega) = \partial\Omega$.

In this case, for every $f \in \mathscr{C}(\partial\Omega)$, the *classical Dirichlet problem*

$$\begin{cases} \Delta u = 0, & u \in \mathscr{C}(\bar{\Omega}) \cap \mathscr{C}^2(\Omega), \\ u_{|\partial\Omega} = f, \end{cases} \tag{2.6.11}$$

has a unique solution, namely H_f (where we continue to denote by H_f the continuous extension of H_f to $\bar{\Omega}$ according to (2.6.10)).

Here $\mathscr{C}^2(\Omega)$ denotes the space of all real-valued continuous functions on Ω which are twice continuously differentiable in Ω.

Examples of regular subsets are furnished by those open bounded subsets Ω of $\mathbb{R}^p$ such that for every $x \in \partial\Omega$ there is a ball B such that $B \cap \Omega = \varnothing$ and $x \in \partial B$ or, if $p \geq 2$, by those open bounded subsets such that for every $x \in \partial\Omega$ there exists a truncated closed solid cone in the complement of Ω with vertex at x. Thus every open convex bounded subset of $\mathbb{R}^p$ is regular. For further details on regular subsets we refer to Helms [1969, Section 8.3].

We can now state two important results.

2.6.2 Proposition. *Let Ω be a bounded open subset of $\mathbb{R}^p$. Set $X = \bar{\Omega}$ and $H(\Omega) := \{u \in \mathscr{C}(X)| u$ is harmonic on $\Omega\}$. Then $\partial^+_{H(\Omega)} X = \partial^{1,+}_{H(\Omega)} X = r(\Omega)$.*

Proof. Fix $x_0 \in \partial^+_{H(\Omega)} X$. First, we show that $x_0 \in \partial\Omega$. Otherwise, if $x_0 \in \Omega$, we can choose $\varepsilon > 0$ such that $B(x_0, \varepsilon) \subset \Omega$. By the averaging principle of harmonic functions (see Helms [1969, Theorems 1.5 and 1.6]), we would have

$$u(x_0) = \frac{1}{\lambda_p(B(x_0, \varepsilon))} \int_{\partial B(x_0, \varepsilon)} u(y) \, d\lambda_p(y) \quad \text{for every } u \in H(\Omega).$$

Hence ε_{x_0} coincides on $H(\Omega)$ with the positive Radon measure $\mu \in \mathscr{M}^+(X)$ defined by

$$\mu(f) = \frac{1}{\lambda_p(B(x_0, \varepsilon))} \int_{\partial B(x_0, \varepsilon)} f(y) \, d\lambda_p(y) \quad (f \in \mathscr{C}(X)).$$

Since $x_0 \in \partial^+_{H(\Omega)} X$, we would have $\mu = \varepsilon_{x_0}$ which is not possible. Therefore $x_0 \in \partial\Omega$.

Now, let $f \in \mathscr{C}(\partial\Omega)$ and take a continuous extension $\tilde{f} \in \mathscr{C}(X)$ of f. If $\varepsilon > 0$, by Theorem 2.5.2, there exist $h, k \in H(\Omega)$ such that $h \leq \tilde{f} \leq k$ and $k(x_0) - h(x_0) \leq \varepsilon$. So $h_{|\partial\Omega} \leq f \leq k_{|\partial\Omega}$ and hence $h \leq H_f \leq k$. Consequently

$$h(x_0) = \liminf_{x \to x_0} h(x) \leq \liminf_{x \to x_0} H_f(x) \leq \limsup_{x \to x_0} H_f(x) \leq \limsup_{x \to x_0} k(x) = k(x_0)$$

and so

$$\limsup_{x \to x_0} k(x) - \liminf_{x \to x_0} h(x) \leq \varepsilon.$$

This implies that $\lim_{x \to x_0} H_f(x) = f(x_0)$ for every $f \in \mathscr{C}(\partial\Omega)$, i.e., x_0 is a regular point.

Conversely, if $x_0 \in r(\Omega)$, by a result of Keldish [1941a] there exists $h \in H(\Omega)$ such that $h(x_0) = 0$ and $h(x_0) > 0$ for every $x \in X$, $x \neq x_0$. By Theorem 2.5.4, $x_0 \in \partial_{H(\Omega)}^+ X$.

Finally $\partial_{H(\Omega)}^+ X = \partial_{H(\Omega)}^{1,+} X$ by (2.6.1).   □

Another important example of Choquet boundary is the set $\partial_e K$ of extreme points of a convex compact set K. In the sequel we shall see, in fact, that this is the most general example of Choquet boundary.

We recall that, if K is a convex compact subset of some locally convex Hausdorff space, then a point $x_0 \in K$ is an *extreme point* of K, if $K \backslash \{x_0\}$ is convex or, equivalently, if for every $x_1, x_2 \in K$ and $\lambda \in \mathbb{R}$, $0 < \lambda < 1$, satisfying $x_0 = \lambda x_1 + (1 - \lambda)x_2$, it necessarily follows that $x_0 = x_1 = x_2$ (see Section 1.4).

In Section 1.4 we have already introduced the linear subspace $A(K)$ of $\mathscr{C}(K)$ consisting of all continuous functions $u \in \mathscr{C}(K)$ which are affine on K, i.e., $u(\lambda x_1 + (1 - \lambda)x_2) = \lambda u(x_1) + (1 - \lambda)u(x_2)$ for every $x_1, x_2 \in K$ and $\lambda \in \mathbb{R}$, $0 \leq \lambda \leq 1$.

2.6.3 Proposition. *Let K be a convex compact subset of some locally convex Hausdorff space. Then $\partial_{A(K)}^+ K = \partial_{A(K)}^{1,+} K = \partial_e K$.*

Proof. Let $x_0 \in \partial_e K$ and consider $\mu \in \mathscr{M}^+(K)$ satisfying $\mu = \varepsilon_{x_0}$ on $A(K)$. Then $\mu(1) = \varepsilon_{x_0}(1) = 1$. So, to show that $\mu = \varepsilon_{x_0}$, it is enough to prove that $\mathrm{Supp}(\mu) \subset \{x_0\}$.

Let $y \in \mathrm{Supp}(\mu)$ and suppose $y \neq x_0$. Then, there exists a convex compact neighborhood U of y in K satisfying $U \cap \{x_0\} = \varnothing$. Take $f \in \mathscr{C}(K)$ such that $0 \leq f \leq 1$, $\mathrm{Supp}(f) \subset U$ and $\mu(f) \neq 0$. By multiplying, if necessary, the function f by a suitable positive real number we can suppose that $\mu(f) < 1$.

Set $\lambda = \mu(f)$ and consider the following positive Radon measures μ_1 and μ_2 on K defined by

$$\mu_1(g) := \frac{1}{\lambda}\mu(f \cdot g)$$

and

$$\mu_2(g) := \frac{1}{1 - \lambda}\mu((1 - f) \cdot g)$$

for every $g \in \mathscr{C}(K)$. Hence $\mu = \lambda\mu_1 + (1 - \lambda)\mu_2$. Moreover, since $\mu_1(\mathbf{1}) = 1 = \mu_2(\mathbf{1})$, we can consider the barycenters x_1, $x_2 \in K$ respectively of μ_1 and μ_2 (see Section 1.5); then for every $u \in A(K)$ we obtain

$$u(x_0) = \mu(u) = \lambda\mu_1(u) + (1 - \lambda)\mu_2(u) = \lambda u(x_1) + (1 - \lambda)u(x_2)$$

$$= u(\lambda x_1 + (1 - \lambda)x_2).$$

Since $A(K)$ separates the points of K, we have $x_0 = \lambda x_1 + (1 - \lambda)x_2$ and hence $x_0 = x_1 = x_2$, while $x_1 \in \overline{\mathrm{co}}(\mathrm{Supp}(\mu_1)) \subset \overline{\mathrm{co}}(\mathrm{Supp}(f)) \subset U$ (see (1.5.2)).

Conversely, take $x_0 \in \partial_{A(K)}^+ K$ and suppose that $x_0 \notin \partial_e K$. Then there exist x_1, $x_2 \in K \backslash \{x_0\}$ and $\lambda \in \mathbb{R}, 0 < \lambda < 1$, such that $x_0 = \lambda x_1 + (1 - \lambda)x_2$.

So, if we consider the Radon measure $\mu = \lambda\varepsilon_{x_1} + (1 - \lambda)\varepsilon_{x_2} \in \mathscr{M}^+(K)$, clearly $\mu = \varepsilon_{x_0}$ on $A(K)$ and hence $\mu = \varepsilon_{x_0}$.

Consequently $\{x_0\} = \mathrm{Supp}(\mu) = \{x_1, x_2\}$ and this leads to a contradiction. $\square$

We shall now give a geometric interpretation of the Choquet boundary in the case where X is compact.

Suppose that X is compact and consider a cofinal subspace H of $\mathscr{C}(X)$ which separates the points of X, i.e., for every $x, y \in X, x \neq y$, there exists $h \in H$ such that $h(x) \neq h(y)$.

Consider the canonical embedding $\Phi: X \to H'$ defined by

$$\Phi(x) := \varepsilon_{x|H} \quad \text{for every } x \in X.$$

Let us denote by H'_+ the cone of all (continuous) positive linear forms on H.

By the extension theorem 1.4.2 and the approximation theorem 1.2.3 it follows that H'_+ coincides with the closed convex cone in H' with vertex in the origin and generated by $\overline{\mathrm{co}}(\Phi(X))$, i.e.,

$$H'_+ = \{v \in H' | \textit{There exists } \mu \in \overline{\mathrm{co}}(\Phi(X)) \textit{ and } t \geq 0 \textit{ such that } v = t\mu\}. \quad (2.6.12)$$

Taking the Remark to Theorem 2.1.4 into account, one easily sees that, since H is cofinal in $\mathscr{C}(X)$, then $H'_+ \cap (-H'_+) = \{0\}$, i.e., the cone H'_+ is pointed.

If $x \in X$, we set

$$\rho(x) := \{t\Phi(x) | t \geq 0\} \subset H'_+. \quad (2.6.13)$$

The set $\rho(x)$ is a ray of H'_+; in the next Proposition we study the case where $\rho(x)$ is an extreme ray, i.e., $H'_+ \backslash \rho(x)$ is convex.

2.6.4 Proposition. *Let X be a compact Hausdorff space and let H be a cofinal subspace of $\mathscr{C}(X)$ which separates the points of X. Given $x_0 \in X$, the following statements are equivalent:*

(i) *$x_0 \in \partial_H^+ X$.*
(ii) *$\rho(x_0)$ is an extreme ray of H'_+.*

Proof. (i) $\Rightarrow$ (ii): Consider μ_1, $\mu_2 \in H'_+ \backslash \rho(x_0)$, $\lambda \in \mathbb{R}$, $0 < \lambda < 1$ and set $\mu :=$ $\lambda\mu_1 + (1 - \lambda)\mu_2$. Suppose that $\mu \in \rho(x_0)$, i.e., $\mu = t\Phi(x_0)$ for some $t \geq 0$.

Clearly $t > 0$, otherwise $\mu_1 = \mu_2 = 0 \in \rho(x_0)$. Choosing two positive extensions $\tilde{\mu}_1$, $\tilde{\mu}_2 \in \mathscr{M}^+(X)$ of μ_1 and μ_2 (see Theorem 1.4.2), put $\upsilon :=$ $\dfrac{1}{t}(\lambda\tilde{\mu}_1 + (1 - \lambda)\tilde{\mu}_2) \in \mathscr{M}^+(X)$.

Then $\upsilon = \varepsilon_{x_0}$ on H and hence $\upsilon = \varepsilon_{x_0}$. Consequently, since

$$\{x_0\} = \mathrm{Supp}(\upsilon) = \mathrm{Supp}(\tilde{\mu}_1) \cup \mathrm{Supp}(\tilde{\mu}_2),$$

there exist α_1, $\alpha_2 \in \mathbb{R}_+$ such that $\tilde{\mu}_1 = \alpha_1\varepsilon_{x_0}$ and $\tilde{\mu}_2 = \alpha_2\varepsilon_{x_0}$, i.e., $\mu_1 = \alpha_1\Phi(x_0) \in \rho(x_0)$ and $\mu_2 = \alpha_2\Phi(x_0) \in \rho(x_0)$, in contradiction to the assumption.

(ii) $\Rightarrow$ (i): Let $\mu \in \mathscr{M}^+(X)$ be such that $\mu = \varepsilon_{x_0}$ on H. Since H contains a strictly positive function, there exists $h \in H$ such that $h(x_0) \neq 0$. So, to show that $\mu = \varepsilon_{x_0}$, it is enough to establish that $\mathrm{Supp}(\mu) \subset \{x_0\}$.

To this end, fix $y \in \mathrm{Supp}(\mu)$ and suppose $y \neq x_0$. Denote by $\mathfrak{B}$ the set of all neighborhoods U of y such that $x_0 \notin U$ and endow it with the partial ordering $\leq$ defined by

$$U \leq V \text{ if } V \subset U.$$

For every $U \in \mathfrak{B}$ there exists $f_U \in \mathscr{C}(X)$ such that

$$0 \leq f_U \leq 1, \quad \mathrm{Supp}(f_U) \subset U, \quad \mu(f_U) > 0. \tag{1}$$

We set

$$\lambda_U := \mu(f_U) \quad \text{and} \quad \lambda_0 := \mu(\mathbf{1}). \tag{2}$$

By multiplying, if necessary, the function f_U by a suitable positive constant, we can also assume that $\lambda_U < \lambda_0$ for every $U \in \mathfrak{B}$.

Consider now the following Radon measures on X:

$$\mu_U(g) := \frac{1}{\lambda_u}\mu(f_U \cdot g) \quad (g \in \mathscr{C}(X)), \tag{3}$$

and

$$v_U(g) := \frac{1}{\lambda_0 - \lambda_U} \mu((1 - f_U) \cdot g) \quad (g \in \mathscr{C}(X)). \tag{4}$$

It is easy to show that $\lim\limits_{U \in \mathfrak{B}}{}_{\leq} \mu_U = \varepsilon_y$ vaguely. Now, for every $U \in \mathfrak{B}$ put

$$\eta_U := \mu_{U|H} \quad \text{and} \quad \theta_U := v_{U|H}. \tag{5}$$

Then $\eta_U, \theta_U \in H'_+$ and $\lim\limits_{U \in \mathfrak{B}}{}_{\leq} \eta_U = \Phi(y)$. Since $\Phi(y) \notin \rho(x_0)$, there exists $U \in \mathfrak{B}$ such that $\eta_U \notin \rho(x_0)$. But

$$\Phi(x_0) = \mu_{|H} = \lambda_U \eta_U + (\lambda_0 - \lambda_U)\theta_U = \frac{\lambda_U}{\lambda_0}(\lambda_0 \eta_U) + \left(1 - \frac{\lambda_U}{\lambda_0}\right)(\lambda_0 \theta_U),$$

which is not possible because $\rho(x_0)$ is an extreme ray. ☐

Note that, under the assumptions of Proposition 2.6.4, if, in particular, $\mathbf{1} \in H$, then $\overline{\mathrm{co}}(\Phi(X)) = \{\mu \in H' | \mu(\mathbf{1}) = 1 = \|\mu\|\}$ (see (1.4.20)) and hence $\overline{\mathrm{co}}(\Phi(X))$ is a base of the cone H'_+. So, the extreme rays of H'_+ are the rays through extreme points of $\overline{\mathrm{co}}(\Phi(X))$. Consequently, we have the following result.

2.6.5 Corollary. *Let H be a linear subspace of $\mathscr{C}(X)$ containing the constant functions and separating the points of X.*

Then a point $x_0 \in X$ belongs to $\partial_H^+ X$ if and only if $\Phi(x_0)$ is an extreme point of $\overline{\mathrm{co}}(\Phi(X))$.

In particular, $\partial_H^+ X \neq \varnothing$ and every $h \in H$ attains its minimum and maximum on $\partial_H^+ X$.

Proof. By Milman's converse to the Krein-Milman theorem (see Theorem 1.4.12, (2)) we have $\partial_e \overline{\mathrm{co}}(\Phi(X)) \subset \Phi(X)$ and hence it follows from Proposition 2.6.4 that $\Phi(\partial_H^+ X) = \partial_e \overline{\mathrm{co}}(\Phi(X))$.

Consequently $\partial_H^+ X \neq \varnothing$ because $\partial_e \overline{\mathrm{co}}(\Phi(X)) \neq \varnothing$ (see Theorem 1.4.10). Moreover, given $h \in H$, it suffices to apply the Bauer maximum principle (see Theorem 1.4.10) to the convex compact subset $\overline{\mathrm{co}}(\Phi(X))$ and to the continuous linear form $\tilde{h} \colon H' \to \mathbb{R}$ defined by

$$\tilde{h}(\mu) := \mu(h) \quad \text{for every } \mu \in H'. \qquad ☐$$

Remarks. 1. From Corollary 2.6.5 and the Krein-Milman theorem it follows that $\partial_H^+ X$ must contain more than one point unless X is a singleton.

2. In general, only assuming that H is cofinal in $\mathscr{C}(X)$ and separates the points of X, one cannot guarantee that $\partial_H^+ X \neq \varnothing$.

For example, if H is the subspace of $\mathscr{C}([0,1])$ generated by the function $h(x) = \exp(-x)$ $(x \in [0,1])$, then $\partial_H^+ [0,1] = \varnothing$ (in fact, for every $x_0 \in [0,1]$, the measure

$$\mu(f) := \frac{\exp(-x_0)}{1 - \exp(-1)} \int_0^1 f(t)\, dt,$$

coincides with ε_{x_0} on H, but $\mu \neq \varepsilon_{x_0}$).

However, if there exists a strictly positive function $h_0 \in H$ such that the subspace $H_0 := \left\{ \dfrac{h}{h_0} \,\middle|\, h \in H \right\}$ separates the points of X, then $\partial_H^+ X \neq \varnothing$.

This follows, in fact, from Corollary 2.6.5 and the equality $\partial_H^+ X = \partial_{H_0}^+ X$.

3. By using the embedding Φ defined by (1.4.19), we may also give a geometric interpretation of peak-points for H, provided H contains the constants and separates the points of X.

To this end, we recall that a *face* of a convex compact subset K is a convex subset F of K such that, if $x, y \in K$ and $\lambda x + (1 - \lambda)y \in F$ for some $\lambda \in \mathbb{R}$, $0 < \lambda < 1$, then necessarily $x, y \in F$.

Every face different from K is clearly contained in the topological boundary ∂K of K.

Moreover a singleton $\{x\}$ is a face of K if and only if $x \in \partial_e K$.

A face F of K is said to be *exposed* (*relatively exposed*, respectively) if there exists a positive continuous affine function u on K such that $u = 0$ on F and $u(x) > 0$ for every $x \in K \backslash F$ (if for every $x \in K \backslash F$ there exists a positive continuous affine function on K such that $u = 0$ on F and $u(x) > 0$, respectively).

An extreme point $x \in K$ is said to be *exposed* (*relatively exposed*, respectively) if $\{x\}$ is an exposed face (a relatively exposed face, respectively).

Clearly every exposed face F is relatively exposed. The converse holds provided F is the intersection of a countable sequence of open subsets (in particular, if K is metrizable).

Furthermore, it is worthy to mention that, if K is a Choquet simplex (see Section 1.5), then every closed face of K is relatively exposed and hence, every extreme point of K is relatively exposed (see, e.g., Alfsen [1971, pp. 119–121]).

Coming back to the situation of Proposition 2.6.5 we observe that a point $x \in X$ is a peak-point for H if and only if $\Phi(x)$ is a relatively exposed point of $\overline{\mathrm{co}}(\Phi(X))$.

Analogously, if $x_1, \ldots, x_n$ are distinct points of X, then condition (2) of Theorem 2.2.6 means that $\mathrm{co}(\Phi(x_1), \ldots, \Phi(x_n))$ is a relatively exposed face of $\overline{\mathrm{co}}(\Phi(X))$.

As a consequence of these geometric representations, if X is metrizable and if $\overline{\text{co}}(\Phi(X))$ is a Choquet simplex, then on account of the above remarks and Theorem 2.2.6 we have the following properties:

(1) The set of peak-points for H coincides with $\partial_H^+ X$.

(2) If $x_1, \ldots, x_n$ are distinct points of $\partial_H^+ X$ such that there exist $h_1, \ldots, h_n \in H$ satisfying $\det(h_i(x_j)) \neq 0$, then H is a D_+-subspace for every $\mu \in C_+(x_1, \ldots, x_n)$.

We also recall that $\overline{\text{co}}(\Phi(X))$ is a Choquet simplex if and only if $\overline{H}$ has the *Riesz interpolation property*, i.e., if f, g_1, $g_2 \in \overline{H}$ and $0 \leq g_1$, $0 \leq g_2$, $0 \leq f \leq g_1 + g_2$ then there exist f_1, $f_2 \in \overline{H}$ such that $0 \leq f_1 \leq g_1$, $0 \leq f_2 \leq g_2$ and $f = f_1 + f_2$ (see Alfsen [1971, Corollary II.3.11]).

Adapting the proof of Proposition 2.6.4 we may furnish a geometric criterion in order that a subspace is a determining subspace for discrete measures having a fixed support with more than one point.

2.6.6 Proposition. *Let X be a compact Hausdorff space and let H be a linear subspace of $\mathscr{C}(X)$ containing the constant functions and separating the points of X.*

Given n distinct points $x_1, \ldots, x_n$ of X, the following statements are equivalent:

(i) *The subspace H is a D_+-subspace for every $\mu \in C_+(x_1, \ldots, x_n)$.*

(ii) (1) $\Phi(X) \cap \text{co}(\Phi(x_1), \ldots, \Phi(x_n)) = \{\Phi(x_1), \ldots, \Phi(x_n)\}$.

(2) $\Phi(x_1), \ldots, \Phi(x_n)$ *are affinely independent in* H' *(i.e., every* $\mu \in \text{co}(\Phi(x_1), \ldots, \Phi(x_n))$ *can be expressed as a unique convex combination of* $\Phi(x_1), \ldots, \Phi(x_n)$).

(3) $\text{co}(\Phi(x_1), \ldots, \Phi(x_n))$ *is a face of* $\overline{\text{co}}(\Phi(X))$.

Proof. (i) $\Rightarrow$ (ii): Statements (1) and (2) are obvious. To prove (3), fix μ, $\upsilon \in \overline{\text{co}}(\Phi(X))$ and suppose that $t\mu + (1 - t)\upsilon \in \text{co}(\Phi(x_1), \ldots, \Phi(x_n))$ for some $t \in \mathbb{R}$, $0 < t < 1$.

Thus there exist $\alpha_1, \ldots, \alpha_n \geq 0$ such that $\sum_{i=1}^{n} \alpha_i = 1$ and $t\mu + (1 - t)\upsilon = \sum_{i=1}^{n} \alpha_i \Phi(x_i)$.

Denote by $\tilde{\mu}$ and $\tilde{\upsilon}$ two positive Radon measures on X which extend μ and υ, respectively. Then, since $t\tilde{\mu} + (1 - t)\tilde{\upsilon} = \sum_{i=1}^{n} \alpha_i \varepsilon_{x_i}$ on H, it follows from (i) that $t\tilde{\mu} + (1 - t)\tilde{\upsilon} = \sum_{i=1}^{n} \alpha_i \varepsilon_{x_i}$, so that the supports of $\tilde{\mu}$ and $\tilde{\upsilon}$ are contained in $\{x_1, \ldots, x_n\}$.

From this and from the fact that $\mu(\mathbf{1}) = \upsilon(\mathbf{1}) = 1$, it follows that μ, $\upsilon \in \text{co}(\Phi(x_1), \ldots, \Phi(x_n))$.

(ii) $\Rightarrow$ (i): Fix $\mu = \sum_{i=1}^{n} \alpha_i \varepsilon_{x_i} \in C_+(x_1, \ldots, x_n)$ and consider $\upsilon \in \mathscr{M}^+(X)$ such that $\upsilon = \mu$ on H.

Since $\|v\| = v(\mathbf{1}) = \mu(\mathbf{1}) = \|\mu\| = \sum_{i=1}^{n} \alpha_i$, we can always suppose that $\|v\| = \sum_{i=1}^{n} \alpha_i = 1$.

We now prove that $\mathrm{Supp}(v) \subset \{x_1, \ldots, x_n\}$.

Take $y \in \mathrm{Supp}(v)$ and suppose that $y \notin \{x_1, \ldots, x_n\}$. Denote by $\mathfrak{B}$ the set of all neighborhoods U of y satisfying $U \cap \{x_1, \ldots, x_n\} = \varnothing$ and endow it with the order relation $U \leq V$ if $V \subset U$.

As in the proof of Proposition 2.6.4, it is possible to construct two nets $(\mu_U)^{\leq}_{U \in \mathfrak{B}}$ and $(v_U)^{\leq}_{U \in \mathfrak{B}}$ of positive Radon measures on X and a net $(\lambda_U)^{\leq}_{U \in \mathfrak{B}}$ in $\mathbb{R}$, with $0 < \lambda_U < 1$ for every $U \in \mathfrak{B}$, such that

$$\lim_{\substack{\leq \\ U \in \mathfrak{B}}} \mu_U = \varepsilon_y \quad \text{vaguely}$$

and

$$v = \lambda_U \mu_U + (1 - \lambda_U) v_U \quad \text{for every } U \in \mathfrak{B}.$$

Thus, for every $U \in \mathfrak{B}$, since $\lambda_U \mu_{U|H} + (1 - \lambda_U) v_{U|H} = v_{|H} = \mu_{|H} \in \mathrm{co}(\Phi(x_1), \ldots, \Phi(x_n))$, we have $\mu_{U|H}, v_{U|H} \in \mathrm{co}(\Phi(x_1), \ldots, \Phi(x_n))$ on account of . (3).

Consequently, $\Phi(y) = \lim_{\substack{\leq \\ U \in \mathfrak{B}}} \mu_{U|H} \in \mathrm{co}(\Phi(x_1), \ldots, \Phi(x_n))$. By virtue of (1), there exists $i = 1, \ldots, n$ such that $\Phi(y) = \Phi(x_i)$, i.e., $y = x_i$ and we have reached a contradiction.

Thus, $\mathrm{Supp}(v) \subset \{x_1, \ldots, x_n\}$ and from (2) and the fact that $v(\mathbf{1}) = 1$ it follows that $\mu = v$. $\qquad \square$

In order to describe the subset $\partial_H^{1,+} X$ in a geometric way, we need a stronger separation property.

We say that a linear subspace H *separates strongly* the points of X if it separates the points of X and, furthermore, if

$$\text{for every } x \in X \text{ there exists } h \in H \text{ such that } h(x) \neq 0. \qquad (2.6.14)$$

2.6.7 Proposition. *Let X be a compact Hausdorff space and let H be a linear subspace of $\mathscr{C}(X)$ which separates strongly the points of X.*

Then a point $x_0 \in X$ belongs to $\partial_H^{1,+} X$ if and only if $\Phi(x_0)$ is an extreme point of $\overline{\mathrm{co}}(\Phi(X) \cup \{0\})$.

Consequently, $\partial_H^{1,+} X$ is non-empty.

Proof. First, suppose $x_0 \in \partial_H^{1,+} X$ and consider $\delta_1, \delta_2 \in \overline{\mathrm{co}}(\Phi(X) \cup \{0\})$ and $\lambda \in \mathbb{R}$, $0 < \lambda < 1$, satisfying $\Phi(x_0) = \lambda \delta_1 + (1 - \lambda) \delta_2$. Then there exists a net $(\eta_i)^{\leq}_{i \in I}$ of linear functionals in $\mathrm{co}(\Phi(X) \cup \{0\})$ such that $\lim_{\substack{\leq \\ i \in I}} \eta_i = \delta_1$ in H'_+.

For each $i \in I$ there exists a discrete Radon measure $\mu_i \in \mathcal{M}^+(X)$ such that $\|\mu_i\| \leq 1$ and $\mu_{i|H} = \eta_i$. So there exists a filter $\mathfrak{F}$ on I finer than the filter of sections on I and $\mu_1 \in \mathcal{M}^+(X)$, $\|\mu_1\| \leq 1$, such that $\lim_{\mathfrak{F}} \mu_i = \mu_1$; hence, $\mu_1 = \delta_1$ $\underset{i \in I}{}$ on H. Similarly, one proves that there exists $\mu_2 \in \mathcal{M}^+(X)$, $\|\mu_2\| \leq 1$ such that $\mu_2 = \delta_2$ on H. Put $\mu = \lambda \mu_1 + (1 - \lambda)\mu_2$. Then $\|\mu\| \leq 1$ and $\mu = \varepsilon_{x_0}$ on H. So $\mu = \varepsilon_{x_0}$ and consequently $\mu_1 = \mu_2 = \varepsilon_{x_0}$, i.e., $\delta_1 = \delta_2 = \Phi(x_0)$.

Before showing the converse, we observe that if $\mu \in \mathcal{M}^+(X)$, $\|\mu\| \leq 1$, then, by the approximation theorem 1.2.3, there exists a net $(\mu_i)_{i \in I}^{\leq}$ of discrete positive Radon measures which converges vaguely to μ and such that $\|\mu_i\| = \|\mu\| \leq 1$ for each $i \in I$. Hence for every $i \in I$ one has $\mu_{i|H} = \mu_{i|H} + (1 - \|\mu_i\|)0 \in$ co$(\Phi(X) \cup \{0\})$.

With this observation in mind, one may show the converse implication by using a reasoning analogous to the one of the proof of (ii) $\Rightarrow$ (i) in Proposition 2.6.4. In this case the functionals $\lambda_0 \eta_U$ and $\lambda_0 \theta_U$ considered in that proof belong to $\overline{\text{co}}(\Phi(X) \cup \{0\})$ (see the above Remark) and, moreover, the hypothesis that H strongly separates the points of X (instead of supposing that H is cofinal) is enough to complete the proof.

Finally, $\partial_H^{1,+} X$ is non-empty, because $\partial_e \overline{\text{co}}(\Phi(X) \cup \{0\})$ is contained in $\Phi(X) \cup \{0\}$ and contains more than one point. $\qquad\qquad\qquad\qquad\qquad\qquad\qquad\qquad\square$

Remark. Under the hypotheses of Proposition 2.6.7 we see that a point $x_0 \in X$ is a weak peak point for H if and only if $\Phi(x_0)$ is a relatively exposed point of $\overline{\text{co}}(\Phi(X) \cup \{0\})$.

When the subspace H is finite dimensional, then the geometric aspects of the previous results become more transparent.

In fact, suppose that H is algebraically generated by $n + 1$ linearly independent functions $h_0, h_1, \ldots, h_n$. Then H' can be identified with $\mathbb{R}^{n+1}$ by means of the mapping

$$\mu \in H' \mapsto (\mu(h_0), \mu(h_1), \ldots, \mu(h_n)).$$

So the canonical embedding Φ defined by (1.4.19) can be identified with the mapping $\Phi^*: X \to \mathbb{R}^{n+1}$ defined by

$$\Phi^*(x) := (h_0(x), h_1(x), \ldots, h_n(x)) \quad (x \in X). \qquad (2.6.15)$$

If $h_0 = 1$, we can further replace Φ^* by the mapping $\Phi_0: X \to \mathbb{R}^n$ defined by

$$\Phi_0(x) := (h_1(x), \ldots, h_n(x)) \quad (x \in X). \qquad (2.6.16)$$

Furthermore, since in a finite dimensional space the convex hull of a compact set is closed, H'_+ can be identified with the convex cone in $\mathbb{R}^{n+1}$ generated by $\overline{\mathrm{co}}(\Phi^*(X))$ and with vertex at the origin. Thus, H is cofinal in $\mathscr{C}(X)$ if and only if $H'_+ \cap (-H'_+) = \{0\}$.

Moreover, if H separates strongly the points of X, then

$$\Phi^*(\partial_H^{1,+} X) = \partial_e \mathrm{co}(\Phi^*(X) \cup \{0\}) \backslash \{0\} \qquad (2.6.17)$$

and, if $h_0 = 1$,

$$\Phi_0(\partial_H^+ X) = \partial_e \mathrm{co}(\Phi_0(X)). \qquad (2.6.18)$$

In particular, if X is a compact subset of $\mathbb{R}^n$ and if we denote by H the linear subspace of $\mathscr{C}(X)$ generated by the function $\mathbf{1}$ and the projections $\mathrm{pr}_1, \ldots, \mathrm{pr}_n \in \mathscr{C}(X)$ defined in (1.2.49), i.e.,

$$\mathrm{pr}_k(x) := x_k \quad \text{for every } x = (x_1, \ldots, x_n) \in X, \quad k = 1, \ldots, n, \qquad (2.6.19)$$

then Φ_0 becomes the natural embedding from X into $\mathbb{R}^n$ and so

$$\partial_H^+ X = \partial_e \mathrm{co}(X). \qquad (2.6.20)$$

If H_1 denotes the subspace of $\mathscr{C}(X)$ generated by $\mathrm{pr}_1, \ldots, \mathrm{pr}_n$ and if $0 \notin X$, then it follows from (2.6.17) that

$$\partial_{H_1}^{1,+} X = \partial_e \mathrm{co}(X \cup \{0\}) \backslash \{0\}. \qquad (2.6.21)$$

We conclude this section by studying some Choquet boundaries for the product of a finite family of compact spaces.

Let $(X_i)_{1 \le i \le p}$ be a finite family of compact Hausdorff spaces and denote by $X = \prod_{i=1}^{p} X_i$ the (compact) product space of the family $(X_i)_{1 \le i \le p}$ endowed with the product topology. As usual, we shall denote by $\mathrm{pr}_i \colon X \to X_i$ the i-th projection from X onto X_i (see (1.2.49)).

We recall that, if $(f_i)_{1 \le i \le p} \in \prod_{i=1}^{p} \mathscr{C}(X_i)$, we denoted by $\bigotimes_{i=1}^{p} f_i \colon X \to \mathbb{R}$ the continuous function defined by

$$\left(\bigotimes_{i=1}^{p} f_i \right)(x) := \prod_{i=1}^{p} f_i(x_i) \quad (x = (x_i)_{1 \le i \le p} \in X) \qquad (2.6.22)$$

$$\left(\text{thus } \bigotimes_{i=1}^{p} f_i = \prod_{i=1}^{p} f_i \circ \mathrm{pr}_i \right) \text{ (see (1.2.51))}.$$

Suppose now that for every $i \in I$ a subspace H_i of $\mathscr{C}(X_i)$ is assigned and suppose that every H_i contains the constant functions and separates the points of X_i.

We set

$$\sum_{i=1}^{p} H_i := \mathscr{L}(\{h_i \circ \mathrm{pr}_i | h_i \in H_i, i = 1, \ldots, p\}) \tag{2.6.23}$$

and

$$\bigotimes_{i=1}^{p} H_i := \mathscr{L}\left(\left\{\bigotimes_{i=1}^{p} h_i | h_i \in H_i, i = 1, \ldots, p\right\}\right). \tag{2.6.24}$$

We also consider the subspace

$$\underset{i=1}{\overset{p}{\mathbf{M}}} H_i := \Big\{ f \in \mathscr{C}(X) | f_{j,\bar{x}} \in H_j \text{ for every } j = 1, \ldots, p$$

$$\text{and} \quad \bar{x} = (\bar{x}_i)_{i \in \{1,\ldots,p\} \setminus \{j\}} \in \prod_{\substack{i=1 \\ i \neq j}}^{p} X_i \Big\}; \tag{2.6.25}$$

here $f_{j,\bar{x}}$ denotes the continuous function on X_j defined by

$$f_{j,\bar{x}}(x_j) := f((y_i)_{1 \le i \le p}) \quad (x_j \in X_j), \tag{2.6.26}$$

where $y_i = \bar{x}_i$ if $i \neq j$ and $y_i = x_j$ if $i = j$.

Clearly,

$$\sum_{i=1}^{p} H_i \subset \bigotimes_{i=1}^{p} H_i \subset \underset{i=1}{\overset{p}{\mathbf{M}}} H_i. \tag{2.6.27}$$

According to the following proposition, we see that, in fact, these three subspaces of $\mathscr{C}(X)$ have the same Choquet boundary.

2.6.8 Proposition. *Under the above assumption one has*

$$\partial^+_{\sum_{i=1}^{p} H_i} X = \partial^+_{\bigotimes_{i=1}^{p} H_i} X = \partial^+_{\mathbf{M}_{i=1}^{p} H_i} X = \prod_{i=1}^{p} \partial^+_{H_i} X_i.$$

Proof. From (2.6.27) it follows that $\partial^+_{\sum_{i=1}^{p} H_i} X \subset \partial^+_{\bigotimes_{i=1}^{p} H_i} X \subset \partial^+_{\mathbf{M}_{i=1}^{p} H_i} X$.

Let $x = (x_i)_{1 \leq i \leq p} \in \partial^+_{\underset{i=1}{\overset{p}{M}} H_i} X$ and fix $j = 1, \ldots, p$ and $\mu_j \in \mathcal{M}^+(X_j)$ such that $\mu_j = \varepsilon_{x_j}$ on H_j. For every $i = 1, \ldots, p$, $i \neq j$, put $\mu_i = \varepsilon_{x_i}$ and consider the product Radon measure $\mu = \bigotimes_{i=1}^{p} \mu_i \in \mathcal{M}^+(X)$. Then for every $h \in \underset{i=1}{\overset{p}{M}} H_i$ we have

$$\mu(h) = \int \ldots \int h(y_1, \ldots, y_j, \ldots, y_p) \, d\mu_1(y_1) \ldots d\mu_p(y_p) = h(x_1, \ldots, x_p) = h(x).$$

Hence $\mu = \varepsilon_x$ and consequently $\mu_j = \mathrm{pr}_j(\mu) = \varepsilon_{x_j}$.

This reasoning shows that $\partial^+_{\underset{i=1}{\overset{p}{M}} H_i} X \subset \prod_{i=1}^{p} \partial^+_{H_i} X_i$.

The proof will be complete if we show that $\prod_{i=1}^{p} \partial^+_{H_i} X_i \subset \partial^+_{\underset{i=1}{\overset{p}{\sum} H_i}} X$. In fact, fix $x = (x_i)_{1 \leq i \leq p} \in \prod_{i=1}^{p} \partial^+_{H_i} X_i$, an open neighborhood U of x and set $\varepsilon > 0$. Then, for every $i = 1, \ldots, p$, there exists an open neighborhood U_i of x_i such that $\prod_{i=1}^{p} U_i \subset U$. For every $i = 1, \ldots, p$, since $x_i \in \partial^+_{H_i} X_i$, by Theorem 2.5.2, there exists $h_i \in H_i$ satisfying $0 \leq h_i$, $1 \leq h_i$ on $X_i \backslash U_i$ and $h_i(x_i) < \dfrac{\varepsilon}{p}$.

Then, if we put $h := \sum_{i=1}^{p} h_i \circ \mathrm{pr}_i \in \sum_{i=1}^{p} H_i$, we have $0 \leq h$, $1 \leq h$ on $X \backslash U$ and $h(x) < \varepsilon$. Again, by Theorem 2.5.2, it follows that $x \in \partial^+_{\underset{i=1}{\overset{p}{\sum} H_i}} X$. $\square$

Example. If $(X_i)_{1 \leq i \leq p}$ is a finite family of convex compact sets and if we consider the subspaces $A(X_i) \subset \mathscr{C}(X_i)$ $(i = 1, \ldots, p)$, then $\underset{i=1}{\overset{p}{M}} A(X_i)$ coincides with the subspace of all continuous multiaffine functions on $X = \prod_{i=1}^{p} X_i$ (i.e., those continuous functions which are affine with respect to each variable). Furthermore, according to Propositions 2.6.3 and 2.6.8,

$$\partial^+_{\underset{i=1}{\overset{p}{M}} A(X_i)} X = \prod_{i=1}^{p} \partial_e X_i. \tag{2.6.28}$$

Remark. In this last Remark we would like to draw the attention of the reader to some subsets that generalize Choquet boundaries and whose study could be of interest.

Indeed, consider a subspace H of $\mathscr{C}_0(X)$ and fix $n \in \mathbb{N}$. Define

$$\partial^+_{H,n} X := \{(x_1, \ldots, x_n) \in X^n \, | \, x_1, \ldots, x_n \text{ are distinct and } H \text{ is a } D_+\text{-subspace}$$

$$\text{for every } \mu \in C_+(x_1, \ldots, x_n)\}.$$

Analogously, one can define $\partial_{H,n}^{1,+}X$. Thus $\partial_{H,1}^{+}X = \partial_H^{+}X$ and $\partial_{H,n}^{+}X \subset (\partial_H^{+}X)^n$.
What properties of the Choquet boundary are extensible to this subset? Under what assumptions is $\partial_{H,n}^{+}X$ non-empty? (For instance, if X is convex and compact and $H = A(X)$, then $\partial_{H,n}^{+}X = \varnothing$ for every $n \geq 2$. On the other hand, if H is a K_+-subspace of order n in $\mathscr{C}_0(X)$, then $\partial_{H,n}^{+}X = \{(x_1,\ldots,x_n) \in X^n | x_1,\ldots,$ x_n are distinct$\}$ (see Theorem 3.4.7)).

Proposition 2.6.6 and Theorems 2.2.2 and 2.2.3 contain several characterizations of the points belonging to $\partial_{H,n}^{+}X$.

Notes and references

The notion of Choquet boundary (sometimes called the *fine boundary*) was introduced by Bauer [1958], [1959] as the set of H-extremal points.

Bishop and de Leeuw [1959] named it explicitly as Choquet boundary.

In the above quoted papers of Bauer and Bishop de Leeuw the Choquet boundary was introduced respectively in connection with the need to frame in a more general and suitable setting some important problems from potential theory (e.g., the Dirichlet problem) as well as to extend Choquet's integral representation theory to arbitrary compact Hausdorff spaces.

Propositions 2.6.2 and 2.6.3 are due to Bauer [1959], [1961], [1966] (see also Bishop and de Leeuw [1959]).

The essence of Proposition 2.6.4 in contained in Choquet [1953]. The proof presented here is taken from Berens and Lorentz [1975] (see also Shashkin [1960], [1962]).

Corollary 2.6.5 is due to Bauer [1958], [1959] (see also Bishop and de Leeuw [1959]).

As a matter of fact, Bauer [1958], [1959] showed that $\overline{\partial_H^{+}X}$ is the smallest closed subset of X on which every $h \in H$ takes its maximum.

The main significance of Corollary 2.6.5 lies in the fact that it allows to translate what it is known about the set of extreme points and the integral representation theory for convex compact sets in the general context of compact Hausdorff spaces.

For instance, we have that $\partial_H^{+}X$ is a Baire space in the relative topology (see D. A. Edwards [1966]). Moreover, if X is metrizable, $\partial_H^{+}X$ is a countable intersection of open subsets of X.

For a metrizable compact space X an integral representation result runs as follows (see Bishop and de Leeuw [1959]):

> For every $x \in X$ there exists $\mu \in \mathscr{M}_1^{+}(X)$ supported by
> $\partial_H^{+}X$ (i.e., $\mu^*(X \backslash \partial_H^{+}X) = 0$) such that $\mu(h) = h(x)$ for each $h \in H$.

In this general setting Bauer [1959], [1961] formulated and studied an abstract *Dirichlet problem on the Choquet boundary*. It consists in extending every continuous function $f \in \mathscr{C}(\overline{\partial_H^{+}X})$ to a function belonging to a suitable subspace of $\mathscr{C}(X)$ (namely, to $K_+(H)$).

The connection between this abstract Dirichlet problem and the classical Dirichlet problem becomes clear if one looks to Proposition 2.6.2 and to Corollary 3.2.6 on account of the fact that $K_+(H(\Omega)) = H(\Omega)$ by virtue of Theorem 4.1.17.

Among other things, Bauer showed that the abstract Dirichlet problem is solvable if and only if $K_+(H)$ is a lattice in its own ordering or, equivalently, if for every $x \in X$ there exists a unique $\mu \in \mathcal{M}_1^+(X)$ supported by $\overline{\partial_H^+ X}$ such that $\mu(h) = h(x)$ for each $h \in H$.

When H is the real part of a (complex) function algebra A of continuous functions on X containing the constants and separating the points of X, then an important description of its Choquet boundary can be given. More precisely, Bishop and de Leeuw [1959] showed that for every $x_0 \in X$ the following statements are equivalent:

(i) $x_0 \in \partial_H^+ X$ and $\{x_0\}$ is a countable intersection of open subsets of X.
(ii) There exists $f \in A$ such that $|f(x)| < |f(x_0)|$ for every $x \in X$, $x \neq x_0$.

Proposition 2.6.6 seems to be new. Propositions 2.6.7 and 2.6.8 can be found in Berens and Lorentz [1975] and Grossman [1965], respectively. However, note that Proposition 2.6.8 still holds if the family of compact spaces is infinite (see Grossman [1970]).

For more complete information and details about Choquet boundaries we refer the reader to Alfsen [1971, Section I.5], Choquet [1969, Section 29] and Phelps [1966, Section 6].

Chapter 3

Korovkin-type theorems
for positive linear operators

The development of Korovkin-type approximation theory is surely motivated by the significant and elegant results obtained on the convergence of nets of positive linear operators and positive linear contractions. The original theorem of Korovkin and the first developments of the theory were directed to these objects. The characterizations obtained in relation with the convergence to the identity operator have subsequently stimulated similar investigations for other classes of linear operators.

In this chapter we shall deal with the problem of the convergence of equicontinuous nets of positive linear operators to a fixed positive linear operator.

In the first two sections we explain the general results for an arbitrary positive linear operator. In the subsequent sections we require additional conditions on the limit operator, namely that it is a projection, a finitely defined operator and, finally, a lattice homomorphism. The most important case of the identity operator deserves a special place in our treatment and will be discussed in the subsequent chapter.

3.1 Korovkin closures and Korovkin subspaces
for positive linear operators

In the present section we shall deal with a basic topic of Korovkin-type approximation theory.

Namely, we fix two locally compact Hausdorff spaces X and Y and a positive linear operator $T: \mathscr{C}_0(X) \to \mathscr{C}_0(Y)$; we are interested in studying the strong convergence of nets of positive operators from $\mathscr{C}_0(X)$ in $\mathscr{C}_0(Y)$ to the operator T.

Unless otherwise stated, in these spaces we shall always consider the supnorm topology.

According to our general aim we try to ensure the convergence of these nets on the whole space $\mathscr{C}_0(X)$, once that they converge toward T on some suitable (small) subsets of $\mathscr{C}_0(X)$.

3.1.1 Definition. *A subset H of $\mathscr{C}_0(X)$ is called a Korovkin subset for T with respect to positive linear operators or, briefly, a K_+-subset for T if it satisfies the following property:*

> *if $(L_i)_{i \in I}^{\leq}$ is an arbitrary net of positive linear operators from $\mathscr{C}_0(X)$ into $\mathscr{C}_0(Y)$ such that $\sup_{i \in I} \|L_i\| < +\infty$ and if*
>
> $$\lim_{i \in I}{}_{\leq} L_i(h) = T(h) \text{ for all } h \in H, \text{ then } \lim_{i \in I}{}_{\leq} L_i(f) = T(f)$$
>
> *for every $f \in \mathscr{C}_0(X)$, i.e., $(L_i)_{i \in I}^{\leq}$ converges strongly to T.*   (3.1.1)

If T is a positive linear contraction (i.e., $\|T\| \leq 1$) and if H satisfies (3.1.1) only for nets $(L_i)_{i \in I}^{\leq}$ of positive linear contractions, then H is called a Korovkin subset for T with respect to positive linear contractions or, briefly, a K_+^1-subset for T.

We observe that H is a Korovkin subset of $\mathscr{C}_0(X)$ if and only if the linear subspace $\mathscr{L}(H)$ generated by H is itself a Korovkin set; thus, our attention might be devoted only to subspaces of $\mathscr{C}_0(X)$ satisfying (3.1.1) and in this case we shall call H a *Korovkin subspace for T with respect to positive linear operators* or, more briefly, a K_+-*subspace for T*.

Analogously, we also define the K_+^1-subspaces for T.

Note that, if X is compact and $\mathbf{1} \in H$, then by replacing, if necessary, the net $(L_i)_{i \in I}^{\leq}$ in (3.1.1) with a suitable family $(L_i)_{i \geq i_0}^{\leq}$ $(i_0 \in I)$, the condition $\sup_{i \in I} \|L_i\| < +\infty$ is satisfied provided $\lim_{i \in I}{}_{\leq} L_i(h) = T(h)$ for every $h \in H$.

The tools we have at our disposal will allow us not only to characterize Korovkin subsets of $\mathscr{C}_0(X)$ but, more generally, to describe, for a given subset H of $\mathscr{C}_0(X)$, the maximal subspace of $\mathscr{C}_0(X)$ where every equicontinuous net of positive linear operators (positive linear contractions, respectively) from $\mathscr{C}_0(X)$ into $\mathscr{C}_0(Y)$ converges to T whenever it converges to T on H.

More precisely, if H is a subset of $\mathscr{C}_0(X)$, we introduce the following subspace of $\mathscr{C}_0(X)$:

$$K_+(H, T) := \left\{ f \in \mathscr{C}_0(X) \,\middle|\, \lim_{i \in I}{}_{\leq} L_i(f) = T(f) \text{ for each net } (L_i)_{i \in I}^{\leq} \text{ of} \right.$$
$$\text{positive linear operators from } \mathscr{C}_0(X) \text{ into } \mathscr{C}_0(Y) \text{ satisfying}$$
$$\left. \sup_{i \in I} \|L_i\| < +\infty \text{ and } \lim_{i \in I}{}_{\leq} L_i(h) = T(h) \text{ for all } h \in H \right\}; \quad (3.1.2)$$

the subspace $K_+(H, T)$ is called the *Korovkin closure of H for T with respect to positive linear operators*, or the K_+-*closure of H for T*.

Clearly, the K_+-closure of a subset H of $\mathscr{C}_0(X)$ coincides with the K_+-closure of the linear subspace generated by H. Furthermore, $K_+(H, T) = \mathscr{C}_0(X)$ exactly when H is a K_+-subset for T.

Analogously, the *Korovkin closure of a subset H of $\mathscr{C}_0(X)$ for T with respect to positive linear contractions*, or the K_+^1-closure of H for T, is defined as in (3.1.2) by considering nets of positive linear contractions rather than positive operators. It will be denoted by $K_+^1(H, T)$.

At this point we can start with some characterizations both of K_+-closures and K_+^1-closures in $\mathscr{C}_0(X)$ for a given positive linear operator $T: \mathscr{C}_0(X) \to \mathscr{C}_0(Y)$. Our first step consists in showing the relationship between these closures and the convergence subspaces associated with suitable Radon measures which depend on T. The connection is achieved by considering, for each $y \in Y$, the positive Radon measure $\mu_y^T: \mathscr{C}_0(X) \to \mathbb{R}$ defined by

$$\mu_y^T(f) := Tf(y) \quad \text{for each } f \in \mathscr{C}_0(X) \tag{3.1.3}$$

(thus, $\mu_y^T = \varepsilon_y \circ T$).

We begin with the case of positive contractions, since this will simplify the analogous subsequent result about positive operators.

3.1.2 Theorem. *Let $T: \mathscr{C}_0(X) \to \mathscr{C}_0(Y)$ be a positive linear contraction and let H be a subspace of $\mathscr{C}_0(X)$. For each $f \in \mathscr{C}_0(X)$ the following statements are equivalent*:
(i) $f \in K_+^1(H, T)$.
(ii) (1) $f \in U_+^1(H, \mu_y^T)$ *(or, equivalently, $f \in D_+^1(H, \mu_y^T)$) for each $y \in Y$.*
   (2) *If $\mu \in \mathscr{M}_b^+(X)$ and $\mu = 0$ on H, then $\mu(f) = 0$.*
In short,

$$K_+^1(H, T) = H^* \cap \bigcap_{y \in Y} U_+^1(H, \mu_y^T) = H^* \cap \bigcap_{y \in Y} D_+^1(H, \mu_y^T). \tag{3.1.4}$$

Moreover, if Y is compact, then statement (i) *is equivalent only to part* (1) *of* (ii) *and*

$$K_+^1(H, T) = \bigcap_{y \in Y} U_+^1(H, \mu_y^T) = \bigcap_{y \in Y} D_+^1(H, \mu_y^T). \tag{3.1.5}$$

Proof. (i) $\Rightarrow$ (ii): We first establish condition (1). Consider $y \in Y$ and $v \in \mathscr{M}_b^+(X)$ such that $\|v\| \leq 1$ and $v = \mu_y^T$ on H, and fix a base $\mathfrak{B}$ of open relatively compact neighborhoods of y in Y. According to Theorem 1.1.2, for each $V \in \mathfrak{B}$, choose a continuous function $g_V: Y \to [0, 1]$ such that $g_V(y) = 1$ and $g_V = 0$ on $Y \backslash V$ and, hence, consider the linear operator $L_V: \mathscr{C}_0(X) \to \mathscr{C}_0(Y)$ defined by $L_V(g) := v(g)g_V + T(g)(1 - g_V)$ for each $g \in \mathscr{C}_0(X)$.

Then each L_V is a positive contraction, because

$$|L_V(g)(z)| \leq |v(g)|g_V(z) + \|T(g)\|(1 - g_V)(z)$$

$$\leq \|g\|g_V(z) + \|g\|(1 - g_V)(z) \leq \|g\|$$

for each $g \in \mathscr{C}_0(X)$ and $z \in Y$.

Consider on $\mathfrak{B}$ the partial ordering $\leq$ defined by $U \leq V$ if $V \subset U$.

If $h \in H$ and $\varepsilon > 0$, there exists $U_0 \in \mathfrak{B}$ such that $|Th(x) - Th(y)| \leq \varepsilon$ for every $x \in U_0$.

Consequently, if $U \in \mathfrak{B}$ and $U_0 \leq U$ (i.e., $U \subset U_0$), then $\|L_V(h) - T(h)\| \leq \varepsilon$ because $L_V(h)(x) - T(h)(x) = (T(h)(y) - T(h)(x))g_V(x)$ for every $x \in X$.

Thus we have proved that $\lim_{V \in \mathfrak{B}}^{\leq} L_V(h) = T(h)$ for every $h \in H$ and therefore, by (i) and (3.1.2), the net $(L_V(f))_{V \in \mathfrak{B}}^{\leq}$ converges to $T(f)$. Hence $\mu_y^T(f) = T(f)(y) = \lim_{V \in \mathfrak{B}}^{\leq} L_V(f)(y) = v(f)$. Thus $f \in U_+^1(H, \mu_y^T)$ or, equivalently, $f \in D_+^1(H, \mu_y^T))$ by virtue of (2.1.21).

Suppose that Y is non compact. To show part (2) of (ii), let $\mu \in \mathscr{M}_b^+(X)$ such that $\mu = 0$ on H. Denote by $\mathfrak{A}$ the set of all compact subsets of Y ordered by inclusion. For every $K \in \mathfrak{A}$, choose a point $x_K \in Y \backslash K$ and a function $\varphi_K \in \mathscr{K}(Y)$ such that $0 \leq \varphi_K \leq 1$, $\varphi_K(x_K) = 1$ and $\varphi_K = 0$ on K. Hence $\|\varphi_K\| = 1$. Furthermore, it is easy to show that $\lim_{K \in \mathfrak{A}}^{\subset} \varphi_K f = 0$ uniformly on Y for every $f \in \mathscr{C}_0(Y)$.

For each $K \in \mathfrak{A}$ consider the positive linear contraction $L_K : \mathscr{C}_0(X) \to \mathscr{C}_0(Y)$ defined by $L_K(g) := \dfrac{\mu(g)}{\|\mu\|} \varphi_K + T(g)(1 - \varphi_K)$ for each $g \in \mathscr{C}_0(X)$.

It is clear that the net $(L_K(h))_{K \in \mathfrak{A}}^{\subset}$ converges to $T(h)$ for each $h \in H$; by (i) and (3.1.2) the net $(L_K(f))_{K \in \mathfrak{A}}^{\subset}$ must converge to $T(f)$. Consequently,

$$\frac{\mu(f)}{\|\mu\|} \varphi_K = L_K(f) - T(f) + T(f)\varphi_K \to 0,$$

and this yields $\mu(f) = 0$.

(ii) $\Rightarrow$ (i): Let $(L_i)_{i \in I}^{\leq}$ be a net of positive linear contractions from $\mathscr{C}_0(X)$ into $\mathscr{C}_0(Y)$, which converges to $T(h)$ for each $h \in H$. If $(L_i(f))_{i \in I}^{\leq}$ does not converge to $T(f)$, we can find $\varepsilon > 0$ and, for each $i \in I$, an index $\alpha(i) \in I$, $\alpha(i) \geq i$, and $y_i \in Y$ such that

$$|L_{\alpha(i)}f(y_i) - Tf(y_i)| \geq \varepsilon. \tag{1}$$

If the net $(y_i)_{i \in I}^{\leq}$ converges to the point at infinity of Y, we have $\lim_{i \in I}^{\leq} \psi(y_i) = 0$ for each $\psi \in \mathscr{C}_0(Y)$. We can consider, for each $i \in I$, the positive contractive Radon measure $\mu_i : \mathscr{C}_0(X) \to \mathbb{R}$ defined by $\mu_i(g) := L_{\alpha(i)}(g)(y_i)$ for each $g \in \mathscr{C}_0(X)$. By the Alaoglu-Bourbaki theorem 1.4.1 there exist a filter $\mathfrak{F}$ on I finer than the filter of sections on I and $\mu \in \mathscr{M}_b^+(X)$ such that $\|\mu\| \leq 1$ and the net $(\mu_i)_{i \in I}^{\mathfrak{F}}$ is vaguely convergent to μ. If $h \in H$ we have

$$|\mu_i(h)| \leq |L_{\alpha(i)}h(y_i) - Th(y_i)| + |Th(y_i)| \leq \|L_{\alpha(i)}(h) - T(h)\| + |Th(y_i)|$$

for each $i \in I$ and hence the net $(\mu_i(h))_{i \in I}^{\mathfrak{F}}$ converges to 0. By part (2) of (ii), it follows that $\lim_{\substack{\mathfrak{F} \\ i \in I}} \mu_i(f) = \mu(f) = 0$, which contradicts (1).

If the net $(y_i)_{i \in I}^{\leq}$ does not converge to the point at infinity of Y, we can find a compact subset K of Y and, for each $i \in I$, an index $\beta(i) \in I$, $\beta(i) \geq i$, such that $y_{\beta(i)} \in K$. Now, we consider the positive contractive Radon measure $\mu_i: \mathscr{C}_0(X) \to \mathbb{R}$ defined by $\mu_i(g) := L_{\alpha(i)}g(y_{\beta(i)})$ for each $g \in \mathscr{C}_0(X)$. We can again apply the Alaoglu-Bourbaki theorem to obtain a filter $\mathfrak{F}_1$ on I finer than the filter of sections on I and $\mu \in \mathscr{M}_b^+(X)$ such that $\|\mu\| \leq 1$ and the net $(\mu_i)_{i \in I}^{\mathfrak{F}_1}$ is vaguely convergent to μ. Since K is compact, we can consider another filter $\mathfrak{F}_2$ on I, finer than $\mathfrak{F}_1$, such that $(y_{\beta(i)})_{i \in I}^{\mathfrak{F}_2}$ converges to an element $y \in Y$. If $h \in H$, for each $i \in I$, we have

$$\mu_i(h) = L_{\alpha(i)}h(y_{\beta(i)}) - Th(y_{\beta(i)}) + Th(y_{\beta(i)}),$$

and

$$|L_{\alpha(i)}h(y_{\beta(i)}) - Th(y_{\beta(i)})| \leq \|L_{\alpha(i)}(h) - T(h)\| \to 0.$$

Therefore $\mu(h) = \lim_{\substack{\mathfrak{F}_2 \\ i \in I}} \mu_i(h) = Th(y) = \mu_y^T(h)$. Since $f \in U_+^1(H, \mu_y^T)$, we also have $\mu(f) = \mu_y^T(f) = Tf(y)$, which is again a contradiction to (1).

Now note that condition (2) of part (ii) means that $f \in U_+(H, 0) = H^*$ (apply Theorem 2.1.6 to the null functional) and therefore we obtain the equality (3.1.4).

Finally, if Y is compact, the case where the net $(y_i)_{i \in I}^{\leq}$ converges to the point at infinity of Y cannot arise. Therefore condition (2) is not needed in the proof of (ii) $\Rightarrow$ (i). $\qquad\qquad\qquad\qquad\qquad\qquad\qquad\qquad\qquad\qquad\square$

Remarks. 1. An inspection of the proof of Theorem 3.1.2 shows that, in fact, in the definition of $K_+^1(H, T)$ we may also consider *parametric nets* of positive linear contractions, i.e., families of nets of positive linear contractions.

More precisely, by using Theorem 3.1.2, it is possible to show that a function $f \in \mathscr{C}_0(X)$ belongs to $K_+^1(H, T)$ if and only if for every family of nets $(L_{i,\lambda})_{i \in I}^{\leq}$, $\lambda \in \Lambda$, of positive linear contractions from $\mathscr{C}_0(X)$ into $\mathscr{C}_0(Y)$ satisfying

$$\lim_{\substack{\leq \\ i \in I}} L_{i,\lambda}(h) = T(h) \quad \text{uniformly in } \lambda \in \Lambda \text{ for every } h \in H,$$

one also has

$$\lim_{\substack{\leq \\ i \in I}} L_{i,\lambda}(f) = T(f) \quad \text{uniformly in } \lambda \in \Lambda.$$

A similar remark also holds for $K_+(H, T)$ if one considers parametric nets of positive linear operators $(L_{i,\lambda})_{i \in I}^{\leq}$, $\lambda \in \Lambda$, such that $\sup_{(i,\lambda) \in I \times \Lambda} \|L_{i,\lambda}\| < +\infty$ (see the proof of Theorem 3.1.3).

These additional properties of $K_+^1(H, T)$ and $K_+(H, T)$ will be important in the study of $\mathscr{A}$-summation processes of positive linear operators (see Section 5.4).

2. If Y is not compact, part (2) of condition (ii) is essential, and (3.1.5) does not hold, as we make clear with the following example.

Let T be the identity operator from the space $\mathscr{C}_0(\mathbb{N})$ into itself and put

$$H := \left\{ f \in \mathscr{C}_0(\mathbb{N}) \,\middle|\, \sum_{n=1}^{\infty} \frac{1}{2^n} f(n) = 0 \right\}.$$

As we have showed in the Remark to Theorem 2.1.10, H satisfies condition (1) in (ii) of Theorem 3.1.2 for each $f \in \mathscr{C}_0(\mathbb{N})$.

On the other hand, the positive Radon measure $\mu_0 \in \mathscr{M}_b^+(\mathbb{N})$ defined by

$$\mu_0(f) := \sum_{n=1}^{\infty} \frac{1}{2^n} f(n) \text{ for each } f \in \mathscr{C}_0(\mathbb{N}) \text{ satisfies } \mu_0(f) = 0 \text{ if and only if } f \in H.$$

Consequently, from part (2) of condition (ii) in Theorem 3.1.2 we obtain

$$K_+^1(H, T) = H \neq \mathscr{C}_0(\mathbb{N}) = \bigcap_{n \in \mathbb{N}} U_+^1(H, \mu_n^T).$$

Now we may easily give the corresponding result for the K_+-closures; however, in this case, we can always drop part (2) of condition (ii).

3.1.3 Theorem. *Let* $T: \mathscr{C}_0(X) \to \mathscr{C}_0(Y)$ *be a positive linear operator and let* H *be a subspace of* $\mathscr{C}_0(X)$. *For each* $f \in \mathscr{C}_0(X)$ *the following statements are equivalent*:
(i) $f \in K_+(H, T)$.
(ii) $f \in U_+(H, \mu_y^T)$ *(or, equivalently,* $f \in D_+(H, \mu_y^T)$*) for each* $y \in Y$.
In other words,

$$K_+(H, T) = \bigcap_{y \in Y} U_+(H, \mu_y^T) = \bigcap_{y \in Y} D_+(H, \mu_y^T). \tag{3.1.6}$$

Proof. We can argue as in the proof of Theorem 3.1.2, once we have shown that part (2) of (ii) in Theorem 3.1.2 is automatically true whenever $f \in \mathscr{C}_0(X)$ satisfies $f \in U_+(H, \mu_y^T)$ for each $y \in Y$. But this is, in fact, a consequence of Theorem 2.1.6. $\qquad\square$

The preceding theorems allow us a direct use of the results in the preceding Chapter 2. In particular, if T is the null operator, we have that

$$K_+(H, 0) = K_+^1(H, 0) = H^*.$$

So, H is a K_+^1-subspace (and a K_+-subspace) for 0 if and only if $H^* = \mathscr{C}_0(X)$.

Next, we shall state other characterizations of $K_+(H, T)$ which are of practical interest in the applications.

Other properties may be found in the subsequent section.

3.1.4 Theorem. *Let X and Y be locally compact Hausdorff spaces, let $T: \mathscr{C}_0(X) \to \mathscr{C}_0(Y)$ be a positive linear operator and let H be a subspace of $\mathscr{C}_0(X)$. For a given $f \in \mathscr{C}_0(X)$ the following statements are equivalent:*

(i) *$f \in K_+(H, T)$.*

(ii) *(Uniqueness property) $\mu(f) = Tf(y)$ for every $\mu \in \mathscr{M}_b^+(X)$ and $y \in Y$ satisfying $\mu(h) = Th(y)$ for all $h \in H$.*

(iii) *(Analytic property)*

$$\inf_{\varepsilon > 0} \left(\sup_{\substack{h \in H \\ h - \varepsilon \leq f}} T(h) \right) = T(f) = \sup_{\varepsilon > 0} \left(\inf_{\substack{k \in H \\ f \leq k + \varepsilon}} T(k) \right).$$

(iv) *For every $\varepsilon > 0$ there exist finitely many functions $h_0, \ldots, h_n \in H$ and $k_0, \ldots, k_n \in H$ such that*

$$\left\| \inf_{0 \leq j \leq n} T(k_j) - \sup_{0 \leq i \leq n} T(h_i) \right\| \leq \varepsilon \quad and \quad \sup_{0 \leq i \leq n} h_i - \varepsilon \leq f \leq \inf_{0 \leq j \leq n} k_j + \varepsilon.$$

(v) *For every $\varepsilon > 0$ there exist finitely many functions $h_0, \ldots, h_n \in H$ and $k_0, \ldots, k_n \in H$ and $u, v \in \mathscr{C}_0^+(X)$ such that $\|u\| \leq \varepsilon$, $\|v\| \leq \varepsilon$ and, further,*

$$\left\| \inf_{0 \leq j \leq n} T(k_j) - \sup_{0 \leq i \leq n} T(h_i) \right\| \leq \varepsilon \quad and \quad \sup_{0 \leq i \leq n} h_i - u \leq f \leq \inf_{0 \leq j \leq n} k_j + v.$$

Proof. The equivalence of statements (i) and (ii) is, in fact, Theorem 3.1.3.

To show that (ii) $\Leftrightarrow$ (iii), we fix $y \in Y$ and observe that the condition $f \in U_+(H, \mu_y^T)$ means that

$$\inf_{\varepsilon > 0} \left(\sup_{\substack{h \in H \\ h - \varepsilon \leq f}} \mu_y^T(h) \right) = \mu_y^T(f) = \sup_{\varepsilon > 0} \left(\inf_{\substack{k \in H \\ f \leq k + \varepsilon}} \mu_y^T(k) \right)$$

(see Theorem 2.1.6 and (2.1.10), (2.1.11)).

Therefore, on account of (3.1.3) and since $y \in Y$ is arbitrary, we have (ii) $\Leftrightarrow$ (iii).

(iii) $\Rightarrow$ (iv): Let $\varepsilon > 0$ and set $\varepsilon' := \dfrac{\varepsilon}{1 + \|T\|}$. By virtue of (iii) we can consider h_0 and k_0 in H such that $h_0 - \dfrac{\varepsilon'}{2} \leq f \leq k_0 + \dfrac{\varepsilon'}{2}$. Since $T(h_0)$ and $T(k_0)$ are in $\mathscr{C}_0(Y)$, there exists a compact set K in Y such that $Tk_0(y) - Th_0(y) \leq \varepsilon'$ for each $y \in Y \backslash K$. If $y \in K$, by (iii) and Theorem 2.1.6 applied to the measure $\mu = \mu_y^T$, we can consider h_y and k_y in H such that $Tk_y(y) - Th_y(y) \leq \varepsilon'$ and $h_y - \dfrac{\varepsilon'}{2} \leq f \leq k_y + \dfrac{\varepsilon'}{2}$. The continuity of $T(h_y)$ and $T(k_y)$ yields an open neighborhood $U(y)$ of y in Y such that $T(k_y) - T(h_y) \leq \varepsilon'$ on $U(y)$. Since y is arbitrary in K, we obtain an open cover $(U(y))_{y \in K}$ of K; the compactness of K implies the existence of a finite subcover $U(y_0), \ldots, U(y_n)$ of K. Put $h_i := h_{y_i}$ and $k_i := k_{y_i}$ for every $i = 0, \ldots, n$. Since

$$\sup_{0 \leq i \leq n} h_i - \frac{\varepsilon'}{2} \leq f \leq \inf_{0 \leq j \leq n} k_j + \frac{\varepsilon'}{2},$$

we obtain

$$\sup_{0 \leq i \leq n} T(h_i) - \frac{\varepsilon'}{2}\|T\| \leq T(f) \leq \inf_{0 \leq j \leq n} T(k_j) + \frac{\varepsilon'}{2}\|T\|$$

and hence

$$\sup_{0 \leq i \leq n} T(h_i) - \inf_{0 \leq j \leq n} T(k_j) \leq \varepsilon'\|T\| \leq \varepsilon.$$

On the other hand we have

$$\inf_{0 \leq j \leq n} T(k_j) - \sup_{0 \leq i \leq n} T(h_i) \leq \varepsilon' \leq \varepsilon,$$

and so the result follows.

(iv) $\Leftrightarrow$ (v): This is a direct consequence of Lemma 2.1.5.

(iv) $\Rightarrow$ (iii): It follows from Theorem 2.1.6 applied to every measure μ_y^T ($y \in Y$). $\quad\square$

Remarks. 1. We observe that Theorem 2.1.6 furnishes further alternative formulations of statement (iii), but for brevity we shall not state them explicitly.

2. The functions $f \in \mathscr{C}_0(X)$ satisfying condition (iii) are often called *almost H_T-affine functions* (see Bauer and Donner [1978] and Grossman [1976a, Section 1]).

At this point, by using the preceding theorem we may immediately give some characterizations of K_+-subspaces for a positive operator T.

To state them in a more complete form, we introduce the notion of *Choquet boundary for a subspace H of $\mathscr{C}_0(X)$ associated with the operator T or T-Choquet boundary for H in Y*. This is, by definition, the subset $\partial_{H,T}^+ Y$ of Y consisting of all elements $y \in Y$ such that H is D_+-subspace for μ_y^T.

Moreover, if in addition T is a contraction, we shall denote by $\partial_{H,T}^{1,+} Y$ the subset of Y consisting of all elements $y \in Y$ such that H is D_+^1-subspace for μ_y^T.

If $X = Y$ and T is the identity operator we obtain the classical Choquet boundary for H already studied in Section 2.6. We are not interested in studying in depth this type of Choquet boundary but we just present a characterization of K_+- and K_+^1-subspaces for T in terms of these subsets according to Theorems 3.1.2 and 3.1.3.

3.1.5 Corollary. *Let X and Y be locally compact Hausdorff spaces and let T: $\mathscr{C}_0(X) \to \mathscr{C}_0(Y)$ be a positive linear operator. Then a subspace H of $\mathscr{C}_0(X)$ is a K_+-subspace for T if and only if one of the statements* (ii)–(v) *in Theorem* 3.1.4 *holds for each $f \in \mathscr{C}_0(X)$ or, equivalently, if and only if $\partial_{H,T}^+ Y = Y$.*

Although Theorem 3.1.4 also holds for X compact, its statements can be arranged in a simplified form in this case. For these reasons we consider it separately.

3.1.6 Theorem. *Suppose X to be compact. Let $T: \mathscr{C}(X) \to \mathscr{C}_0(Y)$ be a positive linear operator and let H be a cofinal subspace of $\mathscr{C}(X)$. For a given $f \in \mathscr{C}(X)$ the following statements are equivalent:*

(i) $f \in K_+(H,T)$.

(ii) (*Uniqueness property*) $\mu(f) = Tf(y)$ *for every $\mu \in \mathscr{M}^+(X)$ and $y \in Y$ satisfying $\mu(h) = Th(y)$ for all $h \in H$.*

(iii) (*Analytic property*)

$$\sup_{\substack{h \in H \\ h \le f}} T(h) = T(f) = \inf_{\substack{k \in H \\ f \le k}} T(k).$$

(iv) *For every $\varepsilon > 0$ there exist finitely many functions $h_0, \ldots, h_n \in H$ and $k_0, \ldots, k_n \in H$ such that*

$$\left\| \inf_{0 \le j \le n} T(k_j) - \sup_{0 \le i \le n} T(h_i) \right\| \le \varepsilon \quad and \quad \sup_{0 \le i \le n} h_i \le f \le \inf_{0 \le j \le n} k_j.$$

Proof. The equivalences from (i) to (iv) can be established by the same arguments of the proof of Theorem 3.1.4 using Theorem 3.1.3 and Theorem 2.1.3 instead of Theorem 2.1.6. □

As an incidental remark note that the functions $f \in \mathscr{C}(X)$ satisfying condition (iii) are called H_T-affine functions (see Bauer [1973] and Grossman [1976a]).

3.1.7 Corollary. *Suppose X to be compact. If $T: \mathscr{C}(X) \to \mathscr{C}_0(Y)$ is a positive linear operator and H an arbitrary subspace of $\mathscr{C}(X)$, then H is a K_+-subspace for T if and only if one of the following equivalent assertions are valid:*
(ii)' $\partial_{H,T}^+ Y = Y$.
(iii)' *H is cofinal in $\mathscr{C}(X)$ and for each $f \in \mathscr{C}(X)$*

$$\sup_{\substack{h \in H \\ h \leq f}} T(h) = T(f) = \inf_{\substack{k \in H \\ f \leq k}} T(k).$$

(iv)' *H is cofinal in $\mathscr{C}(X)$ and for each $f \in \mathscr{C}(X)$ and $\varepsilon > 0$ there exist finitely many functions $h_0, \ldots, h_n \in H$ and $k_0, \ldots, k_n \in H$ such that*

$$\left\| \inf_{0 \leq j \leq n} T(k_j) - \sup_{0 \leq i \leq n} T(h_i) \right\| \leq \varepsilon \quad and \quad \sup_{0 \leq i \leq n} h_i \leq f \leq \inf_{0 \leq j \leq n} k_j.$$

Proof. The result follows from Theorem 3.1.6 and Theorem 2.1.4.   □

The reader may note that, on account of the Remark subsequent to Theorem 2.1.4, the K_+-subspaces for the null operator are just the cofinal subspaces of $\mathscr{C}(X)$.

We now proceed to state some characterizations of Korovkin closures and Korovkin subspaces for T with respect to positive linear contractions.

First, we observe that from (3.1.5), (3.1.6) and (2.1.22) it follows that

$$K_+^1(H, T) = K_+(H, T) \tag{3.1.7}$$

in the case where X and Y are *compact*, $1 \in H$ and $T(1) \leq 1$.

In the general case only the inclusion $K_+(H, T) \subset K_+^1(H, T)$ can be expected, and this follows directly from the definitions.

As a simple counterexample to the converse inclusion, consider the positive contraction $T: \mathscr{C}_0(]0, 1]) \to \mathscr{C}([0, 1])$ defined by assigning to every $f \in \mathscr{C}_0(]0, 1])$ the constant function of constant value $f(\frac{2}{3})$ and denote by H the subspace of $\mathscr{C}_0(]0, 1])$ generated by the functions e_1 and e_2 defined by $e_i(x) := x^i$ ($x \in]0, 1]$, $i = 1, 2$). Then, for each $y \in [0, 1]$, we have $\mu_y^T = \varepsilon_{2/3}$ and therefore, on account of the example following (2.1.21) in Section 2.1 and Theorems 3.1.2 and 3.1.3, we have $K_+^1(H, T) = \mathscr{C}_0(]0, 1])$ while $K_+(H, T) \neq \mathscr{C}_0(]0, 1])$.

In the following result we describe K_+^1-closures from an analytic point of view.

3.1.8 Theorem. *Let X and Y be locally compact Hausdorff spaces, let $T: \mathscr{C}_0(X) \to \mathscr{C}_0(Y)$ be a positive linear contraction and consider a subspace H of $\mathscr{C}_0(X)$. For each $f \in \mathscr{C}_0(X)$ the following statements are equivalent:*

(i) $f \in K_+^1(H, T)$.

(ii) (1) (*Uniqueness property*) $\mu(f) = Tf(y)$ *for every* $\mu \in \mathscr{M}_b^+(X)$, $\|\mu\| \le 1$, *and* $y \in Y$ *satisfying* $\mu(h) = Th(y)$ *for all* $h \in H$.

(2) *If* $\mu \in \mathscr{M}_b^+(X)$ *and* $\mu = 0$ *on* H, *then* $\mu(f) = 0$.

(iii) (*Analytic property*) $f \in H^*$ *and*

$$\sup_{\substack{h \in H, \alpha \ge 0 \\ h - \alpha \le f}} (T(h) - \alpha) = T(f) = \inf_{\substack{k \in H, \beta \ge 0 \\ f \le k + \beta}} (T(k) + \beta).$$

(iv) $f \in H^*$ *and for every* $\varepsilon > 0$ *there exist finitely many functions* $h_0, \ldots, h_n$, $k_0, \ldots, k_n$ *in* H *and* $\alpha_0, \ldots, \alpha_n, \beta_0, \ldots, \beta_n$ *in* $\mathbb{R}_+$ *such that*

$$\left\| \inf_{0 \le j \le n} (T(k_j) + \beta_j) - \sup_{0 \le i \le n} (T(h_i) - \alpha_i) \right\| \le \varepsilon$$

and

$$\sup_{0 \le i \le n} (h_i - \alpha_i) \le f \le \inf_{0 \le j \le n} (k_j + \beta_j).$$

Consequently, H is a K_+^1-subspace for T if and only if one of the statements (i)–(iv) *holds for each $f \in \mathscr{C}_0(X)$ or, equivalently, if and only if*

$$H^* = \mathscr{C}_0(X) \quad \text{and} \quad \partial_{H, T}^{1, +} Y = Y. \tag{3.1.8}$$

Finally, if Y is compact, then condition (2) *in statement* (ii), *the condition $f \in H^*$ in* (iii) *and* (iv) *and the condition $H^* = \mathscr{C}_0(X)$ in* (3.1.8) *can be dropped.*

Proof. The equivalence (i) $\Leftrightarrow$ (ii) is, in fact, Theorem 3.1.2.

Now, observe that condition (2) in (ii) is equivalent to $f \in H^*$ (see (2.1.7) and Theorem 2.1.6 applied to the null functional); so, on account of Theorems 2.1.10 and 3.1.2 (see (2.1.23) and (2.1.24) too) we also obtain the equivalence (ii) $\Leftrightarrow$ (iii).

(iii) $\Rightarrow$ (iv): Let $\varepsilon > 0$; since $f \in H^*$, there exist h_0 and k_0 in H such that $h_0 - \frac{\varepsilon}{3} \le f \le k_0 + \frac{\varepsilon}{3}$; moreover, there exists a compact set K in Y such that

$T(k_0)(y) - T(h_0)(y) < \frac{\varepsilon}{3}$ for each $y \in Y \backslash K$. Put $\alpha_0 := \frac{\varepsilon}{3}$ and $\beta_0 := \frac{\varepsilon}{3}$; then

$Tk_0(y) + \alpha_0 - (Th_0(y) - \beta_0) \le \varepsilon$ for each $y \in Y \backslash K$.

Now, if $y \in K$, by (iii) there exist h_y, k_y in H and $\alpha_y, \beta_y \ge 0$ such that $h_y - \alpha_y \le f \le k_y + \beta_y$ and $|T(k_y) + \beta_y - T(h_y) + \alpha_y| \le \varepsilon$ in a neighborhood $U(y)$ of y in Y.

Since K is compact, we can find $y_1, \ldots, y_n$ such that $U(y_1), \ldots, U(y_n)$ is a cover of K; hence, arguing as in the proof of the implication (iii) $\Rightarrow$ (iv) of Theorem 3.1.4 and taking the hypothesis $\|T\| \leq 1$ into account, it is easy to show that the functions $h_0, h_{y_1}, \ldots, h_{y_n}, k_0, k_{y_1}, \ldots, k_{y_n}$ and the constants $\alpha_0, \alpha_{y_1}, \ldots, \alpha_{y_n}, \beta_0, \beta_{y_1}, \ldots, \beta_{y_n} \geq 0$ satisfy condition (iv).

As the above proof shows, if Y is compact (i.e., $K = Y$), then it is not necessary to suppose $f \in H^*$.

(iv) $\Rightarrow$ (iii): From (iv) it obviously follows that

$$\inf_{\substack{k \in H, \beta \geq 0 \\ f \leq k + \beta}} (T(k) + \beta) \leq \sup_{\substack{h \in H, \alpha \geq 0 \\ h - \alpha \leq f}} (T(h) - \alpha).$$

On the other hand, if $h, k \in H$ and $\alpha, \beta \geq 0$ satisfy $h - \alpha \leq f \leq k + \beta$, then, by Lemma 2.1.5, there exist u and v in $\mathscr{C}_0(X)$, $u, v \geq 0$, such that $h - u \leq f \leq k + v$ and $\|u\| \leq \alpha$, $\|v\| \leq \beta$. Since T is a positive contraction, we have $T(h) - \alpha \leq T(h) - \alpha\|T\| \leq T(h) - \|T(u)\| \leq T(h) - T(u) \leq T(f)$ and, analogously, $T(f) \leq T(k) + \beta$. So

$$\sup_{\substack{h \in H, \alpha \geq 0 \\ h - \alpha \leq f}} (T(h) - \alpha) = T(f) = \inf_{\substack{k \in H, \beta \geq 0 \\ f \leq k + \beta}} (T(k) + \beta). \qquad \square$$

Finally, we observe that property (2) in condition (ii) can be expressed in various ways, as we have pointed out after Corollary 2.1.7. In Corollary 2.1.8 we have considered the specific case where condition (2) in (ii) is satisfied for every $f \in \mathscr{C}_0(X)$.

Moreover, further statements equivalent to those listed in Theorem 3.1.8 can be easily derived from Theorem 2.1.10.

Notes and references

Korovkin closures and Korovkin subsets with respect to positive linear operators and positive linear contractions were introduced and studied first for the identity operator (we refer to the Notes of Section 4.1 for precise historical references). The generalizations to an arbitrary positive linear operator was suggested by Lorentz [1972b] and were considered by Kutateladze and Rubinov [1972] in the context of ordered vector spaces and by Micchelli [1975], Rusk [1975] and Ferguson and Rusk [1976] in the setting of $\mathscr{C}(X)$-spaces, X compact.

Afterwards, other extensions have been considered in other function spaces (see Sections 2.7 and 5.3 of the Subject Index of Appendix D), in the setting of ordered topological (normed) vector spaces (see Section 6.5 of the Subject Index) or in Banach algebras (see Romanelli [1990]).

Other results deal with particular classes of positive linear operators, such as finitely defined operators, lattice homomorphisms and positive projections; we shall have the opportunity of returning with more details to each of these different directions in the notes of the subsequent sections.

Theorem 3.1.2 has been obtained by Altomare [1987c]. Theorem 3.1.3 has been proved by Watanabe [1979b] and, in the compact case, by Micchelli [1975], Rusk [1975] and Ferguson and Rusk [1976]. For an analogous result for adapted spaces see Grossman [1976a].

The additional properties of $K_+^1(H, T)$ and $K_+(H, T)$ dealing with parametric nets of positive linear operators or contractions (see Remark 1 to Theorem 3.1.2) were first considered in a slightly different form by Nishishiraho [1983].

Theorems 3.1.4, 3.1.6 and 3.1.7 seem to be new and they generalize the corresponding results for the identity operator obtained by several authors.

Results similar to Theorem 3.1.6 were established in the setting of adapted spaces by Grossman [1976a].

We observe that, if X is a metrizable locally compact space which is countable at infinity, then in (3.1.1) we may consider only sequences of positive operators rather than nets, as the proof of Theorem 3.1.2 shows.

If X is not metrizable, the distinction is essential (even if X is compact), since the restriction to sequences of positive operators gives rise to the so-called *sequential Korovkin closure* for a positive operator T with respect to positive linear operators which is in general larger than the K_+-closure for T; the case of K_+^1-closures is analogous. An example which emphasizes the difference between the sequential Korovkin closure and the Korovkin closure can be constructed as follows for the identity operator. Take a locally compact, non-compact Hausdorff space X which is countable at infinity and consider

$$H := \{f \in \mathscr{C}(X) | f \text{ is constant on } \beta X \setminus X\},$$

where βX denotes the Stone-Čech compactification of X. Then, H is a K_+-subspace for the identity operator with respect to equicontinuous sequences but not with respect to equicontinuous nets of positive linear operators (see Kitto and Wulbert [1976]).

We also point out another interesting possibility which consists in considering arbitrary nets (not necessarily equicontinuous) of positive linear operators. This Korovkin closure is strictly contained in $K_+(H, T)$, unless X is compact and $1 \in H$.

For example, if $X =]0, 1]$, $H = \mathscr{L}(\{x, x^2, x^3\})$ and T is the identity operator on $\mathscr{C}_0(]0, 1])$, then $K_+(H, T) = \mathscr{C}_0(]0, 1])$ (see Proposition 4.2.4), while the Korovkin closure of H with respect to nets of positive linear operators is the subspace

$$\left\{ f \in \mathscr{C}_0(X) \middle| \sup_{0 < x \leq 1} \frac{|f(x)|}{x} < +\infty \right\}$$

(see Amir and Ziegler [1978] and, for more general results, Altomare [1980]).

More generally, Donner [1975], [1979] showed that a subspace H of $\mathscr{C}_0(X)$ is a Korovkin subspace for the identity operator with respect to nets of positive linear operators (not necessarily equicontinuous) if and only if it is a K_+-subspace for the identity operator and it is cofinal in $\mathscr{C}_0(X)$.

Again in the compact case, Shashkin [1965] defined a positive linear operator T: $\mathscr{C}(X) \to \mathscr{C}(Y)$ to be *uniquely determined on H among the class of positive linear operators* if the equality $S = T$ on H implies $S = T$ for each positive linear operator $S: \mathscr{C}(X) \to \mathscr{C}(Y)$ (i.e., the restriction of T to the subspace H has a unique extension to a positive linear operator on $\mathscr{C}(X)$, namely T itself). Obviously, if H is a K_+-subset of $\mathscr{C}(X)$ for T, then T must be necessarily uniquely determined on H among the class of positive operators. The converse does not hold in general.

For example, if $X = [0, 2\pi]$, $H = \mathscr{L}(\{\mathbf{1}, \sin, \cos\})$ and T is the identity operator on $\mathscr{C}(X)$, then, by virtue of (3.1.6) and the Example to Proposition 2.5.7,

$$K_+(H, T) = \{f \in \mathscr{C}(X) | f(0) = f(2\pi)\}.$$

Hence, the function $e_1(x) = x \ (x \in [0, 2\pi])$ does not belong to $K_+(H, T)$, while it belongs to the subspace

$$\{f \in \mathscr{C}(X) | S(f) = T(f) \text{ for each positive linear operator } S: \mathscr{C}(X) \to \mathscr{C}(X)$$
$$\text{satisfying } S = T \text{ on } H\},$$

as it was shown by Amir and Ziegler [1978, (1.18)].

However, Rusk [1975], [1977] proved that the converse holds when X is a compact real interval or the unit circle and H is a Chebyshev system of odd order.

Papadopoulou [1979] studied conditions under which H is a K_+-subset of $\mathscr{C}(X)$ for the identity operator, provided the identity operator is uniquely determined on H.

We conclude our references with a brief note on the "pointwise" aspect of Korovkin-type approximation theory.

If we replace the uniform convergence with the pointwise convergence in (3.1.1) and (3.1.2), we obtain the corresponding definition of pointwise K_+-subset and pointwise K_+-closure for a positive linear operator T. Analogously, one can define pointwise K_+^1-subsets and pointwise K_+^1-closures for a positive linear contraction T. These closures can be related to suitable convergence subspaces and they also enjoy interesting properties as showed by Flösser [1980a] and [1981] for the identity operator.

However, we observe that in general the pointwise K_+- (or K_+^1-) closure of H for a positive linear operator T is different from the K_+- (or K_+^1-) closure for T; if T is the identity operator, they coincide in the case where H is finite dimensional (see Flösser [1979]) or in the case where X is compact and $\mathbf{1} \in H$ (see Bauer [1973] and Amir and Ziegler [1978]).

Moreover, also in this case the previous definitions can be included in more general settings (see, e.g., Amir and Ziegler [1978] and L. C. Kurtz [1975]).

3.2 Special properties of Korovkin closures

Besides the properties explained in the previous section, here we shall emphasize further properties of Korovkin closures (and hence of Korovkin subsets) that may fruitfully serve in concrete applications to approximation processes.

First we recall that, for given Banach lattices E and F, a *lattice homomorphism* is a linear operator $S: E \to F$ satisfying $|S(f)| = S(|f|)$ for every $f \in E$.

Equivalently, this means that S preserves the (finite) lattice operations, i.e., for every $f_1, \dots, f_n \in E$

$$S\left(\inf_{1 \le i \le n} f_i\right) = \inf_{1 \le i \le n} S(f_i) \quad \text{and} \quad S\left(\sup_{1 \le i \le n} f_i\right) = \sup_{1 \le i \le n} S(f_i).$$

Clearly every lattice homomorphism is positive and hence is continuous.

In the sequel we shall fix two locally compact Hausdorff spaces X and Y.

Our first result tells us that a K_+-subset for a positive operator $T: \mathscr{C}_0(X) \to \mathscr{C}_0(Y)$ is also a K_+-subset for each operator of the form $S \circ T$, where $S: \mathscr{C}_0(Y) \to E$ is an arbitrary lattice homomorphism.

3.2.1 Theorem. *Let* $T: \mathscr{C}_0(X) \to \mathscr{C}_0(Y)$ *be a positive linear operator and let* H *be a subspace of* $\mathscr{C}_0(X)$. *For a given* $f \in \mathscr{C}_0(X)$ *the following statements are equivalent:*

(i) $f \in K_+(H, T)$.

(ii) *(Universal Korovkin-type property with respect to positive linear operators) If* E *is a Banach lattice,* $S: \mathscr{C}_0(Y) \to E$ *is a lattice homomorphism and if* $(L_i)_{i \in I}^{\le}$ *is a net of positive linear operators from* $\mathscr{C}_0(X)$ *into* E *such that* $\sup_{i \in I} \|L_i\| < +\infty$ *and* $\lim_{i \in I}{}_{\le} L_i(h) = S(T(h))$ *for all* $h \in H$, *then*

$$\lim_{i \in I}{}_{\le} L_i(f) = S(T(f)).$$

Proof. Since (ii) $\Rightarrow$ (i) is obvious, we proceed directly to show that (i) $\Rightarrow$ (ii). To this end, let E, S and $(L_i)_{i \in I}^{\le}$ be as in (ii).

Let $\varepsilon > 0$ and consider $\delta > 0$ such that, for each $\varphi \in \mathscr{C}_0(Y)$,

$$\|\varphi\| \le \delta \Rightarrow \|S(\varphi)\| \le \varepsilon. \tag{1}$$

Analogously, since the net $(L_i)_{i \in I}^{\le}$ is equicontinuous, we can find $\eta > 0$, $\eta \le \delta$, such that, for each $g \in \mathscr{C}_0(X)$,

$$\|g\| \le \eta \Rightarrow \|T(g)\| \le \delta \quad \text{and} \quad \|L_i(g)\| \le \varepsilon \quad (i \in I). \tag{2}$$

Since $f \in K_+(H, T)$, by Theorem 3.1.4, (v), there exist $h_0, \ldots, h_n$ and $k_0, \ldots, k_n$ in H and $u, v \in \mathscr{C}_0(X)$ such that $u, v \geq 0$, $\|u\| \leq \eta$, $\|v\| \leq \eta$ and

$$\sup_{0 \leq \alpha \leq n} h_\alpha - u \leq f \leq \inf_{0 \leq \beta \leq n} k_\beta + v, \tag{3}$$

$$\left\| \inf_{0 \leq \beta \leq n} T(k_\beta) - \sup_{0 \leq \alpha \leq n} T(h_\alpha) \right\| \leq \delta. \tag{4}$$

Consequently,

$$\left\| \inf_{0 \leq \beta \leq n} S(T(k_\beta)) - \sup_{0 \leq \alpha \leq n} S(T(h_\alpha)) \right\| \leq \varepsilon. \tag{5}$$

Moreover, by (3) we obtain, for each $i \in I$,

$$\sup_{0 \leq \alpha \leq n} L_i(h_\alpha) - L_i(u) \leq L_i(f) \leq \inf_{0 \leq \beta \leq n} L_i(k_\beta) + L_i(v) \tag{6}$$

and

$$\sup_{0 \leq \alpha \leq n} S(T(h_\alpha)) - S(T(u)) \leq S(T(f)) \leq \inf_{0 \leq \beta \leq n} S(T(k_\beta)) + S(T(v)). \tag{7}$$

By (6) and (7), it follows that, for each $i \in I$,

$$L_i(f) - S(T(f)) \leq \inf_{0 \leq \beta \leq n} L_i(k_\beta) + L_i(v) - \sup_{0 \leq \alpha \leq n} S(T(h_\alpha)) + S(T(u))$$

$$= \inf_{0 \leq \beta \leq n} L_i(k_\beta) - \inf_{0 \leq \beta \leq n} S(T(k_\beta)) + \inf_{0 \leq \beta \leq n} S(T(k_\beta))$$

$$- \sup_{0 \leq \alpha \leq n} S(T(h_\alpha)) + L_i(v) + S(T(u))$$

$$\leq \sum_{\beta=0}^{n} |L_i(k_\beta) - S(T(k_\beta))| + \left| \inf_{0 \leq \beta \leq n} S(T(k_\beta)) - \sup_{0 \leq \alpha \leq n} S(T(h_\alpha)) \right|$$

$$+ |L_i(v)| + |S(T(u))|.$$

Analogously,

$$S(T(f)) - L_i(f) \leq \sum_{\alpha=0}^{n} |L_i(h_\alpha) - S(T(h_\alpha))| + \left| \inf_{0 \leq \beta \leq n} S(T(k_\beta)) - \sup_{0 \leq \alpha \leq n} S(T(h_\alpha)) \right|$$

$$+ |L_i(u)| + |S(T(v))|.$$

Therefore, for each $i \in I$,

$$|S(T(f)) - L_i(f)| \leq \sum_{\alpha=0}^{n} |L_i(h_\alpha) - S(T(h_\alpha))| + \sum_{\beta=0}^{n} |L_i(k_\beta) - S(T(k_\beta))|$$

$$+ \left| \inf_{0 \leq \beta \leq n} S(T(k_\beta)) - \sup_{0 \leq \alpha \leq n} S(T(h_\alpha)) \right| + |L_i(v)| + |S(T(u))|$$

$$+ |L_i(u)| + |S(T(v))|.$$

By (1), (2) and the preceding inequality, we have

$$\|S(T(f)) - L_i(f)\| \leq \sum_{\alpha=0}^{n} \|L_i(h_\alpha) - S(T(h_\alpha))\| + \sum_{\beta=0}^{n} \|L_i(k_\beta) - S(T(k_\beta))\|$$

$$+ \left\| \inf_{0 \leq \beta \leq n} S(T(k_\beta)) - \sup_{0 \leq \alpha \leq n} S(T(h_\alpha)) \right\| + 4\varepsilon. \tag{8}$$

Since the net $(L_i(h_\alpha))_{i \in I}^{\leq}$ converges to $S(T(h_\alpha))$ and the net $(L_i(k_\beta))_{i \in I}^{\leq}$ converges to $S(T(k_\beta))$ for each $\alpha, \beta = 0, \ldots, n$, there exists $i_0 \in I$, such that, for each $i \in I$, $i \geq i_0$, and $\alpha, \beta = 0, \ldots, n$

$$\|L_i(h_\alpha) - S(T(h_\alpha))\| \leq \frac{\varepsilon}{n+1}, \quad \|L_i(k_\beta) - S(T(k_\beta))\| \leq \frac{\varepsilon}{n+1}.$$

Therefore it follows from (8) that $\|S(T(f)) - L_i(f)\| \leq 7\varepsilon$ for each $i \in I, i \geq i_0$. Since $\varepsilon > 0$ is arbitrary, we conclude that $\lim_{\substack{\leq \\ i \in I}} L_i(f) = S(T(f))$. $\qquad \square$

Remarks. 1. As one can verify directly, in the universal Korovkin-type property (ii) with respect to positive linear operators one may require even more. In fact, property (ii) is equivalent to a similar one obtained by considering families $(L_{i,\lambda})_{i \in I}^{\leq}$, $\lambda \in \Lambda$, of nets of positive linear operators from $\mathscr{C}_0(X)$ in E such that $\sup_{(i,\lambda) \in I \times \Lambda} \|L_{i,\lambda}\| < +\infty$ and by requiring that the convergence toward $S \circ T$ is uniform in $\lambda \in \Lambda$ (see also Remark 1 to Theorem 3.1.2).

2. Thus, according to Corollary 3.1.5 and the universal Korovkin-type property (ii) with respect to positive linear operators, if H is a K_+-subspace for T, then H may be used to check the convergence toward T also for nets of positive linear operators which do not necessarily take their values in $\mathscr{C}_0(Y)$. It is enough that their ranges are contained in a suitable Banach lattice E containing $\mathscr{C}_0(Y)$ and with respect to which the same nets are equicontinuous.

In this case one can apply Theorem 3.2.1 to the natural embedding S: $\mathscr{C}_0(Y) \to E$.

In many concrete cases we shall consider the Banach lattice $\mathscr{B}(Y)$ of all real-valued bounded functions on Y endowed with the sup-norm.

In the case where X is compact, we may add to the universal Korovkin-type property with respect to positive linear operators other ones, such as universal Korovkin-type properties with respect to monotone operators and linear contractions.

We recall that an operator (not necessarily linear) $L: E \to F$ acting between two Banach lattices E and F is called *monotone* if $L(f) \le L(g)$ for each $f, g \in E$ such that $f \le g$.

3.2.2 Theorem. *Suppose X to be compact. Let $T: \mathscr{C}(X) \to \mathscr{C}_0(Y)$ be a positive linear operator and let H be a cofinal subspace of $\mathscr{C}(X)$. Then for every $f \in \mathscr{C}(X)$ the following statements are equivalent:*

(i) $f \in K_+(H, T)$.

(ii) (*Universal Korovkin-type property with respect to positive linear operators*) *If E is a Banach lattice, $S: \mathscr{C}_0(Y) \to E$ is a lattice homomorphism and if $(L_i)_{i \in I}^{\le}$ is a net of positive linear operators from $\mathscr{C}(X)$ into E such that $\sup_{i \in I} \|L_i\| < +\infty$ and $\lim_{i \in I}^{\le} L_i(h) = S(T(h))$ for all $h \in H$, then*

$$\lim_{i \in I}^{\le} L_i(f) = S(T(f)).$$

(iii) (*Universal Korovkin-type property with respect to monotone operators*) *If E is a Banach lattice, $S: \mathscr{C}_0(Y) \to E$ is a lattice homomorphism and if $(L_i)_{i \in I}^{\le}$ is a net of monotone operators from $\mathscr{C}(X)$ into E such that $\lim_{i \in I}^{\le} L_i(h) = S(T(h))$ for all $h \in H$, then*

$$\lim_{i \in I}^{\le} L_i(f) = S(T(f)).$$

Moreover, if $\mathbf{1} \in H$, Y is compact and $T(\mathbf{1}) = \mathbf{1}$, then each one of the preceding statements (i)–(iii) is equivalent to the following:

(iv) (*Korovkin-type property with respect to linear contractions*) *If $(L_i)_{i \in I}^{\le}$ is a net of linear contractions from $\mathscr{C}(X)$ into $\mathscr{C}(Y)$ such that $\lim_{i \in I}^{\le} L_i(h) = T(h)$ for all $h \in H$, then*

$$\lim_{i \in I}^{\le} L_i(f) = T(f).$$

Proof. The universal Korovkin-type property with respect to monotone operators clearly implies the universal Korovkin-type property with respect to positive linear operators. Now we show that (i) $\Rightarrow$ (iii); to do this, we follow the proof (i) $\Rightarrow$ (ii) in Theorem 3.2.1.

Let E be a Banach lattice, $S: \mathscr{C}_0(Y) \to E$ be a lattice homomorphism and $(L_i)_{i \in I}^{\leq}$ be a net of monotone operators from $\mathscr{C}(X)$ into E such that $\lim_{\substack{\leq \\ i \in I}} L_i(h) = S(T(h))$ for each $h \in H$.

Let $\varepsilon > 0$ and consider $\delta > 0$ such that, for each $\varphi \in \mathscr{C}_0(Y)$,

$$\|\varphi\| \leq \delta \Rightarrow \|S(\varphi)\| \leq \varepsilon. \tag{1}$$

By Theorem 3.1.6, (iv), there exist $h_1, \ldots, h_n$ and $k_1, \ldots, k_n$ in H such that

$$\sup_{1 \leq \alpha \leq n} h_\alpha \leq f \leq \inf_{1 \leq \beta \leq n} k_\beta \tag{2}$$

and

$$\left\| \inf_{1 \leq \beta \leq n} T(k_\beta) - \sup_{1 \leq \alpha \leq n} T(h_\alpha) \right\| \leq \delta. \tag{3}$$

By (1) and (3) we obtain

$$\left\| \inf_{1 \leq \beta \leq n} S(T(k_\beta)) - \sup_{1 \leq \alpha \leq n} S(T(h_\alpha)) \right\| \leq \varepsilon. \tag{4}$$

Moreover, by (2) and for each $i \in I$ we have

$$\sup_{1 \leq \alpha \leq n} L_i(h_\alpha) \leq L_i(f) \leq \inf_{1 \leq \beta \leq n} L_i(k_\beta) \tag{5}$$

and

$$\sup_{1 \leq \alpha \leq n} S(T(h_\alpha)) \leq S(T(f)) \leq \inf_{1 \leq \beta \leq n} S(T(k_\beta)). \tag{6}$$

From (5) and (6) it follows that

$$L_i(f) - S(T(f)) \leq \inf_{1 \leq \beta \leq n} L_i(k_\beta) - \sup_{1 \leq \alpha \leq n} S(T(h_\alpha))$$

$$= \inf_{1 \leq \beta \leq n} L_i(k_\beta) - \inf_{1 \leq \beta \leq n} S(T(k_\beta)) + \inf_{1 \leq \beta \leq n} S(T(k_\beta))$$

$$- \sup_{1 \leq \alpha \leq n} S(T(h_\alpha))$$

$$\leq \sum_{\beta=1}^{n} |L_i(k_\beta) - S(T(k_\beta))| + \left| \inf_{1 \leq \beta \leq n} S(T(k_\beta)) - \sup_{1 \leq \alpha \leq n} S(T(h_\alpha)) \right|$$

and analogously

$$S(T(f)) - L_i(f) \le \sum_{\alpha=1}^{n} |L_i(h_\alpha) - S(T(h_\alpha))| + \left| \inf_{1 \le \beta \le n} S(T(k_\beta)) - \sup_{1 \le \alpha \le n} S(T(h_\alpha)) \right|.$$

Therefore, for each $i \in I$,

$$|S(T(f)) - L_i(f)| \le \sum_{\alpha=1}^{n} |L_i(h_\alpha) - S(T(h_\alpha))| + \sum_{\beta=1}^{n} |L_i(k_\beta) - S(T(k_\beta))|$$

$$+ \left| \inf_{1 \le \beta \le n} S(T(k_\beta)) - \sup_{1 \le \alpha \le n} S(T(h_\alpha)) \right|.$$

By the preceding inequality, we have

$$\|S(T(f)) - L_i(f)\| \le \sum_{\alpha=1}^{n} \|L_i(h_\alpha) - S(T(h_\alpha))\| + \sum_{\beta=1}^{n} \|L_i(k_\beta) - S(T(k_\beta))\|$$

$$+ \left\| \inf_{1 \le \beta \le n} S(T(k_\beta)) - \sup_{1 \le \alpha \le n} S(T(h_\alpha)) \right\|. \tag{7}$$

Now, there exists $i_0 \in I$, such that, for each $i \in I$, $i \ge i_0$ and $\alpha, \beta = 1, \ldots, n$

$$\|L_i(h_\alpha) - S(T(h_\alpha))\| \le \frac{\varepsilon}{n}, \quad \|L_i(k_\beta) - S(T(k_\beta))\| \le \frac{\varepsilon}{n},$$

and therefore, by (7) and for each $i \in I$, $i \ge i_0$, $\|S(T(f)) - L_i(f)\| \le 3\varepsilon$.

Since $\varepsilon > 0$ is arbitrary, we conclude that $\lim_{\substack{\le \\ i \in I}} L_i(f) = S(T(f))$.

(i) $\Rightarrow$ (iv): Let $(L_i)_{i \in I}^{\le}$ be a net of linear contractions from $\mathscr{C}(X)$ into $\mathscr{C}(Y)$ such that $\lim_{\substack{\le \\ i \in I}} L_i(h) = T(h)$ for each $h \in H$ and suppose that the net $(L_i(f))_{i \in I}^{\le}$ does not converge to $T(f)$.

Then, there exists $\varepsilon > 0$ and, for each $i \in I$, there exists $\alpha(i) \ge i$ and $y_i \in Y$ such that

$$|L_{\alpha(i)}f(y_i) - Tf(y_i)| \ge \varepsilon. \tag{1}$$

Arguing as in the proof of the implication (ii) $\Rightarrow$ (i) of Theorem 3.1.2, one shows that there exist $\mu \in \mathscr{M}(X)$, $\|\mu\| \le 1$, two filters $\mathscr{F}_1$ and $\mathscr{F}_2$ on I, both finer than the filter of sections on I with $\mathscr{F}_1 \subset \mathscr{F}_2$, and $y \in Y$ such that $(y_i)_{i \in I}^{\mathscr{F}_1}$ converges to an element $y \in Y$ and the net $(L_{\alpha(i)}g(y_i))_{i \in I}^{\mathscr{F}_2}$ converges to $\mu(g)$ for every $g \in \mathscr{C}(X)$.

Consequently $\mu = \mu_y^T$ on H. In particular $\mu(1) = 1$ and hence μ is positive.

From (i) it follows $\mu(f) = \mu_y^T(f)$ which contradicts (1).

(iv) $\Rightarrow$ (i): It suffices to repeat the proof of the implication (i) $\Rightarrow$ (ii) of Theorem 3.1.2, since the operators L_V constructed there are contractions too. $\square$

Remarks. 1. A remark similar to Remark 1 to Theorem 3.2.1 holds also in this case. Properties (ii), (iii) and (iv) of Theorem 3.2.2 may be generalized to similar properties obtained by considering families $(L_{i,\lambda})_{i \in I}^{\leq}$, $\lambda \in \Lambda$, of nets of positive linear operators satisfying $\sup_{(i,\lambda) \in I \times \Lambda} \|L_{i,\lambda}\| < +\infty$, or monotone operators or linear contractions, respectively. In this case the convergence must be understood to be uniform in $\lambda \in \Lambda$.

2. We observe that if the Korovkin-type property with respect to linear contractions holds for each $f \in \mathscr{C}(X)$, then necessarily $\|\mu_{y|H}^T\| = 1$ for each $y \in Y$.

Indeed, if $y \in Y$ and $\|\mu_{y|H}^T\| = \alpha < 1$, we can consider $\beta \in \mathbb{R}$ such that $\alpha < \beta \leq 1$ and $\beta \neq \|\mu_y^T\|$. By applying the Hahn-Banach theorem (see Rusk [1975, Lemma B] for details) we obtain an extension $\mu \in \mathscr{M}(X)$ of $\mu_{y|H}^T$ such that $\|\mu\| = \beta$ and hence $\mu \neq \mu_y^T$. Fix an open neighborhood V of y and consider another open neighborhood U of y such that $\overline{U} \subset V$ and a continuous function $\varphi: X \to [0,1]$ such that $\varphi = 1$ on U and $\varphi = 0$ outside of V. Now, we can consider the linear contraction $L_V: \mathscr{C}(X) \to \mathscr{C}(X)$ defined by $L_V f(x) := \varphi(x)\mu(f) + (1 - \varphi(x))f(x)$ for each $f \in \mathscr{C}(X)$ and $x \in X$. At this point we denote by $\mathfrak{B}$ a base of open neighborhoods at y and consider on $\mathfrak{B}$ the partial ordering defined by $U \leq V$ if $V \subset U$ $(U, V \in \mathfrak{B})$. Then the net $(L_V)_{V \in \mathfrak{B}}^{\leq}$ converges to T on H, but not on $\mathscr{C}(X)$ because $L_V(f)(y) = \mu(f)$ for every $V \in \mathfrak{B}$ and $f \in \mathscr{C}(X)$; therefore the Korovkin-type property with respect to linear contractions does not hold for every $f \in \mathscr{C}(X)$.

As a consequence, we also point out that, if H contains the constant functions, then the hypothesis $T(1) = 1$ in the Korovkin-type property with respect to linear contractions of Theorem 3.2.2 is essential, as noted by Ferguson and Rusk [1976, Theorem 6] with the following counterexample (see also Lorentz [1972b, Theorem 8]).

Example. Consider $X = [0,1]$ and let $T: \mathscr{C}(X) \to \mathscr{C}(X)$ be the linear positive operator defined by $Tf(x) := \dfrac{1+x}{3}f(x)$ for each $f \in \mathscr{C}(X)$ and $x \in X$.

If H is the subspace generated by the functions e_i $(i = 0, 1, 2)$ defined by $e_i(x) := x^i$ $(i = 0, 1, 2$ and $x \in [0, 1])$, then H is a K_+-subset for T.

Indeed, let $\mu \in \mathscr{M}^+(X)$ and $x_0 \in X$ satisfying $\mu(e_i) = Te_i(x_0)$ for each $i = 0, 1, 2$. In particular, $\mu((e_1 - x_0)^2) = 0$ and therefore the support $\text{Supp}(\mu)$ must be contained in $\{x_0\}$. Hence, there exists $\lambda \in \mathbb{R}$ such that $\mu = \lambda\varepsilon_{x_0}$; the condition $\mu(e_0) = \lambda\varepsilon_{x_0}(e_0)$ implies $\lambda = \dfrac{1+x_0}{3}$ and therefore $\mu = \dfrac{1+x_0}{3}\varepsilon_{x_0} = \mu_{x_0}^T$.

By the uniqueness property of Theorem 3.1.6 we have $K_+(H, T) = \mathscr{C}(X)$.

However, the Korovkin-type property with respect to linear contractions does not hold in this case since $\|\mu_0^T\| = \frac{1}{3}$.

As regards K_+^1-closures, a statement analogous to the universal Korovkin-type property with respect to positive linear operators will be stated by considering a particular class of Banach lattices.

We recall that a Banach lattice E is an *abstract M-space* (an *AM-space*, in short) if

$$\|x \vee y\| = \|x\| \vee \|y\| \tag{3.2.1}$$

for each $x, y \in E$, $x \geq 0$, $y \geq 0$.

The class of AM-spaces includes $\mathscr{C}_0(X)$-spaces (X locally compact Hausdorff space) and, hence, $\mathscr{C}(X)$-spaces (X compact Hausdorff space). By a well-known result of Kakutani [1941], every AM-space is order-isomorphic to a closed vector sublattice of a suitable $\mathscr{C}(X)$-space with X compact.

If E is a Banach lattice we shall denote by $V(E)$ the set of all real-valued lattice homomorphisms on E and by $V_1(E)$ the set of all $\delta \in V(E)$ satisfying $\|\delta\| = 1$.

For example, if $E = \mathscr{C}_0(X)$, then

$$V_1(E) = \{\varepsilon_x | x \in X\} \tag{3.2.2}$$

and

$$\overline{V_1(E)} = V_1(E) \cup \{0\}, \tag{3.2.3}$$

where $\overline{V_1(E)}$ is the weak*-closure of $V_1(E)$ in E'. If X is compact, then $\overline{V_1(E)} = V_1(E)$.

We recall that a Banach lattice is an AM-space if and only if $V_1(E) \cup \{0\}$ coincides with the set of extreme points of the positive part of the unit ball in E' (see Goullet de Rugy [1972]).

As a consequence, if E is an AM-space, we have

$$\|x\| = \sup_{\delta \in V_1(E)} |\delta(x)| \tag{3.2.4}$$

for each $x \in E$.

3.2.3 Theorem. *Let* $T: \mathscr{C}_0(X) \to \mathscr{C}_0(Y)$ *be a positive linear contraction and consider a subspace H of $\mathscr{C}_0(X)$. Then for every $f \in \mathscr{C}_0(X)$ the following statements are equivalent:*

(i) $f \in K^1_+(H, T)$.

(ii) (*Universal Korovkin-type property with respect to positive linear contractions*) *If E is an AM-space, if $S: \mathscr{C}_0(Y) \to E$ is a lattice homomorphism such that for each $\delta \in V_1(E)$, $\delta \circ S$ either is 0 or has norm 1 and if $(L_i)^{\leq}_{i \in I}$ is a net of positive linear contractions from $\mathscr{C}_0(X)$ into E such that $\lim_{\leq \atop i \in I} L_i(h) = S(T(h))$ for all $h \in H$, then*

$$\lim_{\substack{\leq \\ i \in I}} L_i(f) = S(T(f)).$$

Proof. We have only to show that (i) implies (ii). So suppose that (i) holds. Let E, S, δ and $(L_i)^{\leq}_{i \in I}$ be as in (ii). If the net $(L_i(f))^{\leq}_{i \in I}$ does not converge to $S(T(f))$, we can find $\varepsilon_0 > 0$ and, for every $i \in I$, an index $\alpha(i) \in I$, $\alpha(i) \geq i$, such that

$$\|L_{\alpha(i)}(f) - S(T(f))\| > \varepsilon_0.$$

By (3.2.4), for every $i \in I$ there exists $\delta_i \in V(E)_1$ such that

$$|\delta_i(L_{\alpha(i)}(f)) - \delta_i(S(T(f)))| > \varepsilon_0. \tag{1}$$

By the Alaoglu-Bourbaki theorem 1.4.1, there exists a filter $\mathfrak{F}_1$ finer than the filter of sections on I and $\mu \in \mathcal{M}^+_b(X)$ such that $(\delta_i \circ L_{\alpha(i)})^{\mathfrak{F}_1}_{i \in I}$ converges vaguely to μ. Analogously, there exists a filter $\mathfrak{F}$ on I finer than $\mathfrak{F}_1$ and $\delta \in \overline{V_1(\mathscr{C}_0(Y))}$ such that $(\delta_i \circ S)^{\mathfrak{F}}_{i \in I}$ converges vaguely to δ.

For each $h \in H$ and $i \in I$ we have

$$|\delta_i(L_{\alpha(i)}(h)) - \delta_i(S(T(h)))| \leq \|L_{\alpha(i)}(h) - S(T(h))\| \tag{2}$$

and therefore $\mu(h) = \delta(T(h))$.

So, if $\delta = 0$, then $\mu(f) = 0 = \delta(T(f))$ by (ii), part (2) of Theorem 3.1.8.

If $\delta \neq 0$, then there exists $y \in Y$ such that $\delta = \varepsilon_y$ (see (3.2.3)) and hence $\delta \circ T = \mu^T_y$. From (ii), part (1) of Theorem 3.1.8 it follows that $\mu(f) = T(f)(y) = \delta(T(f))$. In any case we have $\mu(f) = \delta(T(f))$ which contradicts (1). $\qquad\square$

Remark. Also in this case it is possible to show without difficulties that the above property (ii) holds for every family $(L_{i,\lambda})^{\leq}_{i \in I}$, $\lambda \in \Lambda$, of nets of positive linear contractions from $\mathscr{C}_0(X)$ into E, by requiring that the convergence is uniform in $\lambda \in \Lambda$.

Applying Theorems 3.1.3 and 3.2.1 to the identity operator T on $\mathscr{C}_0(X)$ (in this case $\mu^T_y = \varepsilon_x$ for every $x \in X$), we see that

$$\bigcap_{x \in X} U_+(H, \varepsilon_x) \subset K_+(H, S) \tag{3.2.5}$$

for every lattice homomorphism $S: \mathscr{C}_0(X) \to E$.

In fact, it is possible to show a better result and inclusion (3.2.5) can be considerably improved.

For sake of simplicity we shall consider only the compact case.

Thus, from now on we shall keep fixed a compact Hausdorff space X, a Banach lattice E and a lattice homomorphism $S: \mathscr{C}(X) \to E$. If H is a subset of $\mathscr{C}(X)$ we can still consider the Korovkin closure $K_+(H, S)$ of H for S with respect to positive linear operators (briefly, the K_+-closure of H for S) defined as in (3.1.2) provided the operators under consideration are understood to assume their values in E.

We define the *support* of S as the following subset of X:

$$\mathrm{Supp}(S) := \{x \in X \mid f \in \mathscr{C}(X), S(f) = 0 \Rightarrow f(x) = 0\} \left(= \bigcap_{f \in \mathrm{Ker}(S)} f^{-1}(\{0\}) \right),$$

$$(3.2.6)$$

where $\mathrm{Ker}(S) := \{f \in \mathscr{C}(X) \mid S(f) = 0\}$.

The support of S is a closed subset of X which coincides with X if and only if S is injective. On the other hand, $\mathrm{Supp}(S)$ is empty if and only if S is the null operator.

Note that the intersection in (3.2.6) can be obtained by considering only positive functions $f \in \mathrm{Ker}(S)$ satisfying $\|f\| \leq 1$ (indeed, if $f \in \mathrm{Ker}(S)$, then the positive and negative part f^+ and f^- of f are again in $\mathrm{Ker}(S)$). Other properties of the support of S are given in the following Lemma.

3.2.4 Lemma. *Let $S: \mathscr{C}(X) \to E$ be a non-null lattice homomorphism. Then the following statements hold:*
  (i) *If $f \in \mathscr{C}(X)$ vanishes on $\mathrm{Supp}(S)$, then $S(f) = 0$.*
  (ii) *If $f, g \in \mathscr{C}(X)$ and $f = g$ on $\mathrm{Supp}(S)$, then $S(f) = S(g)$.*
  (iii) *If $f \in \mathscr{C}(X)$ is positive on $\mathrm{Supp}(S)$, then $S(f) \geq 0$.*

Proof. (i): Let $f \in \mathscr{C}(X)$ be such that $f = 0$ on $\mathrm{Supp}(S)$. Fix $\varepsilon > 0$ and consider an open neighborhood V of $\mathrm{Supp}(S)$ such that $|f(x)| \leq \varepsilon$ for each $x \in V$. By the compactness of $X \backslash V$, there exist positive functions $g_1, \ldots, g_n \in \mathrm{Ker}(S)$ such that $\bigcap_{i=1}^n g_i^{-1}(\{0\}) \subset V$. We may also suppose that, for each $i = 1, \ldots, n$, g_i is positive and $\|g_i\| \leq 1$. Now, consider the function $g := \dfrac{1}{n}(g_1 + \cdots + g_n)$. Then $g \in \mathrm{Ker}(S)$, $\|g\| \leq 1$ and $g > 0$ on $X \backslash V$; consequently, there exists a constant $c > 0$ such that $g \geq c$ on $X \backslash V$.

For each $x \in X$, we have $|f(x)| \leq \varepsilon + \dfrac{1}{c} \|f\| \cdot g(x)$ and hence $|S(f)| \leq \varepsilon S(1) + \dfrac{1}{c} \|f\| S(g) = \varepsilon S(1)$.

Since $\varepsilon > 0$ is arbitrary, we have $S(f) = 0$.

(ii): This is a simple consequence of (i).

(iii): Apply property (ii) to f and $g = \sup(f, 0)$.   □

We may now prove the following result which also holds for locally compact spaces (see the final notes).

3.2.5 Theorem. *Let $S: \mathscr{C}(X) \to E$ be a non-null lattice homomorphism and let H be a cofinal subspace of $\mathscr{C}(X)$. Then*

$$\bigcap_{x \in \text{Supp}(S)} U_+(H, \varepsilon_x) \subset K_+(H, S). \tag{3.2.7}$$

Accordingly, if H is a determining subspace for every ε_x ($x \in \text{Supp}(S)$), then H is a K_+-subspace for S.

Proof. Let $f \in \bigcap_{x \in \text{Supp}(S)} U_+(H, \varepsilon_x)$ and let $(L_i)_{i \in I}^{\leq}$ be a net of positive operators from $\mathscr{C}(X)$ into E satisfying $\lim_{i \in I}^{\leq} L_i(h) = S(h)$ for each $h \in H$.

Fix $\varepsilon > 0$; according to Theorem 2.1.3, for every $x \in \text{Supp}(S)$ there exist h_x, $k_x \in H$ such that $h_x \leq f \leq k_x$ and $k_x(x) - h_x(x) < \varepsilon$. Hence, an open neighborhood $U(x)$ of x can be found so that

$$k_x(y) - h_x(y) \leq \varepsilon \quad \text{for each } y \in U(x). \tag{1}$$

Since $\text{Supp}(S)$ is compact, we may find finitely many points $x_1, \ldots, x_n \in \text{Supp}(S)$ such that $\text{Supp}(S) \subset \bigcup_{i=1}^{n} U(x_i)$. If we put $h_i = h_{x_i}$ and $k_i = k_{x_i}$ for every $i = 1, \ldots, n$ and if we set $h = \sup_{1 \leq i \leq n} h_i$ and $k = \inf_{1 \leq i \leq n} k_i$, we obviously have $h \leq f \leq k$ and

$$k(x) - h(x) \leq \varepsilon \quad \text{for each } x \in \text{Supp}(S). \tag{2}$$

Then, for each $i \in I$,

$$L_i(h) \leq L_i(f) \leq L_i(k) \tag{3}$$

and

$$S(h) \leq S(f) \leq S(k). \tag{4}$$

From Lemma 3.2.4 and (2) it follows that

$$S(k) - S(h) \leq \varepsilon \cdot S(\mathbf{1}). \tag{5}$$

Hence, by (3)–(5),

$$L_i(h) - S(h) - \varepsilon S(\mathbf{1}) \leq L_i(h) - S(k) \leq L_i(f) - S(f) \leq L_i(k) - S(h)$$

$$\leq L_i(k) - S(k) + \varepsilon S(\mathbf{1}). \qquad (6)$$

For each $j = 1, \ldots, n$ we have $k \leq k_j$ and $h_j \leq h$ and therefore

$$L_i(k) \leq \inf_{1 \leq j \leq n} L_i(k_j), \quad \sup_{1 \leq j \leq n} L_i(h_j) \leq L_i(h).$$

Now observe that, if $j_0 = 1, \ldots, n$, then

$$L_i(k_{j_0}) - S(k_{j_0}) \leq |L_i(k_{j_0}) - S(k_{j_0})| \leq \sum_{j=1}^{n} |L_i(k_j) - S(k_j)|$$

and therefore $L_i(k_{j_0}) \leq \sum_{j=1}^{n} |L_i(k_j) - S(k_j)| + S(k_{j_0})$.

Since $j_0 = 1, \ldots n$ is arbitrary, we obtain

$$L_i(k) \leq \inf_{1 \leq j \leq n} L_i(k_j) \leq \sum_{j=1}^{n} |L_i(k_j) - S(k_j)| + \inf_{1 \leq j \leq n} S(k_j)$$

$$= \sum_{j=1}^{n} |L_i(k_j) - S(k_j)| + S(k),$$

that is $L_i(k) - S(k) \leq \sum_{j=1}^{n} |L_i(k_j) - S(k_j)|$.

Analogously, we have $-\sum_{j=1}^{n} |L_i(h_j) - S(h_j)| \leq L_i(h) - S(h)$. Then we obtain from (6)

$$L_i(f) - S(f) \leq \sum_{j=1}^{n} |L_i(k_j) - S(k_j)| + \varepsilon S(\mathbf{1}),$$

$$S(f) - L_i(f) \leq \sum_{j=1}^{n} |L_i(h_j) - S(h_j)| + \varepsilon S(\mathbf{1})$$

and finally

$$|L_i(f) - S(f)| \leq \sum_{j=1}^{n} |L_i(k_j) - S(k_j)| + \sum_{j=1}^{n} |L_i(h_j) - S(h_j)| + \varepsilon S(\mathbf{1}).$$

At this point, it is enough to utilize the convergence of the nets $(L_i(h_j))_{i \in I}^{\leq}$ and $(L_i(k_j))_{i \in I}^{\leq}$ for $j = 1, \ldots, n$ and the fact that $\varepsilon > 0$ is arbitrarily chosen.   □

In general, the inclusion in (3.2.7) is strict.

Nevertheless, a characterization of the subspace $K_+(H, S)$ for a fixed lattice homomorphism has been actually found by Donner [1982, Section 8] even in the more general case where X is locally compact (see the last part of the subsequent notes).

After having presented general characterizations and properties of Korovkin closures and Korovkin subspaces for a positive linear operator (or for a positive linear contraction) $T: \mathscr{C}_0(X) \to \mathscr{C}_0(Y)$, we can proceed further by dealing with the following problems which arise in a natural way:

1) Under which conditions do there exist non-trivial (i.e., non dense) K_+-subspaces (or K_+^1-subspaces) for T?
2) Describe them explicitly.
3) When are they finite dimensional?

In the next sections we shall present several answers to these questions in the following three cases:

I) $X (= Y)$ is a compact space, and T is a suitable positive projection.
II) T is a finitely defined operator (in particular a lattice homomorphism).
III) T is the identity operator.

In other cases different from I)–III) these problems seem to be completely open.

Notes and references

The properties of Korovkin closures discussed in this section have been previously investigated only for the identity operator and we refer to the final notes of Section 4.1 for suitable references and comments.

Theorems 3.2.1, 3.2.2 and 3.2.3 seem to be new.

However, the equivalence (i) ⇔ (iv) of Theorem 3.2.2 was established by Ferguson and Rusk [1976], who also presented a counterexample showing that the equivalence does not hold if $T(\mathbf{1}) \neq \mathbf{1}$ (see Remark 2 to Theorem 3.2.2).

For these reasons, some authors, as Lorentz [1972b], Rusk [1975] and Ferguson and Rusk [1976], studied, independently, Korovkin closures for T with respect to nets of linear contractions, which are defined as in (3.1.2) by considering only nets of linear contractions.

For the case of the identity operator these kinds of Korovkin closures have been studied also in the setting of general Banach spaces.

But, up to now, a complete and satisfactory characterization of them is not yet available.

However, for some partial results along this direction we refer to the final subject index in Appendix D.

As regards the last part of this section, which is devoted to Korovkin closures with respect to lattice homomorphisms, we point out that the investigation of this case was first carried out by Berens and Lorentz [1973] and, afterwards, it was continued by other authors, such as Gonska [1975] and Donner [1982]. Lemma 3.2.4 and Theorem 3.2.5 are contained in the paper of Berens and Lorentz [1973, Lemma 1, Lemma 2 and Theorem 1] in the case where X is metrizable and in the dissertation of Gonska [1975, 0.43, 0.44 and Theorem 3.1] in the general case.

Finally, we give more details about the characterization of Korovkin closures obtained by Donner [1982, Section 8].

Let X be locally compact and consider a Banach lattice E whose bidual E'' has the bounded positive approximation property (i.e., there exists a net of linear positive finite rank operators defined on E'' converging strongly to the identity operator). Moreover, let $S: \mathscr{C}_0(X) \to E$ be a lattice homomorphism. A subset K of X is an *essential set with respect to S* (or, an *S-essential set*) if, for every downward directed net $(f_i)_{i \in I}^{\leq}$ of positive functions in $\mathscr{C}(X)$ satisfying $\inf_{i \in I} f_i(x) = 0$ for each $x \in K$, we also have $\lim_{\leq} S(f_i) = 0$.

Moreover, if $f \in \mathscr{C}_0(X)$, the *generalized H-bordering set B_f of f* is defined by $B_f :=$

$$\{x \in X \mid \hat{f}(x) = \check{f}(x)\}, \text{ where } \check{f}(x) := \inf_{\varepsilon > 0} \left(\sup_{\substack{h \in H \\ h \leq f + \varepsilon}} h(x) \right) \text{ and } \hat{f}(x) := \sup_{\varepsilon > 0} \left(\inf_{\substack{h \in H \\ h \geq f - \varepsilon}} h(x) \right) \text{ for}$$

every $x \in X$.

Then he proved that, given a subspace H of $\mathscr{C}_0(X)$, the K_+-closure of H for S consists of those functions $f \in \mathscr{C}_0(X)$ such that the generalized H-bordering set B_f is S-essential.

In particular, if $f \in \bigcap_{x \in \mathrm{Supp}(S)} U_+(H, \varepsilon_x)$, then $\mathrm{Supp}(S) \subset B_f$ by Theorem 2.1.6 (note that $\mathrm{Supp}(S)$ can be defined as in (3.2.6) also if X is locally compact). Since $\mathrm{Supp}(S)$ is always an S-essential set, B_f is S-essential too and hence $f \in K_+(H, S)$.

This reasoning shows that Theorem 3.2.5 also holds when X is locally compact and H is an arbitrary subspace of $\mathscr{C}_0(X)$.

Another important consequence of this result is that, if the Choquet boundary $\partial_H^+ X$ of H in X is an S-essential set, then $K_+(H, S) = \mathscr{C}_0(X)$ (see Donner [1982, Theorem 8.9]).

Different examples of S-essential sets are given in Donner [1975, p. 167–168]. One of the most interesting ones deals with the natural embedding $J_p: \mathscr{C}(X) \to L^p(\mu)$, where μ is a bounded positive Radon measure on X and $1 \leq p < +\infty$. In this case a subset K of X is J_p-essential if and only if $X \setminus K$ has inner measure 0 (see (1.2.32)). Consequently, if $X \setminus \partial_H^+ X$ has inner measure 0, then H is a K_+-subspace for J_p.

When X is compact, this last result was proved by Berens and Lorentz [1973].

3.3 Korovkin subspaces for positive projections

In the present section we shall always suppose that X is a compact Hausdorff space. We shall study in more detail how to construct Korovkin subspaces for a positive linear projection $T\colon \mathscr{C}(X) \to \mathscr{C}(X)$.

We shall also present some applications to Bauer simplices and to potential theory. In Chapter 6 we shall see further applications concerning positive approximation processes associated with a positive projection and their connection with some diffusion equations.

So, let $T\colon \mathscr{C}(X) \to \mathscr{C}(X)$ be a positive linear operator and suppose that T is a *projection* on $\mathscr{C}(X)$, i.e., $T \circ T = T$. We shall denote by H the range of T, i.e.,

$$H := T(\mathscr{C}(X)). \tag{3.3.1}$$

We shall always require that H contains the constant functions $\mathbf{1} \in H$ (i.e., $T(\mathbf{1}) = \mathbf{1}$) and separates the points of X.

First, we note that

$$f \in H \Leftrightarrow T(f) = f \tag{3.3.2}$$

for each $f \in \mathscr{C}(X)$. Moreover,

$$h^2 \le T(h^2) \quad \text{for every } h \in H. \tag{3.3.3}$$

Indeed, let $x \in X$ and consider the positive Radon measure $\mu_x^T \in \mathscr{M}^+(X)$ defined by (3.1.3). Then

$$|h(x)| = |\mu_x^T(h)| \le \int |h| \, d\mu_x^T \le \left(\int |h|^2 \, d\mu_x^T \right)^{1/2} \left(\int \mathbf{1} \, d\mu_x^T \right)^{1/2} = (T(h^2)(x))^{1/2},$$

for every $h \in H$ and hence (3.3.3) follows.

Our first result deals with the determination of the Choquet boundary of the range H of T.

3.3.1 Proposition. *Under the above assumptions one obtains*

$$\partial_H^+ X = \{x \in X \,|\, \mu_x^T = \varepsilon_x\} = \{x \in X \,|\, Tf(x) = f(x) \text{ for every } f \in \mathscr{C}(X)\}. \tag{3.3.4}$$

Proof. Clearly, if $x \in \partial_H^+ X$, then $\mu_x^T = \varepsilon_x$ since $\mu_x^T = \varepsilon_x$ on H by (3.3.2).

To prove the converse, fix $x \in X$ such that $\mu_x^T = \varepsilon_x$ and consider $y \in X$, $y \ne x$.

Then there exists $h \in H$ such that $h(y) \ne h(x)$. Thus the function $k = T((h - h(x))^2) \in H$ is positive, vanishes at x and, by (3.3.3), $0 < (h(y) - h(x))^2 \le k(y)$.

From Theorem 2.5.4 it follows that $x \in \partial_H^+ X$. $\qquad\square$

Remark. As a simple consequence of the above result, we notice that, if F is a closed subset of $\partial_H^+ X$, then every $f \in \mathscr{C}(F)$ can be extended to a function in H.

Indeed, denoting by $\tilde{f} \in \mathscr{C}(X)$ an extension of f to X and setting $h :=$ $T(\tilde{f}) \in H$, we have $h(x) = T(\tilde{f})(x) = \tilde{f}(x) = f(x)$ for every $x \in F$ by virtue of Proposition 3.3.1.

In some particular cases, for example if X is metrizable, we can use the next proposition to improve Proposition 3.3.1.

3.3.2 Proposition. *Suppose that there exists a finite or countable family $(h_n)_{n \in \mathbb{N}}$ in H separating the points of X and such that the series $u := \sum_{n=1}^{\infty} h_n^2$ converges uniformly on X. Then $u \leq T(u)$ and*

$$\partial_H^+ X = \{x \in X \mid Tu(x) = u(x)\}. \tag{3.3.5}$$

Proof. Clearly $u \leq T(u)$ by virtue of (3.3.3). On account of Proposition 3.3.1 we have only to show that, if $x \in X$ and $Tu(x) = u(x)$, then $x \in \partial_H^+ X$. To this end, note that, since $\mu_x^T(u) = \varepsilon_x(u)$ and $\mu_x^T(h_n) = \varepsilon_x(h_n)$ for every $n \in \mathbb{N}$, then $\mu_x^T = \varepsilon_x$ by Proposition 2.5.7, part (2). Hence $x \in \partial_H^+ X$ again by Proposition 3.3.1. □

Remark. If X is metrizable, then it is always possible to construct a sequence $(h_n)_{n \in \mathbb{N}}$ in H separating the points of X and such that the series $\sum_{n=1}^{\infty} h_n^2$ converges uniformly on X. Indeed, it suffices to consider a countable dense family $(\ell_n)_{n \in \mathbb{N}}$ of H, which exists since H is separable, and to put $h_n := \dfrac{\ell_n}{2^n(\|\ell_n\| + 1)}$ for every $n \in \mathbb{N}$.

In the sequel, we shall denote by $H[u]$ the subspace of $\mathscr{C}(X)$ generated by H and u.

If $u \in \mathscr{C}(X)$ satisfies the conditions $u \leq T(u)$ and $\partial_H^+ X = \{x \in X \mid Tu(x) = u(x)\}$, then the elements of $H[u]$ are also called *generalized parabola-like functions*. According to the next Corollary 3.3.4, these functions generalize the so-called parabola-like functions that will be studied in Section 4.3 (see (4.3.18)).

Moreover, the above Remark and Proposition 3.3.2 guarantee the existence of subspaces of parabola-like functions. In the next result we shall see that these functions constitute a K_+-subspace for the projection T. Thus, a posteriori, we obtain that every positive projection on a metrizable compact space, whose range contains the constants and separates the points, always possesses a non trivial K_+-subspace.

3.3.3 Theorem. *Let X be a metrizable compact space and let $T: \mathscr{C}(X) \to \mathscr{C}(X)$ be a positive linear projection such that the range $H = T(\mathscr{C}(X))$ of T contains the* *constants and separates the points. Let us consider* $u \in \mathscr{C}(X)$ *such that*

$$u \leq T(u) \quad \text{and} \quad \partial_H^+ X = \{x \in X \,|\, Tu(x) = u(x)\}.$$

Then $H[u] := \mathscr{L}(H \cup \{u\})$ *is a* K_+*-subspace for* T.
More explicitly, this means that if $(L_i)_{i \in I}^{\leq}$ *is a net of positive linear operators (monotone operators, linear contractions, respectively) on* $\mathscr{C}(X)$ *satisfying*

$$\lim_{\substack{\leq \\ i \in I}} L_i(h) = h (= T(h)) \quad \text{for every } h \in H \tag{3.3.6}$$

and

$$\lim_{\substack{\leq \\ i \in I}} L_i(u) = T(u), \tag{3.3.7}$$

then $\lim_{\substack{\leq \\ i \in I}} L_i(f) = T(f)$ *for every* $f \in \mathscr{C}(X)$.

Proof. Let $x \in X$ and $\mu \in \mathscr{M}^+(X)$ be such that $\mu = \mu_x^T$ on $H[u]$ (see (3.1.3)). In particular,

$$\mu(T(u) - u) = \mu_x^T(T(u) - u) = T^2(u)(x) - Tu(x) = 0;$$

since $T(u) - u$ is positive, the support of μ must be contained in $\{y \in X \,|\, Tu(y) - u(y) = 0\} = \partial_H^+ X$ (see Proposition 3.3.2). Hence, by (3.3.4), for each $f \in \mathscr{C}(X)$ the functions f and $T(f)$ coincide on the support of μ and consequently,

$$\mu(f) = \mu(T(f)) = \mu_x^T(T(f)) = T^2(f)(x) = Tf(x) = \mu_x^T(f).$$

Then $\partial_{H[u], T}^+ X = X$ and therefore, by Corollary 3.1.5, we conclude that $H[u]$ is a K_+-subspace for T. $\qquad\square$

Remarks. 1. If X is metrizable and if we take a sequence $(h_n)_{n \in \mathbb{N}}$ in H as in the Remark following Proposition 3.3.2, the subspace generated by $(h_n)_{n \in \mathbb{N}}$ is dense in H and therefore, by Theorem 3.3.3, the set

$$\mathbf{M} := \{\mathbf{1}\} \cup \{h_n \,|\, n \in \mathbb{N}\} \cup \{u\}$$

is a K_+-subset for T. In particular, $H \cup H^2$ is a K_+-subset for T.
    If K is a metrizable convex compact set and $A(K) \subset H$, then a sequence $(h_n)_{n \in \mathbb{N}}$ satisfying the hypotheses of Proposition 3.3.2 can be chosen in $A(K)$ as well (see Remark to Proposition 3.3.2). So, according to Theorem 3.3.3, we have that $H \cup A(K)^2$ is a K_+-subset for T.

2. On account of Theorems 3.3.3 and 3.1.6, (iii), we also obtain a representation of the projection T, namely

$$T(f) = \sup_{\substack{h \in H, \alpha \in \mathbb{R} \\ h + \alpha u \leq f}} (h + \alpha T(u)) = \inf_{\substack{k \in H, \beta \in \mathbb{R} \\ f \leq k + \beta u}} (k + \beta T(u)) \qquad (3.3.8)$$

for every $f \in \mathscr{C}(X)$.

3. As the proof of Theorem 3.3.3 shows, the inclusion $\mathrm{Supp}(\mu_x^T) \subset \partial_H^+ X$ holds for every $x \in X$.

Now, we examine some important examples of positive projections together with some applications of the preceding theorem.

In this first example we consider a metrizable Bauer simplex K, i.e., a metrizable compact convex subset of a locally convex Hausdorff space such that the set $\partial_e K$ of its extreme points is closed and such that, for every $x \in K$, there exists a unique probability Radon measure μ_x on K with barycenter x, which is supported by $\partial_e K$ (see Section 1.5).

We denote by $A(K)$ the subspace of all continuous affine functions on K (see Section 1.4).

We also recall that a function $u: K \to \mathbb{R}$ defined on a convex compact subset K is said to be strictly convex if $u(\lambda x + (1 - \lambda)y) < \lambda u(x) + (1 - \lambda)u(y)$ for each $x, y \in K$, $x \neq y$, and $\lambda \in \,]0, 1[$.

3.3.4 Corollary (*Positive projections associated with Bauer simplices*). *Let K be a metrizable Bauer simplex and consider the canonical positive projection T: $\mathscr{C}(K) \to \mathscr{C}(K)$ defined by $Tf(x) := \mu_x(f)$ for every $f \in \mathscr{C}(K)$ and $x \in K$.*

If $u \in \mathscr{C}(K)$ is a strictly convex function, then $u \leq T(u)$ and $\partial_e K = \{x \in K \,|\, u(x) = Tu(x)\}$, so that $A(K)[u] := \mathscr{L}(A(K) \cup \{u\})$ is a K_+-subspace for T.

Proof. By virtue of Corollary 1.5.9, T is the unique positive projection with range $A(K)$.

Now let $u \in \mathscr{C}(K)$ be a strictly convex continuous function. By the barycenter formula (1.5.3) for each $x \in K$ we have $u(x) \leq \int u \, d\mu_x = \mu_x(u) = Tu(x)$, and therefore $u \leq T(u)$. Taking (1.5.14) into account we obtain

$$\partial_{A(K)}^+ K = \partial_e K = \left\{ x \in K \,|\, u(x) = \inf_{\substack{a \in A(K) \\ u \leq a}} a(x) \right\}$$

$$= \{x \in K \,|\, u(x) = \mu_x(u)\} = \{x \in K \,|\, u(x) = Tu(x)\}.$$

Then, by Theorem 3.3.3, we can conclude that $A(K)[u]$ is a K_+-subspace for T. $\square$

3.3.5 Example (*Positive projection associated with the canonical simplex of* $\mathbb{R}^p$).
Let us consider the *canonical simplex* of $\mathbb{R}^p$ defined by

$$K_p := \left\{ x = (x_i)_{1 \le i \le p} \in \mathbb{R}^p \,|\, x_i \ge 0, i = 1, \ldots, p, \sum_{i=1}^{p} x_i \le 1 \right\}. \qquad (3.3.9)$$

In this case, the positive projection associated with K_p is the operator T_p:
$\mathscr{C}(K_p) \to \mathscr{C}(K_p)$ defined by

$$T_p(f)(x) := \sum_{\substack{h_1 + \cdots + h_p \le 1 \\ h_i \in \mathbb{N}_0}} \alpha_f(h_1, \ldots, h_p) x_1^{h_1} \ldots x_p^{h_p} \left(1 - \sum_{i=1}^{p} x_i \right)^{1 - h_1 - \cdots - h_p} \qquad (3.3.10)$$

for every $f \in \mathscr{C}(K_p)$ and $x = (x_i)_{1 \le i \le p} \in K_p$, where $\alpha_f(h_1, \ldots, h_p) := f(\delta_{h_1 1}, \ldots, \delta_{h_p 1})$,
$\delta_{h_i 1}$ being the Kronecker symbol.
In this case, $A(K_p)$ is generated by $\{\mathbf{1}, \mathrm{pr}_1, \ldots, \mathrm{pr}_p\}$ (where $\mathrm{pr}_h \in \mathscr{C}(K_p)$ denotes
the h-th projection (see (2.6.19)) and the function $u := \sum_{i=1}^{p} \mathrm{pr}_i^2$ is strictly convex
on K_p. So

$$\left\{ \mathbf{1}, \mathrm{pr}_1, \ldots, \mathrm{pr}_p, \sum_{i=1}^{p} \mathrm{pr}_i^2 \right\} \text{ is a } K_+\text{-subset for } T_p. \qquad (3.3.11)$$

When $p = 1$, then $K_1 = [0, 1]$ and the operator T_p in (3.3.10) becomes

$$T_1(f)(x) := (1 - x)f(0) + xf(1) \qquad (3.3.12)$$

for every $f \in \mathscr{C}([0, 1])$ and $x \in [0, 1]$. Furthermore,

$$\{\mathbf{1}, x, x^2\} \text{ is a } K_+\text{-subset for } T_1. \qquad (3.3.13)$$

A simple application of (3.3.13) to Bernstein polynomials will be presented in
the Example 1 to Theorem 4.2.7.
In Section 2.6 we have already shown some connections between the Choquet
boundary of a bounded open subset of $\mathbb{R}^p$ and its regular points. Here we shall
present an interesting example of a positive projection defined in this framework.
We shall fix a regular bounded open subset Ω of $\mathbb{R}^p$ ($p \ge 2$) (see Section 2.6)
and we consider the subspace $H(\Omega) := \{u \in \mathscr{C}(\overline{\Omega}) \,|\, u \text{ is harmonic on } \Omega\}$. By Prop-
osition 2.6.2 and on account of the regularity of Ω, we obtain the equality

$\partial_{H(\Omega)}^+ \overline{\Omega} = \partial\Omega$. Moreover, if $f \in \mathscr{C}(\overline{\Omega})$, the Dirichlet problem

$$\begin{cases} \Delta v = 0 \text{ on } \Omega, \quad v \in \mathscr{C}(\overline{\Omega}) \cap \mathscr{C}^2(\Omega), \\ v_{|\partial\Omega} = f_{|\partial\Omega}, \end{cases} \tag{3.3.14}$$

has a unique solution $H_{f_{|\partial\Omega}}$ (see (2.6.9) and (2.6.10)) and consequently, we may consider the linear positive operator $T: \mathscr{C}(\overline{\Omega}) \to \mathscr{C}(\overline{\Omega})$ defined by

$$T(f) := H_{f_{|\partial\Omega}} \quad \text{for every } f \in \mathscr{C}(\overline{\Omega}). \tag{3.3.15}$$

The operator T is often called the *Dirichlet operator* associated with Ω.

In some cases we can give an explicit expression of the operator T; for example, if $\Omega = B(x_0, r)$ is the open ball of center x_0 and radius $r > 0$ in $\mathbb{R}^p$, then the *Poisson formula* for the solution of the Dirichlet problem for a ball furnishes the representation

$$Tf(x) = \begin{cases} \dfrac{r^2 - \|x - x_0\|^2}{r\sigma_p} \displaystyle\int_{\partial\Omega} \dfrac{f(z)}{\|z - x\|^p} d\sigma(z), & \text{if } \|x - x_0\| < r, \\[2ex] f(x), & \text{if } \|x - x_0\| = r, \end{cases} \tag{3.3.16}$$

for each $f \in \mathscr{C}(\overline{\Omega})$ and $x \in \overline{\Omega}$, where σ_p denotes the surface area of the unit sphere of $\mathbb{R}^p$ and σ is the surface area on $\partial\Omega$.

3.3.6 Corollary (*Dirichlet operators on regular subsets of* $\mathbb{R}^p$). *Let Ω be a regular bounded open subset of $\mathbb{R}^p$. Then the Dirichlet operator $T: \mathscr{C}(\overline{\Omega}) \to \mathscr{C}(\overline{\Omega})$ is a positive projection with range $H(\Omega)$.*

Moreover, if $u \in \mathscr{C}(\overline{\Omega}) \cap \mathscr{C}^2(\Omega)$ satisfies $\Delta u > 0$ on Ω, then $u \leq T(u)$ and $\partial_{H(\Omega)}^+ \overline{\Omega} = \{x \in \overline{\Omega} | u(x) = Tu(x)\}$.

Therefore, $H(\Omega)[u] := \mathscr{L}(H(\Omega) \cup \{u\})$ is a K_+-subspace for T.

Proof. Since $\Delta(T(u) - u) \leq 0$ on Ω, $T(u) - u$ attains its minimum on $\partial\Omega$ where it vanishes. So $u \leq T(u)$ and $T(u) = u$ exactly on $\partial\Omega = \partial_{H(\Omega)}^+ \overline{\Omega}$.

The last part of the statement follows from Theorem 3.3.3. $\square$

Remarks. 1. According to Theorem 3.3.3, in order that a net of positive linear operators on $\mathscr{C}(\overline{\Omega})$ strongly converges to the Dirichlet operator T, it is sufficient that (3.3.6) and (3.3.7) hold.

However, if we assume that Ω is *strictly convex* (i.e., Ω is convex and $\partial\Omega = \partial_e \overline{\Omega}$) and $L_i(H(\Omega)) \subset H(\Omega)$ for every $i \in I$, then we can replace (3.3.6) by the following weaker condition

$$\lim_{i \in I} \leq L_i(h) = h \quad \text{for each } h \in A(\overline{\Omega}) \tag{3.3.17}$$

(see Leha [1977, Bemerkung 3.8]).

2. In particular, from Corollary 3.3.6 it follows that if $S: \mathscr{C}(\bar{\Omega}) \to \mathscr{C}(\bar{\Omega})$ is a monotone operator satisfying $S(h) = h$ for each $h \in H(\Omega)$ and $S(u) = T(u)$ for some $u \in \mathscr{C}(\bar{\Omega}) \cap \mathscr{C}^2(\Omega)$ with $\Delta u > 0$ on Ω (for example, $u(x) = \|x\|^2$), then S coincides with the Dirichlet operator on $\bar{\Omega}$.

This result can be interpreted as a variant of the well-known *Keldish's theorem* (see Keldish [1941b]) which states that every positive linear operator from $\mathscr{C}(\bar{\Omega})$ into $H(\Omega)$ which coincides with T on $H(\Omega)$, is necessarily equal to T.

For more details on this theorem and its generalizations see Kutateladze and Rubinov [1972] and Netuka [1985]. In the same papers one can also find other kinds of Korovkin-type theorems for the Dirichlet operator.

For the rest of this section we shall be interested in the construction of a positive projection on the space of continuous real-valued functions defined on the product of a finite family of compact Hausdorff spaces, provided that a positive projection on the space of continuous real functions on every factor is assigned.

Let $(X_i)_{1 \le i \le p}$ be a finite family of compact Hausdorff spaces and denote by $X := \prod_{i=1}^{p} X_i$ its product space. We shall denote by $\mathrm{pr}_i: X \to X_i$ the i-th projection from X onto X_i $(1 \le i \le p)$.

If $(f_i)_{1 \le i \le p} \in \prod_{i=1}^{p} \mathscr{C}(X_i)$, we denote by $\bigotimes_{i=1}^{p} f_i: X \to \mathbb{R}$ the function defined by (1.2.50).

Moreover, if for every $i = 1, \ldots, p$ a subspace H_i of $\mathscr{C}(X_i)$ is assigned, we shall consider the subspaces $\bigotimes_{i=1}^{p} H_i$ and $\underset{i=1}{\overset{p}{\mathrm{M}}} H_i$ of $\mathscr{C}(X)$ defined by (2.6.24) and (2.6.25).

If, for each $i = 1, \ldots, p$, $L_i: \mathscr{C}(X_i) \to \mathscr{C}(X_i)$ is a positive linear operator, then $\bigotimes_{i=1}^{p} L_i: \mathscr{C}(X) \to \mathscr{C}(X)$ denotes the unique positive linear operator on $\mathscr{C}(X)$ satisfying

$$\left(\bigotimes_{i=1}^{p} L_i \right) \left(\bigotimes_{i=1}^{p} f_i \right) = \prod_{i=1}^{p} L_i(f_i) \qquad (3.3.18)$$

for each $(f_i)_{1 \le i \le p} \in \prod_{i=1}^{p} \mathscr{C}(X_i)$ (see Section 1.2).

In the next proposition we show that the operator $\bigotimes_{i=1}^{p} L_i$ is a projection if every L_i is a positive linear projection.

3.3.7 Proposition. *For each* $i = 1, \ldots, p$ *let* $T_i: \mathscr{C}(X_i) \to \mathscr{C}(X_i)$ *be a positive linear projection on* $\mathscr{C}(X_i)$ *and denote by* $H_i := T_i(\mathscr{C}(X_i))$ *its range.*

Then the positive linear operator $\bigotimes\limits_{i=1}^{p} T_i \colon \mathscr{C}(X) \to \mathscr{C}(X)$ is itself a projection and its range $H := \left(\bigotimes\limits_{i=1}^{p} T_i \right)(\mathscr{C}(X))$ is given by

$$H = \bigotimes_{i=1}^{p} H_i = \overline{\underset{i=1}{\overset{p}{\mathbf{M}}} H_i}. \tag{3.3.19}$$

Consequently, H contains the constant functions and separates the points provided that every H_i satisfies the same properties.

Proof. If $h = \bigotimes\limits_{i=1}^{p} h_i \in \bigotimes\limits_{i=1}^{p} H_i$, by (3.3.18) we obviously have

$$\left(\bigotimes_{i=1}^{p} T_i \right)(h) = \bigotimes_{i=1}^{p} T_i(h_i) = \bigotimes_{i=1}^{p} h_i = h. \tag{1}$$

Moreover, for each $f = \bigotimes\limits_{i=1}^{p} f_i \in \bigotimes\limits_{i=1}^{p} \mathscr{C}(X_i)$,

$$\left(\bigotimes_{i=1}^{p} T_i \right)(f) = \bigotimes_{i=1}^{p} T_i(f_i) \in \bigotimes_{i=1}^{p} H_i \tag{2}$$

and therefore $\left(\bigotimes\limits_{i=1}^{p} T_i \right)^2 (f) = \left(\bigotimes\limits_{i=1}^{p} T_i \right)(f)$.

Since $\bigotimes\limits_{i=1}^{p} \mathscr{C}(X_i)$ is dense in $\mathscr{C}(X)$, by (1) and (2) $\bigotimes\limits_{i=1}^{p} T_i$ is a projection on $\mathscr{C}(X)$ and its range is $\overline{\bigotimes\limits_{i=1}^{p} H_i}$. On the other hand, since $\overline{\underset{i=1}{\overset{p}{\mathbf{M}}} H_i}$ is closed, clearly $\overline{\bigotimes\limits_{i=1}^{p} H_i} \subset \overline{\underset{i=1}{\overset{p}{\mathbf{M}}} H_i}$.

Conversely, fix $f \in \underset{i=1}{\overset{p}{\mathbf{M}}} H_i$; then, for each $x = (x_i)_{1 \le i \le p} \in X$, since $\mu_x^{\overset{p}{\underset{i=1}{\bigotimes}} T_i} = \bigotimes\limits_{i=1}^{p} \mu_{x_i}^{T_i}$, we have

$$\left(\bigotimes_{i=1}^{p} T_i \right)(f)(x) = \int f \, d\mu_x^{\overset{p}{\underset{i=1}{\bigotimes}} T_i}$$

$$= \int f d \bigotimes_{i=1}^{p} \mu_{x_i}^{T_i}$$

$$= \int \dots \int f(y_1, \dots, y_p) \, d\mu_{x_1}^{T_1}(y_1) \dots d\mu_{x_p}^{T_p}(y_p)$$

$$= f(x_1, \dots, x_p).$$

Hence f belongs to the range of $\bigotimes_{i=1}^{p} T_i$, i.e., $f \in \overline{\bigotimes_{i=1}^{p} H_i}$. $\qquad\qquad\square$

Finally, we can state the following Korovkin type result for the product of a finite family of positive linear projections.

3.3.8 Corollary. *For each $i = 1, \ldots, p$ let $T_i \colon \mathscr{C}(X_i) \to \mathscr{C}(X_i)$ be a positive linear projection on $\mathscr{C}(X_i)$ and suppose that its range $H_i := T_i(\mathscr{C}(X_i))$ contains the constants and separates the points.*

Moreover, for every $i = 1, \ldots, p$ let $u_i \in \mathscr{C}(X_i)$ be such that

$$u_i \leq T_i(u_i) \quad and \quad \partial_{H_i}^{+} X_i = \{x_i \in X_i \,|\, T_i u_i(x_i) = u_i(x_i)\}$$

and consider the function $u := \sum_{i=1}^{p} u_i \circ \mathrm{pr}_i \in \mathscr{C}(X)$.
Setting $H := \mathop{M}\limits_{i=1}^{p} H_i$ then we have

$$u \leq \left(\bigotimes_{i=1}^{p} T_i\right)(u) \quad and \quad \partial_H^{+} X = \left\{x \in X \,\middle|\, \left(\bigotimes_{i=1}^{p} T_i\right)(u)(x) = u(x)\right\}.$$

Furthermore, the subspace $H[u] := \mathscr{L}(H \cup \{u\})$ of $\mathscr{C}(X)$ is a K_{+}-subspace for $\bigotimes_{i=1}^{p} T_i$.

Proof. By virtue of the preceding Proposition 3.3.7, the range of $\bigotimes_{i=1}^{p} T_i$ contains the constants and separates the points.

Moreover, if we put $u_{i,j} := \mathbf{1}$ for $i \neq j$ and $u_{i,i} := u_i$, we obtain by (3.3.18)

$$u = \sum_{j=1}^{p} u_j \circ \mathrm{pr}_j = \sum_{j=1}^{p} \bigotimes_{i=1}^{p} u_{j,i} \leq \sum_{j=1}^{p} \bigotimes_{i=1}^{p} T_i u_{j,i}$$

$$= \sum_{j=1}^{p} \left(\bigotimes_{i=1}^{p} T_i\right)\left(\bigotimes_{i=1}^{p} u_{j,i}\right) = \sum_{j=1}^{p} \left(\bigotimes_{i=1}^{p} T_i\right)(u_j \circ \mathrm{pr}_j) = \left(\bigotimes_{i=1}^{p} T_i\right)(u).$$

By the preceding inequalities, we also deduce that, if $x = (x_i)_{1 \leq i \leq p} \in X$, the equality $\left(\bigotimes_{i=1}^{p} T_i\right)(u)(x) = u(x)$ holds if and only if $T_i u_i(x_i) = u_i(x_i)$ for each $i = 1, \ldots, p$. Hence, by Propositions 2.6.8 and 3.3.7,

$$\partial_H^{+} X = \prod_{i=1}^{p} \partial_{H_i}^{+} X_i = \prod_{i=1}^{p} \{x_i \in X_i \,|\, T_i u_i(x_i) = u_i(x_i)\}$$

$$= \left\{x \in X \,\middle|\, \left(\bigotimes_{i=1}^{p} T_i\right)(u)(x) = u(x)\right\}.$$

The last part of the assertion is a consequence of Theorem 3.3.3. $\qquad\qquad\square$

We proceed further to consider an immediate consequence of Corollary 3.3.8 concerning the product of a finite family of Bauer simplices.

Let $(K_i)_{1 \le i \le p}$ be a finite family of Bauer simplices and, for each $i = 1, \ldots, p$, consider the positive linear projection $T_i \colon \mathscr{C}(K_i) \to \mathscr{C}(K_i)$ defined as in Corollary 3.3.4.

Let $K := \prod_{i=1}^{p} K_i$; in this case the range H of the positive linear projection $\bigotimes_{i=1}^{p} T_i \colon \mathscr{C}(K) \to \mathscr{C}(K)$ is the space $\underset{i=1}{\overset{p}{\mathrm{M}}} A(K_i)$ of all continuous functions $h \in \mathscr{C}(K)$ which are affine with respect to each variable.

By applying Corollaries 3.3.4 and 3.3.8 we obtain the following corollary.

3.3.9 Corollary. *Under the above assumptions, consider, for each $i = 1, \ldots, p$, a strictly convex function $u_i \in \mathscr{C}(K_i)$ and the function $u := \sum_{i=1}^{p} u_i \circ \mathrm{pr}_i \in \mathscr{C}(K)$. Then $\underset{i=1}{\overset{p}{\mathrm{M}}} A(K_i)[u]$ is a K_+-subspace for the positive linear projection $\bigotimes_{i=1}^{p} T_i$.*

3.3.10 Example (*Positive projection associated with the hypercube of $\mathbb{R}^p$*). Let $K := [0, 1]$ be the hypercube of $\mathbb{R}^p$. If for every $i = 1, \ldots, p$ we denote by T_i the positive projection T_1 on $\mathscr{C}([0, 1])$ defined by (3.3.12), then $\bigotimes_{i=1}^{p} T_i$ becomes the positive projection $S_p \colon \mathscr{C}(K) \to \mathscr{C}(K)$ defined by

$$S_p(f)(x) := \sum_{h_1, \ldots, h_p = 0}^{1} \alpha_f(h_1, \ldots, h_p) x_1^{h_1}(1 - x_1)^{1-h_1} \ldots x_p^{h_p}(1 - x_p)^{1-h_p} \quad (3.3.20)$$

for every $f \in \mathscr{C}(K)$ and $x = (x_i)_{1 \le i \le p} \in K$, where $\alpha_f(h_1, \ldots, h_p) = f(\delta_{h_1 1}, \ldots, \delta_{h_p 1})$ (see Example 3.3.5).

In this case, $\underset{i=1}{\overset{p}{\mathrm{M}}} A(K_i)$ is the subspace of $\mathscr{C}(K)$ generated by the set $\{\mathbf{1}\} \cup \left\{ \prod_{i \in J} \mathrm{pr}_i \,\middle|\, J \subset \{1, \ldots, p\} \right\}$ and, hence, taking $u_i(x) := x^2$ ($x \in [0, 1], i = 1, \ldots, p$), we infer from Corollary 3.3.9 that

$$\{\mathbf{1}\} \cup \left\{ \prod_{i \in J} \mathrm{pr}_i \,\middle|\, J \subset \{1, \ldots, p\} \right\} \cup \left\{ \sum_{i=1}^{p} \mathrm{pr}_i^2 \right\} \quad \text{is a } K_+\text{-subset for } S_p. \quad (3.3.21)$$

Notes and references

Many of the results of this section as well as the main Theorem 3.3.3 have been established by Altomare [1980b], [1989a].

For other results concerning the existence of parabola-like functions and some generalizations of Theorem 3.3.3 see Rasa [1981], [1985].

We also quote the following result of Nishishiraho [1976], [1987] that may be useful to approximate positive projections in terms of suitable nets of positive linear operators:

Let $T: \mathscr{C}(X) \to \mathscr{C}(X)$ be a positive linear projection whose range contains the constants and separates the points, and let M be a subset of $\mathscr{C}(X)$ which separates the points of X. Fix $p \in \mathbb{R}$, $p > 0$ and for every $h \in M$ consider the continuous function on X

$$\varphi_{T,h,p}(x) := T(|h - h(x) \cdot \mathbf{1}|^p)(x) \quad (x \in X).$$

Then, if $(L_i)_{i \in I}^{\leq}$ is a net of positive linear operators on $\mathscr{C}(X)$ satisfying $L_i \circ T = T$ $(i \in I)$ and if

$$\lim_{i \in I} {}_{\leq} L_i(\varphi_{T,h,p}) = 0 \quad \text{uniformly on } X \tag{$*$}$$

for each $h \in M$, then $\lim_{i \in I} {}_{\leq} L_i(f) = T(f)$ for each $f \in \mathscr{C}(X)$.

Note also that, if p is an even positive integer, then $(*)$ is certainly true if for every $h \in M$ we suppose $T(h^r) = h^r$ $(r = 1, \ldots, p - 1)$ and $\lim_{i \in I} {}_{\leq} L_i(h^p) = T(h^p)$ uniformly on X.

Proposition 3.3.1 was obtained by Rogalski [1968] for a more general class of positive operators that are called Lion's operators (see also Altomare [1977]). Corollary 3.3.4 has been obtained by Micchelli [1975] with different methods (see also Felbecker [1972]).

Corollary 3.3.6 is due to Altomare [1980]. Other kinds of Korovkin-type theorems for the Dirichlet operator were obtained by Kutateladze and Rubinov [1972], Brosowski [1979b], [1981] and Netuka [1985]. For some Korovkin-type results for the solution operator of the Dirichlet problem for partial differential equations other than the Laplace equation, we refer to Flösser and Roth [1979].

Furthermore, in the above quoted papers Brosowski applies his results to the finite difference method for the solution of the Dirichlet problem.

3.4 Korovkin subspaces for finitely defined operators

In the first section of this chapter we pointed out the strong connection between Korovkin subspaces for a given positive linear operator T and determining subspaces for suitable Radon measures which are associated to T. This justifies the detailed study of the convergence of nets of positive Radon measures that we carried out in Chapter 2. In this section we shall now study the convergence of nets of positive linear operators (or positive linear contractions) toward an operator which corresponds in a natural way to the discrete Radon measures that we have considered in Sections 2.2–2.4.

First, we give the definition of the operators that will be at the center of our investigations.

3.4.1 Definition. *Let X and Y be locally compact Hausdorff spaces and let $T: \mathscr{C}_0(X) \to \mathscr{C}_0(Y)$ be a positive linear operator.*

We shall say that T is finitely defined of order n if there exist n mappings $\varphi_1, \ldots, \varphi_n: Y \to X$ and n positive real-valued functions $\psi_1, \ldots, \psi_n: Y \to \mathbb{R}$ such that

$$T(f) = \sum_{i=1}^{n} \psi_i \cdot (f \circ \varphi_i) \quad \text{for each } f \in \mathscr{C}_0(X). \tag{3.4.1}$$

Obviously, it is assumed that the second member in (3.4.1) belongs to $\mathscr{C}_0(Y)$ for each $f \in \mathscr{C}_0(X)$. Of course, this automatically holds in the case where Y is compact and the functions $\varphi_1, \ldots, \varphi_n, \psi_1, \ldots, \psi_n$ are continuous or, if Y is non compact, if the functions $\psi_1, \ldots, \psi_n$ are in $\mathscr{C}_0(Y)$ and $\varphi_1, \ldots, \varphi_n$ are continuous. Another case occurs if the functions $\psi_1, \ldots, \psi_n$ are continuous and bounded and the mappings $\varphi_1, \ldots, \varphi_n$ are continuous proper mappings (see Section 1.2).

In the sequel we shall denote by $\mathscr{F}_n(X, Y)$ the set of all finitely defined operators of order n from $\mathscr{C}_0(X)$ into $\mathscr{C}_0(Y)$ and by $\mathscr{F}_n^1(X, Y)$ the set of all finitely defined operators of order n from $\mathscr{C}_0(X)$ in $\mathscr{C}_0(Y)$ which admit a representation (3.4.1) with $\sum_{i=1}^{n} \psi_i = 1$.

Observe that $\mathscr{F}_n(X, Y) \subset \mathscr{F}_{n+1}(X, Y)$ and $\mathscr{F}_n^1(X, Y) \subset \mathscr{F}_{n+1}^1(X, Y)$ for each $n \in \mathbb{N}$.

Moreover, a finitely defined operator $T: \mathscr{C}_0(X) \to \mathscr{C}_0(Y)$ of order 1 is of the form $T(f) = \psi \cdot (f \circ \varphi)$ for each $f \in \mathscr{C}_0(X)$, where $\varphi: Y \to X$ is a suitable mapping and $\psi: Y \to \mathbb{R}$ is a positive real function; therefore T is finitely defined of order 1 if and only if it is a lattice homomorphism from $\mathscr{C}_0(X)$ into $\mathscr{C}_0(Y)$ (see, e.g., Wolff [1974] and Goullet de Rugy [1972]). By (3.4.1) it also follows that, for each $n > 1$, $\mathscr{F}_n(X, Y)$ contains the subspace of all operators from $\mathscr{C}_0(X)$ into $\mathscr{C}_0(Y)$ which are the sum of n lattice homomorphisms from $\mathscr{C}_0(X)$ into $\mathscr{C}_0(Y)$.

In this connection, note that by virtue of a result of Bernau, Huijsmans and De Pagter [1992], if Y is compact and *extremally disconnected* (i.e., the closure of each open subset of Y is open), then a positive linear operator $T: \mathscr{C}_0(X) \to \mathscr{C}(Y)$ is a sum of n lattices homomorphisms from $\mathscr{C}_0(X)$ into $\mathscr{C}(Y)$ if and only if for every $f_0, f_1, \ldots, f_n \in \mathscr{C}_0(X)$ such that $\inf(|f_i|, |f_j|) = 0$ for all $i, j = 0, \ldots, n$, $i \neq j$, one has $\inf_{0 \leq i \leq n} |T(f_i)| = 0$.

As we have already sketched above, finitely defined operators are strictly related to discrete positive Radon measures; in fact, if $T: \mathscr{C}_0(X) \to \mathscr{C}_0(Y)$ is a finitely defined operator of order n with a representation (3.4.1), we have

$$\mu_y^T(f) = Tf(y) = \sum_{i=1}^n \psi_i(y) \cdot f(\varphi_i(y)) = \sum_{i=1}^n \psi_i(y) \cdot \varepsilon_{\varphi_i(y)}(f) \quad \text{for each } y \in Y \text{ and } f \in$$

$\mathscr{C}_0(X)$, and consequently the measure

$$\mu_y^T = \sum_{i=1}^n \psi_i(y) \cdot \varepsilon_{\varphi_i(y)} \tag{3.4.2}$$

is a discrete positive Radon measure for each $y \in Y$. Thus, by using the same notations as in Section 2.2, we can write

$$\mu_y^T \in C_+(\varphi_1(y), \ldots, \varphi_n(y)) \tag{3.4.3}$$

for each $y \in Y$.

Moreover, if $T \in \mathscr{F}_n^1(X, Y)$, then μ_y^T is also contractive and

$$\mu_y^T \in \tilde{C}_+^1(\varphi_1(y), \ldots, \varphi_n(y)), \tag{3.4.4}$$

for each $y \in Y$.

The previous relations allow us to state the following proposition, which points out the importance of finitely defined operators in Korovkin approximation theory. In fact, they are the only positive linear operators which can possess a finite dimensional Korovkin subspace.

3.4.2 Proposition. *Let* $T: \mathscr{C}_0(X) \to \mathscr{C}_0(Y)$ *be a positive linear operator and consider an n-dimensional subspace H of $\mathscr{C}_0(X)$. Suppose that H separates the points of X and, if X is not compact, that for every $x \in X$ there exists $h \in H$ such that $h(x) \neq 0$.*

Then there exist $n + 1$ mappings $\varphi_1, \ldots, \varphi_{n+1}: Y \to X$ and $n + 1$ positive real functions $\psi_1, \ldots, \psi_{n+1}: Y \to \mathbb{R}$ such that

$$T(h) = \sum_{i=1}^{n+1} \psi_i \cdot (h \circ \varphi_i) \quad \text{for each } h \in H. \tag{3.4.5}$$

Consequently, if H is a K_+-subspace for T (or a K_+^1-subspace, if X is compact and $\|T\| \leq 1$), then T is finitely defined of order $n + 1$.

Finally, if X and Y are both compact and $T(\mathbf{1}) = \mathbf{1}$ and if H is a K_+^1-subspace for T, then $T \in \mathscr{F}_{n+1}^1(X, Y)$.

Proof. For each $y \in Y$, we may apply Proposition 2.2.1 to the bounded positive Radon measure μ_y^T and so there exist $\varphi_1(y), \ldots, \varphi_p(y) \in X$ and a family $\psi_1(y), \ldots,$ $\psi_p(y)$ in $\mathbb{R}_+$, $p \leq n + 1$, such that $\mu_y^T = \sum\limits_{i=1}^{p} \psi_i(y)\varepsilon_{\varphi_i(y)}$ on H. Furthermore $\|\mu_y^T\| = \sum\limits_{i=1}^{n+1} \psi_i(y)$ if X is compact (see Remark 1 to Proposition 2.2.1). We can always suppose $p = n + 1$ by adding, if necessary, $(n + 1) - p$ arbitrary points of X and by defining the corresponding coefficients $\psi_i(y)$ equal to zero. In this way we obtain $n + 1$ mappings $\varphi_1, \ldots, \varphi_{n+1} \colon Y \to X$ and $n + 1$ positive real functions $\psi_1, \ldots, \psi_{n+1} \colon Y \to \mathbb{R}$ such that, for each $y \in Y$,

$$\mu_y^T = \sum_{i=1}^{n+1} \psi_i(y)\varepsilon_{\varphi_i(y)} \quad \text{on } H \tag{1}$$

and

$$\|\mu_y^T\| = \sum_{i=1}^{n+1} \psi_i(y), \quad \text{if } X \text{ is compact.} \tag{2}$$

By (3.1.3) this yields $T(h) = \sum\limits_{i=1}^{n+1} \psi_i \cdot (h \circ \varphi_i)$ for each $h \in H$.

If H is a K_+-subspace for T, then, by Theorem 3.1.3, H is a D_+-subspace for each $\mu_y^T (y \in Y)$; then, equality (1) holds on $\mathscr{C}_0(X)$.

Analogously, if X is compact and $\|T\| \leq 1$ and if H is a K_+^1-subspace for T, then H is a D_+^1-subspace for μ_y^T (see Theorem 3.1.2) and hence, on account of (2), equality (1) holds on $\mathscr{C}(X)$.

In both cases, by (3.1.3) and (3.4.2) we obtain $T \in \mathscr{F}_{n+1}(X, Y)$.

Finally, the last part of the proposition easily follows by equality (2), because in this case $\|\mu_y^T\| = \mu_y^T(\mathbf{1}) = T(\mathbf{1})(y) = 1$ for every $y \in Y$. $\quad\square$

Remark. According to Remark 2 to Proposition 2.2.1, if the one-point compactification X_ω of X is connected (or if X is compact and connected, respectively), then for the representation formula (3.4.5) n functions ψ_i and φ_i are enough.

Now, we turn our attention to the study of K_+- and K_+^1-closures with respect to finitely defined operators.

First we shall deal with some characterizations of the K_+-closure for a *fixed* finitely defined operator $T = \sum\limits_{i=1}^{n} \psi_i(f \circ \varphi_i)$. In this framework, by virtue of

(3.4.2), we can use all the results in Sections 2.2–2.4. For the sake of brevity, we state only the following consequences of Theorems 2.2.2 and 2.4.1 and we leave to the reader the formulation of other analogous applications.

In the sequel, we consider a fixed locally compact Hausdorff space X.

3.4.3 Theorem. *Let Y be a locally compact Hausdorff space and let $T: \mathscr{C}_0(X) \to \mathscr{C}_0(Y)$ be a finitely defined operator of order n with a representation (3.4.1).*

If a subspace H of $\mathscr{C}_0(X)$ satisfies the following conditions for every $y \in Y$:

(1) *For every $\varepsilon > 0$ and for every compact subset K of X satisfying $K \cap \{\varphi_1(y), \ldots, \varphi_n(y)\} = \varnothing$, there exist $k \in H$ and $u \in \mathscr{C}_0^+(X)$ such that $\|u\| < \varepsilon$, $0 \leq k + u$, $1 \leq k + u$ on K and $k(\varphi_i(y)) + u(\varphi_i(y)) < \varepsilon$ for every $i = 1, \ldots, n$.*

(2) *There exist $h_1, \ldots, h_{n(y)} \in H$ such that $\det(h_i(\varphi_{j_k}(y))) \neq 0$, where $\{\varphi_{j_1}(y), \ldots, \varphi_{j_{n(y)}}(y)\}$ is the set of distinct points of $\{\varphi_1(y), \ldots, \varphi_n(y)\}$,*

then H is a K_+-subspace for T.

3.4.4 Theorem. *Let Y be a locally compact Hausdorff space and let $T: \mathscr{C}_0(X) \to \mathscr{C}_0(Y)$ be a finitely defined operator of order n with a representation (3.4.1).*

Let M be a subset of $\mathscr{C}_0(X)$ and suppose that, for each $y \in Y$, there exists $h \in M$ satisfying $h(\varphi_i(y)) \neq 0$ for every $i = 1, \ldots, n$ and, if $n > 1$, $h(\varphi_i(y)) \neq h(\varphi_j(y))$ for every $i, j = 1, \ldots, n$ such that $\varphi_i(y) \neq \varphi_j(y)$.

Moreover, suppose that $f_0 \in \mathscr{C}_0(X)$ is a strictly positive function which is constant on $M(\varphi_i(y))$ for each $i = 1, \ldots, n$ and for each $y \in Y$ (see (2.4.5)).

Then $\{f \in \mathscr{C}_0(X) | f$ is constant on each $M(\varphi_i(y))$, $i = 1, \ldots, n$, $y \in Y\} \subset K_+(Q_n(f_0, M), T)$ (see (2.4.2) and (2.4.5)) and hence $\mathbf{A}(M) \subset K_+(Q_n(f_0, M), T)$, where $\mathbf{A}(M)$ denotes the closed subalgebra generated by M.

Furthermore, if for every $y \in Y$, for every $i = 1, \ldots, n$ and for every $x \in X$ such that $x \neq \varphi_i(y)$ there exists $h \in M$ satisfying $h(x) \neq h(\varphi_i(y))$, then $Q_n(f_0, M)$ is a K_+-subset for T.

On account of equality (3.4.2) and formula (3.4.4), we can also obtain the following partial result concerning the K_+^1-closure for a fixed finitely defined operator. The proof is a direct consequence of Theorems 3.1.2 and 2.2.4.

3.4.5 Theorem. *Let Y be a locally compact Hausdorff space and let $T \in \mathscr{F}_n^1(X, Y)$ be a finitely defined operator of order n with a representation (3.4.1) satisfying $\sum_{i=1}^{n} \psi_i = 1$. Let H be a subspace of $\mathscr{C}_0(X)$ satisfying the condition:*

For every $y \in Y$ there exist $h_1, \ldots, h_{n(y)} \in H$ such that $\det(h_i(\varphi_{j_k}(y))) \neq 0$, where $n(y)$ is defined as in Theorem 3.4.3.

Then the following statements are equivalent:

(i) *H is a K_+^1-subspace for T.*

(ii) (1) *For every $y \in Y$, for every compact subset K of X satisfying $y \notin \bigcup_{i=1}^{n} \varphi_i^{-1}(K)$ and for every $\varepsilon > 0$, there exist $h \in H$ and $u \in \mathscr{C}_0^+(X)$ such that $0 \leq h + u$ on X, $1 \leq h + u$ on K, $\|u\| < \varepsilon + \sum_{i=1}^{n} \psi_i(y) \cdot u(\varphi_i(y))$ and $h(\varphi_i(y)) + u(\varphi_i(y)) < \varepsilon$ for every $i = 1, \ldots, n$.*

(2) *If $\mu \in \mathscr{M}_b^+(X)$ and $\mu = 0$ on H, then $\mu = 0$, i.e., $H^* = \mathscr{C}_0(X)$.*

Furthermore, if Y is compact, then part (2) of condition (ii) can be dropped.

Remark. If X is discrete, we can give the following equivalent formulation of part (1) of (ii) taking the remark following Theorem 2.2.4 into account.

(1)' *For every $y \in Y$, for every $x \in X$ satisfying $y \notin \bigcup_{i=1}^{n} \varphi_i^{-1}(\{x\})$ and for every $\varepsilon > 0$, there exist $h \in H$ and $u \in \mathscr{C}_0^+(X)$ such that $0 \leq h + u$ on X, $1 \leq h(x) + u(x)$, $\|u\| < \varepsilon + \sum_{i=1}^{n} \psi_i(y) \cdot u(\varphi_i(y))$ and $h(\varphi_i(y)) + u(\varphi_i(y)) < \varepsilon$ for every $i = 1, \ldots, n$.*

Until now, we have dealt with a fixed finitely defined operator. A different and important aspect consists in fixing a positive integer n and in studying K_+- and K_+^1-closures for every finitely defined operator of order n.

3.4.6 Definition. *Let X be a locally compact Hausdorff space. A subset H of $\mathscr{C}_0(X)$ is called a K_+-subset (a K_+^1-subset, respectively) of order n in $\mathscr{C}_0(X)$, if it is a K_+-subset for T (a K_+^1-subset for T, respectively) for every locally compact Hausdorff space Y and for every finitely defined operator $T \in \mathscr{F}_n(X, Y)$ $(T \in \mathscr{F}_n^1(X, Y)$, respectively).*

If, in addition, H is a subspace of $\mathscr{C}_0(X)$, we shall call it a K_+-subspace (a K_+^1-subspace, respectively) of order n in $\mathscr{C}_0(X)$.

Clearly, every K_+-subspace of order n in $\mathscr{C}_0(X)$ is a K_+-subspace of order p for every $1 \leq p \leq n$.

In the sequel, we shall always suppose that X possesses more than n points.

3.4.7 Theorem. *Let H be a subspace of $\mathscr{C}_0(X)$. Then the following statements are equivalent:*

(i) *H is a K_+-subspace of order n in $\mathscr{C}_0(X)$.*

(ii) *For every choice of different points $x_1, \ldots, x_n \in X$ and for every $\mu \in C_+(x_1, \ldots, x_n)$ the subspace H is a D_+-subspace for μ.*

(iii) *For every set of distinct points $x_1, \ldots, x_n \in X$, for every compact subset K of X satisfying $K \cap \{x_1, \ldots, x_n\} = \emptyset$ and for every $\varepsilon > 0$ there exist $h \in H$ and $u \in \mathscr{C}_0^+(X)$ such that $\|u\| < \varepsilon$, $0 \leq h + u$ on X, $1 \leq h + u$ on K and $h(x_i) + u(x_i) < \varepsilon$ for every $i = 1, \ldots, n$.*

Proof. (i) $\Rightarrow$ (ii): Let $x_1, \ldots, x_n \in X$ and fix $\mu \in C_+(x_1, \ldots, x_n)$. Then $\mu = \sum_{i=1}^{n} \alpha_i \varepsilon_{x_i}$ with $\alpha_1, \ldots, \alpha_n \in \mathbb{R}_+$. Set $Y := \{1\}$ and, for each $i = 1, \ldots, n$, consider the constant mappings $\varphi_i: Y \to X$ and $\psi_i: Y \to \mathbb{R}$ of constant values x_i and α_i respectively. Then the operator $T: \mathscr{C}_0(X) \to \mathscr{C}(Y)$ defined by $T(f) := \sum_{i=1}^{n} \psi_i(f \circ \varphi_i)$ for each $f \in \mathscr{C}_0(X)$ is finitely defined of order n and consequently, by virtue of (i) and Theorem 3.1.3, H is a D_+-subspace for μ_y^T for each $y \in Y$. But $\mu_y^T = \mu$ and, so, (ii) follows.

(ii) $\Rightarrow$ (iii): It follows directly by Theorem 2.2.2.

(iii) $\Rightarrow$ (ii): Taking Theorem 2.2.2 into account, it is only necessary to show that for every distinct points $x_1, \ldots, x_n$ of X there exist $h_1, \ldots, h_n \in H$ such that $\det(h_i(x_j)) \neq 0$. To this end we shall use Proposition 2.2.5. In fact, given $m = 1, \ldots, n$ and $\varepsilon > 0$, by applying (iii) to the points $x_1, \ldots, x_{m-1}, \bar{x}, x_{m+1}, \ldots, x_n$ and $K = \{x_m\}$ (where $\bar{x}$ is an arbitrary point of X different from $x_1, \ldots, x_n$), we see that there exist $h \in H$ and $u \in \mathscr{C}_0^+(X)$ which exactly satisfy conditions (2.2.2) and hence the proof is complete.

(ii) $\Rightarrow$ (i): This is an immediate consequence of Theorem 3.1.3 and (3.4.3). $\square$

Remark. In the case where X is discrete, the remark following Theorem 2.2.2 furnishes the following different formulation of (iii) of Theorem 3.4.7.

(iii)' *For every set of distinct points $x_1, \ldots, x_n \in X$, for every $x \in X$ different from $x_1, \ldots, x_n$ and for every $\varepsilon > 0$ there exist $h \in H$ and $u \in \mathscr{C}_0^+(X)$ such that $\|u\| < \varepsilon$, $0 \leq h + u$ on X, $1 \leq h(x) + u(x)$ and $h(x_i) + u(x_i) < \varepsilon$ for every $i = 1, \ldots, n$.*

Now we state some consequences of the preceding theorem which are easier to handle in concrete examples.

3.4.8 Corollary. *Let H be a subspace of $\mathscr{C}_0(X)$ satisfying the following condition:*

$$\text{For every choice of different points } x_1, \ldots, x_n \in X$$
$$\text{and for every } x \in X \setminus \{x_1, \ldots, x_n\} \text{ there exists } h \in H, h \geq 0,$$
$$\text{such that } h(x_1) = \cdots = h(x_n) = 0 \text{ and } h(x) > 0. \tag{3.4.6}$$

Then H is a K_+-subspace of order n in $\mathscr{C}_0(X)$.

Proof. As we showed in the proof of Theorem 2.2.6, our hypotheses imply that condition (1) of Theorem 2.2.2 is satisfied for every $x_1, \ldots, x_n \in X$, i.e., condition (iii) of Theorem 3.4.7 holds, and so the result follows. $\square$

Examples. **1.** Consider a compact subset X of $\mathbb{R}^p$. For every $i = 1, \ldots, p$ denote by $\mathrm{pr}_i \in \mathscr{C}(X)$ the i-th projection from X to $\mathbb{R}$. Given $n \in \mathbb{N}$ let us denote by H the linear subspace of $\mathscr{C}(X)$ generated by the functions

$$u^{2r} \cdot \mathrm{pr}_1^{h_{1,r}} \cdots \mathrm{pr}_p^{h_{p,r}},$$

where u is the function $u(x) = \|x\|$ $(x \in X)$, $0 \le r \le n$ and $h_{1,r}, \ldots, h_{p,r} \in \mathbb{N}_0$, $h_{1,r} + \cdots + h_{p,r} \le n - r$.

According to Riordan [1963, pp. 14–15], we have that

$$\dim H = \binom{p+n}{n} + \binom{p+n-1}{n-1}.$$

Furthermore, if $x_1, \ldots, x_n \in X$, then the function

$$h(x) := \prod_{i=1}^{n} \|x - x_i\|^2 \quad (x \in X),$$

belongs to H, is positive and vanishes only on $x_1, \ldots, x_n$.

Consequently, according to Corollary 3.4.8, H is a K_+-subspace of order n in $\mathscr{C}(X)$.

A similar reasoning shows that if $X = \mathbb{R}^p$ and if one considers the above mentioned functions defined on the whole $\mathbb{R}^p$, then the subspace $f_0 \cdot H := \{ f_0 \cdot h \,|\, h \in H \}$ is a K_+-subspace of order n in $\mathscr{C}_0(\mathbb{R}^p)$, where f_0 denotes the function $f_0(x) := \exp(-\|x\|^2)$ $(x \in \mathbb{R}^p)$.

2. (i) Let $X = \,]0,1]$ and let H be the subspace of $\mathscr{C}_0(X)$ generated by the functions $x, x^2, \ldots, x^{2n+1}$.

If $x_1, \ldots, x_n, \bar{x} \in X$ and $\bar{x}$ is different from $x_1, \ldots, x_n$, there exists a polynomial P of degree less than or equal to n such that $P(\bar{x}) = 1$ and $P(x_1) = \cdots = P(x_n) = 0$; then the function $h(x) := x \cdot P(x)^2$ $(x \in \,]0,1])$ satisfies the hypothesis of Corollary 3.4.8. So, H is a K_+-subspace of order n in $\mathscr{C}_0(]0,1])$.

(ii) A similar example may be given by considering $X = [1, +\infty[$ and the subspace H of $\mathscr{C}_0(X)$ generated by the functions $x^{-1}, x^{-2}, \ldots, x^{-2n-1}$.

In this case, if $x_1, \ldots, x_n, \bar{x} \in X$, if $\bar{x}$ is different from $x_1, \ldots, x_n$ and if P is a polynomial of degree less than or equal to n such that $P(1/\bar{x}) = 1$ and $P(1/x_1) = \cdots = P(1/x_n) = 0$, then the function $h(x) := P(1/x)^2/x (x \in [1, +\infty[)$ satisfies again the hypothesis of Corollary 3.4.8.

(iii) Finally, we may also consider $X = [0, +\infty[$ and the subspace H of $\mathscr{C}_0(X)$ generated by the functions $\exp(-\alpha x), \exp(-2\alpha x), \ldots, \exp(-(2n + 1)\alpha x)$, where α is a fixed positive real number.

If $x_1, \ldots, x_n \in X$ and $\bar{x} \in X$ is different from $x_1, \ldots, x_n$, we consider now a polynomial P of degree less than or equal to n such that $P(\exp(-\alpha \bar{x})) = 1$

and $P(\exp(-\alpha x_1)) = \cdots = P(\exp(-\alpha x_n)) = 0$; the function $h \in H$ satisfying the hypothesis of Corollary 3.4.8 is given by $h(x) := \exp(-\alpha x) \cdot P(\exp(-\alpha x))^2$ $(x \in [0, +\infty[)$.

We explicitly consider also the case where X is compact since the more specific conditions that we can state in this case will turn out to be very useful in the subsequent applications. We also refer the reader to the subsequent Theorem 4.5.11 where we present a characterization of those spaces X for which there exists a finite dimensional K_+-subspace of order n in $\mathscr{C}(X)$.

3.4.9 Theorem. *Suppose that X is compact and let H be a subspace of $\mathscr{C}(X)$. Then the following statements are equivalent*:
 (i) *H is a K_+-subspace of order n in $\mathscr{C}(X)$.*
 (ii) *For every set of distinct points $x_1, \ldots, x_n \in X$, for every compact subset K of X satisfying $K \cap \{x_1,\ldots,x_n\} = \varnothing$ and for every $\varepsilon > 0$ there exists $h \in H$ such that $0 \leq h$, $1 \leq h$ on K and $h(x_i) < \varepsilon$ for every $i = 1, \ldots, n$.*
 (iii) *For every set of distinct points $x_1, \ldots, x_n \in X$, for every finite family of open neighborhoods $U_1, \ldots, U_n$ of $x_1, \ldots, x_n$ respectively and for every $\varepsilon > 0$ there exists $h \in H$ such that $0 \leq h$, $1 \leq h$ on $X \setminus \bigcup_{i=1}^{n} U_i$ and $h(x_i) < \varepsilon$ for every $i = 1, \ldots, n$.*
Furthermore, if $1 \in H$, then one can add the further equivalent statement:
 (iv) (1) *H separates the points of X and $\Phi(X) = \partial_e \overline{\mathrm{co}}(\Phi(X))$.*
   (2) *For every set of distinct points $x_1, \ldots, x_n \in X$, $\Phi(x_1), \ldots, \Phi(x_n)$ are affinely independent in H' and $\mathrm{co}(\Phi(x_1),\ldots,\Phi(x_n))$ is a face of $\overline{\mathrm{co}}(\Phi(X))$, where $\Phi: X \to H'$ denotes the canonical embedding (1.4.19).*

Proof. The proof of the equivalences (i) $\Leftrightarrow$ (ii) $\Leftrightarrow$ (iii) follows the same line of the proof of Theorem 3.4.7 by using Theorem 2.2.3 instead of Theorem 2.2.2.

Suppose now that $1 \in H$. If H is a K_+-subspace of order n in $\mathscr{C}(X)$, then H is a K_+-subspace of order 1 and hence H is a D_+-subspace for each ε_x, $x \in X$. This means that $\partial_H^+ X = X$, or, equivalently, on account of Corollary 2.6.5, $\Phi(X) = \partial_e \overline{\mathrm{co}}(\Phi(X))$.

The second part of statement (iv) follows from Theorem 3.4.7 and Proposition 2.6.6.

Conversely, if (iv) holds, then, by virtue of Proposition 2.6.6, H is a D_+-subspace for every $\mu \in C_+(x_1,\ldots,x_n)$ and for every set of distinct points $x_1, \ldots, x_n \in X$. Consequently, H is a K_+-subspace of order n in $\mathscr{C}(X)$ by Theorem 3.4.7. $\square$

Remark. According to Remark 3 to Corollary 2.6.5, we can give an interesting geometric interpretation of condition (3.4.6) in Corollary 3.4.8.

To see this, assume that X is compact and let H be a subspace of $\mathscr{C}(X)$ which contains the constants and satisfies condition (3.4.6) (and hence separates the points of X).

Consider again the canonical embedding $\Phi\colon X \to H'$ defined by (1.4.19).

Then condition (3.4.6) means that for every set of distinct points $x_1, \ldots, x_n \in X$, $\mathrm{co}(\Phi(x_1), \ldots, \Phi(x_n))$ is a relatively exposed face of $\overline{\mathrm{co}}(\Phi(X))$.

A slightly more restrictive condition on H is also of interest. Suppose, indeed, that

> *for every choice of different points $x_1, \ldots, x_n \in X$ there exists*
> *$h \in H, h \geq 0$, satisfying $h(x_1) = \cdots = h(x_n) = 0$ and $h(x) > 0$*
> *whenever $x \in X \setminus \{x_1, \ldots, x_n\}$.*   (3.4.7)

From a geometric point of view, condition (3.4.7) means that for every $x_1, \ldots, x_n \in X$, $\mathrm{co}(\Phi(x_1), \ldots, \Phi(x_n))$ is an exposed face of $\overline{\mathrm{co}}(\Phi(X))$ (see Remark 3 to Corollary 2.6.5).

It is obvious that condition (3.4.7) implies (3.4.6).

Moreover, if X is first countable (i.e., there exists a countable base for the neighborhood system of each point) and H is closed, condition (3.4.6) is equivalent to (3.4.7). This can be proved by using a general result concerning exposed faces (see Remark 3 to Corollary 2.6.5). Here we present a direct proof. Indeed, let $x_1, \ldots, x_n$ be distinct points in X and, for each $i = 1, \ldots, n$, consider a decreasing base $(V_m(x_i))_{m \in \mathbb{N}}$ of open neighborhoods of x_i. By (3.4.6) and the compactness of $X \setminus \bigcup_{i=1}^{n} V_m(x_i)$, we can find for each $m \in \mathbb{N}$ a function $h_m \in H, h_m \geq 0$ such that $h_m(x_1) = \cdots = h_m(x_n) = 0$ and $h_m > 0$ on $X \setminus \bigcup_{i=1}^{n} V_m(x_i)$. Then the function $h := \sum_{m=1}^{\infty} \frac{1}{2^m \|h_m\|} h_m$ belongs to H and satisfies (3.4.7).

As a consequence of Theorem 3.4.7 we can state the following corollary, where we investigate the relations between K_+-subspaces of order n and Chebyshev subspaces (see Section 2.3).

3.4.10 Corollary. *Let X be a real interval $[a, b]$ or the unit circle $\mathbb{T}$. If H is a subspace of $\mathscr{C}(X)$ of dimension $n + 1$, then the following statements hold:*

(1) *If n is even, H is a Chebyshev subspace of order $n + 1$ in $\mathscr{C}(X)$ if and only if H is a K_+-subspace of order $\dfrac{n}{2}$ in $\mathscr{C}(X)$.*

(2) *If n is odd and H is a Chebyshev subspace of order $n + 1$ in $\mathscr{C}(X)$, then H is a K_+-subspace of order $\dfrac{n-1}{2}$ in $\mathscr{C}(X)$.*

Proof. Put $p_n := \dfrac{n}{2}$ if n is even and $p_n := \dfrac{n-1}{2}$ if n is odd. Then for every choice $x_1, \ldots, x_{p_n}$ of different points of X we always have $\sum\limits_{i=1}^{p_n} \omega(x_i) \le n$, and hence, by Theorem 2.3.5, H is a D_+-subspace for every $\mu \in C_+(x_1, \ldots, x_{p_n})$. Consequently, by Theorem 3.4.7, H is a K_+-subspace of order p_n in $\mathscr{C}(X)$.

To conclude the proof, we only have to show that if n is even and if H is a K_+-subspace of order $n/2$ in $\mathscr{C}(X)$, then H is a Chebyshev subspace of order $n + 1$ in $\mathscr{C}(X)$.

Suppose that H is algebraically generated by the linearly independent functions $h_0, \ldots, h_n$. If H is not a Chebyshev subspace of order n in $\mathscr{C}(X)$, there exist $n + 1$ distinct points $x_0, \ldots, x_n \in X$ and $n + 1$ real numbers $\alpha_0, \ldots, \alpha_n$ not simultaneously equal to zero such that $\sum\limits_{i=0}^{n} \alpha_i h_j(x_i) = 0$ for every $j = 0, \ldots, n$. Let $J = \{i = 0, \ldots, n \mid \alpha_i > 0\}$; we may assume $\operatorname{card}(J) \le n/2$, otherwise we multiply every α_i by a factor -1. Consider the positive Radon measures $\mu := \sum\limits_{i \in J} \alpha_i \varepsilon_{x_i}$ (with the convention $\mu = 0$ if $J = \varnothing$) and $\upsilon := -\sum\limits_{i \notin J} \alpha_i \varepsilon_{x_i}$. Then $\mu = \upsilon$ on H and therefore, by Theorem 3.4.7, (ii), $\mu = \upsilon$. This contradicts the fact that the points $x_0, \ldots, x_n$ are distinct. $\qquad\square$

Remark. Note that, in general, the converse of part (2) of the above theorem does not hold. For example, consider the interval $[0, 1]$ and the two convex functions

$$u_2(x) := \begin{cases} \exp(x) - 1, & \text{if } 0 \le x < \tfrac{1}{2}, \\ (1 + \exp(\tfrac{1}{2}))(x - \tfrac{1}{2}) + \exp(\tfrac{1}{2}) - 1, & \text{if } \tfrac{1}{2} \le x \le 1, \end{cases}$$

and

$$u_3(x) := \begin{cases} 0, & \text{if } 0 \le x < \tfrac{1}{2}, \\ \exp(x) - \exp(\tfrac{1}{2}), & \text{if } \tfrac{1}{2} \le x \le 1. \end{cases}$$

Furthermore, denote by e_1 the function $e_1(x) = x$ ($x \in [0, 1]$).

Then, by using the subsequent Theorem 4.3.7, it is easy to show that the subspace H generated by $\{1, e_1, u_2, u_3\}$ is a K_+-subspace of order 1 (see also the Remark to Corollary 4.1.3); on the other hand, H is not a Chebyshev subspace of order 4 since, if one takes four distinct points x_1, x_2, x_3, x_4 in $[0, \tfrac{1}{2}]$, then the determinant

$$\begin{vmatrix} 1 & x_1 & u_2(x_1) & u_3(x_1) \\ 1 & x_2 & u_2(x_2) & u_3(x_2) \\ 1 & x_3 & u_2(x_3) & u_3(x_3) \\ 1 & x_4 & u_2(x_4) & u_3(x_4) \end{vmatrix}$$

is zero.

From the preceding corollary we can draw various examples of K_+-subspaces of order n, for instance, from the list of the Chebyshev subspaces indicated at the end of Section 2.3.

In the case of positive linear contractions, we do not have a result analogous to Theorem 3.4.7. However, the next theorem, which is very similar to Corollary 3.4.8, is enough in concrete examples to establish whether a subspace is a K_+^1-subspace of order n.

3.4.11 Theorem. *Let H be a subspace of $\mathscr{C}_0(X)$ satisfying the following conditions:*

(1) *For every set of distinct points $x_1, \ldots, x_n \in X$ there exist $h_1, \ldots, h_n \in H$, such that $\det(h_i(x_j)) \neq 0$.*

(2) *For every set of distinct points $x_1, \ldots, x_n \in X$ and for every $x \in X \backslash \{x_1, \ldots, x_n\}$ there exists $h \in H + \mathbb{R}_+$, $h \geq 0$ such that $h(x) > 0$ and $h(x_1) = \cdots = h(x_n) = 0$.*

(3) *For every $x \in X$ there exists $h \in H$, $h \geq 0$, such that $h(x) > 0$.*

Then H is a K_+^1-subspace of order n in $\mathscr{C}_0(X)$. Furthermore, if X is compact, condition (3) can be omitted.

Proof. Let Y be a locally compact Hausdorff space and $T \in \mathscr{F}_n^1(X, Y)$. Conditions (1) and (2) ensure that the hypotheses of Theorem 2.2.6 are satisfied and therefore, by (3.4.4), H is a D_+^1-subspace for μ_y^T for every $y \in Y$.

Now, let K be a compact subset of X; by using hypothesis (3) it is easy to find a positive function $h \in H$ such that $1 \leq h$ on K. Then condition (iv) of Corollary 2.1.8 is satisfied for every $\varepsilon > 0$ with $u = 0$. Accordingly, by virtue of (i) of Corollary 2.1.8, H is a D_+-subspace for the null functional.

Thus we have shown that both part (1) and (2) of statement (ii) in Theorem 3.1.2 hold for every $f \in \mathscr{C}_0(X)$ and this, by virtue of the same theorem, completes the proof.

The final part of our statement follows from the last part of Theorem 3.1.2. $\qquad \square$

Another useful result may be stated as follows.

3.4.12 Theorem. *Let M be a subset of $\mathscr{C}_0(X)$ which separates the points of X and such that, for each choice of distinct points $x_1, \ldots, x_n \in X$, there exists $h \in M$ satisfying $h(x_j) \neq 0$ for every $i = 1, \ldots, n$ and, if $n > 1$, $h(x_i) \neq h(x_j)$ for every $i, j = 1, \ldots, n, i \neq j$.*

Then the following statements hold:

(1) *If $f_0 \in \mathscr{L}(M)$ is a strictly positive function, $Q_n(f_0, M)$ is a K_+-subset of order n in $\mathscr{C}_0(X)$.*

(2) *$P_n^*(M)$ is a K_+^1-subset of order n in $\mathscr{C}_0(X)$.*

(3) *If X is compact, $P_n(M)$ is a K_+-subset of order n in $\mathscr{C}_0(X)$.*

Here $Q_n(f_0, M)$, $P_n^*(M)$ and $P_n(M)$ are the subsets defined by (2.4.2), (2.4.3) and (2.4.4), respectively.

Proof. The result follows directly from Theorems 2.4.1, 2.4.2 and 2.4.3. Note that in statement (2), denoting by H the subspace generated by $P_n^*(M)$, we have $H^* = \mathscr{C}_0(X)$ because for every $x \in X$, choosing $h \in M$ such that $h(x) \neq 0$, we have $h^2 \in P_n^*(M) \subset H$ and $h^2(x) > 0$. □

Observe that, if $n > 1$, then statement (3) in Theorem 3.4.12 holds under the weaker assumption that, for each $x_1, \ldots, x_n \in X$, there exists $h \in M$ such that $h(x_n) \neq h(x_i)$ for every $i = 1, \ldots, n - 1$.

3.4.13 Corollary. *Let M be a subset of $\mathscr{C}_0(X)$ satisfying the following conditions:*
(1) *For every set of distinct points $x_1, \ldots, x_n \in X$ and for every $x \in X \backslash \{x_1, \ldots, x_n\}$ there exists $h \in M$ such that $h(x) \neq h(x_i)$ for every $i = 1, \ldots, n$.*
(2) *For every set of distinct points $x_1, \ldots, x_n \in X$ there exists $h \in M$ such that $h(x_i) \neq 0$ for each $i = 1, \ldots, n$ and, if $n > 1$, $h(x_i) \neq h(x_j)$ for each $i, j = 1, \ldots, n$ such that $x_i \neq x_j$.*
If $f_0 \in \mathscr{C}_0(X)$ is a strictly positive function, then the set $\{f_0\} \cup f_0 \cdot M \cup f_0 \cdot M^2 \cup \cdots \cup f_0 \cdot M^{2n}$ is a K_+-subset of order n in $\mathscr{C}_0(X)$ and the set $M \cup M^2 \cup \cdots \cup M^{2n}$ is a K_+^1-subset of order n in $\mathscr{C}_0(X)$.
If X is compact and M only satisfies condition (1), then the subspace H generated by $\{\mathbf{1}\} \cup M \cup M^2 \cup \cdots \cup M^{2n}$ is a K_+-subspace of order n in $\mathscr{C}(X)$.

Proof. If M satisfies conditions (1) and (2), then (2.4.7) and (2.4.8) hold; therefore, by virtue of Proposition 2.4.4, for every set of distinct points $x_1, \ldots, x_n \in X$ the set $\{f_0\} \cup f_0 M \cup \cdots \cup f_0 M^{2n}$ is a D_+-subset for every $\mu \in C_+(x_1, \ldots, x_n)$ and the set $M \cup \cdots \cup M^{2n}$ is a D_+^1-subset for every $\mu \in \tilde{C}_+^1(x_1, \ldots, x_n)$. Furthermore, if H denotes the subspace generated by $M \cup \cdots \cup M^{2n}$, by hypothesis (2), we have $H^* = \mathscr{C}_0(X)$. By virtue of (3.4.3) and (3.4.4) and Theorems 3.1.2 and 3.1.3, the proof of the first part is complete.

Now suppose that X is compact and that M satisfies condition (1). Let $x_1, \ldots, x_n, x \in X$ be distinct points and consider $f \in M$ such that $f(x) \neq f(x_i)$ for every $i = 1, \ldots, n$. Moreover, let P be a polynomial on $\mathbb{R}$ with degree less than or equal to n such that $P(f(x)) = 1$ and $P(f(x_i)) = 0$ for each $i = 1, \ldots, n$. Then the function $h: X \to \mathbb{R}$ defined by $h(y) := P(f(y))^2$ for each $y \in X$ belongs to H and satisfies $h \geq 0$, $h(x) = 1$ and $h(x_i) = 0$ for every $i = 1, \ldots, n$. Hence we can apply Corollary 3.4.8 to obtain the desired result. □

Examples. 1. If X is a compact subset of $\mathbb{R}^2$, then the subset $M := \{pr_1, pr_2, pr_1 + pr_2\}$ satisfies condition (1) of Corollary 3.4.13 for $n = 2$ but not for any $n \geq 3$.

2. Consider $X = \mathbb{R}$ and let $f_1, f_2, f_3 \in \mathscr{C}_0(\mathbb{R})$ such that

f_1 is strictly increasing in $]-\infty, 0]$ and strictly decreasing in $[0, +\infty[$;
f_2 vanishes in $]-\infty, 0]$ and is strictly positive in $]0, +\infty[$;
f_3 is strictly positive in $]-\infty, 0[$ and vanishes in $[0, +\infty[$.

Then, the set $M := \{f_1, f_2, f_3, f_1 + f_2, f_1 + f_3\}$ satisfies hypotheses (1) and (2) of the preceding Corollary for $n = 2$, but not for any $n \geq 3$ (to see, for instance, that (2) is not satisfied, it is sufficient to consider four points x_1, x_2, x_3, x_4 such that $x_1 < 0 < x_2, x_3 < 0 < x_4$ and $f_1(x_3) + f_3(x_3) = f_1(x_1) = f_1(x_2) = f_1(x_4) + f_2(x_4)$).

3. If $f \in \mathscr{C}_0(X)$ is an injective function which does not vanish at any point of X, the preceding Corollary 3.4.13 can be easily applied to the subset $M := \{f\}$ and $f_0 = f$. Thus $\{f, f^2, \ldots, f^{2n+1}\}$ is a K_+-subset of order n in $\mathscr{C}_0(X)$ and $\{f, f^2, \ldots, f^{2n}\}$ is a K_+^1-subset of order n in $\mathscr{C}_0(X)$.

If X is compact, we also obtain that $\{1, f, f^2, \ldots, f^{2n}\}$ is a K_+-subset of order n in $\mathscr{C}(X)$.

Notes and references

Finitely defined operators have been introduced and studied by Shashkin [1965b]. They were considered also by Cavaretta [1973], Micchelli [1973a], Rusk [1975], [1977], Ferguson and Rusk [1976] and Rasa [1981] in the case where the space X is compact and $Y = X$. Several results have been successively extended by Altomare [1987c] to more general function spaces which include the spaces of continuous functions on locally compact spaces dealt with in this section.

As regards the definition of finitely defined operator, we do not require that the representing functions $\varphi_1, \ldots, \varphi_n: Y \to X$ and $\psi_1, \ldots, \psi_n: Y \to \mathbb{R}$ are continuous as in the paper of Altomare; this allow us to apply our results to a larger class of operators and makes available the proof of Proposition 3.4.2 given by Ferguson and Rusk [1976]. A simple example of a (contractive) finitely defined operator which admits no continuous representation can be given by considering $X =]0, 1]$, $Y = [0, 1]$ and $T: \mathscr{C}_0(X) \to \mathscr{C}(Y)$ defined by $T(f) := \psi \cdot (f \circ \varphi)$ for each $f \in \mathscr{C}_0(X)$, where $\psi: Y \to \mathbb{R}$ takes the value $|\sin 1/y|$ at $y \neq 0$ and $\psi(0) = 0$, while $\varphi: Y \to X$ takes the value y at $y \neq 0$ and an arbitrary element of X at $y = 0$.

The connection between finitely defined operators and the existence of finite dimensional Korovkin subspaces has been investigated by Shashkin [1965b] and Cavaretta [1973]; Proposition 3.4.2 has been established in the compact case by Ferguson and Rusk [1976, Corollaries 2.3 and 3.3] (see also Rusk [1975]), although it already appears in Shashkin [1965b] in a slightly different form.

Theorem 3.4.7 is due to Altomare [1987c]. The equivalences (i) ⇔ (ii) ⇔ (iii) of Theorem 3.4.9 are due to Rasa [1981]. The equivalence (i) ⇔ (iv) of the same theorem is due to Shashkin [1965b] in the finite dimensional case and to the authors in the general case. For further investigations concerning those subsets of $\mathbb{R}^n$ which verify a property similar to property (iv), (2), of Theorem 3.4.9 we refer to Shashkin [1974] (see also the final notes of Section 4.5).

Corollary 3.4.8 was first obtained by Shashkin [1965b] and Cavaretta [1973] when X is metrizable and compact; here we have followed the approach of Altomare [1987c, Theorem 2.3].

Some particular examples of subspaces of $\mathscr{C}(X)$ which satisfy condition (iii) in Theorem 3.4.9 can be obtained by considering the subspaces generated by a cofinal cone H in $\mathscr{C}(X)$ which *supremally generates* $\mathscr{C}(X)$ *of order* n, i.e., which satisfies the following condition:

for every $f \in \mathscr{C}(X)$, for every $\varepsilon > 0$ and for every $x_1, \ldots, x_n \in X$ there exists $h \in H$ such that $f \leq h$ and $h(x_i) < f(x_i) + \varepsilon$ for every $i = 1, \ldots, n$.

Supremally generating cones have been studied by Drozhzhin [1989], who has shown that, for a cofinal cone in $\mathscr{C}(X)$, the preceding condition is equivalent to the following property:

For every $x_1, \ldots, x_n \in X$, for every finite family of open pairwise disjoint neighborhoods $U_1, \ldots, U_n$ of $x_1, \ldots, x_n$, respectively, and for every $\varepsilon > 0$ and $\alpha_1, \ldots, \alpha_n \geq 0$ there exists $h \in H$ such that $0 \leq h$ on $X \backslash \bigcup_{i=1}^{n} U_i$, and, for every $i = 1, \ldots, n$, $h \geq -\alpha_i$ on U_i and $h(x_i) < \varepsilon - \alpha_i$.

In the case where $1 \in H$, it is easy to prove that the preceding condition implies the validity of (iii) in Theorem 3.4.9 and, accordingly, that the subspace generated by H is a K_+-subspace of order n in $\mathscr{C}(X)$. A more detailed study of supremally generating cones and their connection with K_+-subspaces of order n can be found in the papers of Kutateladze and Rubinov [1972] and Drozhzhin [1989].

The relations between Chebyshev subspaces and K_+-subspaces of $\mathscr{C}(X)$ have been investigated by Shashkin [1965b], Cavaretta [1973], Micchelli [1973a], Berens and Lorentz [1975], Rusk [1977] and Rasa [1981]. Micchelli [1973a] and Shashkin [1965b] established the equivalence in Corollary 3.4.10, part (1), for a compact real interval and for the unit circle, respectively (see also Rusk [1977] and Rasa [1981]). The counterexample exhibited in the Remark to Corollary 3.4.10 is taken from Rasa [1991].

Theorem 3.4.11 is due to Cavaretta [1973, Theorem 2] in the case X compact and to Altomare [1987c, Theorem 2.4] in the case where X is countable at infinity; also Corollary 3.4.13 and the subsequent Examples can be found in the paper of Altomare [1987c, Theorem 2.2, Corollary 2.5 and Example 2.7.2] (see also Jiménes Poso [1975]).

For other results concerning finitely defined operators we refer to Sections 1.7 and 2.6 of the Subject Classifications of Appendix D.

Here we briefly discuss a generalization of finitely defined operators in locally compact spaces, namely the countably defined operators introduced and studied by Campiti [1987a, Section 2]. A positive linear operator $T: \mathscr{C}_0(X) \to \mathscr{C}_0(Y)$ is called *countably defined* if there exist a sequence $(\varphi_n)_{n \in \mathbb{N}}$ of mappings from Y in X and a sequence $(\psi_n)_{n \in \mathbb{N}}$ of positive real functions such that the series $\sum_{n=1}^{\infty} \psi_n(y)$ is pointwise convergent on Y,

$K \cap \{\varphi_n(y)|n \in \mathbb{N}\}$ is finite for every compact subset K of X and for every $y \in Y$ and
$T(f) = \sum_{n=1}^{\infty} \psi_n(f \circ \varphi_n)$ for each $f \in \mathscr{C}_0(X)$.

Examples of countably defined operators are, for instance, the operators of the form

$$Tf(x) := \sum_{n=1}^{\infty} f(a_n)\psi_n(x),$$

where $(a_n)_{n \in \mathbb{N}}$ is a sequence in X which converges to the point at infinity of X and $(\psi_n)_{n \in \mathbb{N}}$ is a sequence of positive functions in $\mathscr{C}_0(X)$ such that the series $\sum_{n=1}^{\infty} \psi_n$ is uniformly convergent in X.

Countably defined operators have the property that, for each $y \in Y$, the support of μ_y^T is contained in a countable discrete closed subset of Y. This makes available the results described in the notes and references of Section 2.2 and we may deduce the analogue of Theorem 3.4.9 for such operators.

More precisely, we have the following result (see Campiti [1987a, Theorem 2.2]).

If X is a locally compact Hausdorff space and if H is a subspace of $\mathscr{C}_0(X)$, then the following statements are equivalent:

(i) *For every locally compact Hausdorff space Y and for every countably defined operator $T: \mathscr{C}_0(X) \to \mathscr{C}_0(Y)$, H is a K_+-subspace for T.*

(ii) *H is a D_+-subspace for μ for every $\mu \in \mathscr{M}_b^+(X)$ whose support is contained in a countable discrete closed subset of X.*

(iii) *For every sequence $(x_n)_{n \in \mathbb{N}}$ of distinct elements of X which converges to the point at infinity of X, for every $(\alpha_n)_{n \in \mathbb{N}} \in \ell_+^1$ and for each $x \in X \backslash P$ (where $P = \{x_n|n \in \mathbb{N}\}$) there exist $M \in \mathbb{R}_+$ and a neighborhood U of x such that $U \cap P = \varnothing$ and, for every $\varepsilon > 0$ there exist a sequence $(h_n)_{n \in \mathbb{N}}$ in H and a sequence $(u_n)_{n \in \mathbb{N}}$ in $\mathscr{C}_0^+(X)$ satisfying*

$$\|u_n\| \leq \varepsilon, \qquad 0 \leq h_n + u_n,$$

$$1 \leq h_n + u_n \text{ on } U, \quad h_n + u_n \leq \varepsilon \text{ on } \{x_0, \ldots, x_n\}$$

for every $n \in \mathbb{N}$ and

$$\liminf_{n \to \infty} \sum_{p=n+1}^{\infty} \alpha_p(h_n(x_p) + u_n(x_p)) \leq M\varepsilon.$$

(iv) *For every sequence $(x_n)_{n \in \mathbb{N}}$ of distinct elements of X which converges to the point at infinity of X, for every $(\alpha_n)_{n \in \mathbb{N}} \in \ell_+^1$ and for each $x \in P$ there exist $M \in \mathbb{R}_+$ and a neighborhood U of x such that $U \cap (P \backslash \{x\}) = \varnothing$ and, for every $\varepsilon > 0$, there exist a sequence $(h_n)_{n \in \mathbb{N}}$ in H and a sequence $(u_n)_{n \in \mathbb{N}}$ in $\mathscr{C}_0^+(X)$ satisfying*

$$\|u_n\| \leq \varepsilon, \qquad 0 \leq h_n + u_n,$$

$$1 \leq h_n + u_n \text{ on } U, \quad h_n + u_n \leq \varepsilon \text{ on } \{x_0, \ldots, x_n\} \backslash \{x\}$$

for every $n \in \mathbb{N}$ and

$$\liminf_{n \to \infty} \sum_{\substack{p=n+1 \\ x_p \neq x}}^{\infty} \alpha_p(h_n(x_p) + u_n(x_p)) \leq M\varepsilon.$$

Chapter 4

Korovkin-type theorems for the identity operator

With this chapter we enter in the most classical aspects of Korovkin-type approximation theory by dealing with the convergence of nets of positive linear operators toward the identity operator.

Historically, however, the results contained in this chapter were established before those in Chapters 2 and 3 and they may indeed be considered as the fundamental ones of the theory.

Through the next five sections we carry out a study of Korovkin closures and Korovkin sets in many respects. Among other things, the reader will have the opportunity to realize the strong connection between this theory and other areas of functional analysis.

Of course, the most important applications of Korovkin sets are those concerning the approximation of continuous functions by positive linear operators and to this topic we have devoted both Chapters 5 and 6.

4.1 Korovkin closures and Korovkin subspaces for the identity operator

In this section we shall study Korovkin closures for the identity operator, both from an analytic and a geometric point of view.

Our considerations will also lead to a satisfactory characterization of Korovkin subspaces.

However, the classical Korovkin theorem together with some other examples of Korovkin sets will be presented in Section 4.2.

As usual, we shall denote by X a locally compact Hausdorff space and by H a subset of $\mathscr{C}_0(X)$.

The Korovkin closure $K_+(H, I_{\mathscr{C}_0(X)})$ of H for the identity operator on $\mathscr{C}_0(X)$ (see (3.1.2)) will be simply denoted by $K_+(H)$ and it will be called the K_+-*closure of* H.

Thus, a function $f \in \mathscr{C}_0(X)$ belongs to $K_+(H)$ if and only if it satisfies the following property:

For each net $(L_i)_{i \in I}^{\leq}$ of positive linear operators on $\mathscr{C}_0(X)$ satisfying
$$\sup_{i \in I} \|L_i\| < +\infty \text{ and } \lim_{i \in I} {}_{\leq} L_i(h) = h \text{ for all } h \in H, \text{ one has } \lim_{i \in I} {}_{\leq} L_i(f) = f.$$

$$(4.1.1)$$

In a similar way, we shall denote by $K_+^1(H)$ the Korovkin closure $K_+^1(H, I_{\mathscr{C}_0(X)})$ of H for the identity operator with respect to positive linear contractions and we shall call it the K_+^1-*closure of H*.

In other words, a function $f \in \mathscr{C}_0(X)$ lies in $K_+^1(H)$ if and only if it satisfies a property similar to (4.1.1) but considering only nets $(L_i)_{i \in I}^{\leq}$ of positive linear contractions.

The subspaces $K_+(H)$ and $K_+^1(H)$ are closed and, in general, the following inclusions hold:

$$H \subsetneqq K_+(H) \subsetneqq K_+^1(H) \subsetneqq \mathscr{C}_0(X). \qquad (4.1.2)$$

Note however that, if X is compact and $1 \in H$, then we have $K_+(H) = K_+^1(H)$ (see (3.1.7)).

Clearly, the subspaces $K_+(H)$ and $K_+^1(H)$ are the maximal subspaces (with respect to the inclusion relation) on which every equicontinuous net of positive linear operators (of positive linear contractions, respectively) converges strongly to the identity operator whenever it converges on H.

For these reasons, besides devoting out attention to the case when

$$K_+(H) = \mathscr{C}_0(X) \quad (K_+^1(H) = \mathscr{C}_0(X), \text{ respectively}), \qquad (4.1.3)$$

which is the most important for the applications in approximation theory, we shall study these subspaces for their own interest by deriving several characterizations of them.

Moreover, we shall briefly deal even with the extreme case when

$$K_+(H) = H \quad (K_+^1(H) = H, \text{ respectively}). \qquad (4.1.4)$$

The following notion will be at the center of our investigation.

4.1.1 Definition. *A subset H which satisfies (4.1.3) will be called a K_+-subset in $\mathscr{C}_0(X)$ or, a Korovkin subset in $\mathscr{C}_0(X)$ with respect to positive linear operators (a K_+^1-subset in $\mathscr{C}_0(X)$ or, a Korovkin subset in $\mathscr{C}_0(X)$ with respect to positive linear contractions, in the respective case).*

This means that, whenever one considers an equicontinuous net $(L_i)_{i \in I}^{\leq}$ of positive linear operators on $\mathscr{C}_0(X)$ (or, of positive linear contractions, respectively), then the relation

$$\lim_{\substack{i \in I}} {}_{\leq} L_i(h) = h \quad \text{for every } h \in H$$

entails

$$\lim_{\substack{i \in I}} {}_{\leq} L_i(f) = f \quad \text{for every } f \in \mathscr{C}_0(X).$$

In the sequel, the terms "*Korovkin set*" and "*Korovkin closure*" will mean, indifferently, K_+-subset or K_+^1-subset and K_+-closure or K_+^1-closure, respectively.

As explained at the beginning of Section 3.1, the Korovkin closure of a subset H coincides with the Korovkin closure of $\mathscr{L}(H)$. For these reasons, in many situations, we shall restrict ourselves to consider linear subspaces of $\mathscr{C}_0(X)$.

However, if a linear subspace is a Korovkin set, we shall emphasize this situation by calling it a *Korovkin subspace in* $\mathscr{C}_0(X)$.

Before collecting the main characterizations of Korovkin closures, we remark that all the above definitions can be made exactly in the same manner by replacing the space $\mathscr{C}_0(X)$ with an arbitrary Banach lattice E.

On account of Theorems 2.5.1, 3.1.3, 3.1.4 and 3.2.1 and of Corollary 2.1.8, we can immediately state the following characterizations of K_+-closures and K_+-subspaces.

4.1.2 Theorem. *For every $f \in \mathscr{C}_0(X)$ the following statements are equivalent:*
  (i) $f \in K_+(H)$.
 (ii) $f \in D_+(H, \varepsilon_x)$ *for every* $x \in X$.
(iii) (*Uniqueness property*) $\mu(f) = f(x)$ *for every* $\mu \in \mathscr{M}_b^+(X)$ *and* $x \in X$ *satisfying* $\mu(h) = h(x)$ *for all* $h \in H$.
(iv) (*Analytic property*)

$$\inf_{\varepsilon > 0} \left(\sup_{\substack{h \in H \\ h - \varepsilon \leq f}} h \right) = f = \sup_{\varepsilon > 0} \left(\inf_{\substack{k \in H \\ f \leq k + \varepsilon}} k \right).$$

 (v) *For every $\varepsilon > 0$ there exist finitely many functions $h_0, \ldots, h_n \in H$ and $k_0, \ldots, k_n \in H$ such that*

$$\left\| \inf_{0 \leq j \leq n} k_j - \sup_{0 \leq i \leq n} h_i \right\| \leq \varepsilon \quad \text{and} \quad \sup_{0 \leq i \leq n} h_i - \varepsilon \leq f \leq \inf_{0 \leq j \leq n} k_j + \varepsilon.$$

(vi) *For every $\varepsilon > 0$ there exist finitely many functions $h_0, \ldots, h_n \in H$ and $k_0, \ldots, k_n \in H$ and $u, v \in \mathscr{C}_0^+(X)$ such that $\|u\| \leq \varepsilon$, $\|v\| \leq \varepsilon$ and*

$$\left\| \inf_{0 \leq j \leq n} k_j - \sup_{0 \leq i \leq n} h_i \right\| \leq \varepsilon \quad \text{and} \quad \sup_{0 \leq i \leq n} h_i - u \leq f \leq \inf_{0 \leq j \leq n} k_j + v.$$

(vii) (*Universal Korovkin-type property with respect to positive linear operators*)
If E is a Banach lattice, if $S: \mathscr{C}_0(X) \to E$ *is a lattice homomorphism and if*
$(L_i)_{i \in I}^{\leq}$ *is a net of positive linear operators from* $\mathscr{C}_0(X)$ *into E such that*
$\sup_{i \in I} \|L_i\| < +\infty$ *and* $\lim_{\leq_{i \in I}} L_i(h) = S(h)$ *for all* $h \in H$, *then*

$$\lim_{\substack{\leq \\ i \in I}} L_i(f) = S(f).$$

4.1.3 Corollary. *A linear subspace H of* $\mathscr{C}_0(X)$ *is a* K_+-*subspace in* $\mathscr{C}_0(X)$ *if and only if statements* (iii)–(vii) *of Theorem 4.1.2 hold for each* $f \in \mathscr{C}_0(X)$ *or, equivalently, if one of the following further statements is true:*
(viii) $\partial_H^+ X = X$, *i.e.,* $D_+(H, \varepsilon_x) = \mathscr{C}_0(X)$ *for every* $x \in X$.
 (ix) (*Quasi peak-point property*) *H separates strongly the points of X* (*see* (2.6.14)) *and for every* $x \in X$, *for every compact subset K of X which does not contain x and for every* $\varepsilon > 0$ *there exist* $h \in H$ *and* $u \in \mathscr{C}_0(X)$, $u \geq 0$, *such that*

$$\|u\| \leq \varepsilon, \quad 0 \leq h + u, \quad 1 \leq h + u \text{ on } K \quad \text{and} \quad h(x) + u(x) < \varepsilon.$$

Remark. As explained in the introduction of Section 3.4, every lattice homomorphism $T: \mathscr{C}_0(X) \to \mathscr{C}_0(Y)$ is of the form $T(f) = \psi \cdot (f \circ \varphi)$ $(f \in \mathscr{C}_0(X))$, where $\varphi: Y \to X$ and $\psi: Y \to \mathbb{R}$ are suitable functions, i.e., T is finitely defined of order 1.

Consequently, Corollary 4.1.3 tells us that the K_+-subspaces of $\mathscr{C}_0(X)$ are exactly the K_+-subspaces of order 1. Therefore, every K_+-subspace of order n in $\mathscr{C}_0(X)$ is a K_+-subspace of $\mathscr{C}_0(X)$.

Properties (viii) and (ix) are the most useful to determine K_+-subspaces in concrete situations. Property (v) will help us in the next section to investigate some relations between $K_+(H)$ and the Riesz subspaces generated by H. Finally, the universal Korovkin-type property (vii) is particularly helpful in the case when S is the embedding from $\mathscr{C}_0(X)$ in some Banach function lattice E.

Furthermore, note that if H is a K_+-subspace in $\mathscr{C}_0(X)$, then H is a K_+-subspace for every continuous lattice homomorphism $S: \mathscr{C}_0(X) \to E$.

Theorem 4.1.2 applies, of course, also when X is compact. However, in this case and under the assumption that H is cofinal in $\mathscr{C}(X)$, the functions belonging to $K_+(H)$ satisfy some additional interesting properties.

The following result follows directly from Theorem 3.1.6 and 3.2.2.

4.1.4 Theorem. *Suppose that X is compact and consider a cofinal subspace H of* $\mathscr{C}(X)$. *Then, for every* $f \in \mathscr{C}(X)$, *the following statements are equivalent:*
 (i) $f \in K_+(H)$.
(ii) $f \in D_+(H, \varepsilon_x)$ *for every* $x \in X$.

(iii) *(Uniqueness property)* $\mu(f) = f(x)$ *for every* $\mu \in \mathcal{M}^+(X)$ *and* $x \in X$ *satisfying* $\mu(h) = h(x)$ *for all* $h \in H$.

(iv) *(Analytic property)*

$$\sup_{\substack{h \in H \\ h \leq f}} h = f = \inf_{\substack{k \in H \\ f \leq k}} k.$$

(v) *For every* $\varepsilon > 0$, *there exist finitely many functions* $h_0, \ldots, h_n \in H$ *and* $k_0, \ldots, k_n \in H$ *such that*

$$\left\| \inf_{0 \leq j \leq n} k_j - \sup_{0 \leq i \leq n} h_i \right\| \leq \varepsilon \quad and \quad \sup_{0 \leq i \leq n} h_i \leq f \leq \inf_{0 \leq j \leq n} k_j.$$

(vi) *(Universal Korovkin-type property with respect to positive linear operators) If E is a Banach lattice, if* $S: \mathscr{C}(X) \to E$ *is a lattice homomorphism and if* $(L_i)_{i \in I}^{\leq}$ *is a net of positive linear operators from* $\mathscr{C}(X)$ *into E such that* $\lim_{\substack{\leq \\ i \in I}} L_i(h) = S(h)$ *for all* $h \in H$, *then*

$$\lim_{\substack{\leq \\ i \in I}} L_i(f) = S(f).$$

(vii) *(Universal Korovkin-type property with respect to monotone operators) If E is a Banach lattice,* $S: \mathscr{C}(X) \to E$ *is a lattice homomorphism and if* $(L_i)_{i \in I}^{\leq}$ *is a net of monotone operators from* $\mathscr{C}(X)$ *into E such that* $\lim_{\substack{\leq \\ i \in I}} L_i(h) = S(h)$ *for all* $h \in H$, *then*

$$\lim_{\substack{\leq \\ i \in I}} L_i(f) = S(f).$$

Furthermore, if $1 \in H$, *the above statements* (i)–(vii) *are equivalent to the further one:*

(viii) *(Korovkin-type property with respect to linear contractions) For every net* $(L_i)_{i \in I}^{\leq}$ *of linear contractions on* $\mathscr{C}(X)$ *satisfying* $\lim_{\substack{\leq \\ i \in I}} L_i(h) = h$ *for each* $h \in H$, *one also has*

$$\lim_{\substack{\leq \\ i \in I}} L_i(f) = f.$$

Remark. In general, if $1 \notin H$, the Korovkin-type property with respect to linear contractions (viii) is not equivalent to the other properties. As an example, consider the interval $X := [\sqrt{2} - 1, 1]$ and the subspace H generated by the two functions $e_1(x) := x$ and $e_2(x) := x^2$ $(x \in X)$.

Then $K_+(H) = H$ as we shall show in Proposition 4.1.16 while every $f \in \mathscr{C}(X)$ enjoys the Korovkin-type property with respect to linear contractions (see Berens and Lorentz [1975, Example 1, p. 180]).

From Theorem 4.1.4 and from the results of Sections 2.5 and 2.6 we may now derive several characterizations of K_+-subspaces in $\mathscr{C}(X)$ in terms of analytic, topological and geometric properties.

4.1.5 Corollary. *Suppose that X is compact. Then an arbitrary linear subspace H of $\mathscr{C}(X)$ is a K_+-subspace in $\mathscr{C}(X)$ if and only if one of the statements* (iii), (vi) *or* (vii) *of Theorem 4.1.4 holds for every $f \in \mathscr{C}(X)$ or, equivalently, if one of the following further statements is true:*
(iv)′ *H is cofinal in $\mathscr{C}(X)$ and* (iv) *holds for every $f \in \mathscr{C}(X)$.*
 (v)′ *H is cofinal in $\mathscr{C}(X)$ and* (v) *holds for every $f \in \mathscr{C}(X)$.*
 (ix) *$\partial_H^+ X = X$, i.e., $D_+(H, \varepsilon_x) = \mathscr{C}(X)$ for every $x \in X$.*
  (x) *(Quasi peak-point property) H separates strongly the points of X and for every $x \in X$, for every compact subset K of X which does not contain x and for every $\varepsilon > 0$ there exists $h \in H$ such that $0 \leq h$, $1 \leq h$ on K and $h(x) < \varepsilon$.*
 (xi) *H separates strongly the points of X and for every $x \in X$, for every open neighborhood U of x and for every $\varepsilon > 0$ there exists $h \in H$ such that $0 \leq h$, $1 \leq h$ on $X \setminus U$ and $h(x) < \varepsilon$.*
(xii) *(Geometric property) H separates the points of X, the cone H'_+ is pointed and for every $x \in X$, the ray $\rho(x)$ is extreme in H'_+ (see* (2.6.13)).

Furthermore, if $\mathbf{1} \in H$, then the above statements are equivalent to the assertion that statement (viii) *of Theorem 4.1.4 holds for every $f \in \mathscr{C}(X)$ or, equivalently, to the further one:*
(xiii) *(Geometric property) H separates the points of X and $\partial_e \overline{\mathrm{co}}(\Phi(X)) = \Phi(X)$, where Φ is the canonical embedding defined by* (1.4.19).

Proof. By Theorem 2.1.4, if $D_+(H, \varepsilon_x) = \mathscr{C}(X)$ for some $x \in X$, H is necessarily cofinal in $\mathscr{C}(X)$. Consequently, by Theorem 4.1.4, it is clear that (iv)′ holds if and only if $D_+(H, \varepsilon_x) = \mathscr{C}(X)$ for every $x \in X$ or, equivalently, if (ix) is true.

The equivalences (ix) ⇔ (x) and (ix) ⇔ (xi) follow from Theorem 2.5.2.

By Proposition 2.6.4 and Corollary 2.6.5, (ix) is also equivalent to (xii) and (xiii), respectively. This completes the proof. □

Thus, Corollary 4.1.5 tells us that, if H is a K_+-subspace of $\mathscr{C}(X)$ containing the constant functions, then H is a Korovkin subspace not only for (equicontinuous) nets of positive linear operators on $\mathscr{C}(X)$, but also for nets of monotone operators and of linear contractions on $\mathscr{C}(X)$.

In fact, there are other classes of linear operators with respect to which every K_+-subspace is a Korovkin subspace.

An example is furnished by the class of all linear operators L on $\mathscr{C}(X)$ for which there exist subsets X^+ and X^- of X satisfying $X = X^+ \cup X^-$ and such that, for every $f \in \mathscr{C}(X)$, $f \geq 0$, one has

$$L(f) \geq 0 \text{ on } X^+ \quad and \quad L(f) \leq 0 \text{ on } X^-,$$

(see Campiti [1991b], who introduced and studied these operators in the more general setting of spaces of continuous vector-valued functions).

For instance, if $X = [a, b]$ is an interval of $\mathbb{R}$ and if $x_0 \in \,]a, b[$, the operator L on $\mathscr{C}([a, b])$ defined by

$$Lf(x) := \int_{x_0}^{x} f(t)\,dt,$$

belongs to the above mentioned class (and it is not positive).

Moreover, note that these operators can be characterized as the linear operators L on $\mathscr{C}(X)$ satisfying the inequality

$$|L(f)| \leq |L(g)| \quad \text{for every } f, g \in \mathscr{C}(X), |f| \leq g.$$

The operators satisfying this last property were studied independently by Nishishiraho [1991] and were called *quasi positive operators*.

For other classes of non positive linear operators for which a Korovkin-type theorem holds, see also Baskakov [1972], [1973].

On the other hand, there are classes of linear operators with respect to which a K_+-subspace is not a Korovkin subspace.

Example. Set $X := [-\pi, \pi]$ and denote by H the linear subspace generated by the functions $\mathbf{1}$, sin, cos. Then $K_+(H) = \{f \in \mathscr{C}([-\pi, \pi]) | f(-\pi) = f(\pi)\}$ by virtue of Theorem 4.1.2 and the Example to Proposition 2.5.7.

Now consider the *linear* (non positive) operator S_n^* on $\mathscr{C}([-\pi, \pi])$ defined by

$$S_n^*(f)(x) := \frac{1}{2\pi} \int_{-\pi}^{\pi} f(x - t) \frac{\sin(n + 1/2)t}{\sin t/2}\,dt$$

for every $f \in \mathscr{C}([-\pi, \pi])$ and $x \in [-\pi, \pi]$ (*singular integral of Dirichlet*).

As is well-known, if $f \in \mathscr{C}([-\pi, \pi])$ and $f(-\pi) = f(\pi)$, then

$$S_n^*(f)(x) = \frac{a_0}{2} + \sum_{k=1}^{n} a_k \cos kx + b_k \sin kx,$$

where the a_n's and b_n's are the Fourier coefficients of f (see Section 5.4).

Moreover, there exist functions $f \in K_+(H)$ such that the sequence $(S_n^*(f))_{n \in \mathbb{N}}$ does not converge to f (see, for instance, Butzer and Nessel [1971, Section 1.2.2] or Section 5.4). On the other hand,

$$S_n^*(1) = 1, \quad S_n^*(\sin) = \sin, \quad S_n^*(\cos) = \cos,$$

for each $n \geq 1$.

Before studying the corresponding characterizations of the K_+^1-closures, we shall indicate a simple and important consequence of the universal Korovkin-type property (vii) of Theorem 4.1.2.

4.1.6 Proposition. *Let E be a Banach lattice and let $S: \mathscr{C}_0(X) \to E$ be a lattice homomorphism whose range $S(\mathscr{C}_0(X))$ is dense in E. If H is a K_+-subspace in $\mathscr{C}_0(X)$, then $S(H)$ is a K_+-subspace in E.*

Proof. Let $(L_i)_{i \in I}^{\leq}$ be an equicontinuous net of positive linear operators from E into E such that $\lim_{i \in I}^{\leq} L_i(k) = k$ for every $k \in S(H)$.

Consider the net $(L_i \circ S)_{i \in I}^{\leq}$ of positive linear operators from $\mathscr{C}_0(X)$ in E. Since, for every $h \in H$, the net $(L_i(S(h)))_{i \in I}^{\leq}$ converges to $S(h)$, by the universal Korovkin-type property (vii) of Theorem 4.1.2, the net $(L_i(S(f)))_{i \in I}^{\leq}$ converges to $S(f)$ for every $f \in \mathscr{C}_0(X)$, i.e., $\lim_{i \in I}^{\leq} L_i(g) = g$ for every $g \in S(\mathscr{C}_0(X))$.

Since $S(\mathscr{C}_0(X))$ is dense in E, we can conclude that $\lim_{i \in I}^{\leq} L_i(g) = g$ for every $g \in E$. $\qquad \square$

We indicate some consequences of the previous result.

Let E be a Banach lattice which contains $\mathscr{C}_0(X)$ as a dense sublattice and consider the natural embedding $J: \mathscr{C}_0(X) \to E$, which is obviously a lattice homomorphism.

Thus, by virtue of Proposition 4.1.6, every K_+-subspace of $\mathscr{C}_0(X)$ is a K_+-subspace of E.

For example, consider $\mu \in \mathscr{M}_b^+(X)$ and, for every $1 \leq p < +\infty$, consider the Banach lattice $L^p(X, \mu)$ (see Section 1.2).

By applying Proposition 4.1.6 to the natural embedding $J_p: \mathscr{C}_0(X) \to L^p(X, \mu)$ defined by $J_p(f) := f$ $(f \in \mathscr{C}_0(X))$, we obtain that Korovkin subspaces in $\mathscr{C}_0(X)$ are also Korovkin subspaces in $L^p(X, \mu)$.

4.1.7 Corollary. *If $\mu \in \mathscr{M}_b^+(X)$ and if H is a K_+-subspace in $\mathscr{C}_0(X)$, then for every $1 \leq p < +\infty$, H is a K_+-subspace in $L^p(X, \mu)$.*

Some applications of the above corollary will be illustrated in Section 5.3 (see Examples 5.3.7 and 5.3.8). Furthermore we point out that if X is metrizable and

compact and $\mu \in \mathcal{M}^+(X)$, $\mu(1) = 1$, then Corollary 4.1.7 equally holds by considering the more general Banach function spaces $L^p(X, \mu)$ (in the sense of Luxem-
burg) instead of the spaces $L^p(X, \mu)$ (see Berens and Lorentz [1973, n. 4] for
more details).

We are now in a position to present the main characterizations of the
K_+^1-closures.

4.1.8 Theorem. *Given a linear subspace H of $\mathscr{C}_0(X)$ and $f \in \mathscr{C}_0(X)$, then the following statements are equivalent:*
  (i) $f \in K_+^1(H)$.
 (ii) $f \in H^* \cap D_+^1(H, \varepsilon_x)$ *for every* $x \in X$.
(iii) (*Uniqueness property*)
      (1) $\mu(f) = f(x)$ *for every* $\mu \in \mathcal{M}_b^+(X)$, $\|\mu\| \leq 1$, *and for every* $x \in X$ *satisfying*
      $\mu(h) = h(x)$ *for each* $h \in H$.
      (2) $\mu(f) = 0$ *for every* $\mu \in \mathcal{M}_b^+(X)$ *such that* $\mu = 0$ *on* H.
 (iv) (*Analytic property*) $f \in H^*$ *and*

$$\sup_{\substack{h \in H, \alpha \geq 0 \\ h - \alpha \leq f}} (h - \alpha) = f = \inf_{\substack{k \in H, \beta \geq 0 \\ f \leq k + \beta}} (k + \beta).$$

  (v) $f \in H^*$ *and for every* $\varepsilon > 0$ *there exist finitely many functions* $f_0, \ldots, f_n$
  in $H - \mathbb{R}_+$ *and* $g_0, \ldots, g_n$ *in* $H + \mathbb{R}_+$ *such that*

$$\left\| \inf_{0 \leq j \leq n} g_j - \sup_{0 \leq i \leq n} f_i \right\| \leq \varepsilon \quad and \quad \sup_{0 \leq i \leq n} f_i \leq f \leq \inf_{0 \leq j \leq n} g_j.$$

 (vi) (*Universal Korovkin-type property with respect to positive linear contrac-
     tions*) *If E is an AM-space, if $S: \mathscr{C}_0(X) \to E$ is a lattice homomorphism such
     that for every $\delta \in V_1(E)$, $\delta \circ S$ either is zero or has norm 1, and if $(L_i)_{i \in I}$ is a
     net of positive linear contractions from $\mathscr{C}_0(X)$ into E such that $\lim_{i \in I}^{\leq} L_i(h) =
     S(h)$ for all $h \in H$, then*

$$\lim_{i \in I}{}^{\leq} L_i(f) = S(f).$$

*Furthermore, if X is compact, then condition (2) in (iii), and the condition $f \in H^*$
in (ii), (iv), (v) can be dropped.*

Proof. This is a direct consequence of Theorems 3.1.8 and 3.2.3.   □

4.1.9 Corollary *A linear subspace H of $\mathscr{C}_0(X)$ is a K_+^1-subspace in $\mathscr{C}_0(X)$ if and
only if one of statements (iii)–(vi) holds for every $f \in \mathscr{C}_0(X)$ or, equivalently, if one
of the following further statements is true:*

(vii) $H^* = \mathscr{C}_0(X)$ and $\partial_H^{1,+} X = X$, i.e., $D_+^1(H, \varepsilon_x) = \mathscr{C}_0(X)$ for every $x \in X$.
(viii) (*Quasi peak-point property*) $H^* = \mathscr{C}_0(X)$ and for every $x \in X$, for every compact subset K of X which does not contain x and for every $\varepsilon > 0$ there exist $h \in H$ and $u \in \mathscr{C}_0(X)$, $u \geq 0$, such that

$$\|u\| < \varepsilon + u(x), \quad 0 \leq h + u, \quad 1 \leq h + u \text{ on } K \quad \text{and} \quad h(x) + u(x) < \varepsilon.$$

Furthermore, if X is compact, then the condition $H^* = \mathscr{C}(X)$ in both statements (vii) and (viii) can be replaced by the assumption that H separates strongly the points of X. Moreover, one may also add the assertion:
  (ix) (*Geometric property*) H separates strongly the points of X and

$$\Phi(X) \subset \partial_e \overline{\mathrm{co}}(\Phi(X) \cup \{0\}),$$

where, again, Φ denotes the canonical embedding defined by (1.4.19).

Proof. First, note that, if $H^* = \mathscr{C}_0(X)$, then H satisfies (2.6.14) by virtue of Corollary 2.1.8, (v).

So, Theorem 2.5.3 implies that statements (vii) and (viii) are both equivalent to the assertion that statement (ii) holds for every $f \in \mathscr{C}_0(X)$.

Finally, if X is compact, then (vii) is equivalent to (ix) by Proposition 2.6.7. $\qquad\square$

Also for K_+^1-subspaces, in analogy with Proposition 4.1.6, we may state an interesting property involving lattice homomorphisms with dense range. The proof is similar to that of Proposition 4.1.6 and uses the universal Korovkin-type property (vi) in Theorem 4.1.8. For the sake of brevity we omit it.

4.1.10 Proposition. *Let E be an AM-space and let $S: \mathscr{C}_0(X) \to E$ be a lattice homomorphism such that its range $S(\mathscr{C}_0(X))$ is dense in E and for every $\delta \in V_1(E)$, $\delta \circ S$ either is zero or has norm 1. If H is a K_+^1-subspace in $\mathscr{C}_0(X)$, then $S(H)$ is a K_+^1-subspace in E.*

Remarks 1. Theorems 4.1.2 and 4.1.8 and their Corollaries 4.1.3 and 4.1.9 apply, of course, also when X is discrete and, in particular, for the space $c_0 = \mathscr{C}_0(\mathbb{N})$.

In this case, however, we point out that condition (viii) of both Corollaries 4.1.3 and 4.1.9 can be reformulated by considering only singletons of X instead of compact subsets K of X (see Remarks to Theorems 2.2.2 and 2.2.4).

2. The above results make already evident the difference between K_+- and K_+^1-closures (we point out that they coincide when X is compact and $1 \in H$, by virtue of Theorems 4.1.2 and 4.1.4 and (2.1.22)). However, in the next sections, we shall present several examples from which the situation will become clearer. For instance, we shall see that $\mathscr{C}_0(X)$ may possess two-dimensional K_+^1-subspaces (see Proposition 4.2.4) but it cannot contain two-dimensional K_+-subspaces (see Proposition 4.1.16).

However, from Corollaries 4.1.5 and 4.1.9 it follows that, if X is compact and if H is a K_+^1-subspace of $\mathscr{C}(X)$, then the subspace generated by $\{1\}$ and H is a K_+-subspace of $\mathscr{C}(X)$ (for a partial converse of this result, see Corollary 4.5.5).

A useful criterion to establish whether a given subspace is a K_+-subspace or a K_+^1-subspace is indicated in the following Corollary, which is a direct consequence of the previous results and of Theorem 2.5.4 (see also Corollary 3.4.8).

4.1.11 Corollary. *Suppose that a linear subspace H of $\mathscr{C}_0(X)$ satisfies the following condition:*

For every $x, y \in X$, $x \neq y$, there exists $k \in H$ (respectively, $k \in H + \mathbb{R}_+$) such that

$$k \geq 0, k(y) > 0 \text{ and } k(x) = 0. \qquad (4.1.5)$$

Then, H is a K_+-subspace in $\mathscr{C}_0(X)$ (a K_+^1-subspace, respectively, provided that $H^* = \mathscr{C}_0(X)$ or, if X is compact, H separates strongly the points of X).

4.1.12 Corollary. *Let M be a subset of $\mathscr{C}_0(X)$ such that for every $x, y \in X$, $x \neq y$, there exists $h \in \mathscr{L}(M)$, $h \geq 0$, such that $h(x) \neq h(y) \neq 0$. Then, $M \cup M^2 \cup M^3$ is a K_+-subset in $\mathscr{C}_0(X)$.*

Proof. If $x, y \in X$, $x \neq y$, and if h is a function of $\mathscr{L}(M)$ satisfying $h \geq 0$ and $h(x) \neq h(y) \neq 0$, then the function $k := h \cdot (h - h(x))^2 \in \mathscr{L}(M \cup M^2 \cup M^3)$ satisfies (4.1.5). $\qquad \square$

We close this section by explaining some other simple properties of Korovkin closures and by characterizing the cases in which the equality $K_+(H) = H$ occurs.

We begin by studying the Korovkin closure of $f_0 \cdot H$, when $f_0 \in \mathscr{C}_0^+(X)$. We set

$$I(f_0) := \{f \in \mathscr{C}_0(X) | f(x) = 0 \text{ for every } x \in X \text{ such that } f_0(x) = 0\}.$$

Clearly $I(f_0) = \mathscr{C}_0(X)$ if f_0 never vanishes on X.

4.1.13 Proposition *The following statements hold:*
(1) $f_0 \cdot K_+(H) \subset K_+(f_0 \cdot H) \subset I(f_0)$.
(2) *If H is a K_+-subspace in $\mathscr{C}_0(X)$, then $K_+(f_0 \cdot H) = I(f_0)$.*
   Hence $f_0 \cdot H$ is also a K_+-subspace in $\mathscr{C}_0(X)$ if and only if f_0 never vanishes on X.

Proof. Part (1) easily follows by Theorem 4.1.2, (v). Part (2) is a consequence of part (1) and the equality $\overline{f_0 \cdot \mathscr{C}_0(X)} = I(f_0)$.   □

The following result can be useful to construct K_+-subspaces in $\mathscr{C}(X_\omega)$ starting from K_+-subspaces in $\mathscr{C}_0(X)$ and conversely.

As in (2.6.3), we shall set $H_\omega := \mathscr{L}(\{1\} \cup \tilde{H}) \subset \mathscr{C}(X_\omega)$.

4.1.14 Proposition. *Under the above assumption, we have*

$$K_+^1(H)\tilde{} \subset K_+(H_\omega). \tag{4.1.6}$$

Consequently, if H is a K_+^1-subspace in $\mathscr{C}_0(X)$ (in particular, if H is a K_+-subspace in $\mathscr{C}_0(X)$), then H_ω is a K_+-subspace in $\mathscr{C}(X_\omega)$.

Proof. The result follows by a direct application of Theorems 4.1.2 and 4.1.8, by using the procedure of extending a Radon measure $\mu \in \mathscr{M}_b^+(X)$ to the Radon measure $\tilde{\mu} \in \mathscr{M}^+(X_\omega)$ given by (1.2.11).   □

4.1.15 Proposition. *Let H_ω be a K_+-subspace of $\mathscr{C}(X_\omega)$ (not necessarily of the above mentioned form) and let $f_0 \in \mathscr{C}_0(X)$ be a strictly positive function.*

Then the subspace

$$\{f_0 \cdot h_{|X} | h \in H_\omega\} \tag{4.1.7}$$

is a K_+-subspace in $\mathscr{C}_0(X)$.

Proof. Consider $\mu \in \mathscr{M}_b^+(X)$ and $x \in X$ satisfying $\mu(f_0 h_{|X}) = f_0(x)h(x)$ for every $h \in H_\omega$. Then, by passing to the measure $\hat{\mu}$ on X_ω defined by $\hat{\mu}(f) := \mu(f_0 f_{|X})/f_0(x)$ for every $f \in \mathscr{C}(X_\omega)$ and by using the fact that H_ω is a K_+-subspace in $\mathscr{C}(X_\omega)$, we deduce that $\mu(f_0 f) = f_0(x)f(x)$ for every $f \in \mathscr{C}_0(X)$. Hence, $\mu = \varepsilon_x$ since $f_0 \cdot \mathscr{C}_0(X)$ is dense in $\mathscr{C}_0(X)$.   □

Finally, we remark that, if Y is another locally compact Hausdorff space and $\sigma : Y \to X$ is a homeomorphism then, after putting

$$H_\sigma := \{h \circ \sigma | h \in H\}, \tag{4.1.8}$$

it is easy to verify that

$$K_+(H_\sigma) = (K_+(H))_\sigma, \tag{4.1.9}$$

and

$$K_+^1(H_\sigma) = (K_+^1(H))_\sigma. \tag{4.1.10}$$

Consequently, H_σ is a K_+- or a K_+^1-subspace in $\mathscr{C}_0(Y)$ if and only if H is so in $\mathscr{C}_0(X)$.

As regards the extreme case when

$$K_+(H) = H \quad (K_+^1(H) = H, \text{ respectively}), \tag{4.1.11}$$

we proceed by examining some non trivial cases where such an equality holds.

We shall see in the next sections that (4.1.11) is true if H is a closed lattice subspace (or, in particular, a closed subalgebra) of $\mathscr{C}_0(X)$. Another case occurs if X is a convex compact subset of some locally convex Hausdorff space and $H = A(X)$ is the space of all continuous affine functions on X (see (4.3.9)). A more complicated case is furnished by $H = \mathscr{L}(A(X) \cup \{u\})$, where $u \in \mathscr{C}(X)$ is a function neither convex nor concave (see Bauer, Leha and Papadopoulou [1979]).

In general, if E is a Banach lattice and $(L_i)_{i \in I}^{\leq}$ is an equicontinuous net of positive linear operators from $\mathscr{C}_0(X)$ into E, the subspace

$$H := \left\{ f \in \mathscr{C}_0(X) \middle| \lim_{i \in I}{}_{\leq} L_i(f) = f \right\} \tag{4.1.12}$$

satisfies (4.1.11). Similarly, the same holds if H is the range of a positive projection on $\mathscr{C}_0(X)$ or if

$$H := \left\{ f \in \mathscr{C}_0(X) \middle| \lim_{i \in I}{}_{\leq} L_i(f) = 0 \right\}, \tag{4.1.13}$$

$$H := \left\{ f \in \mathscr{C}_0(X) \middle| \lim_{i \in I}{}_{\leq} \mu_i(f) = \alpha f(x_0) \right\}, \tag{4.1.14}$$

$$H := \left\{ f \in \mathscr{C}_0(X) \middle| \lim_{i \in I}{}_{\leq} \mu_i(f) = 0 \right\}, \tag{4.1.15}$$

where $(\mu_i)_{i \in I}^{\leq}$ is an arbitrary equicontinuous net in $\mathscr{M}_b^+(X)$, $x_0 \in X$ and $\alpha \geq 0$.

Similar observations hold for other classes of linear operators according to Theorems 4.1.2, 4.1.4 and 4.1.8. Furthermore, (4.1.11) is satisfied by each subspace which is the intersection on an arbitrary family of subspaces of the above form. So, in particular,

$$K_+(K_+(H)) = K_+(H) \quad (K_+^1(K_+^1(H)) = K_+^1(H), \text{ respectively}). \tag{4.1.16}$$

We also point out the following result.

4.1.16 Proposition. *If* $\dim(H) \leq 2$, *then* $K_+(H) = H$ *and, if* $\dim(H) \leq 1$, *then* $K_+^1(H) = H$.

Proof. As regards the first part of the statement, it is enough to show the result in the case $\dim(H) = 2$.

Thus, suppose that $H = \mathscr{L}(\{f_0, f_1\})$, where $f_0, f_1 \in \mathscr{C}_0(X)$ are linearly independent. Fix $x_1, x_2 \in X$ satisfying $f_0(x_1)f_1(x_2) - f_0(x_2)f_1(x_1) \neq 0$.
Then

$$H = \left\{ f \in \mathscr{C}_0(X) \middle| \begin{vmatrix} f(x) & f(x_1) & f(x_2) \\ f_0(x) & f_0(x_1) & f_0(x_2) \\ f_1(x) & f_1(x_1) & f_1(x_2) \end{vmatrix} = 0 \text{ for every } x \in X \right\}$$

$$= \{f \in \mathscr{C}_0(X) \mid \text{For every } x \in X, \text{ there exist numbers } \alpha(x), \beta(x), \gamma(x)$$
$$\text{such that } \alpha(x)f(x) + \beta(x)f(x_1) + \gamma(x)f(x_2) = 0\}.$$

We may suppose that $\alpha(x) \geq 0$ for every $x \in X$. Put

$$X_1 := \{x \in X \mid \beta(x) \geq 0 \text{ and } \gamma(x) \geq 0\},$$

$$X_2 := \{x \in X \mid \beta(x) \leq 0 \text{ and } \gamma(x) \leq 0\},$$

$$X_3 := \{x \in X \mid \beta(x) \geq 0 \text{ and } \gamma(x) \leq 0\},$$

$$X_4 := \{x \in X \mid \beta(x) \leq 0 \text{ and } \gamma(x) \geq 0\}.$$

Then

$$H = \{f \in \mathscr{C}_0(X) \mid (\alpha(x)\varepsilon_x + \beta(x)\varepsilon_{x_1} + \gamma(x)\varepsilon_{x_2})(f) = 0 \text{ for every } x \in X_1\}$$

$$\cap \{f \in \mathscr{C}_0(X) \mid \alpha(x)\varepsilon_x(f) = (-\beta(x)\varepsilon_{x_1} - \gamma(x)\varepsilon_{x_2})(f) \text{ for every } x \in X_2\}$$

$$\cap \{f \in \mathscr{C}_0(X) \mid (\alpha(x)\varepsilon_x + \beta(x)\varepsilon_{x_1})(f) = -\gamma(x)\varepsilon_{x_2}(f) \text{ for every } x \in X_3\}$$

$$\cap \{f \in \mathscr{C}_0(X) \mid (\alpha(x)\varepsilon_x + \gamma(x)\varepsilon_{x_2})(f) = -\beta(x)\varepsilon_{x_1}(f) \text{ for every } x \in X_4\},$$

and hence the result follows by the above remarks.

Similar (and simpler) reasoning may be used to show the last part of the assertion.   □

Remarks. 1. As Proposition 4.2.4 below shows, it may happen that $\dim(H) = 2$ and $K_+^1(H) = \mathscr{C}_0(X)$.

2. When $X = [a, b]$ Proposition 4.1.16 leads to *Korovkin's third theorem*, i.e., in $\mathscr{C}([a, b])$ there exist no K_+-subsets consisting of two functions.

However all the above results can be obtained from the following theorem where the equality $K_+(H) = H$ is completely characterized.

To state it we recall that a cone P of $\mathscr{C}_0(X)$ is said to be *inf-stable* if

$$\inf(f, g) \in P \quad \text{for every } f, g \in P. \tag{4.1.17}$$

4.1.17 Theorem. *Let H be a linear subspace of $\mathscr{C}_0(X)$. Then the following statements are equivalent:*
(i) *$K_+(H) = H$.*
(ii) *There exists a closed convex inf-stable cone P of $\mathscr{C}_0(X)$ such that $H = P \cap (-P)$.*

Proof. (i) $\Rightarrow$ (ii): This is obvious because the closed convex cone $P :=$ $\{f \in \mathscr{C}_0(X) | \mu(f) \leq f(x) \text{ for every } \mu \in \mathscr{M}_b^+(X) \text{ and } x \in X \text{ satisfying } \mu(h) = h(x) \text{ for every } h \in H\}$ is inf-stable and $K_+(H) = P \cap (-P)$ by virtue of Theorem 4.1.2.
(ii) $\Rightarrow$ (i): Suppose that $H = P \cap (-P)$ for some closed convex inf-stable cone P of $\mathscr{C}_0(X)$. We have to show that $K_+(H) \subset H$ and to this end it is enough to prove the inclusion $K_+(H) \subset P$.

So fix $f \in K_+(H)$. By Theorem 4.1.2, given $\varepsilon > 0$, there exist $h_0, \ldots, h_n$, $k_0, \ldots, k_n \in H \subset P$ such that $\left\| \inf_{0 \leq j \leq n} k_j - \sup_{0 \leq i \leq n} h_i \right\| \leq \varepsilon$ and $\sup_{0 \leq i \leq n} h_i - \varepsilon \leq f \leq \inf_{0 \leq j \leq n} k_j + \varepsilon$.

Hence

$$-2\varepsilon \leq \inf_{0 \leq j \leq n} k_j - f \leq \inf_{0 \leq j \leq n} k_j - \sup_{0 \leq i \leq n} h_i + \sup_{0 \leq i \leq n} h_i - f \leq 2\varepsilon.$$

So $\left\| f - \inf_{0 \leq j \leq n} k_j \right\| \leq 2\varepsilon$. This reasoning shows that $f \in \bar{P} = P$. $\qquad \square$

Notes and references

The notion of Korovkin closure (sometimes called shadow) first appeared in Baskakov [1961] and, later on, in Krasnosel'skii and Lifšic [1969] and in Franchetti [1970]. Implicitly this notion is also contained in Shashkin [1967].

Theorem 4.1.2 is due to Bauer and Donner [1978] and, when $X = \mathbb{N}$ (i.e., $\mathscr{C}_0(X) = c_0$), to Kitto and Wulbert [1976].

Theorem 4.1.4 and its Corollary 4.1.5 collect the results of Bauer [1973], Baskakov [1961], Berens and Lorentz [1973], [1976] (see also Lorentz [1972b]), Shashkin [1972], [1973], Wulbert [1968], Kutateladze and Rubinov [1972].

The enveloping functions considered in part (iv) of Theorem 4.1.4 appear for the first time in Baskakov [1961] and they were extensively studied by Bauer [1973], [1974] and Fakhoury [1974] and, in an abstract setting, by Kutateladze and Rubinov [1971a], [1971b], [1972] and by Donner [1979] (see also Altomare [1979]).

The same enveloping functions have been considered, as an essential tool, by Bauer in [1961] in connection with an abstract Dirichlet problem (see the final notes of Section 2.6). In that paper (see Beispiel 6) it is also shown that, for $X = [0, 1]$ and $H = \mathscr{L}(\{\mathbf{1}, x, x^2\})$, $\bigcap_{x \in X} D_+(H, \varepsilon_x) = \mathscr{C}(X)$, i.e., Korovkin's theorem 4.2.7.

Another quasi-peak point criterion similar to assertion (x) of Corollary 4.1.5 can be found in Brodskii [1961].

The characterizations of K_+-subspaces in terms of Choquet boundaries also appear in an unpublished paper of Arveson [1970] where for the first time appears an universal Korovkin-type property of algebraic nature. This was, perhaps, the starting point for a systematic investigation of Korovkin-type approximation theory in the setting of Banach algebras (see Appendices A and B).

Proposition 4.1.6 is due to Wolff [1977]. Corollary 4.1.7 was established in the compact case by Berens and Lorentz [1973] (see also Kitto and Wulbert [1976]) and, in the general case, by Donner [1982]. In fact, these authors showed a stronger result, namely that, if $\mu \in \mathscr{M}_b^+(X)$, then a subspace H of $\mathscr{C}_0(X)$ is a Korovkin space in $L^p(X, \mu)$ $(1 \le p)$ provided that $X \backslash \partial_H^+ X$ has inner measure 0. More generally, Krasnosel'skii and Lifšic [1968a] showed that, if X is metrizable and compact, then the same result holds if the complement of the set of peak points for H (see Section 2.6) has inner measure 0 (for a proof see Berens and Lorentz [1973, n. 5]).

If X is metrizable and compact and if $X \backslash \partial_H^+ X$ has inner measure 0, then Campiti [1987a] showed that H is a K_+-subspace even in the space of all Riemann μ-integrable functions on X endowed with the so-called Riemann convergence. This kind of convergence introduced by Dickmeis, Mevissen, Nessel and van Wickeren [1988] plays an important role in the approximation of Riemann integrable functions on a bounded real interval.

Theorem 4.1.8 and its Corollary 4.1.9 are due to Bauer and Donner [1986], to Flösser [1981] for the locally compact case and to Berens and Lorentz [1975], [1976] and Scheffold [1973] in the compact case.

Corollary 4.1.12 is due to Fakhoury [1973].

The geometric properties of Corollary 4.1.5 and Corollary 4.1.9 were established by Bauer [1961] and, independently, by Shashkin [1962] in the finite dimensional case (see also Berens and Lorentz [1975]).

The universal Korovkin-type property with respect to monotone operators (Theorem 4.1.4, (vii)) was noted by Bauer [1974] whereas the Korovkin-type property with respect to linear contractions (property (viii) of the same theorem) first appears in Wulbert [1968] and Shashkin [1969].

The universal Korovkin-type properties indicated in Theorems 4.1.2, 4.1.4 and 4.1.8 are due to Bauer and Donner [1978], Berens and Lorentz [1973] and Flösser [1981], respectively. They were the starting point for a systematic treatment of the universal

Korovkin closures in different settings such as Banach lattices, locally convex vector lattices, commutative Banach algebras and C^*-algebras. The interested reader is referred to the papers of Altomare [1979], [1982a], [1982c], [1984a], [1984b], [1986], [1987a], Beckhoff [1985], [1987], [1988], [1990], Donner [1975], [1982], Fakhoury [1974], Flösser [1978], [1979], [1980a], [1981], Flösser, Irmisch and Roth [1981], Pannenberg [1985], [1986a], [1986b], [1987a], [1987b], [1990b], [1991], Pannenberg and Romanelli [1991], Romanelli [1990], Scheffold [1973b], Wolff [1973a], [1973b], [1973c], [1975], [1977], [1978] (see also Appendix A).

In the case when X is compact and H is finite dimensional, then Corollary 4.1.11 was obtained by Freud [1964] (see also Volkov [1958]).

The results presented in the last part of this section were suggested by the paper of Amir and Ziegler [1978] who established Proposition 4.1.13 and 4.1.16 in the compact case.

For the space $\mathscr{C}_0(X)$ and for other classical Banach lattices such as $L^p(X, \mu)$-spaces Proposition 4.1.16 was proved, with a different method, by Donner [1982, Theorem 7.8 and Corollary 7.9].

The problem of showing the equality $K_+(H) = H$ in the space $\mathscr{C}_0(]0, 1])$ for subspaces H generated by the two functions $e_1(x) = x$ and $u \in \mathscr{C}_0(]0, 1])$ led Bauer to prove the existence of so-called u-means and to solve some related classes of functional equations (see Bauer [1986], [1987]).

Theorem 4.1.17 is essentially due to Bauer [1961], [1978b]. The same proof can be used to characterize the closed convex cones of $\mathscr{C}_0(X)$ which are inf-stable (i.e., to obtain a Choquet-Deny-type theorem in $\mathscr{C}_0(X)$).

The same problem was solved by Flösser, Irmisch and Roth [1981] in the context of AM-spaces and dual atomic vector lattices.

4.2 Strict Korovkin subsets. Korovkin's theorems

In the first part of this section we shall briefly discuss some particular Korovkin subsets which are called strict Korovkin subsets. Their usefulness is based on a peak-point type property, which is equivalent to a simple geometric condition in the relative state space.

Furthermore, in concrete situations they are easier to compute than Korovkin subsets.

In the second part of this section we present the classical Korovkin theorems. On account of all the above results, these theorems are obvious corollaries.

However, for the reader's convenience, we shall also present direct proofs.

We begin by introducing the notion of strict Korovkin subset.

4.2.1 Definition. *Let X be a locally compact Hausdorff space. A subspace H of $\mathscr{C}_0(X)$ is called a strict K_+-subspace (a strict K_+^1-subspace, respectively) if it satisfies the following property:*

For every $x \in X$ there exists $k \in H$ ($k \in H + \mathbb{R}_+$, respectively) such that

$$k(x) = 0 \text{ and } k(y) > 0 \text{ for every } y \in X, \, y \neq x. \tag{4.2.1}$$

If this is the case and if $H = \mathscr{L}(M)$ for some subset M of $\mathscr{C}_0(X)$, then M is called a strict K_+- or K_+^1 subset.

Since condition (4.2.1) is stronger than condition (4.1.5) of Corollary 4.1.11, this same corollary implies the next result.

4.2.2 Corollary. *Every strict K_+-subspace of $\mathscr{C}_0(X)$ is a K_+-subspace in $\mathscr{C}_0(X)$. Furthermore, every strict K_+^1-subspace H satisfying $H^* = \mathscr{C}_0(X)$ (or, separating strongly the points of X, if X is compact) is a K_+^1-subspace in $\mathscr{C}_0(X)$.*

For a direct proof of the above assertion see Theorem 4.2.10 and the subsequent Example 3. The converse, in general, is not true as we see in the following example.

Example. Consider the two continuous functions h_1 and h_2 on $[0, 2\pi]$ defined by

$$
h_1(t) := \begin{cases} 1 + \sin t, & 0 \leq t < \dfrac{\pi}{2}, \\[2mm] 2 \sin t, & \dfrac{\pi}{2} \leq t < \dfrac{3}{2}\pi, \\[2mm] -1 + \sin t, & \dfrac{3}{2}\pi \leq t \leq 2\pi, \end{cases}
\qquad
h_2(t) := \begin{cases} -\cos t, & 0 \leq t < \dfrac{\pi}{2}, \\[2mm] -2\cos t, & \dfrac{\pi}{2} \leq t < \dfrac{3}{2}\pi, \\[2mm] -\cos t, & \dfrac{3}{2}\pi \leq t \leq 2\pi. \end{cases}
$$

In this case, denoting by Φ_0 the corresponding embedding (2.6.16), it is easy to show that $\Phi_0([0, 2\pi])$ is the union of the three arcs

$$X_1 := \{(x, y) \in \mathbb{R}^2 | 1 \leq x \leq 2, \quad -1 \leq y \leq 0, \quad (x - 1)^2 + y^2 = 1\},$$

$$X_2 := \{(x, y) \in \mathbb{R}^2 | -2 \leq x \leq 2, \quad 0 \leq y \leq 2, \quad x^2 + y^2 = 4\}$$

and

$$X_3 := \{(x, y) \in \mathbb{R}^2 | -2 \leq x \leq -1, -1 \leq y \leq 0, (x + 1)^2 + y^2 = 1\}.$$

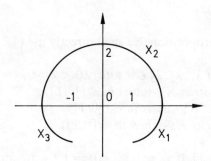

Figure 1

Hence, by Corollary 4.1.5, $\{\mathbf{1}, h_1, h_2\}$ is a K_+-subspace of $\mathscr{C}([0, 2\pi])$ while it is not a strict K_+-subspace, since $(-1, -1)$ (as well as $(1, -1)$) is not an exposed point of $\mathrm{co}(\Phi_0([0, 2\pi]))$ (see Figure 1 and the subsequent remarks).

However, some cases do occur in which every K_+-subspace is also a strict K_+-subspace. This is the case when X is the unit sphere of $\mathbb{R}^n$ ($n \geq 2$) (this has been shown by Volkov [1960] for $n = 3$ and by Berens and Lorentz [1976, Theorem 1] for the general case).

In Section 4.5 we shall also present some results about the existence of finite strict K_+-subsets. Here we present some other important examples.

Note that H is a strict K_+-subspace (a strict K_+^1-subspace, respectively) if and only if the set of peak-points for H (of weak peak-points for H) coincides with X (see Section 2.6).

When X is compact, an equivalent geometric condition can be stated in terms of the embedding $\Phi \colon X \to H'$ defined by (1.4.19).

In fact, if H contains the constants and separates the points of X, then H is a strict K_+-subspace of $\mathscr{C}(X)$ if and only if, for every $x \in X$, $\Phi(x)$ is an exposed point of $\overline{\mathrm{co}}(\Phi(X))$.

Moreover, if H separates strongly the points of X, then H is a strict K_+^1-subspace of $\mathscr{C}(X)$ if and only if, for every $x \in X$, $\Phi(x)$ is an exposed point of $\overline{\mathrm{co}}(\Phi(X) \cup \{0\})$.

When H is finite dimensional, analogous statements hold by considering the embeddings Φ^* and Φ_0 instead of Φ (see (2.6.15) and (2.6.16)).

From these remarks we immediately obtain the following result.

4.2.3 Proposition. *If X is a compact subset of $\mathbb{R}^n$, $n \geq 2$, then*

(1) $\{1, \mathrm{pr}_1, \ldots, \mathrm{pr}_n\}$ *is a strict K_+-subset in $\mathscr{C}(X)$ if and only if each point of X is an exposed point of $\mathrm{co}(X)$.*

Moreover, if $0 \notin X$, then

(2) $\{\mathrm{pr}_1, \ldots, \mathrm{pr}_n\}$ *is a strict K_+^1-subset in $\mathscr{C}(X)$ if and only if each point of X is an exposed point of $\mathrm{co}(X \cup \{0\})$.*

In particular, if X is a closed subset of the unit sphere of $\mathbb{R}^n$, $n \geq 2$, then

(3) $\{1, \mathrm{pr}_1, \ldots, \mathrm{pr}_n\}$ *is a strict K_+-subset in $\mathscr{C}(X)$ and $\{\mathrm{pr}_1, \ldots, \mathrm{pr}_n\}$ is a strict K_+^1-subset in $\mathscr{C}(X)$.*

To show the next proposition, we shall directly use (4.2.1).

4.2.4 Proposition. *Let $\lambda_1, \lambda_2, \lambda_3 \in \mathbb{R}$ satisfy $0 < \lambda_1 < \lambda_2 < \lambda_3$. Then*

(1) $\{x^{\lambda_1}, x^{\lambda_2}, x^{\lambda_3}\}$ *is a strict K_+-subset in $\mathscr{C}_0(]0, 1])$.*

(2) $\{x^{\lambda_1}, x^{\lambda_2}\}$ *is a strict K_+^1-subset in $\mathscr{C}_0(]0, 1])$.*

(3) $\{1, x^{\lambda_1}, x^{\lambda_2}\}$ *is a strict K_+-subset in $\mathscr{C}([0, 1])$.*

Proof. (1) Fix $x_0 \in \,]0, 1]$. If $x_0 = 1$, set $k(t) := t^{\lambda_1} - t^{\lambda_2}$ for $t \in \,]0, 1]$.

If $x_0 < 1$, consider $k(t) := (\lambda_3 - \lambda_2) x_0^{\lambda_3 - \lambda_1} t^{\lambda_1} - (\lambda_3 - \lambda_1) x_0^{\lambda_3 - \lambda_2} t^{\lambda_2} + (\lambda_2 - \lambda_1) t^{\lambda_3}$ for every $t \in \,]0, 1]$.

In any case, k satisfies (4.2.1) as is easy to see by using elementary calculus.

To prove (2), it is enough to consider, for a given $x_0 \in \,]0, 1]$, the functions

$$k(t) := 1 - t^{\lambda_1} \qquad (t \in \,]0, 1])$$

if $x_0 = 1$, and, when $x_0 < 1$,

$$k(t) := (\lambda_2 - \lambda_1) x_0^{\lambda_2} - \lambda_2 x_0^{\lambda_2 - \lambda_1} t^{\lambda_1} + \lambda_1 t^{\lambda_2} \qquad (t \in \,]0, 1]).$$

Of course, denoting by H the subspace generated by $\{x^{\lambda_1}, x^{\lambda_2}\}$, we have $H^* = \mathscr{C}_0(]0, 1])$ since x^{λ_1} is strictly positive on $]0, 1]$.

The same functions can be used to show (3). $\qquad\qquad\square$

Combining Proposition 4.2.4 with Proposition 4.1.6 it is easy to obtain more examples of Korovkin sets in spaces of continuous functions and in spaces of p-th power integrable functions $(1 \leq p < +\infty)$ both in bounded and unbounded intervals.

For Korovkin-type theorems on unbounded intervals we refer to Section 5 of Appendix D.2 and to the survey article of S. M. Eisenberg [1979].

In the next result the symbol ℓ^p $(1 \leq p < +\infty)$ denotes, as usual, the space of all sequences $(\alpha_n)_{n \in \mathbb{N}}$ of real numbers such that $\sum_{n=1}^{\infty} |\alpha_n|^p < +\infty$.

4.2.5 Proposition. *Consider* $0 < \lambda_1 < \lambda_2 < \lambda_3$ *and* $1 \leq p < +\infty$. *Then the following statements hold:*
(1) *The set* $\{1, x^{\lambda_1}, x^{\lambda_2}\}$ *is a* K_+-*subset in* $L^p([0,1])$.
(2) *The sets* $\{\exp(-x), \exp(-x)(1+x)^{\lambda_1}, \exp(-x)(1+x)^{\lambda_2}\}$ *and* $\{\exp(-\lambda_1 x), \exp(-\lambda_2 x), \exp(-\lambda_3 x)\}$ *are* K_+-*subsets both in* $\mathscr{C}_0([0, +\infty[)$ *and in* $L^p([0, +\infty[)$.
(3) *The set* $\{\exp(-x), \exp(-x)x^{\lambda_1}, \exp(-x)x^{\lambda_2}\}$ *is a* K_+-*subset both in* $\mathscr{C}_0([1, +\infty[)$ *and in* $L^p([1, +\infty[)$. *The same statement holds for* $\{x^{-\lambda_1}, x^{-\lambda_2}, x^{-\lambda_3}\}$ *provided* $\frac{1}{p} < \lambda_1$.
(4) *If* $\varphi \colon \mathbb{R} \to]0, 1[$ *is a strictly monotone continuous function satisfying* $\lim_{x \to -\infty} \varphi(x) = 0$ *and* $\lim_{x \to +\infty} \varphi(x) = 1$, *then the set* $\{\exp(-x^2), \exp(-x^2)\varphi(x)^{\lambda_1}, \exp(-x^2)\varphi(x)^{\lambda_2}\}$ *is a* K_+-*subset both in* $\mathscr{C}_0(\mathbb{R})$ *and* $L^p(\mathbb{R})$.
(5) *The set* $\{\exp(-\lambda_1 n), \exp(-\lambda_2 n), \exp(-\lambda_3 n)\}$ *is a* K_+-*subset both in* c_0 *and in* ℓ^p. *The same statement holds for* $\{n^{-\lambda_1}, n^{-\lambda_2}, n^{-\lambda_3}\}$ *provided* $\frac{1}{p} < \lambda_1$.
(6) *Consider the Banach lattice* $E_{\lambda_2} := \left\{ g \in \mathscr{C}([0, +\infty[) \, \middle| \, \dfrac{g(x)}{1 + x^{\lambda_2}} \text{ is convergent as } x \to +\infty \right\}$ *endowed with the norm* $\|f\|_* := \sup_{x \geq 0} \dfrac{|f(x)|}{1 + x^{\lambda_2}}$. *Then* $\{1, x^{\lambda_1}, x^{\lambda_2}\}$ *is a* K_+-*subset in* E_{λ_2}.
(7) *Let* $\mathscr{C}^*([0, +\infty[)$ *be the Banach lattice of all continuous functions on* $[0, +\infty[$ *converging at infinity endowed with the sup-norm. Then* $\{1, \exp(-\lambda_1 x), \exp(-\lambda_2 x)\}$ *is a* K_+-*subset in* $\mathscr{C}^*([0, +\infty[)$.

Proof. Statement (1) follows directly from Proposition 4.2.4, (3), and Corollary 4.1.7.

To prove (2), denote by F_2 either the space $\mathscr{C}_0([0, +\infty[)$ or the space $L^p([0, +\infty[)$ and consider the lattice homomorphism $T_2 \colon \mathscr{C}([0,1]) \to F_2$ and $S_2 \colon \mathscr{C}([0,1]) \to F_2$ defined by

$$T_2(f)(x) := \exp(-x)(1+x)^{\lambda_2} f\left(\frac{1}{1+x}\right)$$

and

$$S_2(f)(x) := \exp(-\lambda_1 x) f(\exp(-x))$$

for each $f \in \mathscr{C}([0,1])$ and $x \in [0, +\infty[$.

Then $\mathscr{K}([0, +\infty[)$ is contained in the ranges of T_2 and S_2 and so they are dense in F_2. Furthermore, T_2 maps the set $\{1, x^{\lambda_2 - \lambda_1}, x^{\lambda_2}\}$ onto the set $\{\exp(-x), \exp(-x)(1 + x)^{\lambda_1}, \exp(-x)(1 + x)^{\lambda_2}\}$, while S_2 maps the set $\{1, x^{\lambda_2 - \lambda_1}, x^{\lambda_3 - \lambda_1}\}$ onto the set $\{\exp(-\lambda_1 x), \exp(-\lambda_2 x), \exp(-\lambda_3 x)\}$. So the result follows from Proposition 4.1.6 and Proposition 4.2.4, (3).

The same reasoning may be used to show (3), (4) and (5) by considering the spaces $F_3 = \mathscr{C}_0([1, +\infty[)$ or $L^p([1, +\infty[)$, $F_4 = \mathscr{C}_0(\mathbb{R})$ or $L^p(\mathbb{R})$ and $F_5 = c_0$ or ℓ^p and the lattice homomorphisms $T_i: \mathscr{C}([0, 1]) \to F_i$ and $S_i: \mathscr{C}([0, 1]) \to F_i$ ($i = 3, 4, 5$) defined by

$$T_3(f)(x) := \exp(-x)x^{\lambda_2} f\left(\frac{1}{x}\right),$$

$$S_3(f)(x) := x^{-\lambda_1} f\left(\frac{1}{x}\right),$$

$$T_4(f)(x) := S_4(f)(x) := \exp(-x^2) f(\varphi(x)),$$

$$T_5(f)(x) := S_2(f)(x),$$

$$S_5(f)(x) := S_3(f)(x)$$

for every $f \in \mathscr{C}([0, 1])$ and $x \in [1, +\infty[$ ($x \in \mathbb{R}$, $x \in \mathbb{N}$, respectively).

To show (6) consider the lattice homomorphism $S: \mathscr{C}([0, 1]) \to E_{\lambda_2}$ defined by

$$Sf(x) := (1 + x^{\lambda_2}) f(\exp(-x))$$

for every $f \in \mathscr{C}([0, 1])$ and $x \in [0, +\infty[$.

The operator S is an isometric lattice isomorphism and its inverse is the operator T defined by

$$Tg(t) := \frac{g(\log 1/t)}{1 + \log^{\lambda_2}(1/t)}$$

for every $f \in E_{\lambda_2}$ and $t \in [0, 1]$ (of course, $Tg(0) := \lim_{t \to 0^+} Tg(t) = \lim_{x \to +\infty} \frac{g(x)}{1 + x^{\lambda_2}}$).

Consider the three functions

$$f_1(t) := \frac{1}{1 + \log^{\lambda_2}(1/t)}, \quad f_2(t) := \frac{\log^{\lambda_1}(1/t)}{1 + \log^{\lambda_2}(1/t)}, \quad f_3(t) := \frac{\log^{\lambda_2}(1/t)}{1 + \log^{\lambda_2}(1/t)}$$
$$(t \in [0, 1])$$

(where $f_k(0) := \lim_{t \to 0^+} f_k(t)$ for every $k = 1, 2, 3$).

Then $\{f_1, f_2, f_3\}$ is a strict K_+-subset in $\mathscr{C}([0,1])$. Indeed, if $t_0 \in {]}0,1]$, then the function

$$k(t) := (\lambda_2 - \lambda_1) \log^{\lambda_2}\left(\frac{1}{t_0}\right) f_1(t) - \lambda_2 \log^{\lambda_2 - \lambda_1}\left(\frac{1}{t_0}\right) f_2(t) + \lambda_1 f_3(t), \quad (t \in [0,1]),$$

is positive and vanishes only at t_0.

At $t_0 = 0$ we can choose f_1 as a positive function which vanishes only at t_0.

Now the result follows because $S(f_1)(x) = 1$, $S(f_2)(x) = x^{\lambda_1}$ and $S(f_3)(x) = x^{\lambda_2}$ $(x \in [0, +\infty[)$.

Statement (7) can be proved by an analogous reasoning considering the isometric lattice isomorphism $S: \mathscr{C}([0,1]) \to \mathscr{C}^*([0,+\infty[)$ defined by $Sf(x) := f(\exp(-x))$ $(f \in \mathscr{C}([0,1]))$ and $x \in [0,+\infty[$ which maps $\mathbf{1}$, x^{λ_1}, x^{λ_2} in $\mathbf{1}$, $\exp(-\lambda_1 x)$, $\exp(-\lambda_2 x)$, respectively. $\qquad\square$

Remark. Statement (5) of the above Proposition 4.2.5 is a particular case of a more general result. In fact, Kitto and Wulbert [1976] proved that if H is a finite dimensional subspace of some ℓ^p $(1 \le p < +\infty)$ and if it contains a strictly positive element, then H is K_+-subspace in ℓ^p if and only if H is a K_+-subspace in c_0 (see also Flösser [1979] and Donner [1982]).

Again, by applying Proposition 4.2.4 and (4.1.10) to the homomorphisms $\sigma_1: [0,+\infty[\to {]}0,1]$ and $\sigma_2: [1,+\infty[\to {]}0,1]$ defined respectively by

$$\sigma_1(x) := \exp(-x) \quad (x \in [0,+\infty[)$$

and

$$\sigma_2(x) := x^{-1} \quad (x \in [1,+\infty[),$$

we obtain more examples of K_+^1-subsets.

4.2.6 Proposition. *If* $0 < \lambda_1 < \lambda_2$, *then*
(1) *the set* $\{\exp(-\lambda_1 x), \exp(-\lambda_2 x)\}$ *is a* K_+^1-*subset in* $\mathscr{C}_0([0,+\infty[)$.
(2) *the set* $\{x^{-\lambda_1}, x^{-\lambda_2}\}$ *is a* K_+^1-*subset in* $\mathscr{C}_0([1,+\infty[)$.
(3) *the sets* $\{\exp(-\lambda_1 n), \exp(-\lambda_2 n)\}$ *and* $\{n^{-\lambda_1}, n^{-\lambda_2}\}$ *are* K_+^1-*subsets in* c_0.

Proof. We have only to show (3). In fact, this follows from (1), (2) and Proposition 4.1.10 by considering the lattice homomorphism $S: \mathscr{C}_0([0,+\infty[) \to c_0$ $(S: \mathscr{C}_0([1,+\infty[) \to c_0$, respectively) defined as $S(f)(n) := f(n)$ $(n \in \mathbb{N})$. $\qquad\square$

For $\lambda_1 = 1$, $\lambda_2 = 2$ and $\lambda_3 = 3$ statements (1) and (2) of Proposition 4.2.4 have been proved by Bauer and Donner [1978], while statement (3) leads to the famous Korovkin's theorem in $\mathscr{C}([0,1])$ (Korovkin [1953]).

4.2.7 Theorem. *The set* $\{1, x, x^2\}$ *is a (strict)* K_+*-subset in* $\mathscr{C}([0, 1])$.

Korovkin's theorem, (often called *Korovkin's first theorem*) has many important applications in the study of positive approximation processes in $\mathscr{C}([0, 1])$ which we shall present in Chapter 5.

However, at moment we recall that, more explicitly, Korovkin's theorem means that, if $(L_i)_{i \in I}^{\leq}$ is a net of positive linear operators (or linear contractions) on $\mathscr{C}([0, 1])$, it is enough to verify that

$$\lim_{i \in I} L_i(e_k) = e_k \quad \text{uniformly on } [0, 1]$$

for each $k = 0, 1, 2$, where $e_k(x) = x^k$ $(x \in [0, 1])$, to obtain

$$\lim_{i \in I} L_i(f) = f \quad \text{uniformly on } [0, 1]$$

for every $f \in \mathscr{C}([0, 1])$.

If each L_i leaves invariant the subspace of all polynomials on $[0, 1]$ of degree ≤ 2, then these last conditions are satisfied if there exist $x_1, x_2, x_3 \in [0, 1]$ such that

$$\lim_{i \in I} L_i(e_k)(x_h) = x_h^k, \quad k = 0, 1, 2, h = 1, 2, 3.$$

Similar remarks hold in $\mathscr{C}_0(]0, 1])$ for the subset $\{x, x^2, x^3\}$. In this case it is necessary to require that the subspace of all polynomials of degree ≤ 3 is invariant under each L_i $(i \in I)$.

Here, we present a first simple application of Korovkin's theorem.

Examples.
1. Bernstein operators. Consider the sequence $(B_n)_{n \in \mathbb{N}}$ of *Bernstein polynomial operators* on $[0, 1]$ defined by (2.5.3). Thus, for every $f \in \mathscr{C}([0, 1])$ and $x \in [0, 1]$,

$$B_n(f)(x) := \sum_{k=0}^{n} f\left(\frac{k}{n}\right)\binom{n}{k} x^k (1 - x)^{n-k}.$$

Consider the positive projection $T_1 \colon \mathscr{C}([0, 1]) \to \mathscr{C}([0, 1])$ defined by (3.3.12), i.e.,

$$T_1(f)(x) := (1 - x)f(0) + xf(1)$$

for every $f \in \mathscr{C}([0, 1])$ and $x \in [0, 1]$ (note that $T_1 = B_1$).

According to (3.3.13) and Theorem 4.2.7, $\{1, x, x^2\}$ is a K_+-subset both for T_1 and the identity operator.

Consequently, taking (2.5.15), (2.5.16) and (2.5.18) into account, for every $f \in \mathscr{C}([0,1])$ we have

(1) $\lim\limits_{n \to \infty} B_n^p(f) = f$ uniformly on $[0,1]$ for every $p \geq 1$.

(2) $\lim\limits_{p \to \infty} B_n^p(f) = T_1(f)$ uniformly on $[0,1]$ for every $n \geq 1$.

(3) If $(k(n))_{n \in \mathbb{N}}$ is a sequence of positive integers, then

$$\lim_{n \to \infty} B_n^{k(n)}(f) = \begin{cases} f & \text{uniformly on } [0,1], \quad \text{if } \lim\limits_{n \to \infty} \dfrac{k(n)}{n} = 0, \\[3em] T_1(f) & \text{uniformly on } [0,1], \quad \text{if } \lim\limits_{n \to \infty} \dfrac{k(n)}{n} = +\infty, \end{cases}$$

(here, B_n^p denotes the p-th power of B_n).

In Section 5.2 and Chapter 6 we shall considerably extend these results to a more general context including multidimensional simplices and hypercubes. There the reader will also find several estimates of the rate of convergence.

The above formula (1) gives in particular a constructive proof of the *classical Weierstrass theorem*, i.e.,

The space of all polynomials is dense in $\mathscr{C}([0,1])$.

On the other hand Korovkin's theorem 4.2.7 can be also derived from the Weierstrass theorem.

Indeed, set $e_p(t) := t^p$ ($t \in [0,1]$, $p \in \mathbb{N}_0$). Then, since

$$|x^p - t^p| = |x - t| |x^{p-1} + x^{p-2}t + \cdots + xt^{p-2} + t^{p-1}| \leq p|x - t|$$

for every $x, t \in [0,1]$ and $p \geq 1$, we obtain

$$|e_p - t^p \cdot 1| \leq p|e_1 - t \cdot 1| \quad \text{for every } t \in [0,1].$$

Accordingly, if $(L_i)_{i \in I}^{\leq}$ is a net of positive linear operators from $\mathscr{C}([0,1])$ into $\mathscr{C}([0,1])$, by using the Cauchy-Schwarz inequality (1.2.4), for every $i \in I$ and $t \in [0,1]$ we obtain

$$|L_i(e_p) - t^p L_i(\mathbf{1})| \leq L_i(|e_p - t^p \cdot 1|) \leq p L_i(|e_1 - t \cdot 1|)$$

$$\leq p\sqrt{L_i(\mathbf{1})}\sqrt{L_i((e_1 - t \cdot 1)^2)}$$

$$= p\sqrt{L_i(\mathbf{1})}\sqrt{L_i(e_2) - 2tL_i(e_1) + t^2 L_i(\mathbf{1})}.$$

In other words

$$|L_i(e_p) - e_p L_i(1)| \leq p\sqrt{L_i(1)}\sqrt{L_i(e_2) - 2e_1 L_i(e_1) + e_2 L_i(1)}.$$

There, if $L_i(e_p) \to e_p$ for $p = 0, 1, 2$, then $L_i(e_p) \to e_p$ for every $p \in \mathbb{N}$ and hence $L_i(f) \to f$ for every $f \in \mathscr{C}([0,1])$ by virtue of Weierstrass theorem and the fact that $\sup_{i \geq i_0} \|L_i\| = \sup_{i \geq i_0} \|L_i(1)\| < +\infty$ for some $i_0 \in I$.

Summing up, the *Weierstrass theorem and Korovkin's first theorem are equivalent*.

In Section 4.4 (see Theorem 4.4.6) we shall see that also Stone's generalization of the Weierstrass theorem is equivalent to a more general Korovkin-type theorem.

2. Bernstein-Stancu operators. Consider the sequence $(B_{n,\alpha,\beta})_{n \in \mathbb{N}}$ of Bernstein-Stancu operators defined by (2.5.2) i.e.,

$$B_{n,\alpha,\beta}(f)(x) := \sum_{k=0}^{n} f\left(\frac{k+\alpha}{n+\beta}\right)\binom{n}{k} x^k (1-x)^{n-k}$$

for every $f \in \mathscr{C}([0,1])$ and $x \in [0,1]$ $(0 \leq \alpha \leq \beta)$.

By (2.5.12)–(2.5.14) or, more generally, by (2.5.15)–(2.5.19) and by virtue of Korovkin's theorem 4.2.7, we obtain

$$\lim_{n \to \infty} B_{n,\alpha,\beta}^p(f) = f \quad \text{uniformly on } [0,1]$$

for every $f \in \mathscr{C}([0,1])$ and $p \in \mathbb{N}$.

Furthermore it is also possible to show that

$$\|B_{n,\alpha,\beta}(f) - f\|$$

$$\leq \begin{cases} \dfrac{3}{2}\omega(f,(n+4\alpha^2)^{-1/2}), & \text{if } 0 \leq \alpha \leq \dfrac{1}{4}, \beta \leq 2\alpha \\[4mm] & \text{or } \dfrac{1}{4} \leq \alpha \leq \dfrac{1}{2}, 4\alpha^2 \leq \beta \leq 2\alpha, \\[4mm] \left(1 + \dfrac{4\alpha^2+1}{2\beta+2}\right)\omega(f,(n+4\alpha^2)^{-1/2}), & \text{if } \dfrac{1}{4} \leq \alpha \leq \dfrac{1}{2}, \alpha \leq \beta \leq 4\alpha^2 \\[4mm] & \text{or } \dfrac{1}{2} \leq \alpha \leq \beta \leq 2\alpha, \\[4mm] \dfrac{3}{2}\omega(f,(n+(\beta-\alpha)^2)^{-1/2}), & \text{if } \alpha \leq \beta - \sqrt{\beta}/2, \beta \leq 1, \\[4mm] \left(1 + \dfrac{4(\beta-\alpha)^2+1}{2\beta+2}\right)\omega(f,(n+(\beta-\alpha)^2)^{-1/2}), & \text{if } \alpha \leq \beta/2, \beta \geq 1, \end{cases}$$

where for a given $f \in \mathscr{C}([0,1])$ and $\delta > 0$

$$\omega(f,\delta) := \sup\{|f(x) - f(y)| \,|\, x, y \in [0,1], |x - y| \le \delta\}$$

denotes the modulus of continuity of f (see Section 5.1 for more details on $\omega(f,\delta)$) (see Stancu [1969a]).

By using Corollary 4.2.2 it is possible to easily derive Korovkin's second theorem.

Set $\mathbb{T} := \{z = (x,y) \in \mathbb{R}^2 \,|\, \|z\|^2 := x^2 + y^2 = 1\}$ and consider the canonical projections $\mathrm{pr}_1, \mathrm{pr}_2 \in \mathscr{C}(\mathbb{T})$ (see (2.6.19)). If $z_0 = (x_0, y_0) \in \mathbb{T}$, then the function $h(z) := 2(1 - x_0\mathrm{pr}_1(z) - y_0\mathrm{pr}_2(z)) = \|z - z_0\|^2$ $(z \in \mathbb{T})$, is positive and vanishes only at z_0.

Thus $\{\mathbf{1}, \mathrm{pr}_1, \mathrm{pr}_2\}$ is a strict K_+-subset in $\mathscr{C}(\mathbb{T})$.

But it is easy to see that the space $\mathscr{C}(\mathbb{T})$ is isometrically (order) isomorphic to the space

$$\mathscr{C}_{2\pi} := \{f \in \mathscr{C}([-\pi, \pi]) \,|\, f(-\pi) = f(\pi)\}.$$

Moreover, under this identification, the set $\{\mathbf{1}, \mathrm{pr}_1, \mathrm{pr}_2\}$ is mapped onto the set $\{\mathbf{1}, \sin, \cos\}$ and hence we arrive to *Korovkin's second theorem*.

4.2.8 Theorem. *The subset $\{\mathbf{1}, \sin, \cos\}$ is a (strict) K_+-subset in $\mathscr{C}_{2\pi}$.*

This last theorem will be used in Section 5.4 in the study of the approximation properties of convolution operators.

Now, for the sake of completeness, we present a direct proof of Korovkin's theorems 4.2.7 and 4.2.8.

We shall use an idea due to Korovkin [1960, pp. 18–19] and which has been subsequently developed by many other authors. Hence we follow the approach used by Nishishiraho [1983]. The setting is sufficiently general to include also a direct proof of many other results scattered in this book.

For another direct proof see Bauer [1978a].

Consider a compact Hausdorff space X and set $X^2 := X \times X$.

We shall denote by $\mathscr{B}(X)$ the Banach lattice of all real-valued bounded functions on X endowed with the sup-norm.

If $\phi: X^2 \to \mathbb{R}$ is a function such that $\phi(\cdot, y) \in \mathscr{C}(X)$ for each $y \in X$ and if $L: \mathscr{C}(X) \to \mathscr{B}(X)$ is a linear operator, we set

$$\sigma(L, \phi) := \sup_{y \in X} |L(\phi(\cdot, y))(y)|. \tag{4.2.2}$$

From now on we fix a positive bounded function $\phi: X^2 \to \mathbb{R}$ such that

$$\phi(\cdot, y) \in \mathscr{C}(X) \quad \text{for each } y \in X \tag{4.2.3}$$

and

$$0 < \inf_{(x,y) \in F} \phi(x,y) \qquad (4.2.4)$$

for every closed subset F of X^2 disjoint from the diagonal $D := \{(t,t) | t \in X\}$ of X^2.

Clearly, if ϕ is continuous and $\phi(x,y) > 0$ for every $x, y \in X$, $x \neq y$, then (4.2.3) and (4.2.4) are fulfilled.

We need the following preliminary result.

4.2.9 Lemma. *Let X be a compact Hausdorff space and let $\phi \colon X^2 \to \mathbb{R}$ be a positive bounded function satisfying (4.2.3) and (4.2.4).*

Consider a family $(L_{i,\lambda})_{i \in I}^{\leq}$, $\lambda \in \Lambda$, of nets of positive linear operators from $\mathscr{C}(X)$ into $\mathscr{B}(X)$ such that

$$\sup_{\lambda \in \Lambda, i \geq i_0} \|L_{i,\lambda}(1)\| < +\infty \quad \text{for some } i_0 \in I \qquad (4.2.5)$$

and

$$\lim_{i \in I} {}^{\leq} \sigma(L_{i,\lambda}, \phi) = 0 \quad \text{uniformly in } \lambda \in \Lambda. \qquad (4.2.6)$$

Then, for every $\psi \in \mathscr{C}(X^2)$ vanishing on the diagonal D of X^2, one also has

$$\lim_{i \in I} {}^{\leq} \sigma(L_{i,\lambda}, \psi) = 0 \quad \text{uniformly in } \lambda \in \Lambda.$$

Proof. Let $\psi \in \mathscr{C}(X^2)$ be such that $\psi(t,t) = 0$ for every $t \in X$.

Given $\varepsilon > 0$, for each $t \in X$ there exists an open neighborhood V_t of (t,t) in X^2 such that $|\psi(x,y)| \leq \varepsilon$ for every $(x,y) \in V_t$. Setting

$$F := X^2 \setminus \bigcup_{t \in X} V_t,$$

$$m := \inf_{(x,y) \in F} \phi(x,y),$$

$$M := \max_{(x,y) \in F} |\psi(x,y)|$$

and by using (4.2.4), one easy obtains that

$$|\psi(x,y)| \leq \varepsilon + \frac{M}{m} \phi(x,y) \quad \text{for each } (x,y) \in X^2.$$

Consequently, for each $\lambda \in \Lambda$, $i \in I$ and $y \in X$

$$|L_{i,\lambda}(\psi(\cdot, y))(y)| \leq \varepsilon L_{i,\lambda}(\mathbf{1})(y) + \frac{M}{m} L_{i,\lambda}(\phi(\cdot, y))(y)$$

and hence

$$\sigma(L_{i,\lambda}, \psi) \leq \varepsilon \|L_{i,\lambda}(\mathbf{1})\| + \frac{M}{m} \sigma(L_{i,\lambda}, \phi).$$

The result now follows from (4.2.5) and (4.2.6).

We may now state the following generalization of Korovkin's theorems.

4.2.10 Theorem. *Let X be a compact Hausdorff space and let $\phi: X^2 \to \mathbb{R}$ be a positive bounded function satisfying (4.2.3) and (4.2.4). Consider a family $(L_{i,\lambda})_{i \in I}^{\leq}$, $\lambda \in \Lambda$, of nets of positive linear operators from $\mathscr{C}(X)$ into $\mathscr{B}(X)$ satisfying (4.2.6), i.e.,*

$$\lim_{\substack{\leq \\ i \in I}} \sigma(L_{i,\lambda}, \phi) = 0 \quad \text{uniformly in } \lambda \in \Lambda.$$

If $\lim_{\substack{\leq \\ i \in I}} L_{i,\lambda}(f_0) = f_0$ uniformly in $\lambda \in \Lambda$ for some strictly positive function $f_0 \in \mathscr{C}(X)$, then, for every $f \in \mathscr{C}(X)$, $\lim_{\substack{\leq \\ i \in I}} L_{i,\lambda}(f) = f$ uniformly in $\lambda \in \Lambda$.

Proof. We set $m := \min_{x \in X} f_0(x)$. Then, since $\mathbf{1} \leq \frac{1}{m} f_0$, we have $L_{i,\lambda}(\mathbf{1}) \leq \frac{1}{m} L_{i,\lambda}(f_0)$ for every $i \in I$, $\lambda \in \Lambda$, so that condition (4.2.5) of Lemma 4.2.9 is satisfied.

Given $f \in \mathscr{C}(X)$ and denoting by $\psi \in \mathscr{C}(X^2)$ the function defined by $\psi(x, y) := f(x) - \frac{f(y)}{f_0(y)} f_0(x)$, $((x, y) \in X^2)$, it follows from Lemma 4.2.9 that $\lim_{\substack{\leq \\ i \in I}} \sigma(L_{i,\lambda}, \psi) = 0$ uniformly in $\lambda \in \Lambda$, and hence

$$\lim_{\substack{\leq \\ i \in I}} L_{i,\lambda}(f) - \frac{f}{f_0} L_{i,\lambda}(f_0) = 0 \quad \text{uniformly in } \lambda \in \Lambda.$$

Now the result follows from our hypothesis on f_0 and the identity

$$L_{i,\lambda}(f) - f = L_{i,\lambda}(f) - \frac{f}{f_0} L_{i,\lambda}(f_0) + \frac{f}{f_0}(L_{i,\lambda}(f_0) - f_0). \qquad \square$$

A simple case where condition (4.2.6) of Lemma 4.2.9 is satisfied occurs if $\phi(t,t) = 0$ for every $t \in X$ and

$$\lim_{\substack{\le \\ i \in I}} L_{i,\lambda}(\phi(\cdot, y)) = \phi(\cdot, y) \quad \text{uniformly in } y \in X \text{ and } \lambda \in \Lambda.$$

Here we briefly indicate some applications of the above result.

Examples. 1. Take $X := [a, b]$ and consider $\phi: X^2 \to \mathbb{R}$ defined by

$$\phi(x, y) := (x - y)^2 \quad ((x, y) \in X^2).$$

In this case (4.2.3) and (4.2.4) are satisfied. If $(L_{i,\lambda})_{i \in I}^{\le}$ $(\lambda \in \Lambda)$ is a family of nets of positive linear operators from $\mathscr{C}(X)$ into $\mathscr{B}(X)$ such that for every $h \in \{1, x, x^2\}$

$$\lim_{\substack{\le \\ i \in I}} L_{i,\lambda}(h) = h \quad \text{uniformly in } \lambda \in \Lambda,$$

then (4.2.6) holds (see the above remark) and hence, by Theorem 4.2.10, we conclude that for every $f \in \mathscr{C}(X)$, $\lim_{\substack{\le \\ i \in I}} L_{i,\lambda}(f) = f$ uniformly in $\lambda \in \Lambda$.

Thus we obtain in particular Korovkin's theorem 4.2.7.

Here we present another more direct proof of Korovkin's theorem which is independent of Theorem 4.2.10. For a given $f \in \mathscr{C}([0,1])$ and $\delta > 0$ consider again the modulus of continuity

$$\omega(f, \delta) := \sup\{|f(x) - f(y)| \, | \, x, y \in [0, 1], |x - y| \le \delta\}.$$

Then, for every $x, y \in [0, 1]$ we have

$$|f(x) - f(y)| \le \frac{2}{\delta} \|f\| (x - y)^2 + \omega(f, \delta).$$

Hence for every $y \in [0, 1]$ and $(i, \lambda) \in I \times \Lambda$,

$$|L_{i,\lambda}(f) - f(y)L_{i,\lambda}(1)| \le \frac{2}{\delta} \|f\| |L_{i,\lambda}(e_2) - 2yL_{i,\lambda}(e_1) + y^2L_{i,\lambda}(1)| + \omega(f, \delta)L_{i,\lambda}(1),$$

where $e_k(x) := x^k$ $(x \in [0, 1], k = 1, 2)$.

Accordingly,

$$\|L_{i,\lambda} - fL_{i,\lambda}(1)\| \le \frac{2}{\delta} \|f\| \|L_{i,\lambda}(e_2) - 2e_1 L_{i,\lambda}(e_1) + e_2 L_{i,\lambda}(1)\| + \omega(f, \delta)\|L_{i,\lambda}(1)\|.$$

Thus the result follows because of the assumptions on the net $(L_{i,\lambda})_{(i,\lambda)\in I\times\Lambda}$ and the fact that $\omega(f,\delta)\to 0$ provided $\delta\to 0$ by the uniform continuity of f (see Lemma 5.1.1, (2)).

2. Let X be a compact Hausdorff space and let $M=\{h_n|n\in\mathbb{N}\}$ be a finite or countable subset of $\mathscr{C}(X)$ which separates the points of X and such that $\sum_{n=1}^{\infty}h_n^2$ is uniformly convergent on X. Consider the function $\phi\in\mathscr{C}(X^2)$ defined by

$$\phi(x,y):=\sum_{n=1}^{\infty}(h_n(x)-h_n(y))^2 \quad ((x,y)\in X^2).$$

Applying Theorem 4.2.10 to this function, one obtains a direct proof of Theorem 4.4.6, (3), below.

If $X=\mathbb{T}$ and $M=\{\mathrm{pr}_1,\mathrm{pr}_2\}$, then one obtains Korovkin's second theorem 4.2.8.

A more direct proof can be given according to the above Example 1.

3. More generally, let $f_1,\ldots,f_n\in\mathscr{C}(X)$, where X is a compact Hausdorff space, and suppose that there exist finitely many functions $a_1,\ldots,a_n\colon X\to\mathbb{R}$ such that the function

$$\phi(x,y):=\sum_{i=1}^{n}a_i(y)f_i(x) \quad ((x,y)\in X^2)$$

is positive, satisfies (4.2.4) and $\phi(t,t)=0$ for each $t\in X$.

In particular, this implies that $\{f_1,\ldots,f_n\}$ is a strict K_+-subset in $\mathscr{C}(X)$ (see (4.2.1)).

If $(L_{i,\lambda})_{i\in I}^{\leq}$, $\lambda\in\Lambda$, is a family of nets of positive linear operators from $\mathscr{C}(X)$ into $\mathscr{B}(X)$ such that for every $j=1,\ldots,n$

$$\lim_{i\in I}{}^{\leq} L_{i,\lambda}(f_j)=f_j \quad \text{uniformly in } \lambda\in\Lambda,$$

then (4.2.6) is fulfilled. Moreover, the further hypothesis of Theorem 4.2.10 is satisfied by $f_0:=\phi(\cdot,y_1)+\phi(\cdot,y_2)$, where y_1 and y_2 are distinct points of X. Hence, from Theorem 4.2.10 we obtain, in a direct way, that $\{f_1,\ldots,f_n\}$ is a K_+-subset in $\mathscr{C}(X)$ (see Corollary 4.2.2).

4. Let (X,d) be a compact metric space and consider a strictly increasing continuous function $u\colon\mathbb{R}_+\to\mathbb{R}$ such that $u(0)=0$. Then the function $\phi\colon X^2\to\mathbb{R}$ defined by

$$\phi(x,y):=u(d(x,y)) \quad ((x,y)\in X^2),$$

satisfies (4.2.3) and (4.2.4).

Notes and references

Strict Korovkin subspaces were first considered by Volkov [1958] and Shashkin [1962] and, subsequently, studied by Lorentz [1972], Berens and Lorentz [1975], [1976], Amir and Ziegler [1978]. The example subsequent to Corollary 4.2.2 is due to Shashkin [1962].

Statement (7) of Proposition 4.2.5 in the case $\lambda_1 = 1$ and $\lambda_2 = 2$ is due to Bojanov and Veselinov [1970].

Korovkin's theorems 4.2.7 and 4.2.8 were obtained by Korovkin in 1953. However, in 1952 Bohman showed that $\{1, x, x^2\}$ is a Korovkin subspace in $\mathscr{C}([0,1])$ with respect to the class of positive linear operators L on $\mathscr{C}([0,1])$ of the form

$$Lf(x) = \sum_{i \in I} f(a_i)\varphi_i(x) \quad (x \in [0,1]),$$

where $(a_i)_{i \in I}$ is a finite family in $[0,1]$ and every φ_i is a continuous function on $[0,1]$. For these reasons, Theorem 4.2.7 is also referred to as the Bohman-Korovkin theorem. However, the germ of this theorem can already be found in a paper by Popoviciu [1951].

The classical Weierstrass approximation theorem appeared in Weierstrass [1885]. An interesting survey on the extensions and ramifications of this theorem, both from a qualitative and a quantitative point of view, can be found in Lubinsky [1993].

Bernstein polynomials were introduced by Bernstein [1912] to give his famous proof of the Weierstrass theorem by means of probabilistic considerations (to this respect see Section 5.2). Qualitative and quantitative properties of them together with several extensions and generalizations will be shown in Section 5.2 and in the whole Chapter 6.

For a comprehensive treatment of Bernstein polynomials and their generalizations see the classical monograph of Lorentz [1986b] and the papers listed in the bibliography of Gonska and Meier-Gonska [1983], [1986].

Results concerning the behavior of the powers of Bernstein polynomials first appeared in Sikkema [1966] and Kelisky and Rivlin [1967] (see also Micchelli [1973b]). For some generalizations we refer to Section 6.2.

Historical notes about Bernstein-Stancu operators can be found in the final notes of Section 2.5.

An extension of Theorem 4.2.8 to the bidimensional case was established by Morozov [1958] and to the general n-dimensional case by Džafarov [1975].

For other extensions to the setting of (locally) compact abelian groups we refer the reader to Altomare [1981], Bloom and Sussich [1980] and Pannenberg [1990a] (see also Appendix A, Theorem A.2.2).

Theorem 4.2.10 is due to Nishishiraho [1983]. In the same paper the interested reader can also find quantitative estimates of the rates of convergence of the approximation processes.

The same theorem can be found in Prolla [1988]. There the proof is based on a Stone-Weierstrass type theorem.

A very similar result was also established by Yoshinaga and Tamura [1976].

Example 3 to Theorem 4.2.10 can be found in Lorentz [1986a, Section 1.3].

Statements (1) and (2) of Proposition 4.2.5, with $\lambda_1 = 1$ and $\lambda_2 = 2$, were first obtained by Dzjadyk [1966] and Wolff [1977], respectively.

4.3 Korovkin closures and state spaces.
Spaces of parabola-like functions

This section is primarily devoted to the description of other properties of Korovkin closures via the state spaces where it is possible to apply Choquet's integral representation theory for convex compact sets. In the last part we shall also determine the Korovkin closures of subspaces of so-called parabola-like functions.

A short review on the needed results from Choquet's integral representation theory can be found in Section 1.5.

Let X be a *compact* Hausdorff space and H a closed subspace of $\mathscr{C}(X)$ which contains the constant functions and separates the points of X.

Consider the canonical embedding $\Phi\colon X \to H'$ defined by (1.4.19) and the state space

$$K = \overline{\mathrm{co}}(\Phi(X)) = \{\mu \in H' \,|\, \mu(\mathbf{1}) = 1 = \|\mu\|\}$$

(see Section 1.4).

We know that K is a convex compact set and, by virtue of Corollary 2.6.5,

$$\Phi(\partial_H^+ X) = \partial_e K \qquad (4.3.1)$$

Hence, by the Krein-Milman theorem, we have

$$K = \overline{\mathrm{co}}(\Phi(\partial_H^+ X)). \qquad (4.3.2)$$

By (1.4.22) we have

$$H = \{a \circ \Phi \,|\, a \in A(K)\};$$

hence, if we set

$$A(K, \Phi(X)) := \{g \in \mathscr{C}(\Phi(X)) \,|\, \text{There exists } a \in A(K) \text{ such that } a_{|\Phi(X)} = g\}, \qquad (4.3.3)$$

then, by (4.1.9), we obtain

$$K_+(H) = \{f \circ \Phi \,|\, f \in K_+(A(K, \Phi(X)))\}. \qquad (4.3.4)$$

On the basis of this equality we now proceed to study the subspace $K_+(A(K, \Phi(X)))$.

For the sake of formal convenience we shall work in a more general setting.

Let K be a convex compact subset of some locally convex Hausdorff space and consider a compact subset K_1 of K such that

$$\partial_e K \subset K_1. \qquad (4.3.5)$$

We set

$$A(K, K_1) := \{g \in \mathscr{C}(K_1) | \text{There exists } a \in A(K) \text{ such that } a_{|K_1} = g\}. \quad (4.3.6)$$

Theorem 4.1.4 implies that

$$K_+(A(K, K_1))$$

$$= \left\{ f \in \mathscr{C}(K_1) \middle| \sup_{\substack{a \in A(K) \\ a \le f \text{ on } K_1}} a(x) = f(x) = \inf_{\substack{b \in A(K) \\ f \le b \text{ on } K_1}} b(x) \quad \text{for every } x \in K_1 \right\}$$

$$= \{f \in \mathscr{C}(K_1) | \text{ If } \mu \in \mathscr{M}_1^+(K_1) \text{ and } x \in K_1 \text{ and } r(\mu) = x, \text{ then } \mu(f) = f(x)\}. \quad (4.3.7)$$

In (4.3.7) $r(\mu)$ denotes the resultant of the canonical extension of μ over K (see (1.2.44)).

The functions in $K_+(A(K, K_1))$ are also called *pseudo-affine functions* on K_1.

If K_2 is a convex compact subset of K_1 and $f \in \mathscr{C}(K_1)$ is pseudo-affine on K_1, then obviously f is affine on K_2. Thus, taking Theorem 1.4.9, (3), into account, we have

$$K_+(A(K, K_1)) = A(K_1) \quad \text{if } K_1 \text{ is convex.} \quad (4.3.8)$$

In particular (since $A(K, K) = A(K)$),

$$K_+(A(K)) = A(K). \quad (4.3.9)$$

On the other hand, if $\partial_e K$ is closed, then by virtue of (4.3.7) and by Proposition 2.6.3 we have

$$K_+(A(K, \partial_e K)) = \mathscr{C}(\partial_e K). \quad (4.3.10)$$

Our first goal will be the characterization of the pseudo-affine functions in terms of (affine) *boundary dependences*, which are, by definition, those measures $\mu \in \mathscr{M}(K)$ such that there exist two measures $\mu_1, \mu_2 \in \mathscr{M}^+(K)$, which are maximal for the Choquet-Meyer ordering (see (1.5.5)) and which satisfy the following properties:

$$\mu = \mu_1 - \mu_2, \quad (4.3.11)$$

$$\|\mu_1\| = \|\mu_2\| \quad \text{and} \quad r(\mu_1) = r(\mu_2). \quad (4.3.12)$$

For every $x \in K$ we shall denote by $\mathscr{D}(x)$ the subset of all boundary depen-
dences $\mu = \mu_1 - \mu_2$, such that $\|\mu_1\| > 0$ (or $\|\mu_2\| > 0$) and $r(\mu_1) = r(\mu_2) = x$.

So, if K is a Choquet simplex, then $\mathscr{D}(x) = \{0\}$ for every $x \in K$.

In the sequel by a common abuse of language every measure on K supported
by $\overline{\partial_e K}$ (i.e., whose support is contained in $\overline{\partial_e K}$) will be also considered as a
measure on K_1.

4.3.1 Theorem. *Under the above assumptions, for a given $f \in \mathscr{C}(\overline{\partial_e K})$, the following statements are equivalent:*

(i) *There exists $g \in K_+(A(K, K_1))$ such that $g_{|\overline{\partial_e K}} = f$.*

(ii) (1) $f \in K_+(A(K, \overline{\partial_e K}))$.

(2) $\mu(f) = 0$ *for every $x \in K_1 \setminus \overline{\partial_e K}$ and $\mu \in \mathscr{D}(x)$.*

Proof. (i) $\Rightarrow$ (ii): Consider the two functions $\bar{f} : K \to \mathbb{R}$ and $\underline{f} : K \to \mathbb{R}$ defined by

$$\bar{f}(x) := \inf_{\substack{b \in A(K) \\ f \leq b \text{ on } \overline{\partial_e K}}} b(x) \tag{4.3.13}$$

and

$$\underline{f}(x) := \sup_{\substack{a \in A(K) \\ a \leq f \text{ on } \overline{\partial_e K}}} a(x) \tag{4.3.14}$$

for every $x \in K$.

Thus, $\bar{f}$ is concave and upper semi-continuous and $\underline{f}$ is convex and lower
semi-continuous. Clearly, on $\overline{\partial_e K}$ we have $\sup\limits_{\substack{a \in A(K) \\ a \leq g \text{ on } K_1}} a \leq \underline{f} \leq \bar{f} \leq \inf\limits_{\substack{b \in A(K) \\ g \leq b \text{ on } K_1}} b$.

Hence, from these relations and from (4.3.7) part (1) of statement (ii) follows.

Part (ii) is also a consequence of (4.3.7) since maximal measures are supported
by $\overline{\partial_e K}$.

(ii) $\Rightarrow$ (i): Fix $x \in K_1 \setminus \overline{\partial_e K}$ and put $\alpha := \underline{f}(x)$ and $\beta := \bar{f}(x)$, where $\bar{f}$ and $\underline{f}$ are
defined by (4.3.13) and (4.3.14), respectively. By a straightforward argument
based on the Hahn-Banach theorem, there exists a probability Radon measure
$v_1 \in \mathscr{M}_1^+(K)$ supported by $\overline{\partial_e K}$ such that $r(v_1) = x$ and $v_1(f) = \alpha$.

Note that, if f_0 is a continuous extension of f to K, and if we consider the
enveloping functions $\hat{f}_0$ and $\check{f}_0$ defined by (1.5.7) and (1.5.8) then, by virtue of
(1.5.10) and (1.5.11), we have $\check{f}_0 \leq \underline{f} \leq \bar{f} \leq \hat{f}_0$ on $\overline{\partial_e K}$.

So, if $\mu \in \mathscr{M}_1^+(K)$ is a maximal measure, then $\mathrm{Supp}(\mu) \subset \overline{\partial_e K}$ and hence by
Theorem 1.5.4

$$\mu(f) = \mu(f_0) = \int \check{f}_0 \, d\mu \leq \int \underline{f} \, d\mu \leq \mu(f) \leq \int \bar{f} \, d\mu = \int \hat{f}_0 \, d\mu = \mu(f_0).$$

Accordingly, $\mu(f) = \int \underline{f}\,d\mu = \int \bar{f}\,d\mu.$

Now, let $\mu_1 \in \mathcal{M}_1^+(K)$ be a maximal measure such that $v_1 < \mu_1$ (where $<$ denotes the Choquet-Meyer ordering). Then $r(\mu_1) = x$ and by (1.5.6) and Theorem 1.5.4 we obtain

$$\alpha = \int f\,dv_1 = \int \bar{f}\,dv_1 \geq \int \bar{f}\,d\mu_1 = \mu_1(f)$$

as well as

$$\alpha = \int f\,dv_1 = \int \underline{f}\,dv_1 \leq \int \underline{f}\,d\mu_1 = \mu_1(f),$$

i.e., $\mu_1(f) = \alpha$.

Analogously, one finds a maximal measure $\mu_2 \in \mathcal{M}_1^+(K)$ such that $r(\mu_2) = x$ and $\mu_2(f) = \beta$. Since $\mu_1 - \mu_2 \in \mathcal{D}(x)$, (2) yields $\alpha = \beta$.

Since $x \in K_1 \backslash \overline{\partial_e K}$ was arbitrarily chosen, we conclude that $\bar{f} = \underline{f}$ on K_1.

Now, if we put $g = \bar{f}_{|K_1}$, then $g \in \mathscr{C}(K_1)$ and $g_{|\overline{\partial_e K}} = f$.

Finally, note that if $a, b \in A(K)$ and $a \leq f \leq b$ on $\overline{\partial_e K}$, then $a \leq \bar{f}$ and $\underline{f} \leq b$ on $\overline{\partial_e K}$. Accordingly, by Theorem 1.4.11, $a \leq \bar{f}$ and $\underline{f} \leq b$ on K, i.e., $a \leq g \leq b$ on K_1. So,

$$\sup_{\substack{a \in A(K) \\ a \leq g \text{ on } K_1}} a = \underline{f} = \bar{f} = \inf_{\substack{b \in A(K) \\ g \leq b \text{ on } K_1}} b,$$

and hence $g \in K_+(A(K, K_1))$. □

For $K_1 = K$ we obtain the following criterion for the solvability of the Dirichlet problem on the extreme points.

4.3.2 Corollary. *If $f \in \mathscr{C}(\overline{\partial_e K})$, the following statements are equivalent:*
(i) *There exists $g \in A(K)$ such that $g_{|\overline{\partial_e K}} = f$.*
(ii) (1) $f \in K_+(A(K, \overline{\partial_e K}))$.
   (2) $\mu(f) = 0$ *for every $x \in K \backslash \overline{\partial_e K}$ and $\mu \in \mathcal{D}(x)$.*

If $\partial_e K$ is closed, part (1) of statements (ii) of Theorem 4.3.1 and Corollary 4.3.2 is redundant by virtue of (4.3.10).

When K is a Bauer simplex, part (2) is obviously fulfilled and hence Corollary 4.3.2 implies the following further theorem on the solvability of the Dirichlet problem on extreme points due to Bauer (see Theorem 1.5.8):

4.3.3 Corollary. *Let K be a Bauer simplex. Then every $f \in \mathscr{C}(\partial_e K)$ admits a continuous affine extension to K.*

In the next corollary we shall see that condition (2) in statement (ii) of Theorem 4.3.1 and Corollary 4.3.2 can be considerably weakened.

For every $x \in K$ we shall denote by F_x the smallest face of K containing x. Clearly,

$$F_x := \{y \in K | \text{There exist } z \in K \text{ and } \lambda \in \mathbb{R}, 0 < \lambda \leq 1, \text{ such that } x = \lambda y + (1-\lambda)z\}.$$
(4.3.15)

Note that, if $\mu \in \mathcal{M}_1^+(K)$ and $r(\mu) = x$, then, choosing a net $(\mu_i)_{i \in I}^{\leq}$ of discrete probability measures converging to μ and having x as a common barycenter (see Proposition 1.5.1), we see that $\mathrm{Supp}(\mu_i) \subset F_x$ for every $i \in I$, and hence

$$\mathrm{Supp}(\mu) \subset F_x.$$
(4.3.16)

The same reasoning shows that a closed convex subset of K is a face if and only if every $\mu \in \mathcal{M}_1^+(K)$ having its barycenter in F must have its support $\mathrm{Supp}(\mu)$ contained in F.

4.3.4 Corollary. *Under the hypotheses of Theorem 4.3.1 consider a subset X of $K_1 \backslash \overline{\partial_e K}$ satisfying*

$$K_1 \backslash \overline{\partial_e K} \subset \bigcup_{x \in X} F_x.$$
(4.3.17)

Then, for every $f \in \mathscr{C}(\overline{\partial_e K})$, the following statements are equivalent:
(i) *There exists $g \in K_+(A(K, K_1))$ such that $g_{|\overline{\partial_e K}} = f$.*
(ii) (1) $f \in K_+(A(K, \overline{\partial_e K}))$.
     (2) $\mu(f) = 0$ *for every $x \in X$ and $\mu \in \mathscr{D}(x)$.*

Proof. On account of Theorem 4.3.1, the result will be demonstrated if we show that for every $x \in K$ we have $\mathscr{D}(x) = \bigcup_{y \in F_x} \mathscr{D}(y)$.

In fact, take $y \in F_x$; then there exist $z \in K$ and $\lambda \in \,]0, 1[$ such that $x = \lambda y + (1 - \lambda)z$. Consider $\mu = \mu_1 - \mu_2 \in \mathscr{D}(y)$ with μ_1 and μ_2 maximal, $\alpha := \|\mu_1\| = \|\mu_2\| > 0$ and $r(\mu_1) = r(\mu_2) = y$; after fixing a maximal measure $\mu \in \mathcal{M}_1^+(K)$ with $r(\mu) = z$, set $v_1 := \dfrac{\lambda}{\alpha}\mu_1 + (1 - \lambda)\mu$, and $v_2 := \dfrac{\lambda}{\alpha}\mu_2 + (1 - \lambda)\mu$.

Then v_1 and v_2 are maximal, $\|v_1\| = v_1(1) = \lambda = \|v_2\|$, $r(v_1) = r(v_2) = x$ and

$$v_1 - v_2 = \frac{\lambda}{\alpha}\mu_1 - \frac{\lambda}{\alpha}\mu_2 = \frac{\lambda}{\alpha}(\mu_1 - \mu_2).$$

Hence $\mu_1 - \mu_2 \in \mathscr{D}(x)$. $\qquad\qquad\square$

Now we are going to characterize the functions belonging to $K_+(A(K, K_1))$ in terms of an extension property.

4.3.5 Proposition. *The following statements hold*:
(1) *If* $f \in K_+(A(K, K_1))$, *then, for every* $x \in K_1$ *such that* F_x *is closed, there exists a unique continuous affine extension of* $f_{|K_1 \cap F_x}$ *to* F_x.
(2) *If* $f \in \mathscr{C}(K_1)$ *and if, for every* $x \in K_1$, *there exists a continuous affine extension of* $f_{|K_1 \cap \overline{F}_x}$ *to* $\overline{F}_x$, *then* $f \in K_+(A(K, K_1))$.

Proof. (1): By Theorem 4.3.1 and Corollary 4.3.4 (with $X = K_1 \backslash \overline{\partial_e K}$) there exists $g \in K_+(A(K, K_1 \cup F_x))$ satisfying $g = f$ on $\overline{\partial_e K}$.

If we set $g_0 := g_{|F_x} \in A(F_x)$, then $g_0 = f$ on $K_1 \cap F_x$. In fact, on $\overline{\partial_e K}$ we have $\bar{g} = g = f = \bar{f}$ (see (4.3.13)), hence $\bar{g} = \bar{f}$ on K_1 by Theorem 1.4.11. Consequently, $g_0 = g = \bar{g} = \bar{f} = f$ on $K_1 \cap F_x$.

Since $\partial_e F_x = \partial_e K \cap F_x \subset K_1 \cap F_x$, we obtain $g_0 = f$ on $\partial_e F_x$ and, hence, g_0 is unique by Bauer's maximum principle (see Theorem 1.4.10).

(2): Set $\mu \in \mathscr{M}^+(K_1)$ and $x \in K_1$ satisfying $r(\mu) = x$. By (4.3.16), $\text{Supp}(\mu) \subset \overline{F}_x$ which implies, by the hypotheses, $\mu(f) = f(x)$, i.e., $f \in K_+(A(K, K_1))$ by (4.3.7). $\qquad \square$

4.3.6 Corollary. *Suppose that, for every* $x \in K$, F_x *is closed. Then, for every* $f \in \mathscr{C}(K_1)$, *the following statements are equivalent*:
(i) $f \in K_+(A(K, K_1))$.
(ii) *For every* $x \in K_1$ *there exists a unique continuous affine extension to* F_x *of* $f_{|K_1 \cap F_x}$.

Note that Corollary 4.3.6 may be applied to convex compact subsets K of $\mathbb{R}^p$. In this case, if $u \in \mathscr{C}(K)$, and if we denote by H the linear subspace generated by $A(K) \cup \{u\}$, then the state space of H can be identified with the convex hull in $\mathbb{R}^{p+1}$ of $\{(x, u(x)) | x \in K\}$, i.e., the convex hull of the graph of u.

So, by using (4.3.4) and Corollaries 4.3.2 and 4.3.6 one may derive some suitable information about $K_+(A(K) \cup \{u\})$.

We shall illustrate this by means of another more general approach.

Consider, in fact, a family $\mathscr{U}$ of convex continuous functions on K and set

$$P(K, \mathscr{U}) := \mathscr{L}(A(K) \cup \mathscr{U}), \qquad (4.3.18)$$

the linear subspace of $\mathscr{C}(K)$ generated by $A(K) \cup \mathscr{U}$. The functions in $P(K, \mathscr{U})$ are also called *parabola-like functions on* K by the analogy with the case $K = [0, 1]$ and $\mathscr{U} = \{e_2\}$, where $e_2(x) = x^2$ ($x \in [0, 1]$).

When $\mathscr{U}$ reduces to a single convex continuous function u, we put

$$P(K, u) := P(K, \{u\}) = \mathscr{L}(A(K) \cup \{u\}). \qquad (4.3.19)$$

We also denote by $\Re(\mathcal{U})$ the set of all convex compact subsets of K on which all the functions of $\mathcal{U}$ are affine.

4.3.7 Theorem. *If $\mathcal{U}$ is a family of convex continuous functions on K, then*

$$K_+(P(K, \mathcal{U})) = \{f \in \mathscr{C}(K)|\, f \text{ is affine on each } K_1 \in \Re(\mathcal{U})\}. \qquad (4.3.20)$$

Hence $P(K, \mathcal{U})$ is a K_+-subspace in $\mathscr{C}(X)$ if and only if $\mathcal{U}$ separates convexely the points of K, i.e.,

For every $x, y, \in K$, $x \neq y$, there exists $u \in \mathcal{U}$ such that $u\left(\dfrac{x+y}{2}\right) < \dfrac{1}{2}(u(x) + u(y))$.

$$(4.3.21)$$

Proof. Suppose $f \in K_+(P(K, \mathcal{U}))$ and set $K_1 \in \Re(\mathcal{U})$. Then, for every $x, y, \in K_1$ and $\lambda \in [0,1]$, after putting $z := \lambda x + (1 - \lambda)y$, we have $\varepsilon_z = \lambda \varepsilon_x + (1 - \lambda)\varepsilon_y$ on $P(K, \mathcal{U})$. Hence $f(z) = \lambda f(x) + (1 - \lambda)f(y)$, i.e., f is affine on K_1.

Conversely, suppose that f is affine on each subset $K_1 \in \Re(\mathcal{U})$. Consider $\mu \in \mathscr{M}^+(K)$ and $x \in K$ and suppose that $\mu = \varepsilon_x$ on $A(K) \cup \mathcal{U}$. In particular $r(\mu) = x$ and $x \in \overline{\text{co}}(\text{Supp}(\mu))$. Moreover, for each $u \in \mathcal{U}$, $\mu(u) = u(x)$ and hence u is affine on $\overline{\text{co}}(\text{Supp}(\mu))$ by Rasa's theorem 1.5.2.

Consequently, f is affine on $\overline{\text{co}}(\text{Supp}(\mu))$ and hence, by Theorem 1.4.9, (3), there exists a net $(a_i)_{i \in I}^{\leq}$ of continuous affine functions on K which converges uniformly to f on $\overline{\text{co}}(\text{Supp}(\mu))$. This implies $\mu(f) = \lim_{\substack{\leq \\ i \in I}} \mu(a_i) = \lim_{\substack{\leq \\ i \in I}} a_i(x) = x$.

The final part of the assertion is a consequence of the fact that, as is easy to see, a convex function $u \in \mathscr{C}(K)$ is not affine on a segment $[x, y]$ if and only if $u((x + y)/2) < (u(x) + u(y))/2$. $\qquad \square$

Remark. According to some results of Rasa [1986] and Campiti and Rasa [1991], it is also possible to describe the Choquet boundary of $P(K, \mathcal{U})$. More precisely, if $x \in K$, then the following statements are equivalent:

(i) $x \in \partial^+_{P(K, \mathcal{U})}K$.
(ii) *For every $y, z \in K$, $y \neq z$, such that $x = \dfrac{y + z}{2}$ there exists $u \in \mathcal{U}$ such that*
$$u(x) < \frac{u(y) + u(z)}{2}.$$
(iii) *There exists $K_1 \in \Re(\mathcal{U})$ which is maximal with respect to the inclusion relation, such that $x \in \partial_e K_1$.*

Theorem 4.3.7 takes a particularly simple form in the case when $\mathcal{U}$ reduces to a single function.

4.3.8 Corollary. *If* $u \in \mathscr{C}(K)$ *is a convex function on* K, *then*

$$K_+(P(K, u)) = \{f \in \mathscr{C}(K) | f \text{ is affine on all convex compact}$$
$$\text{subsets } K_1 \text{ of } K \text{ on which } u \text{ is affine}\}. \qquad (4.3.22)$$

Consequently, $A(K) \cup \{u\}$ *is a* K_+*-set in* $\mathscr{C}(X)$ *if and only if* u *is strictly convex.*

As an example, we shall specialize Corollary 4.3.8 to the particular case when K is a subset of $\mathbb{R}^p$. In this case

$$A(K) = \mathscr{L}(1, \mathrm{pr}_1, \ldots, \mathrm{pr}_p), \qquad (4.3.23)$$

where $\mathrm{pr}_1, \ldots, \mathrm{pr}_p$ denote the canonical projections on K (see (2.6.19)).

4.3.9 Corollary. *Let* K *be a convex compact subset of* $\mathbb{R}^p$ *and let* $u \in \mathscr{C}(K)$ *be a convex function. Then the following assertions hold:*
(1) *If* u *is strictly convex on* K *(in particular, if* $u(x) = \|x\|^2$ $(x \in K))$, *then*

$$\{1, \mathrm{pr}_1, \ldots, \mathrm{pr}_p, u\} \text{ is a } K_+\text{-subset in } \mathscr{C}(K).$$

(2) *If there exists a finite family* $(K_i)_{1 \le i \le n}$ *of convex compact subsets of* K *whose interiors are pairwise disjoint and such that* $K = \bigcup_{i=1}^{n} K_i$ *and* u *is affine on every* K_i, *then*

$$K_+(\{1, \mathrm{pr}_1, \ldots, \mathrm{pr}_p, u\}) = \{f \in \mathscr{C}(K) | f \text{ is affine on } K_i, i = 1, \ldots, n\}. \quad (4.3.24)$$

As a consequence of Corollary 4.3.9 for $p = 1$ we indicate below other simple and useful examples of Korovkin subsets.

Example. Let $[a, b]$ be a real interval. Then the following subsets of $\mathscr{C}([a, b])$ are K_+-subsets in $\mathscr{C}([a, b])$:

$$\{1, x, x^\alpha\}, \quad \alpha > 0, \alpha \ne 1 \text{ (provided } a \ge 0), \qquad (4.3.25)$$

$$\{1, x, \exp(\lambda x)\}, \quad \lambda \ne 0. \qquad (4.3.26)$$

Remarks. 1. We point out that, if u is neither convex nor concave, then $K_+(P(K, u)) = P(K, u)$, as has been shown by Bauer, Leha and Papadopoulou [1979, Proposition 6].

2. For $p = 1$ and $u(x) = x^2$, Corollary 4.3.9 reduces to the classical Korovkin theorem 4.2.7. This adds a geometric significance to those indicated in Corollary 4.1.5. In the next sections we shall also indicate an algebraic and a probabilistic aspect of the same theorem (see Theorem 4.4.6 and Section 5.2).

In the last part of this section we apply some of the previous results to present a characterization of $A(K)$-spaces among all Banach spaces.

4.3.10 Theorem. *Let E be a real Banach space. Then the following statements are equivalent:*

(i) *There exists a convex compact subset K (of some locally convex Hausdorff space) having its set of extreme points closed, such that E is isometrically isomorphic to $A(K)$.*

(ii) *There exist a compact Hausdorff space X and a K_+-subspace H of $\mathscr{C}(X)$ containing 1, such that E is isometrically isomorphic to H.*

(iii) *There exist a weak*-closed face F of the unit ball B' in the dual space E' of E and a point $f_0 \in E$ such that*
   (1) *$\partial_e F$ is weak*-closed.*
   (2) *$\mu(f_0) = 1$ for each $\mu \in F$.*
   (3) *$B' = \mathrm{co}(F \cup (-F))$.*

Proof. (i) $\Rightarrow$ (ii): Without loss of generality, we may assume that $E = A(K)$. Now put $X = \partial_e K$ and $H = A(K, \partial_e K)$ (see (4.3.6)).

Then, by (4.3.10), H is a K_+-subspace in $\mathscr{C}(X)$ and E is isometrically isomorphic to H by Bauer's maximum principle.

(ii) $\Rightarrow$ (iii): For sake of simplicity, assume that $E = H$. Consider the canonical embedding $\Phi: X \to E'$ (see (1.4.19)) and put $F = \overline{\mathrm{co}}(\Phi(X))$. Then F is a weak*-closed face of B' and $\partial_e F = \Phi(X)$ by Corollary 4.1.5, (viii). Hence $\partial_e F$ is weak*-closed. Moreover, (2) is satisfied by $f_0 = 1$.

Note that, if $\mu \in \partial_e B'$, then, by using the Hahn-Banach theorem and the characterization of extreme points of the unit ball of $\mathscr{C}(X)$ (see (1.4.15)), we may choose $x \in X$ such that $\mu = \pm \varepsilon_x$ on H, i.e., $\mu = \pm \Phi(x)$. Thus, $\partial_e B' \subset F \cup (-F)$ and hence, by the Krein-Milman theorem,

$$B' = \overline{\mathrm{co}}^{w^*} \partial_e B' \subset \overline{\mathrm{co}}^{w^*}(F \cup (-F)) = \mathrm{co}(F \cup (-F)) \subset B'$$

(where the above closures are considered with respect to the weak*-topology).

So the result follows.

(iii) $\Rightarrow$ (i): Put $K = F$ and consider the canonical linear operator $\theta: E \to A(K)$ defined by $\theta(f)(x) := f(x)$ for every $f \in E$ and $x \in K$.

From the hypotheses, it follows that $\|\theta(f)\| = \|f\|$, i.e., θ is an isometry.

Finally, we show that θ is onto. In fact, take $g \in A(K)$ and consider the function $\tilde{g}: B' \to \mathbb{R}$ defined by

$$\tilde{g}(\mu) := \alpha g(\mu_1) - (1 - \alpha)g(\mu_2)$$

for every $\mu \in B'$, $\mu = \alpha \mu_1 + (1 - \alpha)(-\mu_2)$ (where $\mu_1, \mu_2 \in F$, $0 \le \alpha \le 1$).

The function $\tilde{g}$ is well defined because, if $\mu = \beta v_1 + (1 - \beta)(-v_2)$ with v_1, $v_2 \in F$ and $0 \leq \beta \leq 1$, then by (2), we infer $2\alpha - 1 = 2\beta - 1$, i.e., $\alpha = \beta$. So, since $\alpha \mu_1 + (1 - \alpha)v_2 = \alpha v_1 + (1 - \alpha)\mu_2$ and $g \in A(K)$, we have

$$\alpha g(\mu_1) + (1 - \alpha)g(v_2) = \alpha g(v_1) + (1 - \alpha)g(\mu_2).$$

This, in turn, implies that

$$\alpha g(\mu_1) - (1 - \alpha)g(\mu_2) = \beta g(v_1) - (1 - \beta)g(v_2).$$

Clearly, endowing B' with the weak*-topology we have that $\tilde{g} \in A(B')$ and $\tilde{g}(0) = 0$ (it suffices to write $0 = \frac{1}{2}\mu - \frac{1}{2}\mu$, where μ is an arbitrary element of F). Consequently, by using (1.4.13), Theorem 1.4.9, (3) and the representation of the weak*-continuous linear functionals on E', it is easy to show that there exists $f \in E$ such that, for every $\mu \in B'$, $\mu(f) = \tilde{g}(\mu)$, i.e., $\theta(f) = g$. $\square$

Notes and references

The results 4.3.1–4.3.6, together with their proofs, are taken from Bauer, Leha and Papadopoulou [1979].

The descriptions of Korovkin closures of parabola-like functions (also called parabolic functions) contained in Theorem 4.3.7 and Corollary 4.3.8 are due to Rasa [1980], [1986].

Corollary 4.3.8 also appears in Bauer, Leha and Papadopoulou [1979] in the case when $K \subset \mathbb{R}^p$ and in Micchelli [1975] in the case when u is smooth.

Extensions and generalizations of Theorem 4.3.7 can be found in a paper of Campiti and Rasa [1991].

Theorem 4.3.10 is due to Amir and Ziegler [1978]. In the same article the reader may find other intrinsic characterizations of Korovkin closures.

4.4 Korovkin closures and Stone-Weierstrass theorems

The aim of this section is the investigation of possible connections between Korovkin closures of subspaces of $\mathscr{C}_0(X)$ and other familiar subspaces such as closed sublattices (or subalgebras, or ideals) generated by suitable subsets.

As a consequence, new proofs of classical Stone-Weierstrass theorems will be presented. Furthermore, it will become more evident as much as these theorems are closely related to Korovkin-type approximation theory.

We come back to the general hypotheses of this chapter by considering a locally compact Hausdorff space X and a subset M of $\mathscr{C}_0(X)$.

We denote by $R(M)$ the *closed vector sublattice generated by M*, i.e., the intersection of all closed vector sublattices of $\mathscr{C}_0(X)$ containing M. From Theorem 4.1.2, (v), it follows that

$$K_+(M) \subset R(M). \tag{4.4.1}$$

In general, this inclusion is strict, even if X is compact.

For example, set $X = [0,1]$ and $M = \{1, e_1\}$, where $e_1(x) = x$ ($x \in [0,1]$). Then $R(M) = \mathscr{C}([0,1])$ by the Stone-Weierstrass theorem, while $K_+(M) = \mathscr{L}(M)$ by Proposition 4.2.6.

However, if one adds to M the function $e_2(x) = x^2$, then $K_+(M \cup \{e_2\}) = \mathscr{C}([0,1]) = R(M)$ (see Theorem 4.2.7).

In fact, this observation also holds for the general setting of our considerations: by adding to M sufficiently many functions, the Korovkin closure of the bigger subspace coincides with $R(M)$.

As an incidental remark, note, however, that in other Banach lattices, such as $L^p(X, \mu)$-spaces, $1 \le p < +\infty$, we always have $R(M) = K_+^1(M)$ (see Berens and Lorentz [1973]).

For every $x \in X$ and $\alpha \ge 0$ we set

$$M(x, \alpha) := \{y \in X \,|\, h(y) = \alpha h(x) \text{ for every } h \in M\}. \tag{4.4.2}$$

Each $M(x, \alpha)$ is closed and, if there exists $h \in M$ such that $h(x) \ne 0$, then $M(x, \alpha) \cap M(x, \beta) = \varnothing$ for every $\alpha, \beta \in \mathbb{R}_+$, $\alpha \ne \beta$.

We also set

$$A(x) := \{\alpha \in \mathbb{R}_+ \,|\, M(x, \alpha) \ne \varnothing\}. \tag{4.4.3}$$

Clearly $1 \in A(x)$ since $x \in M(x, 1)$.

We shall denote by $I(M)$ the *closed ideal generated by M*, i.e., the intersection of all closed ideals of $\mathscr{C}_0(X)$ containing M (we recall that a linear subspace H of a Banach lattice E is said to be an *ideal* (or a *lattice ideal*) of E if for every $x \in H$ and $y \in E$ satisfying $|y| \le |x|$, one has $y \in H$).

By a theorem of Kakutani [1941] (see also Stone [1962, Theorem 7]), it is well-known that

$$I(M) = \begin{cases} \{f \in \mathscr{C}_0(X)| f(x) = 0 \text{ for every } x \in Z(M)\}, & \text{if } Z(M) \neq \varnothing, \\ \mathscr{C}_0(X), & \text{if } Z(M) = \varnothing, \end{cases} \quad (4.4.4)$$

where

$$Z(M) := \{x \in X | h(x) = 0 \text{ for every } h \in M\}. \quad (4.4.5)$$

It is formally convenient to introduce some more notations.
We set

$$|M| := \{|h|\,|\,h \in M\} \quad (4.4.6)$$

and

$$R^*(M) := M \cup |\mathscr{L}(M \cup |M|)|. \quad (4.4.7)$$

The following theorem gives a simple and useful description of the Korovkin closure of $R^*(M)$.

4.4.1 Theorem. *The following equality holds:*

$$K_+(R^*(M)) = I(M) \cap \{f \in \mathscr{C}_0(X)| f(y) = \alpha f(x)$$
$$\textit{for every } x \in X, \alpha \in A(x) \textit{ and } y \in M(x, \alpha)\}. \quad (4.4.8)$$

Proof. One inclusion is rather obvious taking Theorem 4.1.2 into account.

Conversely, fix a function f belonging to the right member of (4.4.8) and consider $\mu \in \mathscr{M}_b^+(X)$ and $x \in X$ satisfying $\mu(h) = h(x)$ for every $h \in R^*(M)$.

First, suppose that $h(x) = 0$ for every $h \in M$, i.e., $x \in Z(M)$. Then, for every $h \in M$ we have $\mu(|h|) = |h(x)| = 0$ and hence $\text{Supp}(\mu) \subset \{y \in X | h(y) = 0$ for each $h \in M\} = Z(M)$.

Since $f \in I(M)$, we obtain $\mu(f) = 0 = f(x)$.

Suppose now that $h(x) \neq 0$ for some $h \in M$. Then, $h(x) = 1$ for some function h of $\mathscr{L}(|M|)$ that we may also suppose positive by replacing it, if necessary, with $|h|$.

Consider $y \in X$, $y \notin \bigcup_{\alpha \in A(x)} M(x, \alpha)$. Then, if $h(y) \notin A(x)$, it follows $M(x, h(y)) = \varnothing$, while, if $h(y) \in A(x)$, we have $y \notin M(x, h(y))$. In any case, there exists $k \in M$ such that $k(y) \neq h(y)k(x)$.

So the function $k_0 := |k - k(x)h| \in R^*(M)$ satisfies the following properties:

$$k_0 \geq 0, \quad k_0(x) = 0 \quad \text{and} \quad k_0(y) > 0.$$

By reasoning as in the proof of Theorem 2.4.1, one easily shows that

$$\text{Supp}(\mu) \subset \bigcup_{\alpha \in A(x)} M(x, \alpha).$$

Consequently, if $f(x) = 0$, then $f = 0$ on each $M(x, \alpha)$ and hence $\mu(f) = 0 = f(x)$.

Finally, if $f(x) \neq 0$, after choosing $h \in M$ satisfying $h(x) \neq 0$, for every $\alpha \in A(x)$ and $y \in M(x, \alpha)$ we obtain $h(y) = \alpha h(x) = \dfrac{h(x)}{f(x)} f(y)$.

This reasoning shows that $h = \dfrac{h(x)}{f(x)} f$ on $\text{Supp}(\mu)$ and so, $h(x) = \mu(h) = \dfrac{h(x)}{f(x)} \mu(f)$, i.e., $\mu(f) = f(x)$, and this completes the proof. $\square$

From Theorem 4.4.1 we obtain the following *lattice generalization of the Weierstrass theorem*, which is due to Kakutani [1941, Theorem 3] (see also Stone [1962, Theorem 3]).

4.4.2 Corollary. *If M is an arbitrary subset of $\mathscr{C}_0(X)$, then*

$$R(M) = K_+(R(M)) = K_+(R^*(M))$$

$$= I(M) \cap \{f \in \mathscr{C}_0(X) | f(y) = \alpha f(x) \text{ for every } x \in X, \alpha \in A(x) \text{ and } y \in M(x, \alpha)\}.$$
$$(4.4.9)$$

Consequently $R(M) = \mathscr{C}_0(X)$ (or, equivalently, $R^*(M)$ is a K_+-subset in $\mathscr{C}_0(X)$) if and only if M satisfies the following two conditions:
(1) For every $x \in X$ there exists $h \in M$ such that $h(x) \neq 0$.
(2) For every $x, y \in X$, $x \neq y$, and for every $\alpha \geq 0$ there exists $h \in M$ such that $h(x) \neq \alpha h(y)$.

Proof. Since $R^*(M) \subset R(M)$, we have $K_+(R^*(M)) \subset K_+(R(M)) \subset R(M)$ by virtue of (4.4.1). From (4.4.8) it follows that $K_+(R^*(M))$ is a closed sublattice of $\mathscr{C}_0(X)$ and it contains M. So, $R(M) \subset K_+(R^*(M))$ and the assertion follows.

Remark. We shall subsume conditions (1) and (2) of the above corollary by saying that the subset M *separates linearly the points of X*.

In the next result we shall also derive the algebraic formulation of the Weierstrass theorem.

We recall that a linear sublattice H of $\mathscr{C}_0(X)$ is called a *Stone sublattice* if

$$\mathbf{1} \wedge f := \inf\{\mathbf{1}, f\} \in H \quad \text{for every } f \in H. \qquad (4.4.10)$$

For example, if $\mu \in \mathcal{M}_b^+(X)$ and $1 \leq p < +\infty$, then $\mathcal{C}_0(X) \cap \mathcal{L}^p(X, \mu)$ is a Stone sublattice. Another simple example in $\mathcal{C}_0(]0, 1])$ is furnished by

$$\{f \in \mathcal{C}_0(]0, 1]) | f \text{ is differentiable at the point } 0\}.$$

The notion of Stone sublattice is fundamental in the theory of Daniell integral and can be introduced for a general subspace of continuous functions on X.

In the context of $\mathcal{C}_0(X)$-spaces, in fact, every closed Stone sublattice is a closed subalgebra and conversely. This appears in a more general context in Nöbeling and Bauer [1955] (see also Muni [1979]). In our setting it will be a particular consequence of the next results.

We shall use the notion of Stone sublattice to obtain a representation of the closed subalgebra generated by a subset M of $\mathcal{C}_0(X)$.

We premise the following general result.

4.4.3 Lemma. *Every closed subalgebra A of $\mathcal{C}_0(X)$ is a Stone sublattice of $\mathcal{C}_0(X)$.*

Proof. We shall make use of the identity

$$t^{1/2} = \sum_{n=0}^{\infty} \binom{1/2}{n} (t - 1)^n = \lim_{n \to \infty} p_n(t),$$

which holds uniformly on $[0, 2]$, where, of course,

$$p_n(t) = \sum_{k=0}^{n} \binom{1/2}{k} (t - 1)^k \qquad (n \geq 1, t \in [0, 2]).$$

Since $\lim_{n \to \infty} p_n(0) = 0$, we have, indeed,

$$t^{1/2} = \lim_{n \to \infty} q_n(t) \qquad \text{uniformly with respect to } t \in [0, 2],$$

where $q_n = p_n - p_n(0)$ $(n \in \mathbb{N})$. So, if $f \in A$, $f \neq 0$, we have

$$|f| = \|f\| \left(\frac{f^2}{\|f\|^2}\right)^{1/2} = \lim_{n \to \infty} \|f\| q_n \left(\frac{f^2}{\|f\|^2}\right) \in \overline{A} = A.$$

Consequently, $|f|^{1/2} \in A$ and, by induction, $|f|^{1/2^n} \in A$ for every $n \geq 1$. Now the proof will be complete if we prove that

$$\inf\{1, f\} = \lim_{n \to \infty} (\inf\{|f|^{1/2^n}, f\}).$$

In fact, set $f_n := \inf\{|f|^{1/2^n}, f\} \in A$ for each $n \geq 1$. Clearly $f_n \to 1 \wedge f$ point-wise on X. But on the compact subset $X_- := \{x \in X \mid f(x) \geq 1\}$, each f_n is equal to $|f|^{1/2^n}$ and hence $(f_n)_{n \in \mathbb{N}}$ is decreasing. Consequently, $f_n \to 1 \;(= 1 \wedge f)$ uniformly on X_- by Dini's theorem.

On the other hand, on the subset $X_+ := \{x \in X \mid f(x) \leq 1\}$, every f_n coincides with $f \;(= 1 \wedge f)$ and this yields the assertion. $\qquad\square$

At this point we need some more notations.

If M is a subset of $\mathscr{C}_0(X)$, we shall denote by $S(M)$ the *closed Stone sublattice generated by* M. In Section 2.4 we have already introduced the symbol $\mathbf{A}(M)$ to denote the closed subalgebra generated by M.

On account of Lemma 4.4.3 the following inclusions hold:

$$R(M) \subset S(M) \subset \mathbf{A}(M) \subset I(M). \tag{4.4.11}$$

We shall also set

$$1 \wedge M := \{1 \wedge h \mid h \in M\} \tag{4.4.12}$$

and

$$S^*(M) := M \cup |\mathscr{L}(M \cup 1 \wedge \mathscr{L}(M \cup |\mathscr{L}(M)|))|. \tag{4.4.13}$$

Finally we recall that, according to (2.4.5), for every $x \in X$ we have defined $M(x)$ as the subset of all $y \in X$ such that $h(x) = h(y)$ for each $h \in M$.

Now we present a description of $\mathbf{A}(M)$ (and of $S(M)$) together with a new proof of the well-known generalization of Weierstrass theorem due to Stone ([1948] and [1962]; see also Shilov [1951]).

4.4.4 Theorem. *If M is a subset of $\mathscr{C}_0(X)$, then*

$$\mathbf{A}(M) = S(M) = K_+(S^*(M))$$

$$= I(M) \cap \{f \in \mathscr{C}_0(X) \mid f \text{ is constant on } M(x) \text{ for each } x \in X\}. \tag{4.4.14}$$

Consequently, $\mathbf{A}(M) = \mathscr{C}_0(X)$ or, equivalently, $S^*(M)$ is a K_+-subspace in $\mathscr{C}_0(X)$, if and only if M separates strongly the points of X.

Proof. First, we shall establish the equality

$$K_+(S^*(M)) = I(M) \cap \{f \in \mathscr{C}_0(X) \mid f \text{ is constant on } M(x) \text{ for each } x \in X\}. \tag{1}$$

One inclusion being obvious, we directly proceed to show the other one. Fix $f \in I(M)$ and suppose that f is constant on each $M(x)$ ($x \in X$).

Consider $\mu \in \mathcal{M}_b^+(X)$ and $x \in X$ satisfying $\mu = \varepsilon_x$ on $S^*(M)$.

There are two possible cases:

1. $h(x) = 0$ for each $h \in M$. Then $f(x) = 0$ and, reasoning as in the proof of Theorem 4.4.1, we have $\mu(f) = 0$.
2. There exists $h \in M$ such that $h(x) \neq 0$. In this case, we have $M(x) \cap Z(M) = \varnothing$.

Note that, if $y \notin M(x) \cup Z(M)$, then one can choose $k \in M$ such that $k(y) \neq k(x)$ and $h_1, h_2 \in \mathcal{L}(M)$ satisfying $h_1(x) = h_2(y) = 1$.

So, if we put

$$h_0 := \mathbf{1} \wedge (\sup(h_1, h_2)) = \mathbf{1} \wedge \left(\frac{h_1 + h_2}{2} + \frac{|h_1 - h_2|}{2} \right),$$

and

$$k_0 := |k - k(x)h_0| \in S^*(M),$$

then

$$k_0 \geq 0, \quad k_0(x) = 0 \quad \text{and} \quad k_0(y) = |k(y) - k(x)| > 0.$$

Again by reasoning as in the proof of Theorem 2.4.1, we conclude that

$$\mathrm{Supp}(\mu) \subset M(x) \cup Z(M).$$

Consequently, if $f(x) = 0$, then $f = 0$ on $M(x) \cup Z(M)$ and hence $\mu(f) = 0 = f(x)$.

If $f(x) \neq 0$, choosing $h \in M$ such that $h(x) \neq 0$, it is easy to show that

$$h = \frac{h(x)}{f(x)} f \quad \text{on } M(x) \cup S(M).$$

This, in turn, yields $h(x) = \mu(h) = \dfrac{h(x)}{f(x)} \mu(f)$, that is $\mu(f) = f(x)$.

Having now proved equality (1), it is clear that $K_+(S^*(M))$ is a closed sub-algebra which contains M and hence $\mathbf{A}(M) \subset K_+(S^*(M))$.

On the other hand, since $S^*(M) \subset S(M)$, it follows from (4.4.1) and (4.4.11) that

$$K_+(S^*(M)) \subset K_+(S(M)) \subset S(M) \subset \mathbf{A}(M),$$

and so the result is proved. $\square$

Next we are going to describe $\mathbf{A}(M)$ in terms of K_+^1-closures. When $\mathscr{L}(M)$ contains a strictly positive function, then the following result may also be derived from Theorem 2.5.6.

4.4.5 Theorem. *If M is a subset of $\mathscr{C}_0(X)$, then*

$$K_+^1(M \cup M^2) = \mathbf{A}(M). \qquad (4.4.15)$$

In particular, $M \cup M^2$ is a K_+^1-subset in $\mathscr{C}_0(X)$ if and only if M separates strongly the points of X.

Furthermore, if $M = \{h_n | n \in \mathbb{N}\}$ is finite or countable and the series $\sum\limits_{n=1}^{\infty} h_n^2$ is uniformly convergent on X, then the same statement holds by replacing M^2 with $\sum\limits_{n=1}^{\infty} h_n^2$.

Proof. The inclusion $K_+^1(M \cup M^2) \subset \mathbf{A}(M)$ easily follows from (4.4.14) and Theorem 4.1.8.

To prove the converse inclusion, set $f \in \mathbf{A}(M)$ and consider a measure $\mu \in \mathscr{M}_b^+(X)$ which vanishes on $M \cup M^2$. Since $\mu(h^2) = 0$ for every $h \in M$, it follows that $\mathrm{Supp}(\mu) \subset Z(M)$ and hence $\mu(f) = 0$, because $f \in I(M)$.

Thus the above reasoning shows that $\mathbf{A}(M) \subset \mathscr{L}(M \cup M^2)^*$.

Fix now $x \in X$ and $\mu \in \mathscr{M}_b^+(X)$, $\|\mu\| \leq 1$, satisfying $\mu = \varepsilon_x$ on $M \cup M^2$. By carefully adapting the proof of Theorem 2.4.2, one shows that $\mathrm{Supp}(\mu) \subset M(x)$.

If $h(x) = 0$ for every $h \in M$, then $\mu(h^2) = h^2(x) = 0$ for $h \in M$ and hence $\mathrm{Supp}(\mu) \subset Z(M)$. Therefore $\mu(f) = 0 = f(x)$.

Otherwise we can take $h \in M$ such that $h(x) \neq 0$. Then, under the additional assumption that $f(x) \neq 0$, we have $h = \dfrac{h(x)}{f(x)} f$ on $M(x)$ and hence $\mu(f) = f(x)$, because $h(x) = \mu(h) = \dfrac{h(x)}{f(x)} \mu(f)$.

We arrive at the same conclusion, even when $f(x) = 0$ because, in this case, $f = 0$ on $M(x)$ and so $\mu(f) = 0$.

Finally, if M is finite or countable, then the assertion follows by arguing as in the proof of Proposition 2.5.7. $\qquad\square$

When the subspace generated by M contains a strictly positive function, we can give a simpler description of $\mathbf{A}(M)$ in terms of K_+-closures (note that in this case $I(M) = \mathscr{C}_0(X)$).

In the following theorem we also present both a compact version of the Stone-Weierstrass theorem and an extension of Korovkin's theorem to arbitrary compact spaces. Furthermore, as for the interval $[0, 1]$, these theorems are, in fact, equivalent.

4.4.6 Theorem. *Given a subset M of $\mathscr{C}_0(X)$, the following statements hold:*
(1) *If the linear subspace generated by M contains a strictly positive function f_0, then*

$$K_+(\{f_0\} \cup f_0 M \cup f_0 M^2)$$

$$= \mathbf{A}(M) = \{f \in \mathscr{C}_0(X) | f \text{ is constant on } M(x) \text{ for each } x \in X\}. \quad (4.4.16)$$

Therefore $\mathbf{A}(M) = \mathscr{C}_0(X)$ or, equivalently, $\{f_0\} \cup f_0 M \cup f_0 M^2$ is a K_+-subset in $\mathscr{C}_0(X)$ if and only if M separates the points of X.
(2) *If X is compact, then*

$$K_+(\{\mathbf{1}\} \cup M \cup M^2) = \mathbf{A}(\{\mathbf{1}\} \cup M)$$

$$= \{f \in \mathscr{C}(X) | f \text{ is constant on } M(x) \text{ for each } x \in X\}. \quad (4.4.17)$$

Hence $\mathbf{A}(\{\mathbf{1}\} \cup M) = \mathscr{C}(X)$ or, equivalently, $\{\mathbf{1}\} \cup M \cup M^2$ is a K_+-subset in $\mathscr{C}(X)$ if and only if M separates the points of X.

(3) *If $M = \{h_n | n \in \mathbb{N}\}$ is finite or countable and the series $\sum\limits_{n=1}^{\infty} h_n^2$ is uniformly convergent on X, then statements (1) and (2) hold by replacing M^2 with the function $\sum\limits_{n=1}^{\infty} h_n^2$.*

Proof. It is a direct consequence of Theorems 4.1.2 and 2.5.6 and Proposition 2.5.7. □

Remark. For a direct proof of part (3) of the above theorem, see Theorem 4.2.10 and the subsequent Example 2.

Next we derive some useful consequences from Theorems 4.4.5 and 4.4.6.

Examples. 1. Consider $X = \mathbb{R}^p$. Then, by applying Theorems 4.4.5 and 4.4.6 to

$$f_0(x) := \exp(-\|x\|^2) \quad \text{and}$$

$$M := \{\exp(-\|x\|^2), x_1 \exp(-\|x\|^2), \ldots, x_p \exp(-\|x\|^2)\},$$

or to

$$f_0(x) := \frac{1}{1 + \|x\|^2} \quad \text{and} \quad M := \left\{\frac{1}{1 + \|x\|^2}, \frac{x_1}{1 + \|x\|^2}, \ldots, \frac{x_p}{1 + \|x\|^2}\right\},$$

we obtain that the following subsets

$$\{\exp(-\|x\|^2), \exp(-2\|x\|^2), x_1 \exp(-2\|x\|^2), \ldots, x_p \exp(-2\|x\|^2),$$

$$(1 + \|x\|^2)\exp(-3\|x\|^2)\} \qquad (4.4.18)$$

and

$$\left\{\frac{1}{1 + \|x\|^2}, \frac{1}{(1 + \|x\|^2)^2}, \frac{x_1}{(1 + \|x\|^2)^2}, \ldots, \frac{x_p}{(1 + \|x\|^2)^2}\right\} \qquad (4.4.19)$$

are K_+-subsets in $\mathscr{C}_0(\mathbb{R}^p)$.

Furthermore, the subsets

$$\{\exp(-\|x\|^2), x_1 \exp(-\|x\|^2), \ldots, x_p \exp(-\|x\|^2), (1 + \|x\|^2)\exp(-2\|x\|^2)\} \qquad (4.4.20)$$

and

$$\left\{\frac{1}{1 + \|x\|^2}, \frac{x_1}{1 + \|x\|^2}, \ldots, \frac{x_p}{1 + \|x\|^2}\right\} \qquad (4.4.21)$$

are K_+^1-subsets in $\mathscr{C}_0(\mathbb{R}^p)$.

2. (*Volkov's theorem*) Let X be a compact subset of $\mathbb{R}^p$. Then, denoting by $M = \{\mathrm{pr}_1, \ldots, \mathrm{pr}_p\}$ the set of the canonical projections on X, we obtain that

$$\left\{1, \mathrm{pr}_1, \ldots, \mathrm{pr}_p, \sum_{i=1}^{p} \mathrm{pr}_i^2\right\} \qquad (4.4.22)$$

is a K_+-subset in $\mathscr{C}(X)$.

In particular, if $X \subset \{x \in \mathbb{R}^p \,|\, \|x\| = 1\}$, then

$$\{1, \mathrm{pr}_1, \ldots, \mathrm{pr}_p\} \qquad (4.4.23)$$

is a K_+-subset in $\mathscr{C}(X)$.

For $p = 2$, we then have that

$$\{1, \mathrm{pr}_1, \mathrm{pr}_2\} \qquad (4.4.24)$$

is a K_+-subset in $\mathscr{C}(\mathbb{T})$, where $\mathbb{T} := \{z \in \mathbb{R}^2 \,|\, \|z\| = 1\}$.

3. If K is a convex compact subset of some locally convex Hausdorff space, then

$$A(K) \cup A(K)^2 \tag{4.4.25}$$

is a K_+-subset in $\mathscr{C}(K)$.

This last result will be extremely useful in the study of the approximation processes associated with positive projections, which will be extensively considered in Chapter 6.

4. If X is locally compact and $f_0 \in \mathscr{C}_0(X)$ is an injective strictly positive function, then $\{f_0, f_0^2, f_0^3\}$ and $\{f_0, f_0^2\}$ are K_+- and K_+^1-subsets in $\mathscr{C}_0(X)$, respectively.

5. If X is compact and $f \in \mathscr{C}(X)$ is injective, then $\{1, f, f^2\}$ is a K_+-subset in $\mathscr{C}(X)$.

6. $\{1, x, x^2\}$ and $\{1, \exp(x), \exp(2x)\}$ are K_+-subsets in $\mathscr{C}([a, b])$.

It should be noticed that Example 5 leads again to Korovkin's theorem 4.2.7.

Furthermore, Example 2 and, in particular, (4.4.24) is in fact Korovkin's second theorem.

Thus both Korovkin's first and second theorems are particular cases of a unique more general result, namely Theorem 4.4.6, (2).

Notes and references

The germ of Stone's generalization of Weierstrass approximation theorem already appears in Stone [1937]. A more complete exposition can be found in Stone [1948], [1962]. The lattice formulation of the Stone-Weierstrass theorem was obtained by Kakutani [1941].

The methods developed in this section to describe the closed sublattice (and the closed subalgebra) generated by a subset of $\mathscr{C}_0(X)$ in terms of Korovkin closures are due to the authors. However, Shashkin [1972] first recognized a strong analogy between Korovkin-type approximation theory and Stone-Weierstrass theorems.

Other results in this direction can be found in Bauer [1974] and Attalienti [1994], who proved among other things a sublattice version of Stone-Weierstrass theorem in the setting of adapted spaces, in Brosowski [1983], where a vector-valued formulation is presented, in Flösser [1981], Scheffold [1977] and Dieckmann [1992].

The last part of statement (2) of Theorem 4.4.6 was obtained by different methods by Freud [1964] when M is finite, and by Schempp [1972] in the general case (see also Grossman [1974], Jiménes Poso [1974] and Yoshinaga and Tamura [1975]).

One of the two equalities which appear in (4.4.17) is due to Amir and Ziegler [1978] and, in the finite dimensional case, to B. Eisenberg [1976] and Franchetti [1969].

Bauer [1973] gave an extension of part (2) of Theorem 4.4.6 to the setting of adapted spaces. For other function spaces see Komornik [1980], Schnabl [1972] and Yoshinaga and Tamura [1975].

Altomare [1982a] established a more general version of (4.4.17) in the context of commutative unital Banach algebras with a symmetric involution (see Appendix A, Theorems A.1.12 and A.1.13).

For the non commutative case (in particular, for C^*-algebras) see Beckhoff [1985], [1987], [1988], [1991], Limaye and Namboodiri [1982], Priestley [1976] and Robertson [1977] (see Appendix B, Section B.2).

Equalities (4.4.17) have been generalized by Campiti and Rasa [1991] by replacing the set M^2 with the set $\{u \circ h | h \in M, u \in \mathscr{U}\}$, where $\mathscr{U}$ is a set of convex real functions on $\mathbb{R}$ which separates convexely the points of $\mathbb{R}$ (see also Altomare [1991]).

The fact that the set described in (4.4.22) is a K_+-subset was first established by Volkov [1957].

4.5 Finite Korovkin sets

The power of Korovkin-type theorems is even more evident when finite Korovkin sets are involved.

This section is devoted to their study. We shall present several characterizations of finite Korovkin sets and we shall characterize those spaces on which such sets exist. We shall see that they are precisely those metrizable (and separable) spaces having finite small inductive dimension. Furthermore, some estimates about their minimal cardinality are also given in terms of the small inductive dimension. As usual, unless otherwise stated, we shall denote by X an arbitrary locally compact Hausdorff space.

4.5.1 Theorem. Let $h_0, \ldots, h_n$ be functions in $\mathscr{C}_0(X)$, $n \geq 2$. Then the following statements are equivalent:

(i) $\{h_0, \ldots, h_n\}$ is a K_+-subset in $\mathscr{C}_0(X)$.

(ii) (1) $\{h_0, \ldots, h_n\}$ separates strongly the points of X.
  (2) For every $m \geq 2$ and for every choice of different points $x_0, \ldots, x_m \in X$, no column of the matrix $(h_i(x_j))_{\substack{0 \leq i \leq n \\ 0 j \leq m}}$ is a positive or negative linear combination of the other columns.

(iii) (1) $\{h_0, \ldots, h_n\}$ separates strongly the points of X.
  (2) For every $x \in X$ and $m = 0, \ldots, n + 2$ and for every choice of $m + 1$ points $x_0, \ldots, x_m \in X$ and $m + 1$ strictly positive real numbers $\alpha_0, \ldots, \alpha_m$ satisfying

$$\sum_{j=0}^{m} \alpha_j h_i(x_j) = h_i(x) \quad \text{for every } i = 0, \ldots, n,$$

it follows that $\sum_{j=0}^{m} \alpha_j = 1$ and $x_j = x$ for every $j = 0, \ldots, m$.

If X is compact, the above statements (i)–(iii) are also equivalent to the following one:

(iv) (1) The subset $\{h_0, \ldots, h_n\}$ separates the points of X.
  (2) The convex cone $V_+(X)$ of $\mathbb{R}^{n+1}$ generated by $\mathrm{co}(\Phi^*(X))$ and having the vertex at the origin is pointed.
  (3) For every $x \in X$ the ray $\rho(x)$ defined by (2.6.13) is an extreme ray in $V_+(X)$, where $\Phi^*: X \to \mathbb{R}^{n+1}$ is the embedding defined by (2.6.15).

Furthermore, if $h_0 = \mathbf{1}$, then we may also add the statement:

(v) The subset $\{\mathbf{1}, h_1, \ldots, h_n\}$ separates the points of X and

$$\Phi_0(X) = \partial_e \mathrm{co}(\Phi_0(X)),$$

where $\Phi_0: X \to \mathbb{R}^n$ is the embedding defined by (2.6.16), i.e., there exists a convex compact subset K of $\mathbb{R}^n$ such that $\Phi_0(X) = \partial_e K$.

Proof. (i) ⇒ (ii): Part (1) is obvious. As regards part (2) suppose that there exist distinct points $x_0, \ldots, x_m \in X$ such that for some $k = 0, \ldots, m$ there exist $\alpha_0, \ldots,$ $\alpha_{k-1}, \alpha_{k+1}, \ldots, \alpha_m \in \mathbb{R}_+$ satisfying $\sum\limits_{\substack{j=0 \\ j \neq k}}^{m} \alpha_j h_i(x_j) = h_i(x_k)$ for every $i = 0, \ldots, n.$

From (i) and Theorem 4.1.2 (take $\mu = \sum\limits_{\substack{j=0 \\ j \neq k}}^{m} \alpha_j \varepsilon_{x_j}$) it follows that $\sum\limits_{\substack{j=0 \\ j \neq k}}^{m} \alpha_j \varepsilon_{x_j} = \varepsilon_{x_k}.$

Hence $x_j = x_k$ for every $j = 0, \ldots, m$ and this contradicts the initial assumptions.

On the other hand, if all the numbers α_j are negative, then we can write

$$2 h_i(x_k) - \sum_{\substack{j=0 \\ j \neq k}}^{m} \alpha_j h_i(x_j) = h_i(x_k) \quad \text{for every } i = 0, \ldots, n,$$

and hence we reach again the contradiction $x_j = x_k$ for each $j = 0, \ldots, m.$

(ii) ⇒ (iii): This is obvious.

(iii) ⇒ (i): We verify statement (iii) of Theorem 4.1.2. Consider $\mu \in \mathcal{M}_b^+(X)$ and $x \in X$ satisfying

$$\mu(h_i) = h_i(x) \quad \text{for every } i = 0, \ldots, n. \tag{1}$$

Given $f \in \mathscr{C}_0(X)$, by applying Proposition 2.2.1 to the subspace generated by $f, h_0, \ldots, h_n$, we may consider a finite family $x_0, \ldots, x_m$ of points of X with $m \leq n + 2$ and (strictly) positive numbers $\alpha_0, \ldots, \alpha_m$ such that

$$\mu(f) = \sum_{j=0}^{m} \alpha_j f(x_j) \tag{2}$$

and

$$\mu(h_i) = \sum_{j=0}^{m} \alpha_j h_i(x_j) \quad \text{for every } i = 0, \ldots, n. \tag{3}$$

Therefore, combining (1) with (3), by virtue of (iii), it follows that $x_j = x$ for every $j = 0, \ldots, m$ and $\sum\limits_{j=0}^{m} \alpha_j = 1$. So (2) implies $\mu(f) = f(x)$.

Since f was arbitrarily chosen we conclude that $\mu = \varepsilon_x$.

The remaining assertions are a direct consequence of Corollary 4.1.5, of the Remarks after Proposition 2.6.7 and of the Krein-Milman theorem. □

Remarks. 1. Note that, if $\{h_0, \ldots, h_n\}$ is a K_+-subspace of $\mathscr{C}_0(X)$, then for every choice of three different points $x_0, x_1, x_2 \in X$, no column of the matrix $(h_i(x_j))_{\substack{0 \leq i \leq n \\ 0 \leq j \leq 2}}$ is a linear combination of the other columns (i.e., the matrix is of rank 3).

In fact, if, for instance, there exist $\alpha_1, \alpha_2 \in \mathbb{R}$ such that

$$h_i(x_0) = \alpha_1 h_i(x_1) + \alpha_2 h_i(x_2), \quad i = 0, \ldots, n,$$

then, by virtue of Theorem 4.5.1, we may restrict ourselves to the case $\alpha_1 \cdot \alpha_2 < 0$. Suppose $\alpha_1 < 0$ and $\alpha_2 > 0$. Then $h_i(x_0) - \alpha_1 h_i(x_1) = \alpha_2 h_i(x_2)$, $i = 0, \ldots, n$. Hence $\varepsilon_{x_0} - \alpha_1 \varepsilon_{x_1} = \alpha_2 \varepsilon_{x_2}$ and this is not possible.

2. If X is compact and if $\{1, h_1, \ldots, h_n\}$ is a K_+-subset in $\mathscr{C}(X)$ consisting of linearly independent functions, then the convex compact subset K of $\mathbb{R}^n$ considered in statement (v) (i.e., $K = \mathrm{co}(\Phi_0(X))$) has non-empty interior.

Otherwise, in fact, K would be contained in a hyperplane of $\mathbb{R}^{n-1}$, so that the functions $1, h_1, \ldots, h_n$ would be linearly dependent.

As a particular case of Theorem 4.5.1 we obtain the following result.

4.5.2 Corollary. *Let X be a compact subset of $\mathbb{R}^n$, $n \geq 2$, and let us consider the projections* $\mathrm{pr}_1, \ldots, \mathrm{pr}_n \in \mathscr{C}(X)$.

Then $\{1, \mathrm{pr}_1, \ldots, \mathrm{pr}_n\}$ is a K_+-subset in $\mathscr{C}(X)$ if and only if $X = \partial_e \mathrm{co}(X)$ or, equivalently, if $X = \partial_e K$ for some convex compact subset K of $\mathbb{R}^n$.

Proof. The result follows from Theorem 4.5.1 since, in this case, the embedding Φ_0 is the natural embedding from X into $\mathbb{R}^n$. ☐

By using the same reasoning as in the proof of Theorem 4.5.1 and taking Theorem 4.1.8 into account instead of Theorem 4.1.2, it is possible to prove the next result without difficulty.

4.5.3 Theorem. *Let $h_0, \ldots, h_n \in \mathscr{C}_0(X)$. Then the following statements are equivalent:*

(i) *$\{h_0, \ldots, h_n\}$ is a K_+^1-subset in $\mathscr{C}_0(X)$.*

(ii) *For every $m \geq 2$ and for every choice of different points $x_0, \ldots, x_m \in X$ the following two properties hold:*

   (1) *no column of the matrix $(h_i(x_j))_{\substack{0 \leq i \leq n \\ 0 \leq j \leq m}}$ is a convex combination of the other columns and of the origin of $\mathbb{R}^n$.*

   (2) *The origin of $\mathbb{R}^n$ cannot be a non-trivial positive linear combination of the columns of the matrix $(h_i(x_j))_{\substack{0 \leq i \leq n \\ 0 \leq j \leq m}}$.*

(iii) (1) *For every $x \in X$ and $m = 0, \ldots, n + 2$, and for every choice of $m + 1$ points $x_0, \ldots, x_m \in X$ and $m + 1$ strictly positive real numbers $\alpha_0, \ldots, \alpha_m$*

satisfying $\sum\limits_{j=0}^{m} \alpha_j \le 1$ *and*

$$\sum_{j=0}^{m} \alpha_j h_i(x_j) = h_i(x) \quad \text{for every } i = 0, \ldots, n,$$

it follows that $\sum\limits_{j=0}^{m} \alpha_j = 1$ *and* $x_j = x$ *for every* $j = 0, \ldots, m$.

(2) *For every* $m \in \mathbb{N}$, $2 \le m \le n + 2$, *and for every choice of different points* $x_0, \ldots, x_m \in X$, *the origin of* $\mathbb{R}^n$ *cannot be a non-trivial positive linear combination of the columns of the matrix* $(h_i(x_j))_{\substack{0 \le i \le n \\ 0 \le j \le m}}$.

Furthermore, if X is compact, condition (2) *in both statements* (ii) *and* (iii) *can be dropped. Moreover, one may also add the further statement*

(iv) *The subset* $\{h_0, \ldots, h_n\}$ *separates strongly the points of X and*

$$\Phi^*(X) \subset \partial_e \mathrm{co}(\Phi^*(X) \cup \{0\}).$$

4.5.4 Corollary. *Let X be a compact subset of* $\mathbb{R}^n$, $n \ge 2$, *and suppose that* $0 \notin X$. *Then* $\{\mathrm{pr}_1, \ldots, \mathrm{pr}_n\}$ *is a* K_+^1-*subset in* $\mathscr{C}(X)$ *if and only if* $X \subset \partial_e \mathrm{co}(X \cup \{0\})$.

Proof. In this case Φ^* coincides with the embedding from X into $\mathbb{R}^n$ and the result follows from Theorem 4.5.3. ∎

4.5.5 Corollary. *Suppose that X is compact and let* $h_1, \ldots, h_n \in \mathscr{C}(X)$ *be functions such that* $\{1, h_1, \ldots, h_n\}$ *is a* K_+-*subset in* $\mathscr{C}(X)$.

If $0 \notin \Phi_0(X)$ *and if for every* $y_0 \in \Phi_0(X)$ *the line through* 0 *and* y_0 *intersects* $\Phi_0(X)$ *only in* y_0, *then* $\{h_1, \ldots, h_n\}$ *is a* K_+^1-*subset in* $\mathscr{C}(X)$.

Proof. Clearly $\{h_1, \ldots, h_n\}$ separates strongly the points of X. Furthermore the embedding Φ^* corresponding to $\{h_1, \ldots, h_n\}$ coincides with the embedding Φ_0 corresponding to $\{1, h_1, \ldots, h_n\}$.

From the hypothesis it follows that $\partial_e \mathrm{co}(\Phi_0(X)) \subset \partial_e \mathrm{co}(\Phi_0(X) \cup \{0\})$, and hence the result follows from Theorems 4.5.1 and 4.5.3. ∎

Remark. As the proof shows, in statement (ii) of both Theorems 4.5.1 and 4.5.3 it is sufficient to require that $m \le n + 2$.

Moreover, if the one-point compactification X_ω of X is connected (or if X is compact and connected, respectively), then we need only to require that statements (ii) and (iii) of both Theorems 4.5.1 and 4.5.3 hold for $m \le n + 1$.

Here we indicate another geometric characterization of finite K_+-subsets.

To see this we first recall the following definition due to Borsuk [1957].

Consider $m \in \mathbb{N}$, $m \geq 2$ and $k \in \mathbb{N}$, $2 \leq k \leq m$. We say that a subset Q of $\mathbb{R}^m$ is *k-independent* if no point of Q is a non-trivial linear combination of k other points of Q.

Here a trivial linear combination $\sum_{i=1}^{k} \alpha_i y_i$ of k points $y_1, \ldots, y_k$ means that one of the coefficients α_i is equal to 1 and all others are 0.

Thus Q is k-independent if and only if every k-dimensional subspace of $\mathbb{R}^m$ contains no more than k points of Q.

The set Q is called *k-positively independent* if no point of Q is a non-trivial positive linear combination of k other points of Q.

Finally, for $1 \leq k \leq m$ we say that Q is *k-regular* if no points of Q is a non trivial linear combination, with sum of coefficients equal to 1, of some k points of Q, i.e., every $(k-1)$-dimensional subspace of $\mathbb{R}^m$ contains no more than k points of Q.

If, in this last case, we also require that the coefficients are positive, then Q is said to be *k-convexely regular*.

The logical relations among these notions are as follows

$$(k\text{-independent}) \rightarrow \quad (k\text{-regular}) \quad \rightarrow (k\text{-convexely regular})$$

$$\searrow \qquad\qquad \nearrow$$

$$(k\text{-positively independent})$$

We may now translate the preceding Theorems 4.5.1 and 4.5.3 and their respective remarks in the following result.

4.5.6 Corollary. *Suppose that X is compact and consider $h_0, \ldots, h_n \in \mathscr{C}(X)$, $n \geq 2$. Then the following assertions hold:*

(1) *$\{h_0, \ldots, h_n\}$ is a K_+-subset in $\mathscr{C}(X)$ if and only if $\{h_0, \ldots, h_n\}$ separates the points of X, the cone $V_+(X)$ is pointed and $\Phi^*(X)$ is k-positively independent in $\mathbb{R}^{n+1}$ for every $2 \leq k \leq n+2$.*

Furthermore, if this is the case, then $\Phi^*(X)$ also is 2-independent in $\mathbb{R}^{n+1}$ and hence 2-regular.

(2) *$\{h_0, \ldots, h_n\}$ is a K_+^1-subset in $\mathscr{C}(X)$ if and only if $\{h_0, \ldots, h_n\}$ separates strongly the points of X and $\Phi^*(X)$ is k-convexely regular in $\mathbb{R}^{n+1}$ for every $2 \leq k \leq n+2$.*

(3) *If $h_0 = 1$, then $\{1, h_1, \ldots, h_n\}$ is a K_+-subset in $\mathscr{C}(X)$ if and only if $\Phi_0(X)$ is k-convexely regular in $\mathbb{R}^n$ for every $2 \leq k \leq n+2$.*

Remark. Some of the above results can be easily extended to finite K_+-subsets of order n, $n \geq 2$.

For instance, given $h_0, \ldots, h_p \in \mathscr{C}_0(X)$, $p \geq 2$, and $n \geq 1$, then it is possible to show that $\{h_0, \ldots, h_p\}$ is a K_+-subset of order n in $\mathscr{C}_0(X)$ if and only if for every $r \leq n$ and $m \geq n+1$ and for every choice of different points $x_0, \ldots, x_m \in$

X, no positive linear combination of r columns of the matrix $(h_i(x_j))_{\substack{0 \le i \le p \\ 0 \le j \le m}}$ is a positive or negative linear combination of the other $m - r$ columns.

If this is the case and if X is compact, then $\Phi^*(X)$ (or $\Phi_0(X)$ if $h_0 = 1$) is $2n$-regular.

When X is a compact real interval or the unit circle, we may easily show the close connection between K_+-subspaces and Chebyshev subspaces.

4.5.7 Theorem. *Let X be a compact real interval or the unit circle and let H be a subspace of $\mathscr{C}(X)$ of dimension $n + 1$, $n \ge 2$. Then the following statements hold:*

(1) *If $n = 2$, H is a Chebyshev subspace of order 3 in $\mathscr{C}(X)$ if and only if H is a K_+-subspace in $\mathscr{C}(X)$.*

(2) *If $n \ge 3$ and if H is a Chebyshev subspace of order $n + 1$ in $\mathscr{C}(X)$, then H is a K_+-subspace in $\mathscr{C}(X)$.*

Proof. This is a direct consequence of Corollary 3.4.10 and of the fact that the K_+-subspaces of order 1 in $\mathscr{C}(X)$ are exactly the K_+-subspaces in $\mathscr{C}(X)$. $\qquad\square$

Remarks. 1. As it has been shown by Shashkin [1962, Remark to Theorem 10], if X is a non connected closed subset of the unit circle containing more than three points, then there exists a Chebyshev subspace of order 3 of $\mathscr{C}(X)$ which is not a K_+-subspace.

On the other hand, the only compact connected subsets of the unit circle containing more than one point are the circle itself and the intervals.

So, on account of the Mairhuber-Curtis theorem (see the discussion after Proposition 2.3.2), all compact spaces which admit a Chebyshev system and for which the equivalence (1) holds for every subspace H, are homeomorphic to a real compact interval or to the unit circle.

2. In the Remark to Corollary 3.4.10 an example of a four dimensional K_+-subspace which is not a Chebyshev subspace is presented.

More generally, Rasa [1991, Section 9] showed that, if X is a real compact interval, then for every $n \ge 3$ there exists a strict K_+-subset of $\mathscr{C}(X)$ which is not a Chebyshev system and such that no proper subset of it is a K_+-subset of $\mathscr{C}(X)$.

K_+-subsets which possess this last property are also called *minimal K_+-subsets*. For more information on these subsets see Amir and Ziegler [1978, Section 7].

We now proceed to study those (locally) compact spaces X such that $\mathscr{C}_0(X)$ admits finite-dimensional K_+- or K_+^1-subspaces. In this case, we will try to obtain some information about the minimal dimension of such Korovkin subspaces.

In the sequel we shall set

$$m_+(X) := \min\,\{n \geq 3 | \text{there exists a } n\text{-dimensional } K_+\text{-subspace in } \mathscr{C}_0(X)\},$$
(4.5.1)

$$m_+^1(X) := \min\{n \geq 2 | \text{there exists a } n\text{-dimensional } K_+^1\text{-subspace in } \mathscr{C}_0(X)\},$$
(4.5.2)

$$r(X) := \min\{n \geq 2 | X \text{ is homeomorphic to a subset of the unit sphere in } \mathbb{R}^n\},$$
(4.5.3)

$$c(X) := \min\{n \geq 2 | X \text{ is homeomorphic to the set of extreme points of}$$
$$\text{a convex compact subset of } \mathbb{R}^n \text{ which has non-empty interior}\}$$
(4.5.4)

provided that the subsets indicated on the right are non-empty.

By virtue of Proposition 4.1.16 we know that

$$3 \leq m_+(X),$$
(4.5.5)

$$2 \leq m_+^1(X) \leq m_+(X).$$
(4.5.6)

Furthermore, if X is a compact real interval or the unit circle, then on account of Proposition 4.1.16, Theorem 4.2.7 and Examples 2 and 4 to Theorem 4.4.6, we have

$$m_+(X) = 3 \quad \text{and} \quad m_+^1(X) = 2.$$
(4.5.7)

In addition we denote by $\text{ind}(X)$ the *Menger-Urysohn dimension*, or the *small inductive dimension*, of the space X.

This number $\text{ind}(X)$ is defined as follows:

(1) $\text{ind}(X) := -1$ if and only if $X = \emptyset$.

(2) $\text{ind}(X) \leq n$, where $n \in \mathbb{N}_0$, if for every $x \in X$ and for every neighborhood V of x there exists an open set U such that $x \in U \subset V$ and $\text{ind}(\partial U) \leq n - 1$ (here, $\partial U = \overline{U} \cap \overline{X \setminus U}$ is the boundary of U).

(3) $\text{ind}(X) := n$, where $n \in \mathbb{N}_0$, if $\text{ind}(X) \leq n$ and the property $\text{ind}(X) \leq n - 1$ is false.

(4) $\text{ind}(X) := \infty$ if for every $n \in \mathbb{N}_0$ it is not true that $\text{ind}(X) \leq n$.

For further details and properties of $\text{ind}(X)$ we refer to Hurewicz and Wallman [1948] and Engelking [1978], [1989].

We also point out that it is possible to assign to X other numbers, namely the *large inductive dimension* $\text{Ind}(X)$ and the *covering dimension* $\dim(X)$, all of which have the characteristics of dimension (see Engelking [1989, Section 7.1]; see also Section A.2).

However, by a consequence of the Katětov-Morita theorem, if X is metrizable and separable (the only case of interest for us), then $\text{ind}(X) = \text{Ind}(X) = \dim(X)$ (see Engelking [1989, Theorems 7.3.2 and 7.3.3]).

We begin by considering the compact case.

4.5.8 Theorem. *Let X be a compact Hausdorff space and fix $n \geq 3$. Then, the following statements are equivalent:*

(i)   *There exists an n-dimensional K_+-subspace in $\mathscr{C}(X)$.*

(ii) *There exists an n-dimensional K_+-subspace in $\mathscr{C}(X)$ containing the function $\mathbf{1}$.*

(iii) *X is homeomorphic to the set of extreme points of a convex compact subset of $\mathbb{R}^{n-1}$ which has non-empty interior.*

(iv) *X is homeomorphic to a subset of the unit sphere of $\mathbb{R}^{n-1}$.*

(v)   *There exists an n-dimensional strict K_+-subspace in $\mathscr{C}(X)$ containing the function $\mathbf{1}$.*

(vi) *There exists an $(n-1)$-dimensional strict K_+^1-subspace in $\mathscr{C}(X)$.*

(vii) *There exists an $(n-1)$-dimensional K_+^1-subspace in $\mathscr{C}(X)$.*

Furthermore, if one of the statements (i)–(vii) hold, then X is metrizable (and hence separable), it has finite small inductive dimension and

$$\text{ind}(X) \leq n - 2. \tag{4.5.8}$$

Proof. (i) $\Rightarrow$ (ii): Suppose that H is an n-dimensional K_+-subspace of $\mathscr{C}(X)$. By Corollary 4.1.5, H is cofinal in $\mathscr{C}(X)$, so that there exists $h_0 \in H$ such that $h_0(x) > 0$ for every $x \in X$.

By Proposition 4.1.13, $\dfrac{1}{h_0} \cdot H$ is a K_+-subspace in $\mathscr{C}(X)$, it contains the function $\mathbf{1}$ and has dimension n.

(ii) $\Rightarrow$ (iii): By hypothesis there exist $h_1, \ldots, h_{n-1} \in \mathscr{C}(X)$ such that $\mathbf{1}, h_1, \ldots, h_{n-1}$ are linearly independent and form a K_+-subset. Then the result follows from Theorem 4.5.1, (v) (see also Remark 2 to the same theorem).

(iii) $\Rightarrow$ (iv): Suppose that X is homeomorphic to the set of extreme points of a convex compact subset K of $\mathbb{R}^{n-1}$ having an interior point $x_0 \in K$.

Let Q be a suitable sphere having x_0 as center and contained in the interior of K. Let us denote by $\varphi \colon \partial_e K \to Q$ the mapping which assigns to every $x_0 \in \partial_e K$ the (unique) point belonging to Q and to the line through x_0 and x. In this way, $\partial_e K$ (and hence K) is homeomorphically embedded into Q.

(iv) $\Rightarrow$ (v): Suppose that X is homeomorphic to a (closed) subset Q of the unit sphere on $\mathbb{R}^{n-1}$. The subset $\{\mathbf{1}, \text{pr}_1, \ldots, \text{pr}_{n-1}\}$ is a strict K_+-subset in $\mathscr{C}(Q)$ (see Proposition 4.2.3) and hence a strict K_+-subset of the form $\{\mathbf{1}, h_1, \ldots, h_{n-1}\}$ exists in $\mathscr{C}(X)$ too.

(v) $\Rightarrow$ (i): This is obvious.

(iv) $\Rightarrow$ (vi): Suppose that X is homeomorphic to a closed subset Q of the unit sphere of $\mathbb{R}^{n-1}$. Since $\mathscr{C}(Q)$ possesses an $(n-1)$-dimensional strict K_+^1-subspace, namely the subspace generated by $\{\mathrm{pr}_1, \ldots, \mathrm{pr}_{n-1}\}$ (see Proposition 4.2.3), it follows that $\mathscr{C}(X)$ possesses an $(n-1)$-dimensional strict K_+^1-subspace.

(vi) $\Rightarrow$ (vii): This is obvious.

(vii) $\Rightarrow$ (i): According to Remark 2 to Proposition 4.1.10, if $\{h_1, \ldots, h_{n-1}\}$ is a K_+^1-subset in $\mathscr{C}(X)$, then $\{\mathbf{1}, h_1, \ldots, h_{n-1}\}$ is a K_+-subset in $\mathscr{C}(X)$.

Finally, if (iv) holds, then clearly X is metrizable and hence separable. Furthermore, since the small inductive dimension is a topological invariant and the unit sphere $\mathbb{S}_{n-2}$ of $\mathbb{R}^{n-1}$ has finite dimension, X has finite small inductive dimension and $\mathrm{ind}(X) \leq \mathrm{ind}(S_{n-2}) = n - 2$ (see Hurewicz and Wallman [1948, Theorems III.1 and IV.1]).   □

4.5.9 Corollary. *If X is a compact Hausdorff space, then $\mathscr{C}(X)$ possesses a three dimensional K_+-subspace or, equivalently, it possesses a two dimensional K_+^1-subspace if and only if X is homeomorphic to a subset of the unit circle.*

Thus if X is also connected, then $\mathscr{C}(X)$ possesses a three dimensional K_+-subspace (or a two dimensional K_+^1-subspace) if and only if X is homeomorphic to the unit circle or to a compact real interval.

From Theorem 4.5.8 we can easily obtain a complete characterization of those compact Hausdorff spaces X for which $\mathscr{C}(X)$ possesses a finite K_+-subspace.

4.5.10 Corollary. *Let X be a compact Hausdorff space. Then the following statements are equivalent:*

(i) *$\mathscr{C}(X)$ possesses a finite dimensional K_+-subspace (or, a finite dimensional strict K_+-subspace containing the constants).*

(ii) *$\mathscr{C}(X)$ possesses a finite dimensional K_+^1-subspace (or, a finite dimensional strict K_+^1-subspace).*

(iii) *X is homeomorphic to the set of extreme points of a convex compact subset of $\mathbb{R}^n$ with non-empty interior, for some $n \geq 2$.*

(iv) *X is homeomorphic to a subset of the unit sphere of $\mathbb{R}^n$, for some $n \geq 2$.*

(v) *X is metrizable and has finite small inductive dimension.*

Furthermore, if one of statements (i)–(v) holds, then

$$\mathrm{ind}(X) + 2 \leq m_+(X) = r(X) + 1 = c(X) + 1 \leq 2\,\mathrm{ind}(X) + 3 \qquad (4.5.9)$$

and

$$m_+^1(X) = m_+(X) - 1 = r(X) = c(X). \qquad (4.5.10)$$

Proof. Clearly the implications (i) $\Rightarrow$ (ii) $\Rightarrow$ (iii) $\Rightarrow$ (iv) $\Rightarrow$ (v) follow from Theorem 4.5.8.

Moreover, on account of (4.5.8), we also have

$$\text{ind}(X) + 2 \leq m_+(X) = r(X) + 1 = c(X) + 1.$$

To show the implication (v) $\Rightarrow$ (iv), suppose that X is metrizable (and hence separable) and $\text{ind}(X) = n$ for some $n \in \mathbb{N}_0$.

By the *Menger-Nöbeling theorem* X is homeomorphic to a subset of $\mathbb{R}^{2n+1}$, i.e., to a subset of the unit sphere of $\mathbb{R}^{2n+2}$ (see Hurewicz and Wallman [1948, Theorem V.2]). So (iv) holds and $r(X) \leq 2n + 2 = 2\,\text{ind}(X) + 2$. Hence (4.5.9) follows.

Finally the implication (iv) $\Rightarrow$ (i) and (4.5.10) are again a consequence of Theorem 4.5.8. $\qquad\qquad\square$

Examples. 1. In general, inequalities (4.5.9) may be strict. For instance, we have $m_+(X) < 2\,\text{ind}(x) + 3$ if X is a compact real interval or the unit circle, since we have $\text{ind}(X) = 1$ and $m_+(X) = 3$.

The inequality $\text{ind}(X) + 2 < m_+(X)$ occurs, for instance, if $\text{ind}(X) = 0$, i.e., X is totally disconnected (i.e., no connected subset of X contains more than one point (see Hurewicz and Wallman [1948, Chapter II, §4]).

2. If $\mathbb{S}_{n-1}$ denotes the unit sphere of $\mathbb{R}^n$, $n \geq 2$, then $r(\mathbb{S}_{n-1}) = n$, so that, by (4.5.9) and (4.5.10), we have

$$m_+(\mathbb{S}_{n-1}) = n + 1, \quad m_+^1(\mathbb{S}_{n-1}) = n. \qquad (4.5.11)$$

3. Let us consider the n-dimensional torus $\mathbb{T}^n = \{(z_1, \dots, z_n) \in (\mathbb{R}^2)^n | z_i \in \mathbb{T}, i = 1, \dots, n\}$. In this case $r(\mathbb{T}^n) = n + 1$ (see Shashkin [1962, Section 4.4] and, for $n = 2$, Morozov [1958]). Hence

$$m_+(\mathbb{T}^n) = n + 2, \quad m_+^1(\mathbb{T}^n) = n + 1. \qquad (4.5.12)$$

4. If X is a compact subset of $\mathbb{R}^n$ having non-empty interior and $n \geq 2$, then $\text{ind}(X) = n$ (see Hurewicz and Wallman [1948, Theorem IV.3]). Hence, by (4.5.9),

$$m_+(X) = n + 2, \quad m_+^1(X) = n + 1. \qquad (4.5.13)$$

5. If X is the set of extreme points of a strictly convex compact subset of $\mathbb{R}^n$, then $\text{ind}(X) = n - 1$ (see Hurewicz and Wallman [1948, Corollary 2 to Theorem IV.3]). On the other hand, it follows from Corollary 4.5.2 that $m_+(X) \leq n + 1$. So, by (4.5.9), we have

$$m_+(X) = n + 1, \quad m_+^1(X) = n. \qquad (4.5.14)$$

6. Volkov [1960] showed that, if X is a closed orientable surface of $\mathbb{R}^3$ not homeomorphic to a sphere, then $m_+(X) = 5$.

We now proceed to present a characterization of those compact spaces X such that $\mathscr{C}(X)$ possesses a finite dimensional K_+-subspace of order n, n being a fixed integer greater than 1 (see Definition 3.4.6).

Surprisingly, the answer is independent of n. More precisely, we shall see that $\mathscr{C}(X)$ possesses a finite dimensional K_+-subspace of order n for every $n \geq 1$ if and only if X is metrizable and has finite small inductive dimension.

Given $n \geq 1$, we set

$$m_{n,+}(X) := \min\{p \in \mathbb{N} \mid \text{There exists a } p\text{-dimensional } K_+\text{-subspace}$$
$$\text{of order } n \text{ in } \mathscr{C}_0(X)\}. \tag{4.5.15}$$

4.5.11 Theorem. *Let X be a compact Hausdorff space and let $n \geq 1$ be fixed. Then the following statements are equivalent:*

(i) *There exists a finite dimensional K_+-subspace of order n in $\mathscr{C}(X)$.*

(ii) *There exists $p \geq 2$ such that X is homeomorphic to the set of extreme points of a convex compact subset K of $\mathbb{R}^p$ having non-empty interior and such that for every choice of n distinct points $x_1, \ldots, x_n \in \partial_e K$, $x_1, \ldots, x_n$ are affinely independent and $\mathrm{co}(x_1, \ldots, x_n)$ is a face of K.*

(iii) *X is metrizable and has finite small inductive dimension.*

Furthermore, if one of the statements (i)–(iii) holds and if we denote by $c_n(X)$ the minimum of the integers $p \geq 2$ satisfying (ii), then we have

$$\mathrm{ind}(X) + 2n \leq m_{n,+}(X) = c_n(X) + 1 \leq \binom{2\,\mathrm{ind}(X) + n + 1}{n} + \binom{2\,\mathrm{ind}(X) + n}{n-1}.$$
$$\tag{4.5.16}$$

Proof. (i) $\Rightarrow$ (ii): Let $f_0, \ldots, f_p$ be linearly independent functions in $\mathscr{C}(X)$ such that the subspace H generated by them is a K_+-subspace of order n in $\mathscr{C}(X)$. Since H is a K_+-subspace, it contains a strictly positive function g_0. Then the subspace $\dfrac{1}{g_0} \cdot H$ is again a (p-dimensional) K_+-subspace of order n and it contains the constants. So the result follows from Theorem 3.4.9 by considering the embedding $\Phi_0 : X \to \mathbb{R}^p$ defined by (2.6.16) instead of the embedding Φ.

This reasoning also shows that $c_n(X) + 1 \leq m_{n,+}(X)$.

(ii) $\Rightarrow$ (i): According to Theorem 3.4.9, $\{1, \mathrm{pr}_1, \ldots, \mathrm{pr}_p\}$ is a K_+-subset of order n in $\mathscr{C}(\partial_e K)$. Thus there exists a $(p+1)$-dimensional K_+-subspace of order n in $\mathscr{C}(X)$ and

$$m_{n,+}(X) \leq c_n(X) + 1.$$

(i) $\Rightarrow$ (iii): Suppose that $\{f_0, \ldots, f_p\}$ is a K_+-subset of order n in $\mathscr{C}(X)$. We may always suppose that $f_0 = \mathbf{1}$. By virtue of Theorem 3.4.9 we see that X is homeomorphic to a subset of $\mathbb{R}^p$ (namely, to $\Phi_0(X)$), so that X is metrizable and has finite small inductive dimension.

Furthermore, since $\Phi_0(X)$ is $2n$-regular (see Remark to Corollary 4.5.6) then by a theorem of Borsuk [1957], for every open set U containing at least $2n - 1$ points of $\Phi_0(X)$, the difference $\Phi_0(X) \setminus U$ is homeomorphic to a subset of $\mathbb{R}^{p-2n+1}$. This shows that $\mathrm{ind}(X) \leq p - 2n + 1$, and hence $\mathrm{ind}(X) + 2n \leq m_{n,+}(X)$.

(iii) $\Rightarrow$ (i): Set $p := \mathrm{ind}(X)$. By the Menger-Nöbeling theorem, X is homeomorphic to a subset Q of $\mathbb{R}^{2p+1}$. On account of Example 1 to Corollary 3.4.8, we see that there exists a K_+-subspace H of order n in $\mathscr{C}(X)$ and

$$\dim(H) = \binom{2p + n + 1}{n} + \binom{2p + n}{n - 1}.$$

This also shows the second inequality of (4.5.16).   □

Example. Let $\mathbb{T}$ be the unit circle. Then, for every $n \geq 1$,

$$m_{n,+}(\mathbb{T}) = 2n + 1. \tag{4.5.17}$$

In fact, by (4.5.16) we have $2n + 1 \leq m_{n,+}(\mathbb{T})$. On the other hand, after setting $p = 2n + 1$, we know that $\mathbf{1}, \mathscr{R}e\, z, \ldots, \mathscr{R}e\, z^p, \mathscr{I}m\, z, \ldots, \mathscr{I}m\, z^p$ is a Chebyshev system of order $2p + 1$ on $\mathbb{T}$ (see Example 2 to Theorem 2.3.5).

By Corollary 3.4.10, the subset $\{\mathbf{1}, \mathscr{R}e\, z, \ldots, \mathscr{R}e\, z^p, \mathscr{I}m\, z, \ldots, \mathscr{I}m\, z^p\}$ is a K_+-subset of order $p = 2n + 1$ in $\mathscr{C}(\mathbb{T})$. Hence $m_{n,+}(\mathbb{T}) = 2n + 1$.

We now deal with similar problems in the case where X is locally compact. We begin by stating the following preliminary result.

4.5.12 Lemma. *Let X be a locally compact Hausdorff space. Given $n \geq 2$, let us consider the following statements:*
(i) *There exists an n-dimensional K_+^1-subspace in $\mathscr{C}_0(X)$.*
(ii) *There exists a convex compact subset K of $\mathbb{R}^n$ having non-empty interior and containing 0 as an extreme point such that X is homeomorphic to $\partial_e K \setminus \{0\}$.*
(iii) *There exists a compact subset Q of the unit sphere of $\mathbb{R}^n$ and a point $q_0 \in Q$ such that X is homeomorphic to $Q \setminus \{q_0\}$.*
(iv) *There exists an $(n + 1)$-dimensional strict K_+-subspace in $\mathscr{C}_0(X)$.*
(v) *There exists an $(n + 1)$-dimensional K_+-subspace in $\mathscr{C}_0(X)$.*
   Then the implications (i) $\Rightarrow$ (ii) $\Rightarrow$ (iii) $\Rightarrow$ (iv) $\Rightarrow$ (v) hold.

Furthermore, if (i) holds, then X is metrizable and separable, it has finite small inductive dimension and

$$\mathrm{ind}(X) \leq n - 1. \tag{4.5.18}$$

Proof. (i) $\Rightarrow$ (ii): Let $h_1, \ldots, h_n$ be linearly independent functions in $\mathscr{C}_0(X)$ such that $\{h_1, \ldots, h_n\}$ is a K_+^1-subset in $\mathscr{C}_0(X)$. Consider the one-point compactification $X_\omega = X \cup \{\omega\}$ of X and denote by $\tilde{h}_1, \ldots, \tilde{h}_n$ the continuous extensions to X_ω of $h_1, \ldots, h_n$ (see (1.1.7)).

By Proposition 4.1.14, $\{1, \tilde{h}_1, \ldots, \tilde{h}_n\}$ is a K_+-subset in $\mathscr{C}(X_\omega)$. Consequently, denoting by $\Phi_0: X_\omega \to \mathbb{R}^n$ the mapping

$$\Phi_0(x) := (\tilde{h}_1(x), \ldots, \tilde{h}_n(x)) \quad (x \in X_\omega),$$

we have by Theorem 4.5.1 that $\Phi_0(X_\omega) = \partial_e \mathrm{co}(\Phi_0(X_\omega))$ and Φ_0 is a homeomorphism from X_ω to $\Phi_0(X_\omega)$. Since $\tilde{h}_1, \ldots, \tilde{h}_n$ are linearly independent, the subset $K := \mathrm{co}(\Phi_0(X_\omega))$ has non-empty interior and $0 = \Phi_0(\omega) \in \partial_e K$.

Finally, $X = X_\omega \backslash \{\omega\}$ is homeomorphic to $\partial_e K \backslash \{0\}$.

(ii) $\Rightarrow$ (iii): For the proof of this implication one can argue as in the proof of the implication (iv) $\Rightarrow$ (v) of Theorem 4.5.8.

(iii) $\Rightarrow$ (iv): It follows from (iii) that X_ω is homeomorphic to Q. By Theorem 4.5.8, $\mathscr{C}(X_\omega)$ possesses an n-dimensional strict K_+^1-subspace H_ω. On the other hand, from (iii) it also follows that X is countable at infinity, so that there exists a strictly positive function $f_0 \in \mathscr{C}_0(X)$.

Consequently, the subspace generated by $\{f_0\}$ and $\{f_0 \cdot h_{|X} | h \in H_\omega\}$ is a strict K_+-subspace in $\mathscr{C}_0(X)$.

(iv) $\Rightarrow$ (v): This is obvious.

The last part of the statement follows as in the final part of the proof of Theorem 4.5.8. $\qquad\square$

We may now prove the following result.

4.5.13 Theorem. *Let X be a locally compact, non-compact Hausdorff space. Then the following statements are equivalent:*

(i) *$\mathscr{C}_0(X)$ possesses a finite dimensional K_+-subspace (or, a finite dimensional strict K_+-subspace).*

(ii) *$\mathscr{C}_0(X)$ possesses a finite dimensional K_+^1-subspace (or, a finite dimensional strict K_+^1-subspace).*

(iii) *There exists a convex compact subset K of $\mathbb{R}^n$ for some $n \geq 2$, having non-empty interior and containing 0 as an extreme point such that X is homeomorphic to $\partial_e K \backslash \{0\}$.*

(iv) *There exist $n \geq 2$, a compact subset Q of the unit sphere of $\mathbb{R}^n$ and a point $q_0 \in Q$ such that X is homeomorphic to $Q \backslash \{q_0\}$.*

(v) *X is metrizable and separable and has finite small inductive dimension.*

Furthermore, if one of the statements (i)–(v) *holds, then*

$$1 + \mathrm{ind}(X) \leq m_+^1(X) \leq m_+(X) \leq m_+^1(X) + 1 \tag{4.5.19}$$

and

$$m_+(X) \leq 2\,\mathrm{ind}(X) + 3. \tag{4.5.20}$$

Proof. The implication (i) $\Rightarrow$ (ii) is obvious. From Lemma 4.5.12, the implications (ii) $\Rightarrow$ (iii) $\Rightarrow$ (iv) $\Rightarrow$ (v) follow with their respective counterparts as well as the inequalities (4.5.19).

To prove the implication (v) $\Rightarrow$ (i), suppose that $\mathrm{ind}(X) = n$ for some $n \in \mathbb{N}_0$. Then $\dim(X_\omega) \leq n$ by the sum theorems for dimensions (see Hurewicz and Wallman [1948, Corollary 2 to Theorem III.2]). So, by the Menger-Nöbeling theorem X_ω is homeomorphic to a closed subset of the unit sphere of $\mathbb{R}^{2n+2}$.

By Theorem 4.5.8 there exists a $(2n + 3)$-dimensional K_+-subspace in $\mathscr{C}(X_\omega)$ and hence in $\mathscr{C}_0(X)$ (see Proposition 4.1.15).

This reasoning also shows that $m_+(X) \leq 2\,\mathrm{ind}(X) + 3$. $\square$

Remark. We were not able to show that $m_+(X) = m_+^1(X) + 1$ when X is a (non compact) locally compact Hausdorff space X and we leave this as an open problem.

It would also be interesting to find some relations between $m_+(X)$ and $m_+(X_\omega)$.

Notes and references

This section is largely based on the papers of Shashkin [1962] and Berens and Lorentz [1975], although if it also contains some new results.

Theorem 4.5.1, in the case where X is compact, was announced by Shashkin in [1960] and appeared in Shashkin [1962] (see also Berens and Lorentz [1975] and Amir and Ziegler [1978]).

Its extension to the locally compact case was obtained by Flösser [1979] and Donner [1982]. More generally, Flösser [1979] studied the Korovkin closures of finite sets in M-spaces and in dual atomic locally convex vector lattices. Finite K_+-subspaces were also studied in ℓ^p-spaces by Kitto and Wulbert [1976] as well as in $L^p(X,\mu)$-spaces by Donner [1982].

For X compact, Theorem 4.5.3 is due, for the case $n = 2$, to Korovkin [1951] and, for $n \geq 3$, to Berens and Lorentz [1975].

Corollary 4.5.6 is due to Shashkin [1962] (see also Berens and Lorentz [1975]). However, the fact that every finite K_+-subset of $\mathscr{C}(X)$ is 2-independent was first discovered by Volkov [1958].

The notion of k-regularity for subsets of Euclidean spaces was introduced and studied by Borsuk [1957].

Further investigations have been carried out by Boltiansky, Ryshkov and Shashkin [1960], Gustin [1947], Haupt and Künneth [1967], Ryshkov [1959], Shashkin [1965a], [1974].

We also quote two further interesting characterizations of finite K_+-subsets. The first one is due to Baskakov [1961]:

Let $h_1, \ldots, h_n$ be in $\mathscr{C}([a,b])$ such that the linear subspace generated by them contains a positive function. Then $\{h_1, \ldots, h_n\}$ is a K_+-subset in $\mathscr{C}([a,b])$ if and only if for every $f \in \mathscr{C}([a,b])$ and for every $x_0 \in [a,b]$, $f(x_0)$ is the only number $\alpha \in \mathbb{R}$ which satisfies the following property:

$$\text{For every } r_1, \ldots, r_{n+1} \in \mathbb{R} \text{ such that } \sum_{i=1}^{n} r_i h_i + r_{n+1} f \geq 0$$

$$\text{it follows that } \sum_{i=1}^{n} r_i h_i(x_0) + r_{n+1}\alpha \geq 0.$$

The second result is due to Garkavi [1972].

Let X be a compact Hausdorff space. For every $f \in \mathscr{C}(X)$, we shall set $Z(f) := \{x \in X \mid f(x) = 0\}$. If $h_1, \ldots, h_n$ are linearly independent functions in $\mathscr{C}(X)$, then $\{h_1, \ldots, h_n\}$ is a K_+-subset in $\mathscr{C}(X)$ if and only if for every $x_0 \in X$ there exist finitely many functions f_1, $\ldots, f_m$, $m < n$, belonging to the subspace generated by $h_1, \ldots, h_n$, such that $\bigcap_{i=1}^{m} Z(f_i) = \{x_0\}$, $f_1 \geq 0$ and $f_i \geq 0$ on $\bigcap_{j=1}^{i-1} Z(f_j)$ for each $i = 2, \ldots, m$.

Theorem 4.5.8 and Corollary 4.5.10 are essentially due to Berens and Lorentz [1975] and Shashkin [1962].

(4.5.9) was obtained by Shashkin [1965a] in the case where X is a one-dimensional locally connected continuum or a two-dimensional closed manifold without an edge. For the general case, (4.5.9) is due to Berens and Lorentz [1975].

The remainder of the section seems to be new, except Theorem 4.5.11 whose essence is due to Shashkin [1965b] and which, however, is presented here in a more complete form.

According to Theorem 4.5.11, the topological properties of those compact Hausdorff spaces admitting a K_+-subspace of order $n \geq 1$ can be studied by investigating the analogous ones of the convex compact subsets of $\mathbb{R}^p$ ($p \geq n$) having non-empty interior and such that the convex hull of n arbitrary extreme points of it is a face.

These subsets are called K_n^p subsets and they are of independent interest.

A first example of them was exhibited by Carathéodory [1911] in connection with the study of the positive harmonic functions on the unit disk.

Further investigation of the properties of these subsets has been carried out by Gale [1956], Grünbaum [1967], [1970] and Shashkin [1974], who studied in particular the topological properties of the set of their extreme points.

The existence of finite K_+-subsets or finite K_+^1-subsets was investigated also in other settings such as Banach lattices and Banach algebras.

In his papers [1973a], [1973b], [1975], [1977], Wolff characterized in a satisfactory way those Banach lattices which possess finite K_+-subsets (see also Scheffold [1973b]). A similar problem was considered by Pannenberg for commutative Banach algebras ([1985], [1987], [1988], [1990b], [1990c], [1992b]) (see Appendix A) and by Beckhoff [1987] for some particular C^*-algebras.

Chapter 5

Applications to positive approximation processes on real intervals

Korovkin-type theorems stated in the previous chapters furnish a powerful tool in studying the convergence toward the identity operator of a great variety of (generalized) sequences of positive linear operators that, because of this property, are also called positive approximation processes.

More precisely, given two function spaces E and F defined on the same locally compact Hausdorff space and such that $E \subset F$, a *positive approximation process on E with respect to F* is a net $(L_i)_{i \in I}^{\leq}$ of positive linear operators from E into F such that for every $f \in E$

$$\lim_{\substack{i \in I}}^{\leq} L_i(f) = f,$$

the convergence being understood with respect to a suitable topology on F.

In this chapter we describe many of these sequences of operators which act on spaces of continuous functions defined on real intervals (bounded or not), and we show their convergence by means of Korovkin-type theorems.

However, it is not the purpose of this chapter to give a complete and systematic treatment of the theory of approximation by positive linear operators.

More simply, we only wish to illustrate some classical and more recent positive approximation processes in order to appreciate better the usefulness of Korovkin-type approximation theory.

Throughout the sections we describe different methods of defining them. In each case we also give concrete examples, often enriched with more specific properties.

Since in many cases we shall also give some quantitative estimates of the convergence, in the preliminary Section 5.1 we include some basic properties of moduli of continuity and some general estimates. In the same section we also treat some extensions of moduli of continuity to a more general context.

In the second section we consider several sequences of positive linear operators which can be defined by using well-known concepts from probability theory. This is, in fact, an additional method to construct positive approximation processes for bounded continuous functions on an arbitrary real interval. We study the convergence of these sequences by means of Feller's theorem, which is equivalent to Korovkin's first theorem in the case of a compact interval.

In Sections 5.3 and 5.4 we briefly discuss the most classical applications of Korovkin-type approximation theory concerning the convergence of discrete-type operators and convolution operators.

In fact, these interesting positive approximation processes have been the subjects of many investigations.

Attempting to limit the size of the book, we were forced to make a drastic choice among them, so that we selected those examples that have particularly stimulated our imagination.

However other examples and references are indicated in the final notes of the sections.

As regards the material selected in Sections 5.3 and 5.4, we mention in particular several generalizations of Bernstein operators, including their integral modifications such as Kantorovich and Bernstein-Durrmeyer operators, Szász-Mirakjan and Baskakov operators, Cheney-Sharma and Meyer-König and Zeller operators, Fejér, Jackson, Fejér-Korovkin and Abel-Poisson convolution operators together with some general summation methods.

In general our results are concerned with approximation processes which converge uniformly on the whole interval or on compact subsets of it. However in some cases we also investigate their convergence in L^p-spaces or in suitable weighted function spaces by using, also in this case, Korovkin-type theorems.

5.1 Moduli of continuity and degree of approximation by positive linear operators

This section has a preparatory character. We collect here some properties concerning classical moduli of continuity and smoothness, as well as some of their extensions to the setting of convex compact sets. Furthermore, we use them to give quantitative error estimates for the (pointwise and uniform) approximation by positive linear operators.

However the estimates we obtain are not the sharpest possible. They have the specific purpose to describe in some respect the rate of convergence of the approximation processes we shall consider in the next sections. So we shall not give an up-to-date survey on quantitative estimates of rates of convergence by positive approximation processes. More complete details in this direction can be found, for instance, in the books of Butzer and Nessel [1971], De Vore [1972], Ditzian and Totik [1987], Korovkin [1960], Lorentz [1986a], Meinardus [1967], Sendov and Popov [1988].

We begin with the classical modulus of continuity of a bounded continuous function on a real interval.

If I is a real interval, we denote by $\mathscr{C}_b(I)$ ($\mathscr{UC}_b(I)$, respectively) the space of all bounded continuous (uniformly continuous, respectively) real functions on I.

If $f \in \mathscr{C}_b(I)$, for every $x \in I$, $h \in \mathbb{R}$, $h \neq 0$ and $k \in \mathbb{N}$, the *k-th difference* $\Delta_h^k f(x)$ of f *with step* h *at the point* x *is given by*

$$\Delta_h^k f(x) := \sum_{m=0}^{k} (-1)^{m+k} \binom{k}{m} f(x + mh), \qquad (5.1.1)$$

provided that $x + kh \in I$. For the sake of brevity we shall set $\Delta_h f(x) := \Delta_h^1 f(x) = f(x + h) - f(x)$.

The $(k + 1)$-th difference can be expressed in terms of the k-th difference by means of the following identity:

$$\Delta_h^{k+1} f(x) = \Delta_h^k f(x + h) - \Delta_h^k f(x). \qquad (5.1.2)$$

Now, if $f: I \to \mathbb{R}$ is a bounded real function and if $\delta > 0$, the *modulus of continuity* $\omega(f, \delta)$ of f with argument δ is defined by

$$\omega(f, \delta) := \sup_{\substack{x, y \in I \\ |x-y| \leq \delta}} |f(x) - f(y)| \left(= \sup_{\substack{|h| \leq \delta \\ x, x+h \in I}} |\Delta_h f(x)| \right). \qquad (5.1.3)$$

The modulus of continuity of f is also called the *first modulus of smoothness of* f.

More generally, if $f: I \to \mathbb{R}$ is a bounded real function and if $\delta > 0$ and $k \in \mathbb{N}$, we can define the *k-th modulus of smoothness* $\omega_k(f, \delta)$ of f by

$$\omega_k(f, \delta) := \sup_{\substack{|h| \leq \delta \\ x, x+kh \in I}} |\Delta_h^k f(x)|. \qquad (5.1.4)$$

In the following lemma we state some elementary properties of moduli of smoothness (and, in particular, of the modulus of continuity).

5.1.1 Lemma. *Let I be a real interval and fix $f \in \mathscr{C}_b(I)$. Then, for every $k \in \mathbb{N}$, the following statements hold:*
(1) *If $0 < \delta_1 \leq \delta_2$, then $\omega_k(f, \delta_1) \leq \omega_k(f, \delta_2)$.*
(2) $\lim_{\delta \to 0^+} \omega_k(f, \delta) = 0$ *provided $f \in \mathscr{UC}_b(I)$.*
(3) *For every $\delta > 0$, $\omega_{k+1}(f, \delta) \leq 2 \omega_k(f, \delta)$.*
(4) *If f is differentiable and $f' \in \mathscr{C}_b(I)$, then $\omega_{k+1}(f, \delta) \leq \delta \, \omega_k(f', \delta)$ for every $\delta > 0$.*
(5) *For every $\delta > 0$ and $n \in \mathbb{N}$, $\omega_k(f, n\delta) \leq n^k \omega_k(f, \delta)$.*
(6) *For every $\delta > 0$ and $\lambda > 0$, $\omega_k(f, \lambda\delta) \leq (1 + [\lambda])^k \omega_k(f, \delta)$, where $[\lambda]$ denotes the integer part of λ.*

Proof. Properties (1) and (2) easily follow from the definition. Property (3) is a direct consequence of (5.1.2) and (5.1.3). As regards property (4), we observe that, for every $h \in \mathbb{R}$, $|h| \leq \delta$, and $x \in I$ such that $x + kh \in I$, we have

$$|\Delta_h^{k+1} f(x)| = |\Delta_h^k f(x + h) - \Delta_h^k f(x)|$$

$$= \left| \sum_{m=0}^{k} (-1)^{m+k} \binom{k}{m} (f(x + (m + 1)h) - f(x + mh)) \right|$$

$$= \left| \sum_{m=0}^{k} (-1)^{m+k} \binom{k}{m} \int_{mh}^{(m+1)h} f'(x + t) \, dt \right|$$

$$= \left| \sum_{m=0}^{k} (-1)^{m+k} \binom{k}{m} \int_{0}^{h} f'(x + mh + t) \, dt \right|$$

$$= \left| \int_{0}^{h} \sum_{m=0}^{k} (-1)^{m+k} \binom{k}{m} f'(x + mh + t) \, dt \right| = \left| \int_{0}^{h} \Delta_h^k f'(x + t) \, dt \right|$$

$$\leq \left| \int_{0}^{h} |\Delta_h^k f'(x + t)| \, dt \right| \leq \left| \int_{0}^{h} \omega_k(f', \delta) \, dt \right| \leq \delta \omega_k(f', \delta).$$

Consequently, property (4) follows from (5.1.4).

If $k = 0$ or $k = 1$, property (5) is obviously true. So assume $k \geq 2$. Given $\varepsilon > 0$, by (5.1.4) there exist $h \in \mathbb{R}$, $|h| \leq \delta$, and $x \in I$ such that $x + n(k + 1)h \in I$ and $\omega_k(f, n\delta) \leq |\Delta_{nh}^k f(x)| + \varepsilon$.

On account of (5.1.2) we obtain

$$\Delta_{nh}^k f(x) = \Delta_{nh}^{k-1} f(x + nh) - \Delta_{nh}^{k-1} f(x)$$

$$= \sum_{i_1=0}^{n-1} \sum_{i_2=0}^{n-1} \cdots \sum_{i_{k-1}=0}^{n-1} \Delta_h^{k-1} \left(\sum_{i_k=0}^{n-1} \Delta_h f(x + i_1 h + \cdots + i_k h) \right)$$

$$= \sum_{i_1=0}^{n-1} \sum_{i_2=0}^{n-1} \cdots \sum_{i_k=0}^{n-1} \Delta_h^k f(x + i_1 h + \cdots + i_k h),$$

and hence

$$\omega_k(f, n\delta) \leq |\Delta_{nh}^k f(x)| + \varepsilon \leq \sum_{i_1=0}^{n-1} \sum_{i_2=0}^{n-1} \cdots \sum_{i_k=0}^{n-1} |\Delta_h^k f(x + i_1 h + \cdots + i_k h)| + \varepsilon$$

$$\leq n^k \omega_k(f, \delta) + \varepsilon.$$

Finally, property (6) is a simple consequence of (1) and (5).   □

We now state a general estimate in terms of the modulus of continuity.

5.1.2 Theorem. *Let I be a real interval and consider a vector sublattice E of $\mathscr{C}(I)$ containing $\mathscr{C}_b(I)$ as well as the functions $e_1(t) := t$ and $e_2(t) := t^2$ $(t \in I)$.*

For every $x \in I$ denote by $\psi_x: I \to \mathbb{R}$ the function defined by $\psi_x(t) := t - x$ for every $t \in I$.

If $L: E \to \mathscr{B}(I)$ is a positive linear operator, then, for every $f \in \mathscr{C}_b(I)$, $\delta > 0$ and $x \in I$, we have

$$|Lf(x) - f(x)| \leq |f(x)||L(1)(x) - 1|$$

$$+ \left(L(1)(x) + \frac{1}{\delta}\sqrt{L(\psi_x^2)(x)}\sqrt{L(1)(x)} \right)\omega(f, \delta). \quad (5.1.5)$$

Moreover, if f is differentiable on I and $f' \in \mathscr{C}_b(I)$, we also have

$$|Lf(x) - f(x)| \leq |f(x)||L(1)(x) - 1| + |f'(x)||L(\psi_x)(x)|$$

$$+ \sqrt{L(\psi_x^2)(x)}\left(\sqrt{L(1)(x)} + \frac{1}{\delta}\sqrt{L(\psi_x^2)(x)} \right)\omega(f', \delta). \quad (5.1.6)$$

Proof. Fix $f \in \mathscr{C}_b(I)$, $x \in I$ and $\delta > 0$. By Lemma 5.1.1, (6), we have for every $t \in I$

$$|f(x) - f(t)| \leq \omega(f, |t - x|) = \omega\left(f, \frac{1}{\delta}|t - x|\delta \right) \leq \left(1 + \frac{|t - x|}{\delta} \right)\omega(f, \delta) \quad (1)$$

and therefore, since the operator L is positive,

$$L(|f - f(x) \cdot 1|)(x) \leq \left(L(1)(x) + \frac{1}{\delta}L(|\psi_x|)(x) \right)\omega(f, \delta).$$

Hence the Cauchy-Schwarz inequality applied to the positive linear form $\mu_x(g) = Lg(x)$ $(g \in E)$ (see (1.2.4)) implies

$$|Lf(x) - f(x)| \leq |L(f(x) \cdot 1)(x) - f(x)| + L(|f - f(x) \cdot 1|)(x)$$

$$\leq |f(x)||L(1)(x) - 1| + \left(L(1)(x) + \frac{1}{\delta}L(|\psi_x|)(x) \right)\omega(f, \delta)$$

$$\leq |f(x)||L(1)(x) - 1| + \left(L(1)(x) \right.$$

$$\left. + \frac{1}{\delta}\sqrt{L(\psi_x^2)(x)}\sqrt{L(1)(x)} \right)\omega(f, \delta).$$

Now assume that f is differentiable and $f' \in \mathscr{C}_b(I)$. Then for every $x, t \in I$ we can write

$$f(t) = f(x) + f'(x)(t - x) + f(t) - f(x) - f'(x)(t - x),$$

and consequently

$$|Lf(x) - f(x)| \leq |f(x)||L(\mathbf{1})(x) - 1| + |f'(x)||L(\psi_x)(x)|$$

$$+ L(|f - f(x)\mathbf{1} - f'(x)\psi_x|)(x).$$

By the mean value theorem we have as in (1)

$$|f(t) - f(x) - f'(x)(t - x)| \leq |t - x|\left(1 + \frac{|t - x|}{\delta}\right)\omega(f', \delta)$$

for every $t \in I$ and hence, by the Cauchy-Schwarz inequality,

$$L(|f - f(x)\mathbf{1} - f'(x)\psi_x|)(x) \leq \left(L(|\psi_x|)(x) + \frac{1}{\delta}L(\psi_x^2)(x)\right)\omega(f', \delta)$$

$$\leq \left(\sqrt{L(\psi_x^2)(x)L(\mathbf{1})(x)} + \frac{1}{\delta}L(\psi_x^2)(x)\right)\omega(f', \delta)$$

$$\leq \sqrt{L(\psi_x^2)(x)}\left(\sqrt{L(\mathbf{1})(x)} + \frac{1}{\delta}\sqrt{L(\psi_x^2)(x)}\right)\omega(f', \delta).$$

$\square$

Remark. Theorem 5.1.2 applied to $\delta = \sqrt{L(\psi_x^2)(x)}$ shows in particular that

$$|Lf(x) - f(x)| \leq |f(x)||L(\mathbf{1})(x) - 1| + \left(L(\mathbf{1})(x) + \sqrt{L(\mathbf{1})(x)}\right)\omega(f, \sqrt{L(\psi_x^2)(x)}),$$

(respectively,

$$|Lf(x) - f(x)| \leq |f(x)||L(\mathbf{1})(x) - 1| + |f'(x)||L(\psi_x)(x)|$$

$$+ \left(1 + \sqrt{L(\mathbf{1})(x)}\right)\sqrt{L(\psi_x^2)(x)}\,\omega(f', \sqrt{L(\psi_x^2)(x)}))$$

provided $f \in \mathscr{C}_b(I)$ (respectively, if f is differentiable on I and $f' \in \mathscr{C}_b(I)$).

The above defined moduli of smoothness involve bounded continuous functions defined on a real interval. For our purposes it is useful to consider an extension of them in order to obtain quantitative estimates of the rates of convergence for positive approximation processes on spaces of real continuous functions defined on a general convex compact set.

So let E be a real locally convex Hausdorff space and consider a convex compact subset K of E. We denote by E'_K the space of all restrictions to K of continuous linear functionals on E, i.e.,

$$E'_K := \{h \in \mathscr{C}(K) | \text{There exists } \varphi \in E' \text{ such that } h = \varphi_{|K}\}. \qquad (5.1.7)$$

The space E'_K is endowed with the sup-norm induced by $\mathscr{C}(K)$.
If $h_1, \ldots, h_m \in E'_K$ and $\delta > 0$, we set

$$K(h_1, \ldots, h_m, \delta) := \left\{ (x, y) \in K \times K \left| \sum_{j=1}^{m} (h_j(x) - h_j(y))^2 \le \delta^2 \right. \right\}. \qquad (5.1.8)$$

We define the *modulus of continuity of a function* $f \in \mathscr{B}(K)$ *with respect to* $h_1, \ldots, h_m$ as

$$\omega(f; h_1, \ldots, h_m, \delta) := \sup\{|f(x) - f(y)| \,|\, (x, y) \in K(h_1, \ldots, h_m, \delta)\}. \qquad (5.1.9)$$

Moreover the *(total) modulus of continuity of* f is by definition

$$\Omega(f, \delta) := \inf \left\{ \omega(f; h_1, \ldots, h_m, \delta) \,\middle|\, m \in \mathbb{N}, h_1, \ldots, h_m \in E'_K, \left\| \sum_{j=1}^{m} h_j^2 \right\| = 1 \right\}. \qquad (5.1.10)$$

It is easy to see that in fact

$$\Omega(f, \delta) = \inf \left\{ \omega(f; h_1, \ldots, h_m, 1) \,\middle|\, m \in \mathbb{N}, h_1, \ldots, h_m \in E'_K, \left\| \sum_{j=1}^{m} h_j^2 \right\| = \frac{1}{\delta^2} \right\}. \qquad (5.1.11)$$

If E is a normed space, according to (5.1.3), we shall also define a *modulus of continuity* of $f \in \mathscr{B}(K)$ by

$$\omega(f, \delta) := \sup\{|f(x) - f(y)| \,|\, \|x - y\| \le \delta\} \quad (\delta > 0). \qquad (5.1.12)$$

If $E = \mathbb{R}^p$ and if we denote by $\mathrm{pr}_i \colon \mathbb{R}^p \to \mathbb{R}$ the i-th projection, clearly

$$\omega(f; \mathrm{pr}_1, \ldots, \mathrm{pr}_p, \delta) = \omega(f, \delta). \qquad (5.1.13)$$

Consequently, setting $\rho(K) := \sqrt{\max\{\|x\|^2 \mid x \in K\}}$, we obtain

$$\Omega(f, \delta) \leq \omega(f, \delta\rho(K)) \qquad (5.1.14)$$

because

$$\Omega(f, \delta) \leq \omega\left(f; \frac{\mathrm{pr}_1}{\rho(K)}, \ldots, \frac{\mathrm{pr}_p}{\rho(K)}, \delta\right) = \omega(f; \mathrm{pr}_1, \ldots, \mathrm{pr}_p, \delta\rho(K)) = \omega(f, \delta\rho(K)).$$

If $p = 1$ and $K = [a, b]$, then equality holds in (5.1.14).

We now state the analogue of Lemma 5.1.1 in this general setting.

5.1.3 Lemma. *Let $h_1, \ldots, h_m \in E'_K$ and $f \in \mathscr{C}(K)$. Then the following statements hold:*
(1) *If $0 < \delta_1 \leq \delta_2$, then $\omega(f; h_1, \ldots, h_m, \delta_1) \leq \omega(f; h_1, \ldots, h_m, \delta_2)$ ($\Omega(f, \delta_1) \leq \Omega(f, \delta_2)$, respectively).*
(2) $\lim\limits_{\delta \to 0^+} \Omega(f, \delta) = 0.$
(3) *For every $\delta > 0$ and $\lambda > 0$, $\omega(f; h_1, \ldots, h_m, \lambda\delta) \leq (1 + [\lambda])\omega(f; h_1, \ldots, h_m, \delta)$, ($\Omega(f, \lambda\delta) \leq (1 + [\lambda])\,\Omega(f, \delta)$, respectively), where $[\lambda]$ denotes the integer part of λ.*

Proof. Property (1) follows directly from the definitions.

To show (2), note first that the topology on K coincides with the coarsest topology on K with respect to which each $\varphi \in E'$ is continuous. Furthermore, this last topology is the uniform topology defined by the saturated family $\{d_{\varphi_1, \ldots, \varphi_m} \mid m \geq 1, \varphi_1, \ldots, \varphi_m \in E'\}$ of pseudo-metrics on K defined by

$$d_{\varphi_1, \ldots, \varphi_m}(x, y) := \sup_{1 \leq i \leq m} |\varphi_i(x) - \varphi_i(y)|$$

for every $x, y \in K$ and $\varphi_1, \ldots, \varphi_m \in E'$ (see Choquet [1969, Section 5] for more details on uniform spaces).

Now, since K is compact, f is uniformly continuous with respect to this uniform topology. So, if $\varepsilon > 0$, then there exist $\delta > 0$ and $\varphi_1, \ldots, \varphi_m \in E'$ such that $|f(x) - f(y)| \leq \varepsilon$ for every $x, y \in X$ satisfying $d_{\varphi_1, \ldots, \varphi_m}(x, y) \leq \delta$.

Thus, if we set $h_i = \varphi_{i|K}$, $i = 1, \ldots, m$, we obtain $\omega(f; h_1, \ldots, h_m, \delta) \leq \varepsilon$ and hence $\Omega(f, \delta) \leq \varepsilon$.

As regards property (3), let $\lambda > 0$ and $\varepsilon > 0$. By (5.1.9) we can consider $(x, y) \in K(h_1, \ldots, h_m, \lambda\delta)$ such that $\omega(f; h_1, \ldots, h_m, \lambda\delta) \leq |f(y) - f(x)| + \varepsilon$.

Now define $x_k := \left(1 - \dfrac{k}{1+[\lambda]}\right)x + \dfrac{k}{1+[\lambda]}y, k = 0,\ldots,1+[\lambda]$. Then, since every $h_i\ (i = 1,\ldots,m)$ is linear, for every $k = 0,\ldots,[\lambda]$ we have

$$\sum_{j=1}^{m}(h_j(x_{k+1}) - h_j(x_k))^2 = \sum_{j=1}^{m}h_j^2\left(\frac{1}{1+[\lambda]}(y-x)\right) \le \frac{1}{(1+[\lambda])^2}\lambda^2\delta^2 \le \delta^2,$$

and therefore $(x_k, x_{k+1}) \in K(h_1,\ldots,h_m,\delta)$. Hence we obtain

$$\omega(f;h_1,\ldots,h_m,\lambda\delta) \le |f(y) - f(x)| + \varepsilon = \left|\sum_{k=0}^{[\lambda]}f(x_{k+1}) - f(x_k)\right| + \varepsilon$$

$$\le \sum_{k=0}^{[\lambda]}\left|f(x_{k+1}) - f(x_k)\right| + \varepsilon$$

$$\le (1 + [\lambda])\omega(f;h_1,\ldots,h_m,\delta) + \varepsilon.$$

Since $\varepsilon > 0$ was arbitrarily chosen, we conclude

$$\omega(f;h_1,\ldots,h_m,\lambda\delta) \le (1 + [\lambda])\omega(f;h_1,\ldots,h_m,\delta).$$

Finally, the remaining part follows from (5.1.10). □

If L is a positive linear operator from $\mathscr{C}(K)$ into $\mathscr{B}(K)$, then for every $f \in \mathscr{C}(K)$ and $x \in K$, we put

$$\mu(x; L, f) := L((f - f(x)\mathbf{1})^2)(x). \tag{5.1.15}$$

The following preliminary result will enable us to proceed further.

5.1.4 Proposition. *Let* $L: \mathscr{C}(K) \to \mathscr{B}(K)$ *be a positive linear operator and let* $f \in \mathscr{C}(K)$ *and* $x \in K$.
Then for every $h_1,\ldots,h_m \in E_K'$ *and* $\delta > 0$ *we have*

$$|Lf(x) - f(x)L(\mathbf{1})(x)| \le \left(L(\mathbf{1})(x) + \frac{1}{\delta^2}\sum_{j=1}^{m}\mu(x; L, h_j)\right)\omega(f;h_1,\ldots,h_m,\delta). \tag{5.1.16}$$

In particular, if $L(h) = h$ *for every* $h \in A(K)$, *then*

$$|Lf(x) - f(x)| \le (1 + \tau(\delta, x))\Omega(f,\delta), \tag{5.1.17}$$

where

$$\tau(\delta, x) := \sup \left\{ \sum_{j=1}^{m} \mu(x; L, h_j) | m \geq 1, h_1, \ldots, h_m \in E'_K \text{ and } \left\| \sum_{j=1}^{m} h_j^2 \right\| = \frac{1}{\delta^2} \right\}. \quad (5.1.18)$$

Proof. Given $y \in K$, we note that, if $\sum_{j=1}^{m} (h_j(y) - h_j(x))^2 \geq \delta^2$, then, by Lemma 5.1.3, (3),

$$|f(y) - f(x)| \leq \omega \left(f; h_1, \ldots, h_m, \sqrt{\sum_{j=1}^{m} (h_j(y) - h_j(x))^2} \right)$$

$$\leq \left(1 + \frac{1}{\delta} \sqrt{\sum_{j=1}^{m} (h_j(y) - h_j(x))^2} \right) \omega(f; h_1, \ldots, h_m, \delta)$$

$$\leq \left(1 + \frac{1}{\delta^2} \sum_{j=1}^{m} (h_j(y) - h_j(x))^2 \right) \omega(f; h_1, \ldots, h_m, \delta).$$

On the other hand, if $\sum_{j=1}^{m} (h_j(x) - h_j(y))^2 \leq \delta^2$, we obviously have

$$|f(y) - f(x)| \leq \omega(f; h_1, \ldots, h_m, \delta)$$

$$\leq \left(1 + \frac{1}{\delta^2} \sum_{j=1}^{m} (h_j(y) - h_j(x))^2 \right) \omega(f; h_1, \ldots, h_m, \delta).$$

Thus we have proved that $|f - f(x)\mathbf{1}| \leq \left(1 + \frac{1}{\delta^2} \sum_{j=1}^{m} (h_j - h_j(x)\mathbf{1})^2 \right)$ $\omega(f; h_1, \ldots, h_m, \delta)$ and hence, since L is a positive linear operator,

$$|Lf - f(x)L(\mathbf{1})| \leq \left(L(\mathbf{1}) + \frac{1}{\delta^2} \sum_{j=1}^{m} L(h_j - h_j(x)\mathbf{1})^2 \right) \omega(f; h_1, \ldots, h_m, \delta).$$

By evaluating the preceding expression at $x \in K$, we obtain (5.1.16). □

Remark. If $E = \mathbb{R}^p$ and if $L(h) = h$ for every $h \in A(K)$, then (5.1.16) furnishes a quantitative estimate in terms of the modulus of continuity $\omega(f, \delta)$ defined by (5.1.12). Indeed, denoting by $\mathrm{pr}_i : \mathbb{R}^p \to \mathbb{R}$ the canonical projections, we have $\omega(f, \delta) = \omega(f; \mathrm{pr}_1, \ldots, \mathrm{pr}_p, \delta)$ (see (5.1.13)) and by (5.1.15)

$$\sum_{j=1}^{p} \mu(x; L, \mathrm{pr}_j) = \sum_{j=1}^{p} L((\mathrm{pr}_j - \mathrm{pr}_j(x) \cdot 1)^2)(x)$$

$$= \sum_{j=1}^{p} (L(\mathrm{pr}_j^2)(x) - 2\,\mathrm{pr}_j(x) \cdot L(\mathrm{pr}_j)(x) + \mathrm{pr}_j^2(x))$$

$$= \sum_{j=1}^{p} (L(\mathrm{pr}_j^2)(x) - \mathrm{pr}_j^2(x)) = L\left(\sum_{j=1}^{p} \mathrm{pr}_j^2\right)(x) - \sum_{j=1}^{p} \mathrm{pr}_j^2(x)$$

for every $x \in K$.

So, if we consider the function $e: K \to \mathbb{R}$ defined by

$$e(t) := \|t\|^2 \quad \text{for every } t \in K, \tag{5.1.19}$$

by (5.1.16) we obtain

$$|Lf(x) - f(x)| \le \left(1 + \frac{1}{\delta^2}(L(e)(x) - e(x))\right)\omega(f, \delta). \tag{5.1.20}$$

In the next result we show that (5.1.20) holds more generally in the case where E is a real inner-product space.

5.1.5 Proposition. *Let K be a convex compact subset of a real inner-product space $(E, \langle \cdot, \cdot \rangle)$ and let $L: \mathscr{C}(K) \to \mathscr{B}(K)$ be a positive linear operator satisfying $L(h) = h$ for every $h \in A(K)$.*

Then for every $f \in \mathscr{C}(K)$, $x \in K$ and $\delta > 0$ we have

$$|Lf(x) - f(x)| \le \left(1 + \frac{1}{\delta^2}(L(e)(x) - e(x))\right)\omega(f, \delta). \tag{5.1.21}$$

where the function e is defined as in (5.1.19).

Proof. Fix $f \in \mathscr{C}(K)$, $x \in K$ and $\delta > 0$. For every $t \in K$ we obviously have

$$|f(y) - f(x)| \le \left(1 + \frac{1}{\delta}\|x - y\|\right)\omega(f, \delta) \le \left(1 + \frac{1}{\delta^2}\|x - y\|^2\right)\omega(f, \delta)$$

$$= \left(1 + \frac{1}{\delta^2}(\|x\|^2 - 2\langle x|y \rangle + \|y\|^2)\right)\omega(f, \delta)$$

provided $\|x - y\| \ge \delta$. Clearly the same inequality also holds if $\|x - y\| \le \delta$.

Therefore, by applying the operator L to both sides of the preceding inequality and by evaluating them at the point $x \in K$, we obtain (5.1.21). $\qquad \square$

We now return to the general case where E is a real locally convex Hausdorff space and K is a convex compact subset of E.

Our next aim is to give a general quantitative estimate for the convergence of a parametric net $(L_{i,\lambda})_{i \in I}^{\leq}$, $\lambda \in \Lambda$, of linear positive operators from $\mathscr{C}(K)$ into $\mathscr{B}(K)$, uniformly with respect to $\lambda \in \Lambda$. These estimates will be used in Section 5.4 to study $\mathscr{A}$-summation processes.

We shall assume that the parametric net $(L_{i,\lambda})_{i \in I}^{\leq}$, $\lambda \in \Lambda$, satisfies the condition

$$\sup_{\lambda \in \Lambda} \|L_{i,\lambda}\| < +\infty \tag{5.1.22}$$

for every $i \in I$.

Moreover, for every $h_1, \ldots, h_m \in E'_K$ and $i \in I$, we define (see (5.1.15)),

$$\mu_i(h_1, \ldots, h_m) := \sup_{\lambda \in \Lambda} \left\| \sum_{j=1}^{m} \mu(\cdot; L_{i,\lambda}, h_j) \right\| \left(= \sup_{\lambda \in \Lambda} \sup_{x \in K} \left| \sum_{j=1}^{m} L_{i,\lambda}((h_j - h_j(x)\mathbf{1})^2)(x) \right| \right). \tag{5.1.23}$$

Furthermore for every $f \in \mathscr{C}(K)$ we set

$$\omega_i(f) := \inf\{\omega(f; h_1, \ldots, h_m, \sqrt{\mu_i(h_1, \ldots, h_m)}) | m \in \mathbb{N}, h_1, \ldots, h_m \in E'_K$$

$$\text{and } \mu_i(h_1, \ldots, h_m) > 0\}. \tag{5.1.24}$$

5.1.6 Theorem. *Let $(L_{i,\lambda})_{i \in I}^{\leq}$, $\lambda \in \Lambda$, be a parametric net of positive linear operators from $\mathscr{C}(K)$ into $\mathscr{B}(K)$ satisfying (5.1.22).*

Then, for every $f \in \mathscr{C}(K)$ and $i \in I$, we have

$$\sup_{\lambda \in \Lambda} \|L_{i,\lambda}(f) - f\| \leq \|f\| \sup_{\lambda \in \Lambda} \|L_{i,\lambda}(\mathbf{1}) - \mathbf{1}\| + C_i \omega_i(f), \tag{5.1.25}$$

where

$$C_i := \sup_{\lambda \in \Lambda} \|L_{i,\lambda}(\mathbf{1}) + \mathbf{1}\|. \tag{5.1.26}$$

In particular, if $L_{i,\lambda}(\mathbf{1}) = \mathbf{1}$ for every $\lambda \in \Lambda$, we obtain

$$\sup_{\lambda \in \Lambda} \|L_{i,\lambda}(f) - f\| \leq 2\,\omega_i(f). \tag{5.1.27}$$

Proof. We first remark that

$$\sup_{\lambda \in \Lambda} \|L_{i,\lambda}(f) - f\| \leq \|f\| \sup_{\lambda \in \Lambda} \|L_{i,\lambda}(\mathbf{1}) - \mathbf{1}\| + K_i(f),$$

where

$$K_i(f) := \sup_{\lambda \in \Lambda} \|L_{i,\lambda}(f) - fL_{i,\lambda}(\mathbf{1})\|.$$

Now, let $m \in \mathbb{N}$ and $h_1, \dots, h_m \in E'_K$ be such that $\mu_i(h_1, \dots, h_m) > 0$. Recalling Proposition 5.1.4, we obtain

$$\|L_{i,\lambda}(f) - fL_{i,\lambda}(\mathbf{1})\| \leq \left\| L_{i,\lambda}(\mathbf{1}) + \frac{1}{\delta^2} \mu_i(h_1, \dots, h_m) \cdot \mathbf{1} \right\| \omega(f; h_1, \dots, h_m, \delta)$$

for every $\lambda \in \Lambda$ and $\delta > 0$ (see also (5.1.23)).

In particular, setting $\delta = \sqrt{\mu_i(h_1, \dots, h_m)}$, the preceding formula yields

$$\|L_{i,\lambda}(f) - fL_{i,\lambda}(\mathbf{1})\| \leq \|L_{i,\lambda}(\mathbf{1}) + \mathbf{1}\| \omega(f; h_1, \dots, h_m, \sqrt{\mu_i(h_1, \dots, h_m)}).$$

Since

$$\|L_{i,\lambda}(\mathbf{1}) + \mathbf{1}\| \omega(f; h_1, \dots, h_m, \sqrt{\mu_i(h_1, \dots, h_m)}) \leq C_i \omega_i(f),$$

we conclude that $K_i(f) \leq C_i \omega_i(f)$ and this completes the proof. $\qquad\square$

As a simple consequence of the preceding theorem, we obtain the following result which will be very useful in the applications.

5.1.7 Corollary. *Let $(L_{i,\lambda})_{i \in I}^{\leq}$, $\lambda \in \Lambda$, be a parametric net of positive linear operators from $\mathscr{C}(K)$ into $\mathscr{B}(K)$ satisfying (5.1.22) and assume that $L_{i,\lambda}(h) = h$ for every $i \in I$, $\lambda \in \Lambda$ and $h \in A(K)$.*

Moreover suppose that there exist a constant $C > 0$ and a net $(\xi_{i,\lambda})_{i \in I}^{\leq}$, $\lambda \in \Lambda$, such that

$$\lim_{\substack{\leq \\ i \in I}} \xi_{i,\lambda} = 0 \quad \text{uniformly in } \lambda \in \Lambda$$

and

$$\left\| \sum_{j=1}^{m} (L_{i,\lambda}(h_j^2) - h_j^2) \right\| \leq C\xi_{i,\lambda} \left\| \sum_{j=1}^{m} h_j^2 \right\|$$

for every $i \in I$, $\lambda \in \Lambda$ and $h_1, \dots, h_m \in E'_K$.

Then, for every $f \in \mathscr{C}(K)$ and $i \in I$, we have

$$\sup_{\lambda \in \Lambda} \|L_{i,\lambda}(f) - f\| \leq 2\Omega(f, \sqrt{C\xi_i}), \tag{5.1.28}$$

where

$$\xi_i = \sup_{\lambda \in \Lambda} \xi_{i,\lambda}. \qquad (5.1.29)$$

Remark. Incidentally in the proof of Theorem 5.1.6 we also showed that

$$\sup_{\lambda \in \Lambda} \|L_{i,\lambda}(f) - f\| \le \|f\| \sup_{\lambda \in \Lambda} \|L_{i,\lambda}(\mathbf{1}) - \mathbf{1}\|$$

$$+ \sup_{\lambda \in \Lambda} \left\| L_{i,\lambda}(\mathbf{1}) + \frac{1}{\delta^2} \left(\sum_{j=1}^{m} \mu(\cdot; L_{i,\lambda}, h_j) \right) \cdot \mathbf{1} \right\|$$

$$\times \omega(f; h_1, \ldots, h_m, \delta), \qquad (5.1.30)$$

provided that $h_1, \ldots, h_m \in E_K'$.

In particular, if $E = \mathbb{R}^p$ and $L_{i,\lambda}(h) = h$ for every $h \in A(K)$ and $\lambda \in \Lambda$, then on account of (5.1.23) (see also (5.1.19)), we have $\omega(f; \mathrm{pr}_1, \ldots, \mathrm{pr}_p, \delta) = \omega(f, \delta)$ and $\sum_{j=1}^{m} \mu(\cdot; L_{i,\lambda}, \mathrm{pr}_j) = L_{i,\lambda}(e) - e$.

Therefore (5.1.30) becomes

$$\sup_{\lambda \in \Lambda} \|L_{i,\lambda}(f) - f\| \le \sup_{\lambda \in \Lambda} \left\| \mathbf{1} + \frac{1}{\delta^2}(L_{i,\lambda}(e) - e) \right\| \omega(f, \delta). \qquad (5.1.31)$$

In fact we can obtain a similar estimate in the more general case where E is a real inner-product space.

5.1.8 Theorem. *Let K be a compact convex subset of a real inner-product space $(E, \langle \cdot, \cdot \rangle)$ and let $(L_{i,\lambda})_{i \in I}^{\le}$, $\lambda \in \Lambda$, be a parametric net of positive linear operators from $\mathscr{C}(K)$ into $\mathscr{B}(K)$ satisfying (5.1.22).*

Then, for every $f \in \mathscr{C}(K)$, $i \in I$ and $\delta > 0$, we have

$$\sup_{\lambda \in \Lambda} \|L_{i,\lambda}(f) - f\| \le \|f\| \sup_{\lambda \in \Lambda} \|L_{i,\lambda}(\mathbf{1}) - \mathbf{1}\| + C_i(\delta)\omega(f, \delta\mu_i), \quad (5.1.32)$$

where

$$C_i(\delta) := \sup_{\lambda \in \Lambda} \left\| L_{i,\lambda}(\mathbf{1}) + \frac{1}{\delta^2}\mathbf{1} \right\| \qquad (5.1.33)$$

and

$$\mu_i := \sup_{\lambda \in \Lambda} \sqrt{\sup_{x \in K} L_{i,\lambda}(\| \cdot - x\|^2)(x)}. \qquad (5.1.34)$$

In particular, if $L_{i,\lambda}(\mathbf{1}) = \mathbf{1}$ for every $\lambda \in \Lambda$, we obtain

$$\sup_{\lambda \in \Lambda} \|L_{i,\lambda}(f) - f\| \leq \left(1 + \frac{1}{\delta^2}\right) \omega(f, \delta\mu_i). \qquad (5.1.35)$$

Proof. By reasoning as in the proof of Theorem 5.1.6, we can obtain

$$\sup_{\lambda \in \Lambda} \|L_{i,\lambda}(f) - f\| \leq \|f\| \sup_{\lambda \in \Lambda} \|L_{i,\lambda}(\mathbf{1}) - \mathbf{1}\| + K_i(f),$$

where

$$K_i(f) := \sup_{\lambda \in \Lambda} \|L_{i,\lambda}(f) - fL_{i,\lambda}(\mathbf{1})\|.$$

Now, for every $\lambda \in \Lambda$ and $x \in K$, we have as in the proof of Proposition 5.1.5 (see also (5.1.34))

$$|L_{i,\lambda}(f)(x) - f(x)L_{i,\lambda}(\mathbf{1})(x)| \leq \left(L_{i,\lambda}(\mathbf{1})(x) + \frac{1}{\delta^2\mu_i^2} L_{i,\lambda}(\|\cdot - x\|^2)(x)\right) \omega(f, \delta\mu_i)$$

$$\leq \left(L_{i,\lambda}(\mathbf{1})(x) + \frac{1}{\delta^2}\right) \omega(f, \delta\mu_i).$$

Consequently (see (5.1.33)) $\|L_{i,\lambda}(f) - fL_{i,\lambda}(\mathbf{1})\| \leq C_i(\delta)\omega(f, \delta\mu_i)$, and taking the supremum over $\lambda \in \Lambda$, we obtain (5.1.32).

Finally, (5.1.35) is a simple consequence of (5.1.32). □

The above estimates take a particularly simple form for nets $(L_i)_{i \in I}^{\leq}$ of positive linear operators from $\mathscr{C}(K)$ into $\mathscr{B}(K)$. In this case (5.1.25)–(5.1.29) yield

$$\|L_i(f) - f\| \leq \|f\| \|L_i(\mathbf{1}) - \mathbf{1}\| + \|L_i(\mathbf{1}) + \mathbf{1}\| \omega_i(f), \qquad (5.1.36)$$

for every $f \in \mathscr{C}(K)$ and $i \in I$ and if in particular $L_i(\mathbf{1}) = \mathbf{1}$,

$$\|L_i(f) - f\| \leq 2\omega_i(f). \qquad (5.1.37)$$

Moreover, by virtue of Corollary 5.1.7, assuming that

$$L_i(h) = h \quad \text{for every } i \in I \text{ and } h \in A(K)$$

and that there exist a constant $C > 0$ and a net $(\xi_i)_{i \in I}^{\leq}$ in $\mathbb{R}_+$ converging to 0 and such that

$$\left\|\sum_{j=1}^m (L_i(h_j^2) - h_j^2)\right\| \leq C\xi_i \left\|\sum_{j=1}^m h_j^2\right\|$$

for every $i \in I$ and $h_1, \ldots, h_m \in E'_K$, then we have

$$\| L_i(f) - f \| \le 2\Omega(f, \sqrt{C\xi_i}) \tag{5.1.38}$$

for every $f \in \mathscr{C}(K)$ and $i \in I$.

Under several circumstances it will be useful to estimate the norm $\| L_i(f) - T(f) \|$ as well, where $T \colon \mathscr{C}(K) \to \mathscr{C}(K)$ is a positive projection on $\mathscr{C}(K)$ different from the identity operator and $(L_i)_{i \in I}^{\le}$ is a net approximating T.

Along these lines we state the following result.

5.1.9 Theorem. *Let $T \colon \mathscr{C}(K) \to \mathscr{C}(K)$ be a positive projection on $\mathscr{C}(K)$ different from the identity operator such that $T(\mathbf{1}) = \mathbf{1}$ and let $(L_i)_{i \in I}^{\le}$ be a net of positive linear operators from $\mathscr{C}(K)$ into itself such that $L_i \circ T = T$ for every $i \in I$.*

Then, for every $f \in \mathscr{C}(K)$ and $i \in I$, we have

$$\| L_i(f) - T(f) \| \le \inf \left\{ c_\varepsilon \omega \left(f; h_1, \ldots, h_m, \varepsilon \sqrt{\left\| \sum_{j=1}^{m} L_i(\mu(\cdot; T, h_j)) \right\|} \right) \middle| \varepsilon > 0, m \ge 1, \right.$$

$$\left. h_1, \ldots, h_m \in E'_K, \left\| \sum_{j=1}^{m} L_i(\mu(\cdot; T, h_j)) \right\| > 0 \right\}, \tag{5.1.39}$$

where

$$c_e := 1 + \min \left\{ \frac{1}{\varepsilon^2}, \frac{1}{\varepsilon} \right\}. \tag{5.1.40}$$

Moreover, if $A(K)$ is contained in the range of the projection T (i.e., $Th = h$ for each $h \in A(X)$), we can replace the quantity $\left\| \sum_{j=1}^{m} L_i(\mu(\cdot; T, h_j)) \right\|$ in (5.1.39) by the term $\sum_{j=1}^{m} \| L_i(h_j^2) - T(h_j^2) \|$.

Proof. An argument similar to the one used in the proof of Proposition 5.1.4 shows that, for every $m \in \mathbb{N}$, $\delta > 0$ and $h_1, \ldots, h_m \in E'_K$ such that $\left\| \sum_{j=1}^{m} \mu(\cdot; T, h_j) \right\| > 0$, one has

$$| Tf(x) - f(x) | \le (1 + \alpha)\omega(f; h_1, \ldots, h_m, \delta),$$

where x is an arbitrary point lying in K and

$$\alpha := \min \left\{ \frac{1}{\delta} \sqrt{\sum_{j=1}^{m} \mu(x; T, h_j)}, \frac{1}{\delta^2} \sum_{j=1}^{m} \mu(x; T, h_j) \right\}.$$

Now, since $L_i \circ T = T$, by the preceding inequality we obtain

$$|Tf(x) - L_i(f)(x)| \leq (1 + M)\omega(f; h_1, \ldots, h_m, \delta),$$

where

$$\mathrm{M} := \min\left\{\frac{1}{\delta}\sqrt{\left\|\sum_{j=1}^{m} L_i(\mu(\,\cdot\,; T, h_j))\right\|}, \frac{1}{\delta^2}\left\|\sum_{j=1}^{m} L_i(\mu(\,\cdot\,; T, h_j))\right\|\right\}.$$

Then (5.1.39) follows by setting $\delta = \varepsilon\sqrt{\left\|\sum_{j=1}^{m} L_i(\mu(\,\cdot\,; T, h_j))\right\|}$ and taking the supremum over $x \in K$.

The last part of the theorem is obvious because, if the range of T contains the space $A(K)$, then $\left\|\sum_{j=1}^{m} L_i(\mu(\,\cdot\,; T, h_j))\right\| \leq \sum_{j=1}^{m} \|L_i(\mu(\,\cdot\,; T, h_j))\| \leq$

$\sum_{j=1}^{m} \|L_i(h_j^2) - T(h_j^2)\|.$ ☐

In the last part of this section we want to obtain some general estimates for the convergence of the powers of a net of positive linear operators.

To this end for every $f \in \mathscr{C}(K)$ and $\delta > 0$ we define

$$\Psi(f, \delta) := \inf\left\{c_\varepsilon \omega\left(f; h_1, \ldots, h_m, \delta\varepsilon\sqrt{\left\|\sum_{j=1}^{m} (T(h_j^2) - h_j^2)\right\|}\right)\middle| \varepsilon > 0, m \geq 1, \right.$$

$$\left. h_1, \ldots, h_m \in E_K', \left\|\sum_{j=1}^{m} T(h_j^2) - h_j^2\right\| > 0\right\}, \tag{5.1.41}$$

where c_ε is defined by (5.1.40) and $T: \mathscr{C}(K) \to \mathscr{C}(K)$ is a projection which is different from the identity operator.

Consider a net $(L_i)_{i \in I}^{\leq}$ of positive linear operators from $\mathscr{C}(K)$ into itself. Suppose that

$$L_i \circ T = T \quad \text{for every } i \in I \tag{5.1.42}$$

and there exists $\lambda_i \in \,]0, 1[$ such that

$$L_i(h^2) = h^2 + \lambda_i(T(h^2) - h^2)(=(1 - \lambda_i)h^2 + \lambda_i T(h^2)), \tag{5.1.43}$$

whenever $h \in A(K)$.

The last conditions simplify the study of the powers of L_i. In particular by induction on $n \in \mathbb{N}$, one easily obtains that

$$L_i^n(h^2) = (1 - \lambda_i)^n h^2 + (1 - (1 - \lambda_i)^n) T(h^2) \qquad (5.1.44)$$

for every $h \in A(K)$.

Taking the preceding formula into account, by Theorems 5.1.6 and 5.1.9 we obtain the following result.

5.1.10 Theorem. *Under the above assumptions, let* $(k_i)_{i \in I}^{\leq}$ *be a net of positive integers.*

Then, for every $f \in \mathscr{C}(K)$ *and* $i \in I$, *we have*

$$\|L_i^{k_i}(f) - f\| \leq \Psi(f, \sqrt{1 - (1 - \lambda_i)^{k_i}}) \leq \Psi(f, \sqrt{k_i \lambda_i}) \qquad (5.1.45)$$

and

$$\|L_i^{k_i}(f) - T(f)\| \leq \Psi(f, \sqrt{(1 - \lambda_i)^{k_i}}). \qquad (5.1.46)$$

Until now we have considered estimates which are valid for every continuous function. When we add more properties to a function $f \in \mathscr{C}(K)$, we can obtain more precise estimates of the norm $\|L(f) - f\|$. In what follows we shall present a simple result concerning Lipschitz functions.

Let (X, d) be a compact metric space. For a given $M > 0$ we shall denote by $\mathrm{Lip}_M 1$ the subset of all Lipschitz functions $f \in \mathscr{C}(X)$ having M as a Lipschitz constant, i.e., satisfying

$$|f(x) - f(y)| \leq M d(x, y), \quad \text{for every } x, y \in X. \qquad (5.1.47)$$

If $(L_i)_{i \in I}^{\leq}$ is a net of positive linear operators from $\mathscr{C}(X)$ into itself, we consider for every $i \in I$ the continuous functions $\alpha_i \colon X \to \mathbb{R}$ and $\beta_i \colon X \to \mathbb{R}$ defined by

$$\alpha_i(x) := L_i(d(\cdot, x))(x) \quad \text{and} \quad \beta_i(x) := L_i(d^2(\cdot, x))(x) \qquad (5.1.48)$$

for every $x \in X$.

With these notations our result runs as follows.

5.1.11 Theorem. *Let* $(L_i)_{i \in I}^{\leq}$ *be a net of positive linear operators from* $\mathscr{C}(X)$ *into itself and let* $f \in \mathrm{Lip}_M 1$ *with* $M > 0$.

Then, for every $i \in I$, *we have*

$$\|L_i(f) - f \cdot L_i(1)\| \leq M \|\alpha_i\|. \qquad (5.1.49)$$

Moreover, if $L_i(\mathbf{1}) = \mathbf{1}$ for every $i \in I$, then

$$\|L_i(f) - f\| \le M \sqrt{\|\beta_i\|}. \tag{5.1.50}$$

Proof. Since $|f(x) - f(y)| \le Md(x, y)$ for every $x, y \in X$, we can write

$$-Md(\cdot, x) \le f - f(x)\mathbf{1} \le Md(\cdot, x).$$

For every $i \in I$ the operator L_i is positive and therefore, by (5.1.48), $|L_i(f)(x) - f(x)L_i(\mathbf{1})(x)| \le M\alpha_i(x)$. Hence (5.1.49) follows.

Finally (5.1.50) is a consequence of (5.1.48) and the Cauchy-Schwarz inequality. □

Remark. Assume that K is a compact subset of a real inner-product space $(E, \langle \cdot, \cdot \rangle)$ and, for every $x \in K$, consider the function $e_x \colon K \to \mathbb{R}$ defined by $e_x(t) := \langle x, t \rangle$ for every $t \in K$. Moreover let $e \colon K \to \mathbb{R}$ be defined as in (5.1.19).

If $(L_i)_{i \in I}^{\le}$ is a net of positive linear operators from $\mathscr{C}(K)$ into itself satisfying $L_i(\mathbf{1}) = \mathbf{1}$ for every $i \in I$, then, for each $i \in I$ and for every Lipschitz function $f \in \operatorname{Lip}_M 1$ we have

$$\|L_i(f) - f\| \le M \sqrt{\|a_i - 2b_i\|}, \tag{5.1.51}$$

where $a_i := L_i(e) - e$ and $b_i(x) = L_i(e_x)(x) - e(x)$ for every $x \in K$.

Indeed, by the equality $\|t - x\|^2 = e(t) - 2e_x(t) + e(x)$, $(t, x \in K)$, we obtain

$$\beta_i(x) = L_i(e)(x) - 2L_i(e_x)(x) + e(x) = L_i(e)(x) - e(x) - 2(L_i(e_x)(x) - e(x))$$

for every $i \in I$ and $x \in K$ and therefore (5.1.51) is a consequence of (5.1.50).

Notes and references

The properties of moduli of smoothness are classical and are treated in many books in a more complete form (see, for instance, De Vore [1972] and Sendov and Popov [1988]).

Theorem 5.1.2 is taken from De Vore [1972]. In particular, the estimates given in the Remark to Theorem 5.1.2 are due to Shisha and Mond [1968] (and to Mamedov [1962] provided $L(\mathbf{1}) = \mathbf{1}$). Further extensions and generalizations can be found in Censor [1971], Gonska [1975] and Jiménez Poso [1980], [1981].

The definition of modulus of continuity in the setting of compact convex sets or in more general contexts was given by Nishishiraho in a series of papers (see Nishishiraho [1977], [1982b], [1988a]). From the same papers we took the results 5.1.4–5.1.10. There the reader can also find other estimates of the rate of convergence for positive linear operators and their powers.

The final estimates described in Theorem 5.1.11 are taken from Andrica and Mustata [1989].

5.2 Probabilistic methods and positive approximation processes

In this section we begin the description of several positive approximation processes by first considering those that may be defined by using some well-known concepts of probability theory.

As a matter of fact, we shall see that every positive approximation process on a compact interval arises in this way and, under this point of view, Korovkin's theorem has a simple and natural probabilistic interpretation.

Apart from that, the aim of this section is also to describe in some respect those approximation processes which arise by considering some basic probability distributions.

Besides showing their approximation properties, we shall also give estimates of the rate of convergence in terms of the associated probabilistic parameters.

According to the general aim of this chapter in the present section the reader will find only few examples and we refer to the final notes for further references.

In the sequel we shall assume that the reader is acquainted with a solid background in classical probability theory. Some of these prerequisites are surveyed in Section 1.3.

However the reader who knows little or no probability theory can equally derive the approximation properties of the sequences studied in this section by directly using (for compact intervals) Korovkin's theorem 4.2.7 (see Theorem 5.2.3 and the subsequent Remark 1).

Consider a real interval I and denote by $\mathscr{B}(I)$ the Banach lattice of all bounded real functions on I endowed with the sup-norm (1.1.5). We recall that the symbol $\mathscr{C}_b(I)$ ($\mathscr{UC}_b(I)$, respectively) denotes the subspace of $\mathscr{B}(I)$ of all continuous (uniformly continuous, respectively) real functions on I.

Clearly every $f \in \mathscr{UC}_b(I)$ can be extended to a function in $\mathscr{UC}_b(\mathbb{R})$.

From now on we fix a probability space $(\Omega, \mathscr{F}, P)$ and we denote by $M_2(\Omega)$ the space of all real square-integrable random variables on Ω.

A *random scheme on I* is, by definition, a mapping $Z \colon \mathbb{N} \times I \to M_2(\Omega)$.

In the sequel, given a random scheme $Z \colon \mathbb{N} \times I \to M_2(\Omega)$, we set

$$\alpha_{n,x} := E(Z(n,x)) \tag{5.2.1}$$

and

$$\sigma_{n,x} := \sigma(Z(n,x)) \tag{5.2.2}$$

for every $(n,x) \in \mathbb{N} \times I$.

Furthermore for every $n \in \mathbb{N}$ and $f \in \mathscr{C}_b(\mathbb{R})$ we consider the function $P_n(f) \colon I \to \mathbb{R}$ defined by

$$P_n(f)(x) := E(f \circ Z(n,x)) = \int_\Omega f \circ Z(n,x)\, dP = \int_{-\infty}^{+\infty} f\, dP_{Z(n,x)} \tag{5.2.3}$$

for every $x \in I$.

Note that $P_n(f)$ is a bounded function and clearly satisfies $\|P_n(f)\| \le \|f\|$.

Thus by means of the random scheme Z we have constructed a sequence $(P_n)_{n \in \mathbb{N}}$ of positive linear contractions from $\mathscr{C}_b(\mathbb{R})$ into $\mathscr{B}(I)$.

The question now arises under which conditions this sequence is an approximation process.

However note that, if we require

$$P\{Z(n, x) \in I\} = 1 \quad \text{for every } (n, x) \in \mathbb{N} \times I, \tag{5.2.4}$$

then we can restrict the operators P_n to a suitable function space on I. More precisely set

$$\mathscr{C}_b^*(I) := \begin{cases} \mathscr{C}_b(I), & \text{if } I \text{ is closed,} \\ \{f \in \mathscr{C}_b(I) | f \text{ is convergent} & \text{if } I \ne \mathbb{R} \text{ and } I \text{ is not closed.} \\ \text{at every point } \bar{x} \in \partial I \cap \mathbb{R}\}, \end{cases} \tag{5.2.5}$$

Moreover for every $n \in \mathbb{N}$ consider the positive linear contraction P_n^*: $\mathscr{C}_b^*(I) \to \mathscr{B}(I)$ defined by

$$P_n^*(f)(x) := P_n(f^*)(x) = \int_{-\infty}^{+\infty} f^* \, dP_{Z(n, x)} = \int_I f \, dP_{Z(n, x)} \tag{5.2.6}$$

for every $f \in \mathscr{C}_b^*(I)$ and $x \in I$, where $f^* \in \mathscr{C}_b(\mathbb{R})$ is an arbitrary bounded continuous extension of f over $\mathbb{R}$ such that $\|f^*\| = \|f\|$.

The definition of $P_n^*(f)$ is independent of the extension f^* because, by virtue of (5.2.4), $\text{Supp}(P_{Z(n, x)}) \subset \bar{I}$ for every $(n, x) \in \mathbb{N} \times I$ (see Section 1.3).

Clearly the approximation properties of $(P_n^*)_{n \in \mathbb{N}}$ will follow from the analogous ones that will be established for $(P_n)_{n \in \mathbb{N}}$.

As regards some qualitative properties of the functions $P_n(f)$ and $P_n^*(f)$, let us remark that, if for every $n \in \mathbb{N}$ the mapping $x \mapsto P_{Z(n, x)}$ is continuous from I into $\mathscr{M}_1^+(\mathbb{R})$ with respect to the weak topology, then $P_n(f) \in \mathscr{C}_b(I)$ for every $f \in \mathscr{C}_b(\mathbb{R})$ $(P_n^*(f) \in \mathscr{C}_b(I)$ if $f \in \mathscr{C}_b^*(I))$, respectively).

If I is not compact, we can ask whether the operators P_n map $\mathscr{C}_0(\mathbb{R})$ into $\mathscr{C}_0(I)$ or whether the operators P_n^* map $\mathscr{C}_0(I)$ into $\mathscr{C}_0(I)$.

By the above remark we have first to require that the function $x \mapsto P_{Z(n, x)}$ is continuous for every $n \in \mathbb{N}$.

Now assume that $I = [a, b[(-\infty < a < b \le +\infty)$ and fix $n \in \mathbb{N}$; for every $x \in I$, denote by $F_{n, x} \colon \mathbb{R} \to \mathbb{R}$ the distribution function of $Z(n, x)$.

If

$$\lim_{x \to b^-} F_{n, x} = 0 \quad \text{pointwise on } \mathbb{R}, \tag{5.2.7}$$

then $P_n(f) \in \mathscr{C}_0(I)$ for every $f \in \mathscr{C}_0(\mathbb{R})$ (respectively, $P_n^*(f) \in \mathscr{C}_0(I)$ for every $f \in \mathscr{C}_0(I)$ provided (5.2.4) holds).

Indeed, fix $f \in \mathscr{C}_0(\mathbb{R})$ and $\varepsilon > 0$. So there exists $t_1 \in \mathbb{R}$ such that $|f(t)| \le \varepsilon/2$ for every $t \ge t_1$. By (5.2.7) there exists $x_0 \in [a, b[$ such that $F_{n,x}(t_1) \le \dfrac{\varepsilon}{2(\|f\| + 1)}$ for every $x \in [x_0, b[$.

Accordingly

$$|P_n(f)(x)| \le \int_{-\infty}^{t_1} |f| \, dP_{Z(n,x)} + \int_{t_1}^{+\infty} |f| \, dP_{Z(n,x)} \le \|f\| F_{n,x}(t_1) + \frac{\varepsilon}{2} \le \varepsilon.$$

Under the assumption (5.2.4), note first that $\mathscr{C}_0(I) \subset \mathscr{C}_b^*(I)$ and for every $f \in \mathscr{C}_0(I)$ one can choose an extension $f^* \in \mathscr{C}_0(\mathbb{R})$ of f and hence $P_n^*(f) = P_n(f^*) \in \mathscr{C}_0(I)$.

In a similar way, if $I =]a, b]$ $(-\infty \le a < b < +\infty)$ and if

$$\lim_{x \to a^+} F_{n,x} = 0 \quad \text{pointwise on } \mathbb{R}, \tag{5.2.8}$$

then $P_n(f) \in \mathscr{C}_0(I)$ for every $f \in \mathscr{C}_0(\mathbb{R})$ (respectively, $P_n^*(f) \in \mathscr{C}_0(I)$ for every $f \in \mathscr{C}_0(I)$ provided (5.2.4) holds).

Finally, if $I =]a, b[$ $(-\infty < a < b \le +\infty)$, we have to require both conditions (5.2.7) and (5.2.8).

We now proceed to show the approximation properties of the sequence $(P_n)_{n \in \mathbb{N}}$ and, more precisely, to find conditions under which

$$P_n(f) \to f \tag{5.2.9}$$

for every $f \in \mathscr{C}_b(\mathbb{R})$ with respect to some topology on $\mathscr{B}(I)$.

If for every $x \in I$ we denote by $Z_\infty(x)$ the random variable on Ω which is P-almost surely equal to x, then (5.2.9) holds pointwise on I if and only if

$$\lim_{n \to \infty} Z(n, x) = Z_\infty(x) \quad \text{in distribution for every } x \in I. \tag{5.2.10}$$

To verify this last property we have at our disposal several powerful results from probability theory. For instance each of the next conditions ensures the validity of (5.2.10) (and hence of (5.2.9) with respect to the pointwise convergence):

(i) The sequence $(F_{n,x})_{n \in \mathbb{N}}$ of the distribution functions of $(Z(n, x))_{n \in \mathbb{N}}$ converges pointwise to $F_{Z_\infty(x)}$ for every $x \in I$.

(ii) The sequence $(\hat{P}_{Z(n,x)})_{n \in \mathbb{N}}$ of the Fourier transforms of $(P_{Z(n,x)})_{n \in \mathbb{N}}$ converges pointwise to $\hat{P}_{Z_\infty(x)}$ for every $x \in I$.

(iii) The sequence $(Z(n,x))_{n \in \mathbb{N}}$ converges P-almost surely to $Z_\infty(x)$ for every $x \in I$.

(iv) The sequence $(Z(n,x))_{n \in \mathbb{N}}$ converges stochastically to $Z_\infty(x)$ for every $x \in I$.

In Theorem 5.2.2 below we shall give some simple conditions ensuring that the sequence $(P_n(f))_{n \in \mathbb{N}}$ converges to f uniformly on I or on compact subintervals of I.

We need the following lemma which also includes the classical Chebyshev inequality.

5.2.1 Lemma. *Given a measure space $(\Omega, \mathscr{F}, \mu)$, for every measurable numerical function $f \colon \Omega \to \tilde{\mathbb{R}}$ and every $\alpha > 0$, $p > 0$ we have $\mu(\{|f| \geq \alpha\}) \leq \dfrac{1}{\alpha^p} \displaystyle\int_\Omega |f|^p \, d\mu$ (Chebyshev-Markov inequality).*

In particular, if $(\Omega, \mathscr{F}, P)$ is a probability space and $Z \in M_2(\Omega)$, then

$$P\{|Z - E(Z)| \geq \alpha\} \leq \frac{1}{\alpha^2} \operatorname{Var}(Z) \qquad (5.2.11)$$

(Chebyshev inequality).

Proof. Set $A := \{|f| \geq \alpha\} \in \mathscr{F}$. Then $\alpha^p \mathbf{1}_A \leq |f|^p$ (where $\mathbf{1}_A$ denotes the characteristic function of A) and hence the result follows. $\qquad\square$

We are now able to show the next theorem that will be referred as *Feller's theorem* (as a matter of fact, only (5.2.15) was obtained by Feller).

5.2.2 Theorem. *Let $(\Omega, \mathscr{F}, P)$ be a probability space, I a real interval and $Z \colon \mathbb{N} \times I \to M_2(\Omega)$ a random scheme on I. Furthermore consider the sequences of positive linear operators $(P_n)_{n \in \mathbb{N}}$ and $(P_n^*)_{n \in \mathbb{N}}$ defined by (5.2.3) and (5.2.6), respectively. If*

$$\lim_{n \to \infty} \alpha_{n,x} = x \quad \text{uniformly in } x \in I \qquad (5.2.12)$$

and

$$\lim_{n \to \infty} \sigma_{n,x}^2 = 0 \quad \text{uniformly in } x \in I, \qquad (5.2.13)$$

then, for every $f \in \mathscr{C}_b(\mathbb{R})$,

$$\lim_{n \to \infty} P_n(f) = f \quad \text{uniformly on compact subintervals of } I. \qquad (5.2.14)$$

Moreover, if $f \in \mathcal{U}\mathcal{C}_b(\mathbb{R})$, then

$$\lim_{n \to \infty} P_n(f) = f \quad \text{uniformly on } I. \tag{5.2.15}$$

Finally, if in addition (5.2.4) holds, then

$$\lim_{n \to \infty} P_n^*(f) = f \quad \text{uniformly on compact subintervals of } I \tag{5.2.16}$$

for every $f \in \mathcal{C}_b^*(I)$ (uniformly on I for every $f \in \mathcal{U}\mathcal{C}_b(I)$, respectively).

Proof. First we show (5.2.15). In fact, given $f \in \mathcal{U}\mathcal{C}_b(\mathbb{R})$ and $\varepsilon > 0$, there exists $\delta > 0$ such that $|f(s) - f(t)| \le \dfrac{\varepsilon}{2}$ for every $s, t \in \mathbb{R}$, $|s - t| \le \delta$. On account of (5.2.13), there exists $\upsilon \in \mathbb{N}$ such that $\sigma_{n,x}^2 \le \dfrac{\delta^2 \varepsilon}{4(\|f\| + 1)}$ for every $n \ge \upsilon$ and $x \in I$. Then

$$|P_n(f)(x) - f(\alpha_{n,x})| = \left| \int_{-\infty}^{+\infty} (f(t) - f(\alpha_{n,x}))\, dP_{Z(n,x)}(t) \right|$$

$$\le \int_{|t - \alpha_{n,x}| < \delta} |f(t) - f(\alpha_{n,x})|\, dP_{Z(n,x)}(t)$$

$$+ \int_{|t - \alpha_{n,x}| \ge \delta} |f(t) - f(\alpha_{n,x})|\, dP_{Z(n,x)}(t)$$

$$\le \frac{\varepsilon}{2} + 2\|f\| P\{|Z(n,x) - \alpha_{n,x}| \ge \delta\} \le \frac{\varepsilon}{2} + \frac{2\|f\|}{\delta^2} \sigma_{n,x}^2$$

$$\le \varepsilon.$$

Since $\varepsilon > 0$ is arbitrary, we conclude that $\lim\limits_{n \to \infty} (P_n(f)(x) - f(\alpha_{n,x})) = 0$ uniformly in $x \in I$. Moreover, by (5.2.12) and the uniform continuity of f, we have $\lim\limits_{n \to \infty} f(x) - f(\alpha_{n,x}) = 0$ uniformly in $x \in I$, and this implies (5.2.15).

In order to show (5.2.14), let $f \in \mathcal{C}_b(\mathbb{R})$ and fix a compact subinterval J of I and $\varepsilon < 6(\|f\| + 1)$. Choose $g \in \mathcal{K}(\mathbb{R})$ such that $0 \le g \le 1$ and $1 - \dfrac{\varepsilon}{6(\|f\| + 1)} \le g$ on J.

Since $f \cdot g \in \mathcal{K}(\mathbb{R}) \subset \mathcal{U}\mathcal{C}_b(\mathbb{R})$ as well as $1 - g \in \mathcal{U}\mathcal{C}_b(\mathbb{R})$, there exists $\upsilon \in \mathbb{N}$ such that for each $n \ge \upsilon$ and $x \in I$,

$$|P_n(1 - g)(x) - (1 - g(x))| \leq \frac{\varepsilon}{6(\|f\| + 1)}, \tag{1}$$

$$|P_n(f \cdot g)(x) - f(x) \cdot g(x)| \leq \frac{\varepsilon}{3}. \tag{2}$$

From (1) it also follows that, for every $x \in J$,

$$|P_n(1 - g)(x)| \leq \frac{\varepsilon}{6(\|f\| + 1)} + 1 - g(x) \leq \frac{\varepsilon}{3(\|f\| + 1)}.$$

Consequently, for $n \geq v$ and $x \in J$,

$$|P_n(f)(x) - f(x)| \leq |P_n(f)(x) - P_n(f \cdot g)(x)| + |P_n(f \cdot g)(x) - f(x)g(x)|$$

$$+ |f(x)g(x) - f(x)| \leq \int_{-\infty}^{+\infty} |f|(1 - g) \, dP_{Z(n, x)} + \frac{\varepsilon}{3}$$

$$+ \frac{\varepsilon}{3(\|f\| + 1)} \|f\| \leq \|f\| P_n(1 - g)(x) + \frac{2}{3}\varepsilon \leq \varepsilon. \qquad \square$$

Remarks. 1. If (5.2.12) and (5.2.13) hold uniformly only on compact subintervals of I, then (5.2.15) holds uniformly on compact subintervals of I.

2. As the above proof shows, (5.2.14) and (5.2.15) (and hence (5.2.16)) hold under the assumption (5.2.12) and the following condition which is weaker than (5.2.13):

$$\lim_{n \to \infty} (Z(n, x) - \alpha_{n, x}) = 0 \quad \text{stochastically uniformly in } x \in I, \tag{5.2.17}$$

i.e.,

$$\lim_{n \to \infty} \sup_{x \in I} P\{|Z(n, x) - \alpha_{n, x}| \geq \delta\} = 0 \quad \text{for every } \delta > 0. \tag{5.2.18}$$

The preceding theorem constitutes a powerful tool in the study of the convergence of many sequences of positive linear operators, as we shall see in the sequel by means of various examples. Here we show how Feller's theorem is actually equivalent to Korovkin's first theorem in the case where I is a compact real interval.

5.2.3 Theorem. *The following statements hold and are equivalent:*
(i) *Given a probability space* $(\Omega, \mathscr{F}, P)$, *if* $Z\colon \mathbb{N} \times [a, b] \to M_2(\Omega)$ *is a random scheme on* $[a, b]$ *satisfying* (5.2.12), (5.2.13) *and* (5.2.4) *and if* $(P_n^*)_{n \in \mathbb{N}}$ *is the sequence of positive linear operators defined by* (5.2.6), *then for every* $f \in \mathscr{C}([a, b])$

$$\lim_{n \to \infty} P_n^*(f) = f \text{ uniformly on } [a, b].$$

(ii) *If* $(L_n)_{n \in \mathbb{N}}$ *is a sequence of positive linear operators from* $\mathscr{C}([a, b])$ *into* $\mathscr{B}([a, b])$ *satisfying* $\lim_{n \to \infty} L_n(e_i) = e_i$, $i = 0, 1, 2$, *where* $e_i(x) = x^i$ *for every* $i = 0, 1, 2$ *and* $x \in [a, b]$, *then for every* $f \in \mathscr{C}([a, b])$

$$\lim_{n \to \infty} L_n(f) = f \quad \text{uniformly on } [a, b].$$

Proof. We have only to show that statements (i) and (ii) are equivalent.

(i) $\Rightarrow$ (ii): Let $(L_n)_{n \in \mathbb{N}}$ be a sequence of positive linear operators from $\mathscr{C}([a, b])$ in $\mathscr{B}([a, b])$ such that $\lim_{n \to \infty} L_n(e_i) = e_i$, $i = 0, 1, 2$.

Since $\lim_{n \to \infty} L_n(1) = 1$, we can always assume that $L_n(1) = 1$ for every $n \in \mathbb{N}$.

On $[a, b]$ we consider the σ-algebra $\mathfrak{B}([a, b])$ of Borel sets. For every $(n, x) \in \mathbb{N} \times [a, b]$ let $\mu_{n, x}\colon \mathscr{C}([a, b]) \to \mathbb{R}$ be the positive Radon measure defined by $\mu_{n, x}(f) := L_n(f)(x)$ for every $f \in \mathscr{C}([a, b])$. By the Riesz representation theorem 1.2.4, there exists a unique probability Borel measure $\tilde{\mu}_{n, x}\colon \mathfrak{B}([a, b]) \to \mathbb{R}$ such that $\mu_{n, x}(f) = \displaystyle\int f \, d\tilde{\mu}_{n, x}$ for every $f \in \mathscr{C}([a, b])$. Consequently, there exists a probability space $(\Omega, \mathscr{F}, P)$ and a family $(Z_{n, x})_{n \in \mathbb{N}, x \in [a, b]}$ of independent random variables such that $P_{Z_{n,x}} = \tilde{\mu}_{n, x}$ for every $n \in \mathbb{N}$ and $x \in [a, b]$ (see Section 1.3). Clearly every $Z_{n, x}$ is square-integrable because $E(Z_{n, x}^2) = \mu_{n, x}(e_2)$.

Now consider the random scheme $Z\colon \mathbb{N} \times [a, b] \to M_2(\Omega)$ on $[a, b]$ defined by $Z(n, x) := Z_{n, x}$ for every $(n, x) \in \mathbb{N} \times [a, b]$. In this case (5.2.4) is satisfied and hence, for every $f \in \mathscr{C}([a, b])$ and $x \in [a, b]$,

$$P_n^*(f)(x) = \int_{-\infty}^{+\infty} f^* \, dP_{Z(n, x)} = \int_a^b f \, d\tilde{\mu}_{n, x} = \mu_{n, x}(f) = L_n(f)(x),$$

i.e., $P_n^*(f) = L_n(f)$ for every $f \in \mathscr{C}([a, b])$.

Moreover the hypothesis $\lim_{n \to \infty} L_n(e_1) = e_1$ yields

$$\lim_{n \to \infty} \alpha_{n, x} = \lim_{n \to \infty} \int_a^b t \, dP_{Z(n, x)}(t) = \lim_{n \to \infty} L_n(e_1)(x) = x \quad \text{uniformly in } x \in [a, b].$$

At the same time, by the hypothesis $\lim_{n \to \infty} L_n(e_2) = e_2$, we obtain

$$\lim_{n \to \infty} \sigma_{n,x}^2 = \lim_{n \to \infty} \int_a^b (t - \alpha_{n,x})^2 \, dP_{Z(n,x)}(t)$$

$$= \lim_{n \to \infty} (L_n(e_2)(x) - \alpha_{n,x}^2) = 0 \quad \text{uniformly in } x \in [a, b].$$

So, by Feller's theorem 5.2.2, we conclude that the sequence $(L_n)_{n \in \mathbb{N}}$ converges strongly to the identity operator.

(ii) $\Rightarrow$ (i): Let $Z: \mathbb{N} \times [a, b] \to M_2(\Omega)$ be a random scheme on $[a, b]$ satisfying (5.2.4) and consider the sequence $(P_n^*)_{n \in \mathbb{N}}$ of positive linear operators on $\mathscr{C}([a, b])$ defined by (5.2.6). Suppose that

$$\lim_{n \to \infty} \alpha_{n,x} = x \quad \text{and} \quad \lim_{n \to \infty} \sigma_{n,x}^2 = 0 \quad \text{uniformly in } x \in [a, b].$$

As we have showed in the proof of (i) $\Rightarrow$ (ii), these assumptions mean that $\lim_{n \to \infty} P_n^*(e_i) = e_i$, $i = 1, 2$. Since $P_n(1) = 1$ for every $n \in \mathbb{N}$, we may conclude that $\lim_{n \to \infty} P_n^*(f) = f$ for every $f \in \mathscr{C}([a, b])$. $\square$

Remarks. 1. The reader will observe that assertion (ii) in Theorem 5.2.3 is a slight generalization of Korovkin's first theorem (see also Example 6 to Theorem 4.4.6) because every L_n takes values in $\mathscr{B}([a, b])$). On the other hand, Korovkin's first theorem and the universal Korovkin-type property (vi) in Theorem 4.1.4 clearly imply assertion (ii) of Theorem 5.2.3.

2. The proof of the implication (i) $\Rightarrow$ (ii) of the above theorem also shows that, if $(L_n)_{n \in \mathbb{N}}$ is a sequence of positive linear operators from $\mathscr{C}([a, b])$ into $\mathscr{B}([a, b])$ satisfying $L_n(1) = 1$ for every $n \in \mathbb{N}$, then there exists a random scheme $Z: \mathbb{N} \times [a, b] \to M_2(\Omega)$ satisfying (5.2.4) and such that the sequence $(L_n)_{n \in \mathbb{N}}$ coincides with the sequence $(P_n^*)_{n \in \mathbb{N}}$ of positive linear operators on $\mathscr{C}([a, b])$ associated with Z.

Moreover, for every $n \in \mathbb{N}$ and $x \in [a, b]$, we have

$$L_n(e_1)(x) = \alpha_{n,x},$$

and

$$L_n(e_2)(x) = \sigma_{n,x}^2 + \alpha_{n,x}^2.$$

Before starting with some applications of Feller's theorem, we investigate some quantitative aspects of the convergence described in (5.2.16).

5.2.4 Theorem. *Let $Z: \mathbb{N} \times I \to M_2(\Omega)$ be a random scheme on a real interval I satisfying (5.2.4) and consider the sequence $(P_n^*)_{n \in \mathbb{N}}$ of positive linear operators defined by (5.2.6). If $f \in \mathscr{C}_b(I)$ and $x \in I$, then for every $n \geq 1$ and $\delta > 0$ we have*

$$|P_n^*(f)(x) - f(x)| \leq \left(1 + \frac{1}{\delta}\sqrt{(\alpha_{n,x} - x)^2 + \sigma_{n,x}^2}\right)\omega(f,\delta). \qquad (5.2.19)$$

In particular, if we take $\delta = n^{-1/2}$, we obtain

$$|P_n^*(f)(x) - f(x)| \leq (1 + \sqrt{n((\alpha_{n,x} - x)^2 + \sigma_{n,x}^2)})\,\omega(f,n^{-1/2}). \qquad (5.2.20)$$

In the case where f is differentiable and $f' \in \mathscr{C}_b(I)$, we also have

$$|P_n^*(f)(x) - f(x)| \leq \sqrt{\sigma_{n,x}^2 + (\alpha_{n,x} - x)^2}\left(1 + \frac{1}{\delta}\sqrt{\sigma_{n,x}^2 + (\alpha_{n,x} - x)^2}\right)\omega(f',\delta)$$

$$+ |f'(x)||\alpha_{n,x} - x| \qquad (5.2.21)$$

and, if $\delta = n^{-1/2}$ and $\alpha_{n,x} = x$,

$$|P_n^*(f)(x) - f(x)| \leq \sigma_{n,x}(1 + \sqrt{n}\,\sigma_{n,x})\omega(f',n^{-1/2}). \qquad (5.2.22)$$

Proof. This is a direct consequence of Theorem 5.1.2. Indeed $P_n^*(1) = 1$ for every $n \in \mathbb{N}$ and also, by considering the function $\psi_x: I \to \mathbb{R}$ defined by $\psi_x(t) = t - x$ $(x, t \in I)$, we obtain

$$P_n^*(\psi_x^2)(x) = \int_I (x - t)^2\, dP_{Z(n,x)}(t) \leq \int_{-\infty}^{+\infty} (x - t)^2\, dP_{Z(n,x)}(t)$$

$$= \int_{-\infty}^{+\infty} ((x - \alpha_{n,x}) + (\alpha_{n,x} - t))^2\, dP_{Z(n,x)}(t)$$

$$= \int_{-\infty}^{+\infty} (x - \alpha_{n,x})^2\, dP_{Z(n,x)}(t) + 2\int_{-\infty}^{+\infty} (x - \alpha_{n,x})(\alpha_{n,x} - t)\, dP_{Z(n,x)}(t)$$

$$+ \int_{-\infty}^{+\infty} (\alpha_{n,x} - t)^2\, dP_{Z(n,x)}(t) = (\alpha_{n,x} - x)^2$$

$$+ \int_{-\infty}^{+\infty} (\alpha_{n,x} - t)^2\, dP_{Z(n,x)}(t) = (\alpha_{n,x} - x)^2 + \sigma_{n,x}^2. \qquad \square$$

Remarks. 1. Under the hypotheses of Theorem 5.2.2, i.e.,

$$\lim_{n\to\infty} \alpha_{n,x} = x \quad \text{and} \quad \lim_{n\to\infty} \sigma_{n,x}^2 = 0 \quad \text{uniformly in } x \in I,$$

if we set

$$\beta_n := \sup_{x\in I} \sqrt{(\alpha_{n,x} - x)^2 + \sigma_{n,x}^2}, \tag{5.2.23}$$

we obviously have

$$\lim_{n\to\infty} \beta_n = 0.$$

Moreover, if we put $\delta = \gamma\beta_n$ in (5.2.19) with $\gamma > 0$, we obtain

$$\|P_n^*(f) - f\| \leq \left(1 + \frac{1}{\gamma}\right)\omega(f, \gamma\beta_n) \tag{5.2.24}$$

for every $n \in \mathbb{N}$ and for every bounded continuous function $f: I \to \mathbb{R}$. Thus we reobtain Theorem 5.2.2.

Finally, we observe that the hypothesis $\alpha_{n,x} = x$ $((n,x) \in \mathbb{N} \times I)$ in (5.2.22) is satisfied in many concrete examples; under this assumption, if we consider a differentiable function f satisfying $f' \in \mathscr{C}_b(I)$, Theorem 5.2.4 yields the following estimate

$$\|P_n^*(f) - f\| \leq \beta_n(1 + \beta_n\sqrt{n})\,\omega(f', n^{-1/2}) \tag{5.2.25}$$

for every $n \in \mathbb{N}$.

2. In the subsequent examples we shall consider some approximation processes on a real interval I which are associated with a random scheme $Z: \mathbb{N} \times I \to M_2(\Omega)$ defined as

$$Z(n,x) := \frac{1}{n}\sum_{k=1}^{n} Y(k,x), \quad ((n,x) \in \mathbb{N} \times I), \tag{5.2.26}$$

where $Y: \mathbb{N} \times I \to M_2(\Omega)$ is another random scheme on I.

In this case, if $(Y(n,x))_{n\in\mathbb{N}}$ obeys the *weak law of large numbers* uniformly in $x \in I$, i.e., for every $\delta > 0$

$$\lim_{n\to\infty} \sup_{x\in I} P\left\{\left|\frac{1}{n}\sum_{k=1}^{n}(Y(k,x) - E(Y(k,x)))\right| \geq \delta\right\} = 0, \tag{5.2.27}$$

then (5.2.18) holds. Consequently, if in addition

$$\lim_{n \to \infty} \frac{1}{n} \sum_{k=1}^{n} E(Y(k,x)) = x \quad \text{uniformly in } x \in I, \qquad (5.2.28)$$

then the conclusion of Theorem 5.2.2 remains true for the random scheme Z.

Note that condition (5.2.27) is satisfied if for every $x \in I$ the random variables $Z(n,x)$ $(n \in \mathbb{N})$ are pairwise uncorrelated and

$$\lim_{n \to \infty} \frac{1}{n^2} \sum_{k=1}^{n} \text{Var}(Y(k,x)) = 0 \quad \text{uniformly in } x \in I. \qquad (5.2.29)$$

Furthermore Z satisfies (5.2.4) if Y satisfies the same condition.

Summing up, if the random scheme Y satisfies (5.2.27) (or (5.2.29) and (5.2.28), respectively), then the sequence $(P_n)_{n \in \mathbb{N}}$ verifies (5.2.14) and (5.2.15) (as well as $(P_n^*)_{n \in \mathbb{N}}$ verifies (5.2.16) if Y satisfies (5.2.4)).

In this particular context, according to some results of Butzer and Hahn [1978, Theorem 4] and Hahn [1982, Corollary 1], it is possible to give estimates of the order of approximation in terms of the second modulus of smoothness $\omega_2(f, \delta)$ (see Section 5.1).

More precisely, under the assumptions (5.2.4), (5.2.27) and (5.2.28) for every $f \in \mathcal{UC}_b(I)$ and for every $n \in \mathbb{N}$ and $x \in I$, we have

$$|P_n^*(f)(x) - f(\alpha_{n,x})| \leq M\omega_2\left(f, \left(\frac{1}{n^2} \sum_{k=1}^{n} \text{Var}(Y(k,x))\right)^{1/2}\right), \qquad (5.2.30)$$

where the constant M is independent of f and n and where we continue to denote by f its extension to $\bar{I}$ (note that by formula (1.5.2), $\alpha_{n,x} \in \overline{\text{co}}(\text{Supp}(P_{Z(n,x)})) \subset \bar{I})$.

Furthermore in many circumstances we shall require that for every $x \in I$ the random variables $Y(n,x)$ are identically distributed. In this case, if in addition $E(Y(n,x)) = x$, we have

$$|P_n^*(f)(x) - f(x)| \leq M\omega_2(f, \sigma(x)n^{-1/2}), \qquad (5.2.31)$$

for every $f \in \mathcal{UC}_b(I)$, $n \in \mathbb{N}$ and $x \in I$, where M and $\sigma(x) := \sigma(Y(n,x))$ are constants independent of f and n.

In the remaining part of this section we shall apply the above results to some sequences of linear positive operators which have particular interest from a probabilistic point of view. In all the following examples the operators under consideration map continuous bounded functions into continuous (bounded) functions. Furthermore (5.2.4) also holds.

The first examples will concern the case where the random variables $Z(n, x)$ are discretely distributed, i.e., for every $(n, x) \in \mathbb{N} \times I$ there exist a sequence $(\lambda_{n,x,k})_{k \in \mathbb{N}_0}$ of positive real numbers satisfying $\sum\limits_{k=0}^{\infty} \lambda_{n,x,k} = 1$ and a sequence $(a_{n,x,k})_{k \in \mathbb{N}_0}$ of real numbers, such that the distribution $P_{Z(n, x)}$ of $Z(n, x)$ is given by

$$P_{Z(n, x)} = \sum_{k=0}^{\infty} \lambda_{n,x,k} \varepsilon_{a_{n,x,k}}. \qquad (5.2.32)$$

In this case, for every $n \in \mathbb{N}$, $f \in \mathscr{C}_b(\mathbb{R})$ and $x \in I$,

$$P_n(f)(x) = \sum_{k=0}^{\infty} f(a_{n,x,k}) \lambda_{n,x,k}, \qquad (5.2.33)$$

or, if $a_{n,x,k} \in I$ for every $k \in \mathbb{N}_0$,

$$P_n^*(f)(x) = \sum_{k=0}^{\infty} f(a_{n,x,k}) \lambda_{n,x,k} \qquad (5.2.34)$$

for every $f \in \mathscr{C}_b^*(I)$.

Moreover we observe that, if the functions $x \mapsto \lambda_{n,x,k}$ and $x \mapsto a_{n,x,k}$ are continuous on I for every $k \geq 0, n \geq 1$ and if

$$\lim_{k \to \infty} a_{n,x,k} = +\infty \quad \text{uniformly in } x \in I,$$

provided there are infinitely many $\lambda_{n,x,k}$ different from zero, then the mapping $x \mapsto P_{Z(n, x)}$ is continuous for every $n \in \mathbb{N}$.

The same conclusion also holds when the function $x \mapsto \lambda_{n,x,k}$ is continuous on I and the function $x \mapsto a_{n,x,k}$ is constant, i.e., there exists $(a_{n,k})_{k \in \mathbb{N}_0}$ such that

$$a_{n,x,k} = a_{n,x} \quad \text{for every } x \in I \text{ and } k \in \mathbb{N}_0$$

(see Section 1.3).

Our first two examples are connected with a sequence of Bernoulli trials.

5.2.5 Bernstein-King operators.

For every $x \in [0, 1]$ we consider a sequence of Bernoulli experiments. We assume that for every $n \in \mathbb{N}$ a success occurs at the n-th performance of the experiment with probability $p_n(x)$ and that the function p_n is continuous on $[0, 1]$. Now, for every $n \in \mathbb{N}$ we consider the random variable $Y(n, x)$ (defined on a suitable probability space $(\Omega, \mathscr{F}, P)$) which assumes only the values 1 or 0 and specifies whether a success does or does not occur at the n-th performance. Thus we obtain a sequence $(Y(n, x))_{n \in \mathbb{N}}$ of independent Bernoulli random variables with parameters 1 and $p_n(x)$, i.e., $P\{Y(n, x) = 1\} = p_n(x)$ and $P\{Y(n, x) = 0\} = 1 - p_n(x)$. Let $Z: \mathbb{N} \times [0, 1] \to M_2(\Omega)$ be the random scheme defined as in (5.2.26); then, for every $x \in [0, 1]$ and $n \in \mathbb{N}$, $Z(n, x)$ is a discretely distributed random variable which represents the average numbers of successes in the first n performances. Furthermore (5.2.4) is satisfied.

If we consider the matrix $(a_{nk})_{(n,k) \in \mathbb{N}_0 \times \mathbb{N}_0}$ defined by

$$a_{00}(x) = 1, \quad a_{0k}(x) = 0, \quad \prod_{i=1}^{n} (p_i(x)t + 1 - p_i(x)) = \sum_{k=0}^{n} a_{nk}(x)t^k \quad (t \in \mathbb{R}),$$

we obtain, for every $n \geq 1, k \geq 0$, $P\left\{\sum_{i=0}^{n} Y(i, x) = k\right\} = a_{nk}(x)$, and therefore $P\left\{Z(n, x) = \dfrac{k}{n}\right\} = a_{nk}(x)$.

We denote by $L_n: \mathscr{C}([0, 1]) \to \mathscr{C}([0, 1])$ the linear positive operator defined by (5.2.6) and associated with Z. Then, for every $f \in \mathscr{C}([0, 1])$ and $x \in [0, 1]$,

$$L_n(f)(x) = \sum_{k=0}^{n} f\left(\frac{k}{n}\right) a_{nk}(x). \tag{5.2.35}$$

The operators L_n constitute a probabilistic generalization of Bernstein operators (2.5.3). In fact, if we put $p_i(x) = x$ for every $i \in \mathbb{N}$ and $x \in [0, 1]$, the random variable $\sum_{k=0}^{n} Y(k, x)$ becomes a Bernoulli random variable with parameters n and x, i.e.,

$$P\left\{\sum_{i=0}^{n} Y(i, x) = k\right\} = \binom{n}{k} x^k(1 - x)^{n-k}.$$

Thus its distribution is given by $\sum_{k=0}^{n} \binom{n}{k} x^k(1 - x)^{n-k}\varepsilon_k$ and therefore the distribution of $Z(n, x)$ is $\sum_{k=0}^{n} \binom{n}{k} x^k(1 - x)^{n-k}\varepsilon_{k/n}$.

In this case the operators L_n become the Bernstein operators B_n already encountered in Sections 2.5 and 4.2 and defined by

$$B_n(f)(x) = \sum_{k=0}^{n} f\left(\frac{k}{n}\right)\binom{n}{k} x^k(1 - x)^{n-k} \tag{5.2.36}$$

for every $f \in \mathscr{C}([0, 1])$ and $x \in [0, 1]$.

In order to study the convergence of the sequence $(L_n)_{n \in \mathbb{N}}$, we point out that from formula (1.3.21), it follows that

$$\alpha_{n,x} = \frac{1}{n} \sum_{i=1}^{n} E(Y(i,x)) = \frac{1}{n} \sum_{i=1}^{n} p_i(x)$$

and

$$\sigma_{n,x}^2 = \frac{1}{n^2} \mathrm{Var}\left(\sum_{i=1}^{n} Y(i,x) \right) = \frac{1}{n^2} \sum_{i=1}^{n} \mathrm{Var}(Y(i,x)) = \frac{1}{n^2} \left(\sum_{i=1}^{n} p_i(x) - \sum_{i=1}^{n} p_i(x)^2 \right).$$

Then, if

$$\lim_{n \to \infty} \frac{1}{n} \sum_{i=1}^{n} p_i(x) = x \quad \text{uniformly in } x \in [0,1]$$

both conditions (5.2.12) and (5.2.13) are satisfied and therefore, by Theorem 5.2.2, we conclude that

$$\lim_{n \to \infty} L_n(f) = f \quad \text{uniformly on } [0,1], \tag{5.2.37}$$

for every $f \in \mathscr{C}([0,1])$.

Moreover, by recalling (5.2.20) we obtain

$$|L_n(f)(x) - f(x)| \le \left(1 + \sqrt{\frac{1}{n}\left(\left(\sum_{i=1}^{n} p_i(x) - nx \right)^2 + \sum_{i=1}^{n} p_i(x)(1 - p_i(x)) \right)} \right)$$

$$\times \omega(f, n^{-1/2}) \tag{5.2.38}$$

for every $f \in \mathscr{C}([0,1])$ and $x \in [0,1]$.

In the particular case of Bernstein operators we obtain

$$\alpha_{n,x} = \frac{1}{n} \sum_{i=1}^{n} x = x, \quad \sigma_{n,x}^2 = \frac{1}{n^2} \sum_{i=1}^{n} x(1-x) = \frac{x(1-x)}{n}$$

and therefore $\lim_{n \to \infty} B_n(f) = f$ for every $f \in \mathscr{C}([0,1])$.

Moreover, by (5.2.38),

$$|B_n(f)(x) - f(x)| \le (1 + \sqrt{x(1-x)})\,\omega(f, n^{-1/2}) \tag{5.2.39}$$

for every $f \in \mathscr{C}([0,1])$ and $x \in [0,1]$. Furthermore, since $x(1-x) \le 1/4$, it follows that

$$\|B_n(f) - f\| \leq 3/2\omega(f, n^{-1/2}).$$ (5.2.40)

This last formula is due to Popoviciu [1935]. Different authors have considered the problem to find the best constant M such that $\|B_n(f) - f\| \leq M\omega(f, n^{-1/2})$ for every $f \in \mathscr{C}([0, 1])$. In 1953 Lorentz (see Lorentz [1986b, Theorem 1.6.1]) showed the estimate with $M = 5/4$, but the best value was found by Sikkema [1961] (see also Sikkema [1970b]) and is given by

$$M_0 := \frac{4306 + 837\sqrt{6}}{5832}.$$ (5.2.41)

If in addition $f \in \mathscr{C}^1([0, 1])$, then we can apply (5.2.22) to obtain

$$|B_n(f)(x) - f(x)| \leq \sqrt{\frac{x(1-x)}{n}}(1 + \sqrt{x(1-x)})\omega(f', n^{-1/2}).$$ (5.2.42)

Finally we also observe that by (5.2.31) we have an estimate in terms of the second modulus of smoothness given by

$$|B_n(f)(x) - f(x)| \leq M\omega_2\left(f, \sqrt{\frac{x(1-x)}{n}}\right)$$ (5.2.43)

($f \in \mathscr{C}([0, 1])$, $x \in [0, 1]$) which, apart the constant M, is the best possible.

Furthermore, as it has been shown by Berens and Lorentz [1972] and De Vore [1972], the inequality

$$|B_n(f)(x) - f(x)| \leq M\left(\frac{x(1-x)}{n}\right)^{\alpha/2}$$

holds for some $\alpha \leq 2$ if and only if $\omega_2(f, \delta) = O(\delta^\alpha)$, while, according to Ditzian [1979], we can have

$$\|B_n(f) - f\| = O(n^{-\alpha}), \quad n \to \infty,$$

for a given $\alpha \leq 1$ if and only if $\omega_\varphi^2(f, \delta)_\infty = O(\delta^{2\alpha})$, $\delta \to 0^+$, $\omega_\varphi^2(f, \delta)_\infty$ being the Ditzian-Totik modulus of smoothness (Ditzian and Totik [1987]) and $\varphi(x) := (x(1-x))^{1/2}$ ($0 \leq x \leq 1$).

Finally Totik [1991] showed that there exist $M_1, M_2 > 0$ such that

$$M_1\omega_\varphi^2\left(f, \frac{1}{\sqrt{n}}\right)_\infty \leq \|B_n(f) - f\| \leq M_2\omega_\varphi^2\left(f, \frac{1}{\sqrt{n}}\right)_\infty$$

for every $n \geq 1$ and $f \in \mathscr{C}([0, 1])$.

For multidimensional and infinite dimensional extensions of Bernstein operators see the subsequent Examples 5.2.11 and 5.2.12 and Chapter 6. Further
generalizations can be found in Section 5.3.

The construction of the next approximation process is essentially based on
negative binomial-type random variables.

5.2.6 Baskakov-King operators. In analogy to the preceding example, for every
$x \in [0, +\infty[$ consider a sequence of Bernoulli experiments and assume that a
success occurs at the n-th performance of the experiment with probability
$p_n(x) > 0$. We shall suppose that the functions p_n are continuous on $[0, +\infty[$.

For every $n \in \mathbb{N}$ and $x \geq 0$ let $U(n, x)$ be the random variable which counts
the number of performances after the $(n - 1)$-th success before obtaining the n-th
success.

Accordingly, the random variable $\sum_{i=1}^{n} U(i, x)$ gives the waiting time for the
n-th success, i.e., $\sum_{i=1}^{n} U(i, x) = p$ for some $p \geq n$ means that we have had $n - 1$
successes in the preceding p performances and the n-th occurs at the $(p + 1)$-th
performance.

Then we know that the probability-generating function $g_{n,x}$ of the random
variable $\sum_{i=1}^{n} U(i, x)$ is given by

$$g_{n,x}(t) = \prod_{i=1}^{n} E(t^{U(i,x)}) = \prod_{i=1}^{n} \sum_{k=0}^{\infty} q_i(x)^k p_i(x) t^k = \prod_{i=1}^{n} \frac{p_i(x)}{1 - q_i(x)t}, \quad (t \in [-1, 1]),$$

where $q_n(x) = 1 - p_n(x)$ for every $n \in \mathbb{N}$ and $x \geq 0$.

Hence, as in the preceding example, we can consider the matrix $(b_{nk})_{(n,k) \in \mathbb{N} \times \mathbb{N}_0}$
defined by the relations

$$\prod_{i=1}^{n} \frac{p_i(x)}{1 - q_i(x)t} = \sum_{k=0}^{\infty} b_{nk}(x) t^k,$$

and we obtain $P\left\{\frac{1}{n} \sum_{i=1}^{n} U(i, x) = \frac{k}{n}\right\} = P\left\{\sum_{i=1}^{n} U(i, x) = k\right\} = b_{nk}(x)$ for every
$(n, k) \in \mathbb{N} \times \mathbb{N}_0$.

Let $Z(n, x)$ be the random scheme defined as in (5.2.26). Clearly Z satisfies
(5.2.4). For every $n \in \mathbb{N}$ denote by $T_n \colon \mathscr{C}_b([0, +\infty[) \to \mathscr{C}_b([0, +\infty[)$ the linear
positive operator defined by (5.2.6). On account of (5.2.34) we have

$$T_n(f)(x) = \sum_{k=0}^{\infty} f\left(\frac{k}{n}\right) b_{nk}(x), \qquad (5.2.44)$$

whenever $f \in \mathscr{C}_b([0, +\infty[)$ and $x \in [0, +\infty[$.

In the particular case where $p_n(x) = \dfrac{1}{1+x}$, then $\displaystyle\prod_{i=1}^{n} \dfrac{p_i(x)}{1 - q_i(x)t} =$

$(1 + x - xt)^{-n} = \displaystyle\sum_{k=0}^{\infty} \binom{n+k-1}{k} \dfrac{x^k}{(1+x)^{n+k}} \cdot t^k$ and the random variables

$\displaystyle\sum_{i=1}^{n} U(i, x)$ are also called *negative binomial random variables*. The corresponding operators are the classical *Baskakov operators* $(A_n)_{n \in \mathbb{N}}$ defined by

$$A_n(f)(x) = \sum_{k=0}^{\infty} f\left(\frac{k}{n}\right) \binom{n+k-1}{k} \frac{x^k}{(1+x)^{n+k}} \tag{5.2.45}$$

for every $f \in \mathscr{C}_b([0, +\infty[)$ and $x \in [0, +\infty[$.

In order to study the convergence of the sequence $(T_n)_{n \in \mathbb{N}}$, we observe that

$$g_{n,x}'(t) = g_{n,x}(t) \sum_{i=1}^{n} \frac{h_i(x)}{1 + h_i(x) - h_i(x)t},$$

where $h_i(x) = \dfrac{q_i(x)}{1 - q_i(x)}$, and

$$g_{n,x}''(t) = g_{n,x}(t) \left(\sum_{i=1}^{n} \frac{h_i(x)^2}{1 + h_i(x) - h_i(x)t} + \left(\sum_{i=1}^{n} \frac{h_i(x)}{1 + h_i(x) - h_i(x)t} \right)^2 \right);$$

so, on account of formulas (1.3.10) and (1.3.11), we obtain

$$\alpha_{n,x} = \frac{1}{n} E\left(\sum_{i=1}^{n} U(i, x) \right) = \frac{1}{n} g_{n,x}'(1) = \frac{1}{n} \sum_{i=1}^{n} h_i(x)$$

and

$$\sigma_{n,x}^2 = \frac{1}{n^2} \operatorname{Var}\left(\sum_{i=1}^{n} U(i, x) \right) = \frac{1}{n^2} (g_{n,x}''(1) + g_{n,x}'(1) - g_{n,x}'(1)^2)$$

$$= \frac{1}{n^2} \left(\sum_{i=1}^{n} h_i(x)^2 + \sum_{i=1}^{n} h_i(x) \right).$$

As we shall see, the uniform convergence of $(T_n(f))_{n \in \mathbb{N}}$ on a fixed interval $[0, b]$ depends on a boundedness assumption on the functions q_n. In fact, we shall assume that there exists $\delta \in \mathbb{R}$ such that $0 < q_n(x) \le \delta < 1$ for every $x \in [0, b]$ and $n \in \mathbb{N}$. This yields $h_n(x) \le \dfrac{\delta}{1 - \delta}$ for every $x \in [0, b]$ and $n \in \mathbb{N}$, and therefore, if

$$\lim_{n \to \infty} \frac{1}{n} \sum_{i=1}^{n} h_i(x) = x \quad \text{uniformly in } x \in [0, b],$$

conditions (5.2.12) and (5.2.13) are satisfied uniformly in the interval $[0, b]$. In this case, we conclude by Theorem 5.2.2 that

$$\lim_{n \to \infty} T_n(f) = f, \tag{5.2.46}$$

uniformly in the interval $[0, b]$ for every function $f \in \mathscr{C}_b([0, +\infty[)$.

Consequently, if $0 < q_n(x) \le \delta < 1$ for every $x \in [0, +\infty[$ and $n \in \mathbb{N}$ and if

$$\lim_{n \to \infty} \frac{1}{n} \sum_{i=1}^{n} h_i(x) = x \quad \text{uniformly in } x \in [0, +\infty[,$$

then $\lim_{n \to \infty} T_n(f) = f$ uniformly on compact subintervals of $[0, +\infty[$ for every $f \in \mathscr{C}_b([0, +\infty[)$ (uniformly on $[0, +\infty[$ for every $f \in \mathscr{U}\mathscr{C}_b([0, +\infty[))$.

By taking the above values of $\alpha_{n,x}$ and $\sigma_{n,x}^2$ into account and by using Theorem 5.2.4, we can obtain estimates of the rate of convergence of generalized Baskakov operators in terms of the modulus of continuity of f and f', respectively.

For the sake of simplicity we state some estimates in the particular case of Baskakov operators where $p_i(x) = \dfrac{1}{1+x}$ and $h_i(x) = x$ ($n \ge 1, i = 1, \ldots, n$). Since

$$\alpha_{n,x} = \frac{1}{n} \sum_{i=1}^{n} x = x$$

and

$$\sigma_{n,x}^2 = \frac{1}{n^2} \left(\sum_{i=1}^{n} x^2 + \sum_{i=1}^{n} x \right) = \frac{x(1+x)}{n},$$

we obtain by (5.2.20)

$$|A_n(f)(x) - f(x)| \le (1 + \sqrt{x(1+x)})\omega(f, n^{-1/2}) \tag{5.2.47}$$

for every $f \in \mathscr{C}_b([0, +\infty[)$ and $x \in [0, +\infty[$. Moreover, if f is differentiable and $f' \in \mathscr{C}_b([0, +\infty[)$, we can apply (5.2.22) to conclude

$$|A_n(f)(x) - f(x)| \le \sqrt{\frac{x(1+x)}{n}}(1 + \sqrt{x(1+x)})\omega(f', n^{-1/2}). \tag{5.2.48}$$

Finally, since $\text{Var}(U(n, x)) = x(1 + x)$, by virtue of (5.2.31), we also have

$$|A_n(f)(x) - f(x)| \leq M\omega_2\left(f, \sqrt{\frac{x(1 + x)}{n}}\right) \tag{5.2.49}$$

for every $f \in \mathscr{U}\mathscr{C}_b([0, +\infty[)$ and $x \in [0, +\infty[$.

For another generalization of Baskakov operators we refer to Section 5.3.

The next example constitutes a wide generalization of Bernstein operators in the interval $[0, 1]$.

5.2.7 Stancu operators. We fix a sequence $(a_n)_{n \in \mathbb{N}}$ of positive real numbers. The random scheme Z is defined by considering, for every $x \in [0, 1]$ and $n \in \mathbb{N}$, a random variable $Z(n, x)$ having the following distribution function:

$$F_{Z(n, x)}(t) := \begin{cases} 0 & \text{if } t < 0 \\ \displaystyle\sum_{0 \leq k \leq [nt]} \omega_{n,k}(x, a_n) & \text{if } 0 \leq t \leq 1, \\ 1 & \text{if } t > 1, \end{cases} \tag{5.2.50}$$

where

$$\omega_{n,k}(x, a_n)$$

$$:= \binom{n}{k} \frac{x(x + a_n)\cdots(x + (k-1)a_n)(1 - x)(1 - x + a_n)\cdots(1 - x + (n - k - 1)a_n)}{(1 + a_n)(1 + 2a_n)\cdots(1 + (n-1)a_n)}. \tag{5.2.51}$$

A probabilistic interpretation may be deduced by considering the *Markov-Pólya urn scheme* (see, e.g., Feller [1966, Chapter V, Example (2.c)]): an urn contains N balls, a of which are white and b black; a ball is drawn at random, its colour noted and it is replaced along with c_n balls of the same colour. This procedure is repeated n times ($n \geq 1$). For every $h = 1, \ldots, n$, consider the random variable $Y(h, x)$ which takes the value 1 if the result of the h-th drawing is a white ball, otherwise it takes the value 0. The probability $P(k; n, a, b, c_n)$ that the total number of white balls is k ($0 \leq k \leq n$), i.e., that the random variable $\sum_{i=1}^{n} Y(i, x)$ takes the value k, depends on a, b and c_n and is given by (see, e.g. Feller [1966, Chapter V, Example (2.c)])

$$P(k; n, a, b, c_n)$$

$$= \binom{n}{k} \frac{a(a + c_n)\cdots(a + (k-1)c_n)b(b + c_n)\cdots(b + (n - k - 1)c_n)}{N(N + c_n)\cdots(N + (n-1)c_n)}.$$

If x is a rational number between 0 and 1, and if a, b, c_n and N satisfy $x = \dfrac{a}{N}$, $a_n = \dfrac{c_n}{N}$ and $1 - x = \dfrac{b}{N}$, then the random variable $Z(n, x) = \dfrac{1}{n} \sum\limits_{i=1}^{n} Y(i, x)$ takes the value $\dfrac{k}{n}$ $(0 \le k \le n)$ with probability $\omega_{n,k}(x, a_n)$.

For every $n \in \mathbb{N}$ we consider the positive linear operator $S_{n, a_n} : \mathscr{C}([0,1]) \to \mathscr{C}([0,1])$ associated with the random scheme Z according to (5.2.6) and call it the *n-th Stancu operator*. In fact, taking (5.2.50) into account, we have

$$S_{n, a_n}(f)(x) = \sum_{k=0}^{n} f\left(\frac{k}{n}\right) \omega_{n,k}(x, a_n) \tag{5.2.52}$$

for every $f \in \mathscr{C}([0,1])$ and $x \in [0,1]$.

If the sequence $(a_n)_{n \in \mathbb{N}}$ is fixed and no confusion arises, we shall use the simpler notation S_n instead of S_{n, a_n}.

Clearly, if $a_n = 0$ for each $n \in \mathbb{N}$, S_n is the *n*-th Bernstein operator B_n.

In the sequel, it will be useful to have a different representation of the operators S_n in terms of finite differences at the initial point 0.

To this purpose, we first consider the *Newton interpolation polynomial* $N_f : \mathbb{R} \to \mathbb{R}$ with respect to the nodes $\dfrac{k}{n}$ $(k = 0, \ldots, n)$, defined by

$$N_f(t) := f(0) + \sum_{k=1}^{n} \binom{nt}{k} \Delta_{1/n}^k f(0) \quad (t \in \mathbb{R}). \tag{5.2.53}$$

By recalling that for every $n \ge 1$

$$\int_{-\infty}^{+\infty} N_f\left(\frac{t}{n}\right) dP_{Z(n, x)}(t) = f(0) + \sum_{j=1}^{n} \frac{m_{n,x,j}}{j!} \Delta_{1/n}^j f(0),$$

where

$$m_{n,x,j} := \int_{-\infty}^{+\infty} t(t-1) \cdots (t-j+1) \, dP_{Z(n, x)}(t)$$

is the *j-th factorial moment* of $Z(n, x)$ (see (1.3.13)), we can write

$$S_n(f)(x) = \sum_{k=0}^{n} f\left(\frac{k}{n}\right) \omega_{n,k}(x, a_n) = \sum_{k=0}^{n} N_f\left(\frac{k}{n}\right) \omega_{n,k}(x, a_n)$$

$$= f(0) + \sum_{j=1}^{n} \Delta_{1/n}^j f(0) \frac{m_{n,x,j}}{j!}. \tag{5.2.54}$$

Now consider the probability generating function $g_{n,x}: [-1, 1] \to \mathbb{R}$ of $Z(n, x)$. Then

$$g_{n,x}(t) = \sum_{k=0}^{n} \omega_{n,k}(x, a_n)t^k \quad (t \in [-1, 1]), \tag{5.2.55}$$

and

$$m_{n,x,j} = g_{n,x}^{(j)}(1) = \sum_{h=0}^{n} h(h - 1) \cdots (h - j + 1)\omega_{n,h}(x, a_n)$$

$$= \sum_{k=j}^{n} k(k - 1) \cdots (k - j + 1)\omega_{n,k}(x, a_n). \tag{5.2.56}$$

Setting

$$N_n := (1 + a_n)(1 + 2a_n) \cdots (1 + (n - 1)a_n),$$

we obtain

$$m_{n,x,j} = \frac{1}{N_n} \sum_{k=j}^{n} k(k - 1) \cdots (k - j + 1)\frac{n(n - 1) \cdots (n - k + 1)}{k!}$$

$$\times \prod_{p=0}^{k-1} (x + pa_n) \prod_{q=0}^{n-k-1} (1 - x + qa_n)$$

$$= \frac{1}{N_n} \sum_{k=j}^{n} \frac{n(n - 1) \cdots (n - k + 1)}{(k - j)!} \prod_{p=0}^{k-1} (x + pa_n) \prod_{q=0}^{n-k-1} (1 - x + qa_n)$$

$$= \frac{1}{N_n} \sum_{s=0}^{n-j} \frac{n(n - 1) \cdots (n - j - s + 1)}{s!} \prod_{p=0}^{j+s-1} (x + pa_n) \prod_{q=0}^{n-j-s-1} (1 - x + qa_n).$$

Moreover, the equality $\prod_{p=0}^{j+s-1} (x + pa_n) = \prod_{p_1=0}^{j-1} (x + p_1 a_n) \prod_{p_2=j}^{j+s-1} (x + p_2 a_n)$ yields

$$m_{n,x,j} = \frac{n!}{N_n} \prod_{p_1=0}^{j-1} (x + p_1 a_n) \sum_{s=0}^{n-j} \left(\frac{1}{s!} \prod_{p_2=j}^{j+s-1} (x + p_2 a_n) \right)$$

$$\times \left(\frac{1}{(n - j - s)!} \prod_{q=0}^{n-j-s-1} (1 - x + qa_n) \right)$$

$$= n(n-1)\cdots(n-j+1)\frac{x(x+a_n)\cdots(x+(j-1)a_n)}{(1+a_n)\cdots(1+(n-1)a_n)}\sum_{s=0}^{n-j}\binom{n-j}{s}$$

$$\times \prod_{p_2=j}^{j+s-1}(x+p_2a_n)\prod_{q=0}^{n-j-s-1}(1-x+qa_n).$$

At this point we recall the *Vandermonde formula*

$$\prod_{k=0}^{p-1}(a+b+kc) = \sum_{s=0}^{p}\binom{p}{s}\prod_{k=0}^{s-1}(a+kc)\prod_{h=0}^{p-s-1}(b+hc),$$

which holds for $a,b,c \in \mathbb{R}$ and $p \geq 1$ $\left(\text{where } \prod_{k=0}^{-1}(a+kc) = \prod_{h=0}^{-1}(b+hc) = 1 \text{ by}\right.$ convention$\left.\right)$.

By applying this formula to $a = x + ja_n$, $b = 1 - x$, $c = a_n$ and $p = n - j$, we obtain

$$\sum_{s=0}^{n-j}\binom{n-j}{s}\prod_{p_2=j}^{j+s-1}(x+p_2a_n)\prod_{q=0}^{n-j-s-1}(1-x+qa_n)$$

$$= (1+ja_n)(1+(j+1)a_n)\cdots(1+(n-1)a_n)$$

and hence

$$m_{n,x,j} = n(n-1)\cdots(n-j+1)\frac{x(x+a_n)\cdots(x+(j-1)a_n)}{(1+a_n)\cdots(1+(j-1)a_n)}.$$

So, by (5.2.54), we have

$$S_n(f)(x) = f(0) + \sum_{j=1}^{n}\binom{n}{j}\Delta_{1/n}^j f(0)\frac{x(x+a_n)\cdots(x+(j-1)a_n)}{(1+a_n)\cdots(1+(j-1)a_n)} \quad (5.2.57)$$

for every $f \in \mathscr{C}([0,1])$ and $x \in [0,1]$.

In particular, if we take $a_n = 0$ for all $n \in \mathbb{N}$, we obtain the following representation of Bernstein polynomials in terms of finite differences at the initial point 0:

$$B_n(f)(x) = f(0) + \sum_{j=1}^{n}\binom{n}{j}\Delta_{1/n}^j f(0)x^j \quad (f \in \mathscr{C}([0,1]), x \in [0,1]). \quad (5.2.58)$$

In this case, for every $n \in \mathbb{N}$ and $x \in [0,1]$, the factorial moments are given by

$$m_{n,x,j} = g_{n,x}^{(j)}(1) = n(n-1)\cdots(n-j+1)x^j$$

for every $j = 1,\ldots,n$.

We shall use the identity (5.2.57) to evaluate $\alpha_{n,x}$ and $\sigma_{n,x}^2$. Denote by e_i, $i = 1, 2$, the function $e_i(x) = x^i$ ($x \in [0, 1]$). Then $\Delta_{1/n} e_1(0) = 1/n$, $\Delta_{1/n} e_2(0) = 1/n^2$ and $\Delta_{1/n}^2 e_2(0) = 2/n^2$ and therefore, by (5.2.57), for every $n \in \mathbb{N}$ and $x \in [0, 1]$,

$$\alpha_{n,x} = S_n(e_1)(x) = \binom{n}{1} \frac{x}{n} = x$$

and

$$\sigma_{n,x}^2 = S_n(e_2)(x) - x^2 = \binom{n}{1} \frac{x}{n^2} + \binom{n}{2} \frac{2x(x + a_n)}{n^2(1 + a_n)} - x^2 = \frac{1 + na_n}{n(1 + a_n)} x(1 - x).$$

Therefore, if we assume $\lim_{n \to \infty} a_n = 0$, we can apply Theorem 5.2.2 to obtain

$$\lim_{n \to \infty} S_n(f) = f \quad \text{uniformly on } [0, 1] \tag{5.2.59}$$

for every $f \in \mathscr{C}([0, 1])$.

Moreover, by Theorem 5.2.4, for every $n \in \mathbb{N}$, $f \in \mathscr{C}([0, 1])$ and $x \in [0, 1]$, we obtain

$$|S_n(f)(x) - f(x)| \leq \left(1 + \sqrt{\frac{1 + na_n}{1 + a_n} x(1 - x)}\right) \omega(f, n^{-1/2}) \tag{5.2.60}$$

and consequently

$$\|S_n(f) - f\| \leq \left(1 + \frac{1}{2} \sqrt{\frac{1 + na_n}{1 + a_n}}\right) \omega(f, n^{-1/2}). \tag{5.2.61}$$

Since in this case (see (5.2.23))

$$\beta_n = \sup_{x \in [0, 1]} \sqrt{\frac{1 + na_n}{n(1 + a_n)} x(1 - x)} = \frac{1}{2} \sqrt{\frac{1 + na_n}{n(1 + a_n)}},$$

we obtain by (5.2.24) with $\gamma = 2$ the following further estimate

$$\|S_n(f) - f\| \leq \frac{3}{2} \omega\left(f, \sqrt{\frac{1 + na_n}{n(1 + a_n)}}\right). \tag{5.2.62}$$

In both cases, if we take $a_n = 0$, we achieve (5.2.40).

Furthermore (5.2.25) yields

$$\|S_n(f) - f\| \leq \frac{1}{2}\sqrt{\frac{1 + na_n}{n(1 + a_n)}}\left(1 + \frac{1}{2}\sqrt{\frac{1 + na_n}{1 + a_n}}\right)\omega(f', n^{-1/2}) \qquad (5.2.63)$$

provided $f \in \mathscr{C}^1([0, 1])$.

For multidimensional and infinite dimensional extensions of Stancu operators, see Chapter 6.

As in Example 5.2.6, we now consider an approximation process acting on continuous functions on $[0, +\infty[$.

5.2.8 Bleimann-Butzer-Hahn operators. For every $x \in [0, +\infty[$ let $(Y(n, x))_{n \in \mathbb{N}}$ be a sequence of independent and identically distributed Bernoulli random variables with parameters 1 and $\dfrac{x}{1 + x}$. Thus $P\{Y(n, x) = 1\} = \dfrac{x}{1 + x}$ and $P\{Y(n, x) = 0\} = \dfrac{1}{1 + x}$ for every $n \in \mathbb{N}$.

Consequently the sum $\sum\limits_{j=1}^{n} Y(j, x)$ has a Bernoulli distribution with parameters n and $\dfrac{x}{1 + x}$, and therefore, for every $k = 0, 1, \ldots, n$,

$$P\left\{\sum_{j=1}^{n} Y(j, x) = k\right\} = \binom{n}{k}\left(\frac{x}{1 + x}\right)^k\left(\frac{1}{1 + x}\right)^{n-k}.$$

Define a random scheme $Z\colon \mathbb{N} \times [0, +\infty[\to M_2(\Omega)$ on $[0, +\infty[$ by

$$Z(n, x) := \frac{\sum\limits_{j=1}^{n} Y(j, x)}{n + 1 - \sum\limits_{j=1}^{n} Y(j, x)} \qquad (5.2.64)$$

for every $(n, x) \in \mathbb{N} \times [0, +\infty[$.

For every $n \in \mathbb{N}$ the positive linear operator $H_n\colon \mathscr{C}_b([0, +\infty[) \to \mathscr{C}_b([0, +\infty[)$ associated with the random scheme Z is called the *n-th Bleimann-Butzer-Hahn operator*. On account of (5.2.34) we have

$$H_n(f)(x) = \frac{1}{(1 + x)^n}\sum_{k=0}^{n} f\left(\frac{k}{n - k + 1}\right)\binom{n}{k}x^k \qquad (5.2.65)$$

for every $f \in \mathscr{C}_b([0, +\infty[)$ and $x \in [0, +\infty[$.

As in the above examples, we shall study the convergence of the sequence $(H_n(f))_{n \in \mathbb{N}}$ together with some quantitative estimates. First, we consider the quantities $\alpha_{n,x}$ and $\sigma_{n,x}^2$ and then we apply Theorems 5.2.2 and 5.2.4.

For a fixed $n \in \mathbb{N}$ and $x \in [0, +\infty[$ we obviously have

$$\alpha_{n,x} = x - \frac{x^{n+1}}{(1 + x)^n}.$$

To evaluate $\sigma_{n,x}^2$ we begin by setting $p := \dfrac{x}{1 + x}$ and $q := 1 - p = \dfrac{1}{1 + x}$. By the definition of $Z(n, x)$ we have

$$E(Z(n,x)^2) = \sum_{k=0}^{n} \frac{k^2}{(n - k + 1)^2} P\left\{ \sum_{j=1}^{n} Y(j,x) = k \right\}$$

$$= \sum_{k=1}^{n} \frac{k}{(n - k + 1)(k - 1)!} \frac{n!}{(n - k + 1)!} p^k q^{n-k}$$

$$= \sum_{j=0}^{n-1} \frac{(j + 1)}{(n - j)j!} \frac{n!}{(n - j)!} p^{j+1} q^{n-j-1}$$

$$= \sum_{j=1}^{n-1} \frac{n!}{(n - j)(j - 1)!(n - j)!} p^{j+1} q^{n-j-1}$$

$$+ \sum_{j=0}^{n-1} \frac{n!}{(n - j)j!(n - j)!} p^{j+1} q^{n-j-1} = \frac{pq^{n-1}}{n}$$

$$+ \sum_{j=1}^{n-1} \frac{n! p^{j+1} q^{n-j-1}}{(n - j)(j - 1)!(n - j)!} + \sum_{j=1}^{n-1} \frac{n! p^{j+1} q^{n-j-1}}{(n - j)j!(n - j)!}$$

$$= \frac{x}{n(1 + x)^n} + \sum_{k=0}^{n-2} \binom{n}{k} p^{k+2} q^{n-k-2} \left(\frac{n - k}{n - k - 1} + \frac{n - k}{(k + 1)(n - k - 1)} \right)$$

$$= \frac{x}{n(1 + x)^n} + \sum_{k=0}^{n-2} \binom{n}{k} p^{k+2} q^{n-k-2} \left(1 + \frac{n + 1}{(n - k - 1)(k + 1)} \right)$$

$$= \frac{x}{n(1 + x)^n} + \sum_{k=0}^{n-2} \binom{n}{k} p^{k+2} q^{n-k-2} + \sum_{k=0}^{n-2} \frac{(n + 1)! p^{k+2} q^{n-k-2}}{(k + 1)!(n - k)!(n - k - 1)}$$

$$= \frac{x}{n(1 + x)^n} + \sum_{k=0}^{n} \binom{n}{k} p^{k+2} q^{n-k-2} - p^{n+2} q^{-2} - np^{n+1} q^{-1}$$

$$+ \sum_{k=0}^{n-2} \frac{(n + 1)! p^{k+2} q^{n-k-2}}{(k + 1)!(n - k)!(n - k - 1)}$$

$$= \frac{x}{n(1 + x)^n} + x^2 - x^2 \left(\frac{x}{1 + x} \right)^n - nx \left(\frac{x}{1 + x} \right)^n + R,$$

where

$$R := \sum_{k=0}^{n-2} \frac{(n + 1)! p^{k+2} q^{n-k-2}}{(k + 1)!(n - k)!(n - k - 1)}.$$

Since $(n - j)^{-1} \le 3/(n - j + 2)$ for $j = 1, \ldots, n - 1$, we obtain

$$R = \sum_{j=1}^{n-1} \binom{n + 1}{j} \frac{p^{j+1} q^{n-j-1}}{n - j} \le 3 \sum_{j=1}^{n-1} \binom{n + 1}{j} \frac{p^{j+1} q^{n-j-1}}{n - j + 2}$$

$$\le \frac{3p}{q^2} \sum_{j=0}^{n+1} \binom{n + 1}{j} \frac{p^j q^{n+1-j}}{n - j + 2} = 3x(1 + x) \sum_{j=0}^{n+1} \binom{n + 1}{j} \frac{p^j q^{n+1-j}}{n - j + 2}.$$

On the other hand, we remark that $\sum_{j=0}^{n+1} \binom{n + 1}{j} \frac{p^j q^{n+1-j}}{n - j + 2}$ can be interpreted as the expected value $E \left(\frac{1}{n + 2 - U} \right)$ of the random variable $\frac{1}{n + 2 - U}$, where U is a Bernoulli random variable with parameters $n + 1$ and p. Consequently $n + 1 - U$ is a Bernoulli random variable with parameters $n + 1$ and $1 - p = q$.

So, according to a result of Chao and Strawderman [1972, (3.4)],

$$E \left(\frac{1}{n + 2 - U} \right) = E \left(\frac{1}{(n + 1 - U) + 1} \right) = \frac{1 - p^{n+2}}{(n + 2)q}.$$

Consequently $R \le 3x(1 + x) \frac{1 - p^{n+2}}{(n + 2)q} \le \frac{3x(1 + x)^2}{(n + 2)}.$

Therefore we obtain

$$\sigma_{n,x}^2 = E(Z(n,x)^2) - \alpha_{n,x}^2 = \frac{x}{n(1+x)^n} + x^2 - x^2 \left(\frac{x}{1+x}\right)^n - nx \left(\frac{x}{1+x}\right)^n$$

$$+ R - \left(x - \frac{x^{n+1}}{(1+x)^n}\right)^2 = \frac{x}{n(1+x)^n} + x^2 \left(\frac{x}{1+x}\right)^n - nx \left(\frac{x}{1+x}\right)^n$$

$$+ R - \frac{x^{2(n+1)}}{(1+x)^{2n}} \le \frac{x}{n(1+x)^n} + x^2 \left(\frac{x}{1+x}\right)^n - nx \left(\frac{x}{1+x}\right)^n$$

$$+ \frac{3x(1+x)^2}{(n+2)} - \frac{x^{2(n+1)}}{(1+x)^{2n}} \le \frac{x}{n} + \frac{x^2(1+x)}{n} + \frac{3x(1+x)^2}{n} - \frac{x^{2(n+1)}}{(1+x)^{2n}}$$

$$= \frac{4x(1+x)^2}{n} - \frac{x^{2(n+1)}}{(1+x)^{2n}},$$

where we have used the inequality $\left(\dfrac{x}{1+x}\right)^n \le \dfrac{1+x}{n}$.

Now, for every fixed $b > 0$, we obviously have

$$\lim_{n \to \infty} \alpha_{n,x} = x \quad \text{and} \quad \lim_{n \to \infty} \sigma_{n,x}^2 = 0$$

uniformly in $x \in [0, b]$, and therefore, by Theorem 5.2.2 and the subsequent Remark 1, for every $f \in \mathscr{C}_b([0, +\infty[)$, we obtain

$$\lim_{n \to \infty} H_n(f)(x) = f(x) \quad \text{uniformly in } x \in [0, b]. \tag{5.2.66}$$

We now try to estimate the rate of convergence. Take $f \in \mathscr{C}_b([0, +\infty[)$ and $\delta > 0$. By virtue of (5.2.19) we have

$$|H_n(f)(x) - f(x)| \le \left(1 + \frac{1}{\delta} \sqrt{\frac{x^{2(n+1)}}{(1+x)^{2n}} + \sigma_{n,x}^2}\right) \omega(f, \delta)$$

$$\le \left(1 + \frac{1}{\delta} \sqrt{\frac{4x(1+x)^2}{n}}\right) \omega(f, \delta)$$

for every $x \in [0, +\infty[$.

Letting $\delta = \sqrt{\dfrac{x(1 + x)^2}{n}}$, we obtain

$$|H_n(f)(x) - f(x)| \leq 3\omega\left(f, \sqrt{\frac{x(1 + x)^2}{n}}\right). \tag{5.2.67}$$

Finally, according to R. A. Khan [1988], it is possible to show that

$$|H_n(f)(x) - f(x)| \leq M\left(\omega_2\left(f, \sqrt{\frac{x(1 + x)^2}{n}}\right) + \frac{x(1 + x)^2}{n}\|f\|\right). \tag{5.2.68}$$

We proceed now by giving some other examples of positive approximation processes associated with a random scheme $Z: \mathbb{N} \times I \to M_2(\Omega)$ such that every $Z(n, x)$ is a Lebesgue-continuous random variable. If for every $(n, x) \in \mathbb{N} \times I$ we denote by $f_{n,x}: \mathbb{R} \to \mathbb{R}$ the probability density of $Z(n, x)$, i.e., $f_{n,x}$ is a positive measurable function on $\mathbb{R}$ and

$$P_{Z(n,x)} = f_{n,x} \cdot \lambda_1, \tag{5.2.69}$$

then, for every $n \in \mathbb{N}$, $f \in \mathscr{C}_b(\mathbb{R})$ and $x \in I$ we have

$$P_n(f)(x) = \int_{-\infty}^{+\infty} f(t) \cdot f_{n,x}(t)\,dt, \tag{5.2.70}$$

or, if (5.2.4) is satisfied,

$$P_n^*(f)(x) = \int_{-\infty}^{+\infty} f^*(t) \cdot f_{n,x}(t)\,dt = \int_I f(t) \cdot f_{n,x}(t)\,dt. \tag{5.2.71}$$

In this case, for all $n \in \mathbb{N}$ the mapping $x \mapsto P_{Z(n,x)}$ is continuous if for every $x_0 \in I$

$$\lim_{x \to x_0} f_{n,x} = f_{n,x_0} \quad \text{pointwise on } I \tag{5.2.72}$$

(see Section 1.3).

Our first example is classical and its construction is based on Gaussian random variables.

5.2.9 Gauss-Weierstrass convolution operators.

Given an arbitrary sequence $(\varepsilon_n)_{n \in \mathbb{N}}$ of strictly positive real numbers converging to 0, for every $x \in \mathbb{R}$ consider a sequence $(Y(n, x))_{n \in \mathbb{N}}$ of Gaussian random variables with parameters nx and $2n^2\varepsilon_n$ (see Section 1.3).

Furthermore set

$$Z(n, x) := \frac{1}{n} Y(n, x), \quad (n, x) \in \mathbb{N} \times \mathbb{R}. \tag{5.2.73}$$

Since the probability density of $Y(n, x)$ is

$$g_{nx, 2n^2\varepsilon_n}(t) := \frac{1}{n\sqrt{4\pi\varepsilon_n}} \exp\left(-\frac{1}{2}\left(\frac{t - nx}{n\sqrt{2\varepsilon_n}}\right)^2\right), \quad (t \in \mathbb{R}),$$

then $Z(n, x)$ is Lebesgue-continuous and its probability density is the function

$$f_{n, x}(t) := n g_{nx, 2n^2\varepsilon_n}(nt) = \frac{1}{\sqrt{4\pi\varepsilon_n}} \exp\left(-\frac{(t - x)^2}{4\varepsilon_n}\right) \quad (t \in \mathbb{R}). \tag{5.2.74}$$

The sequence of positive linear operators which are associated with Z according to (5.2.70) will be denoted by $(W_n)_{n \in \mathbb{N}}$. Thus W_n is a positive linear operator from $\mathscr{C}_b(\mathbb{R})$ into itself and for every $f \in \mathscr{C}_b(\mathbb{R})$ and $x \in \mathbb{R}$

$$W_n(f)(x) := \frac{1}{\sqrt{4\pi\varepsilon_n}} \int_{-\infty}^{+\infty} f(y) \exp\left(-\frac{(y - x)^2}{4\varepsilon_n}\right) dy. \tag{5.2.75}$$

Since $\alpha_{n, x} = x$ and $\sigma_{n, x}^2 = 2\varepsilon_n$, by virtue of Theorem 5.2.2 we obtain

$$\lim_{n \to \infty} W_n(f) = f \tag{5.2.76}$$

uniformly on compact subsets of $\mathbb{R}$ for every $f \in \mathscr{C}_b(\mathbb{R})$ and uniformly on $\mathbb{R}$ for every $f \in \mathscr{U}\mathscr{C}_b(\mathbb{R})$ (in particular for every $f \in \mathscr{C}_0(\mathbb{R})$).

Furthermore from (5.2.20) and (5.2.30) it also follows that

$$\|W_n(f) - f\| \leq (1 + \sqrt{2n\varepsilon_n})\omega(f, n^{-1/2}). \tag{5.2.77}$$

The above results allow us to derive the approximation properties of the *convolution operators of Gauss-Weierstrass* which are defined by

$$W_t(f)(x) := \frac{1}{\sqrt{4\pi t}} \int_{-\infty}^{+\infty} f(y) \exp\left(-\frac{(y - x)^2}{4t}\right) dy \tag{5.2.78}$$

for every $f \in \mathscr{C}_b(\mathbb{R})$, $t > 0$ and $x \in \mathbb{R}$.

In fact, since the above formula (5.2.76) holds for an arbitrary sequence $(\varepsilon_n)_{n \in \mathbb{N}}$ converging to zero, we obtain

$$\lim_{t \to 0^+} W_t(f) = f, \tag{5.2.79}$$

uniformly on compact subsets of $\mathbb{R}$ for every $f \in \mathscr{C}_b(\mathbb{R})$ (uniformly on $\mathbb{R}$ for every $f \in \mathscr{UC}_b(\mathbb{R})$, respectively).

Among other things, the significance of these operators is based on the fact that the function $u(x, t) := W_t(f)(x)$ $(x \in \mathbb{R}, t > 0)$ is the solution of the initial value problem of the *heat equation* for an infinite rod with prescribed initial temperature distribution $f \in \mathscr{UC}_b(\mathbb{R})$, namely

$$\begin{cases} \dfrac{\partial u}{\partial t}(x, t) = \dfrac{\partial^2 u}{\partial x^2}(x, t), & x \in \mathbb{R}, t > 0, \\[3mm] \lim_{t \to 0} u(\cdot, t) = f & \text{uniformly on } \mathbb{R}. \end{cases} \tag{5.2.80}$$

The subsequent example is still based on Gaussian distributions that, in this case, are identically distributed.

5.2.10 Cismasiu operators. Consider the interval $I = \,]0, +\infty[$ and, for every $x \in \,]0, +\infty[$, let $(Y(n, x))_{n \in \mathbb{N}}$ be a sequence of independent random variables having the same Gaussian distribution with parameters 0 and x.

Then the sum $\sum_{k=1}^{n} Y(k, x)^2$ is a Chi-squared random variable $\chi_{n, x}^2$ with n degrees of freedom and parameter x (see (1.3.40)). Its probability density function $g_{n, x}$ is given by

$$g_{n, x}(t) := \begin{cases} 0, & \text{if } t < 0, \\[3mm] \dfrac{t^{n/2 - 1} \exp(-t/(2x))}{(2x)^{n/2} \Gamma\!\left(\dfrac{n}{2}\right)}, & \text{if } t \geq 0, \end{cases}$$

where Γ denotes the gamma function.

The random scheme $Z \colon \mathbb{N} \times I \to M_2(\Omega)$ we shall consider is defined by

$$Z(n, x) := \frac{1}{n} \sum_{k=1}^{n} Y(k, x)^2$$

for every $(n, x) \in \mathbb{N} \times I$.

The n-th operator $C_n: \mathscr{C}_b^*(]0, +\infty[) \to \mathscr{C}_b(]0, +\infty[)$ which is associated with the random scheme Z is called the n-th *Cismasiu operator*. By (5.2.6), for every $f \in \mathscr{C}_b^*(]0, +\infty[)$ and $x \in]0, +\infty[$, we have

$$C_n(f)(x) = \frac{1}{(2x)^{n/2}\Gamma\left(\dfrac{n}{2}\right)} \int_0^{+\infty} t^{n/2-1} \exp\left(-\frac{t}{2x}\right) f\left(\frac{t}{n}\right) dt. \qquad (5.2.81)$$

In this case, we know that $\alpha_{n,x} = x$ and $\sigma_{n,x}^2 = \dfrac{2}{n}x^2$ (see (1.3.41)).

Therefore, by Theorem 5.2.2, we obtain

$$\lim_{n \to \infty} C_n(f) = f \quad \text{uniformly on every compact interval of }]0, +\infty[\quad (5.2.82)$$

for every $f \in \mathscr{C}_b^*(]0, +\infty[)$.

Moreover (5.2.20) yields the estimate

$$|C_n(f)(x) - f(x)| \leq (1 + \sqrt{2}\,x)\omega(f, n^{-1/2}), \qquad (5.2.83)$$

which holds for every $f \in \mathscr{C}_b(]0, +\infty[)$ and $x \in]0, +\infty[$.

For further positive approximation processes which can be constructed by probabilistic methods we refer the reader to Adell, Badía and de la Cal [1993], Adell, Badía, de la Cal and Plo [1992], Adell and de la Cal [1992], [1993a], Arató and Rényi [1957], Butzer and Berens [1967], de la Cal [1994], de la Cal and Luquin [1991a], [1991b], [1992], Cismasiu [1985], Feller [1957], J. A. Goldstein [1975], Hahn [1982], Hille and Phillips [1957], Kahn [1980], [1988], Lupaş, and Müller [1967], Müller [1967], [1968], Rényi [1959], Stancu [1969b] and the references quoted therein.

Furthermore, in Sections 5.3 and 6.3 we shall discuss Schurer-Szász-Mirakjan operators and Lototsky-Schnabl operators which also have an interesting probabilistic interpretation.

In the last part of this section we briefly explain how Theorems 5.2.2 and 5.2.3 can be extended *mutatis mutandis* to a multidimensional setting.

Let $(\Omega, \mathscr{F}, P)$ be a probability space and, for a given $p \geq 2$, denote by $M_{2,p}(\Omega)$ the space of all random variables $Z: \Omega \to \mathbb{R}^p$ which have finite variance (see (1.3.8)).

If $Z \in M_{2,p}(\Omega)$, then from the Chebyshev-Markov inequality (see Lemma 5.2.1) we can also derive the *multidimensional Chebyshev inequality*

$$P\{\|Z - E(Z)\| \geq \alpha\} \leq \frac{1}{\alpha^2}\text{Var}(Z). \qquad (5.2.84)$$

Consider now a Borel subset Q of $\mathbb{R}^p$ and an arbitrary mapping $Z\colon \mathbb{N} \times Q \to M_{2,p}(\Omega)$ that we shall call a *p-dimensional random scheme on Q*.

As in (5.2.3) and (5.2.6), we can define a sequence of positive linear operators $P_n\colon \mathscr{C}_b(\mathbb{R}^p) \to \mathscr{B}(Q)$ (respectively, $P_n^*\colon \mathscr{C}_b^*(Q) \to \mathscr{B}(Q)$ if (5.2.4) holds) where in this case

$$\mathscr{C}_b^*(Q) := \{ f \in \mathscr{C}_b(Q) \,|\, \text{There exists } f^* \in \mathscr{C}_b(\mathbb{R}^p) \text{ such that } f^* = f \text{ on } Q \}.$$

Furthermore, every P_n maps $\mathscr{C}_b(\mathbb{R}^p)$ in $\mathscr{C}_b(Q)$ if the mapping $x \mapsto P_{Z(n,x)}$ is continuous from Q into $\mathscr{M}_1^+(\mathbb{R}^p)$ with respect to the weak topology.

After these preliminaries, by using (5.2.84) and the proof of Theorem 5.2.2, it is easy to show that, if

$$\lim_{n\to\infty} \alpha_{n,x} = x \quad \text{uniformly in } x \in Q \tag{5.2.85}$$

and

$$\lim_{n\to\infty} \sigma_{n,x}^2 = 0 \quad \text{uniformly in } x \in Q, \tag{5.2.86}$$

where $\alpha_{n,x} = E(Z(n,x))$ and $\sigma_{n,x}^2 = \text{Var}(Z(n,x))$ (see Section 1.3), then $\lim_{n\to\infty} P_n(f) = f$ uniformly on compact subsets of Q for every $f \in \mathscr{C}_b(\mathbb{R}^p)$ (uniformly on Q for every $f \in \mathscr{U}\mathscr{C}_b(\mathbb{R}^p)$, respectively).

Furthermore, if (5.2.4) holds, then $\lim_{n\to\infty} P_n^*(f) = f$ uniformly on compact subsets of Q for every $f \in \mathscr{C}_b^*(Q)$ (uniformly on Q for every $f \in \mathscr{C}_b^*(Q)$ which admit an extension in $\mathscr{U}\mathscr{C}_b(\mathbb{R}^p)$, respectively).

If Q is compact, then the above result is equivalent to the Volkov extension of Korovkin's theorem (see Example 2 to Theorem 4.4.6). This can be easily seen by using the proof of Theorem 5.2.3.

Furthermore, if $(L_n)_{n \in \mathbb{N}}$ is a sequence of positive linear operators from $\mathscr{C}(Q)$ into $\mathscr{B}(Q)$, and if we denote by Z the random scheme on Q satisfying (5.2.4) and whose associated sequence of positive linear operators coincides with $(L_n)_{n \in \mathbb{N}}$, then for every $n \in \mathbb{N}$ and $x \in Q$ the following equalities hold:

$$L_n(\text{pr}_i)(x) = \text{pr}_i(\alpha_{n,x}) \tag{5.2.87}$$

and

$$L_n\left(\sum_{i=1}^{p} \text{pr}_i^2 \right)(x) = \sigma_{n,x}^2 + \sum_{i=1}^{p} \text{pr}_i^2(\alpha_{n,x}), \tag{5.2.88}$$

where the pr_i's are the canonical projections on Q (see (2.6.19)).

In what follows we present the most well-known examples of multidimensional approximation processes.

5.2.11 Bernstein operators on the p-dimensional simplex. Let us consider the canonical simplex K_p of $\mathbb{R}^p$ (see Example 3.3.5). Given $x = (x_1, \ldots, x_p) \in K_p$ and $n \in \mathbb{N}$, denote by $Y(n, x)$ a multinomial random variable of order $p + 1$ with parameters $n, x_1, \ldots, x_p$ defined on a suitable probability space $(\Omega, \mathcal{F}, P)$ (see (1.3.26)).

So, if $h = (h_1, \ldots, h_p) \in \mathbb{N}^p$, $h_1 + \cdots + h_p \leq n$, then

$$P\{Y(n, x) = h\}$$

$$= \frac{n!}{h_1! \cdots h_p!(n - h_1 - \cdots - h_p)!} x_1^{h_1} \cdots x_p^{h_p} (1 - x_1 - \cdots - x_p)^{n - h_1 - \cdots - h_p}.$$

Accordingly, consider the random scheme $Z: \mathbb{N} \times K_p \to M_{2,p}(\Omega)$ defined as

$$Z(n, x) := \frac{1}{n} Y(n, x), \quad (n, x) \in \mathbb{N} \times K_p,$$

and denote by $B_n: \mathscr{C}(K_p) \to \mathscr{C}(K_p)$ the sequence of the associated positive linear operators. Thus, for every $f \in \mathscr{C}(K_p)$ and $x = (x_1, \ldots, x_p) \in K_p$,

$$B_n(f)(x) = \int_{K_p} f \, dP_{Z(n, x)}$$

$$= \sum_{\substack{h_1, \ldots, h_p \in \mathbb{N}_0 \\ h_1 + \cdots + h_p \leq n}} f\left(\frac{h_1}{n}, \ldots, \frac{h_p}{n}\right) \frac{n!}{h_1! \ldots h_p!(n - h_1 - \cdots - h_p)!} x_1^{h_1} \ldots x_p^{h_p}$$

$$\times (1 - x_1 - \cdots - x_p)^{n - h_1 - \cdots - h_p}. \tag{5.2.89}$$

The operator B_n is called the *n-th Bernstein operator on the p-dimensional simplex*.

Since $\alpha_{n,x} = (x_1, \ldots, x_p) = x$ and

$$\sigma_{n,x}^2 = \frac{1}{n^2} \operatorname{Var}(Y(n, x)) = \frac{1}{n} \sum_{i=1}^{p} x_i(1 - x_i) \leq \frac{p}{4n} \to 0,$$

we have

$$\lim_{n \to \infty} B_n(f) = f \quad \text{uniformly on } K_p \tag{5.2.90}$$

for every $f \in \mathscr{C}(K_p)$.

We point out again that to show (5.2.90) without probabilistic backgrounds, it suffices to apply Volkov's theorem (see Example 2 to Theorem 4.4.6), taking formulas (5.2.87) and (5.2.88) into account.

Finally, by a result of Stancu [1960] we have

$$\|B_n(f) - f\| \leq \left(1 + \frac{p}{2}\right)\omega(f, n^{-1/2}) \tag{5.2.91}$$

for every $f \in \mathscr{C}(K_p)$ and $n \geq 1$. However in Section 6.1 we shall see that, in fact,

$$\|B_n(f) - f\| \leq 2\omega(f, n^{-1/2}). \tag{5.2.92}$$

5.2.12 Bernstein operators on the hypercube of $\mathbb{R}^p$. Let $X = [0, 1]^p$ be the hypercube of $\mathbb{R}^p$. For every $x = (x_1, \ldots, x_p) \in X$ and $i = 1, \ldots, p$ let $(Y_i(n, x))_{n \in \mathbb{N}}$ be an independent sequence of Bernoulli random variables with parameters 1 and x_i. Suppose that for every fixed $(n, x) \in \mathbb{N} \times X$ the family $(Y_i(n, x)_{1 \leq i \leq p}$ is independent too. We set

$$Z(n, x) := \left(\frac{1}{n}\sum_{h=1}^{p} Y_1(h, x), \ldots, \frac{1}{n}\sum_{h=1}^{p} Y_p(h, x)\right), \quad (n, x) \in \mathbb{N} \times X.$$

Then for every $h_1, \ldots, h_p \in \mathbb{N}, 0 \leq h_i \leq n$,

$$P\left\{Z(n, x) = \left(\frac{h_1}{n}, \ldots, \frac{h_p}{n}\right)\right\} = \binom{n}{h_1} \cdots \binom{n}{h_p} x_1^{h_1}(1 - x_1)^{n - h_1} \ldots x_p^{h_p}(1 - x_p)^{n - h_p}.$$

So, if we denote by $B_n: \mathscr{C}(X) \to \mathscr{C}(X)$ $(n \in \mathbb{N})$ the positive linear operators associated with Z, we obtain

$$B_n(f)(x) = \int_X f \, dP_{Z(n, x)}$$

$$= \sum_{h_1, \ldots, h_p = 0}^{n} f\left(\frac{h_1}{n}, \ldots, \frac{h_p}{n}\right)\binom{n}{h_1} \cdots \binom{n}{h_p} x_1^{h_1}(1 - x_1)^{n - h_1} \ldots x_p^{h_p}(1 - x_p)^{n - h_p}$$

$$\tag{5.2.93}$$

for every $f \in \mathscr{C}(X)$ and $x = (x_1, \ldots, x_p) \in X$.

The operator B_n is called the *n-th Bernstein operator on the p-dimensional hypercube*.

Also in this case $\alpha_{n, x} = E(Z(n, x)) = (x_1, \ldots, x_p) = x$ and

$$\sigma_{n, x}^2 = \text{Var}(Z(n, x)) = \frac{1}{n}\sum_{i=1}^{p} x_i(1 - x_i) \leq \frac{p}{4n} \to 0.$$

Therefore

$$\lim_{n \to \infty} B_n(f) = f \quad \text{uniformly on } X \tag{5.2.94}$$

for every $f \in \mathscr{C}(X)$.

According to the general results we shall establish in Section 6.1, estimate (5.2.92) holds for these operators as well.

Notes and references

The term 'random scheme' was introduced by de la Cal and Luquin [1991b] who studied the associated positive approximation process in the general setting of metric spaces.

The proof of Lemma 5.2.1 is taken from Bauer [1981, Lemma 2.11.1]. Formula (5.2.14) of Theorem 5.2.2 seems to be new. Formula (5.2.15) of Theorem 5.2.2 first appeared in Feller [1966]. A quantitative version of it appeared in Stancu [1969b]. Similar proofs can be found also in J.A. Goldstein [1975] and King [1975], where estimates in terms of the modulus of continuity of the function or of its derivative are discussed (see also Stancu [1969b, §3, Theorem 3.1 and the subsequent Remark] and King [1980b]).

The estimates in terms of the second modulus of smoothness (5.2.31) appear in the paper of Butzer and Hahn [1978, Theorem 4] and Hahn [1982, Corollary 1]. More generally they prove that (5.2.30) holds for functions which are the sum of a function in $\mathscr{UC}_b(I)$ and a function whose second derivative belongs to $\mathscr{UC}_b(I)$.

Moreover, in the paper of Hahn [1982] the reader may find other estimates in terms of the third modulus of smoothness.

The probabilistic interpretation of Korovkin's theorem described within Theorem 5.2.4 can be found in King [1980b] (see also Tihomirov [1973] and Kubo and Takashima [1985]).

For all the preceding questions the reader is also referred to the papers of J.A. Goldstein [1975] and Walk [1980]. Further developments and extensions of these results can be found in many recent papers, e.g., de la Cal and Luquin [1991b], [1992].

Bernstein-King and Baskakov-King operators are discussed in King [1980b]. As a matter of fact, Bernstein-King operators were previously introduced by King [1966] from a different point of view. They were associated with a generalized Lototsky summation matrix and he studied their convergence by a direct method. For some historical references concerning Bernstein polynomials see the final notes of Section 4.2. Moreover, for further qualitative and quantitative properties of Bernstein polynomials we refer to Ditzian and Totik [1987], Lorentz [1986a], [1986b], Sendov and Popov [1988] and to our Chapter 6.

Baskakov operators were introduced by Baskakov [1957]. These operators have been extensively studied by many authors (see for instance Ditzian and Totik [1987], Sendov and Popov [1988], Totik [1983d], [1984], Della Vecchia [1987], [1988], [1989] and the references quoted therein). In the book of Ditzian and Totik one can also find a Kantorovich-type modification of these operators (see also Ditzian [1985] and Lenze [1990] and the references quoted therein).

A different proof of the convergence of Baskakov-King operators can be found in Swetits and Wood [1973b].

Stancu operators are extensively treated in the papers of Stancu [1968a], [1968b], [1969b]. In those papers a representation in terms of both finite differences and divided differences at the initial point 0 is also given. In his paper [1968a, Section 4], Stancu shows the approximation properties of $(S_n)_{n \in \mathbb{N}}$ by a direct application of Korovkin's theorem and gives a representation of the remainder $R_n(f)(x) = f(x) - S_n(f)(x)$. Furthermore, some estimates of $R_n(f)(x)$ are established in terms of the first and second divided difference of f in $[0, 1]$ together with an asymptotic estimate in the case where f is twice differentiable at x.

For further properties of Stancu operators see also de la Cal [1994], Della Vecchia [1988], [1989], Mastroianni [1980a], Mastroianni and Occorsio [1978a], [1978b] and Mühlbach [1969], [1970]. For multidimensional versions see Stancu [1970], [1972b] (see also Subsection 6.3.3, formula (6.3.9)). A more general extension was obtained by Felbecker [1972], [1973] and Campiti [1991c] (see Sections 6.1 and 6.3).

Finally we point-out that Stancu operators (5.2.52) and Bernstein-Stancu operators (2.5.2) are particular cases of a more general class of positive linear operators introduced by Stancu [1972a] (see also Gonska and Meier-Gonska [1984]).

As regards Bleimann-Butzer-Hahn operators, we refer to the paper of Bleimann, Butzer and Hahn [1980] for more details; there the convergence is established by using a Jackson-type inequality, which gives, for a function $f \in \mathcal{UC}_b([0, +\infty[)$ with first and second derivative belonging to $\mathcal{UC}_b([0, +\infty[)$, the following estimate:

$$|H_n(f)(x) - f(x)| \leq \frac{2x(1 + x)^2}{n + 2}(\|f\| + \|f'\| + \|f''\|)$$

for every $x \in [0, +\infty[$ and $n \geq 24(1 + x)$. Here we followed the approach of R.A. Khan [1988], where the approximation property of $(H_n)_{n \in \mathbb{N}}$ is proved by means of Feller's theorem. From the same paper we have taken the evaluation of $\alpha_{n,x}$ and $\sigma_{n,x}^2$ and the subsequent estimates (5.2.67) and (5.2.68) in terms of the first and second modulus of smoothness. Further properties such as monotonicity and preservation of Lipschitz classes can be found in M.K. Khan [1989], [1991]. In the paper of Totik [1984] the reader can find a saturation class theorem stating necessary and sufficient conditions on f (depending on f' and f'' too) for which $\sup_{x \in \mathbb{R}_+} |H_n(f)(x) - f(x)| = O(n^{-1})$.

For a generalization of Bleimann-Butzer-Hahn operators see Adell, de la Cal and San Miguel [1994].

The classical convolution operators of Gauss-Weierstrass were introduced by Weierstrass in 1885. For more details about these operators see Butzer and Nessel [1971].

Cismasiu's operators have been introduced and studied in the papers of Cismasiu [1984], [1987a].

In the multidimensional case the random schemes and the associated approximation processes have been treated by de la Cal and Luquin [1991b], J.A. Goldstein [1975] and B. Eisenberg [1976] (see also Tihomirov [1973]). In these papers several other examples and applications can be found. The examples concerning Bernstein polynomials on the p-dimensional simplices and p-dimensional hypercubes are given in J.A. Goldstein

[1975] and de la Cal and Luquin [1991b], respectively. Bernstein polynomials on p-dimensional simplices first appeared in Dinghas [1951] and Lorentz [1953] (see Lorentz [1986b]) (see also Stancu [1959], [1960]). Their approximation behavior has been recently studied by Berens and Xu [1991c] also in term of a new weighted K-modulus. Bernstein polynomials on p-dimensional hypercube were first studied by Hildebrandt and Schoenberg [1933] and Butzer [1953].

A generalization of Bleimann-Butzer-Hahn operators in the case of two variables has been introduced and studied in Ciupa [1986].

5.3 Discrete-type approximation processes

In this section we present several other examples of positive approximation processes. For the sake of brevity we shall show only their approximation properties together with some estimates of the rate of convergence while we shall avoid the treatment of other important questions which are related to them (see the final Notes and the references quoted therein).

All the linear operators we shall consider in this section are of *discrete type*, i.e., they are of the form

$$Lf(x) = \sum_{k=0}^{\infty} \mu_k(f)\varphi_k(x),$$

where the positive functions φ_k and the positive Radon measures μ_k are suitably chosen. In many cases, however, the measures μ_k are Dirac measures.

In fact, in Section 5.2 we have already encountered these kinds of operators and we have discussed them by means of probabilistic methods. Here we give a direct application of Korovkin-type theorems to other classes of discrete-type operators. We point out that some of the next examples provide a general method for constructing sequences of positive linear operators.

First we consider some examples on bounded intervals.

5.3.1 Bernstein-Schurer operators. Among the many different generalizations of Bernstein operators we point-out a simple one where Korovkin's theorem can be applied easily.

For fixed $p \in \mathbb{N}_0$ and $n \in \mathbb{N}$ consider the operator $B_{n,p} : \mathscr{C}([0, 1 + p]) \to \mathscr{C}([0, 1])$ defined by

$$B_{n,p}(f)(x) := \sum_{k=0}^{n+p} f\left(\frac{k}{n}\right)\binom{n+p}{k}x^k(1-x)^{n+p-k} \qquad (5.3.1)$$

for every $f \in \mathscr{C}([0, 1 + p])$ and $x \in [0, 1]$.

Obviously every B_n is positive and linear; moreover, recalling formulas (2.5.8)–(2.5.10), we obtain

$$B_{n,p}(1)(x) = 1,$$

$$B_{n,p}(e_1)(x) = \sum_{k=0}^{n+p} \frac{k}{n}\binom{n+p}{k}x^k(1-x)^{n+p-k}$$

$$= \frac{n+p}{n}\sum_{k=0}^{n+p} \frac{k}{n+p}\binom{n+p}{k}x^k(1-x)^{n+p-k} = \left(1 + \frac{p}{n}\right)x,$$

$$B_{n,p}(e_2)(x) = \sum_{k=0}^{n+p} \frac{k^2}{n^2} \binom{n+p}{k} x^k (1-x)^{n+p-k}$$

$$= \frac{(n+p)^2}{n^2} \sum_{k=0}^{n+p} \frac{k^2}{(n+p)^2} \binom{n+p}{k} x^k (1-x)^{n+p-k}$$

$$= \frac{(n+p)^2}{n^2} \left(\frac{n+p-1}{n+p} x^2 + \frac{1}{n+p} x \right)$$

$$= \left(\frac{n+p}{n} \right)^2 x^2 + \frac{n+p}{n^2} (x - x^2),$$

where $e_k(x) = x^k$ ($k = 1, 2, x \in [0, 1+p]$).

Then we have

$$\lim_{n\to\infty} B_{n,p}(1) = 1, \quad \lim_{n\to\infty} B_{n,p}(e_1) = e_1, \quad \lim_{n\to\infty} B_{n,p}(e_2) = e_2$$

uniformly on $[0,1]$ and therefore, on account of Korovkin's theorem (see Example 6 to Theorem 4.4.6) and the universal Korovkin-type property (vi) of Theorem 4.1.4, we infer that

$$\lim_{n\to\infty} B_{n,p}(f) = f \quad \text{uniformly on } [0,1] \tag{5.3.2}$$

for every $f \in \mathscr{C}([0, 1+p])$.

According to Theorem 5.1.2, quantitative estimates of the rate of convergence are available by evaluating the operators $B_{n,p}$ on the function $\psi_x(t) = t - x$ ($x, t \in [0, 1+p]$).

Actually,

$$B_{n,p}(\psi_x)(x) = B_{n,p}(e_1 - x \cdot 1)(x) = \frac{p}{n} x$$

and

$$B_{n,p}(\psi_x^2)(x) = B_{n,p}(e_2)(x) - 2x B_{n,p}(e_1)(x) + x^2 B_{n,p}(1)(x)$$

$$= \left(\frac{n+p}{n} \right)^2 x^2 + \frac{n+p}{n^2}(x - x^2) - 2x^2 - \frac{2p}{n} x^2 + x^2$$

$$= \frac{p^2}{n^2} x^2 + \frac{n+p}{n^2} x(1-x),$$

so that, if we take $\delta = \sqrt{\dfrac{p^2 x^2 + (n + p)x(1 - x)}{n^2}}$ in (5.1.5), we obtain

$$|B_{n,p}(f)(x) - f(x)| \leq 2\omega\left(f, \sqrt{\frac{p^2 x^2 + (n + p)x(1 - x)}{n^2}}\right) \qquad (5.3.3)$$

for every $f \in \mathscr{C}([0, 1 + p])$ and $x \in [0, 1]$.

Finally, if f is differentiable and $f' \in \mathscr{C}([0, 1 + p])$, then

$$|B_{n,p}(f)(x) - f(x)|$$

$$\leq \frac{p}{n}|f'(x)|x + 2\sqrt{\frac{p^2 x^2 + (n + p)x(1 - x)}{n^2}}\,\omega\left(f', \sqrt{\frac{p^2 x^2 + (n + p)x(1 - x)}{n^2}}\right).$$
$$(5.3.4)$$

The next sequence of operators originates in the identity

$$(x + y + nt)^n = \sum_{k=0}^{n} \binom{n}{k} x(x + kt)^{k-1}(y + (n - k)t)^{n-k}, \qquad (5.3.5)$$

which holds for every $x, y, t \in \mathbb{R}$ and $n \geq 1$. It is due to Jensen [1902] and generalizes the binomial theorem.

In particular, if we set $y = 1 - x$ in (5.3.5), we obtain

$$(1 + nt)^n = \sum_{k=0}^{n} \binom{n}{k} x(x + kt)^{k-1}(1 - x + (n - k)t)^{n-k}. \qquad (5.3.6)$$

5.3.2 Bernstein-Cheney-Sharma operators. Consider a sequence $(t_n)_{n \in \mathbb{N}}$ of positive real numbers and for every $n \in \mathbb{N}$ define the operator $G_n: \mathscr{C}([0, 1]) \to \mathscr{C}([0, 1])$ by

$$G_n(f)(x) := (1 + nt_n)^{-n} \sum_{k=0}^{n} f\left(\frac{k}{n}\right)\binom{n}{k} x(x + kt_n)^{k-1}(1 - x + (n - k)t_n)^{n-k} \quad (5.3.7)$$

for every $f \in \mathscr{C}([0, 1])$ and $x \in [0, 1]$.

The operator G_n is called the *n-th Bernstein-Cheney-Sharma operator* and constitutes a different generalization of the Bernstein operator, which can be obtained by taking $t_n = 0$.

In order to study the convergence of the sequence $(G_n)_{n \in \mathbb{N}}$, it is useful to consider the quantities

$$S(j, n, p, x, y) := \sum_{k=0}^{n} \binom{n}{k}(x + kt_p)^{k+j-1}(y + (n - k)t_p)^{n-k}, \qquad (5.3.8)$$

where $x, y \in [0, 1]$, $p \in \mathbb{N}$ and $j = 0, \ldots, n$.

The following identity is an easy consequence of (5.3.8):

$$S(j, n, p, x, y) = xS(j - 1, n, p, x, y) + nt_p S(j, n - 1, p, x + t_p, y), \quad (5.3.9)$$

and it provides us with a reduction formula for the numbers $S(j, n, p, x, y)$.
The Jensen equality (5.3.5) yields

$$xS(0, n, p, x, y) = (x + y + nt_p)^n, \quad (5.3.10)$$

and we can therefore use the reduction formula (5.3.9) to obtain

$$S(1, n, p, x, y) = \sum_{k=0}^{n} \binom{n}{k} k! \, t_p^k (x + y + nt_p)^{n-k}. \quad (5.3.11)$$

Finally, since $k! = \int_0^{+\infty} s^k \exp(-s) \, ds$, by using the binomial theorem, we obtain

$$S(1, n, p, x, y) = \int_0^{+\infty} \exp(-s)(x + y + nt_p + st_p)^n \, ds. \quad (5.3.12)$$

It is possible to obtain in a similar way

$$S(2, n, p, x, y) = \sum_{k=0}^{n} (x + kt_p) \binom{n}{k} k! \, t_p^k S(1, n - k, p, x + kt_p, y) \quad (5.3.13)$$

and hence

$$S(2, n, p, x, y) = \int_0^{+\infty} \exp(-s) \, ds \int_0^{+\infty} \exp(-u) \Big(x(x + y + nt_p + st_p + ut_p)^n$$

$$+ nt_p^2 u(x + y + nt_p + st_p + ut_p)^{n-1} \Big) du. \quad (5.3.14)$$

Now we are in a position to formulate and prove the main result.

5.3.3 Theorem. *If $(t_n)_{n \in \mathbb{N}}$ is a sequence of positive real numbers such that* $\lim_{n \to \infty} n \cdot t_n = 0$, *then, for every* $f \in \mathscr{C}([0, 1])$,

$$\lim_{n \to \infty} G_n(f) = f \quad \text{uniformly on } [0, 1].$$

Proof. Clearly every G_n is positive and $G_n(1) = 1$ by virtue of (5.3.5). Now, for every $x \in [0,1]$,

$$G_n(e_1)(x) = (1 + nt_n)^{-n} \sum_{k=0}^{n} \binom{n}{k} x(x + kt_n)^{k-1}(1 - x + (n - k)t_n)^{n-k}\frac{k}{n}$$

$$= x(1 + nt_n)^{-n} \sum_{h=0}^{n-1} \binom{n-1}{h}(x + t_n + ht_n)^{h}(1 - x + (n - 1 - h)t_n)^{n-1-h}$$

$$= x(1 + nt_n)^{-n}S(1, n - 1, n, x + t_n, 1 - x),$$

and consequently, by (5.3.12),

$$G_n(e_1)(x) = x(1 + nt_n)^{-n} \int_0^{+\infty} \exp(-s)(1 + nt_n + st_n)^{n-1}\, ds$$

$$= \frac{x}{1 + nt_n} \int_0^{+\infty} \exp(-s)\left(1 + \frac{st_n}{1 + nt_n}\right)^{n-1} ds.$$

On the other hand it follows from the inequality $(1 + u)^{n-1} \le \exp((n - 1)u)$ that

$$G_n(e_1)(x) = \frac{x}{1 + nt_n} \int_0^{+\infty} \exp(-s)\left(1 + \frac{st_n}{1 + nt_n}\right)^{n-1} ds$$

$$\le \frac{x}{1 + nt_n} \int_0^{+\infty} \exp\left(-\frac{1 + t_n}{1 + nt_n}s\right) ds \le \frac{x}{1 + nt_n}\frac{1 + nt_n}{1 + t_n} = \frac{x}{1 + t_n},$$

and

$$\frac{x}{1 + nt_n} \le \frac{x}{1 + nt_n} \int_0^{+\infty} \exp(-s)\left(1 + \frac{st_n}{1 + nt_n}\right)^{n-1} ds = G_n(e_1)(x),$$

so that $\dfrac{x}{1 + nt_n} \le G_n(e_1)(x) \le \dfrac{x}{1 + t_n}$.

Hence, since $\lim_{n \to \infty} n \cdot t_n = 0$, we conclude that $\lim_{n \to \infty} G_n(e_1) = e_1$ uniformly on $[0, 1]$.

Finally, consider the function e_2. On account of (5.3.7) we have

$$G_n(e_2)(x) = (1 + nt_n)^{-n} \sum_{k=0}^{n} \binom{n}{k} x(x + kt_n)^{k-1}(1 - x + (n - k)t_n)^{n-k} \frac{k^2}{n^2}$$

$$= (1 + nt_n)^{-n} \sum_{k=0}^{n} \left(\frac{(n-1)}{n} \frac{k(k-1)}{n(n-1)} + \frac{k}{n^2} \right) \binom{n}{k} x(x + kt_n)^{k-1}$$

$$\times (1 - x + (n - k)t_n)^{n-k}$$

$$= \frac{n-1}{n}(1 + nt_n)^{-n}xS(2, n - 2, n, x + 2t_n, 1 - x) + \frac{1}{n}G_n(e_1)(x).$$

On the one hand, $\frac{1}{n}G_n(e_1)(x)$ tends uniformly to 0 and, on the other hand, (5.3.14) yields

$$\frac{n-1}{n}(1 + nt_n)^{-n}xS(2, n - 2, n, x + 2t_n, 1 - x)$$

$$= \frac{n-1}{n}(1 + nt_n)^{-n}x(x + 2t_n) \int_0^{+\infty} \exp(-s)\, ds \int_0^{+\infty} \exp(-u)$$

$$\times (1 + nt_n + st_n + ut_n)^n\, du + (n - 1)(1 + nt_n)^{-n}xt_n^2$$

$$\times \int_0^{+\infty} \exp(-s)\, ds \int_0^{+\infty} u\exp(-u)(1 + nt_n + st_n + ut_n)^{n-1}\, du$$

$$=: \alpha(n, x) + \beta(n, x).$$

Again by using the inequality $(1 + u)^{n-1} \leq \exp((n - 1)u)$ we see that

$$\beta(n, x) \leq \frac{x(n-1)t_n^2}{1 + nt_n} \int_0^{+\infty} \exp(-s)\, ds \int_0^{+\infty} u\exp(-u)\exp\left(\frac{st_n + ut_n}{1 + nt_n}(n - 1)\right) du$$

$$= \frac{x(n-1)t_n^2}{1 + nt_n}\left(\frac{1 + nt_n}{1 + t_n}\right)^4;$$

so, condition $\lim_{n \to \infty} n \cdot t_n = 0$ ensures that $\beta(n, x) \to 0$ uniformly in $x \in [0, 1]$. Moreover, by the inequalities $\exp(nu)(1 - nu^2) \leq (1 + u)^n \leq \exp(nu)$, we obtain

$$\frac{n-1}{n}(1 + nt_n)^2 x(x + 2t_n)(1 - 6nt_n^2) \leq \alpha(n, x) \leq \frac{n-1}{n}x(x + 2t_n)(1 + nt_n)^2$$

and consequently $\alpha(n, x) \to x^2$ uniformly in $x \in [0, 1]$. Summing up we have proved that

$$\lim_{n \to \infty} G_n(e_2) = e_2 \quad \text{uniformly on } [0, 1],$$

and hence an application of Korovkin's theorem 4.2.7 completes the proof. $\square$

Finally, we observe that on the analogy of the construction of the Bernstein-Cheney-Sharma operators which is based on the identity (5.3.5), we can start with alternative identities to consider different sequences of linear operators.

For example, consider the following further equality of Jensen

$$(x + y)(x + y + nt_n)^{n-1} = \sum_{k=0}^{n} \binom{n}{k} x(x + kt_n)^{k-1} y(y + (n - k)t_n)^{n-k-1}, \quad (5.3.15)$$

which holds for every $x, y, t \in \mathbb{R}$ and $n \geq 1$. Given a sequence $(t_n)_{n \in \mathbb{N}}$ of positive real numbers, for every $n \in \mathbb{N}$ we can define the operator $G_n^*: \mathscr{C}([0, 1]) \to \mathscr{C}([0, 1])$ by

$$G_n^*(f)(x) := (1 + nt_n)^{1-n} \sum_{k=0}^{n} f\left(\frac{k}{n}\right) \binom{n}{k} x(x + kt_n)^{k-1}(1 - x)$$

$$\times (1 - x + (n - k)t_n)^{n-k-1} \qquad (5.3.16)$$

for every $f \in \mathscr{C}([0, 1])$ and $x \in [0, 1]$.

The same arguments we used to establish the convergence of the sequence $(G_n)_{n \in \mathbb{N}}$ can be also applied in studying the convergence of $(G_n^*)_{n \in \mathbb{N}}$. As a matter of fact, it is possible to show that if the sequence $(n \cdot t_n)_{n \in \mathbb{N}}$ converges to 0, then $\lim_{n \to \infty} G_n^*(f) = f$ for every $f \in \mathscr{C}([0, 1])$ (see Cheney and Sharma [1964a] for more details).

A much more complicated (but interesting) example of positive approximation process is the next one which is also due to Cheney and Sharma and is constructed by means of Laguerre polynomials.

5.3.4 Cheney-Sharma operators. Let $\alpha > -1$ and consider the weight function $v: \mathbb{R} \to \mathbb{R}$ which vanishes on $]-\infty, 0]$ and takes the value $v(x) = x^\alpha e^{-x}$ at $x > 0$. The sequence $(\lambda_n^{(\alpha)})_{n \in \mathbb{N}_0}$ of *Laguerre polynomials* is defined as the orthogonal polynomial sequence with respect to the weight function v and such that for every $n \in \mathbb{N}_0$ the leading coefficient of $\lambda_n^{(\alpha)}$ (i.e., the coefficient of x^n) is $(-1)^n/n!$.

Thus every $\lambda_n^{(\alpha)}$ is a polynomial of degree n and for every $n, m \in \mathbb{N}_0$,

$$\int_0^{+\infty} \lambda_n^{(\alpha)}(x)\lambda_m^{(\alpha)}(x)x^\alpha e^{-x}\,dx = \begin{cases} 0, & \text{if } n \neq m, \\ \dfrac{\Gamma(n+\alpha+1)}{n!}, & \text{if } n = m. \end{cases} \quad (5.3.17)$$

For more information regarding the construction of the sequence $(\lambda_n^{(\alpha)})_{n \in \mathbb{N}_0}$ we refer, e.g., to Chihara [1978, p. 9]. Here we simply recall some properties of $(\lambda_n^{(\alpha)})_{n \in \mathbb{N}_0}$ which are useful in the sequel. First we recall the following representation formula:

$$\lambda_n^{(\alpha)}(x) = \frac{x^{-\alpha}\exp(x)}{n!}\frac{d^n}{dx^n}(\exp(-x)x^{n+\alpha}), \quad x \geq 0. \quad (5.3.18)$$

Moreover the next identities can be easily proved (see, e.g., Chihara [1978, pp. 43–44 and p. 155]):
(1) For every $n \in \mathbb{N}_0$ and $x \geq 0$

$$(n+1)\lambda_{n+1}^{(\alpha)}(x) = (n+\alpha+1)\lambda_n^{(\alpha)}(x) - x\lambda_n^{(\alpha+1)}(x). \quad (5.3.19)$$

(2) For every $n \geq 2$ and $x \geq 0$

$$n\lambda_n^{(\alpha)}(x) = (-x+2n+\alpha-1)\lambda_{n-1}^{(\alpha)}(x) - (n+\alpha-1)\lambda_{n-2}^{(\alpha)}(x). \quad (5.3.20)$$

(3) For every $\alpha > -1, 0 \leq x < 1$ and $t \in \mathbb{R}$

$$(1-x)^{\alpha+1}\exp\left(\frac{tx}{1-x}\right)\sum_{k=0}^{\infty} x^k \lambda_k^{(\alpha)}(t) = 1. \quad (5.3.21)$$

The preceding equality is the starting point for defining the sequence of positive linear operators which we shall study in this example. We fix $a \in \,]0, 1[$ and we consider a sequence $(t_n)_{n \in \mathbb{N}}$ of negative real numbers. Given $n \in \mathbb{N}$, we define the operator $Q_n \colon \mathscr{C}([0,1]) \to \mathscr{C}([0,a])$ by

$$Q_n(f)(x) := (1-x)^{n+1}\exp\left(\frac{t_n x}{1-x}\right)\sum_{k=0}^{\infty} \lambda_k^{(n)}(t_n)f\left(\frac{k}{n+k}\right)x^k \quad (5.3.22)$$

for every $f \in \mathscr{C}([0,1])$ and $x \in [0,a]$.

The operator Q_n is called the *n-th Cheney-Sharma operator*. Since the Laguerre polynomial is positive on $]-\infty, 0]$, the Cheney-Sharma operators are obviously positive.

Conditions ensuring the convergence of the sequence $(Q_n)_{n \in \mathbb{N}}$ are stated in the next result.

5.3.5 Theorem. *If* $(t_n)_{n \in \mathbb{N}}$ *is a sequence of negative real numbers such that* $\lim\limits_{n \to \infty} \dfrac{t_n}{n} = 0$, *then, for every* $f \in \mathscr{C}([0,1])$,

$$\lim_{n \to \infty} Q_n(f) = f \quad \textit{uniformly on } [0,a]. \tag{5.3.23}$$

Proof. We shall apply Korovkin's theorem 4.2.7 and the universal Korovkin-type property with respect to positive linear operators (see Theorem 4.1.4, (vi)).
On account of (5.3.21), we clearly have

$$Q_n(1) = 1$$

for every $n \geq 1$. By (5.3.19) we obtain, for given $x \in [0,a]$ and $k \geq 1$,

$$k\lambda_k^{(n)}(t_n) = (k + n)\lambda_{k-1}^{(n)}(t_n) - t_n\lambda_{k-1}^{(n+1)}(t_n)$$

and consequently (see (5.3.22))

$$Q_n(e_1)(x) = (1 - x)^{n+1} \exp\left(\frac{t_n x}{1 - x}\right) \sum_{k=1}^{\infty} \frac{k}{n + k} x^k \lambda_k^{(n)}(t_n)$$

$$= (1 - x)^{n+1} \exp\left(\frac{t_n x}{1 - x}\right) \sum_{k=1}^{\infty} x^k \left(\lambda_{k-1}^{(n)}(t_n) - \frac{t_n}{n + k}\lambda_{k-1}^{(n+1)}(t_n)\right)$$

$$= (1 - x)^{n+1} \exp\left(\frac{t_n x}{1 - x}\right)$$

$$\times \left(x \sum_{k=0}^{\infty} x^k \lambda_k^{(n)}(t_n) - t_n x \sum_{k=0}^{\infty} x^k \frac{1}{n + k + 1}\lambda_k^{(n+1)}(t_n)\right)$$

$$= x - \frac{t_n x}{(n + 1)(1 - x)}.$$

Hence $\lim\limits_{n \to \infty} Q_n(e_1) = e_1$ uniformly on $[0,a]$, because $\lim\limits_{n \to \infty} \dfrac{t_n}{n} = 0$.
To evaluate $Q_n(e_2)$ it is convenient to observe that a repeated application of the equality (5.3.19) yields for $k \geq 2$

$$\left(\frac{k}{n + k}\right)^2 \lambda_k^{(n)}(t_n) = \frac{n + k - 1}{n + k}\lambda_{k-2}^{(n)}(t_n) - \frac{t_n}{n + k}\lambda_{k-2}^{(n+1)}(t_n) - \frac{kt_n}{(n + k)^2}\lambda_{k-1}^{(n+1)}(t_n).$$

By (5.3.22), we have

$$Q_n(e_2)(x) = (1-x)^{n+1} \exp\left(\frac{t_n x}{1-x}\right) \sum_{k=1}^{\infty} \left(\frac{k}{n+k}\right)^2 x^k \lambda_k^{(n)}(t_n)$$

$$=: \alpha(n,x) - \beta(n,x) - \gamma(n,x) + \delta(n,x),$$

where

$$\alpha(n,x) := (1-x)^{n+1} \exp\left(\frac{t_n x}{1-x}\right) \sum_{k=2}^{\infty} x^k \frac{n+k-1}{n+k} \lambda_{k-2}^{(n)}(t_n),$$

$$\beta(n,x) := (1-x)^{n+1} \exp\left(\frac{t_n x}{1-x}\right) \sum_{k=2}^{\infty} x^k \frac{t_n}{n+k} \lambda_{k-2}^{(n+1)}(t_n),$$

$$\gamma(n,x) := (1-x)^{n+1} \exp\left(\frac{t_n x}{1-x}\right) \sum_{k=2}^{\infty} x^k \frac{kt_n}{(n+k)^2} \lambda_{k-1}^{(n+1)}(t_n),$$

$$\delta(n,x) := (1-x)^{n+1} \exp\left(\frac{t_n x}{1-x}\right) \frac{1}{(n+1)^2} x \lambda_1^{(n)}(t_n).$$

Using (5.3.21) we can majorize $\gamma(n,x)$ as follows:

$$|\gamma(n,x)| = |t_n| x (1-x)^{n+1} \exp\left(\frac{t_n x}{1-x}\right) \sum_{k=0}^{\infty} x^k \frac{k+1}{(n+k+1)^2} \lambda_k^{(n+1)}(t_n)$$

$$\leq \frac{|t_n| x}{n(1-x)} (1-x)^{n+2} \exp\left(\frac{t_n x}{1-x}\right) \sum_{k=0}^{\infty} x^k \lambda_k^{(n+1)}(t_n) = \frac{|t_n| x}{n(1-x)}.$$

In a similar way we obtain

$$|\beta(n,x)| \leq \frac{|t_n| x^2}{n(1-x)} (1-x)^{n+2} \exp\left(\frac{t_n x}{1-x}\right) \sum_{k=0}^{\infty} x^k \lambda_k^{(n+1)}(t_n) = \frac{|t_n| x^2}{n(1-x)}$$

and $|\delta(n,x)| \leq \dfrac{1}{(n+1)^2}$.

Finally, since $1 - \dfrac{1}{n} \leq \dfrac{n+k-1}{n+k} \leq 1$, the inequalities $x^2\left(1 - \dfrac{1}{n}\right) \leq \alpha(n,x) \leq$
x^2 also hold and therefore we conclude that $\lim_{n\to\infty} Q_n(e_2) = e_2$ uniformly on $[0,a]$.
Then we have only to apply Theorem 4.2.7 to complete the proof. $\square$

From the expression for $Q_n(e_1)$ we deduce that Cheney-Sharma operators map linear functions into linear functions if and only if $t_n = 0$ for every $n \in \mathbb{N}$. This case is of particular interest and the corresponding operators are called the *Meyer-König and Zeller operators*. We shall denote them by R_n. Thus every operator $R_n \colon \mathscr{C}([0, 1]) \to \mathscr{C}([0, 1])$ is defined by

$$R_n(f)(x) := (1 - x)^{n+1} \sum_{k=0}^{\infty} f\left(\frac{k}{n+k}\right)\binom{n+k}{k} x^k \qquad (5.3.24)$$

for every $f \in \mathscr{C}([0, 1])$ and $x \in [0, 1]$.

We can see directly that $(R_n)_{n \in \mathbb{N}}$ is an approximation process. Indeed

$$R_n(1) = 1,$$

$$R_n(e_1)(x) = x,$$

$$R_n(e_2)(x) = x^2 + \eta_n(x)$$

for every $x \in [0, 1]$, where

$$\eta_n(x) := x(1 - x)^{n+1} \sum_{k=0}^{\infty} \binom{n+k-1}{k} \frac{x^k}{n+k+1} \leq \frac{x(1 - x)}{n+1}. \qquad (5.3.25)$$

Hence Theorem 4.2.7 yields

$$\lim_{n \to \infty} R_n(f) = f \quad \text{uniformly on } [0, 1] \qquad (5.3.26)$$

for every $f \in \mathscr{C}([0, 1])$.

In this particular case, it is easy to obtain a quantitative estimate for the rate of convergence. Indeed, considering the function $\psi_x(t) = t - x$ $(x, t \in [0, 1])$, we have

$$R_n(\psi_x)(x) = R_n(e_1 - x \cdot 1)(x) = 0$$

and

$$R_n(\psi_x^2)(x) = R_n(e_2)(x) - 2xR_n(e_1)(x) + x^2 R_n(1)(x)$$

$$= x^2 + \eta_n(x) - 2x^2 + x^2 = \eta_n(x).$$

Hence, applying (5.1.5) with $\delta = \sqrt{\eta_n(x)}$, we obtain

$$|R_n(f)(x) - f(x)| \leq 2\omega(f, \sqrt{\eta_n(x)}) \qquad (5.3.27)$$

for every $n \in \mathbb{N}$ and $f \in \mathscr{C}([0, 1])$.

Moreover, if f is differentiable on $[0,1]$ and $f' \in \mathscr{C}([0,1])$, then

$$|R_n(f)(x) - f(x)| \leq 2\sqrt{\eta_n(x)}\omega(f', \sqrt{\eta_n(x)}) \qquad (5.3.28)$$

on account of (5.1.6) applied again to $\delta = \sqrt{\eta_n(x)}$).

Finally, we point out the following bound for the numbers $\eta_n(x)$, which is given in Sikkema [1970b]:

$$\eta_n(x) \leq \frac{x(1-x)^2}{n+1} + \frac{x^2(1-x)(2-x)}{(n+1)^2}.$$

Another interesting sequence of positive linear operators is that of Fejér-Hermite operators that are constructed by means of Chebyshev polynomials. Among other things, the main significance of these operators rests on the fact that the Hermite interpolation procedure (see below) furnishes approximations to all continuous functions.

5.3.6 Fejér-Hermite operators. For a given $n \in \mathbb{N}$ consider the *Chebyshev polynomial* $T_n(x) := \cos(n\arccos x)$ $(x \in [-1,1])$ and, for every $k = 1, \ldots, n$, denote by $x_{k,n} := \cos\left(\dfrac{2k-1}{2n}\pi\right)$ the zeros of T_n. The *n-th Fejér-Hermite operator* H_n^*: $\mathscr{C}([-1,1]) \to \mathscr{C}([-1,1])$ is defined by

$$H_n^*(f)(x) := \sum_{k=1}^{n} f(x_{k,n})(1-x_{k,n}x)\left(\frac{T_n(x)}{n(x-x_{k,n})}\right)^2 \qquad (5.3.29)$$

for every $f \in \mathscr{C}([-1,1])$ and $x \in [-1,1]$.

Thus, $H_n^*(f)$ is a polynomial of degree $2n-1$. It interpolates the values of f at each $x_{k,n}$ and has derivative equal to zero at these points.

We have

$$H_n^*(1) = 1$$

and, for every $x \in [-1,1]$,

$$H_n^*(e_1)(x) - x = \sum_{k=1}^{n} (x_{k,n} - x)(1-x_{k,n}x)\left(\frac{T_n(x)}{n(x-x_{k,n})}\right)^2$$

$$= -\sum_{k=1}^{n} ((1-x^2)(x-x_{k,n}) + x(x-x_{k,n})^2)\left(\frac{T_n(x)}{n(x-x_{k,n})}\right)^2$$

$$= -\frac{1}{n}xT_n^2(x) - \frac{1}{n^2}(1-x^2)T_n^2(x)\sum_{k=1}^{n}\frac{1}{x-x_{k,n}}$$

$$= -\frac{1}{n^2}(1-x^2)T_n(x)T_n'(x) - \frac{xT_n^2(x)}{n}.$$

Since

$$|(1 - x)T_n'(x)| = |\sqrt{1 - x^2}n\sin(n\arccos x)| \leq n,$$

we deduce that $\lim_{n\to\infty} H_n^*(e_1) = e_1$ uniformly on $[-1, 1]$.

Finally, since the points $x_{k,n}$ are symmetric about the origin, we have, for every $x \in [-1, 1]$,

$$H_n^*(\psi_x^2)(x) = \sum_{k=1}^{n} (x_{k,n} - x)^2(1 - x_{k,n}x)\left(\frac{T_n(x)}{n(x - x_{k,n})}\right)^2$$

$$= \frac{1}{n^2}T_n^2(x) \sum_{k=1}^{n} (1 - x_{k,n}x) = \frac{1}{n}T_n^2(x),$$

and therefore, on account of Theorem 4.4.6, Example 6, we conclude

$$\lim_{n\to\infty} H_n^*(f) = f \quad \text{for every } f \in \mathscr{C}([-1, 1]). \tag{5.3.30}$$

Moreover, by using the estimates of the Remark to Theorem 5.1.2, we have that

$$|H_n^*(f)(x) - f(x)| \leq 2\omega\left(f, \frac{|T_n(x)|}{\sqrt{n}}\right) \tag{5.3.31}$$

for every $n \in \mathbb{N}$, $f \in \mathscr{C}([-1, 1])$ and $x \in [-1, 1]$; furthermore, if f is differentiable with $f' \in \mathscr{C}([-1, 1])$, then

$$|H_n^*(f)(x) - f(x)| \leq 2|T_n(x)|\left(\frac{|f'(x)|}{n} + \frac{1}{\sqrt{n}}\omega\left(f', \frac{|T_n(x)|}{\sqrt{n}}\right)\right). \tag{5.3.32}$$

Finally we point out the following estimate due to Bojanic [1969]:

$$\|H_n^*(f) - f\| \leq \frac{C}{n} \sum_{k=1}^{n} \omega\left(f, \frac{1}{k}\right), \tag{5.3.33}$$

where C is a suitable constant.

The next example of operators originates from the need to modify classical Bernstein operators (which are not suitable for approximating discontinuous functions) in order to obtain an approximation process in spaces of integrable functions.

5.3.7 Kantorovich operators. For every $n \in \mathbb{N}$ we define the *n-th Kantorovich operator* $U_n: L^1([0,1]) \to \mathscr{C}([0,1])$ by

$$U_n(f)(x) := \sum_{k=0}^{n} (n+1) \left(\int_{k/(n+1)}^{(k+1)/(n+1)} f(t) \, dt \right) \binom{n}{k} x^k (1-x)^{n-k} \quad (5.3.34)$$

for every $f \in L^1([0,1])$ and $x \in [0,1]$.

Note in particular that, if $f \in \mathscr{C}^1([0,1])$, then

$$U_n(f') = (B_{n+1}(f))', \quad (5.3.35)$$

(see Lorentz [1986b, p. 30]), where B_{n+1} is the $(n+1)$-th Bernstein operator.

Also in this case, we obtain a sequence $(U_n)_{n \in \mathbb{N}}$ of positive linear operators and we can study its convergence by evaluating them at the functions $\mathbf{1}$, e_1 and e_2.

In fact for every $n \in \mathbb{N}$ and $x \in [0,1]$ (see also (2.5.8)–(2.5.10)) we have

$$U_n(\mathbf{1})(x) = \sum_{k=0}^{n} \binom{n}{k} x^k (1-x)^{n-k} = 1,$$

$$U_n(e_1)(x) = \sum_{k=0}^{n} \binom{n}{k} x^k (1-x)^{n-k} \frac{2k+1}{2(n+1)} = \frac{n}{n+1} x + \frac{1}{2(n+1)},$$

and finally

$$U_n(e_2)(x) = \sum_{k=0}^{n} \binom{n}{k} x^k (1-x)^{n-k} \frac{3k^2 + 3k + 1}{3(n+1)^2}$$

$$= \sum_{k=0}^{n} \binom{n}{k} x^k (1-x)^{n-k} \frac{k^2}{(n+1)^2} + \sum_{k=0}^{n} \binom{n}{k} x^k (1-x)^{n-k} \frac{k}{(n+1)^2}$$

$$+ \sum_{k=0}^{n} \binom{n}{k} x^k (1-x)^{n-k} \frac{1}{3(n+1)^2}$$

$$= \frac{n(n-1)x^2 + nx}{(n+1)^2} + \frac{nx}{(n+1)^2} + \frac{1}{3(n+1)^2}$$

$$= \frac{n(n-1)}{(n+1)^2} x^2 + 2 \frac{n}{(n+1)^2} x + \frac{1}{3(n+1)^2}.$$

By the above formulas, we easily obtain

$$\lim_{n \to \infty} U_n(\mathbf{1}) = \mathbf{1}, \quad \lim_{n \to \infty} U_n(e_1) = e_1, \quad \lim_{n \to \infty} U_n(e_2) = e_2,$$

in $\mathscr{C}([0,1])$ and hence in $L^p([0,1])$ for every $p \geq 1$. Therefore, by Korovkin's theorem 4.2.7 and Corollary 4.1.7, we obtain

$$\lim_{n\to\infty} U_n(f) = f \quad \text{uniformly on } [0,1] \tag{5.3.36}$$

for every $f \in \mathscr{C}([0,1])$ as well as

$$\lim_{n\to\infty} U_n(f) = f \quad \text{in } L^p([0,1]) \tag{5.3.37}$$

for every $f \in L^p([0,1])$ and $p \geq 1$.

As in the preceding examples, we can give a quantitative estimate for the rate of convergence by evaluating, for every $n \in \mathbb{N}$ and $x \in [0,1]$, the quantities

$$U_n(\psi_x)(x) = U_n(e_1 - x \cdot 1)(x) = \frac{1}{2(n+1)} - \frac{1}{n+1}x$$

and

$$U_n(\psi_x^2)(x) = U_n(e_2)(x) - 2xU_n(e_1)(x) + x^2 U_n(1)(x)$$

$$= \frac{n(n-1)}{(n+1)^2}x^2 + 2\frac{n}{(n+1)^2}x - 2\frac{n}{n+1}x^2 - \frac{1}{n+1}x + x^2$$

$$= \frac{n-1}{(n+1)^2}x(1-x).$$

Thus, if we set $\delta = \dfrac{\sqrt{(n-1)x(1-x)}}{n+1}$ in (5.1.5), we have, for every $f \in \mathscr{C}([0,1])$ and $x \in [0,1]$,

$$|U_n(f)(x) - f(x)| \leq 2\omega\left(f, \frac{\sqrt{(n-1)x(1-x)}}{n+1}\right). \tag{5.3.38}$$

Moreover, if f is differentiable and $f' \in \mathscr{C}([0,1])$, then

$$|U_n(f)(x) - f(x)| \leq \frac{|1-2x|}{n+1}|f'(x)|$$

$$+ 2\frac{\sqrt{(n-1)x(1-x)}}{n+1}\omega\left(f', \frac{\sqrt{(n-1)x(1-x)}}{n+1}\right). \tag{5.3.39}$$

For another finer estimate we refer to Ditzian [1985] who proved that

$$|U_n(f)(x) - f(x)| \le M\omega_2\left(f, \sqrt{\frac{x(1-x)}{n} + \frac{L}{n^2}}\right) + \omega\left(f, \frac{|1 - 2x|}{n}\right), \quad (5.3.40)$$

where L and M are suitable constants.

Finally, by using a general result by Sendov and Popov [1988, Theorem 4.3], one can obtain the estimate

$$\|U_n(f) - f\|_p \le 748\tau(f, \sqrt{1/(n+1)})_p \qquad (5.3.41)$$

for every $f \in \mathcal{B}_0([0, 1])$ and $1 \le p < +\infty$, where

$$\tau(f, \delta)_p := \left(\int_0^1 \omega(f, x, \delta)^p \, dx\right)^{1/p} \qquad (5.3.42)$$

is the *averaged modulus of smoothness and*

$$\omega(f, x, \delta) := \sup\{|f(t + h) - f(t)| \, | t, t + h \in [x - \delta/2, x + \delta/2] \cap [0, 1]\}. \qquad (5.3.43)$$

For the sake of brevity we omit the details.

Further L^p-estimates can be found in Totik [1983b].

Another interesting integral modification of Bernstein polynomials was proposed by Durrmeyer [1967]. These modified operators were first systematically studied by Derrienic [1981] and continue to draw the attention of several mathematicians (see the final Notes).

5.3.8 Bernstein-Durrmeyer operators. Given $n \ge 1$, define for every $f \in L^1([0, 1])$ and $x \in [0, 1]$

$$D_n(f)(x) := (n + 1) \sum_{k=0}^{n} \left(\int_0^1 \binom{n}{k} t^k(1 - t)^{n-k} f(t) \, dt\right) \binom{n}{k} x^k (1 - x)^{n-k}. \quad (5.3.44)$$

Then D_n is a positive linear operator from $L^1([0, 1])$ into $\mathscr{C}([0, 1])$ and is called the *n-th Bernstein-Durrmeyer operator.*

The operators D_n enjoy several interesting properties. Among other things, we point out that they are self-adjoint operators on $L^2([0, 1])$; furthermore, if for every $m \in \mathbb{N}_0$ we set

$$\lambda_{n,m} := \begin{cases} \dfrac{(n+1)!n!}{(n+m+1)!(n-m)!} & \text{if } m \le n, \\[4mm] 0, & \text{if } m > n, \end{cases} \qquad (5.3.45)$$

then every $\lambda_{n,m}$ is an eigenvalue for D_n with corresponding eigenfunction the Legendre polynomial Q_m of degree m. Consequently

$$D_n(f) = \sum_{m=0}^{n} \lambda_{n,m} \left(\int_0^1 f(x) Q_m(x) \, dx \right) Q_m \qquad (5.3.46)$$

(see Derrienic [1981] for more details).

To study the behavior of the sequence $(D_n)_{n \in \mathbb{N}}$ on the functions $\mathbf{1}$, e_1 and e_2 we need the following identities

$$\int_0^1 \binom{n}{k} t^k (1-t)^{n-k} t^m \, dt = \frac{(k+m)!}{k!} \frac{n!}{(n+m+1)!} \qquad (5.3.47)$$

$(n, m, k \in \mathbb{N}_0, 0 \leq k \leq n)$ which follow from the elementary properties of the *beta function*

$$B(u, v) := \int_0^1 t^{u-1}(1-t)^{v-1} \, dt = \frac{\Gamma(u)\Gamma(v)}{\Gamma(u+v)} \qquad (5.3.48)$$

$(u > 0, v > 0)$.

By using (5.3.46) and the identities (2.5.9) and (2.5.10), we obtain for every $x \in [0, 1]$ and $n \geq 1$

$$D_n(\mathbf{1})(x) = (n+1) \sum_{k=0}^{n} \binom{n}{k} x^k (1-x)^{n-k} \frac{1}{n+1} = 1,$$

$$D_n(e_1)(x) = (n+1) \sum_{k=0}^{n} \binom{n}{k} x^k (1-x)^{n-k} \frac{k+1}{(n+1)(n+2)}$$

$$= \frac{n}{n+2} x + \frac{1}{n+2},$$

and finally

$$D_n(e_2)(x) = (n+1) \sum_{k=0}^{n} \binom{n}{k} x^k (1-x)^{n-k} \frac{(k+1)(k+2)}{(n+1)(n+2)(n+3)}$$

$$= \frac{1}{(n+2)(n+3)} \sum_{k=0}^{n} (k^2 + 3k + 2) \binom{n}{k} x^k (1-x)^{n-k}$$

$$= \frac{1}{(n+2)(n+3)} \left(n^2 \left(\frac{n-1}{n} x^2 + \frac{x}{n} \right) + 3nx + 2 \right)$$

$$= \frac{n(n-1)}{(n+2)(n+3)} x^2 + \frac{4n}{(n+2)(n+3)} x + \frac{2}{(n+2)(n+3)}.$$

Hence we see that $D_n(1) \to 1$, $D_n(e_1) \to e_1$ and $D_n(e_2) \to e_2$ both in $\mathscr{C}([0,1])$ and in $L^p([0,1])$, $p \geq 1$.

According to Theorem 4.2.7 and Corollary 4.1.7 we have that

$$\lim_{n \to \infty} D_n(f) = f \quad \text{uniformly on } [0,1] \tag{5.3.49}$$

if $f \in \mathscr{C}([0,1])$, or

$$\lim_{n \to \infty} D_n(f) = f \quad \text{in } L^p([0,1]) \tag{5.3.50}$$

provided $f \in L^p([0,1])$, $p \geq 1$.

Estimates of the rate of the above convergence can be obtained from Theorem 5.1.2 and the subsequent remark.

In fact, by using the preceding formulas, we have for every $x \in [0,1]$ and $n \geq 3$

$$D_n(\psi_x^2)(x) = \frac{2(n-3)x(1-x)+2}{(n+2)(n+3)} \leq \frac{n+1}{2(n+2)(n+3)} \leq \frac{1}{n},$$

so that, by setting $\delta = \sqrt{D_n(\psi_x^2)(x)}$ in (5.1.5), we obtain

$$|D_n(f)(x) - f(x)| \leq 2\omega\left(f, \sqrt{\frac{2(n-3)x(1-x)+2}{(n+2)(n+3)}}\right) \tag{5.3.51}$$

or, respectively,

$$\|D_n(f) - f\| \leq 2\omega(f, \sqrt{1/n}) \tag{5.3.52}$$

for every $f \in \mathscr{C}([0,1])$.

If $f \in \mathscr{B}_0([0,1])$ and $1 \leq p < +\infty$, by using an estimate of Sendov and Popov [1988, Theorem 4.3] one can obtain the same estimate as in (5.3.41), i.e.,

$$\|D_n(f) - f\|_p \leq 748\tau(f, \sqrt{1/(n+1)})_p. \tag{5.3.53}$$

For further L^p-estimates see also Ditzian and Ivanov [1989].

The second part of this section will be devoted to some examples of approximation processes on unbounded intervals.

In this respect in Section 5.2 we have already furnished some examples such as Examples 5.2.6, 5.2.8, 5.2.9 and 5.2.10. There we used Feller's theorem as main tool.

Here we shall apply Korovkin-type results for unbounded intervals, namely Proposition 4.2.5.

We consider the Banach lattice

$$E_2 := \left\{ f \in \mathscr{C}([0, +\infty[) \,\middle|\, \frac{f(x)}{1+x^2} \text{ is convergent as } x \to \infty \right\} \quad (5.3.54)$$

endowed with the norm

$$\|f\|_* := \sup_{x \geq 0} \frac{|f(x)|}{1+x^2}. \quad (5.3.55)$$

In this space the set $\{1, e_1, e_2\}$ is a K_+-subset (see Proposition 4.2.5, (6)). Note also that in the same Proposition 4.2.5 we showed that E_2 is isomorphic to $\mathscr{C}([0, 1])$.

5.3.9 Schurer-Szász-Mirakjan operators. Fix $p \in \mathbb{N}_0$ and consider the linear operator $M_{n,p} \colon E_2 \to \mathscr{C}([0, +\infty[)$ defined by

$$M_{n,p}(f)(x) := \exp(-(n+p)x) \sum_{k=0}^{\infty} f\left(\frac{k}{n}\right) \frac{(n+p)^k x^k}{k!} \quad (5.3.56)$$

for every $f \in E_2$, $x \in [0, +\infty[$ and $n \geq 1$.

The operator $M_{n,p}$ is called the *n-th Schurer-Szász-Mirakjan operator* and it is a generalization of the classical *n-th Szász-Mirakjan operator* M_n which can be obtained by taking $p = 0$, i.e.,

$$M_n(f)(x) := \exp(-nx) \sum_{k=0}^{\infty} f\left(\frac{k}{n}\right) \frac{n^k x^k}{k!} \quad (5.3.57)$$

for $f \in E_2$ and $x \geq 0$.

Note that the series on the right-hand side of (5.3.56) (and (5.3.57)) is absolutely convergent because for every $f \in E_2$ we have

$$\sum_{k=0}^{\infty} \frac{(n+p)^k x^k}{k!} \left| f\left(\frac{k}{n}\right) \right|$$

$$\leq \|f\|_* \sum_{k=0}^{\infty} \frac{(n+p)^k x^k}{k!} \left(1 + \frac{k^2}{n^2} \right)$$

$$= \|f\|_* \left(\exp((n+p)x) + \frac{1}{n^2} \exp((n+p)x)((n+p)^2 x^2 + (n+p)x) \right). \quad (5.3.58)$$

Furthermore we also point out that every $M_{n,p}$ maps $\mathscr{C}_b([0, +\infty[)$ (respectively, $\mathscr{C}_0([0, +\infty[))$ into itself.

Now, fix $b > 0$ and consider the lattice homomorphism $T_b\colon \mathscr{C}([0, +\infty[) \to \mathscr{C}([0, b])$ defined by

$$T_b(f) := f_{|[0,b]} \qquad (5.3.59)$$

for every $f \in \mathscr{C}([0, +\infty[)$.

Since for every $x \geq 0$

$$M_{n,p}(\mathbf{1})(x) = 1,$$

$$M_{n,p}(e_1)(x) = x + \frac{p}{n}x,$$

$$M_{n,p}(e_2)(x) = \left(\frac{n+p}{n}\right)^2 x^2 + \frac{n+p}{n^2}x$$

we see that $T_b(M_{n,p}(\mathbf{1})) \to T_b(\mathbf{1})$, $T_b(M_{n,p}(e_1)) \to T_b(e_1)$ and $T_b(M_{n,p}(e_2)) \to T_b(e_2)$ uniformly on $[0, b]$. Hence the above mentioned Korovkin-type theorem in E_2 and the universal Korovkin-type property (vi) of Theorem 4.1.4 imply that

$$\lim_{n \to \infty} M_{n,p}(f) = f \quad \text{uniformly on } [0, b] \qquad (5.3.60)$$

provided $f \in E_2$ and $b > 0$.

Moreover for every $x \in [0, b]$ one has

$$M_{n,p}(\psi_x)(x) = M_{n,p}(e_1 - x \cdot \mathbf{1})(x) = \frac{p}{n}x$$

and

$$M_{n,p}(\psi_x^2)(x) = M_{n,p}(e_2)(x) - 2xM_{n,p}(e_1)(x) + x^2 M_{n,p}(\mathbf{1})(x)$$

$$= \left(\frac{n+p}{n}\right)^2 x^2 + \frac{n+p}{n^2}x - 2x^2 - \frac{2p}{n}x^2 + x^2$$

$$= \frac{p^2}{n^2}x^2 + \frac{n+p}{n^2}x.$$

Setting $\delta = \sqrt{\dfrac{p^2x^2 + (n + p)x}{n^2}}$ in (5.1.5), we obtain

$$|M_{n,p}(f)(x) - f(x)| \le 2\omega\left(f, \sqrt{\frac{p^2x^2 + (n + p)x}{n^2}}\right) \tag{5.3.61}$$

for every $f \in \mathscr{C}_b([0, +\infty[)$; if $p = 0$, then (5.3.61) reduces to the following estimate

$$|M_n(f)(x) - f(x)| \le 2\omega(f, \sqrt{x/n}), \tag{5.3.62}$$

which was first obtained by Stancu [1969b].

Moreover, if f is differentiable on $[0, +\infty[$ and $f' \in \mathscr{C}_b([0 +\infty[)$, then for every $x \ge 0$ we obtain $\left(\text{see (5.1.6) with } \delta = \sqrt{\dfrac{p^2x^2 + (n + p)x}{n^2}}\right)$

$$|M_{n,p}(f)(x) - f(x)| \le \frac{p}{n}|f'(x)|x + 2\sqrt{\frac{p^2x^2 + (n + p)x}{n^2}}$$

$$\times \omega\left(f', \sqrt{\frac{p^2x^2 + (n + p)x}{n^2}}\right). \tag{5.3.63}$$

In order to have results about uniform convergence on $\mathbb{R}_+$ of the sequence $(M_{n,p}(f))_{n \in \mathbb{N}}$, we have to restrict to a subspace of E_2.

For this purpose it is sufficient to consider the subspace

$$\mathscr{C}^*([0, +\infty[) := \{f \in \mathscr{C}([0, +\infty[) | f(x) \text{ is convergent as } x \to +\infty\} \tag{5.3.64}$$

endowed with the sup-norm.

For a given $\lambda > 0$ consider the function $f_\lambda(x) := \exp(-\lambda x)$ $(x \ge 0)$. Then for every $n \ge 1$ and $x \ge 0$

$$M_{n,p}(f_\lambda)(x) = \exp(-(n + p)x) \sum_{k=0}^{\infty} \frac{(n + p)^k x^k}{k!} \exp\left(\frac{-\lambda k}{n}\right)$$

$$= \exp(-(n + p)x) \sum_{k=0}^{\infty} \frac{((n + p)x \exp(-\lambda/n))^k}{k!}$$

$$= \exp\left(-\lambda x \frac{1 - \exp(-\lambda/n)}{\lambda/(n + p)}\right).$$

Accordingly, $M_{n,p}(f_\lambda) \to f_\lambda$ uniformly on $[0, +\infty[$. Then we can apply Proposition 4.2.5, (7), to obtain that, for every $f \in \mathscr{C}^*([0, +\infty[)$,

$$\lim_{n \to \infty} M_{n,p}(f) = f \quad \text{uniformly on } [0, +\infty[. \tag{5.3.65}$$

As a matter of fact, for $p = 0$ Totik [1984] showed that (5.3.65) holds for a given function $f \in \mathscr{C}_b([0, +\infty[)$ if and only if the function $f(x^2)$ $(x \geq 0)$ is uniformly continuous on $[0, +\infty[$.

Finally, we point out that the operators $M_{n,p}$ have a natural probabilistic interpretation. In fact they are the positive operators associated with a random scheme $Z: \mathbb{N} \times [0, +\infty[\to M_2(\Omega)$ where $Z(n,x) := \dfrac{1}{n} Y_p(n,x)$ and $Y_p(n,x)$ is a Poisson random variable with parameter $(n + p)x$ $(x \geq 0, n \geq 1)$.

We have already encountered Baskakov operators in Section 5.2 (Example 5.2.6). Here we want to consider a different generalization of these operators which depends on a parameter $p \in \mathbb{N}_0$.

5.3.10 Baskakov-Schurer operators. Given $p \in \mathbb{N}_0$, we define the operator $A_{n,p}$: $E_2 \to \mathscr{C}([0, +\infty[)$ by

$$A_{n,p}(f)(x) := (1 + x)^{-n-p} \sum_{k=0}^{\infty} f\left(\frac{k}{n}\right)\binom{n + p + k - 1}{k}\left(\frac{x}{1 + x}\right)^k$$

$$= \sum_{k=0}^{\infty} f\left(\frac{k}{n}\right)\binom{-(n + p)}{k}(-x)^k(1 + x)^{-(n+p)-k}, \tag{5.3.66}$$

for every $f \in E_2$, $x \in [0, +\infty[$ and $n \geq 1$, where E_2 is defined in (5.3.54).

The operator $A_{n,p}$ is called the *n-th Baskakov-Schurer operator* (the classical *n*-th Baskakov operator can be obtained by taking $p = 0$).

Furthermore, $A_{n,p}$ maps $\mathscr{C}_b([0, +\infty[)$ in $\mathscr{C}_b([0, +\infty[)$ because of the identity

$$\sum_{k=0}^{\infty}\binom{-(n + p)}{k}(-x)^k(1 + x)^{-(n+p)-k} = (-x + 1 + x)^{-(n+p)} = 1. \tag{5.3.67}$$

We also observe that for every $f \in E_2$ and $x \geq 0$ we have

$$\sum_{k=0}^{\infty}\left|f\left(\frac{k}{n}\right)\right|\binom{-(n + p)}{k}(-x)^k(1 + x)^{-(n+p)-k}$$

$$\leq \|f\|_* \sum_{k=0}^{\infty}\binom{-(n + p)}{k}(-x)^k(1 + x)^{-(n+p)-k}\left(1 + \frac{k^2}{n^2}\right)$$

$$= \| f \|_* \left(1 + \frac{n+p}{n^2} \sum_{k=1}^{\infty} \binom{-(n+p)}{k} \frac{k^2}{n+p} (-x)^k (1+x)^{-(n+p)-k} \right)$$

$$= \| f \|_* \left(1 + \frac{n+p}{n^2} x \sum_{k=1}^{\infty} \binom{-(n+p)}{k} \frac{k^2}{-(n+p)} k(-x)^{k-1} (1+x)^{-(n+p)-k} \right)$$

$$= \| f \|_* \left(1 + \frac{n+p}{n^2} x \sum_{k=1}^{\infty} \binom{-(n+1+p)}{k-1} k(-x)^{k-1} (1+x)^{-(n+p)-k} \right)$$

$$= \| f \|_* \left(1 + \frac{n+p}{n^2} x \sum_{h=0}^{\infty} \binom{-(n+1+p)}{h} (h+1)(-x)^h (1+x)^{-(n+1+p)-h} \right)$$

$$= \| f \|_* \left(1 + \frac{n+p}{n^2} x + \frac{n+p}{n^2} x \right.$$

$$\left. \times \sum_{h=0}^{\infty} \binom{-(n+1+p)}{h} h(-x)^h (1+x)^{-(n+1+p)-h} \right)$$

$$= \| f \|_* \left(1 + \frac{n+p}{n^2} x + \frac{(n+p)(n+1+p)}{n^2} x^2 \right.$$

$$\left. \times \sum_{h=1}^{\infty} \binom{-(n+2+p)}{h-1} (-x)^{h-1} (1+x)^{-(n+1+p)-h} \right)$$

$$= \| f \|_* \left(1 + \frac{n+p}{n^2} x + \frac{(n+p)(n+1+p)}{n^2} x^2 \right).$$

Consequently, the last member of (5.3.66) is absolutely convergent. By using a similar reasoning one can easily show that

$$A_{n,p}(1)(x) = 1,$$

$$A_{n,p}(e_1)(x) = x + \frac{p}{n} x,$$

and

$$A_{n,p}(e_2)(x) = \left(\frac{n+p}{n} \right)^2 x^2 + \frac{n+p}{n^2} x(1+x).$$

Accordingly, by reasoning as in the above Example 5.3.9, for every $f \in E_2$ we obtain

$$\lim_{n \to \infty} A_{n,p}(f) = f \tag{5.3.68}$$

uniformly on compact subsets of $[0, +\infty[$.

The following identities

$$A_{n,p}(\psi_x)(x) = A_{n,p}(e_1 - x \cdot 1)(x) = \frac{p}{n}x$$

and

$$A_{n,p}(\psi_x^2)(x) = A_{n,p}(e_2)(x) - 2xA_{n,p}(e_1)(x) + x^2 A_{n,p}(1)(x)$$

$$= \left(\frac{n+p}{n}\right)^2 x^2 + \frac{n+p}{n^2}x(1+x) - 2x^2 - \frac{2p}{n}x^2 + x^2$$

$$= \frac{p^2}{n^2}x^2 + \frac{n+p}{n^2}x(1+x)$$

and formula (5.1.5) with $\delta = \sqrt{\dfrac{p^2x^2 + (n+p)x(1+x)}{n^2}}$ allow us to obtain the estimate

$$|A_{n,p}(f)(x) - f(x)| \leq 2\omega\left(f, \sqrt{\frac{p^2x^2 + (n+p)x(1+x)}{n^2}}\right), \tag{5.3.69}$$

which holds for every $n \in \mathbb{N}$, $f \in \mathscr{C}_b([0, +\infty[)$ and $x \geq 0$; if $p = 0$, (5.3.69) becomes

$$|A_n(f)(x) - f(x)| \leq 2\omega\left(f, \sqrt{\frac{x(1+x)}{n}}\right). \tag{5.3.70}$$

Finally, if f is differentiable on $[0, +\infty[$ and $f' \in \mathscr{C}_b([0, +\infty[)$, then

$$|A_{n,p}(f)(x) - f(x)| \leq \frac{p}{n}|f'(x)|x + 2\sqrt{\frac{p^2x^2 + (n+p)x(1+x)}{n^2}}$$

$$\times \omega\left(f', \sqrt{\frac{p^2x^2 + (n+p)x(1+x)}{n^2}}\right). \tag{5.3.71}$$

As in the last part of the preceding Example 5.3.9 we shall also establish some results about the uniform convergence on $\mathbb{R}_+$ of $(A_{n,p}(f))_{n \in \mathbb{N}}$.

As a function space we shall use $\mathscr{C}^*([0, +\infty[)$ (see (5.3.64)) and the test functions $f_\lambda(x) := \exp(-\lambda x)$ $(\lambda > 0, x \geq 0)$.

Since for every $n \geq 1$ and $x \geq 0$

$$A_{n,p}(f_\lambda)(x) = \sum_{k=0}^{\infty} \binom{-(n+p)}{k} (-x \exp(-\lambda/n))^k (1+x)^{-(n+p)-k}$$

$$= (-x \exp(-\lambda/n) + 1 + x)^{-(n+p)}$$

$$= (1 + x(1 - \exp(-\lambda/n)))^{-(n+p)},$$

we see that $A_{n,p}(f_\lambda) \to f_\lambda$ uniformly on $[0, +\infty[$.

Hence for every $f \in \mathscr{C}^*([0, +\infty[)$

$$\lim_{n\to\infty} A_{n,p}(f) = f \quad \text{uniformly on } [0, +\infty[. \tag{5.3.72}$$

More generally, for $p = 0$ Totik [1984] showed that, for a given function $f \in \mathscr{C}_b([0, +\infty[)$, (5.3.72) holds if and only if the function $f(\exp x)$ is uniformly continuous on $[0, +\infty[$.

A general method to construct positive approximation processes on the interval $[0, +\infty[$ is indicated below.

5.3.11 Mastroianni operators. We start with a sequence $(\phi_n)_{n \in \mathbb{N}}$ of real functions on $[0, +\infty[$ which are infinitely differentiable and strictly monotone on $[0, +\infty[$ and which satisfy the following additional conditions:

(i) $\phi_n(0) = 1$ for every $n \in \mathbb{N}$;

(ii) $(-1)^k \phi_n^{(k)}(x) \geq 0$ for every $n \in \mathbb{N}$, $x \in [0, +\infty[$ and $k \in \mathbb{N}_0$;

(iii) for every $(n, k) \in \mathbb{N} \times \mathbb{N}_0$ there exists a positive integer $p(n, k) \in \mathbb{N}$ and a real function $\alpha_{n,k}: [0, +\infty[\to \mathbb{R}$ such that

$$\phi_n^{(i+k)}(x) = (-1)^k \phi_{p(n,k)}^{(i)}(x) \alpha_{n,k}(x) \tag{5.3.73}$$

for every $i \in \mathbb{N}_0$ and $x \in [0, +\infty[$ and

$$\lim_{n\to\infty} \frac{n}{p(n,k)} = \lim_{n\to\infty} \frac{\alpha_{n,k}(x)}{n^k} = 1. \tag{5.3.74}$$

First, we observe that, by multiplying the two members of (5.3.73) by $(-1)^{i+k}$ and by applying condition (ii), we obtain $\alpha_{n,k}(x) \geq 0$. Moreover, by condition (i) and (5.3.73), we obtain by induction on $k \in \mathbb{N}_0$ that

$$\lim_{n\to\infty} \frac{\phi_n^{(k)}(0)}{n^k} = \lim_{n\to\infty} \frac{(-1)^k \alpha_{n,k}(0)}{n^k} = (-1)^k \tag{5.3.75}$$

for every $k \in \mathbb{N}_0$.

Furthermore, for a given $x \geq 0$, by using the Taylor series centered at x of the function ϕ_n as well as of its first and second derivatives and by evaluating them at 0, we obtain

$$\sum_{k=0}^{\infty} (-1)^k x^k \frac{\phi_n^{(k)}(x)}{k!} = \phi_n(0) = 1, \tag{5.3.76}$$

$$\sum_{k=0}^{\infty} k(-1)^k x^k \frac{\phi_n^{(k)}(x)}{k!} = -x\phi_n'(0) \tag{5.3.77}$$

and

$$\sum_{k=0}^{\infty} k^2(-1)^k x^k \frac{\phi_n^{(k)}(x)}{k!} = x^2\phi_n''(0) - x\phi_n'(0). \tag{5.3.78}$$

To the sequence $(\phi_n)_{n \in \mathbb{N}}$ we can associate a sequence $(V_n)_{n \in \mathbb{N}}$ of positive linear operators from E_2 into $\mathscr{C}([0, +\infty[)$ by putting, for every $n \in \mathbb{N}$, $f \in E_2$ and $x \in [0, +\infty[$,

$$V_n(f)(x) := \sum_{k=0}^{\infty} (-1)^k f\left(\frac{k}{n}\right) x^k \frac{\phi_n^{(k)}(x)}{k!}, \tag{5.3.79}$$

the last series being absolutely convergent by virtue of (5.3.76) and (5.3.78).

We can also obtain a representation of the operators V_n in terms of finite differences at the initial point 0. To this end, by reasoning by induction on $i \in \mathbb{N}_0$, we first obtain

$$V_n(f)^{(i)}(x) = \frac{i!}{n^i} \sum_{k=0}^{\infty} (-1)^{k+i} \Delta_{1/n}^i f\left(\frac{k}{n}\right) x^k \frac{\phi_n^{(k+i)}(x)}{k!} \tag{5.3.80}$$

and therefore, by expanding $V_n(f)$ in its MacLaurin series, we obtain

$$V_n(f)(x) = \sum_{k=0}^{\infty} (-1)^k \frac{\phi_n^{(k)}(0)}{k!} \Delta_{1/n}^k f(0) x^k. \tag{5.3.81}$$

From (5.3.76)–(5.3.78) it follows that

$$V_n(1) = 1$$

and, for every $x \geq 0$,

$$V_n(e_1)(x) = -\frac{\phi_n'(0)}{n} x,$$

$$V_n(e_2)(x) = \frac{\phi_n''(0)}{n^2} x^2 - \frac{\phi_n'(0)}{n^2} x.$$

Consequently, as in the preceding examples, we obtain

$$\lim_{n \to \infty} V_n(f) = f \qquad (5.3.82)$$

uniformly on compact subsets of $[0, +\infty[$ for every $f \in E_2$. Furthermore, by using the equalities

$$V_n(\psi_x)(x) = V_n(e_1 - x \cdot \mathbf{1})(x) = -\left(\frac{\phi'_n(0)}{n} + 1\right)x$$

and

$$V_n(\psi_x^2)(x) = V_n(e_2)(x) - 2x V_n(e_1)(x) + x^2 V_n(\mathbf{1})(x)$$

$$= \frac{\phi''_n(0)}{n^2}x^2 - \frac{\phi'_n(0)}{n^2}x + 2\frac{\phi'_n(0)}{n}x^2 + x^2$$

$$= \left(1 + 2\frac{\phi'_n(0)}{n} + \frac{\phi''_n(0)}{n^2}\right)x^2 - \frac{\phi'_n(0)}{n^2}x$$

and by setting

$$\delta_n(x) := \sqrt{\left(1 + 2\frac{\phi'_n(0)}{n} + \frac{\phi''_n(0)}{n^2}\right)x^2 - \frac{\phi'_n(0)}{n^2}x}$$

in (5.1.5), we obtain the estimate

$$|V_n(f)(x) - f(x)| \leq 2\omega(f, \delta_n(x)), \qquad (5.3.83)$$

which holds for every $f \in \mathscr{C}([0, +\infty[)$ and $x \geq 0$; if f is differentiable on $[0, +\infty[$ and $f' \in \mathscr{C}_b([0 +\infty[)$, we also have (see (5.1.6))

$$|V_n(f)(x) - f(x)| \leq |f'(x)|\left|\frac{\phi'_n(0)}{n} + 1\right|x + 2\delta_n(x)\omega(f', \delta_n(x)). \qquad (5.3.84)$$

Another useful tool to study approximation processes on unbounded intervals is the so-called technique of *multiplier enlargement* which was introduced by Hsu [1961] (see also Hsu [1964], Hsu and Wang [1964] and S. M. Eisenberg [1979]). Here we present an example.

5.3.12 Bernstein-Chlodovsky operators. Let $(b_n)_{n\in\mathbb{N}}$ be a sequence of strictly positive real numbers and suppose that

$$\lim_{n\to\infty} b_n = +\infty. \tag{5.3.85}$$

For every $n \in \mathbb{N}$ consider the linear operator $C_n^*: \mathscr{C}([0,+\infty[) \to \mathscr{C}([0,+\infty[)$ defined by

$$C_n^*(f)(x) := \begin{cases} \sum_{k=0}^{n} f\left(\frac{b_n k}{n}\right)\binom{n}{k}\left(\frac{x}{b_n}\right)^k\left(1-\frac{x}{b_n}\right)^{n-k}, & \text{if } 0 \le x \le b_n, \\ f(x), & \text{if } x > b_n \end{cases} \tag{5.3.86}$$

for every $f \in \mathscr{C}([0,+\infty[)$ and $x \in [0,+\infty[$.

The operator C_n^* is called the *n-th Bernstein-Chlodovsky operator*. It is a positive operator and it maps the function spaces E_2, $\mathscr{C}^*([0,+\infty[)$ and $\mathscr{C}_0([0,+\infty[)$ into themselves.

Furthermore for every $n \ge 1$, $x \ge 0$ and $\lambda > 0$ we have

$$C_n^*(1)(x) = 1,$$

$$C_n^*(e_1)(x) = x$$

and

$$C_n^*(e_2)(x) = x^2 - \frac{1}{n}x^2 + \frac{b_n}{n}x$$

for $x \le b_n$, so that (see (5.3.55))

$$\|C_n^*(e_2) - e_2\|_* \le \frac{b_n}{n} \tag{5.3.87}$$

and, finally, if $x \le b_n$,

$$C_n^*(f_\lambda)(x) = \left(\exp(-\lambda b_n/n)\frac{x}{b_n} + 1 - \frac{x}{b_n}\right)^n$$

$$= \left(1 - \lambda x\left(\frac{1 - \exp(-\lambda b_n/n)}{\lambda b_n}\right)\right)^n. \tag{5.3.88}$$

Again by using Proposition 4.2.5, (6) and (7), we obtain as in the preceding examples the following result.

5.3.13 Theorem. *If besides* (5.3.85) *the sequence* $(b_n)_{n \in \mathbb{N}}$ *of strictly positive real numbers satisfies the condition* $\lim\limits_{n \to \infty} \dfrac{b_n}{n} = 0$, *then, for every* $f \in E_2$,

$$\lim_{n \to \infty} C_n^*(f) = f \quad \text{in } E_2 \text{ (and hence uniformly on compact subsets of } [0, +\infty[).$$

$$(5.3.89)$$

Furthermore, if $f \in \mathscr{C}^*([0, +\infty[)$,

$$\lim_{n \to \infty} C_n^*(f) = f \quad \text{uniformly on } [0, +\infty[. \tag{5.3.90}$$

Finally, on account of (5.1.5) and (5.1.6), we obtain

$$|C_n^*(f)(x) - f(x)| \leq 2\omega\left(f, \sqrt{\frac{b_n x - x^2}{n}}\right) \tag{5.3.91}$$

for every $f \in \mathscr{C}_b([0, +\infty[)$ and $0 \leq x \leq b_n$, as well as

$$|C_n^*(f)(x) - f(x)| \leq 2\sqrt{\frac{b_n x - x^2}{n}}\,\omega\left(f', \sqrt{\frac{b_n x - x^2}{n}}\right) \tag{5.3.92}$$

provided f is differentiable in $[0, +\infty[$ and $f' \in \mathscr{C}_b[0 +\infty[)$.

Notes and references

Bernstein-Schurer operators 5.3.1 were introduced by Schurer [1962], [1965] (see also Sikkema [1970] and Pethe [1983]).

Bernstein-Cheney-Sharma operators were introduced and studied in Cheney and Sharma [1964a] from which we took the proof of Theorem 5.3.3. In the same paper, a Voronovskaja type formula for the operators G_n is established. Adapting an idea of Kantorovich, these operators have been modified by Müller [1989] (see also Habib and Umar [1980] and Wolik [1985]) in order to obtain a positive approximation process in $L^p([0, 1])$.

Cheney-Sharma operators were introduced and studied in Cheney and Sharma [1964b]. Among other properties, the authors investigate in detail the particular case when $t_n = 0$ for every $n \in \mathbb{N}$, where these operators coincide with the Meyer-König and Zeller operators, introduced in Meyer-König and Zeller [1960].

As a matter of fact Meyer-König and Zeller operators were originally defined as

$$R_n(f)(x) = (1 - x)^n \sum_{k=0}^{\infty} f\left(\frac{k}{n + k}\right)\binom{n - 1 + k}{k} x^k.$$

The slight modification (5.3.24) proposed by Cheney and Sharma is more appropriate in some respects because these modified operators preserves linear functions.

In the above mentioned paper of Cheney and Sharma in particular, it is shown that Meyer-König and Zeller operators R_n are the unique operators of the form

$$T_n(f)(x) = \frac{1}{h_n(t_n, x)} \sum_{k=0}^{\infty} c_{nk}(t_n) f\left(\frac{k}{n+k}\right) x^k \quad (c_{nk}(t_n) > 0),$$

which map linear functions into linear functions. In the same paper it is shown that, if f is convex, the sequence $(R_n(f))_{n \in \mathbb{N}}$ is decreasing and it is constant if and only if f is linear. Finally the authors show an interesting application of these operators to the approximation of the solution of the initial value problem

$$\begin{cases} y' = f(x, y), \\ y(0) = y_0, \end{cases}$$

where $f : [0, 1[\times \mathbb{R} \to \mathbb{R}$ satisfies suitable assumptions. Della Vecchia [1989] studied the preservation property of Hölder continuous functions by Cheney-Sharma operators. The convergence of the powers of Meyer-König and Zeller operators are studied in Kocić and Stanković [1984].

A different generalization of Meyer-König and Zeller operators has been studied by Swetits and Wood [1973b].

Further generalizations of both Cheney-Sharma operators and Meyer-König and Zeller operators were carried out by Lehnhoff [1979]. A Kantorovich-type modification of the Meyer-König and Zeller operators has been proposed by Müller [1978] (see also Maier, Müller and Swetits [1981] and Totik [1983a], [1983d]).

A detailed list of papers dealing with Meyer-König and Zeller operators as well as with Cheney-Sharma operators and their generalizations can be found in Stark [1984].

As regards Fejér-Hermite operators, they were introduced by Fejér [1916] who proved in [1930] the famous approximation formula (5.3.30). Here we followed the approach of Korovkin [1960, Section 7.4] and De Vore [1972, Section 2.8].

A generalization of these operators to the space of continuous functions on the real line can be found in Hsu [1964] and Eisenberg and Wood [1976]. The estimates (5.3.31) and (5.3.32) were given by Popoviciu [1951].

Kantorovich operators have been introduced and studied in Kantorovich [1930] and they also appear in Lorentz [1937], [1986b]. These operators possess the phenomenon of saturation that has been investigated by Maier [1978] and Riemenschneider [1978]. Further properties and generalizations can be found in Becker and Nessel [1981], Cao [1989], Ditzian and May [1976], Marlewsky [1980], [1984], J. Nagel [1982], Totik [1983c], [1983f].

Bernstein-Durrmeyer operators first appeared in Durrmeyer [1967] and were extensively studied by Derrienic [1981].

Saturation properties of these operators were discussed by Heilmann [1988], while Ditzian and Ivanov [1989] studied their rate of convergence and their derivatives by means of the Ditzian-Totik modulus of smoothness (see also Chen and Ditzian [1991]). Other inverse results have been obtained by Zhou [1990b], [1992b].

An interesting generalization of these operators which involves Jacobi weights has been recently proposed by Berens and Xu [1991a], [1991b]. Multidimensional weighted Bernstein-Durrmeyer operators can be found in Sauer [1992a] (see also [1993]).

Multidimensional Bernstein-Durrmeyer operators have been introduced and studied by Derrienic [1985]. For further investigations see Berens, Schmidt and Xu [1992] and Zhou [1990a], [1992a].

Szász-Mirakjan operators have been introduced by Mirakjan [1941] and were studied by different authors (see, e.g., Favard [1944], Szász [1950]. The generalization considered in Example 5.3.9 has been discussed in Schurer [1962], [1965] (see also Sikkema [1970] and Pethe [1983]). To show (5.3.65) we followed S. M. Eisenberg [1979]. In Cheney and Sharma [1964b] it is shown that if f is convex, the sequence $(S_n(f))_{n \in \mathbb{N}}$ is decreasing; a converse property has been established in Horová [1968]. The same author investigated in [1982] the convergence of their first derivatives.

Further properties can be also found, for instance, in Della Vecchia [1987], [1988], [1989], Ditzian and Totik [1987], Sendov and Popov [1988], Totik [1983e], [1984]. Generalizations and modifications of these operators have been considered by Guo and Zhou [1994], Lenze [1990], Mastroianni [1980a], Stancu [1972a].

Baskakov-Schurer operators were introduced and studied by Schurer [1962], [1965] (see also Sikkema [1970] and Pethe [1983]).

The operators considered in Example 5.3.10 have been considered by Mastroianni [1979]; in this paper the behavior of the sequence $(V_n(f))_{n \in \mathbb{N}}$ is also investigated in the case when f is a Lipschitz function or in the case when f is convex. Moreover the convergence of the derivatives together with some quantitative estimates is studied.

The characterization of the convergence of the powers of Mastroianni operators is obtained in the same paper of Mastroianni [1979] by considering the numbers $\gamma_{n,i} = \phi_n^{(i)}(0)/n^2$.

Indeed, it is shown that the condition $\gamma_{n,2} < 1$ is equivalent to the relation

$$\lim_{k \to \infty} V_n^{(k)}(f)(x) = f(0) + n(1 - \gamma_{n,2}) \, (f(n(1 - \gamma_{n,2})) - f(0))x$$

for every $n \in \mathbb{N}$, $f \in \mathscr{C}([0,b])$ and $x \in [0,b]$.

Moreover, if we consider the operators $V_{n,k} = I_{\mathscr{C}([0,b])} - (I_{\mathscr{C}([0,b])} - V_n)^k = \sum_{j=1}^{k} (-1)^{j-1} \binom{k}{j} V_n^j$

$(n, k \in \mathbb{N})$, then it is possible to show that $\lim_{k \to \infty} V_{n,k}(f)(x) = f(0) + \sum_{i=1}^{\infty} x\left(x - \frac{1}{n}\right) \cdots$
$\left(x - \frac{i-1}{n}\right) \Delta_{1/n}^{i-1} f(0)$ for every $n \in \mathbb{N}$, $f \in \mathscr{C}([0,b])$ and $x \in [0,b]$ (provided that the series is convergent) if and only if $\gamma_{n,i} < 2$ for every $i \geq 2$.

We also point out the papers of Mastroianni [1980a], [1980b] for the construction of other examples of positive linear operators which are obtained starting with different hypotheses on the sequence $(\phi_n)_{n \in \mathbb{N}}$ (see also Della Vecchia [1989]). However, we have to note that other authors have given examples of linear positive operators constructed in a similar way (see, for instance, Baskakov [1957], Jakimovski and Leviatan [1969], Lehnhoff [1981], Lupaş [1967a], [1967b], Schurer [1966], Wood [1985]).

Indeed, the method takes its origin in a paper of Baskakov [1957] (see also Schurer [1966]), where the author considers a sequence $(\phi_n)_{n \in \mathbb{N}}$ of real infinitely differentiable functions on an interval $[0,b]$ which satisfy the following conditions:

(1) $\phi_n(0) = 1$ for every $n \in \mathbb{N}$;

(2) $(-1)^k \phi_n^{(k)}(x) \geq 0$ for every $n \in \mathbb{N}$, $x \in [0, b]$ and $k \in \mathbb{N}_0$;

(3) there exists a positive integer $c \in \mathbb{N}$ such that $-\phi_n^{(k)}(x) = n\phi_{n-c}^{(k-1)}(x)$ for every $n \geq c$, $x \in [0, b]$ and $k \in \mathbb{N}$.

Then, for every $n \in \mathbb{N}$, a linear operator $L_n : E_2 \to \mathscr{C}([0, b])$ can be defined by

$$L_n(f)(x) := \sum_{k=0}^{\infty} f\left(\frac{k}{n}\right) \frac{(-1)^k \phi_n^{(k)}(x) x^k}{k!}$$

for every $f \in E_2$ and $x \in [0, b]$.

The above operators are positive and an application of a Korovkin-type theorem yields the convergence of the sequence $(L_n(f))_{n \in \mathbb{N}}$ toward f uniformly on $[0, b]$ for every $f \in E_2$. In the paper of Schurer [1966] some examples of sequences $(\phi_n)_{n \in \mathbb{N}}$ are considered.

Bernstein-Chlodovsky operators were introduced by Chlodovsky [1937] and can be found in the book of Lorentz [1986b, p. 36]. The convergence of Bernstein-Chlodovsky operators has been investigated also by Gadzhiev [1976] who proved (5.3.96) and by Eisenberg and Wood [1970] who showed that $C_n^*(f) \to f$ uniformly on compact subsets of $[0, +\infty[$ for every continuous functions f on $[0, +\infty[$ satisfying $f(x) \leq M \exp(\alpha x)$ for every $x \geq 0$ (where $M > 0$ and $\alpha > 0$ are suitable constants) (see also S. M. Eisenberg [1979]).

Finally, we have to mention the paper of Lupaş and Lupaş [1987], where a different method to construct sequences of positive linear operators is indicated. Here we briefly describe some details.

Let $\mathscr{A}$ be the algebra of real polynomials; a *delta operator* $Q : \mathscr{A} \to \mathscr{A}$ is a linear operator which maps $\mathbf{1}$ in a constant function and, further, is shift-invariant, i.e., $Q \circ E_a = E_a \circ Q$, for every $a \in \mathbb{R}$, where $E_a : \mathscr{A} \to \mathscr{A}$ is defined by $E_a(p)(x) := p(x + a)$ for every $p \in \mathscr{A}$ and $x \in \mathbb{R}$. The authors give many examples of delta operators, such as the differentiation operator, the backward difference operator, the Touchard operator, the Laguerre operator and so on. One of the most interesting properties of delta operators rests primarily on the fact that to every delta operator it is possible to associate a unique sequence $(p_n)_{n \in \mathbb{N}_0}$ of polynomials satisfying

$$p_0 = \mathbf{1}, \ p_n(0) = 0 \quad \text{and} \quad Q(p_n) = np_{n-1} \quad \text{for every } n \in \mathbb{N}.$$

The polynomials p_n are called the *basic polynomials for the delta operator Q*. Now they define an operator $L_{n,Q} : \mathscr{C}([0, 1]) \to \mathscr{C}([0, 1])$ by

$$L_{n,Q}(f)(x) := \frac{1}{p_n(n)} \sum_{k=0}^{n} f\left(\frac{k}{n}\right) \binom{n}{k} p_k(nx) p_{n-k}(n - nx)$$

for every $f \in \mathscr{C}([0, 1])$ and $x \in [0, 1]$. Korovkin's theorem is then used to show the convergence of the sequence $(L_{n,Q})_{n \in \mathbb{N}}$; furthermore estimates of the rate of convergence by means of the first and the second modulus of smoothness are established as well.

5.4 Convolution operators and summation processes

Approximation processes by means of convolution operators constitute another wide area of interest in approximation theory and Fourier analysis. As a consequence of the large amount of researches devoted to this subject, we have today a satisfactory knowledge which is moreover deepened by some monographs on this type of approximation. Among them we refer to the excellent books of Butzer and Nessel [1971] and De Vore [1972] where the reader can find the main results in this field.

Here we content ourselves with presenting only a few selected results, which are closer related to Korovkin-type approximation theory; we also restrict the large field of applications to the most classical ones which concern the convergence of some summation methods of Fourier series.

In the sequel we shall denote by $\mathscr{C}_{2\pi}$ the space of all 2π-periodic continuous real-valued functions on $\mathbb{R}$.

Furthermore, if $1 \leq p < +\infty$, we shall denote by $L^p_{2\pi}$ the space of all functions on $\mathbb{R}$ which are Lebesgue integrable to the p-th power over $[-\pi, \pi]$ and which satisfy $f(x + 2\pi) = f(x)$ a.e. on $\mathbb{R}$.

The spaces $\mathscr{C}_{2\pi}$ and $L^p_{2\pi}$ ($1 \leq p < +\infty$) will be endowed with the norms

$$\|f\| := \sup_{|x| \leq \pi} |f(x)|, \quad f \in \mathscr{C}_{2\pi},$$

and

$$\|f\|_p := \left(\frac{1}{2\pi} \int_{-\pi}^{\pi} |f(t)|^p \, dt \right)^{1/p}, \quad f \in L^p_{2\pi},$$

respectively.

To make the exposition simpler we shall denote by $(\mathbb{X}_{2\pi}, \|\cdot\|_{\mathbb{X}_{2\pi}})$ indifferently one of the spaces $\mathscr{C}_{2\pi}$ or $L^p_{2\pi}$.

In the sequel we consider nets of convolution operators defined by means of so-called kernels. More precisely, we say that a net $(\chi_i)_{i \in I}$ of functions of $L^1_{2\pi}$ is a (*periodic*) *kernel* if

$$\lim_{i \in I} \frac{1}{2\pi} \int_{-\pi}^{\pi} \chi_i(t) \, dt = 1. \tag{5.4.1}$$

In fact, we shall see that many kernels are *normalized*, i.e., the equality $\int_{-\pi}^{\pi} \chi_i(t) \, dt = 2\pi$ holds for every $i \in I$.

A kernel $(\chi_i)_{i \in I}^{\leq}$ will be called an *approximate identity* if there exists $i_0 \in I$ such that $(\chi_i)_{i \geq i_0}$ is bounded in $L^1_{2\pi}$ and, moreover,

$$\lim_{i \in I} \left(\int_{-\pi}^{-\delta} |\chi_i(t)| \, dt + \int_{\delta}^{\pi} |\chi_i(t)| \, dt \right) = 0 \tag{5.4.2}$$

for every $\delta \in \mathbb{R}$, $0 < \delta < \pi$.

Note that if every χ_i is positive, then condition (5.4.1) implies that the family $(\chi_i)_{i \geq i_0}$ is bounded in $L^1_{2\pi}$ for some $i_0 \in I$.

It is customary to attach to the kernel $(\chi_i)_{i \in I}^{\leq}$ the properties of the χ_i's. So we shall say that the kernel $(\chi_i)_{i \in I}^{\leq}$ is continuous (positive, bounded and so on, respectively) if, for every $i \in I$, χ_i is continuous (positive, bounded a.e. and so on, respectively).

Moreover, in many circumstances the index set I will be a real interval with endpoints a and b $(-\infty \leq a < b \leq +\infty)$or the set $\mathbb{N}$; in these cases the limit must be understood at one of the points a or b or at the point $+\infty$.

From now on we fix a periodic kernel $(\chi_i)_{i \in I}^{\leq}$. Given $i \in I$ and $f \in \mathbb{X}_{2\pi}$, we can consider the convolution $f * \chi_i$ of f and χ_i defined by

$$f * \chi_i(x) := \frac{1}{2\pi} \int_{-\pi}^{\pi} f(x - t)\chi_i(t)\, dt = \frac{1}{2\pi} \int_{-\pi}^{\pi} f(t)\chi_i(x - t)\, dt \qquad (5.4.3)$$

for every $x \in \mathbb{R}$.

By using Fubini's theorem and Hölder's inequality, it is easy to show that $f * \chi_i \in \mathbb{X}_{2\pi}$ and

$$\| f * \chi_i \|_{\mathbb{X}_{2\pi}} \leq \| \chi_i \|_1 \| f \|_{\mathbb{X}_{2\pi}}. \qquad (5.4.4)$$

Accordingly we may consider the linear operator $L_i: \mathbb{X}_{2\pi} \to \mathbb{X}_{2\pi}$ defined by

$$L_i(f) := f * \chi_i \qquad (5.4.5)$$

for every $f \in \mathbb{X}_{2\pi}$.

The operator L_i is called the *convolution operator* associated with χ_i.

Next, we are interested in the approximation properties of the net $(L_i)_{i \in I}^{\leq}$.

Obviously, if the kernel $(\chi_i)_{i \in I}^{\leq}$ is positive, then the corresponding operators $L_i (i \in I)$ are positive; moreover, every L_i is a linear contraction on $\mathbb{X}_{2\pi}$ provided

$$\int_{-\pi}^{\pi} |\chi_i(t)|\, dt \leq 2\pi, \quad \text{i.e.,} \quad \| \chi_i \|_1 \leq 1. \qquad (5.4.6)$$

It is convenient to point out that the function $L_i(1)$ assumes the constant value $\frac{1}{2\pi} \int_{-\pi}^{\pi} \chi_i(t)\, dt$. Furthermore, if for every $x \in \mathbb{R}$ we set $\psi_x(t) := t - x$ $(t \in \mathbb{R})$, then $\sin^2(\frac{1}{2}\psi_x) \in \mathscr{C}_{2\pi}$ and for each $i \in I$ the quantity

$$\beta_i := L_i(\sin^2(\tfrac{1}{2}\psi_x))(x)$$

$$= \frac{1}{2\pi} \int_{-\pi}^{\pi} \chi_i(t) \sin^2 \frac{t}{2}\, dt$$

$$= \frac{1}{2\pi} \int_{-\pi}^{\pi} \chi_i(u - x) \sin^2 \frac{u - x}{2}\, du$$

$$= \frac{1}{2}(1 - \cos x L_i(\cos)(x) - \sin x L_i(\sin)(x)) \tag{5.4.7}$$

is independent of $x \in \mathbb{R}$.

The numbers β_i will play an important role in deriving both qualitative and quantitative properties of the approximation process $(L_i)_{i \in I}^{\leq}$.

5.4.1 Theorem. *Assume that* $(\chi_i)_{i \in I}^{\leq}$ *is a positive kernel and consider the net* $(L_i)_{i \in I}^{\leq}$ *of the associated convolution operators defined by (5.4.5).*

Then the following statements are equivalent:

(i) $\lim_{\substack{\leq \\ i \in I}} L_i(f) = f$ *in* $\mathbb{X}_{2\pi}$ *for every* $f \in \mathbb{X}_{2\pi}$.

(ii) $\lim_{\substack{\leq \\ i \in I}} L_i(\cos) = \cos$ *and* $\lim_{\substack{\leq \\ i \in I}} L_i(\sin) = \sin$ *in* $\mathbb{X}_{2\pi}$.

(iii) $\lim_{\substack{\leq \\ i \in I}} L_i(\cos)(x_0) = \cos(x_0)$ *and* $\lim_{\substack{\leq \\ i \in I}} L_i(\sin)(x_0) = \sin(x_0)$ *for every* $x_0 \in \mathbb{R}$.

(iv) $\lim_{\substack{\leq \\ i \in I}} L_i(\cos)(x_0) = \cos(x_0)$ *and* $\lim_{\substack{\leq \\ i \in I}} L_i(\sin)(x_0) = \sin(x_0)$ *for some* $x_0 \in \mathbb{R}$.

(v) $\lim_{\substack{\leq \\ i \in I}} \beta_i = 0$.

(vi) *The kernel* $(\chi_i)_{i \in I}^{\leq}$ *is an approximate identity.*

Proof. The implications (i) $\Rightarrow$ (ii) and (iii) $\Rightarrow$ (iv) are obvious. To show that (ii) $\Rightarrow$ (iii), note first that for an arbitrary $x \in \mathbb{R}$ we have

$$\frac{1}{2\pi} \int_{-\pi}^{\pi} \sin(t)\chi_i(t)\, dt = \frac{1}{2\pi} \int_{x-\pi}^{x+\pi} \sin(x - u)\chi_i(x - u)\, du$$

$$= \frac{\sin x}{2\pi} \int_{-\pi}^{\pi} \cos(x - t)\chi_i(t)\, dt - \frac{\cos x}{2\pi} \int_{-\pi}^{\pi} \sin(x - t)\chi_i(t)\, dt$$

$$= \sin x L_i(\cos)(x) - \cos x L_i(\sin)(x) \to 0$$

by (ii).

Analogously,

$$\frac{1}{2\pi} \int_{-\pi}^{\pi} (\cos t - 1)\chi_i(t)\, dt = 1 - \cos x L_i(\cos)(x) - \sin x L_i(\sin)(x) \to 0.$$

Hence

$$L_i(\cos)(x_0) - \cos x_0 = \frac{1}{2\pi} \int_{-\pi}^{\pi} (\cos(x_0 - t) - \cos x_0)\chi_i(t)\, dt$$

$$+ \cos x_0 \left(\frac{1}{2\pi} \int_{-\pi}^{\pi} \chi_i(t)\, dt - 1 \right)$$

$$= \frac{\cos x_0}{2\pi} \int_{-\pi}^{\pi} (\cos t - 1)\chi_i(t)\, dt + \frac{\sin x_0}{2\pi} \int_{-\pi}^{\pi} \sin(t)\chi_i(t)\, dt$$

$$+ \cos x_0 \left(\frac{1}{2\pi} \int_{-\pi}^{\pi} \chi_i(t)\, dt - 1 \right) \to 0.$$

A similar reasoning shows that $L_i(\sin)(x_0) - \sin x_0 \to 0$.

Statement (v) is a consequence of (iv) because of (5.4.7) applied to $x = x_0$.

To show that (v) $\Rightarrow$ (vi), note first that, since the kernel $(\chi_i)_{i \in I}^{\le}$ is positive, there exists $i_0 \in I$ such that $(\chi_i)_{i \ge i_0}$ is bounded in $L^1_{2\pi}$. Furthermore for every $0 < \delta < \pi$

$$\frac{\sin^2(\delta/2)}{2\pi} \left(\int_{-\pi}^{-\delta} \chi_i(t)\, dt + \int_{\delta}^{\pi} \chi_i(t)\, dt \right)$$

$$\le \frac{1}{2\pi} \left(\int_{-\pi}^{-\delta} \chi_i(t) \sin^2 \frac{t}{2}\, dt + \int_{\delta}^{\pi} \chi_i(t) \sin^2 \frac{t}{2}\, dt \right)$$

$$\le \beta_i,$$

and hence (5.4.2) holds.

Now, we proceed to show that (vi) $\Rightarrow$ (ii). First we choose $j_0 \in I$ such that

$$M := \sup_{i \ge j_0} \frac{1}{2\pi} \int_{-\pi}^{\pi} \chi_i(t)\, dt < +\infty.$$

Given $\varepsilon > 0$, consider $0 < \delta < \pi$ such that $|1 - \cos t| \le \dfrac{\varepsilon}{6(M + 1)}$ and $|\sin t| \le \dfrac{\varepsilon}{6(M + 1)}$ for every $t \in \mathbb{R}$, $|t| \le \delta$. Moreover, by (5.4.1) and (5.4.2), there exists $i_0 \in I$, $i_0 \ge j_0$, such that

$$\left| \frac{1}{2\pi} \int_{-\pi}^{\pi} \chi_i(t)\, dt - 1 \right| \le \frac{\varepsilon}{3} \quad \text{and} \quad \int_{-\pi}^{-\delta} \chi_i(t)\, dt + \int_{\delta}^{\pi} \chi_i(t)\, dt \le \frac{\varepsilon}{6}$$

for every $i \geq i_0$. Accordingly, if $i \geq i_0$ and $x \in \mathbb{R}$, then

$$|L_i(\sin)(x) - \sin x| \leq \frac{1}{2\pi} \left| \int_{-\pi}^{\pi} (\sin(x-t) - \sin x)\chi_i(t)\, dt \right|$$

$$+ \left| \left(\frac{1}{2\pi} \int_{-\pi}^{\pi} \chi_i(t)\, dt - 1 \right) \sin x \right|$$

$$\leq \frac{1}{2\pi} \int_{\delta \leq |t| \leq \pi} |\sin(x-t) - \sin x| \chi_i(t)\, dt$$

$$+ \frac{1}{2\pi} \int_{|t| < \delta} |(\cos t - 1)\sin x - \cos x \sin t| \chi_i(t)\, dt + \frac{\varepsilon}{3}$$

$$\leq \frac{1}{\pi} \int_{\delta \leq |t| \leq \pi} \chi_i(t)\, dt + \frac{\varepsilon}{3(M+1)} \frac{1}{2\pi} \int_{-\pi}^{\pi} \chi_i(t)\, dt + \frac{\varepsilon}{3} \leq \varepsilon.$$

So $\|L_i(\sin) - \sin\|_{\mathbb{X}_{2\pi}} \leq \|L_i(\sin) - \sin\|_{\mathscr{C}_{2\pi}} \leq \varepsilon$.

This reasoning shows that $\lim_{\substack{i \in I}}^{\leq} L_i(\sin) = \sin$ in $\mathbb{X}_{2\pi}$. In a similar way it can be shown that $\lim_{\substack{i \in I}}^{\leq} L_i(\cos) = \cos$ in $\mathbb{X}_{2\pi}$.

Finally the implication (ii) $\Rightarrow$ (i) is a consequence of Korovkin's second theorem 4.2.8 if $\mathbb{X}_{2\pi} = \mathscr{C}_{2\pi}$ and of Corollary 4.1.7 if $\mathbb{X}_{2\pi} = L^p_{2\pi} \cong L^p(\mathbb{T})$, taking into account that $\lim_{\substack{i \in I}}^{\leq} L_i(1) = 1$ in $\mathbb{X}_{2\pi}$ by virtue of (5.4.1). $\qquad\square$

Remarks. 1. In Theorem 5.4.1 we have required that the kernel $(\chi_i)_{i \in I}^{\leq}$ is positive and this condition makes possible the application of Korovkin's second theorem.

On the other hand, if the kernel $(\chi_i)_{i \in I}^{\leq}$ satisfies (5.4.6) for every $i \in I$, then the operators L_i $(i \in I)$ are linear contractions and therefore, by Theorem 4.1.4, (viii), we can still apply Korovkin's second theorem to the net $(L_i)_{i \in I}^{\leq}$. Consequently, we conclude that statements (i) and (ii) of Theorem 5.4.1 remain equivalent in this case too when $\mathbb{X}_{2\pi} = \mathscr{C}_{2\pi}$.

2. If the kernel $(\chi_i)_{i \in I}^{\leq}$ is not positive only a part of Theorem 5.4.1 remains true. In fact, it can be shown that, if $(\chi_i)_{i \in I}^{\leq}$ is an arbitrary approximate identity and if $(L_i)_{i \in I}^{\leq}$ is defined as in (5.4.5), then, for every $f \in \mathbb{X}_{2\pi}$, we have $\lim_{\substack{i \in I}}^{\leq} \|L_i(f) - f\|_{\mathbb{X}_{2\pi}} = 0$ (see Butzer and Nessel [1971, Theorem 1.1.5]).

Thus, according to Theorem 5.4.1, the convergence of the net $(L_i)_{i \in I}^{\leq}$ can be checked by using only the numbers β_i. In the next result we give a quantitative estimate of the convergence by means of these numbers and of the modulus of continuity $\omega(f, \delta)$ defined in Section 5.1.

5.4.2 Theorem. *Assume that $(\chi_i)_{i \in I}^{\le}$ is a positive kernel and consider the net $(L_i)_{i \in I}^{\le}$ of positive convolution operators defined by (5.4.5). Then, given $f \in \mathscr{C}_{2\pi}$ and $x \in \mathbb{R}$, for every $i \in I$ and $\delta > 0$ we have*

$$|L_i(f)(x) - f(x)| \le |f(x)||L_i(\mathbf{1}) - 1| + \left(L_i(\mathbf{1}) + \frac{\pi}{\delta}\sqrt{\beta_i}\sqrt{L_i(\mathbf{1})} \right)\omega(f, \delta). \quad (5.4.8)$$

In particular,

$$|L_i(f)(x) - f(x)| \le |f(x)||L_i(\mathbf{1}) - 1| + (L_i(\mathbf{1}) + \pi\sqrt{L_i(\mathbf{1})})\omega(f, \sqrt{\beta_i}). \quad (5.4.9)$$

Moreover, if f is differentiable and $f' \in \mathscr{C}_{2\pi}$, then

$$|L_i(f)(x) - f(x)| \le |f(x)||L_i(\mathbf{1}) - 1| + \pi|f'(x)|\sqrt{\beta_i L_i(\mathbf{1})}$$

$$+ \pi\left(\sqrt{L_i(\mathbf{1})} + \frac{\pi}{\delta}\sqrt{\beta_i} \right)\sqrt{\beta_i}\,\omega(f', \delta) \quad (5.4.10)$$

and, in particular,

$$|L_i(f)(x) - f(x)| \le |f(x)||L_i(\mathbf{1}) - 1| + \pi|f'(x)|\sqrt{\beta_i L_i(\mathbf{1})}$$

$$+ \pi(\pi + \sqrt{L_i(\mathbf{1})})\sqrt{\beta_i}\,\omega(f', \sqrt{\beta_i}). \quad (5.4.11)$$

Proof. Since $|t| \le \dfrac{\pi}{2}|\sin t|$ provided $|t| \le \dfrac{\pi}{2}$, we obtain for every $f \in \mathscr{C}_{2\pi}$, $i \in I$ and $x \in \mathbb{R}$

$$|L_i(f)(x) - L_i(f(x)\mathbf{1})| \le \frac{1}{2\pi}\int_{-\pi}^{\pi} |f(x-t) - f(x)|\chi_i(t)\,dt$$

$$\le \frac{1}{2\pi}\int_{-\pi}^{\pi} \omega(f, |t|)\chi_i(t)\,dt$$

$$\le \frac{1}{2\pi}\int_{-\pi}^{\pi} \left(1 + \frac{|t|}{\delta} \right)\omega(f, \delta)\chi_i(t)\,dt$$

$$= \left(L_i(\mathbf{1}) + \frac{1}{\pi\delta}\int_{-\pi}^{\pi} \frac{|t|}{2}\chi_i(t)\,dt \right)\omega(f, \delta)$$

$$\le \left(L_i(\mathbf{1}) + \frac{1}{\pi\delta}\int_{-\pi}^{\pi} \frac{\pi}{2}\left|\sin\frac{t}{2}\right|\chi_i(t)\,dt \right)\omega(f, \delta)$$

$$\leq \left(L_i(1) + \frac{\pi}{\delta} \left(\frac{1}{2\pi} \int_{-\pi}^{\pi} \sin^2 \frac{t}{2} \chi_i(t)\, dt \right)^{1/2} \right.$$

$$\left. \times \left(\frac{1}{2\pi} \int_{-\pi}^{\pi} \chi_i(t)\, dt \right)^{1/2} \right) \omega(f, \delta),$$

$$= \left(L_i(1) + \frac{\pi}{\delta} \sqrt{\beta_i} \sqrt{L_i(1)} \right) \omega(f, \delta),$$

where in the penultimate conclusion we used the Cauchy-Schwarz inequality.
   If we take into account the inequality

$$|L_i(f)(x) - f(x)| \leq |L_i(f)(x) - L_i(f(x)\mathbf{1})| + |f(x)||L_i(\mathbf{1}) - 1|,$$

we then obtain (5.4.8).
   Now assume that f is differentiable and $f' \in \mathscr{C}_{2\pi}$. For every $x \in \mathbb{R}$ and $t \in [-\pi, \pi]$ we use the mean value theorem to obtain $f(x - t) - f(x) = f'(\xi(t))t$, where $\xi(t) \in \mathbb{R}$ satisfies $|\xi(t) - x| \leq |t|$.
   Therefore

$$|f(t - x) - f(x)| = |f'(\xi(t))||t|$$

$$\leq |f'(\xi(t)) - f'(x)||t| + |f'(x)||t|$$

$$\leq \omega(f', |t|)|t| + |f'(x)||t|$$

$$\leq \left(1 + \frac{|t|}{\delta} \right) |t| \omega(f', \delta) + |f'(x)||t|$$

$$= (\omega(f', \delta) + |f'(x)|)|t| + \frac{t^2}{\delta} \omega(f', \delta).$$

By reasoning as above we obtain

$$|L_i(f)(x) - L_i(f(x)\mathbf{1})| \leq \frac{1}{2\pi} \int_{-\pi}^{\pi} |f(x - t) - f(x)| \chi_i(t)\, dt$$

$$\leq (\omega(f', \delta) + |f'(x)|) \frac{1}{2\pi} \int_{-\pi}^{\pi} |t| \chi_i(t)\, dt$$

$$+ \frac{\omega(f', \delta)}{\delta} \frac{1}{2\pi} \int_{-\pi}^{\pi} t^2 \chi_i(t)\, dt$$

$$\leq (\omega(f',\delta) + |f'(x)|)\pi\sqrt{\beta_i}\sqrt{L_i(1)}$$

$$+ 4\frac{\omega(f',\delta)}{\delta}\frac{1}{2\pi}\int_{-\pi}^{\pi}\frac{\pi^2}{4}\sin^2\frac{t}{2}\chi_i(t)\,dt$$

$$= (\omega(f',\delta) + |f'(x)|)\pi\sqrt{\beta_i}\sqrt{L_i(1)} + \pi^2\frac{\omega(f',\delta)}{\delta}\beta_i,$$

and hence inequality (5.4.10) follows.

The particular cases (5.4.9) and (5.4.11) follow by applying (5.4.8) and (5.4.10) respectively, to $\delta = \sqrt{\beta_i}$. □

Remark. If the kernel $(\chi_i)_{i\in I}^{\leq}$ is normalized, formulas (5.4.8) and (5.4.9) become

$$\|L_i(f) - f\| \leq \left(1 + \frac{\pi}{\delta}\sqrt{\beta_i}\right)\omega(f,\delta) \tag{5.4.12}$$

and, respectively,

$$\|L_i(f) - f\| \leq (1 + \pi)\omega(f,\sqrt{\beta_i}). \tag{5.4.13}$$

Furthermore, if f is differentiable and $f' \in \mathscr{C}_{2\pi}$, by (5.4.10) and (5.4.11) we obtain

$$\|L_i(f) - f\| \leq \pi\sqrt{\beta_i}\left(\|f'\| + \left(1 + \frac{\pi}{\delta}\sqrt{\beta_i}\right)\omega(f',\delta)\right) \tag{5.4.14}$$

and, respectively,

$$\|L_i(f) - f\| \leq \pi\sqrt{\beta_i}(\|f'\| + (\pi + 1)\omega(f',\sqrt{\beta_i})). \tag{5.4.15}$$

To state some estimates of the rate of convergence of the net $(L_i(f))_{i\in I}^{\leq}$ towards f in $L_{2\pi}^p$-spaces as well, we shall introduce the *modulus of continuity of order p of a function f* which is defined as

$$\omega^{(p)}(f,\delta) := \sup_{|h|\leq\delta} \|f(\cdot + h) - f\|_p, \tag{5.4.16}$$

for every $f \in L_{2\pi}^p$ and $\delta > 0$.

As it is easy to see, $\omega^{(p)}(f,\delta)$ satisfies properties (1), (2) and (6) (with $k = 1$) of Lemma 5.1.1.

The following result is the analogue of (5.4.12) and (5.4.13).

5.4.3 Theorem. *Let $(\chi_i)_{i \in I}^{\le}$ be a positive normalized kernel and consider the net of positive convolution operators $(L_i)_{i \in I}^{\le}$ defined by (5.4.5). For every $1 \le p < +\infty$, $f \in L_{2\pi}^p$, $\delta > 0$ and $i \in I$ one has*

$$\|L_i(f) - f\|_p \le \left(1 + \frac{\pi}{\delta}\sqrt{\beta_i}\right)\omega^{(p)}(f, \delta). \tag{5.4.17}$$

In particular

$$\|L_i(f) - f\|_p \le (1 + \pi)\omega^{(p)}(f, \sqrt{\beta_i}). \tag{5.4.18}$$

Proof. By using the generalized Minkowski inequality (see Butzer and Nessel [1971, Proposition 0.1.7]), we directly obtain

$$\|L_i(f) - f\|_p = \left\| \frac{1}{2\pi} \int_{-\pi}^{\pi} (f(\cdot - t) - f)\chi_i(t)\,dt \right\|_p$$

$$\le \frac{1}{2\pi} \int_{-\pi}^{\pi} \|f(\cdot - t) - f\|_p \chi_i(t)\,dt$$

$$\le \frac{1}{2\pi} \int_{-\pi}^{\pi} \omega^{(p)}(f, |t|)\chi_i(t)\,dt$$

$$\le \frac{1}{2\pi} \int_{-\pi}^{\pi} \left(1 + \frac{|t|}{\delta}\right)\omega^{(p)}(f, \delta)\chi_i(t)\,dt.$$

Furthermore, by Hölder's inequality,

$$\frac{1}{2\pi} \int_{-\pi}^{\pi} |t|\chi_i(t)\,dt \le \frac{1}{2\pi}\left(\int_{-\pi}^{\pi} t^2 \chi_i(t)\,dt\right)^{1/2}\left(\int_{-\pi}^{\pi} \chi_i(t)\,dt\right)^{1/2}$$

$$= \left(\frac{1}{2\pi}\int_{-\pi}^{\pi} t^2 \chi_i(t)\,dt\right)^{1/2}$$

$$\le \left(\frac{1}{2\pi}\int_{-\pi}^{\pi} \pi^2 \sin^2\frac{t}{2}\chi_i(t)\,dt\right)^{1/2} = \pi\sqrt{\beta_i},$$

and hence (5.4.17) follows.

Clearly (5.4.18) follows from (5.4.17) by setting $\delta = \sqrt{\beta_i}$.   □

Remark. Summing up, from (5.4.13) and (5.4.18) it follows that, if $(\chi_i)_{i \in I}^{\leq}$ is a positive normalized kernel, then for every $x_0 \in \mathbb{R}$, $f \in \mathbb{X}_{2\pi}$ and $i \in I$

$$\|L_i(f) - f\|_{\mathbb{X}_{2\pi}} \leq (1 + \pi)\omega^*(f, \sqrt{\beta_i})$$

$$= (1 + \pi)\omega^*\left(f, \sqrt{\frac{1 - \cos x_0 L_i(\cos)(x_0) - \sin x_0 L_i(\sin)(x_0)}{2}}\right),$$

$$(5.4.19)$$

where $\omega^*(f, \delta) := \omega(f, \delta)$ if $\mathbb{X}_{2\pi} = \mathscr{C}_{2\pi}$ and $\omega^*(f, \delta) := \omega^{(p)}(f, \delta)$ if $\mathbb{X}_{2\pi} = L_{2\pi}^p$.

Approximate identities ensure nice approximating properties for the corresponding nets of convolution operators. Among other things, they are intimately connected with several methods of summation of Fourier series. It is this last aspect that we want to emphasize with the next example and, however, we refer the reader to the book of Butzer and Nessel [1971] for a more complete treatment of this topic.

We recall that a *trigonometric polynomial of degree n*, $n \in \mathbb{N}$, is a real function of the form

$$u_n(x) = \frac{1}{2}a_0 + \sum_{k=1}^{n} (a_k \cos kx + b_k \sin kx), \quad x \in \mathbb{R}, \tag{5.4.20}$$

where $a_0, a_1, \ldots, a_n$ and $b_1, \ldots, b_n$ are real numbers.

A series of the form

$$u(x) = \frac{1}{2}a_0 + \sum_{k=1}^{\infty} (a_k \cos kx + b_k \sin kx), \quad x \in \mathbb{R}, \tag{5.4.21}$$

will be also called a *trigonometric series*.

Since every trigonometric polynomial belongs to $\mathscr{C}_{2\pi}$, a sequence $(u_n)_{n \in \mathbb{N}}$ of trigonometric polynomials may uniformly converge only to a function $f \in \mathscr{C}_{2\pi}$. If this is the case, then the coefficients a_0 and a_n, b_n are given by

$$a_0 = a_0(f) = \frac{1}{\pi} \int_{-\pi}^{\pi} f(t)\, dt, \tag{5.4.22}$$

$$a_n = a_n(f) = \frac{1}{\pi} \int_{-\pi}^{\pi} f(t) \cos nt\, dt, \quad n \geq 1, \tag{5.4.23}$$

$$b_n = b_n(f) = \frac{1}{\pi} \int_{-\pi}^{\pi} f(t) \sin nt\, dt, \quad n \geq 1. \tag{5.4.24}$$

The preceding coefficients can be considered for every function $f \in X_{2\pi}$, and they are called the *real Fourier coefficients* of f. The corresponding trigonometric series (5.4.21) is called the *Fourier series of* f.

For a given $n \in \mathbb{N}$ we can define the linear operator $S_n^*: \mathscr{C}_{2\pi} \to \mathscr{C}_{2\pi}$ by

$$S_n^*(f)(x) := \frac{1}{2}a_0(f) + \sum_{k=1}^{n} (a_k(f)\cos kx + b_k(f)\sin kx) \tag{5.4.25}$$

for every $f \in \mathscr{C}_{2\pi}$ and $x \in \mathbb{R}$.

Thus the trigonometric polynomial $S_n^*(f)$ is the n-th *partial sum* of the Fourier series of f.

By using (5.4.22)–(5.4.24) we can also write

$$S_n^*(f)(x) = \frac{1}{2\pi} \int_{-\pi}^{\pi} f(x-t)D_n(t)\,dt, \tag{5.4.26}$$

where

$$D_n(t) := 1 + 2\sum_{k=1}^{n} \cos kt. \tag{5.4.27}$$

Clearly, the sequence $(D_n)_{n \in \mathbb{N}}$ is a kernel and the corresponding convolution operators are just the partial sum of the Fourier series (see (5.4.5)).

The kernel $(D_n)_{n \in \mathbb{N}}$ is called the *Dirichlet kernel* and is of fundamental interest in Fourier analysis; by multiplying (5.4.27) by $\sin(t/2)$ it is easy to check that, for every $x \in \mathbb{R}$,

$$D_n(t) = \begin{cases} \dfrac{\sin((2n+1)t/2)}{\sin(t/2)}, & \text{if } t \text{ is not a multiple of } 2\pi, \\[2ex] 2n+1 & \text{if } t \text{ is a multiple of } 2\pi. \end{cases} \tag{5.4.28}$$

From this it is possible to deduce that the Dirichlet kernel is not an approximate identity since it is not a bounded sequence in $L_{2\pi}^1$; to be more precise, we have

$$\left| \|D_n\|_1 - \frac{4}{\pi^2}\log n \right| \le M \tag{5.4.29}$$

for every $n \in \mathbb{N}$, where M is a suitable constant (see Butzer and Nessel [1971, Proposition 1.2.3]).

On the other hand it is not difficult to show that $\|S_n^*\| = \|D_n\|_1$ (see, e.g., Butzer and Nessel [1971, Proposition 1.3.1]) so that the sequence $(S_n^*)_{n \in \mathbb{N}}$ is not bounded as well. Consequently the uniform boundedness principle ensures the existence of a continuous function $f \in \mathscr{C}_{2\pi}$ such that the sequence $(S_n^*(f))_{n \in \mathbb{N}}$ does not converge uniformly to f.

Although the Fourier series does not converge for every $f \in \mathscr{C}_{2\pi}$, we may hope to use some other summation methods to obtain trigonometric series which converge to f for every $f \in \mathscr{C}_{2\pi}$.

We begin by considering the Cesaro method of summation.

5.4.4 Fejér Convolution operators.
In this first example we shall deal with the arithmetic means of the partial sum of the Fourier series. Thus, for every $n \in \mathbb{N}$ and $x \in \mathbb{R}$, we define

$$\varphi_n(x) := \frac{1}{n+1} \sum_{k=0}^{n} D_k(x), \tag{5.4.30}$$

where $D_0 := 1$.

On account of (5.4.27), it is easy to show that

$$\varphi_n(x) = 1 + 2 \sum_{k=1}^{n} \left(1 - \frac{k}{n+1}\right) \cos kx. \tag{5.4.31}$$

Moreover, a different representation of φ_n can be obtained by multiplying the right side of (5.4.30) by $\sin(x/2)$; this yields, for every $x \in \mathbb{R}$,

$$\varphi_n(x) = \begin{cases} \dfrac{\sin^2((n+1)x/2)}{(n+1)\sin^2(x/2)}, & \text{if } x \text{ is not a multiple of } 2\pi, \\[3mm] n+1, & \text{if } x \text{ is a multiple of } 2\pi. \end{cases} \tag{5.4.32}$$

By (5.4.32), it follows that the sequence $(\varphi_n)_{n \in \mathbb{N}}$ is a kernel which will be called the *Fejér kernel*. Furthermore, it is positive and on account of (5.4.32) we have

$$\sup_{\delta \le |x| \le \pi} |\varphi_n(x)| \le \frac{1}{(n+1)\sin^2(\delta/2)} \tag{5.4.33}$$

for every $0 < \delta < \pi$.

Consequently the Fejér kernel $(\varphi_n)_{n \in \mathbb{N}}$ is an approximate identity.

Now, for every $n \in \mathbb{N}$, consider the operator $F_n \colon \mathbb{X}_{2\pi} \to \mathbb{X}_{2\pi}$ defined as in (5.4.5); thus

$$F_n(f)(x) := \frac{1}{2\pi} \int_{-\pi}^{\pi} f(x-t) \frac{\sin^2(n+1)t/2}{(n+1)\sin^2 t/2} \, dt \tag{5.4.34}$$

for every $f \in \mathbb{X}_{2\pi}$ and $x \in \mathbb{R}$.

The positive linear operator F_n is called the *n-th Fejér convolution operator* or the *n-th singular integral of Fejér*.

By using (5.4.26) and (5.4.31), we also obtain

$$F_n(f)(x) = \frac{1}{n+1} \sum_{k=0}^{n} S_k^*(f)(x), \qquad (5.4.35)$$

where $S_0^*(f) := \frac{1}{2} a_0(f)\mathbf{1}$, i.e., $F_n(f)$ is the arithmetic mean of the first n partial sums of the Fourier series of f, and hence it is a trigonometric polynomial.

On account of Theorem 5.4.1 we obtain that for every $f \in \mathbb{X}_{2\pi}$

$$\lim_{n \to \infty} F_n(f) = f \quad \text{in } \mathbb{X}_{2\pi}. \qquad (5.4.36)$$

Among other things (5.4.36) furnishes a constructive proof of the classical *Weierstrass approximation theorem* for periodic functions:

The space of all trigonometric polynomials is dense in $\mathbb{X}_{2\pi}$. (5.4.37)

By virtue of Theorems 5.4.2 and 5.4.3 we can also give a quantitative estimate of the rate of convergence described in (5.4.36). First, we observe that from (5.4.7) and (5.4.32) it follows that

$$\beta_n := F_n\left(\sin^2 \frac{\psi_x}{2} \right)(x) = \frac{1}{2\pi} \int_{-\pi}^{\pi} \frac{\sin^2((n+1)t/2)}{n+1}\, dt = \frac{1}{2(n+1)}$$

for every $n \geq 1$.

Consequently, since the kernel $(\varphi_n)_{n \in \mathbb{N}}$ is normalized, (5.4.12) and (5.4.17) yield the following estimate

$$\| F_n(f) - f \|_{\mathbb{X}_{2\pi}} \leq \left(1 + \frac{\pi}{\delta\sqrt{2(n+1)}} \right) \omega^*(f, \delta), \qquad (5.4.38)$$

which holds for every $f \in \mathbb{X}_{2\pi}$, $n \geq 1$ and $\delta > 0$, where $\omega^*(f, \delta)$ is defined as in the Remark to Theorem 5.4.3.

In particular, if we set $\delta = (n+1)^{-1/2}$, we have

$$\| F_n(f) - f \|_{\mathbb{X}_{2\pi}} \leq (1 + \pi/\sqrt{2})\omega^*(f, (n+1)^{-1/2}). \qquad (5.4.39)$$

Moreover, if f is differentiable and $f' \in \mathscr{C}_{2\pi}$, we can apply (5.4.14) to obtain

$$\| F_n(f) - f \| \leq \frac{\pi}{\sqrt{2(n+1)}} \left(\| f' \| + \left(1 + \frac{\pi}{\delta\sqrt{2(n+1)}} \right) \omega(f', \delta) \right) \qquad (5.4.40)$$

and in particular, setting $\delta = (n + 1)^{-1/2}$,

$$\|F_n(f) - f\| \leq \frac{\pi}{\sqrt{2(n + 1)}}(\|f'\| + (1 + \pi/\sqrt{2})\omega(f', (n + 1)^{-1/2})). \quad (5.4.41)$$

Another important example of a convolution approximation process was exhibited by Jackson [1930].

5.4.5 Jackson Convolution operators.
In this example we consider the sequence $(j_n)_{n \in \mathbb{N}}$ in $L^1_{2\pi}$ defined by

$$j_n(x) := a_n \varphi_n(x)^2 \quad (5.4.42)$$

for every $n \in \mathbb{N}$ and $x \in \mathbb{R}$, where φ_n is the n-th Fejér kernel and $a_n :=$
$\left(\dfrac{1}{2\pi} \displaystyle\int_{-\pi}^{\pi} \varphi_n(t)^2 \, dt\right)^{-1}$.

Thus $(j_n)_{n \in \mathbb{N}}$ is a positive normalized kernel which is called the *Jackson kernel*. The corresponding positive linear operator J_n defined according to (5.4.5) is called the *n-th Jackson convolution operator*.

On account of (5.4.31), it is easy to obtain $a_n = \dfrac{3(n + 1)}{2(n + 1)^2 + 1}$, so that

$$J_n(f)(x) = \frac{3}{2\pi(n + 1)(2(n + 1)^2 + 1)} \int_{-\pi}^{\pi} f(x - t) \frac{\sin^4(n + 1)t/2}{\sin^4 t/2} \, dt \quad (5.4.43)$$

for every $f \in \mathbb{X}_{2\pi}$ and $x \in \mathbb{R}$.

By using the same expansion of $\varphi_n(t)^2$, we obtain

$$\beta_n := \frac{1}{2\pi} \int_{-\pi}^{\pi} j_n(t) \sin^2 \frac{t}{2} \, dt$$

$$= \frac{a_n}{4\pi} \int_{-\pi}^{\pi} (1 - \cos t)\varphi_n(t)^2 \, dt$$

$$= \frac{a_n}{2(n + 1)} = \frac{3/2}{2(n + 1)^2 + 1}$$

so that $\beta_n \to 0$ as $n \to \infty$.

Therefore, according to Theorem 5.4.1, for every $f \in \mathbb{X}_{2\pi}$ we have

$$\lim_{n \to \infty} J_n(f) = f \quad \text{in } \mathbb{X}_{2\pi}. \quad (5.4.44)$$

Furthermore, since $\beta_n \leq \dfrac{1}{n^2}$, by (5.4.13), (5.4.15) and (5.4.18) we obtain

$$\|J_n(f) - f\|_{\mathbb{X}_{2\pi}} \leq (1 + \pi)\omega^*\left(f, \frac{1}{n}\right), \tag{5.4.45}$$

and, if f is differentiable with $f' \in \mathscr{C}_{2\pi}$,

$$\|J_n(f) - f\| \leq \frac{\pi}{n}\left(\|f'\| + (\pi + 1)\omega\left(f, \frac{1}{n}\right)\right). \tag{5.4.46}$$

As it can be seen, the operators J_n correspond to a suitable summation method of the Fourier series of f.

By using them, Jackson in 1912 obtained the celebrated estimates of the best trigonometric approximation $E_n(f)$ of an arbitrary $f \in \mathbb{X}_{2\pi}$ by trigonometric polynomials of degree n. More precisely he proved that

$$E_n(f) \leq c\omega^*\left(f, \frac{1}{n}\right), \tag{5.4.47}$$

where $E_n(f) := \min\{\|u - f\|_{\mathbb{X}_{2\pi}} | u \text{ trigonometric polynomial of degree } n\}$ and c is a suitable constant independent of f and n (see Jackson [1930] or Butzer and Nessel [1971, Theorem 2.2.1]). In fact, in 1962 Korneichuk showed that $E_n(f) \leq \omega(f, \pi/(n + 1))$ for every $f \in \mathscr{C}_{2\pi}$ and $n \geq 1$ and the coefficient 1 of $\omega(f, \pi/(n + 1))$ is the best possible one independent of f and n (see Korneichuk [1991] or Cheney [1982, p. 144]).

To illustrate some other summation methods of Fourier series we return to the sequence $(F_n)_{n \in \mathbb{N}}$ defined by (5.4.34) and deduce the identity

$$F_n(f)(x) = \frac{1}{n + 1}\sum_{k=0}^{n} S_k^*(f)(x)$$

$$= \frac{1}{2}a_0(f) + \frac{n}{n + 1}(a_1(f)\cos x + b_1(f)\sin x) + \cdots$$

$$+ \frac{1}{n + 1}(a_n(f)\cos nx + b_n(f)\sin nx)$$

$$= \frac{1}{2}a_0(f) + \sum_{k=1}^{n}\frac{n - k + 1}{n + 1}(a_k(f)\cos kx + b_k(f)\sin kx),$$

which holds for every $f \in \mathbb{X}_{2\pi}$, $n \geq 1$ and $x \in \mathbb{R}$.

If we compare this representation formula with (5.4.25), we realize that the factor $\rho_{n,k} := (n + 1 - k)/(n + 1)$ in the preceding sum substantially modifies the convergence properties of the Fourier series.

So, we can ask whether we can find some general condition on a factor $\rho_{n,k}$ ($n, k \in \mathbb{N}$) which furnish satisfactory convergence properties.

More generally, we shall consider a net $(\theta_i)_{i \in I}^{\leq}$ of real sequences satisfying the assumptions

$$\theta_i \in \ell^1 \left(\text{i.e., } \sum_{k=0}^{\infty} |\theta_i(k)| < +\infty \right) \quad \text{and} \quad \theta_i(0) = 1 \quad \text{for every } i \in I. \quad (5.4.48)$$

Such a family will be called a *factor*.

For every $f \in L^1_{2\pi}$ we see that the series

$$\frac{1}{2} a_0(f) + \sum_{k=1}^{\infty} \theta_i(k)(a_k(f) \cos kx + b_k(f) \sin kx),$$

is uniformly convergent to a 2π-periodic continuous real function because

$$\sum_{k=1}^{\infty} |\theta_i(k)(a_k(f) \cos kx + b_k(f) \sin kx)| \leq 4\|f\|_1 \sum_{k=1}^{\infty} |\theta_i(k)| < +\infty.$$

So we can define the linear operator $L_i^\theta: L^1_{2\pi} \to \mathscr{C}_{2\pi}$ by

$$L_i^\theta(f)(x) := \frac{1}{2} a_0(f) + \sum_{k=1}^{\infty} \theta_i(k)(a_k(f) \cos kx + b_k(f) \sin kx) \quad (5.4.49)$$

for every $f \in L^1_{2\pi}$ and $x \in \mathbb{R}$.

The function $L_i^\theta(f)$ is usually referred to as the *i-th θ-mean of the Fourier series of f*. If $(L_i^\theta(f))_{i \in I}^{\leq}$ converges in $\mathbb{X}_{2\pi}$ for every $f \in \mathbb{X}_{2\pi}$, we say that $(L_i^\theta)_{i \in I}^{\leq}$ is a *summation process* or a *summation method* in $\mathbb{X}_{2\pi}$.

If $\theta = (\theta_i)_{i \in I}^{\leq}$ is a factor, by (5.4.22)–(5.4.24) and the uniform convergence of the series in (5.4.49), we obtain the following representation of each operator L_i:

$$L_i^\theta(f)(x) = \frac{1}{2\pi} \int_{-\pi}^{\pi} f(t) \chi_i^\theta(t - x) \, dt = \frac{1}{2\pi} \int_{-\pi}^{\pi} \chi_i^\theta(x) f(t - x) \, dt \quad (5.4.50)$$

for every $f \in L^1_{2\pi}$ and $x \in \mathbb{R}$, where

$$\chi_i^\theta(x) := 1 + 2 \sum_{k=1}^{\infty} \theta_i(k) \cos kx. \quad (5.4.51)$$

Hence

$$\int_{-\pi}^{\pi} \chi_i^\theta(t)\, dt = 2\pi, \tag{5.4.52}$$

so that the net $(\chi_i^\theta)_{i \in I}^{\le}$ is a normalized (even) kernel and the operators L_i^θ are, in fact, the convolution operators associated with $(\chi_i^\theta)_{i \in I}^{\le}$.

To study the convergence of the net $(L_i^\theta)_{i \in I}^{\le}$ we therefore apply Theorems 5.4.1, 5.4.2 and 5.4.3.

5.4.6 Theorem. *Let* $\theta = (\theta_i)_{i \in I}^{\le}$ *be a factor and consider the net* $(L_i^\theta)_{i \in I}^{\le}$ *of linear operators defined by* (5.4.49).

Assume that, for every $i \in I$ *and* $x \in \mathbb{R}$,

$$0 \le 1 + 2 \sum_{k=1}^{\infty} \theta_i(k) \cos kx \tag{5.4.53}$$

and

$$\lim_{i \in I}^{\le} \theta_i(1) = 1. \tag{5.4.54}$$

Then $\lim_{i \in I}^{\le} \| L_i^\theta(f) - f \|_{X_{2\pi}} = 0$ *for every* $f \in X_{2\pi}$.

Moreover, given $f \in X_{2\pi}$, $i \in I$ *and* $\delta > 0$, *we have*

$$\| L_i^\theta(f) - f \|_{X_{2\pi}} \le \left(1 + \frac{\pi}{\delta} \sqrt{\frac{1 - \theta_i(1)}{2}} \right) \omega^*(f, \delta) \tag{5.4.55}$$

and, in particular,

$$\| L_i^\theta(f) - f \|_{X_{2\pi}} \le (1 + \pi) \omega^* \left(f, \sqrt{\frac{1 - \theta_i(1)}{2}} \right), \tag{5.4.56}$$

where $\omega^*(f, \delta)$ *is defined as in the* **Remark** *to Theorem 5.4.2.*

Finally, if $f \in \mathscr{C}_{2\pi}$ *is differentiable and* $f' \in \mathscr{C}_{2\pi}$, *then*

$$\| L_i^\theta(f) - f \| \le \pi \sqrt{\frac{1 - \theta_i(1)}{2}} \left(\| f' \| + \left(1 + \frac{\pi}{\delta} \sqrt{\frac{1 - \theta_i(1)}{2}} \right) \omega(f', \delta) \right) \tag{5.4.57}$$

and

$$\| L_i^\theta(f) - f \| \le \pi \sqrt{\frac{1 - \theta_i(1)}{2}} \left(\| f' \| + (\pi + 1) \omega \left(f', \sqrt{\frac{1 - \theta_i(1)}{2}} \right) \right). \tag{5.4.58}$$

Proof. Condition (5.4.53) ensures that the kernel $(\chi_i^\theta)_{i \in I}^{\leq}$ defined by (5.4.51) is positive. Moreover a simple calculation shows that

$$\beta_i = \frac{1 - \theta_i(1)}{2}$$

for every $i \in I$ (see also (5.4.7), (5.4.50) and (5.4.51)).

Hence, the result follows from Theorem 5.4.1 and formulas (5.4.12)–(5.4.18).

$\square$

Fejér's and Jackson's convolution operators are examples of summation processes associated to suitable factors. Here we give other important examples.

5.4.7 Fejér-Korovkin convolution operators. We observe that the degree of the trigonometric polynomial J_n is $2n - 2$. In the next example we briefly describe a kernel whose corresponding n-th convolution operators is a trigonometric polynomial of degree at most n.

For every $n \in \mathbb{N}$ and $x \in \mathbb{R}$, we define

$k_n(x)$

$$:= \begin{cases} \dfrac{2\sin^2 \pi/(n+2)}{n+2}\left(\dfrac{\cos(n+2)x/2}{\cos \pi/(n+2) - \cos x}\right)^2, & \text{if } x \neq \pm\dfrac{\pi}{n+2} + 2j\pi,\, j \in \mathbb{Z}, \\[4mm] \dfrac{n+2}{2}, & \text{if } x = \pm\dfrac{\pi}{n+2} + 2j\pi \text{ for some } j \in \mathbb{Z}. \end{cases}$$

$$(5.4.59)$$

It can be easily shown that the sequence $(k_n)_{n \in \mathbb{N}}$ is a positive normalized kernel, which is called the *Fejér-Korovkin kernel*. For every $n \in \mathbb{N}$, the positive linear operator K_n defined according to (5.4.5), i.e.,

$$K_n(f)(x) := \frac{1}{2\pi}\int_{-\pi}^{\pi} f(x - t)k_n(t)\,dt, \qquad (5.4.60)$$

is called the *n-th Fejér-Korovkin convolution operator*.

An argument similar to the one used in Examples 5.4.4 and 5.4.5 gives the following representation of the Fejér-Korovkin kernel:

$$k_n(x) = 1 + 2\sum_{k=1}^{n} \theta_n(k)\cos kx, \qquad (5.4.61)$$

where

$$\theta_n(k) = \frac{1}{2(n+2)\sin \pi/(n+2)}\left((n-k+3)\sin \frac{k+1}{n+2}\pi - (n-k+1)\sin\frac{k-1}{n+2}\pi\right).$$

$$(5.4.62)$$

Since

$$\theta_n(1) = \frac{1}{2(n+2)\sin \pi/(n+2)}(n+2)\sin \frac{2\pi}{n+2} = \cos\frac{\pi}{n+2} \to 1$$

as $n \to \infty$, we have by Theorem 5.4.6

$$\lim_{n\to\infty} K_n(f) = f \quad \text{in } \mathbb{X}_{2\pi} \qquad (5.4.63)$$

for every $f \in \mathbb{X}_{2\pi}$.

Finally, since $\dfrac{1-\theta_i(1)}{2} \le \dfrac{\pi^2}{2(n+2)^2} \le \dfrac{1}{n^2}$, we obtain by (5.4.56) for every $n \in \mathbb{N}$ and $f \in \mathbb{X}_{2\pi}$

$$\|K_n(f) - f\|_{\mathbb{X}_{2\pi}} \le (1+\pi)\omega^*\left(f, \frac{1}{n}\right) \qquad (5.4.64)$$

and by (5.4.58), if f is differentiable with $f' \in \mathscr{C}_{2\pi}$

$$\|K_n(f) - f\| \le \frac{\pi}{n}\left(\|f'\| + (\pi+1)\omega\left(f', \frac{1}{n}\right)\right). \qquad (5.4.65)$$

5.4.8 Abel-Poisson convolution operators. In the next example we shall consider a net of convolution operators whose index set is the interval $[0, 1[$ endowed with the usual ordering $\le$.

More precisely, given $r \in [0, 1[$ and $k \in \mathbb{N}_0$, set

$$\theta_r(k) := r^k, \qquad (5.4.66)$$

where $0^0 = 1$ by convention.

The family $(\theta_r)_{0\le r<1}$ is obviously a factor because

$$\sum_{k=0}^{\infty} r^k = \frac{1}{1-r} < +\infty \quad \text{and} \quad \theta_r(0) = 1$$

for every $r \in [0, 1[$. It is called the *Abel-Poisson factor*.

The kernel $(p_r)_{0 \le r < 1}$ corresponding to the Abel-Poisson factor is called the *Abel-Poisson kernel* and, for every $f \in L_{2\pi}^1$, the corresponding means are called the *Abel-Poisson means* of the Fourier series of f.

By (5.4.51), we have the following explicit expression of the Abel-Poisson kernel:

$$p_r(x) = 1 + 2 \sum_{k=1}^{\infty} r^k \cos kx = \frac{1 - r^2}{1 - 2r \cos x + r^2} \qquad (5.4.67)$$

where the last equality follows by setting $z = re^{ix}$ in the equality

$$1 + 2 \sum_{k=1}^{\infty} z^k = \frac{1 + z}{1 - z} \qquad (z \in \mathbb{C}, |z| < 1)$$

and by taking the real part of both sides.

Denote by $(P_r)_{0 \le r < 1}$ the net of the corresponding linear operators defined by (5.4.49) (or by (5.4.50)).

Then, P_r is called the *r-th Abel-Poisson convolution operator*. More explicitly, for every $f \in L_{2\pi}^1$ and $x \in \mathbb{R}$, we have

$$P_r(f)(x) = \frac{1}{2} a_0(f) + \sum_{k=1}^{\infty} r^k(a_k(f) \cos kx + b_k(f) \sin kx)$$

$$= \frac{1 - r^2}{2\pi} \int_{-\pi}^{\pi} \frac{f(t - x)}{1 - 2r \cos t + r^2} dt. \qquad (5.4.68)$$

It is easy to show that the hypotheses of Theorem 5.4.6 are satisfied and hence we conclude that for every $f \in \mathbb{X}_{2\pi}$

$$\lim_{r \to 1^-} P_r(f) = f \quad \text{in } \mathbb{X}_{2\pi}. \qquad (5.4.69)$$

Direct estimates can be obtained for arbitrary $f \in \mathbb{X}_{2\pi}$ and $r \in [0, 1[$ by taking $\delta = \sqrt{1 - r}$ in (5.4.55) and (5.4.57). In fact we have

$$\|P_r(f) - f\|_{\mathbb{X}_{2\pi}} \le \left(1 + \frac{\pi}{\sqrt{2}}\right) \omega^*(f, \sqrt{1 - r}) \qquad (5.4.70)$$

and, if f is differentiable and $f' \in \mathscr{C}_{2\pi}$

$$\|P_r(f) - f\| \le \pi \sqrt{\frac{1 - r}{2}} \left(\|f'\| + (1 + \pi/\sqrt{2})\omega(f', \sqrt{1 - r})\right). \qquad (5.4.71)$$

As a final remark we emphasize the fact that the Abel-Poisson convolution operators are intimately connected with the Dirichlet problem on the unit disc. In fact it can be shown that, given $f \in X_{2\pi}$, the function $u(r, x) := P_r(f)(x)$ ($|x| \leq \pi$, $0 \leq r < 1$) is the solution of the *Dirichlet problem* (in the polar coordinates)

$$
\begin{cases}
\dfrac{\partial^2 u}{\partial r^2}(r, x) + \dfrac{1}{r}\dfrac{\partial u}{\partial r}(r, x) + \dfrac{1}{r^2}\dfrac{\partial^2 u}{\partial x^2}(r, x) = 0, & \text{for } |x| \leq \pi, 0 \leq r < 1, \\[2mm]
\lim_{r \to 1^-} u(\cdot, r) = f, & \text{in } X_{2\pi}
\end{cases}
\tag{5.4.72}
$$

(see Butzer and Nessel [1971, Sections 1.2.4 and 2.5.1] for more details).

In the last part of this section we briefly indicate more general summation methods of linear operators which extend the preceding ones.

To this end we follow the unifying approach suggested by Nishishiraho [1981] who introduced the notion of $\mathscr{A}$-summation process.

Given a compact Hausdorff space X, we consider a family $\mathscr{A} = (A^{(\lambda)})_{\lambda \in \Lambda}$ of infinite real matrices. To fix the ideas suppose that

$$
A^{(\lambda)} = (a^{(\lambda)}_{n,m})_{(n,m) \in \mathbb{N}_0 \times \mathbb{N}_0}
$$

for every $\lambda \in \Lambda$.

A sequence $(T_n)_{n \in \mathbb{N}}$ of bounded linear operators from $\mathscr{C}(X)$ into $\mathscr{B}(X)$ is called an $\mathscr{A}$-*summation process on* $\mathscr{C}(X)$ if, for every $f \in \mathscr{C}(X)$, $\lambda \in \Lambda$ and $n \in \mathbb{N}$, the series $\sum_{m=0}^{\infty} a^{(\lambda)}_{n,m} T_m(f)$ converges uniformly and, further,

$$
\lim_{n \to \infty} \left\| \sum_{m=0}^{\infty} a^{(\lambda)}_{n,m} T_m(f) - f \right\| = 0 \quad \text{uniformly in } \lambda \in \Lambda.
\tag{5.4.73}
$$

As the following examples show, there is a wide variety of families $\mathscr{A} = (A^{(\lambda)})_{\lambda \in \Lambda}$ of particular interest which cover several summation methods scattered in the literature.

Examples. 1. If $A = (a_{n,m})_{(n,m) \in \mathbb{N}_0 \times \mathbb{N}_0}$ is an infinite matrix and $A^{(\lambda)} = A$ for every $\lambda \in \Lambda$, we obtain the classical *matrix summability* by A, i.e., a sequence $(L_n)_{n \in \mathbb{N}_0}$ is an A-summation process on $\mathscr{C}(X)$ if, for every $f \in \mathscr{C}(X)$ and $n \in \mathbb{N}_0$, the series $\sum_{m=0}^{\infty} a_{n,m} L_m(f)$ converges uniformly to the function f.

In particular the summation processes associated with factors satisfying (5.4.53) and (5.4.54) fall within this case.

2. If $\Lambda = \mathbb{N}$, we obtain the summation method $\mathscr{A} = (A^{(p)})_{p \in \mathbb{N}}$ introduced by Bell [1973].

In particular, if $A^{(p)} = (a^{(p)}_{n,m})_{(n,m) \in \mathbb{N}_0 \times \mathbb{N}_0}$ is given by

$$a^{(p)}_{n,m} := \begin{cases} \dfrac{1}{n}, & \text{if } p + 1 \leq m \leq p + n, \\[2mm] 0, & \text{otherwise}, \end{cases}$$

we obtain the notion of *almost convergence* introduced by Lorentz [1948].

3. For every $\lambda \in \Lambda$ we can consider a sequence $q^{(\lambda)} = (q^{(\lambda)}_n)_{n \in \mathbb{N}_0}$ of positive real numbers satisfying $q^{(\lambda)}_0 + q^{(\lambda)}_1 + \cdots + q^{(\lambda)}_n > 0$ for every $n \in \mathbb{N}_0$, and we can define the matrix $A^{(\lambda)} = (a^{(\lambda)}_{n,m})_{(n,m) \in \mathbb{N}_0 \times \mathbb{N}_0}$ by

$$a^{(\lambda)}_{n,m} := \frac{q^{(\lambda)}_{n-m}}{q^{(\lambda)}_0 + q^{(\lambda)}_1 + \cdots + q^{(\lambda)}_n}, \quad \text{if } m \leq n, \quad \text{and} \quad a^{(\lambda)}_{n,m} := 0, \quad \text{if } m > n.$$

This kind of summability is called *Nörlund summability* in the case where $q^{(\lambda)} = (q_n)_{n \in \mathbb{N}_0}$ is a fixed sequence of positive real numbers satisfying $q_0 > 0$.

Another case of interest is that in which Λ is a set of positive real numbers, $\beta > 0$ and, for every $\lambda \in \Lambda$ and $n \in \mathbb{N}_0$, $q^{(\lambda)}_n = C^{(\lambda+\beta-1)}_n$, where $C^{(\tau)}_n := \binom{n + \tau}{n}(\tau > -1)$.

In particular, if $\Lambda = \{0\}$, we have $q^{(0)}_n = \binom{n + \beta - 1}{n}$ and we obtain the *Cesaro summability* of order β.

4. In the following examples we only briefly indicate the set Λ and the generic entry $a^{(\lambda)}_{n,m}$ of the corresponding matrix $A^{(\lambda)}$.

(i) $\Lambda \subset \,]0, +\infty[$, $\beta > -1$ and

$$a^{(\lambda)}_{n,m} := \frac{C^{(\lambda-1)}_{n-m} C^{(\beta)}_m}{C^{(\beta+\lambda)}_n}, \quad \text{if } m \leq n, \quad \text{and} \quad a^{(\lambda)}_{n,m} := 0, \quad \text{if } m > n,$$

where $C^{(\tau)}_n$ is defined as in the preceding Example 3.

(ii) $\Lambda \subset [0, 1]$ and

$$a^{(\lambda)}_{n,m} := \binom{n}{m} \lambda^m (1 - \lambda)^{n-m}, \quad \text{if } m \leq n, \quad \text{and} \quad a^{(\lambda)}_{n,m} := 0, \quad \text{if } m > n.$$

(iii) $\Lambda \subset [0, +\infty[$ and $a^{(\lambda)}_{n,m} := \dfrac{\exp(-n\lambda)(n\lambda)^m}{m!}$ for every $m, n \in \mathbb{N}_0$.

(iv) $\Lambda \subset [0, 1[$ and $a_{n,m}^{(\lambda)} := \binom{n+m}{m} \lambda^m (1-\lambda)^{n+1}$ for every $m, n \in \mathbb{N}_0$.

(v) $\Lambda \subset [0, +\infty[$ and $a_{n,m}^{(\lambda)} := \binom{n+m-1}{m} \lambda^m (1+\lambda)^{-n-m}$ for every $m, n \in \mathbb{N}_0$.

In all the above Examples 2–4 the matrices $A^{(\lambda)} = (a_{n,m}^{(\lambda)})_{(n,m) \in \mathbb{N}_0 \times \mathbb{N}_0}$ are sto-chastic, i.e., $a_{n,m}^{(\lambda)} \geq 0$ and $\sum_{m=0}^{\infty} a_{n,m}^{(\lambda)} = 1$ for every $n, m \in \mathbb{N}_0$.

The question now arises under which conditions a given sequence of bounded linear operators is an $\mathscr{A}$-summation process.

In this respect a qualitative result on the convergence of $\mathscr{A}$-summation processes can be obtained by applying some Korovkin-type theorems established in Chapter 4 to the family $(B_{n,\lambda})_{n \in \mathbb{N}}$ $(\lambda \in \Lambda)$ of sequences of linear operators from $\mathscr{C}(X)$ into $\mathscr{B}(X)$ defined by

$$B_{n,\lambda}(f) := \sum_{m=0}^{\infty} a_{n,m}^{(\lambda)} T_m(f). \tag{5.4.74}$$

Here we indicate a simple criterion in the framework of convex compact sets.

5.4.9 Theorem. *Let K be a convex compact set of some locally convex Hausdorff space E and let $\mathscr{A} = (A^{(\lambda)})_{\lambda \in \Lambda}$ be a family of infinite stochastic matrices $A^{(\lambda)} = (a_{n,m}^{(\lambda)})_{(n,m) \in \mathbb{N}_0 \times \mathbb{N}_0}$.*

Moreover, let $(T_n)_{n \in \mathbb{N}_0}$ be a sequence of positive linear operators from $\mathscr{C}(K)$ into $\mathscr{B}(K)$ such that $T_n(h) = h$ for every $n \in \mathbb{N}_0$ and $h \in A(K)$.

If there exist a constant $C > 0$ and a sequence $(\xi_n)_{n \in \mathbb{N}_0}$ of positive real numbers such that

$$\lim_{n \to \infty} \sum_{m=0}^{\infty} a_{n,m}^{(\lambda)} \xi_m = 0 \quad \text{uniformly in } \lambda \in \Lambda$$

and

$$\left\| \sum_{j=1}^{m} (T_n(h_j^2) - h_j^2) \right\| \leq C \xi_n \left\| \sum_{j=1}^{m} h_j^2 \right\|$$

for every $n, m \in \mathbb{N}_0$ and $h_1, \ldots, h_m \in E_K'$ (see (5.1.7)), then $(T_n)_{n \in \mathbb{N}}$ is an $\mathscr{A}$-summation process.

Furthermore, for every $f \in \mathscr{C}(K)$ and $n \geq 1$

$$\sup_{\lambda \in \Lambda} \left\| \sum_{m=0}^{\infty} a_{n,m}^{(\lambda)} T_m(f) - f \right\| \leq 2\Omega(f, \sqrt{C\gamma_n}). \tag{5.4.75}$$

where

$$\gamma_n := \sup_{\lambda \in \Lambda} \sum_{m=0}^{\infty} a_{n,m}^{(\lambda)} \xi_m$$

and $\Omega(f, \delta)$ is defined by (5.1.10).

Proof. We apply Theorem 4.4.6, (2), to the parametric sequences $(B_{n,\lambda})_{n \in \mathbb{N}}$, $\lambda \in \Lambda$, defined by (5.4.74) and to the subset $M = E'_K$ which separates the points of K by virtue of the Hahn-Banach theorem. In fact, if $h \in M$, clearly $B_{n,\lambda}(h) = h$ and

$$\| B_{n,\lambda}(h^2) - h^2 \| \leq \sum_{m=0}^{\infty} a_{n,m}^{(\lambda)} \| T_m(h^2) - h^2 \| \leq C \sum_{m=0}^{\infty} a_{n,m}^{(\lambda)} \xi_m.$$

Accordingly $\lim_{n \to \infty} L_{n,\lambda}(h^2) = h^2$ in $\mathcal{B}(K)$ uniformly in $\lambda \in \Lambda$ and hence $(T_n)_{n \in \mathbb{N}_0}$ is an $\mathcal{A}$-summation process.

Finally, estimate (5.4.75) is a direct consequence of Corollary 5.1.7. □

Notes and references

The main part of the material in this section is classical and is taken from the books of Butzer and Nessel [1971], De Vore [1972] and Korovkin [1960].

We refer the reader to the above mentioned books for further aspects concerning convolution operators, such as direct and inverse theorems and saturation properties.

The proofs of Theorems 5.4.1 and 5.4.3 are the same as in Butzer and Nessel [1971, Theorems 1.1.5, 1.3.7, 1.5.10].

In Campiti [1987a, Proposition 3.1] one can also find an extension of Theorem 5.4.1 which involves the space of 2π-periodic Riemann integrable functions equipped with the so-called Riemann sequential convergence.

Theorem 5.4.2 can be found in De Vore [1972] in a slightly different form.

The estimates of the rate of convergence in this section are not the best possible. Further sharper estimates can be found in the monographs quoted at beginning.

Fejér's operators were introduced by Fejér [1900], [1904] who first showed (5.4.36).

Jackson's operators were introduced by Jackson [1930].

Theorem 5.4.6 is due to Korovkin [1960]. Perhaps Abel-Poisson operators first appeared in Abel [1881].

Although the Fejér-Korovkin kernels (5.4.59) were first considered by Fejér [1916a], it was Korovkin [1958] who used the convolution operators K_n in the approximation by trigonometric operators.

Further detailed results concerning summation methods of Fourier series can be found in any book on the subject and, in particular, in the fine treatises of Bari [1964], R. E. Edwards [1967], Hardy and Rogosinsky [1944], Sz-Nagy [1950] and Zygmund [1959].

The notion of $\mathscr{A}$-summation process was introduced by Nishishiraho ([1981], [1982a], [1982b], [1983]).

Quantitative aspects of approximation processes of convolution operators in arbitrary Banach spaces have been also treated by Nishishiraho [1981].

In the particular case where $\Lambda = \mathbb{N}$, the concept of $\mathscr{A}$-summability was introduced by Bell [1973] and includes, besides the almost convergence introduced by Lorentz [1948], also the summation methods studied by Mazhar and Siddiqi [1967] and Jurkat and Peyerimhoff [1971a], [1971b]. Examples 1–4 regarding matrices of $\mathscr{A}$-summation processes are described in Nishishiraho [1983] (further ones can be found in Nishishiraho [1989]).

For further developments on $\mathscr{A}$-summation processes and, in particular, for the convergence of iterations of $\mathscr{A}$-summation processes, we refer again to Nishishiraho [1987], [1990].

Chapter 6

Applications to positive approximation processes on convex compact sets

In this final chapter we carry out a detailed analysis of further sequences of positive linear operators that have been studied recently and that seem to play a non-negligible role not only in approximation theory but also in the theory of partial differential equations and Markov processes.

These operators are defined in the framework of convex compact sets and their construction essentially depends on a positive projection.

This general approach has the advantage of unifying the presentation of various well-known approximation processes and, at the same time, of providing new ones both in univariate and multivariate settings and in the infinite dimensional case.

In Section 6.1 we first consider the approximation properties of these operators and then we proceed to show their monotonicity properties and a related maximum principle and, finally, the property of preserving the spaces of Hölder continuous functions.

In the subsequent sections we point out the existence of a (uniquely determined) Feller semigroup that can be represented in terms of powers of them.

The generator of that semigroup is explicitly determined in a core of its domain and, in the finite dimensional case, it is, in fact, an elliptic second order differential operator which degenerates on the Choquet boundary of the range of the projection.

As a consequence, by means of these operators we can study the qualitative properties of the Cauchy problems associated with some degenerate diffusion equations.

The last section is devoted to stressing some further applications in particular concrete settings where we illustrate in detail both the approximation processes and the corresponding differential operators.

There the reader will better frame both the results previously obtained and some new ones within the existing literature of other apparently unrelated fields of researches.

We finally point out that some results (especially those concerning the interval $[0, 1]$) also admit probabilistic interpretations that we show in some respects. In particular, we describe the transition function of the Markov process whose existence is guaranteed by the general theory (see Section 1.6), and we discuss some connections with a stochastic model from genetics.

6.1 Positive approximation processes associated with positive projections

In this section we define the sequences of positive linear operators we shall study throughout the whole chapter.

Their construction is based on a positive projection acting on the space of continuous functions defined on a convex compact set.

In this respect the examples and the results we stated in Section 3.3 are essential. In particular, in each of the settings considered there, our construction will furnish concrete and significant examples of positive approximation processes.

However, to simplify our exposition, we prefer to develop all the main results first and then to present the various examples in the last section.

We begin by fixing a metrizable convex compact subset K of a locally convex space E.

For every $f \in \mathscr{C}(K)$, $z \in K$ and $\alpha \in [0, 1]$ we denote by $f_{z,\alpha} \in \mathscr{C}(K)$ the function defined by

$$f_{z,\alpha}(x) := f(\alpha x + (1 - \alpha)z) \quad \text{for every } x \in K. \tag{6.1.1}$$

From now on we shall also fix a positive projection $T \colon \mathscr{C}(K) \to \mathscr{C}(K)$ with range

$$H := T(\mathscr{C}(K)). \tag{6.1.2}$$

To avoid triviality, we shall suppose that T *is different from the identity operator on* $\mathscr{C}(K)$.

We also assume that $T(h) = h$ for every affine function $h \in A(K)$, i.e.,

$$A(K) \subset H, \tag{6.1.3}$$

and that

$$h_{z,\alpha} \in H \quad \text{for every } z \in K, \alpha \in [0, 1] \text{ and } h \in H. \tag{6.1.4}$$

As a consequence of (6.1.3) and Proposition 2.6.3, we have that H contains the constants and separates the points of K and

$$\partial_e K \subset \partial_H^+ K. \tag{6.1.5}$$

We shall show that assumption (6.1.4) guarantees that $\partial_H^+ K$ is the union of faces of K.

To do so we first recall that for every $x \in K$ we have denoted by F_x the face of K generated by x, i.e., the smallest face of K containing x. More precisely we have

$$F_x = \{y \in K \,|\, \text{there exist } z \in K \text{ and } \alpha \in \,]0,1] \text{ such that } x = \alpha y + (1 - \alpha)z\},$$

(see Section 4.3, (4.3.15)).

After these preparations we can state the following result.

6.1.1 Proposition. *Let* $T: \mathscr{C}(K) \to \mathscr{C}(K)$ *be a positive projection satisfying conditions (6.1.3) and (6.1.4). Then the Choquet boundary of H is the union of faces of K, i.e.,*

$$\partial_H^+ K = \bigcup_{x \in \partial_H^+ K} F_x. \tag{6.1.6}$$

Accordingly, $\partial_H^+ K$ *is contained in the topological boundary* ∂K *of K.*

Proof. We have to show that $F_x \subset \partial_H^+ K$ for every $x \in \partial_H^+ K$. So let $x \in \partial_H^+ K$ and $y \in F_x$, and write $x = \alpha y + (1 - \alpha)z$ for some $z \in K$ and $a \in \,]0,1]$. Let Y: $\{\alpha u + (1 - \alpha)z \,|\, u \in K\}$ and define the homeomorphism $j: K \to Y$ by $j(u) := \alpha u + (1 - \alpha)z$ for every $u \in K$.

Given $\mu \in \mathscr{M}^+(K)$ such that $\mu = \varepsilon_y$ on H, consider the Radon measure $\upsilon \in \mathscr{M}^+(K)$ defined by $\upsilon(f) := \mu(f \circ j)$ $(f \in \mathscr{C}(K))$; then, for every $h \in H$, we have $h \circ j = h_{z,\alpha} \in H$ and consequently

$$\upsilon(h) = \mu(h \circ j) = \varepsilon_y(h \circ j) = h(\alpha y + (1 - \alpha)z) = h(x) = \varepsilon_x(h).$$

Since $x \in \partial_H^+ K$, we obtain $\upsilon = \varepsilon_x$. Accordingly, we also have $\mu = \varepsilon_y$ because, if $f \in \mathscr{C}(K)$ and if we denote by $g \in \mathscr{C}(K)$ a continuous extension of $f \circ j^{-1}$, then

$$\mu(f) = \mu(g \circ j) = \upsilon(g) = \varepsilon_x(g) = g(x) = f(j^{-1}(x)) = f(y) = \varepsilon_y(f).$$

This reasoning shows that $y \in \partial_H^+ K$ as well.

The last part of the statement follows from Proposition 3.3.1. Indeed, since T is different from the identity operator, then $\partial_H^+ K \neq K$; therefore every face F_x $(x \in \partial_H^+ K)$ is different from K and hence is contained in ∂K. □

Remark. We observe that a closed face F of K is contained in $\partial_H^+ K$ if and only if $H_{|F} = \mathscr{C}(F)$, where $H_{|F} := \{h_{|F} \,|\, h \in H\}$.

Indeed in the Remark to Proposition 3.3.1 we have already noticed that $\mathscr{C}(F) = H_{|F}$ provided F is a closed subset of $\partial_H^+ K$. Conversely, if $\mathscr{C}(F) = H_{|F}$ and if we consider $x \in F$ and $\mu \in \mathscr{M}^+(K)$ such that $\mu = \varepsilon_x$ on H, then x is the barycenter of μ by virtue of (6.1.3). Therefore, we have $\mathrm{Supp}(\mu) \subset F$ (see (4.3.16)).

Hence it follows that $\mu = \varepsilon_x$ because $\mu(f) = \mu(h) = h(x) = f(x)$ for every $f \in \mathscr{C}(K)$ and $h \in H$ such that $h = f$ on F.

Now, for every $x \in K$ we consider the probability Radon measure $\mu_x^T \in \mathscr{M}^+(K)$ defined by (3.1.3), i.e., for every $f \in \mathscr{C}(K)$

$$\mu_x^T(f) := Tf(x). \tag{6.1.7}$$

By assumption (6.1.3), the barycenter of μ_x^T is the point x.
Furthermore, according to Remark 3 to Theorem 3.3.3, we have

$$\operatorname{Supp}(\mu_x^T) \subset \partial_H^+ K \tag{6.1.8}$$

for every $x \in K$.

At this point we are in a position to introduce the first sequence of linear operators which can be associated with the positive projection T. These operators generalize both the classical Bernstein operators defined on the interval $[0, 1]$ and their extensions to the canonical simplex and the hypercube of $\mathbb{R}^p$ (see Section 6.3).

Bernstein-Schnabl operators. We fix an infinite lower triangular stochastic matrix $P = (p_{nj})_{n \geq 1, j \geq 1}$, i.e., P is an infinite matrix of positive real numbers satisfying

$$p_{nj} = 0 \quad \text{for every } j > n \quad \text{and} \quad \sum_{j=1}^{\infty} p_{nj} = \sum_{j=1}^{n} p_{nj} = 1 \quad \text{for every } n \geq 1. \tag{6.1.9}$$

For every $n \geq 1$, we define the positive linear operator $B_{P,n} : \mathscr{C}(K) \to \mathscr{C}(K)$ by

$$B_{P,n}(f)(x) := \int_{K^n} f\left(\sum_{j=1}^{n} p_{nj} x_j \right) d\left(\bigotimes_{i=1}^{n} \mu_{x,i}^T \right)(x_1, \ldots, x_n)$$

$$= \int_K \cdots \int_K f\left(\sum_{j=1}^{n} p_{nj} x_j \right) d\mu_x^T(x_1) \ldots d\mu_x^T(x_n) \tag{6.1.10}$$

for every $f \in \mathscr{C}(K)$ and $x \in K$, where $\mu_{x,i}^T := \mu_x^T$ for every $i = 1, \ldots, n$.

The operator $B_{P,n}$ is called the *n-th Bernstein-Schnabl operator associated with the projection T and the matrix P*. In case of need we shall also write $B_{p_{n1}, \ldots, p_{nn}}$ instead of $B_{P,n}$.

A case of particular interest is when we consider the *arithmetic mean Toeplitz matrix* $P_0 = (p_{nj})_{n \geq 1, j \geq 1}$, which is defined by

$$p_{ni} := \frac{1}{n} \quad \text{if } i = 1, \ldots, n \quad \text{and} \quad p_{ni} := 0 \quad \text{if } i > n. \tag{6.1.11}$$

In this case the operator $B_{P,n}$ is denoted simply by B_n and is called the *n-th Bernstein-Schnabl operator associated with the projection T*.
Thus

$$B_n(f)(x) = \int_K \cdots \int_K f\left(\frac{x_1 + \cdots + x_n}{n}\right) d\mu_x^T(x_1)\ldots d\mu_x^T(x_n) \qquad (6.1.12)$$

for every $f \in \mathcal{C}(K)$ and $x \in K$.

As we shall see in the sequel, the general matrix P considered in (6.1.10) will allow us to represent other sequences of linear operators in terms of Bernstein-Schnabl operators.

Now, we proceed to state some identities which turn out to be useful in studying the convergence of Bernstein-Schnabl operators and their powers.

We first point out that, by Proposition 3.3.1, $B_{P,n}(f)$ interpolates the value of f at each $x \in \partial_H^+ K$ (and hence at every $x \in \partial_e K$), i.e.,

$$B_{P,n}(f)(x) = f(x) \quad \text{for every } x \in \partial_H^+ K \text{ and } f \in \mathcal{C}(K). \qquad (6.1.13)$$

Other simple properties are stated below.

6.1.2 Lemma. *Let $P = (p_{nj})_{n \geq 1, j \geq 1}$ be an infinite lower triangular stochastic matrix and consider the sequence $(B_{P,n})_{n \in \mathbb{N}}$ of Bernstein-Schnabl operators associated with the projection T and the matrix P.*

Given $n \in \mathbb{N}$ and $m \in \mathbb{N}$, then for every $h \in H$,

$$B_{P,n}^m(h) = h \, (= T(h)) \qquad (6.1.14)$$

and, if $h \in A(K)$,

$$B_{P,n}^m(h^2) = \left(1 - \left(1 - \sum_{i=1}^n p_{ni}^2\right)^m\right) T(h^2) + \left(1 - \sum_{i=1}^n p_{ni}^2\right)^m h^2. \qquad (6.1.15)$$

In particular, if we consider the arithmetic mean Toeplitz matrix, then for every $h \in A(K)$ we have

$$B_n^m(h^2) = \left(1 - \left(\frac{n-1}{n}\right)^m\right) T(h^2) + \left(\frac{n-1}{n}\right)^m h^2$$

$$= h^2 + \left(1 - \left(\frac{n-1}{n}\right)^m\right)(T(h^2) - h^2). \qquad (6.1.16)$$

Proof. If $h \in H$, then $T(h) = h$ and therefore $\mu_x^T(h) = h(x)$ for every $x \in K$ (see (6.1.7)). Then by (6.1.10) and (6.1.9) we have

$$B_{P,n}(h)(x) = \int_K \cdots \int_K h\left(\sum_{j=1}^n p_{nj} x_j\right) d\mu_x^T(x_1)\ldots d\mu_x^T(x_n) = h\left(\sum_{j=1}^n p_{nj} x\right) = h(x),$$

where in the second equality we used (6.1.4). Hence (6.1.14) easily follows by induction on $m \in \mathbb{N}$.

Now, let $h \in A(K)$. For every $(x_1,\ldots,x_n) \in K^n$ we have

$$h^2\left(\sum_{i=1}^n p_{ni} x_i\right) = \left(\sum_{i=1}^n p_{ni} h(x_i)\right)^2 = \sum_{i=1}^n p_{ni}^2 h^2(x_i) + 2 \sum_{1 \le i < j \le n} p_{ni} p_{nj} h(x_i) h(x_j)$$

and consequently, by (6.1.10) for every $x \in K$,

$$B_{P,n}(h^2)(x) = \int_K \cdots \int_K h^2\left(\sum_{j=1}^n p_{nj} x_j\right) d\mu_x^T(x_1)\ldots d\mu_x^T(x_n)$$

$$= \sum_{i=1}^n p_{ni}^2 \int_K h^2 \, d\mu_x^T + 2 \sum_{1 \le i < j \le n} p_{ni} p_{nj} h^2(x)$$

$$= \sum_{i=1}^n p_{ni}^2 T(h^2)(x) + \left(1 - \sum_{i=1}^n p_{ni}^2\right) h^2(x).$$

Again identity (6.1.15) can be easily obtained by induction on $m \ge 1$.

Finally, when we consider the arithmetic mean Toeplitz matrix, we have $\sum_{i=1}^n p_{ni}^2 = \dfrac{1}{n}$, $(n \ge 1)$, and therefore (6.1.16) follows directly from (6.1.15). $\qquad\Box$

For the sake of simplicity, from now on we limit ourselves to the case of the arithmetic mean Toeplitz matrix.

As a consequence of formulas (6.1.14)–(6.1.16) and by using some Korovkin-type theorems from Sections 3.3 and 4.4, we can establish the approximation properties of Bernstein-Schnabl operators.

6.1.3 Theorem. *For every $f \in \mathscr{C}(K)$ one has*
(1) $\lim_{n \to \infty} B_n^m(f) = f$ *uniformly on K for every $m \in \mathbb{N}$.*

In particular, $\lim_{n \to \infty} B_n(f) = f$ *uniformly on K.*

(2) $\lim_{m \to \infty} B_n^m(f) = T(f)$ *uniformly on K for every $n \ge 1$.*

(3) *If* $(k(n))_{n \in \mathbb{N}}$ *is a sequence of positive integers, then*

$$\lim_{n \to \infty} B_n^{k(n)}(f) = \begin{cases} f & \text{uniformly on } K, \quad \text{if } \lim_{n \to \infty} \dfrac{k(n)}{n} = 0, \\[2ex] T(f) & \text{uniformly on } K, \quad \text{if } \lim_{n \to \infty} \dfrac{k(n)}{n} = +\infty. \end{cases}$$

Proof. Since $H \cup A(K)^2$ is a K_+-subset both for the identity operator and the projection T (see Example 3 to Theorem 4.4.6 and Remark 1 to Theorem 3.3.3), it is enough to show that properties (1)–(3) hold for every $f \in H \cup A(K)^2$. On the other hand, this is an immediate consequence of (6.1.14) and (6.1.16), because

$$\left(\frac{n-1}{n} \right)^{k(n)} = \exp\left(k(n) \log\left(1 - \frac{1}{n} \right) \right). \qquad \Box$$

Remarks. 1. An argument similar to that used in the preceding proof shows that, if one considers the sequence $(B_{P,n})_{n \in \mathbb{N}}$ of Bernstein-Schnabl operators associated with an arbitrary stochastic matrix $P = (p_{nj})_{n \geq 1, j \geq 1}$, then (2) also holds for this sequence while (1) and (3) hold provided $\lim_{n \to \infty} \sum_{i=1}^n p_{ni}^2 = 0$ and $\lim_{n \to \infty} k(n) \sum_{i=1}^n p_{ni}^2 = 0$ or $= +\infty$, respectively.

2. In Section 6.2 it will become clear what happens when, under the hypotheses of Theorem 6.1.3, $\lim_{n \to \infty} \dfrac{k(n)}{n} = t$ for some $t \in \mathbb{R}, t > 0$.

Now we give some quantitative estimates of the rate of convergence of the operators B_n and their powers by means of the moduli of continuity $\Omega(f, \delta)$ and $\Psi(f, \delta)$ defined in (5.1.11) and (5.1.41), respectively.

6.1.4 Theorem. *For every $n \geq 1, m \geq 1$ and $f \in \mathscr{C}(K)$ we have*

$$\| B_n(f) - f \| \leq 2\Omega(f, \sqrt{1/n}), \tag{6.1.17}$$

$$\| B_n^m(f) - f \| \leq \Psi\left(f, \sqrt{1 - \left(\frac{n-1}{n} \right)^m} \right) \leq \Psi(f, \sqrt{m/n}) \tag{6.1.18}$$

and, finally,

$$\| B_n^m(f) - T(f) \| \leq \Psi\left(f, \sqrt{\left(\frac{n-1}{n} \right)^m} \right). \tag{6.1.19}$$

Proof. Given $h \in E'_K$ (see (5.1.7)) and $x \in K$, we obtain from (6.1.16)

$$B_n((h - h(x)1)^2)(x) = h^2(x) + \frac{1}{n}(T(h^2)(x) - h^2(x)) - 2h^2(x) + h^2(x)$$

$$= \frac{1}{n}(T(h^2)(x) - h^2(x)) \le \frac{1}{n}T(h^2)(x),$$

and consequently, for every $\delta > 0$ and $h_1, \ldots, h_m \in E'_K$ satisfying $\left\| \sum_{j=1}^{m} h_j^2 \right\| = \frac{1}{\delta^2}$, we obtain

$$\sum_{j=1}^{m} B_n((h_j - h_j(x)1)^2)(x) \le \frac{1}{n}T\left(\sum_{j=1}^{m} h_j^2\right)(x) \le \frac{1}{n}\left\| \sum_{j=1}^{m} h_j^2 \right\| = \frac{1}{n}\frac{1}{\delta^2}.$$

An application of Formula (5.1.17) gives

$$\|B_n(f) - f\| \le \left(1 + \frac{1}{n\delta^2}\right)\Omega(f, \delta)$$

and hence (6.1.17) follows by setting $\delta := \sqrt{1/n}$.

To show estimates (6.1.18) and (6.1.19) it is enough to apply Theorem 5.1.10 by recalling that, in the present case, (5.1.42) and (5.1.43) are satisfied because of (6.1.14) and (6.1.16). □

Remarks. 1. The proof of Theorem 6.1.4 is based on equalities (6.1.14) and (6.1.15) which hold more generally for every $h \in A(K)$. As a consequence, (6.1.17) remains true if we consider $A(K)$ instead of E'_K in the definition of $\Omega(f, \delta)$ and $\tau(\delta, x)$ respectively (see (5.1.11) and (5.1.18)).

2. By using the same reasoning as in the proof of Theorem 6.1.4, if we consider the sequence $(B_{P,n})_{n \in \mathbb{N}}$ of Bernstein-Schnabl operators associated with a stochastic matrix $P = (p_{nj})_{n \ge 1, j \ge 1}$, then we have the following more general estimates:

$$\|B_{P,n}(f) - f\| \le 2\Omega\left(f, \sqrt{\sum_{i=1}^{n} p_{ni}^2}\right),$$

$$\|B_{P,n}^m(f) - f\| \le \Psi\left(f, \sqrt{1 - \left(1 - \sum_{i=1}^{n} p_{ni}^2\right)^m}\right) \le \Psi\left(f, \sqrt{m \sum_{i=1}^{n} p_{ni}^2}\right),$$

and, finally,

$$\|B_{P,n}^m(f) - T(f)\| \le \Psi\left(f, \sqrt{\left(1 - \sum_{i=1}^{n} p_{ni}^2\right)^m}\right).$$

We now consider two different generalizations of Bernstein-Schnabl operators.

Although these new sequences of operators can be also constructed by means of a general infinite lower triangular stochastic matrix, for the sake of simplicity we shall consider only the arithmetic mean Toeplitz matrix.

Stancu-Schnabl operators. We denote by $p_n: \mathbb{R} \to \mathbb{R}$ the real function defined by

$$p_n(a) := \prod_{j=0}^{n-1} (1 + ja) \quad (a \in \mathbb{R}). \qquad (6.1.20)$$

Moreover, we shall use the convention to write $|v|_k = n$ to indicate an element $v = (v_1, \ldots, v_k) \in \mathbb{N}^k$ satisfying

$$v_1, \ldots, v_k \geq 1 \quad \text{and} \quad \sum_{i=1}^{k} v_i = n. \qquad (6.1.21)$$

Given a sequence $(a_n)_{n \in \mathbb{N}}$ of real numbers satisfying the conditions

$$a_n \neq -\frac{1}{j} \quad \text{for every } n \geq 2 \text{ and } j = 1, \ldots, n-1, \qquad (6.1.22)$$

the *n-th Stancu-Schnabl operator associated with the projection T and the sequence $(a_n)_{n \in \mathbb{N}}$* is, by definition, the operator $S_{n,a_n}: \mathscr{C}(K) \to \mathscr{C}(K)$ defined by

$$S_{n,a_n}(f)(x) := \frac{1}{p_n(a_n)} \sum_{k=1}^{n} \frac{n!}{k!} a_n^{n-k} \sum_{|v|_k=n} \frac{1}{v_1 \ldots v_k}$$

$$\times \int_K \cdots \int_K f\left(\frac{v_1 x_1 + \cdots + v_k x_k}{n}\right) d\mu_x^T(x_1) \ldots d\mu_x^T(x_k) \qquad (6.1.23)$$

for each $f \in \mathscr{C}(K)$ and $x \in K$.

By comparing (6.1.10) and (6.1.16), we also obtain that

$$S_{n,a_n} = \frac{1}{p_n(a_n)} \sum_{k=1}^{n} \frac{n!}{k!} a_n^{n-k} \sum_{|v|_k=n} \frac{1}{v_1 \ldots v_k} B_{v_1/n, \ldots, v_k/n}. \qquad (6.1.24)$$

Moreover it is clear that $S_{n,a_n} = B_n$ if $a_n = 0$.

In the sequel we always assume that the sequence $(a_n)_{n \in \mathbb{N}}$ is positive so that every S_{n,a_n} is a positive operator.

The study of the convergence of the sequence $(S_{n,a_n})_{n \in \mathbb{N}}$ can be carried out by considering some classical combinatorial identities.

We shall denote by $s(n, k)$ the coefficient of a^{n-k} of the polynomial $p_n(a)$ (see (6.1.20)), i.e.,

$$p_n(a) = \sum_{k=1}^{n} s(n, k) a^{n-k}. \tag{6.1.25}$$

More explicitly, for every $k = 1, \ldots, n$, we have

$$s(n, k) = \frac{n!}{k!} \sum_{|v|_k = n} \frac{1}{v_1 \ldots v_k}. \tag{6.1.26}$$

In particular, by comparing the preceding formulas with (6.1.23) we see that

$$S_{n,a_n}(f)(x) = f(x) \quad \text{for each } x \in \partial_H^+ K \text{ and } f \in \mathscr{C}(K). \tag{6.1.27}$$

Moreover, for every $n \in \mathbb{N}_0$, $k \in \mathbb{N}_0$ and $p \geq 1$, it is useful to consider the numbers

$$r(n, k, p) := \begin{cases} 1, & \text{if } n = 0 \text{ and } k = 0, \\ 0, & \text{if } n \geq 1 \text{ and } k = 0, \\ \displaystyle\sum_{|v|_k = n} \frac{v_1^p + \cdots + v_k^p}{v_1 \ldots v_k}, & \text{if } n \geq 1 \text{ and } k = 1, \ldots, n, \\ 0, & \text{if } k > n. \end{cases} \tag{6.1.28}$$

We collect here some combinatorial properties of the numbers $s(n, k)$ and $r(n, k, p)$ which will also be applied in the next section to develop further properties of the sequence $(S_{n,a_n})_{n \in \mathbb{N}}$.

6.1.5 Proposition. *For every $n \geq 1$, $k = 1, \ldots, n$ and $a \in \mathbb{R}$ the following properties hold:*
(1) *Recurrence formulas for the coefficients $s(n, k)$:*

$$s(n, k) = \frac{(n-1)!}{k!} r(n, k, 1), \tag{6.1.29}$$

$$s(n+1, k+1) = s(n, k) + n s(n, k+1). \tag{6.1.30}$$

(2) *Combinatorial identities:*

$$s(n + 1, k + 1) = \sum_{j=k}^{n} \frac{n!}{j!} s(j, k), \qquad (6.1.31)$$

$$\sum_{k=1}^{n} \frac{(n-1)!}{k!} a^{n-k} r(n, k, 1) = p_n(a), \qquad (6.1.32)$$

$$p_2(a) \sum_{k=1}^{n} \frac{(n-1)!}{k!} a^{n-k} r(n, k, 2) = p_{n+1}(a), \qquad (6.1.33)$$

$$p_3(a) \sum_{k=1}^{n} \frac{(n-1)!}{k!} a^{n-k} r(n, k, 3) = (1 + 2na)p_{n+1}(a), \qquad (6.1.34)$$

and, if $n \geq 2$,

$$p_4(a) \sum_{k=1}^{n} \frac{(n-1)!}{k!} a^{n-k} \sum_{|v|_k=n} \frac{1}{v_1 \ldots v_k} \sum_{\substack{i,j=1 \\ i \neq j}}^{k} v_i^2 v_j^2 = (n-1)p_{n+2}(a). \quad (6.1.35)$$

(3) *Analytic estimates:*
   For a given $p \geq 1$ set

$$N_k(p) := \{(i_1, \ldots, i_p) \in \{1, \ldots, k\}^p | i_r \neq i_s \text{ for } r \neq s\}. \qquad (6.1.36)$$

Then for every $(v_1, \ldots, v_k) \in \mathbb{N}^k$ satisfying (6.1.21) we have

$$\sum_{(i_1, \ldots, i_p) \in N_k(p)} v_{i_1}^2 v_{i_2} \ldots v_{i_p} = n^{p-1} \sum_{i=1}^{k} v_i^2 + U_n(v_1, \ldots, v_k; p) \qquad (6.1.37)$$

with

$$|U_n(v_1, \ldots, v_k; p)| \leq u_{1p} n^{p-2} \sum_{i=1}^{k} v_i^3 + u_{2p} n^{p-3} \sum_{(i_1, i_2) \in N_k(2)} v_{i_1}^2 v_{i_2}^2, \quad (6.1.38)$$

where u_{1p} and u_{2p} are real constants depending on p.
   Finally, if $p \geq 2$, we also have

$$\sum_{(i_1, \ldots, i_p) \in N_k(p)} v_{i_1} \ldots v_{i_p} = n^p - n^{p-2} \frac{p(p-1)}{2} \sum_{i=1}^{k} v_i^2 + W_n(v_1, \ldots, v_k; p) \quad (6.1.39)$$

with

$$|W_n(v_1,\ldots,v_k;p)| \le w_{1p}n^{p-3}\sum_{i=1}^{k}v_i^3 + w_{2p}n^{p-4}\sum_{(i_1,i_2)\in N_k(2)}v_{i_1}^2 v_{i_2}^2, \quad (6.1.40)$$

where w_{1p} and w_{2p} are real constants depending on p.

Proof. Identities (6.1.29)–(6.1.32) follow from the definitions.
   To prove (6.1.33), we use (6.1.29) to write

$$\frac{(n-1)!}{k!}r(n,k,2) = \frac{(n-1)!}{k!}\sum_{|v|_k=n}\frac{v_1^2+\cdots+v_k^2}{v_1\ldots v_k}$$

$$= \frac{(n-1)!}{k!}k\sum_{|v|_k=n}\frac{v_k^2}{v_1\ldots v_k} = \frac{(n-1)!}{(k-1)!}\sum_{|v|_k=n}\frac{v_k}{v_1\ldots v_{k-1}}$$

$$= \frac{(n-1)!}{(k-1)!}\left(\sum_{|v|_{k-1}=n-1}\frac{1}{v_1\ldots v_{k-1}} + 2\sum_{|v|_{k-1}=n-2}\frac{1}{v_1\ldots v_{k-1}} + \cdots \right.$$

$$\left. + (n-k+1)\sum_{|v|_{k-1}=k-1}\frac{1}{v_1\ldots v_{k-1}}\right)$$

$$= \frac{(n-1)!}{(k-1)!}\sum_{j=k-1}^{n-1}\frac{n-j}{j}r(j,k-1,1)$$

$$= \sum_{j=k-1}^{n-1}\frac{(n-1)!}{j!}(n-j)s(j,k-1). \tag{1}$$

Therefore, since $r(n,n,2) = r(n,1,2) = n$, by using (6.1.30) and (6.1.31) we obtain

$$p_2(a)\sum_{k=1}^{n}\frac{(n-1)!}{k!}a^{n-k}r(n,k,2)$$

$$= (1+a)\sum_{k=1}^{n}\frac{(n-1)!}{k!}a^{n-k}r(n,k,2)$$

$$= \sum_{k=1}^{n}\frac{(n-1)!}{k!}a^{n-k}r(n,k,2) + \sum_{k=1}^{n}\frac{(n-1)!}{k!}a^{n+1-k}r(n,k,2)$$

$$= 1 + \sum_{k=1}^{n-1} a^{n-k} \left(\sum_{j=k-1}^{n-1} \frac{(n-1)!}{j!} (n-j)s(j, k-1) \right)$$

$$+ \sum_{k=2}^{n} a^{n+1-k} \left(\sum_{j=k-1}^{n-1} \frac{(n-1)!}{j!} (n-j)s(j, k-1) \right) + n!a^n$$

$$= \sum_{k=2}^{n} a^{n+1-k} \left(\sum_{j=k-2}^{n-1} \frac{(n-1)!}{j!} (n-j)s(j, k-2) \right.$$

$$\left. + \sum_{j=k-1}^{n-1} \frac{(n-1)!}{j!} (n-j)s(j, k-1) \right) + 1 + n!a^n$$

$$= \sum_{k=2}^{n} a^{n+1-k} \left(\sum_{j=k-2}^{n-1} \frac{(n-1)!}{j!} (n-j)(s(j+1, k-1) - js(j, k+1)) \right.$$

$$\left. + \sum_{j=k-1}^{n-1} \frac{(n-1)!}{j!} (n-j)s(j, k-1) \right) + 1 + n!a^n$$

$$= \sum_{k=2}^{n} a^{n+1-k} \left(\sum_{j=k-1}^{n-1} \frac{(n-1)!}{j!} ((n-j+1)j - (n-j)j + (n-j)) \right.$$

$$\left. \times (s(j, k-1) + s(n, k-1)) \right) + 1 + n!a^n$$

$$= \sum_{k=2}^{n} a^{n+1-k} \left(n \sum_{j=k-1}^{n-1} \frac{(n-1)!}{j!} s(j, k-1) + s(n, k-1) \right) + 1 + n!a^n$$

$$= \sum_{k=2}^{n} a^{n+1-k} (ns(n, k) + s(n, k-1)) + 1 + n!a^n$$

$$= \sum_{k=2}^{n} a^{n+1-k} s(n+1, k) + 1 + n!a^n = \sum_{k=1}^{n+1} a^{n+1-k} s(n+1, k) = p_{n+1}(a),$$

and hence (6.1.33) follows.

In order to prove (6.1.34), we preliminarily point out the identity

$$(1 + a) \sum_{j=1}^{n} \frac{(n-1)!}{(j-1)!} p_j(a) a^{n-j} = p_{n+1}(a), \tag{2}$$

which can be established by induction on n.

Moreover, we also need the identity

$$(1 + a) \sum_{k=1}^{n} \frac{(n-1)!}{k!} a^{n-1-k} \sum_{j=k}^{n-1} r(j,k,1) = (n-1)p_n(a), \tag{3}$$

which holds for every $n \geq 1$.

Indeed, if $n = 1$ or $n = 2$, equality (3) becomes trivial. Now, if (3) holds for some $n \geq 1$, then by using (6.1.32) we obtain

$$(1 + a) \sum_{k=1}^{n+1} \frac{n!}{k!} a^{n-k} \sum_{j=k}^{n} r(j,k,1)$$

$$= (1 + a) \sum_{k=1}^{n} \frac{n!}{k!} a^{n-k} \sum_{j=k}^{n} r(j,k,1)$$

$$= na(1 + a) \sum_{k=1}^{n} \frac{(n-1)!}{k!} a^{n-1-k} \sum_{j=k}^{n} r(j,k,1)$$

$$= n(1 + a) \sum_{k=1}^{n} \frac{(n-1)!}{k!} a^{n-k} \sum_{j=k}^{n} r(j,k,1) + n(n-1)ap_n(a)$$

$$= n(1 + a)p_n(a) + n(n-1)ap_n(a) = np_{n+1}(a),$$

and hence (3) holds for the integer $n + 1$. By induction, the result follows.

Besides identity (3), we shall also use the next one:

$$p_3(a) \sum_{k=1}^{n-1} \frac{(n-1)!}{k!} a^{n-1-k} \sum_{j=k}^{n-1} (n-j)r(j,k,1) = (n-1)p_{n+1}(a) \quad (n \geq 1), \tag{4}$$

whose proof is again based on a reasoning by induction.

Indeed, (4) is obviously satisfied if $n = 1$ or $n = 2$. Assume that (4) holds for some $n \geq 1$; by using (2), (3) and (6.1.32) we obtain

$$p_3(a) \sum_{k=1}^{n} \frac{n!}{k!} a^{n-k} \sum_{j=k}^{n} (n+1-j)r(j,k,1)$$

$$= p_3(a) \sum_{k=1}^{n} \frac{n!}{k!} a^{n-k} \sum_{j=k}^{n-1} (n+1-j)r(j,k,1) + p_3(a) \sum_{k=1}^{n} \frac{n!}{k!} a^{n-k} r(n,k,1)$$

$$= p_3(a) \sum_{k=1}^{n} \frac{n!}{k!} a^{n-k} \sum_{j=k}^{n-1} (n-j)r(j,k,1) + p_3(a) \sum_{k=1}^{n} \frac{n!}{k!} a^{n-k} \sum_{j=k}^{n-1} r(j,k,1)$$

$$+ p_3(a) \sum_{k=1}^{n} \frac{n!}{k!} a^{n-k} r(n,k,1)$$

$$= na(n-1)p_{n+1}(a) + na(1+2a)(1+a) \sum_{k=1}^{n} \frac{(n-1)!}{k!} a^{n-1-k} \sum_{j=k}^{n-1} r(j,k,1)$$

$$+ np_3(a)p_n(a)$$

$$= na(n-1)(1+na)p_n(a) + n(n-1)a(1+2a)p_n(a) + np_3(a)p_n(a)$$

$$= np_{n+2}(a).$$

Hence the proof of (4) is complete.

At this point, we observe that, as in (1), we can write

$$r(n,k,3) = \sum_{|v|_k=n} \frac{v_1^3 + \cdots + v_k^3}{v_1 \ldots v_k} = k \sum_{|v|_k=n} \frac{v_k^3}{v_1 \ldots v_k} = k \sum_{|v|_k=n} \frac{v_k^2}{v_1 \ldots v_{k-1}}$$

$$= k\left(\frac{1}{n-1}r(n-1,k-1,1) + \frac{4}{n-2}r(n-2,k-1,1)\right.$$

$$+ \frac{9}{n-3}r(n-3,k-1,1) + \cdots + \frac{(n-k+1)^2}{k-1}r(k-1,k-1,1)\bigg)$$

$$= k \sum_{j=k-1}^{n-1} \frac{(n-j)^2}{j} r(j,k-1,1)$$

$$= kn \sum_{j=k-1}^{n-1} \frac{n-j}{j} r(j,k-1,1) - k \sum_{j=k-1}^{n-1} (n-j)r(j,k-1,1)$$

$$= nr(n,k,2) - k \sum_{j=k-1}^{n-1} (n-j)r(j,k-1,1).$$

Finally, we can use (6.1.33) and (4) and the equality $r(n,1,3) = nr(n,1,2)$ to obtain

$$p_3(a) \sum_{k=1}^{n} \frac{(n-1)!}{k!} a^{n-k} r(n,k,3)$$

$$= n(1+2a)p_2(a) \sum_{k=1}^{n} \frac{(n-1)!}{k!} a^{n-k} r(n,k,2)$$

$$- p_3(a) \sum_{k=2}^{n} \frac{(n-1)!}{(k-1)!} a^{n-k} \sum_{j=k-1}^{n-1} (n-j)r(j,k-1,1)$$

$$= n(1 + 2a)p_{n+1}(a) - p_3(a) \sum_{k=1}^{n-1} \frac{(n-1)!}{k!} a^{n-1-k} \sum_{j=k}^{n-1} (n-j)r(j, k, 1)$$

$$= n(1 + 2a)p_{n+1}(a) - (n-1)p_{n+1}(a) = (1 + 2na)p_{n+1}(a),$$

and this completes the proof of (6.1.34).

Finally we proceed to show (6.1.35). We use the equality

$$(1 + 2a)(1 + 3a) \sum_{j=1}^{n-1} \frac{(n-2)!}{(j-1)!}(n-j)p_{j+1}(a)a^{n-j-1} = p_{n+2}(a) \quad (n \geq 2), \quad (5)$$

which can be shown by induction. Moreover, we observe that

$$\sum_{|v|_k = n} \frac{1}{v_1 \cdots v_k} \sum_{\substack{i,j=1 \\ i \neq j}}^{k} v_i^2 v_j^2 = k(r(n-1, k-1, 2) + 2r(n-2, k-1, 2) + \cdots$$

$$+ (n-k+1)r(k-1, k-1, 2))$$

$$= k \sum_{j=k-1}^{n-1} (n-j)r(j, k-1, 2),$$

and consequently, by using (4) and (5),

$$p_4(a) \sum_{k=1}^{n} \frac{(n-1)!}{k!} a^{n-k} \sum_{|v|_k = n} \frac{1}{v_1 \cdots v_k} \sum_{\substack{i,j=1 \\ i \neq j}}^{k} v_i^2 v_j^2$$

$$= p_4(a) \sum_{k=1}^{n} \frac{(n-1)!}{(k-1)!} a^{n-k} \sum_{j=k-1}^{n-1} (n-j)r(j, k-1, 2)$$

$$= p_4(a) \sum_{k=1}^{n-1} \sum_{j=1}^{n-1} \frac{(n-1)!}{k!} a^{n-1-k}(n-j)r(j, k, 2)$$

$$= p_4(a) \sum_{j=1}^{n-1} \frac{(n-1)!}{(j-1)!}(n-j)a^{n-j-1} \sum_{k=1}^{j} \frac{(j-1)!}{k!} a^{j-k}r(j, k, 2)$$

$$= (1 + 2a)(1 + 3a) \sum_{j=1}^{n-1} \frac{(n-1)!}{(j-1)!}(n-j)p_{j+1}(a)a^{n-j-1} = (n-1)p_{n+2}(a).$$

We shall now establish (6.1.37) by reasoning by induction on p. If $p = 1$, (6.1.37) holds with $u_{11} = u_{21} = 0$. If (6.1.37) holds for some $p \in \mathbb{N}$, then

$$n \sum_{(i_1,\ldots,i_p) \in N_k(p)} v_{i_1}^2 v_{i_2} \ldots v_{i_p} - \sum_{(i_1,\ldots,i_{p+1}) \in N_k(p+1)} v_{i_1}^2 v_{i_2} \ldots v_{i_{p+1}}$$

$$= n \sum_{(i_1,\ldots,i_p) \in N_k(p)} v_{i_1}^2 v_{i_2} \ldots v_{i_p} - \sum_{(i_1,\ldots,i_p) \in N_k(p)} \sum_{\substack{i=1 \\ i \neq i_1,\ldots,i_p}}^{k} v_{i_1}^2 v_{i_2} \ldots v_{i_p} v_i$$

$$= \sum_{(i_1,\ldots,i_p) \in N_k(p)} \left(n - \sum_{\substack{i=1 \\ i \neq i_1,\ldots,i_p}}^{k} v_i \right) v_{i_1}^2 v_{i_2} \ldots v_{i_p}$$

$$= \sum_{(i_1,\ldots,i_p) \in N_k(p)} v_{i_1}^3 v_{i_2} \ldots v_{i_p} + (p-1) \sum_{(i_1,\ldots,i_p) \in N_k(p)} v_{i_1}^2 v_{i_2}^2 v_{i_3} \ldots v_{i_p},$$

and hence

$$\sum_{(i_1,\ldots,i_{p+1}) \in N_k(p+1)} v_{i_1}^2 v_{i_2} \ldots v_{i_{p+1}}$$

$$= n \left(n^{p-1} \sum_{i=1}^{k} v_i^2 + U_n(v_1,\ldots,v_k; p) \right)$$

$$- \sum_{(i_1,\ldots,i_p) \in N_k(p)} v_{i_1}^3 v_{i_2} \ldots v_{i_p} - (p-1) \sum_{(i_1,\ldots,i_p) \in N_k(p)} v_{i_1}^2 v_{i_2}^2 v_{i_3} \ldots v_{i_p}$$

$$= n^p \sum_{i=1}^{k} v_i^2 + U_n(v_1,\ldots,v_k; p+1),$$

with

$$|U_n(v_1,\ldots,v_k; p+1)| \leq n \left(u_{1p} n^{p-2} \sum_{i=1}^{k} v_i^3 + u_{2p} n^{p-3} \sum_{(i_1,i_2) \in N_k(2)} v_{i_1}^2 v_{i_2}^2 \right)$$

$$+ n^{p-1} \sum_{i=1}^{k} v_i^3 + (p-1) n^{p-2} \sum_{(i_1,i_2) \in N_k(2)} v_{i_1}^2 v_{i_2}^2.$$

Thus (6.1.37) holds for $p+1$ with $u_{1,p+1} = u_{1p} + 1$ and $u_{2,p+1} = u_{2p} + p - 1$ and the proof is complete.

Finally, we show (6.1.39) again by reasoning by induction. If $p = 2$, by (6.1.21), (6.1.36) and the equality

$$\left(\sum_{i=1}^{k} v_i \right)^2 = \sum_{i=1}^{k} v_i^2 + \sum_{(i_1,i_2) \in N_k(2)} v_{i_1} v_{i_2}$$

we obtain (6.1.39) with $w_{12} = w_{22} = 0$. If (6.1.39) holds for some $p \geq 2$, then we have

$$
n \sum_{(i_1,\ldots,i_p) \in N_k(p)} v_{i_1} v_{i_2} \cdots v_{i_p} - \sum_{(i_1,\ldots,i_{p+1}) \in N_k(p+1)} v_{i_1} v_{i_2} \cdots v_{i_{p+1}}
$$

$$
= \sum_{(i_1,\ldots,i_p) \in N_k(p)} \left(n - \sum_{\substack{i=1 \\ i \neq i_1,\ldots,i_p}}^{k} v_i \right) v_{i_1} v_{i_2} \cdots v_{i_p}
$$

$$
= \sum_{(i_1,\ldots,i_p) \in N_k(p)} (v_{i_1} + v_{i_2} + \cdots + v_{i_p}) v_{i_1} v_{i_2} \cdots v_{i_p}
$$

$$
= p \sum_{(i_1,\ldots,i_p) \in N_k(p)} v_{i_1}^2 v_{i_2} \cdots v_{i_p}
$$

and hence (see (6.1.37))

$$
\sum_{(i_1,\ldots,i_{p+1}) \in N_k(p+1)} v_{i_1} v_{i_2} \cdots v_{i_{p+1}}
$$

$$
= n \sum_{(i_1,\ldots,i_p) \in N_k(p)} v_{i_1} v_{i_2} \cdots v_{i_p} - p \sum_{(i_1,\ldots,i_p) \in N_k(p)} v_{i_1}^2 v_{i_2} \cdots v_{i_{p+1}}
$$

$$
= n \left(n^p - n^{p-2} \frac{p(p-1)}{2} \sum_{i=1}^{k} v_i^2 + W_n(v_1,\ldots,v_k;p) \right)
$$

$$
- p \left(n^{p-1} \sum_{i=1}^{k} v_i^2 + U_n(v_1,\ldots,v_k;p) \right)
$$

$$
= n^{p+1} - n^{p-1} \frac{p(p+1)}{2} \sum_{i=1}^{k} v_i^2 + W_n(v_1,\ldots,v_k;p+1)
$$

with

$$
|W_n(v_1,\ldots,v_k;p+1)| \leq n \left(w_{1p} n^{p-3} \sum_{i=1}^{k} v_i^3 + w_{2p} n^{p-4} \sum_{(i_1,i_2) \in N_k(2)} v_{i_1}^2 v_{i_2}^2 \right)
$$

$$
+ p \left(u_{1p} n^{p-2} \sum_{i=1}^{k} v_i^3 + u_{2p} n^{p-3} \sum_{(i_1,i_2) \in N_k(2)} v_{i_1}^2 v_{i_2}^2 \right).
$$

Accordingly (6.1.39) holds for $p + 1$ with $w_{1,p+1} = w_{1p} + p u_{1p}$ and $w_{2,p+1} = w_{2p} + p u_{2p}$ and this completes the proof. $\qquad \square$

At this point, we proceed as in the case of Bernstein-Schnabl operators. In the following lemma we evaluate $S_{n,a_n}(h)$ and $S_{n,a_n}(k^2)$ for every $h \in H$ and $k \in A(K)$, respectively.

6.1.6 Lemma. *Let $(a_n)_{n \in \mathbb{N}}$ be a sequence of positive real numbers and consider the corresponding sequence $(S_{n,a_n})_{n \in \mathbb{N}}$ of Stancu-Schnabl operators.*
   Given $n \in \mathbb{N}$ and $m \in \mathbb{N}$, then for every $f \in H$,

$$S_{n,a_n}^m(h) = h \; (= T(h)), \tag{6.1.41}$$

and, if $h \in A(K)$,

$$S_{n,a_n}^m(h^2) = \left(1 - \left(1 - \frac{1 + na_n}{n(1 + a_n)}\right)^m\right) T(h^2) + \left(1 - \frac{1 + na_n}{n(1 + a_n)}\right)^m h^2$$

$$= h^2 + \left(1 - \left(\frac{n-1}{n}\,\frac{1}{1+a_n}\right)^m\right)(T(h^2) - h^2). \tag{6.1.42}$$

Proof. It is enough to show (6.1.41) for $m = 1$. But for arbitrary $h \in H$ and $x \in K$ we have

$$S_{n,a_n}(h)(x) = \frac{1}{p_n(a_n)} \sum_{k=1}^{n} \frac{n!}{k!} a_n^{n-k} \sum_{|v|_k=n} \frac{1}{v_1 \dots v_k} h\left(\frac{v_1 x + \dots + v_k x}{n}\right)$$

$$= h(x) \frac{1}{p_n(a_n)} \sum_{k=1}^{n} \frac{n!}{k!} a_n^{n-k} \sum_{|v|_k=n} \frac{1}{v_1 \dots v_k}$$

$$= h(x) \frac{1}{p_n(a_n)} \sum_{k=1}^{n} s(n,k) a_n^{n-k} = h(x),$$

where again we applied (6.1.4).
   To show (6.1.42), let $h \in A(K)$ and consider $n \geq 1$ and $k = 1, \dots, n$; for every $(x_1, \dots, x_k) \in K^k$ and $v = (v_1, \dots, v_k)$, $|v|_k = n$, we have

$$h^2\left(\frac{v_1 x_1 + \dots + v_k x_k}{n}\right) = \left(\frac{1}{n} \sum_{i=1}^{k} v_i h(x_i)\right)^2$$

$$= \frac{1}{n^2}\left(\sum_{i=1}^{k} v_i^2 h^2(x_i) + 2 \sum_{1 \leq i < j \leq k} v_i v_j h(x_i) h(x_j)\right).$$

Hence, by (6.1.23) and Proposition 6.1.5, we obtain for every $x \in K$

$$S_{n,a_n}(h^2)(x)$$

$$= \frac{1}{p_n(a_n)} \sum_{k=1}^{n} \frac{n!}{k!} a_n^{n-k} \sum_{|v|_k=n} \frac{1}{v_1 \ldots v_k}$$

$$\times \left(\frac{v_1^2 + \cdots + v_k^2}{n^2} \int_K h^2 \, d\mu_x^T + 2 \sum_{1 \le i < j \le n} v_i v_j h^2(x) \right)$$

$$= \frac{1}{n^2} \left(\frac{1}{p_n(a_n)} \sum_{k=1}^{n} \frac{n!}{k!} a_n^{n-k} \sum_{|v|_k=n} \frac{v_1^2 + \cdots + v_k^2}{v_1 \ldots v_k} \right) T(h^2)(x)$$

$$+ \left(\frac{1}{p_n(a_n)} \sum_{k=1}^{n} \frac{n!}{k!} a_n^{n-k} \sum_{|v|_k=n} \frac{1}{v_1 \ldots v_k} \left(1 - \frac{v_1^2 + \cdots + v_k^2}{n^2} \right) \right) h^2(x)$$

$$= \frac{1}{n} \left(\frac{1}{p_n(a_n)} \sum_{k=1}^{n} \frac{(n-1)!}{k!} a_n^{n-k} \sum_{|v|_k=n} \frac{v_1^2 + \cdots + v_k^2}{v_1 \ldots v_k} \right) T(h^2)(x)$$

$$+ \frac{1}{p_n(a_n)} \left(\sum_{k=1}^{n} \frac{n!}{k!} a_n^{n-k} \sum_{|v|_k=n} \frac{1}{v_1 \ldots v_k} \right.$$

$$\left. - \frac{1}{n} \left(\sum_{k=1}^{n} \frac{(n-1)!}{k!} a_n^{n-k} \sum_{|v|_k=n} \frac{v_1^2 + \cdots + v_k^2}{v_1 \ldots v_k} \right) \right) h^2(x)$$

$$= \frac{1}{n} \left(\frac{p_{n+1}(a_n)}{p_n(a_n) p_2(a_n)} \right) T(h^2)(x) + \left(1 - \frac{1}{n} \left(\frac{p_{n+1}(a_n)}{p_n(a_n) p_2(a_n)} \right) \right) h^2(x)$$

$$= \frac{1 + na_n}{n(1 + a_n)} T(h^2)(x) + \left(1 - \frac{1 + na_n}{n(1 + a_n)} \right) h^2(x)$$

$$= \left(1 - \left(1 - \frac{1 + na_n}{n(1 + a_n)} \right) \right) T(h^2)(x) + \left(1 - \frac{1 + na_n}{n(1 + a_n)} \right) h^2(x).$$

This shows (6.1.42) for $m = 1$. The general case can be easily obtained by induction. □

Formulas (6.1.41) and (6.1.42) yield the following convergence results for Stancu-Schnabl operators.

6.1.7 Theorem. *Consider the sequence $(S_{n,a_n})_{n \in \mathbb{N}}$ of Stancu-Schnabl operators associated with the projection T and a sequence $(a_n)_{n \in \mathbb{N}}$ of positive real numbers.*

For every $f \in \mathscr{C}(K)$ the following statements hold:
(1) *If the sequence $(a_n)_{n \in \mathbb{N}}$ converges to 0, then*

$$\lim_{n \to \infty} S^m_{n,a_n}(f) = f \quad \text{uniformly on } K \text{ for every } m \geq 1.$$

In particular

$$\lim_{n \to \infty} S_{n,a_n}(f) = f \quad \text{uniformly on } K.$$

(2) $\lim_{m \to \infty} S^m_{n,a_n}(f) = T(f)$ *uniformly on K for every $n \geq 1$.*

(3) *If the sequence $(a_n)_{n \in \mathbb{N}}$ converges to 0 and if $(k(n))_{n \in \mathbb{N}}$ is a sequence of positive integers, then*

$$\lim_{n \to \infty} S^{k(n)}_{n,a_n}(f)$$

$$= \begin{cases} f, & \text{uniformly on } K \quad \text{if } \lim_{n \to \infty} \dfrac{k(n)}{n} = 0 \text{ and } \lim_{n \to \infty} k(n)a_n = 0, \\[2em] T(f), & \text{uniformly on } K \quad \text{if } \lim_{n \to \infty} \dfrac{k(n)}{n} = 0 \text{ and } \lim_{n \to \infty} k(n)a_n = +\infty, \\[2em] & \qquad\qquad\quad \text{or } \lim_{n \to \infty} \dfrac{k(n)}{n} = +\infty. \end{cases}$$

Proof. The proof is analogous to that of Theorem 6.1.3 so that, according to the Korovkin-type theorems quoted there, we only have to check the desired convergence on $H \cup A(K)^2$. This aim can be easily achieved by means of (6.1.26) and (6.1.33) and by using, in case (3), the familiar formulas

$$\left(\frac{n-1}{n(1+a_n)}\right)^{k(n)} = \exp\left(\frac{k(n)}{n} \frac{\log(1-1/n)}{1/n}\right) \exp\left(-k(n)a_n \frac{\log(1+a_n)}{a_n}\right),$$

and, respectively,

$$\left(\frac{n-1}{n(1+a_n)}\right)^{k(n)} \leq \left(\left(1-\frac{1}{n}\right)^n\right)^{k(n)/n}. \qquad \square$$

Remarks. 1. In statement (1) of the above theorem the assumption that the sequence $(a_n)_{n \in \mathbb{N}}$ converges to 0 is necessary as well. As a matter of fact we have that, if $\lim\limits_{n\to\infty} S_{n,a_n}^m(f) = f$ pointwise on K for every $f \in \mathscr{C}(K)$ and for some $m \geq 1$, then necessarily $\lim\limits_{n\to\infty} a_n = 0$.

Indeed, by Remark 1 to Theorem 3.3.3 there exists $h \in A(K)$ such that $T(h^2) \neq h^2$. Since $\lim\limits_{n\to\infty} S_{n,a_n}^m(h^2)(x) = h^2(x)$ for every $x \in K$, from (6.1.42) it follows that

$$\lim_{n\to\infty} \left(\frac{n-1}{n}\frac{1}{1+a_n}\right)^m = 1 \text{ and hence } a_n \to 0 \text{ as } n \to \infty.$$

2. As a special case of Stancu-Schnabl operators, consider the sequence $(a_n(t))_{n\in\mathbb{N}}$ defined by $a_n(t) := t/n$ for every $n \in \mathbb{N}$, t being a fixed positive real number. Moreover, denote by $S_{n,t}$ the corresponding n-th Stancu-Schnabl operator. Since for every $n \geq 1$ and $m \geq 1$

$$\lim_{t\to+\infty} \left(1 - \frac{1+na_n(t)}{n(1+a_n(t))}\right)^m = \lim_{t\to+\infty} \left(1 - \frac{t+1}{t+n}\right)^m = 0,$$

by recalling (6.1.42) we have

$$\lim_{t\to+\infty} S_{n,t}^m(h^2) = T(h^2) \quad \text{for every } h \in A(K).$$

Thus, according to Remark 1 to Theorem 3.3.3, we obtain

$$\lim_{t\to+\infty} S_{n,t}^m(f) = T(f) \quad \text{for every } f \in \mathscr{C}(K). \tag{6.1.43}$$

At this point we establish some quantitative estimates of the rate of convergence of Stancu-Schnabl operators which generalize those obtained in Theorem 6.1.4 for Bernstein-Schnabl operators.

To this end we shall proceed as in the proof of Theorem 6.1.4.

In fact we estimate for each $h \in E'_K$ and $x \in K$ (see (6.1.33))

$$S_{n,a_n}((h - h(x)1)^2)(x) = \left(1 - \frac{n-1}{n}\frac{1}{1+a_n}\right)(T(h^2)(x) - h^2(x))$$

$$\leq \frac{1+na_n}{n(1+a_n)}T(h^2)(x),$$

and hence we obtain the following result.

6.1.8 Theorem. *For every $n \geq 1$, $m \geq 1$ and $f \in \mathscr{C}(K)$, we have*

$$\|S_{n,a_n}(f) - f\| \leq \left(1 + \frac{1 + na_n}{1 + a_n}\right)\Omega(f, \sqrt{1/n}), \tag{6.1.44}$$

$$\|S_{n,a_n}^m(f) - f\| \leq \Psi\left(f, \sqrt{1 - \left(\frac{n-1}{n(1+a_n)}\right)^m}\right) \leq \Psi\left(f, \sqrt{m\frac{1 + na_n}{n(1+a_n)}}\right), \tag{6.1.45}$$

and

$$\|S_{n,a_n}^m(f) - T(f)\| \leq \Psi\left(f, \sqrt{\left(\frac{n-1}{n(1+a_n)}\right)^m}\right). \tag{6.1.46}$$

The third sequence of positive operators we shall study in this section is obtained by considering a function $\lambda \in \mathscr{C}(K)$ taking its values in the interval $[0, 1]$. Under these assumptions and for every $x \in K$ we shall denote by v_x^T the probability Radon measure defined by

$$v_x^T := \lambda(x)\mu_x^T + (1 - \lambda(x))\varepsilon_x \tag{6.1.47}$$

(see (6.1.7)). The notation $v_{x,\lambda}^T$ is also used to show the dependence on the function λ.

Note that the barycenter of v_x^T is x and

$$\text{Supp}(v_x^T) = \begin{cases} \text{Supp}(\mu_x^T), & \text{if } \lambda(x) = 1, \\ \text{Supp}(\mu_x^T) \cup \{x\}, & \text{if } 0 < \lambda(x) < 1, \\ \{x\}, & \text{if } \lambda(x) = 0. \end{cases} \tag{6.1.48}$$

Lototsky-Schnabl operators. Under the above assumptions for every $n \in \mathbb{N}$ we define the operator $L_{n,\lambda}: \mathscr{C}(K) \to \mathscr{C}(K)$ by

$$L_{n,\lambda}(f)(x) := \int_K \cdots \int_K f\left(\frac{x_1 + \cdots + x_n}{n}\right) dv_x^T(x_1) \ldots dv_x^T(x_n) \tag{6.1.49}$$

for every $f \in \mathscr{C}(K)$ and $x \in K$.

The operator $L_{n,\lambda}$ is called the *n-th Lototsky-Schnabl operator associated with the projection T and the function λ.*

We explicitly observe that the operator $L_{n,\lambda}$ coincides with B_n or the identity operator provided $\lambda = 1$ or $\lambda = 0$, respectively.

More generally, if $x \in K$, we have

$$L_{n,\lambda}(f)(x) = \begin{cases} B_n(f)(x), & \text{if } \lambda(x) = 1, \\ f(x), & \text{if } \lambda(x) = 0. \end{cases} \tag{6.1.50}$$

In any case

$$L_{n,\lambda}(f)(x) = f(x) \quad \text{for every } x \in \partial_H^+ K \text{ and } f \in \mathscr{C}(K). \tag{6.1.51}$$

By using a simple combinatorial argument, we can write

$$L_{n,\lambda}(f)(x) = \sum_{k=0}^{n} \binom{n}{k} \lambda(x)^k (1 - \lambda(x))^{n-k} B_k(f_{x,k/n})(x) \tag{6.1.52}$$

for every $f \in \mathscr{C}(K)$ and $x \in K$, where the function $f_{x,k/n}$ is defined as in (6.1.1), i.e.,

$$f_{x,k/n}(y) := f\left(\frac{k}{n}y + \left(1 - \frac{k}{n}\right)x\right) \tag{6.1.53}$$

for every $y \in K$.

In (6.1.52) B_0 is by convention the identity operator on $\mathscr{C}(K)$.

The representation of Stancu-Schnabl and Lototsky-Schnabl operators in terms of Bernstein-Schnabl operators we gave in (6.1.24) and (6.1.52) will be useful to explicitly describe these operators in the concrete examples of Section 6.3.

In the next results we want to discuss the asymptotic behavior of the sequence $(L_{n,\lambda})_{n \in \mathbb{N}}$ of Lototsky-Schnabl operators.

In this case we do not have an explicit expression for the powers of $L_{n,\lambda}(h^2)$ when $h \in A(K)$ and so we obtain only partial results.

6.1.9 Lemma. *Let* $\lambda \in \mathscr{C}(K)$ *be such that* $0 \le \lambda \le 1$ *and consider the sequence* $(L_{n,\lambda})_{n \in \mathbb{N}}$ *of Lototsky-Schnabl operators associated with the projection T and the function λ.*

Then for every $n, m \in \mathbb{N}$ *and* $h \in H$ *one has*

$$L_{n,\lambda}^m(h) = h \ (= T(h)). \tag{6.1.54}$$

Moreover, if $h \in A(K)$, *then*

$$L_{n,\lambda}(h^2) = \frac{1}{n}\lambda T(h^2) + \left(1 - \frac{1}{n}\lambda\right)h^2 = h^2 + \frac{1}{n}\lambda(T(h^2) - h^2) \tag{6.1.55}$$

and

$$L_{n,\lambda}^m(h^2) = \left(1 - \left(1 - \frac{\lambda}{n}\right)^m\right)T(h^2) + \left(1 - \frac{\lambda}{n}\right)^m h^2 \tag{6.1.56}$$

provided λ is constant.

Proof. As in the proof of Lemmas 6.1.2 and 6.1.6, if $h \in H$ we have $v_x^T(h) = h(x)$ for every $x \in K$ and consequently, by (6.1.49), $L_{n, \lambda}(h)(x) = h(x)$. The equality (6.1.54) now follows by induction on $m \in \mathbb{N}$.

As far as (6.1.55) is concerned, for every $(x_1, \ldots, x_n) \in K^n$ and $h \in A(K)$ we have

$$h^2 \left(\frac{x_1 + \cdots + x_n}{n} \right) = \left(\frac{1}{n} \sum_{i=1}^{n} h(x_i) \right)^2 = \frac{1}{n^2} \left(\sum_{i=1}^{n} h^2(x_i) + 2 \sum_{1 \le i < j \le n} h(x_i) h(x_j) \right),$$

and hence, by (6.1.47) and (6.1.49), for every $x \in K$ we obtain

$$
\begin{aligned}
L_{n, \lambda}(h^2)(x) &= \int_K \cdots \int_K h^2 \left(\frac{x_1 + \cdots + x_n}{n} \right) dv_x^T(x_1) \ldots dv_x^T(x_n) \\
&= \frac{1}{n^2} \left(\sum_{i=1}^{n} \int_K h^2 \, dv_x^T \right) + \frac{1}{n^2} n(n-1) h^2(x) \\
&= \frac{1}{n} (\lambda(x) T(h^2)(x) + (1 - \lambda(x)) h^2(x)) + \left(1 - \frac{1}{n} \right) h^2(x) \\
&= \frac{1}{n} \lambda(x) T(h^2)(x) + \left(1 - \frac{1}{n} \lambda(x) \right) h^2(x).
\end{aligned}
$$

Finally, identity (6.1.56) can be proved by induction. □

Also in this case, we can apply the Korovkin-type theorems of Sections 3.3 and 4.4 to establish the convergence properties of Lototsky-Schnabl operators. For the sake of brevity we omit the details of the proof.

6.1.10 Theorem. *Let $\lambda \in \mathscr{C}(K)$ be such that $0 \le \lambda \le 1$ and consider the sequence $(L_{n, \lambda})_{n \in \mathbb{N}}$ of Lototsky-Schnabl operators associated with the projection T and the function λ.*

For every $f \in \mathscr{C}(K)$ the following statements hold:
(1) $\lim_{n \to \infty} L_{n, \lambda}(f) = f$ *uniformly on K.*

Furthermore, if $\lambda > 0$ is constant, then
(2) $\lim_{n \to \infty} L_{n, \lambda}^m(f) = f$ *uniformly on K for every $m \ge 1$.*

(3) $\lim_{n \to \infty} L_{n, \lambda}^m(f) = T(f)$ *uniformly on K for every $n \ge 1$.*

(4) *If $(k(n))_{n \in \mathbb{N}}$ is a sequence of positive integers, then*

$$
\lim_{n \to \infty} L_{n, \lambda}^{k(n)}(f) =
\begin{cases}
f, & \text{uniformly on } K \quad \text{if } \lim_{n \to \infty} \dfrac{k(n)}{n} = 0, \\[4mm]
T(f), & \text{uniformly on } K \quad \text{if } \lim_{n \to \infty} \dfrac{k(n)}{n} = +\infty.
\end{cases}
$$

Remark. As a simple application of Korovkin-type theorem 4.4.6 (see also (4.4.25)) we can also describe the behavior of the family $(L_{n,\lambda}^m)_{0<\lambda\leq 1}$ when we keep fixed m and n and vary the parameter $\lambda \in]0, 1]$.

In fact, on account of (6.1.54) and (6.1.56), we obtain

$$\lim_{\lambda \to 0^+} L_{n,\lambda}^m(f) = f \quad \text{uniformly on } K \tag{6.1.57}$$

for every $f \in \mathscr{C}(K)$ and $n, m \geq 1$.

In the next result we give some quantitative estimates of the rate of convergence described in Theorem 6.1.10.

6.1.11 Theorem. *Under the hypotheses of Theorem 6.1.10, for every $n \geq 1$, $f \in \mathscr{C}(K)$ and $x \in K$, we have*

$$|L_{n,\lambda}(f)(x) - f(x)| \leq (1 + \lambda(x))\Omega(f, \sqrt{1/n}). \tag{6.1.58}$$

Moreover, if λ is constant, then for every $m \geq 1$, we also have

$$\|L_{n,\lambda}^m(f) - f\| \leq \Psi\left(f, \sqrt{1 - \left(1 - \frac{\lambda}{n}\right)^m}\right) \leq \Psi\left(f, \sqrt{m\lambda/n}\right) \tag{6.1.59}$$

and

$$\|L_{n,\lambda}^m(f) - T(f)\| \leq \Psi\left(f, \sqrt{\left(1 - \frac{\lambda}{n}\right)^m}\right). \tag{6.1.60}$$

Proof. As in the proof of Theorem 6.1.4, by (6.1.54) and (6.1.55), we obtain for every $h \in E_K'$ and $x \in K$,

$$L_{n,\lambda}((h - h(x)\mathbf{1})^2)(x) = \frac{1}{n}\lambda(x)(T(h^2)(x) - h^2(x)) \leq \frac{1}{n}\lambda(x)T(h^2)(x)$$

and consequently, for every $\delta > 0$ and $h_1, \ldots, h_m \in E_K'$ satisfying $\left\|\sum_{j=1}^{m} h_j^2\right\| = \frac{1}{\delta^2}$,

$$\sum_{j=1}^{m} L_{n,\lambda}((h_j - h_j(x)\mathbf{1})^2)(x) \leq \frac{1}{n}\lambda(x)T\left(\sum_{j=1}^{m} h_j^2\right)(x) \leq \frac{1}{n}\lambda(x)\left\|\sum_{j=1}^{m} h_j^2\right\| = \frac{\lambda(x)}{n}\frac{1}{\delta^2}.$$

Thus, in this case formula (5.1.17) yields

$$|L_{n,\lambda}(f)(x) - f(x)| \leq \left(1 + \frac{\lambda(x)}{n}\frac{1}{\delta^2}\right)\Omega(f, \delta),$$

and (6.1.58) follows by setting $\delta := \sqrt{1/n}$.

Finally, (6.1.59) and (6.1.60) follow by (6.1.56) and Theorem 5.1.10, formulas (5.1.45) and (5.1.46), respectively.   □

In addition to the above estimates next we shall show some further ones holding only for special functions, e.g., Lipschitz functions.

To this end we shall use Theorem 5.1.11 and the subsequent remark.

6.1.12 Proposition. *Assume that K is a convex compact subset of a real inner-product space $(E, \langle \cdot, \cdot \rangle)$ and consider the function $e \in \mathscr{C}(K)$ defined by $e(x) := \|x\|^2$ $(x \in K)$.*

Let $T : \mathscr{C}(K) \to \mathscr{C}(K)$ be a positive projection satisfying (6.1.3) and (6.1.4) and consider the sequences of Bernstein-Schnabl, Stancu-Schnabl and Lototsky-Schnabl operators defined by (6.1.12), (6.1.23) and (6.1.49), respectively.

Then for every $n \geq 1$ and for every Lipschitz function $f \in \mathrm{Lip}_M 1$ (see (5.1.47)) we have

$$\|B_n(f) - f\| \leq M \sqrt{\frac{1}{n}} \|T(e) - e\|, \tag{6.1.61}$$

$$\|S_{n,a_n}(f) - f\| \leq M \sqrt{\frac{1 + na_n}{n(1 + a_n)}} \|T(e) - e\| \tag{6.1.62}$$

and

$$\|L_{n,\lambda}(f) - f\| \leq M \sqrt{\frac{1}{n}} \|\lambda(T(e) - e)\|. \tag{6.1.63}$$

Proof. First note that, given $x \in K$, the function $e_x(y) := \langle x, y \rangle$, $(y \in K)$ is affine on K so that

$$B_n(e_x)(x) = S_{n,a_n}(e_x)(x) = L_{n,\lambda}(e_x)(x) = \langle x, x \rangle = e_x(x) \tag{1}$$

by virtue of (6.1.14), (6.1.41) and (6.1.54).

On the other hand for every $x_1, \ldots, x_n \in K$

$$e\left(\frac{x_1 + \cdots + x_n}{n}\right) = \left\langle \frac{x_1 + \cdots + x_n}{n}, \frac{x_1 + \cdots + x_n}{n} \right\rangle$$

$$= \frac{1}{n^2}\left(\sum_{i=1}^{n} \langle x_i, x_i \rangle + 2 \sum_{1 \leq i < j \leq n} \langle x_i, x_j \rangle\right)$$

$$= \frac{1}{n^2}\left(\sum_{i=1}^{n} e(x_i) + 2 \sum_{1 \leq i < j \leq n} \langle x_i, x_j \rangle\right).$$

Accordingly,

$$S_{n,a_n}(e)(x) = \frac{1 + na_n}{n(1 + a_n)}(Te(x) - e(x)) + e(x)$$

as well as

$$L_{n,\lambda}(e)(x) = \frac{1}{n^2}(n(\lambda(x)Te(x) + (1 - \lambda(x))e(x)) + n(n - 1)e(x))$$

$$= \frac{\lambda(x)}{n}(Te(x) - e(x)) + e(x).$$

The results now follow by applying formula (5.1.51).   □

In the last part of this section we investigate two important qualitative properties of Lototsky-Schnabl (and hence of Bernstein-Schnabl) operators.

The first one is concerned with the monotonicity properties of this sequence.

As a matter of fact a similar property established for classical Bernstein operators, revealed itself to be important in the mathematical development of computer-aided geometric design (see, for instance, Bézier [1972], Forrest [1971], Gordon and Riesenfeld [1974]).

Moreover, the same property has been investigated not only for classical Bernstein polynomials but also for other sequences of positive linear operators. Some references to this subject can be found in the final notes of this section as well as in Section 6.3 where further results will be illustrated in the particular settings we treat there.

We begin by introducing some useful definitions.

We say that a function $f \in \mathscr{C}(K)$ is *T-convex* if

$$f_{z,\alpha} \leq T(f_{z,\alpha}) \quad \text{for every } z \in K \text{ and } \alpha \in [0,1] \tag{6.1.64}$$

where $f_{z,\alpha}$ is defined as in (6.1.1).

Setting $\alpha = 1$ in (6.1.64) we obtain $f \leq T(f)$.

We recall that, if $f \in \mathscr{C}(K)$ is convex, then, by virtue of (1.5.4) and the fact that every $x \in K$ is the barycenter of μ_x^T, we have $f \leq T(f)$.

On the other hand every $f_{z,\alpha}$ is convex too and hence $f_{z,\alpha} \leq T(f_{z,\alpha})$.

In other words, every convex function on K is *T*-convex as well.

T-convex functions ensure the monotonicity of the family $(L_{n,\lambda}(f))$ with respect to the parameter $\lambda \in \mathscr{C}(K)$, as the next result shows.

6.1.13 Theorem. *Let* $f \in \mathscr{C}(K)$ *be a T-convex function and* $\alpha, \lambda \in \mathscr{C}(K)$ *satisfying* $0 \le \alpha \le \lambda \le 1$. *Then for every* $n \ge 1$ *we have*

$$L_{n,\alpha}(f) \le L_{n,\lambda}(f) \tag{6.1.65}$$

and in particular

$$f \le L_{n,\lambda}(f) \le B_n(f) \le T(f). \tag{6.1.66}$$

Proof. First we show that

$$f\left(x_0 + \frac{x}{n}\right) \le \int f\left(x_0 + \frac{u}{n}\right) d\mu_x^T(u) \tag{1}$$

for every $n \ge 1$, $x_0 \in \dfrac{n-1}{n}K$ and $x \in K$.

Indeed, (1) is certainly true if $n = 1$ because $x_0 = 0$ and $f \le T(f)$. Assume $n > 1$ and write $x_0 = \dfrac{n-1}{n}z_0$ for some $z_0 \in K$. Then

$$f\left(x_0 + \frac{x}{n}\right) = f_{z_0, 1/n}(x) \le \int f_{z_0, 1/n}(u) \, d\mu_x^T(u) = \int f\left(x_0 + \frac{u}{n}\right) d\mu_x^T(u).$$

Inequality (1) enables us to conclude (6.1.65) because we obtain

$$\int f\left(x_0 + \frac{u}{n}\right) dv_{x,\alpha}^T(u)$$

$$= \alpha(x) \int f\left(x_0 + \frac{u}{n}\right) d\mu_x^T(u) + (1 - \alpha(x)) f\left(x_0 + \frac{x}{n}\right)$$

$$= \alpha(x) \int f\left(x_0 + \frac{u}{n}\right) d\mu_x^T(u) + (\lambda(x) - \alpha(x)) f\left(x_0 + \frac{x}{n}\right)$$

$$+ (1 - \lambda(x)) f\left(x_0 + \frac{x}{n}\right)$$

$$\leq \alpha(x) \int f\left(x_0 + \frac{u}{n}\right) d\mu_x^T(u) + (\lambda(x) - \alpha(x)) \int f\left(x_0 + \frac{u}{n}\right) d\mu_x^T(u)$$

$$+ (1 - \lambda(x)) f\left(x_0 + \frac{x}{n}\right)$$

$$\leq \lambda(x) \int f\left(x_0 + \frac{u}{n}\right) d\mu_x^T(u) + (1 - \lambda(x)) f\left(x_0 + \frac{x}{n}\right)$$

$$= \int f\left(x_0 + \frac{u}{n}\right) dv_{x,\lambda}^T(u)$$

and hence

$L_{n,\alpha}(f)(x)$

$$= \int_K \cdots \int_K \left(\int_K f\left(\frac{x_1 + \cdots + x_{n-1}}{n} + \frac{x_n}{n}\right) dv_{x,\alpha}^T(x_n)\right) dv_{x,\alpha}^T(x_1) \ldots dv_{x,\alpha}^T(x_{n-1})$$

$$\leq \int_K \cdots \int_K \left(\int_K f\left(\frac{x_1 + \cdots + x_{n-1}}{n} + \frac{x_n}{n}\right) dv_{x,\lambda}^T(x_n)\right) dv_{x,\alpha}^T(x_1) \ldots dv_{x,\alpha}^T(x_{n-1})$$

$$\leq \cdots \leq \int_K \cdots \int_K f\left(\frac{x_1 + \cdots + x_n}{n}\right) dv_{x,\lambda}^T(x_1) \ldots dv_{x,\lambda}^T(x_n) = L_{n,\lambda}(f)(x).$$

As regards inequalities (6.1.66), the first two follow from (6.1.65) and the last one is a consequence of the fact that, since $f \leq T(f)$, then $B_n(f) \leq B_n(T(f)) = T(f)$ by virtue of (6.1.14). □

To obtain the monotonicity of the sequence $(L_{n,\lambda}(f))_{n \in \mathbb{N}}$, one has to impose some further hypotheses on the T-convex function $f \in \mathscr{C}(K)$.

In this respect we shall consider another type of convexity.

For a given $\lambda \in \mathscr{C}(K)$ and $x \in K$ we set

$$D_{x,\lambda} := \{(u,v) \in K^2 | \text{There exist } \beta \geq 0 \text{ and } p, q \in \text{Supp}(v_{x,\lambda}^T)$$

$$\text{such that } u - v = \beta(p - q)\}, \qquad (6.1.67)$$

where, as explained after formula (6.1.47), $v_{x,\lambda}^T$ denotes the measure v_x^T.

A function $f \in \mathscr{C}(K)$ is called (T,λ)-convex if

$$f(\alpha u + (1 - \alpha)v) \leq \alpha f(u) + (1 - \alpha)f(v) \qquad (6.1.68)$$

for every $x \in K$, $(u,v) \in D_{x,\lambda}$ and $0 \leq \alpha \leq 1$.

A $(T, 1)$-convex function will be simply called an *axially convex function*.

Clearly every convex function is (T, λ)-convex. We shall see that, if λ is strictly positive, then every (T, λ)-convex function is T-convex. The converse holds in some particular settings such as Bauer simplices (see Theorem 6.3.2).

In the next theorem we collect the main results on the monotonicity of Lototsky-Schnabl and Bernstein-Schnabl operators.

6.1.14 Theorem. *For a given $f \in \mathscr{C}(K)$ and $n \geq 1$, the following statements hold:*
(1) If f is (T, λ)-convex for some $\lambda \in \mathscr{C}(K), 0 \leq \lambda \leq 1$, then

$$f \leq L_{n+1, \lambda}(f) \leq L_{n, \lambda}(f). \tag{6.1.69}$$

(2) If f is axially convex, then

$$f \leq B_{n+1}(f) \leq B_n(f). \tag{6.1.70}$$

Proof. In order to show statement (1), we consider for every $i = 1, \ldots, n + 1$ the mapping $\sigma_i \colon K^{n+1} \to K$ defined by

$$\sigma_i(x_1, \ldots, x_{n+1}) := \frac{1}{n} \sum_{\substack{j=1 \\ j \neq i}}^{n+1} x_j \in K \quad \text{for every } (x_1, \ldots, x_{n+1}) \in K^{n+1}. \tag{1}$$

Clearly we have

$$\sum_{i=1}^{n+1} \sigma_i(x_1, \ldots, x_{n+1}) = \frac{1}{n} \sum_{i=1}^{n+1} \left(\left(\sum_{j=1}^{n+1} x_j \right) - x_i \right) = \sum_{i=1}^{n+1} x_i. \tag{2}$$

Moreover

$$\int_{K^{n+1}} f(\sigma_i(x_1, \ldots, x_{n+1})) \, dv_x^T(x_1) \ldots dv_x^T(x_{n+1})$$
$$= \int_{K^n} f\left(\frac{y_1 + \cdots + y_n}{n} \right) dv_x^T(y_1) \ldots dv_x^T(y_n). \tag{3}$$

Therefore, if $x \in K$, we have

$$L_{n, \lambda}(f)(x) - L_{n+1, \lambda}(f)(x)$$

$$= \int_K \cdots \int_K \left(f\left(\frac{x_1 + \cdots + x_n}{n} \right) - f\left(\frac{x_1 + \cdots + x_{n+1}}{n + 1} \right) \right) dv_x^T(x_1) \ldots dv_x^T(x_{n+1})$$

$$= \int_K \cdots \int_K \left(\left(\frac{1}{n + 1} \sum_{i=1}^{n+1} f(\sigma_i(x_1, \ldots, x_{n+1})) \right) \right.$$

$$\left. - f\left(\frac{1}{n + 1} \sum_{i=1}^{n+1} \sigma_i(x_1, \ldots, x_{n+1}) \right) \right) dv_x^T(x_1) \ldots dv_x^T(x_{n+1}). \tag{4}$$

To conclude the proof we need to introduce some other auxiliary mappings which are defined in terms of the above mappings σ_i. For every $i = 0, \ldots, n$ and $j = i + 1, \ldots, n + 1$ we set

$$
\sigma_{ij} := \begin{cases}
\sigma_j, & \text{if } i = 0, \\[2mm]
\dfrac{\sigma_1 + n\sigma_j}{n + 1}, & \text{if } i = 1, \\[2mm]
\dfrac{\sigma_{i-1,i} + (n - i + 1)\sigma_{i-1,j}}{n - i + 2}, & \text{if } i > 1.
\end{cases}
\tag{5}
$$

Then

$$
\sigma_{n,n+1} = \frac{1}{2}(\sigma_{n-1,n} + \sigma_{n-1,n+1}) = \frac{1}{3}(\sigma_{n-2,n-1} + \sigma_{n-2,n} + \sigma_{n-2,n+1}) = \cdots
$$

$$
= \frac{1}{n + 1}(\sigma_{0,1} + \sigma_{0,2} + \cdots + \sigma_{0,n+1}) = \frac{1}{n + 1}\sum_{i=1}^{n+1} \sigma_i.
$$

On the other hand, if $x \in K$ and $(x_1, \ldots, x_{n+1}) \in (\mathrm{Supp}(v_x^T))^{n+1}$, then

$$
(\sigma_{i,j}(x_1, \ldots, x_{n+1}), \sigma_{i,k}(x_1, \ldots, x_{n+1})) \in D_{x,\lambda}
$$

for each $i = 0, \ldots, n$ and $j, k = i + 1, \ldots, n + 1$, so that

$$
f\left(\frac{1}{n + 1}\sum_{i=1}^{n+1} \sigma_i(x_1, \ldots, x_{n+1})\right) = f(\sigma_{n,n+1}(x_1, \ldots, x_{n+1}))
$$

$$
\leq \frac{1}{2}(f(\sigma_{n-1,n}(x_1, \ldots, x_{n+1})) + f(\sigma_{n-1,n+1}(x_1, \ldots, x_{n+1}))
$$

$$
\leq \frac{1}{3}(f(\sigma_{n-2,n-1}(x_1, \ldots, x_{n+1})) + f(\sigma_{n-2,n}(x_1, \ldots, x_{n+1}))
$$

$$
+ f(\sigma_{n-2,n+1}(x_1, \ldots, x_{n+1})))
$$

$$
\leq \cdots \leq \frac{1}{n + 1}(f(\sigma_{0,1}(x_1, \ldots, x_{n+1})) + \cdots + f(\sigma_{0,n+1}(x_1, \ldots, x_{n+1})))
$$

$$
= \frac{1}{n + 1}\sum_{i=1}^{n+1} f(\sigma_i(x_1, \ldots, x_{n+1})).
$$

By using (4) together with the above inequality and the fact that

$$\mathrm{Supp}(v_x^T \otimes \cdots \otimes v_x^T) = (\mathrm{Supp}(v_x^T))^{n+1} \tag{6}$$

(see Proposition 1.2.5), we obtain $L_{n+1,\lambda}(f) \leq L_{n,\lambda}(f)$.

From the last inequality we obtain by induction $L_{n+p,\lambda}(f) \leq L_{n,\lambda}(f)$ for every $p \geq 1$.

Letting $p \to +\infty$ and taking Theorem 6.1.10, (1), into account, we conclude $f \leq L_{n,\lambda}(f)$.

Statement (2) is an obvious consequence of (1). $\qquad\square$

Remark. From the above theorem it also follows that, if a function $f \in \mathscr{C}(K)$ is (T, λ)-convex for some strictly positive function $\lambda \in \mathscr{C}(K)$ (in particular, if f is axially convex), then f is T-convex.

Indeed, applying (6.1.69) to $n = 1$, we obtain $f \leq L_{1,\lambda}(f) = \lambda T(f) + (1 - \lambda)f$ and so $f \leq T(f)$. On the other hand, for each $z \in K$ and $0 \leq \alpha \leq 1$, $f_{z,\alpha}$ is (T, λ)-convex as well and hence $f_{z,\alpha} \leq T(f_{z,\alpha})$.

The following corollary is worth mentioning separately.

6.1.15 Corollary. *Let $f \in \mathscr{C}(K)$ be a convex function. Then for every $\lambda \in \mathscr{C}(K)$, $0 \leq \lambda \leq 1$ and $n \geq 1$ we have*

$$f \leq L_{n+1,\lambda}(f) \leq L_{n,\lambda}(f) \leq \lambda T(f) + (1 - \lambda)f \leq T(f) \tag{6.1.71}$$

and in particular

$$f \leq B_{n+1}(f) \leq B_n(f) \leq T(f). \tag{6.1.72}$$

Furthermore, for every $x \in K$ the following statements are equivalent:
(i) *$L_{n+1,\lambda}(f)(x) = L_{n,\lambda}(f)(x)$ for every $n \geq 1$ and for every $\lambda \in \mathscr{C}(K), 0 \leq \lambda \leq 1$ satisfying $\lambda(x) > 0$.*
(ii) *$L_{n+1,\lambda}(f)(x) = L_{n,\lambda}(f)(x)$ for every $n \geq 1$ and for some $\lambda \in \mathscr{C}(K), 0 \leq \lambda \leq 1$ satisfying $\lambda(x) > 0$.*
(iii) *$B_{n+1}(f)(x) = B_n(f)(x)$ for every $n \geq 1$.*
(iv) *$Tf(x) = f(x)$.*
(v) *f is affine on $\overline{\mathrm{co}}(\mathrm{Supp}(\mu_x^T))$.*

Proof. As far as (6.1.71) is concerned, we have only to prove that $L_{n,\lambda}(f) \leq \lambda T(f) + (1 - \lambda)f \leq T(f)$. Indeed, since f is convex, we know that $f \leq T(f)$ and hence for every $x \in K$

$$L_{n,\lambda}(f)(x) \leq \int_K \cdots \int_K \frac{f(x_1) + \cdots + f(x_n)}{n} \, dv_x^T(x_1) \ldots dv_x^T(x_n) = v_x^T(f)$$

$$= \lambda(x) Tf(x) + (1 - \lambda(x))f(x) \leq Tf(x).$$

To show the second part of our statement it suffices to prove the implications (ii) $\Rightarrow$ (iv), (iii) $\Rightarrow$ (iv) and (iv) $\Rightarrow$ (v) $\Rightarrow$ (i).

Assume that (ii) holds. Then for every $p \geq 1$ and $n \geq 1$ we obtain $L_{n,\lambda}(f)(x) = L_{n+p,\lambda}(f)(x)$.

Setting $n = 1$ and letting $p \to \infty$, we have $v_x^T(f) = f(x)$ hence $Tf(x) = \mu_x^T(f) = f(x)$, because $\lambda(x) > 0$.

A similar reasoning shows that (iii) implies (iv).

The implication (iv) $\Rightarrow$ (v) directly follows from Rasa's theorem 1.5.2.

Finally note that formulas (1.5.2) and (6.1.48) yield the equality

$$\overline{\mathrm{co}}(\mathrm{Supp}(v_x^T)) = \overline{\mathrm{co}}(\mathrm{Supp}(\mu_x^T)). \tag{1}$$

Accordingly, if (v) holds, then f is affine on $\overline{\mathrm{co}}(\mathrm{Supp}(v_x^T))$ and hence $L_{n+1,\lambda}(f)(x) = L_{n,\lambda}(f)(x)$ by virtue of the equalities (4) and (6) of the proof of Theorem 6.1.14. $\square$

Remark. A close look at the above proof shows that, in fact, conditions (iv) and (v) are equivalent to the apparently more restrictive conditions

(i)' $L_{n,\lambda}(f)(x) = f(x)$ *for every* $n \geq 1$ *and for every* $\lambda \in \mathscr{C}(K)$, $0 \leq \lambda \leq 1$ *satisfying* $\lambda(x) > 0$.

(ii)' $L_{n,\lambda}(f)(x) = f(x)$ *for every* $n \geq 1$ *and for some* $\lambda \in \mathscr{C}(K)$, $0 \leq \lambda \leq 1$ *satisfying* $\lambda(x) > 0$.

(iii)' $B_n(f)(x) = f(x)$ *for every* $n \geq 1$.

According to Theorem 6.1.14 and Corollary 6.1.15, the property

$$f \leq L_{n,\lambda}(f) \quad \text{for every } n \geq 1 \text{ and for some } \lambda \in \mathscr{C}(K), 0 \leq \lambda \leq 1, \tag{6.1.73}$$

is satisfied by very special functions $f \in \mathscr{C}(K)$.

In Section 6.3 (see Corollary 6.3.8) we shall see that, when K is the unit interval and T is the projection (3.3.12) (with respect to which the Bernstein-Schnabl operators are the classical Bernstein operators), then property (6.1.73) implies that f is convex.

In the general context of this section we shall see that, if a function $f \in \mathscr{C}(K)$ satisfies (6.1.73), then f achieves its maximum on $\partial_H^+ K$.

The following lemma will enable us to show this result quickly.

Given a topological vector space E over $\mathbb{K}$ and an arbitrary subset B of E we set

$$m(B) := \left\{ \frac{x_1 + \cdots + x_p}{p} \,\middle|\, p \geq 1, x_1, \ldots, x_p \in B \right\}. \tag{6.1.74}$$

6.1.16 Lemma. *Under the above assumptions, the equality* $\overline{m(B)} = \overline{\mathrm{co}}(B)$ *holds.*

Proof. We have only to show the inclusion $\overline{\text{co}}(B) \subset \overline{m(B)}$. Let $x_0 \in \overline{\text{co}}(B)$ and consider a neighborhood U of 0. After choosing a neighborhood V of 0 such that $V + V \subset U$, we can take $x = \sum\limits_{i=1}^{n} \lambda_i x_i \in \text{co}(B)$ such that $x - x_0 \in V$, where $x_i \in B$, $\lambda_i \geq 0$, $i = 1, \ldots, n$ and $\sum\limits_{i=1}^{n} \lambda_i = 1$.

Obviously we may suppose $n \geq 2$.

Let W be a suitable neighborhood of 0 such that $W + \cdots + W \subset V$, where the left-hand side of the above inclusion is the sum of n copies of W.

Furthermore, choose $\varepsilon_0 > 0$ such that

$$\xi x_i \in W \quad \text{for every } \xi \in \mathbb{K}, |\xi| \leq \varepsilon_0 \text{ and } i = 1, \ldots, n.$$

Then for every $i = 1, \ldots, n - 1$ there exist $h_i, p_i \in \mathbb{N}$ such that $0 \leq \lambda_i - \dfrac{h_i}{p_i}$
$\leq \dfrac{\varepsilon_0}{n - 1}$.

So, setting $q_n := 1 - \sum\limits_{i=1}^{n-1} \dfrac{h_i}{p_i} \in \mathbb{Q}$, we can write $q_n = \dfrac{h_n}{p_n}$ for some $h_n, p_n \in \mathbb{N}$ and

$$|q_n - \lambda_n| \leq \sum\limits_{i=1}^{n-1} \left| \dfrac{h_i}{p_i} - \lambda_i \right| \leq \varepsilon_0.$$

Since $\sum\limits_{i=1}^{n} \dfrac{h_i}{p_i} = 1$, if we write $\dfrac{h_i}{p_i} = \dfrac{k_i}{p}$ for suitable $k_i, p \in \mathbb{N}$ and $i = 1, \ldots, n$,

clearly we have $\sum\limits_{i=1}^{n} k_i = p$, so that the point $y := \sum\limits_{i=1}^{n} \dfrac{k_i x_i}{p}$ belongs to $m(B)$. Finally

$$y - x_0 = y - x + x - x_0$$

$$= \sum\limits_{i=1}^{n} \left(\dfrac{h_i}{p_i} - \lambda_i \right) x_i + x - x_0 \in W + \cdots + W + V \subset V + V \subset U.$$

Since U was arbitrarily chosen, we can conclude that $x_0 \in \overline{m(B)}$. $\square$

After these preliminaries we can show the aforementioned maximum principle.

6.1.17 Theorem. *Let $f \in \mathscr{C}(K)$ be a function satisfying (6.1.73) and set $M := \max\{f(x) | x \in K\}$. Suppose that λ is not identically zero on $\{x \in K | f(x) = M\}$. Then for each $x_0 \in K$ satisfying $f(x_0) = M$ and $\lambda(x_0) > 0$ we also have*

$$f(x) = M \quad \text{for every } x \in \overline{\text{co}}(\text{Supp}(\mu_{x_0}^T)).$$

In particular, f takes its maximum on $\partial_H^+ K$.

Proof. Since

$$0 \le L_{n,\lambda}(f)(x_0) - f(x_0)$$

$$= \int_K \cdots \int_K \left(f\left(\frac{x_1 + \cdots + x_n}{n} \right) - f(x_0) \right) dv_x^T(x_1) \ldots dv_x^T(x_n) \le 0,$$

we obtain that

$$f\left(\frac{x_1 + \cdots + x_n}{n} \right) = f(x_0),$$

for each $(x_1, \ldots, x_n) \in \mathrm{Supp}(v_{x_0}^T \otimes \cdots \otimes v_{x_0}^T) = (\mathrm{Supp}(v_{x_0}^T))^n$.
Consequently, f takes the value $f(x_0)$ on

$$\overline{m}(\mathrm{Supp}(v_{x_0}^T)) = \overline{\mathrm{co}}(\mathrm{Supp}(v_{x_0}^T)) = \overline{\mathrm{co}}(\mathrm{Supp}(\mu_{x_0}^T)),$$

by Lemma 6.1.16 and equality (1) of the proof of Corollary 6.1.15.
The last part of the statement follows from (6.1.8). □

A simple consequence of Theorem 6.1.17 is indicated next.

6.1.18 Corollary. *If a function $f \in \mathscr{C}(K)$ satisfies (6.1.73), then at each point $x_0 \in K$ where f takes on its maximum we have $L_{n,\lambda}(f)(x_0) = f(x_0)$ for each $n \ge 1$ and $\lambda \in \mathscr{C}(K), 0 \le \lambda \le 1$.*

We close this section by discussing another qualitative property of Lototsky-Schnabl operators which concerns the invariance of the space of Hölder continuous functions.

This property can be framed into the problem of asking which global smoothness properties of a function f are retained by the approximants $L_{n,\lambda}(f)$.

Along these lines several results have been established for classical Bernstein operators and for other positive approximation processes as well (for some references see the final notes and Section 6.3).

In our general setting the property of preserving Hölder continuity will be fruitfully used to derive a regularity result for the solutions of the degenerate diffusion equations associated with the Lototsky-Schnabl operators (see Sections 6.2 and 6.3).

From now on we shall fix a metric d on K whose associated topology coincides with the given one on K.

The modulus of continuity of $f \in \mathscr{C}(K)$ is given by

$$\omega(f, \delta) := \sup\{|f(x) - f(y)| \,|\, x, y \in K, d(x, y) \leq \delta\} \tag{6.1.75}$$

$(\delta > 0)$.

Given $M > 0$ and $0 \leq \alpha \leq 1$, we shall denote by $\mathrm{Lip}_M \alpha$ the subset of all Hölder continuous functions f on K with exponent α and constant M, i.e.,

$$|f(x) - f(y)| \leq M d(x, y)^\alpha \quad \text{for every } x, y \in K. \tag{6.1.76}$$

We also set

$$\mathrm{Lip}(K) := \bigcup_{M > 0} \mathrm{Lip}_M 1, \tag{6.1.77}$$

endowed with the seminorm

$$\|f\|_{\mathrm{Lip}} := \sup\left\{\frac{|f(x) - f(y)|}{d(x, y)} \,\middle|\, x, y \in K, x \neq y\right\}. \tag{6.1.78}$$

To our purpose we need to introduce the *least concave majorant* of a function $f \in \mathscr{C}(K)$ which is defined by

$$\tilde{\omega}(f, \delta) := \begin{cases} \sup\left\{\dfrac{(\delta - s)\omega(f, t) + (t - \delta)\omega(f, s)}{t - s} \,\middle|\, 0 \leq s \leq \delta \leq t \leq \delta(K), s \neq t\right\}, \\ \hspace{6cm} \text{if } 0 \leq \delta \leq \delta(K), \\ \omega(f, \delta(K)), \quad \text{if } \delta > \delta(K), \end{cases}$$

$$\tag{6.1.79}$$

where $\delta(K) := \sup\{d(x, y) \,|\, x, y \in K\}$.

In fact, it turns out that

$$\tilde{\omega}(f, \delta) = \inf\{\alpha(\delta) \,|\, \alpha \text{ is concave in } [0, +\infty[\text{ and } \omega(f, t) \leq \alpha(t) \text{ for every } t \geq 0\}.$$

The equality

$$\tilde{\omega}(f, \delta) = 2 K_0\left(f, \frac{\delta}{2}\right) \tag{6.1.80}$$

relates the majorant $\tilde{\omega}(f, \delta)$ with the *Peetre K_0-functional* $K_0(f, \delta)$ of f with respect to $\mathrm{Lip}(K)$, which is defined by

$$K_0(f, \delta) := \inf\{\|f - g\| + \delta\|g\|_{\mathrm{Lip}} \,|\, g \in \mathrm{Lip}(K)\} \tag{6.1.81}$$

(for a proof of (6.1.80), see Mitjagin and Semenov [1977] or Peetre [1963]).

The next result shows under which assumptions a linear operator preserves in some respects the smoothness of a function.

6.1.19 Proposition. *Under the above assumptions, consider a non-null bounded linear operator* $L: \mathscr{C}(K) \to \mathscr{C}(K)$ *such that* $L(\mathrm{Lip}(K)) \subset \mathrm{Lip}(K)$ *and* $\|L(g)\|_{\mathrm{Lip}} \leq c\|g\|_{\mathrm{Lip}}$ *for every* $g \in \mathrm{Lip}(K)$, *c being a suitable positive constant which depends only on* L.

Then for every $f \in \mathscr{C}(K)$ *and* $\delta > 0$ *we have* $\omega(L(f), \delta) \leq \|L\| \tilde{\omega}\left(f, \dfrac{c\delta}{\|L\|}\right)$.

Proof. To achieve the assertion we use (6.1.80). In fact, if $g \in \mathrm{Lip}(K)$, then

$$\omega(L(g), \delta) \leq \delta \|L(g)\|_{\mathrm{Lip}} \leq c\delta \|g\|_{\mathrm{Lip}}$$

and hence for every $f \in \mathscr{C}(K)$

$$\omega(L(f), \delta) \leq \omega(L(f - g), \delta) + \omega(L(g), \delta)$$

$$\leq 2\|L(f - g)\| + c\delta \|g\|_{\mathrm{Lip}} \leq 2\|L\| \left(\|f - g\| + \frac{c\delta}{2\|L\|} \|g\|_{\mathrm{Lip}}\right).$$

By taking the infimum with respect to $g \in \mathrm{Lip}(K)$, the result follows. □

From the above result we can derive some interesting consequence provided we assume that the modulus of continuity $\omega(f, \delta)$ satisfies the inequality (which is very familiar in normed spaces)

$$\omega(f, t\delta) \leq (1 + t)\omega(f, \delta) \tag{6.1.82}$$

for every $f \in \mathscr{C}(K)$ and $\delta, t > 0$.

This assumption is satisfied if the metric d is induced by a metric d_0 on E satisfying $d_0(x + z, y + z) = d_0(x, y)$ and $d_0(\alpha x, 0) \leq \alpha d_0(x, 0)$ for every $x, y, z \in E$ and $0 \leq \alpha \leq 1$ (see Nishishiraho [1983]).

Other examples are those spaces which have a coefficient of convex deformation equal to one (Jiménez Poso [1980]).

6.1.20 Corollary. *Let* $L: \mathscr{C}(K) \to \mathscr{C}(K)$ *be a non-null bounded linear operator such that* $L(\mathrm{Lip}(K)) \subset \mathrm{Lip}(K)$ *and* $\|L(g)\|_{\mathrm{Lip}} \leq c\|g\|_{\mathrm{Lip}}$ *for every* $g \in \mathrm{Lip}(K)$ ($c \geq 0$ *being independent of* g).

Furthermore assume that (6.1.82) *holds.*

Then the following statements hold:
(1) *For every* $f \in \mathscr{C}(K)$ *and* $\delta > 0$

$$\omega(L(f), \delta) \leq (\|L\| + c)\omega(f, \delta).$$

(2) *If $f \in \operatorname{Lip}_M \alpha$ for some $M > 0$ and $0 < \alpha \le 1$, then $L(f) \in \operatorname{Lip}_N \alpha$ where $N :=$
$Mc^\alpha \|L\|^{1-\alpha}$.*

Proof. On account of Proposition 6.1.19, to prove statement (1) it is enough
to show that $\tilde\omega(f, \xi\delta) \le (1 + \xi)\omega(f, \delta)$ for every $\delta, \xi > 0$.
   In fact, if $\xi\delta \le \delta(K)$, then for every $0 \le s \le \xi\delta \le t \le \delta(K)$, $s \ne t$, we obtain

$$\frac{(\xi\delta - s)\omega(f, t) + (t - \xi\delta)\omega(f, s)}{t - s}$$

$$\le \frac{\xi\delta - s}{t - s}\left(1 + \frac{t}{\delta}\right)\omega(f, \delta) + \frac{t - \xi\delta}{t - s}\left(1 + \frac{s}{\delta}\right)\omega(f, \delta) = (1 + \xi)\omega(f, \delta).$$

If $\xi\delta > \delta(K)$, then

$$\tilde\omega(f, \xi\delta) = \omega(f, \delta(K)) \le \left(1 + \frac{\delta(K)}{\delta}\right)\omega(f, \delta) \le (1 + \xi)\omega(f, \delta),$$

and hence the proof is complete.
   Assume now that $f \in \operatorname{Lip}_M \alpha$ for some $M > 0$ and $0 < \alpha \le 1$.
   Then $\omega(f, \delta) \le M\delta^\alpha$ for every $\delta > 0$. Since the real function $\delta \mapsto \delta^\alpha$ is
concave, we obtain $\tilde\omega(f, \delta) \le M\delta^\alpha$ $(\delta > 0)$ and this yields the assertion because,
again by virtue of Proposition 6.1.19, we have $\omega(L(f), \delta) \le \|L\|\tilde\omega\left(f, \frac{c\delta}{\|L\|}\right) \le$
$M\|L\|^{1-\alpha}c^\alpha\delta^\alpha$. $\square$

We shall now connect these results with the above considerations on
Lototsky-Schnabl operators.
   From now on we shall assume that there exists $c \ge 1$ such that

$$T(f) \in \operatorname{Lip}_c 1 \quad \text{for every } f \in \operatorname{Lip}_1 1. \tag{6.1.83}$$

This condition is crucial for what follows and in Section 6.3 we shall see that
it is not always satisfied in concrete examples. Note however that from (6.1.83) it
follows that

$$T(f) \in \operatorname{Lip}_{cM} 1 \quad \text{for every } f \in \operatorname{Lip}_M 1. \tag{6.1.84}$$

Moreover we shall fix

$$\lambda \in \operatorname{Lip}_N 1 \quad \text{such that } 0 \le \lambda \le 1. \tag{6.1.85}$$

Under these assumptions we want to study the property of preserving spaces of Hölder continuous functions by Lototsky-Schnabl operators.

We first observe that the operator $U := \lambda T + (1 - \lambda)I_{\mathscr{C}(K)}$, where $I_{\mathscr{C}(K)}$ denotes the identity operator, preserves the Lipschitz functions; more precisely, we have

$$U(f) \in \text{Lip}_{cM+N\|T(f)-f\|} 1 \quad \text{for every } f \in \text{Lip}_M 1. \tag{6.1.86}$$

Indeed, if $x, y \in K$ and $f \in \mathscr{C}(K)$, then

$$|Uf(x) - Uf(y)| = |\lambda(x)(Tf(x) - Tf(y)) + (1 - \lambda(x))(f(x) - f(y))$$

$$+ (\lambda(x) - \lambda(y))(Tf(y) - f(y))|$$

$$\leq (cM + N\|T(f) - f\|)\,d(x, y).$$

We can now state the following result.

6.1.21 Theorem. *Under assumptions* (6.1.83) *and* (6.1.85), *for every* $f \in \text{Lip}_M 1$ *and* $n \in \mathbb{N}$ *we have*

$$L_{n,\lambda}(f) \in \text{Lip}_{cM+nN\|T(f)-f\|} 1. \tag{6.1.87}$$

Moreover, if λ *is constant, then*

$$L_{n,\lambda}(f) \in \text{Lip}_{cM} 1. \tag{6.1.88}$$

In particular, for $\lambda = 1$, *we have*

$$B_n(f) \in \text{Lip}_{cM} 1. \tag{6.1.89}$$

Proof. If $n = 1$ the operator $L_{1,\lambda}$ coincides with $U := \lambda T + (1 - \lambda)I_{\mathscr{C}(K)}$ and therefore in this case (6.1.87) is a consequence of (6.1.86).

Suppose $n \geq 2$ and let $f \in \text{Lip}_M 1$ and $x \in K$. For every $x_1, \ldots, x_{n-1} \in K$ we define the function $f^x_{x_1,\ldots,x_{n-1}} : K \to \mathbb{R}$ by

$$f^x_{x_1,\ldots,x_{n-1}}(t) := f\left(\frac{x_1 + \cdots + x_{n-1} + t}{n}\right) \quad \text{for every } t \in K.$$

Moreover, for every $k = 2, \ldots, n - 1$ we consider the functions on K which are defined recursively by

$$f^x_{x_1,\ldots,x_{n-k}}(t) := U(f^x_{x_1,\ldots,x_{n-k},t})(x) \quad (t \in K).$$

Finally we define $f^x(t) := U(f_t^x)(x)$ for every $t \in K$.
By finite induction we obtain

$$\|f_{x_1,\ldots,x_{k-1},u}^x - f_{x_1,\ldots,x_{k-1},v}^x\| \leq \frac{M}{n} d(u,v) \quad \text{for every } u, v \in K \text{ and } k = 1, \ldots, n-1$$

(1)

and consequently $f_{x_1,\ldots,x_k}^x \in \mathrm{Lip}_{M/n} 1$.
Moreover, from (1) it follows that $f^x \in \mathrm{Lip}_{M/n} 1$.
Now let $y \in K$ and observe that for every $k = 1, \ldots, n-1$

$$U(f_{x_1,\ldots,x_k}^y)(x) \leq U(f_{x_1,\ldots,x_k}^y)(y) + \left(\frac{cM}{n} + N\|T(f) - f\|\right) d(x,y)$$

and

$$U(f^y)(x) \leq U(f^y)(y) + \left(\frac{cM}{n} + N\|T(f) - f\|\right) d(x,y).$$

Since $f_{x_1,\ldots,x_{n-1}}^y = f_{x_1,\ldots,x_{n-1}}^x$, we obtain

$L_{n,\lambda}(f)(x)$

$$= \int_K \cdots \int_K f\left(\frac{x_1 + \cdots + x_n}{n}\right) dv_x^T(x_1) \ldots dv_x^T(x_n)$$

$$= \int_K \cdots \int_K U(f_{x_1,\ldots,x_{n-1}}^x)(x) dv_x^T(x_1) \ldots dv_x^T(x_{n-1})$$

$$= \int_K \cdots \int_K U(f_{x_1,\ldots,x_{n-1}}^y)(x) dv_x^T(x_1) \ldots dv_x^T(x_{n-1})$$

$$\leq \int_K \cdots \int_K U(f_{x_1,\ldots,x_{n-1}}^y)(y) dv_x^T(x_1) \ldots dv_x^T(x_{n-1})$$

$$+ \left(\frac{cM}{n} + N\|T(f) - f\|\right) d(x,y)$$

$$\leq \cdots \leq \int_K U(f_{x_1}^y)(y) dv_x^T(x_1) + (n-1)\left(\frac{cM}{n} + N\|T(f) - f\|\right) d(x,y)$$

$$= U(f^y)(x) + (n-1)\left(\frac{cM}{n} + N\|T(f) - f\|\right) d(x,y)$$

$$\leq U(f^y)(y) + (cM + nN\|T(f) - f\|) d(x,y).$$

On the other hand,

$$L_{n,\lambda}(f)(y) = \int_K \cdots \int_K f\left(\frac{x_1 + \cdots + x_n}{n}\right) dv_y^T(x_1) \ldots dv_y^T(x_n)$$

$$= \int_K \cdots \int_K U(f_{x_1,\ldots,x_{n-1}}^y)(y) \, dv_y^T(x_1) \ldots dv_y^T(x_{n-1})$$

$$= \cdots = \int_K U(f_{x_1}^y)(y) \, dv_y^T(x_1) = U(f^y)(y).$$

Accordingly,

$$|L_{n,\lambda}(f)(x) - L_{n,\lambda}(f)(y)| \le (cM + nN\|T(f) - f\|)\, d(x,y),$$

and, since x and y are arbitrary in K, the proof of (6.1.87) is complete.
If λ is constant, we have $N = 0$ and therefore we obtain (6.1.88).   □

Since $\|L_{n,\lambda}\| = 1$, by the above result and Corollary 6.1.20 we immediately have the following corollary.

6.1.22 Corollary. *Assume that* (6.1.82) *holds. If* $T(\mathrm{Lip}_1 1) \subset \mathrm{Lip}_c 1$ *for some* $c > 1$, *then for every* $\lambda \in \,]0,1]$ *and* $n \ge 1$ *we have*

$$\omega(L_{n,\lambda}(f),\delta) \le (1 + c)\omega(f,\delta) \quad \text{for every } f \in \mathscr{C}(K) \text{ and } \delta > 0 \quad (6.1.90)$$

and

$$L_{n,\lambda}(\mathrm{Lip}_M \alpha) \subset \mathrm{Lip}_{c^\alpha M} \alpha \quad \text{for every } M > 0 \text{ and } 0 < \alpha \le 1. \quad (6.1.91)$$

In particular, if $T(\mathrm{Lip}_1 1) \subset \mathrm{Lip}_1 1$, *then*

$$\omega(L_{n,\lambda}(f),\delta) \le 2\omega(f,\delta) \quad \text{for every } f \in \mathscr{C}(K) \text{ and } \delta > 0 \quad (6.1.92)$$

and

$$L_{n,\lambda}(\mathrm{Lip}_M \alpha) \subset \mathrm{Lip}_M \alpha \quad \text{for every } M > 0 \text{ and } 0 < \alpha \le 1. \quad (6.1.93)$$

Remark. By applying Corollary 6.1.22 to $\lambda = 1$, we obtain that, if $T(\mathrm{Lip}_1 1) \subset \mathrm{Lip}_c 1$ (respectively, $T(\mathrm{Lip}_1 1) \subset \mathrm{Lip}_1 1$), then

$$\omega(B_n(f),\delta) \le (1 + c)\omega(f,\delta) \quad \text{for every } f \in \mathscr{C}(K) \text{ and } \delta > 0 \quad (6.1.94)$$

and

$$B_n(\text{Lip}_M \alpha) \subset \text{Lip}_{c^\alpha_M} \alpha \quad \text{for every } M > 0 \text{ and } 0 < \alpha \leq 1 \qquad (6.1.95)$$

(respectively,

$$\omega(B_n(f), \delta) \leq 2\omega(f, \delta) \quad \text{for every } f \in \mathscr{C}(K) \text{ and } \delta > 0 \qquad (6.1.96)$$

and

$$B_n(\text{Lip}_M \alpha) \subset \text{Lip}_M \alpha \quad \text{for every } M > 0 \text{ and } 0 < \alpha \leq 1). \qquad (6.1.97)$$

Notes and references

Bernstein-Schnabl operators were first introduced by Schnabl [1968] (see also [1969a], [1969b], [1972]) in the context of the sets of all probability Radon measures on compact Hausdorff spaces, i.e., the Bauer simplices (Alfsen [1971, Corollary II.4.2]), and with respect to the arithmetic mean Toeplitz matrix.

The extension to an arbitrary infinite lower triangular stochastic matrix was performed in the same setting by Felbecker and Schempp [1971].

A more general method of constructing Bernstein-Schnabl operators on an arbitrary convex compact set was proposed by Grossman [1974].

The approach we followed here is due to Altomare [1989a] (see also [1989c], [1991b]).

Contrary to the case considered by Grossman, when the Bernstein-Schnabl operators are associated with a positive projection, they satisfy many additional interesting properties as we shall see in the next section.

In the particular case of Bauer simplices (where one has at his disposal a natural projection (see Corollary 3.3.4)) Bernstein-Schnabl operators have been extensively studied by Nishishiraho ([1974], [1976], [1977], [1978], [1983], [1987]) as well.

Other results can be found in Blümlinger [1987] where a detailed analysis of these operators is carried out on the unit ball of $\mathbb{R}^p$ with respect to the Dirichlet projection (3.3.16), in Andrica and Mustata [1989] and Rasa [1988a], [1988b], [1991], [1993].

Stancu-Schnabl operators were introduced and studied by Felbecker [1972], [1973] in the setting of the sets of probability Radon measures on a compact space and by Campiti [1991c], [1992c] in the general framework of positive projections.

However, note that in those papers Stancu-Schnabl operators were called Stancu-Mühlbach operators.

Here we preferred to name them Stancu-Schnabl operators for the sake of more pertinence because these operators become Stancu operators (5.2.52) in the case of unit interval and, secondly, they are constructed according to the idea of Schnabl.

Lototsky-Schnabl operators were first introduced by Schempp [1971] and, in a more general form, by Grossman [1976b].

In the context of Bauer simplices they were investigated by Nishishiraho [1982b], [1988a].

In the framework of positive projections recent investigations have been carried out by Altomare [1991b], [1991c], [1992a], [1992b], [1994], Altomare and Romanelli [1992], Rasa [1991], [1993].

Proposition 6.1.1 and the subsequent remark are due to Rasa [1992]. Theorem 6.1.3 is due to Altomare [1989]. However part (1), with $m = 1$, of the same theorem as well as part (1) of Theorem 6.1.9 have been previously obtained by Grossman [1974]. In the case of Bauer simplices, in fact, Theorem 6.1.3 is due to Nishishiraho [1976].

Theorem 6.1.3 extends similar results concerning the classical Bernstein operators obtained by Da Silva [1985], Karlin and Ziegler [1970], Keliski and Rivlin [1967], Marlewski [1982], Micchelli [1973b], J. Nagel [1980] and Sikkema [1966].

The estimates of Theorems 6.1.4 and 6.1.11 are taken from Nishishiraho [1982b], [1988a]. In the same articles the reader can also find a more general definition of Lototsky-Schnabl operators.

The combinatorial identities and the estimates contained in Proposition 6.1.5 are taken from Comtet [1970, pp. 49–50] and Felbecker [1972, pp. 14–16].

Theorems 6.1.7 and 6.1.8 are due to Campiti [1991c]. They generalize the analogous results obtained by Felbecker [1972], [1973] in the particular setting of the convex compact set of probability Radon measures on a compact space. However, in the interval [0, 1] the powers of Stancu operators were first studied by Mastroianni and Occorsio [1978b].

Theorem 6.1.10 is a particular case of a more general result obtained by Altomare and Romanelli [1992].

Proposition 6.1.12 is essentially due to Andrica and Mustata [1989]. There further estimates for more general Bernstein-Schnabl and Lototsky-Schnabl operators can be found.

We mention here that, as in the case of the unit interval, for the sequences of operators we considered in this section, a saturation phenomenon occurs.

To be more precise we recall that, if $(P_n)_{n \in \mathbb{N}}$ is a sequence of bounded linear operators on $\mathscr{C}(K)$ which strongly converges to the identity, then the trivial class of $(P_n)_{n \in \mathbb{N}}$ is defined by $\mathrm{Tr}[P_n] := \{f \in \mathscr{C}(K) | P_n(f) = f \text{ for every } n \in \mathbb{N}\}$.

Moreover, if $(\eta_n)_{n \in \mathbb{N}}$ is a sequence of positive real numbers converging to zero, we set $S[P_n, \eta_n] := \{f \in \mathscr{C}(K) | \|P_n(f) - f\| = O(\eta_n)\}$.

We say that $(P_n)_{n \in \mathbb{N}}$ is *saturated* in $\mathscr{C}(K)$ with order η_n if $\|P_n(f) - f\| = o(\eta_n) \Rightarrow f \in \mathrm{Tr}[P_n]$ and $S[P_n, \eta_n] \backslash \mathrm{Tr}[P_n] \neq \varnothing$.

In this case we call $S[P_n, \eta_n]$ the *saturation class* of $(P_n)_{n \in \mathbb{N}}$ (see Butzer and Nessel [1971], De Vore [1972], Lorentz [1986a] and Nishishiraho [1987]).

Now, for the sake of simplicity, set for every $n \geq 1$ and $\lambda > 0$

$$P_{n,\lambda} := \begin{cases} L_{n,\lambda}, & \text{if } \lambda < 1, \\ B_n, & \text{if } \lambda = 1, \\ S_{n,(\lambda-1)/n}, & \text{if } \lambda > 1, \end{cases} \quad \text{and} \quad \xi_{n,\lambda} := = \begin{cases} \dfrac{\lambda}{n}, & \text{if } \lambda \leq 1, \\ \dfrac{\lambda}{n + \lambda - 1}, & \text{if } \lambda > 1. \end{cases}$$

Then it is possible to show the following results:

(1) *For every $\lambda > 0$ and $m \geq 1$ the sequence $(P_{n,\lambda}^m)_{n \in \mathbb{N}}$ is saturated in $\mathscr{C}(K)$ with order $(1 - (1 - \xi_{n,\lambda})^m)$, or equivalently with order $m\xi_{n,\lambda}$ and*

$$\mathrm{Tr}[P_{n,\lambda}^m] = \{ f \in \mathscr{C}(K) | \, \|P_{n,\lambda}^m(f) - f\| = o(1 - (1 - \xi_{n,\lambda})^m) \}$$

$$= \{ f \in \mathscr{C}(K) | \, \|P_{n,\lambda}^m(f) - f\| = o(m\xi_{n,\lambda}) \} = H.$$

(2) *For every $\lambda > 0$ and for every sequence $(k(n))_{n \in \mathbb{N}}$ of positive integers such that*
$$\lim_{n \to \infty} \frac{k(n)}{n} = 0, \text{ the sequence } (P_{n,\lambda}^{k(n)})_{n \in \mathbb{N}} \text{ is saturated with order } 1 - (1 - \xi_{n,\lambda})^{k(n)}, \text{ or equiv-}$$
alently, with order $k(n)\xi_{n,\lambda}$, and

$$\mathrm{Tr}[P_{n,\lambda}^{k(n)}] = \{ f \in \mathscr{C}(K) | \, \|P_{n,\lambda}^{k(n)}(f) - f\| = o(1 - (1 - \xi_{n,\lambda})^{k(n)}) \}$$

$$= \{ f \in \mathscr{C}(K) | \, \|P_{n,\lambda}^{k(n)}(f) - f\| = o(k(n)\xi_{n,\lambda}) \} = H.$$

Furthermore, the saturation classes of $(P_{n,\lambda}^m)_{n \in \mathbb{N}}$ and $(P_{n,\lambda}^{k(n)})_{n \in \mathbb{N}}$ can be characterized in terms of the so-called *relative completion* but we omit the details (see Nishishiraho [1976], [1978], [1987] and Altomare [1991b]).

For further results about the saturation problem we refer to Blümlinger [1987], Butzer and Nessel [1971], Da Silva [1985], De Vore [1972], Felbecker [1972], Lorentz [1964], [1986a], Lorentz and Schumaker [1972], Micchelli [1973b], Mühlbach [1969] and Schnabl [1968], [1969a].

The results which concern the monotonicity properties and the preservation of Hölder continuous functions by (Bernstein-) Lototsky-Schnabl operators are due to Rasa [1988a], [1988b], [1993].

The same author introduced the notions of *T*-convexity and (T, λ)-convexity. The last one generalizes the definition of axially convexity for simplices of $\mathbb{R}^p$ which appears in Dahmen [1991] and Sauer [1991] and which seems to have been first introduced in an unpublished manuscript by Schmid in 1975.

Theorems 6.1.13–6.1.15 are taken from the above mentioned papers of Rasa.

They extend several results obtained in different contexts. Particular cases of it are, indeed, the results obtained by Chang and Davis [1984] for two-dimensional simplices, Dahmen [1991] and Sauer [1991] for simplices of $\mathbb{R}^p$, by Dahmen and Micchelli [1990] for the product of a finite number of simplices and by Temple [1954] and Aramă [1957] for the unit interval (see also Pólya and Schoenberg [1958]).

The monotonicity of the sequence of Bernstein-Schnabl operators associated with an infinite lower triangular stochastic matrix has been recently investigated by Della Vecchia and Rasa [1993].

For further references on the monotonicity of positive approximation processes on real intervals see Section 6.3 and the corresponding notes.

The maximum principle 6.1.17 is a joint result by Altomare and Rasa and it also appears in a paper of Rasa [1994].

This result unifies several maximum principles obtained by different and more complicated methods (see, for instance, Chang and Zhang [1990] for the canonical simplex of

$\mathbb{R}^2$, Dahmen and Micchelli [1990] for the finite products of simplices, Sauer [1992a] for the canonical simplex of $\mathbb{R}^p$), and it contains, as a particular case, the maximum principle for subharmonic functions as well (see Subsection 6.3.9)).

The Peetre K_0-functional was introduced by Peetre [1963] and it has many applications in approximation theory.

The results contained in Proposition 6.1.19 and Corollary 6.1.20 are taken from Anastassiou, Cottin and Gonska [1991a], who also proved (6.1.92) and Corollary 6.1.22 in [1991b] in the context of simplices of $\mathbb{R}^p$ and for Bernstein operators (see also Feng [1990] and Khan and Peters [1989]).

Theorem 6.1.21 and Corollary 6.1.22 have been proved in their full generality by Rasa [1993].

The corresponding results for Bernstein operators on the unit interval have been obtained by Hájek [1965] and Lindvall [1982] and, with a simpler proof, by Brown, Elliot and Paget [1987], as well as by Kratz and Stadtmüller [1988].

More references along these lines can be found in Section 6.3.

6.2 Positive projections and their associated Feller semigroups

The operators considered in the previous section can be also used in representing the solutions of certain diffusion equations. Indeed, we shall see that to every sequence of Bernstein-Schnabl, Stancu-Schnabl or Lototsky-Schnabl operators there corresponds a uniquely determined Feller semigroup that can be represented in terms of their powers.

The generator of the semigroup is explicitly determined in a core of its domain and, in the finite dimensional case, it is an elliptic second order differential operator which degenerates on the Choquet boundary of the range of the projection.

These results allow us to represent the solutions of the Cauchy problems associated with some degenerate diffusion equations in terms of the powers of the operators we are dealing with and, more particularly, to derive qualitative properties of the solutions, e.g., their asymptotic behavior and some regularity results.

By using the general theory of Markov processes that we reviewed in Section 1.6, we also describe the probabilistic meaning of our results, by giving the transition function corresponding to the Markov process associated with the above mentioned Feller semigroup.

In the subsequent Section 6.3 we carry out a further analysis in some concrete examples.

We keep fixed the notation of Section 6.1. So, K will denote a compact convex metrizable set of a locally convex space E and $T: \mathscr{C}(K) \to \mathscr{C}(K)$ will be a positive projection whose range $H := T(\mathscr{C}(K))$ satisfies conditions (6.1.3) and (6.1.4).

Almost all properties we shall establish in this section are concerned with the three kinds of approximation processes which we have dealt with in Section 6.1.

To avoid unnecessary repetitions we shall make use of the following convenient notation.

First we set

$$B_1^+(\mathscr{C}(K)) := \{\lambda \in \mathscr{C}(K) | 0 \leq \lambda \leq 1\}. \tag{6.2.1}$$

Moreover, given $n \in \mathbb{N}$ and $\lambda \in [1, +\infty[\cup B_1^+(\mathscr{C}(K))$ we set

$$P_{n,\lambda} := S_{n,a_n} \quad \text{if } \lambda \in \mathbb{R}, \lambda \geq 1, \tag{6.2.2}$$

where S_{n,a_n} denotes the n-th Stancu-Schnabl operator associated with T and an *arbitrary* sequence $(a_n)_{n \in \mathbb{N}}$ of positive real numbers such that $\lim_{n \to \infty} na_n = \lambda - 1$, whereas we set

$$P_{n,\lambda} := L_{n,\lambda} \quad \text{if } \lambda \in B_1^+(\mathscr{C}(K)), \tag{6.2.3}$$

$L_{n,\lambda}$ being the n-th Lototsky-Schnabl operator associated with T and λ.

Thus, in particular we have

$$P_{n,1} = B_n. \tag{6.2.4}$$

According to this notation, from now on *it is understood that whenever we shall establish a result for the sequence* $(P_{n,\lambda})_{n \in \mathbb{N}}$ *with* $\lambda \geq 1$, *this means that the result in question holds for the sequence of Stancu-Schnabl operators associated with an arbitrary sequence* $(a_n)_{n \in \mathbb{N}}$ *of positive real numbers such that* $\lim\limits_{n \to \infty} na_n = \lambda - 1$.

We shall now introduce an operator that will be essential for the sequel and that, in the finite dimensional case, is in fact a differential operator.

For every $m \geq 1$ we denote by $A_m(K)$ the linear subspace generated by all products of m affine functions, i.e.,

$$A_m(K) := \mathcal{L}\left(\left\{\prod_{i=1}^{m} h_i \,\middle|\, h_1, \ldots, h_m \in A(K)\right\}\right). \tag{6.2.5}$$

The sequence $(A_m(K))_{m \geq 1}$ is increasing and therefore

$$A_\infty(K) := \bigcup_{m=1}^{\infty} A_m(K) \tag{6.2.6}$$

is a subspace of $\mathscr{C}(K)$. It is obvious that $A_\infty(K)$ is a subalgebra of $\mathscr{C}(K)$ separating the points of K and so it is dense in $\mathscr{C}(K)$ by the Stone-Weierstrass theorem.

For every $h_1, \ldots, h_m \in A(K)$ we set

$$L_T\left(\prod_{i=1}^{m} h_i\right) := \begin{cases} 0, & m = 1, \\ T(h_1 h_2) - h_1 h_2, & m = 2, \\ \displaystyle\sum_{1 \leq i < j \leq m} (T(h_i h_j) - h_i h_j) \prod_{\substack{r=1 \\ r \neq i,j}}^{m} h_r, & m \geq 3. \end{cases} \tag{6.2.7}$$

As a first step, we state an abstract Voronovskaja-type formula for Bernstein-Schnabl, Stancu-Schnabl and Lototsky-Schnabl operators.

6.2.1 Theorem. *Given* $\lambda \in [1, +\infty[\,\cup\, B_1^+(\mathscr{C}(K))$, *the sequence* $(n(P_{n,\lambda}(f) - f))_{n \in \mathbb{N}}$ *converges uniformly to a function* $Z_\lambda^*(f) \in \mathscr{C}(K)$ *for every* $f \in A_\infty(K)$.

Moreover, if $h_1, \ldots, h_m \in A(K)$ *then*

$$Z_\lambda^*\left(\prod_{i=1}^{m} h_i\right) = \lambda L_T\left(\prod_{i=1}^{m} h_i\right). \tag{6.2.8}$$

Proof. Let $f \in A_\infty(K)$. Without loss of generality, we can suppose that $f = \prod\limits_{i=1}^{m} h_i$ where $h_1, \ldots, h_m \in A(K)$. First suppose that $\lambda \in \mathbb{R}$, $\lambda \geq 1$. Let $(a_n)_{n \in \mathbb{N}}$ be a sequence of positive real numbers such that $\lim\limits_{n \to \infty} n \cdot a_n = \lambda - 1$ and consider the corresponding sequence $(S_{n,a_n})_{n \in \mathbb{N}}$ of Stancu-Schnabl operators. For every $n \geq 1$, $k = 1, \ldots, n$, $(x_1, \ldots, x_k) \in K^k$ and $(v_1, \ldots, v_k) \in \mathbb{N}^k$ satisfying (6.1.21), we have (see (6.1.36))

$$f\left(\frac{v_1 x_1 + \cdots + v_k x_k}{n}\right)$$

$$= \prod_{j=1}^{m} h_j\left(\frac{v_1 x_1 + \cdots + v_k x_k}{n}\right)$$

$$= \prod_{j=1}^{m} \frac{1}{n} \sum_{i=1}^{k} v_i h_j(x_i)$$

$$= \frac{1}{n^m} \sum_{i_1=1}^{k} \cdots \sum_{i_m=1}^{k} v_{i_1} \ldots v_{i_m} h_1(x_{i_1}) \ldots h_m(x_{i_m})$$

$$= \frac{1}{n^m}\left(\sum_{i \in N_k(1)} v_i^m (h_1 \ldots h_m)(x_i) + \sum_{(i_1,i_2) \in N_k(2)} v_{i_1}^{m-1} v_{i_2} (h_1 \ldots h_{m-1})(x_{i_1}) h_m(x_{i_2})\right.$$

$$+ \sum_{(i_1,i_2) \in N_k(2)} v_{i_1}^{m-1} v_{i_2} (h_1 \ldots h_{m-2} h_m)(x_{i_1}) h_{m-1}(x_{i_2}) + \cdots$$

$$+ \sum_{(i_1,i_2) \in N_k(2)} v_{i_1}^{m-1} v_{i_2} (h_2 \ldots h_m)(x_{i_1}) h_1(x_{i_2}) + \cdots$$

$$+ \sum_{(i_1,i_2) \in N_k(2)} v_{i_1}^{m-2} v_{i_2}^2 (h_1 \ldots h_{m-2})(x_{i_1}) (h_{m-1} h_m)(x_{i_2}) + \cdots$$

$$+ \sum_{(i_1,i_2) \in N_k(2)} v_{i_1}^{m-2} v_{i_2}^2 (h_3 \ldots h_m)(x_{i_1}) (h_1 h_2)(x_{i_2}) + \cdots$$

$$+ \sum_{(i_1,\ldots,i_{m-1}) \in N_k(m-1)} v_{i_1}^2 v_{i_2} \ldots v_{i_{m-1}} (h_1 h_2)(x_{i_1}) h_3(x_{i_2}) \ldots h_m(x_{i_{m-1}}) + \cdots$$

$$+ \sum_{(i_1,\ldots,i_{m-1}) \in N_k(m-1)} v_{i_1}^2 v_{i_2} \ldots v_{i_{m-1}} (h_{m-1} h_m)(x_{i_1}) h_1(x_{i_2}) \ldots h_{m-2}(x_{i_{m-1}})$$

$$+ \left.\sum_{(i_1,\ldots,i_m) \in N_k(m)} v_{i_1} \ldots v_{i_m} h_1(x_{i_1}) \ldots h_m(x_{i_m})\right)$$

and therefore, for each $x \in K$,

$$\int_K \cdots \int_K f\left(\frac{v_1 x_1 + \cdots + v_k x_k}{n}\right) d\mu_x^T(x_1) \ldots d\mu_x^T(x_k)$$

$$= \frac{1}{n^m}\left(\sum_{i \in N_k(1)} v_i^m T(h_1 \ldots h_m)(x) + \sum_{(i_1, i_2) \in N_k(2)} v_{i_1}^{m-1} v_{i_2} T(h_1 \ldots h_{m-1})(x) T(h_m)(x)\right.$$

$$+ \sum_{(i_1, i_2) \in N_k(2)} v_{i_1}^{m-1} v_{i_2} T(h_1 \ldots h_{m-2} h_m)(x) T(h_{m-1})(x) + \cdots$$

$$+ \sum_{(i_1, i_2) \in N_k(2)} v_{i_1}^{m-1} v_{i_2} T(h_2 \ldots h_m)(x) T(h_1)(x) + \cdots$$

$$+ \sum_{(i_1, i_2) \in N_k(2)} v_{i_1}^{m-2} v_{i_2}^2 (h_1 \ldots h_{m-2})(x)(h_{m-1} h_m)(x) + \cdots$$

$$+ \sum_{(i_1, i_2) \in N_k(2)} v_{i_1}^{m-2} v_{i_2}^2 (h_3 \ldots h_m)(x)(h_1 h_2)(x) + \cdots$$

$$+ \left(\sum_{(i_1, \ldots, i_{m-1}) \in N_k(m-1)} v_{i_1}^2 v_{i_2} \ldots v_{i_{m-1}}\right) \sum_{1 \le i < j \le m} T(h_i h_j)(x) \prod_{\substack{r=1 \\ r \ne i,j}}^m T(h_r)(x)$$

$$+ \left.\left(\sum_{(i_1, \ldots, i_m) \in N_k(m)} v_{i_1} \ldots v_{i_m}\right) T(h_1)(x) \ldots T(h_m)(x)\right).$$

Now, by formulas (6.1.37) and (6.1.39) of Proposition 6.1.5 we obtain

$$\int_K \cdots \int_K f\left(\frac{v_1 x_1 + \cdots + v_k x_k}{n}\right) d\mu_x^T(x_1) \ldots d\mu_x^T(x_k)$$

$$= \frac{1}{n^m}\left(\sum_{i=1}^k v_i^m T(h_1 \ldots h_m)(x)\right.$$

$$+ \left(n^{m-2} \sum_{i=1}^k v_i^2 + U_n(v_1, \ldots, v_k; m-1)\right) \sum_{1 \le i < j \le m} T(h_i h_j)(x) \prod_{\substack{r=1 \\ r \ne i,j}}^m T(h_r)(x)$$

$$+ \left.\left(n^m - n^{m-2} \frac{m(m-1)}{2} \sum_{i=1}^k v_i^2 + W_n(v_1, \ldots, v_k; m)\right) T(h_1)(x) \ldots T(h_m)(x)\right)$$

$$= h_1 \ldots h_m(x) + \frac{1}{n^2}\left(\sum_{i=1}^k v_i^2\right) \sum_{1 \le i < j \le m} (T(h_i h_j)(x) - h_i h_j(x)) \prod_{\substack{r=1 \\ r \ne i,j}}^m h_r(x)$$

$$+ \sum_{i=1}^{s(m)} R_i(v_1, \ldots, v_k) L_i(h_1 \ldots h_m)(x),$$

where $s(m)$ is a suitable integer greater than 1 and, for every $i = 1, \ldots, s(m)$, $L_i(h_1 \ldots h_m)$ is a suitable function belonging to the subspace generated by

$$\{h_1 \ldots h_m, T(h_1 h_2)h_3 \ldots h_m, \ldots, T(h_1 h_2 h_3)h_4 \ldots h_m, \ldots, T(h_1 \ldots h_m)\},$$

and

$$|R_i(v_1, \ldots, v_k)| \le \frac{1}{n^3} c_i \sum_{j=1}^{k} v_j^3 + \frac{1}{n^4} d_i \sum_{(j_1, j_2) \in N_k(2)} v_{j_1}^2 v_{j_2}^2$$

for some $c_i, d_i \in \mathbb{R}$.

By (6.2.7) and (6.1.23), we have

$$S_{n, a_n}(f) = \frac{1}{p_n(a_n)} \sum_{k=1}^{n} \frac{(n-1)!}{k!} a_n^{n-k} \left(\sum_{|v|_k = n} \frac{n}{v_1 \ldots v_k} h_1 \ldots h_m \right.$$

$$+ \sum_{|v|_k = n} \frac{v_1^2 + \cdots + v_k^2}{v_1 \ldots v_k} \frac{1}{n} L_T(h_1 \ldots h_m)$$

$$\left. + \sum_{|v|_k = n} \frac{n}{v_1 \ldots v_k} \sum_{i=1}^{s(m)} R_i(v_1, \ldots, v_k) L_i(h_1 \ldots h_m) \right)$$

$$= h_1 \ldots h_m + \frac{1}{n} \frac{1 + na_n}{1 + a_n} L_T(h_1 \ldots h_m)$$

$$+ \frac{1}{p_n(a_n)} \sum_{k=1}^{n} \frac{(n-1)!}{k!} a_n^{n-k} \sum_{|v|_k = n} \frac{n}{v_1 \ldots v_k} \sum_{i=1}^{s(m)} R_i(v_1, \ldots, v_k) L_i(h_1 \ldots h_m).$$

At this point, we obtain by (6.1.34) and (6.1.35)

$$\|n(S_{n, a_n}(f) - f) - \lambda L_T(f)\|$$

$$\le \left\| n(S_{n, a_n}(f) - f) - \frac{1 + na_n}{1 + a_n} L_T(f) \right\| + \left| \frac{1 + na_n}{1 + a_n} - \lambda \right| \|L_T(f)\|$$

$$\le \sum_{i=1}^{s(m)} \frac{1}{p_n(a_n)} \sum_{k=1}^{n} \frac{(n-1)!}{k!} a_n^{n-k} \left(\frac{1}{n} c_i \sum_{|v|_k = n} \frac{v_1^3 + \cdots + v_k^3}{v_1 \ldots v_k} \right.$$

$$\left. + \frac{1}{n^2} d_i \sum_{|v|_k = n} \frac{1}{v_1 \ldots v_k} \sum_{\substack{i,j=1 \\ i \ne j}}^{n} v_i^2 v_j^2 \right) \|L_i(h_1 \ldots h_m)\|$$

$$+ \left| \frac{1 + na_n}{1 + a_n} - \lambda \right| \|L_T(h_1 \ldots h_m)\|$$

$$\leq \sum_{i=1}^{s(m)} \frac{1}{p_n(a_n)} \left(\frac{1}{n} c_i (1 + 2na_n) \frac{p_{n+1}(a_n)}{p_3(a_n)} + \frac{1}{n^2} d_i(n-1) \frac{p_{n+2}(a_n)}{p_4(a_n)} \right) \| L_i(h_1 \ldots h_m) \|$$

$$+ \left| \frac{1 + na_n}{1 + a_n} - \lambda \right| \| L_T(h_1 \ldots h_m) \|$$

$$\leq \frac{1}{n} \sum_{i=1}^{s(m)} \left(c_i \frac{(1 + 2na_n)(1 + na_n)}{(1 + a_n)(1 + 2a_n)} \right.$$

$$\left. + d_i \frac{(n-1)(1 + na_n)(1 + (n+1)a_n)}{n(1 + a_n)(1 + 2a_n)(1 + 3a_n)} \right) \| L_i(h_1 \ldots h_m) \|$$

$$+ \left| \frac{1 + na_n}{1 + a_n} - \lambda \right| \| L_T(h_1 \ldots h_m) \|.$$

Since $\lim\limits_{n \to \infty} na_n = \lambda - 1$, we conclude that $\lim\limits_{n \to \infty} \| n(S_{n,a_n}(f) - f) - \lambda L_T(f) \| = 0$.

Assume now $\lambda \in \mathscr{C}(K)$ and $0 \leq \lambda \leq 1$ and consider the sequence $(L_{n,\lambda})_{n \in \mathbb{N}}$ of Lototsky-Schnabl operators associated with the projection T and the function λ. We can apply the same reasoning as for identity (2) (take $k = n$ and $v_1 = \cdots = v_n = 1$), and we obtain, for every $n \geq 1$ and $x \in K$,

$$\int_K \cdots \int_K f \left(\frac{x_1 + \cdots + x_n}{n} \right) dv_x^T(x_1) \ldots dv_x^T(x_n)$$

$$= \left(h_1 \ldots h_m(x) + \frac{1}{n} \lambda(x) \sum_{1 \leq i < j \leq m} (T(h_i h_j)(x) - h_i h_j(x)) \prod_{\substack{r=1 \\ r \neq i,j}}^{m} h_r(x) \right.$$

$$\left. + \sum_{i=1}^{s(m)} R_i L_i(h_1 \ldots h_m)(x) \right), \tag{3}$$

where $s(m) \in \mathbb{N}$ and for every $i = 1, \ldots, s(m)$, $L_i(h_1 \ldots h_m)$ belongs to the linear subspace generated by

$$\{ h_1 \ldots h_m, T(h_1 h_2) h_3 \ldots h_m, \ldots, T(h_1 h_2 h_3) h_4 \ldots h_m, \ldots, T(h_1 \ldots h_m) \},$$

and

$$|R_i| \leq \frac{1}{n^2} c_i + \frac{n-1}{n^3} d_i$$

for some $c_i, d_i \in \mathbb{R}$.

Hence we have

$$n(L_{n,\lambda}(f) - f) = n\left(\frac{1}{n}\lambda L_T(f) + \sum_{i=1}^{s(m)} R_i L_i(h_1 \ldots h_m)\right)$$

$$= \lambda L_T(f) + \sum_{i=1}^{s(m)} R_i L_i(h_1 \ldots h_m),$$

and consequently, we obtain

$$\|n(L_{n,\lambda}(f) - f) - \lambda L_T(f)\| \leq \sum_{i=1}^{s(m)} |R_i| \|L_i(h_1 \ldots h_m)\|.$$

This yields the result.   □

Remark. By virtue of Theorem 6.2.1 the mapping L_T defined by (6.2.7) can be extended to a linear operator from $A_\infty(K)$ into $\mathscr{C}(K)$. By an abuse of notation, we shall continue to denote this extension by L_T.

Notice that, if K is a convex compact subset of $\mathbb{R}^p$ then $A_\infty(K)$ is in fact the subalgebra of all polynomials on K.

In this case we shall see next that a Voronovskaja-type formula (6.2.8) holds for functions belonging to the bigger subalgebra $\mathscr{C}^2(K)$.

To achieve this result we need some further preliminaries.

6.2.2 Lemma. *Let $x \in K$ and $h_1, h_2 \in A(K)$ be such that $h_1(x) = h_2(x) = 0$. Then for every $\lambda \in [1, +\infty[\cup B_1^+(\mathscr{C}(K))$ there exist suitable constants C and D such that*

$$nP_{n,\lambda}(h_1^2 \cdot h_2^2)(x) \leq \frac{Cn + D}{n^2}\|h_1\|^2 \cdot \|h_2\|^2 \quad \textit{for every } n \geq 1. \tag{6.2.9}$$

Proof. First, we consider the case $\lambda \geq 1$. We observe preliminarily that $L_T(h_1^2 \cdot h_2^2)(x) = 0$ (see (6.2.7)). Now, in analogy with formulas (1) and (2) in the proof of the preceding Theorem 6.2.1, we obtain

$$S_{n,a_n}(h_1^2 \cdot h_2^2)(x)$$

$$= \frac{1}{p_n(a_n)} \sum_{k=1}^{n} \frac{(n-1)!}{k!} a_n^{n-k} \sum_{|v|_k=n} \frac{n}{v_1 \ldots v_k} \sum_{i=1}^{s} R_i(v_1, \ldots, v_k) T(h_1^2 \cdot h_2^2)(x),$$

where s is a suitable natural number and for every $i = 1, \ldots, s$, there exist c_i,

$d_i \in \mathbb{R}$ such that

$$|R_i(v_1,\ldots,v_k)| \le \frac{1}{n^3} c_i \sum_{j=1}^{k} v_j^3 + \frac{1}{n^4} d_i \sum_{(j_1,j_2)\in N_k(2)} v_{j_1}^2 v_{j_2}^2.$$

At this point, choosing $M \in \mathbb{R}$ such that $na_n \le M$ for every $n \ge 1$, by (6.1.34) and (6.1.35) we have

$nS_{n,a_n}(h_1^2 \cdot h_2^2)(x)$

$$\le \sum_{i=1}^{s} \frac{1}{p_n(a_n)} \sum_{k=1}^{n} \frac{(n-1)!}{k!} a_n^{n-k} \left(\frac{1}{n} c_i \sum_{|v|_k=n} \frac{v_1^3 + \cdots + v_k^3}{v_1 \cdots v_k} \right.$$

$$\left. + \frac{1}{n^2} d_i \sum_{|v|_k=n} \frac{1}{v_1 \cdots v_k} \sum_{\substack{i,j=1 \\ i\ne j}}^{n} v_i^2 v_j^2 \right) T(h_1^2 \cdot h_2^2)(x)$$

$$\le \sum_{i=1}^{s} \frac{1}{p_n(a_n)} \left(\frac{1}{n} c_i(1+2na_n)\frac{p_{n+1}(a_n)}{p_3(a_n)} + \frac{1}{n^2} d_i(n-1)\frac{p_{n+2}(a_n)}{p_4(a_n)} \right) T(h_1^2 \cdot h_2^2)(x)$$

$$\le \frac{1}{n} \sum_{i=1}^{s} \left(c_i \frac{(1+2na_n)(1+na_n)}{(1+a_n)(1+2a_n)} \right.$$

$$\left. + d_i \frac{(n-1)(1+na_n)(1+(n+1)a_n)}{n(1+a_n)(1+2a_n)(1+3a_n)} \right) \| T(h_1^2 \cdot h_2^2)\|$$

$$\le \frac{1}{n} \sum_{i=1}^{s} \left(c_i(1+2M)(1+M) + d_i\frac{(n-1)(1+M)(1+2M)}{n} \right) \|h_1\|^2 \cdot \|h_2\|^2$$

$$= \frac{1}{n} \sum_{i=1}^{s} \left(c_i(1+2M)(1+M) + d_i(1+M)(1+2M) \right.$$

$$\left. - d_i\frac{(1+M)(1+2M)}{n} \right) \|h_1\|^2 \cdot \|h_2\|^2$$

$$= \frac{Cn+D}{n^2} \|h_1\|^2 \cdot \|h_2\|^2,$$

where $C := \sum_{i=1}^{s} (c_i(1+2M)(1+M) + d_i(1+M)(1+2M))$ and

$$D := -\sum_{i=1}^{s} d_i(1+M)(1+2M).$$

In the case where $\lambda \in B_1^+(\mathscr{C}(K))$ the proof proceeds in a similar way. By using formulas (1) and (3) from the proof of Theorem 6.2.1, we can write

$$L_{n,\lambda}(h_1^2 \cdot h_2^2)(x) = \sum_{i=1}^{s} R_i T(h_1^2 \cdot h_2^2)(x),$$

where $s \in \mathbb{N}$ is suitably chosen and for every $i = 1, \ldots, s$, there exist $c_i, d_i \in \mathbb{R}$ such that

$$|R_i| \le \frac{1}{n^2} c_i + \frac{n-1}{n^3} d_i.$$

Hence, in this case we have

$$nL_{n,\lambda}(h_1^2 \cdot h_2^2)(x) \le \left(\frac{1}{n} \sum_{i=1}^{s} (c_i + d_i) - \frac{1}{n^2} \sum_{i=1}^{s} d_i\right) T(h_1^2 \cdot h_2^2)(x)$$

$$= \frac{Cn + D}{n^2} \|h_1\|^2 \cdot \|h_2\|^2,$$

where $C := \sum_{i=1}^{s} (c_i + d_i)$ and $D := -\sum_{i=1}^{s} d_i$ and this completes the proof. □

Remark. We point out that when $\lambda \in B_1^+(\mathscr{C}(K))$, the corresponding formulas (6.2.8) and (6.2.9) can be proved with a direct and simpler method, without making use of the combinatorial tools which are necessary for Stancu-Schnabl operators (see Altomare [1992a] for details).

From the preceding lemma we derive the following property which will be useful to obtain a more detailed description of the generator of the Feller semigroup in the case where K is contained in some finite dimensional space.

6.2.3 Proposition. *Let K be a convex compact subset of $\mathbb{R}^p$ and for every $x \in K$ denote by $\sigma_x : K \to \mathbb{R}$ the continuous mapping defined by $\sigma_x(y) := \|y - x\|$ for every $y \in K$. Then for every $\lambda \in [1, +\infty[\cup B_1^+(\mathscr{C}(K))$ we have*

$$\lim_{n \to \infty} nP_{n,\lambda}(\sigma_x^4)(x) = 0 \quad \text{uniformly with respect to } x \in K. \tag{6.2.10}$$

Proof. For every $x = (x_1, \ldots, x_p) \in K$, we have $\sigma_x^4 = \sum_{i,j=1}^{p} (pr_i - x_i)^2 (pr_j - x_j)^2$, and therefore by Lemma 6.2.2 there exist $C, D \in \mathbb{R}$ such that

$$nP_{n,\lambda}(\sigma_x^4)(x) = \sum_{i,j=1}^{p} nP_{n,\lambda}((pr_i - x_i)^2 (pr_j - x_j)^2)(x) \le \frac{Cn + D}{n^2} p^2 \delta(K)^4$$

for every $n \ge 1$, where $\delta(K)$ denotes the diameter of K (see (6.1.79)).
Hence we obtain (6.2.10). □

We are now in a position to establish the promised extension of the Voronovskaja-type formula.

To this end we introduce an elliptic second order differential operator that can be constructed by means of the projection T.

We shall assume that K is a convex compact subset of $\mathbb{R}^p$ having non-empty interior. We denote by $\mathscr{C}^2(K)$ the space of all real-valued continuous functions on K which are twice continuously differentiable in $\mathring{K}$ and whose partial derivatives of order ≤ 2 can be continuously extended to K.

For $u \in \mathscr{C}^2(K)$ and $i, j = 1, \ldots, p$ we shall continue to denote by $\dfrac{\partial u}{\partial x_i}$ and $\dfrac{\partial^2 u}{\partial x_i \partial x_j}$ the continuous extension to K of the partial derivatives $\dfrac{\partial u}{\partial x_i}$ and $\dfrac{\partial^2 u}{\partial x_i \partial x_j}$ defined on $\mathring{K}$.

The differential operator we are dealing with is the operator $W_T : \mathscr{C}^2(K) \to \mathscr{C}(K)$ defined by

$$W_T(u)(x) := \frac{1}{2} \sum_{i,j=1}^{p} a_{ij}(x) \frac{\partial^2 u(x)}{\partial x_i \partial x_j} \tag{6.2.11}$$

for every $u \in \mathscr{C}^2(K)$ and $x \in K$, where

$$a_{ij}(x) := T((pr_i - x_i)(pr_j - x_j))(x) = T(pr_i pr_j)(x) - x_i x_j \tag{6.2.12}$$

for every $i, j = 1, \ldots, p$, and $x = (x_1, \ldots, x_p) \in K$. Here the functions pr_i denote the canonical projections on K (see (2.6.19)).

The operator W_T is elliptic and degenerates on the Choquet boundary $\partial_H^+ K$ and, in particular, on the set of extreme points of K.

Indeed, for every $x \in K$ and $(\xi_1, \ldots, \xi_p) \in \mathbb{R}^p$ we have

$$\sum_{i,j=1}^{p} a_{ij}(x)\xi_i\xi_j = T\left(\sum_{i,j=1}^{p} \xi_i \xi_j (pr_i - x_i)(pr_j - x_j) \right)(x)$$

$$= T\left(\left(\sum_{i=1}^{p} \xi_i (pr_i - x_i) \right)^2 \right)(x) \geq 0,$$

and, if $x \in \partial_H^+ K$, then by Proposition 3.3.1 applied to $f = \displaystyle\sum_{i,j=1}^{p} \xi_i \xi_j (pr_i - x_i)(pr_j - x_j)$ we have

$$\sum_{i,j=1}^{p} a_{ij}(x)\xi_i\xi_j = 0.$$

The operator W_T will be called the *elliptic second order differential operator associated with the projection T*.

In the next Section 6.3 we shall give the explicit expression of W_T in some concrete cases such as the unit interval and the canonical simplex of $\mathbb{R}^p$, the hypercube and the balls of $\mathbb{R}^p$.

Next we shall see that, in fact, the operator W_T and some multiplicative perturbations of it are closable and their closures generate a Feller semigroup that can be described in terms of the approximation processes associated with the projection T.

Here we need the following further identities whose proofs are all similar to those of formulas (6.1.42) and (6.1.55), and therefore we omit the details.

6.2.4 Lemma. *Under the general assumptions of Section 6.1, given a sequence* $(a_n)_{n \in \mathbb{N}}$ *of positive real numbers and a function* $\lambda \in \mathscr{C}(K), 0 \leq \lambda \leq 1$, *then for every* $h_1, h_2 \in A(K)$ *and* $n \geq 1$ *we have*

$$S_{n,a_n}(h_1 h_2) = h_1 h_2 + \frac{1}{n} \frac{1 + na_n}{1 + a_n}(T(h_1 h_2) - h_1 h_2), \tag{6.2.13}$$

and

$$L_{n,\lambda}(h_1 h_2) = h_1 \cdot h_2 + \frac{1}{n}\lambda(T(h_1 h_2) - h_1 h_2). \tag{6.2.14}$$

We now proceed to show an extension of Theorem 6.2.1.

6.2.5 Theorem. *Let* K *be a convex compact subset of* $\mathbb{R}^p$ *having non-empty interior and consider a projection* $T : \mathscr{C}(K) \to \mathscr{C}(K)$ *satisfying conditions* (6.1.3) *and* (6.1.4).

Given $\lambda \in [1, +\infty[\cup B_1^+(\mathscr{C}(K))$, *then for every* $u \in \mathscr{C}^2(K)$

$$\lim_{n \to \infty} n(P_{n,\lambda}(u) - u) = \lambda W_T(u) \quad \text{uniformly on } K. \tag{6.2.15}$$

In particular, one has

$$W_T = L_T \quad \text{on } A_\infty(K) \tag{6.2.16}$$

(see (6.2.7) *and* Remark *to Theorem 6.2.1).*

Proof. For a given $u \in \mathscr{C}^2(K)$ we can use its Taylor expansion to obtain

$$u(y) = u(x) + \sum_{i=1}^{p} \frac{\partial u(x)}{\partial x_i}(y_i - x_i) + \frac{1}{2} \sum_{i,j=1}^{p} \frac{\partial^2 u(x)}{\partial x_i \partial x_j}(y_i - x_i)(y_j - x_j)$$

$$+ \omega(x,y)\|y - x\|^2 \tag{1}$$

for every $x, y \in K$, where $\omega \colon K^2 \to \mathbb{R}$ satisfies the conditions

$$|\omega(x, y)| \leq M \quad \text{for every } (x, y) \in K^2 \text{ and some } M \in \mathbb{R}_+ \tag{2}$$

and

$$\lim_{y \to x} \omega(x, y) = 0 \quad \text{uniformly with respect to } x \in K. \tag{3}$$

Now, for every $x = (x_1, \ldots, x_p) \in K$, we consider the following continuous functions on K

$$f_{i,x} := pr_i - x_i, \quad i = 1, \ldots, p,$$

$$f_x := \sum_{i=1}^{p} \frac{\partial u(x)}{\partial x_i} f_{i,x},$$

$$g_x := \frac{1}{2} \sum_{i,j=1}^{p} \frac{\partial^2 u(x)}{\partial x_i \partial x_j} f_{i,x} f_{j,x},$$

$$\sigma_x := \|\cdot - x\|.$$

From (1) it follows that the functions $\omega(x, \cdot) \sigma_x^2$ are continuous. We show preliminarily that

$$\lim_{n \to \infty} n P_{n, \lambda}(\omega(x, \cdot) \sigma_x^2)(x) = 0 \quad \text{uniformly in } x \in K. \tag{4}$$

Indeed, since $\sigma_x^2 = \sum\limits_{i=1}^{p} f_{i,x}^2$ and $f_{i,x} \in A(K)$ for every $i = 1, \ldots, p$, by (6.1.42) and (6.1.55), there exists $c > 0$ such that

$$nP_{n, \lambda}(\sigma_x^2)(x) = \sum_{i=1}^{p} nP_{n, \lambda}(f_{i,x}^2)(x) = \sum_{i=1}^{p} n\left(f_{i,x}^2(x) + \frac{c}{n}(T(f_{i,x}^2)(x) - f_{i,x}^2(x)) \right)$$

$$= c \sum_{i=1}^{p} T(f_{i,x}^2)(x) \leq c \sum_{i=1}^{p} \|f_{i,x}^2\| \leq cp\delta(K)^2. \tag{5}$$

For a given $\varepsilon > 0$, by (3) there exists $\delta > 0$ such that

$$|\omega(x, y)| \leq \frac{\varepsilon}{2cp\delta(K)^2} \quad \text{for every } (x, y) \in K^2 \text{ satisfying } \|x - y\| \leq \delta.$$

Then, for every $(x, y) \in K^2$ satisfying $\|x - y\| \leq \delta$, we obtain

$$|\omega(x, y)\sigma_x^2(y)| \leq \frac{\varepsilon}{2cp\delta(K)^2}\sigma_x^2(y) \leq \frac{\varepsilon}{2cp\delta(K)^2}\sigma_x^2(y) + \frac{M\delta(K)^2}{\delta^4}\sigma_x^4(y);$$

on the other hand, if $\|x - y\| > \delta$, we obtain by (2)

$$|\omega(x, y)\sigma_x^2(y)| \leq M\delta(K)^2 \leq \frac{M\delta(K)^2}{\delta^4}\sigma_x^4(y) \leq \frac{\varepsilon}{2cp\delta(K)^2}\sigma_x^2(y) + \frac{M\delta(K)^2}{\delta^4}\sigma_x^4(y);$$

hence, for every $x \in K$ we have

$$|\omega(x, \cdot)\sigma_x^2| \leq \frac{\varepsilon}{2cp\delta(K)^2}\sigma_x^2 + \frac{M\delta(K)^2}{\delta^4}\sigma_x^4. \tag{6}$$

Moreover, by Proposition 6.2.3, there exists $v \in \mathbb{N}$ such that

$$nP_{n,\lambda}(\sigma_x^4)(x) \leq \frac{\varepsilon\delta^4}{2M\delta(K)^2} \quad \text{for every } n \geq v \text{ and } x \in K. \tag{7}$$

Consequently, by (6), (5) and (7), for every $n \geq v$ and $x \in K$, we obtain

$$|nP_{n,\lambda}(\omega(x, \cdot)\sigma_x^2)(x)| \leq nP_{n,\lambda}(|\omega(x, \cdot)|\sigma_x^2)(x)$$

$$\leq \frac{\varepsilon}{2cp\delta(K)^2}nP_{n,\lambda}(\sigma_x^2)(x) + \frac{M\delta(K)^2}{\delta^4}nP_{n,\lambda}(\sigma_x^4)(x)$$

$$\leq \frac{\varepsilon}{2cp\delta(K)^2}cp\delta(K)^2 + \frac{M\delta(K)^2}{\delta^4}\frac{\varepsilon\delta^4}{2M\delta(K)^2} = \varepsilon,$$

and hence property (4) follows.

Now, for every $x \in K$, by (1) we can write $u = u(x) \cdot 1 + f_x + g_x + \omega(x, \cdot)\sigma_x^2$.
Since $f_x \in A(K)$, we have $P_{n,\lambda}(f_x) = f_x$ for every $n \geq 1$ (see (6.1.41) and (6.1.54)) and therefore

$$P_{n,\lambda}(u) = u(x) \cdot 1 + f_x + P_{n,\lambda}(g_x) + P_{n,\lambda}(\omega(x, \cdot)\sigma_x^2);$$

hence we obtain

$$n(P_{n,\lambda}(u)(x) - u(x)) = nP_{n,\lambda}(g_x)(x) + nP_{n,\lambda}(\omega(x, \cdot)\sigma_x^2)(x).$$

By (4), the proof of (6.2.15) will be complete if we show that

$$\lim_{n \to \infty} nP_{n,\lambda}(g_x)(x) = \lambda W_T(u)(x)$$

uniformly with respect to $x \in K$. But this clearly follows from Lemma 6.2.4. Finally, equality (6.2.16) follows from (6.2.15) and (6.2.8). □

Among other things, the above formula (6.2.15) tells us that, even if we impose further regularity assumptions on a function $f \in \mathscr{C}(K)$, the order of convergence of $(P_{n,\lambda}(f))_{n \in \mathbb{N}}$ toward f does not improve over $O\left(\dfrac{1}{n}\right)$.

However, the main significance of (6.2.8) and (6.2.15) rests primarily on the fact that they lead to the existence of a Feller semigroup in the light of Trotter's theorem 1.6.7 (respectively, Schnabl's theorem 1.6.8).

As a consequence of our result, we shall see that such a semigroup exists whenever K is a Bauer simplex or K is a finite dimensional convex compact set having non-empty interior and the projection T satisfies additional assumptions (that are satisfied in all the concrete examples we shall consider in Section 6.3).

6.2.6 Theorem. *Assume that either*

$$T(A_2(K)) \subset A(K), \quad i.e., \ T(h_1 h_2) \in A(K) \text{ for each } h_1, h_2 \in A(K), \quad (6.2.17)$$

or

$$K \subset \mathbb{R}^p, \quad \mathring{K} \neq \varnothing \quad and \quad T(A_m(K)) \subset A_m(K) \text{ for every } m \geq 1. \quad (6.2.18)$$

Then for every given $\lambda \geq 1$ *or* $\lambda \in B_1^+(\mathscr{C}(K))$, λ *strictly positive, there exists a Feller semigroup* $(T_\lambda(t))_{t \geq 0}$ *such that the following statements hold:*
(1) *If* $t \geq 0$ *and if* $(k(n))_{n \in \mathbb{N}}$ *is a sequence of positive integers satisfying* $\lim\limits_{n \to \infty} \dfrac{k(n)}{n} = t$, *then*

$$\lim_{n \to \infty} P_{n,\lambda}^{k(n)} = T_\lambda(t) \quad \text{strongly on } \mathscr{C}(K). \quad (6.2.19)$$

(2) *We have*

$$\lim_{t \to \infty} T_\lambda(t) = T \quad \text{strongly on } \mathscr{C}(K), \quad (6.2.20)$$

and hence for a given $f \in \mathscr{C}(K)$ *we have*

$$\lim_{t \to \infty} T_\lambda(t)(f) = 0 \quad \text{uniformly on } K \text{ if and only if } f = 0 \text{ on } \partial_H^+ K. \quad (6.2.21)$$

(3) *The generator A_λ of the semigroup $(T_\lambda(t))_{t\geq 0}$ is the closure of the linear opera-*
tor $Z_\lambda\colon D(Z_\lambda) \to \mathscr{C}(K)$ defined by

$$Z_\lambda(f) := \lim_{n\to\infty} n(P_{n,\lambda}(f) - f) \tag{6.2.22}$$

for every $f \in D(Z_\lambda)$, where

$$D(Z_\lambda) := \left\{ g \in \mathscr{C}(K) \mid \lim_{n\to\infty} n(P_{n,\lambda}(g) - g) \text{ exists in } \mathscr{C}(K) \right\}. \tag{6.2.23}$$

(4) $A_\infty(K) \subset D(Z_\lambda)$, $A_\infty(K)$ *is a core for* A_λ *and*

$$A_\lambda = \lambda \cdot L_T \quad \text{on } A_\infty(K) \tag{6.2.24}$$

where L_T is defined by (6.2.7) (see also Remark to Theorem 6.2.1).
(5) $D(A_\lambda) = D(A_1)$ *and*

$$A_\lambda = \lambda \cdot A_1. \tag{6.2.25}$$

(6) *If $K \subset \mathbb{R}^p$ has non-empty interior, then $\mathscr{C}^2(K) \subset D(Z_\lambda)$, $\mathscr{C}^2(K)$ is a core for A_λ*
and

$$A_\lambda = \lambda W_T \quad \text{on } \mathscr{C}^2(K), \tag{6.2.26}$$

where W_T is the elliptic second order differential operator associated with the
projection T (see (6.2.11)).

Proof. First, we suppose that (6.2.17) holds. Let us consider the linear operator
$Z_\lambda\colon D(Z_\lambda) \to \mathscr{C}(K)$ defined in (6.2.22). By Theorem 6.2.1 we have $A_\infty(K) \subset D(Z_\lambda)$
and $Z_\lambda = \lambda L_T$ on $A_\infty(K)$. In particular, $D(Z_\lambda)$ is dense in $\mathscr{C}(K)$.

We now show the existence of the semigroup $(T_\lambda(t))_{t\geq 0}$ provided $\lambda \in \mathbb{R}$, $\lambda \geq 1$,
i.e., the operators $P_{n,\lambda}$ are the Stancu-Schnabl operators associated with an arbi-
trary sequence $(a_n)_{n\in\mathbb{N}}$ of positive real numbers such that $\lim_{n\to\infty} na_n = \lambda - 1$.

A first step consists in showing that for an arbitrary $\xi > 0$, the range
$R(\xi I_{D(Z_\lambda)} - Z_\lambda)$ of $\xi I_{D(Z_\lambda)} - Z_\lambda$ is dense in $\mathscr{C}(K)$ and, to this end, it is enough to
prove that

$$(\xi I_{D(Z_\lambda)} - Z_\lambda)(A_\infty(K)) \quad \text{is dense in } \mathscr{C}(K). \tag{1}$$

Indeed consider a measure $\mu \in \mathscr{M}(K)$ vanishing on $(\xi I_{D(Z_\lambda)} - Z_\lambda)(A_\infty(K))$, i.e.,
$\mu(f) = \dfrac{1}{\xi}\mu(Z_\lambda(f))$ for every $f \in A_\infty(K)$. Then, if $f \in A_1(K)$, by (6.2.7) we have

$\mu(f) = 0$. Analogously, for every $f \in A_2(K)$ we have $\mu(f) = \lambda \frac{1}{\xi}\mu(T(f)) - \lambda \frac{1}{\xi}\mu(f) = -\lambda \frac{1}{\xi}\mu(f)$ and therefore $\mu(f) = 0$.

Suppose now that $\mu = 0$ on $A_m(K)$ with $m \geq 2$ and let $f = \prod_{i=1}^{m+1} h_i \in A_{m+1}(K)$. Then by (6.2.17) $T(h_i h_j) \prod_{r \neq i,j} h_r \in A_m(K)$ and consequently,

$$\mu(f) = \frac{1}{\xi}\mu(Z_\lambda(f)) = \lambda \frac{1}{\xi}\mu\left(\sum_{1 \leq i < j \leq m+1} T(h_i h_j) \prod_{r \neq i,j} h_r - \binom{m+1}{2}f\right)$$

$$= -\lambda \frac{1}{\xi}\frac{m(m+1)}{2}\mu(f);$$

thus, $\mu(f) = 0$. This implies that $\mu = 0$ on $A_{m+1}(K)$.

By induction on m, we have $\mu = 0$ on $A_\infty(K)$ and hence $\mu = 0$.

Thus, having proved that $R(\xi I_{D(Z_\lambda)} - Z_\lambda)$ is dense in $\mathscr{C}(K)$ for every $\xi > 0$, we can apply Trotter's theorem 1.6.7 $\left(\text{take } M = 1, \rho_n = \frac{1}{n} \text{ and } \omega = 1\right)$, and we obtain that Z_λ is closable and its closure A_λ is the generator of a contraction semigroup $(T_\lambda(t))_{t \geq 0}$ which satisfies (6.2.19). In particular, every $T_\lambda(t)$ is positive.

Clearly $A_\lambda = Z_\lambda = \lambda L_T$ on $A_\infty(K)$ and hence $A_\infty(K)$ is a core for A_λ because $(I_{D(A_\lambda)} - A_\lambda)(A_\infty(K)) = (I_{D(Z_\lambda)} - Z_\lambda)(A_\infty(K))$ is dense in $\mathscr{C}(K)$ (see the remarks before Theorem 1.6.1).

On the other hand, the generator A_1 which corresponds to $\lambda = 1$, i.e., to the Bernstein-Schnabl operators, coincides with L_T on $A_\infty(K)$ and hence $A_\lambda = \lambda A_1$ on the core $A_\infty(K)$. Consequently $A_\lambda = \lambda A_1$.

Now, to show (6.2.20), fix $t \geq 0$ and choose a sequence $(k(n))_{n \in \mathbb{N}}$ of integers satisfying $\lim_{n \to \infty} \frac{k(n)}{n} = t$ (for example, $k(n) := [nt]$, the integer part of nt). Hence, for every $h \in H$ we have (see (6.2.19) and (6.1.41))

$$T_\lambda(t)(h) = \lim_{n \to \infty} S_{n,a_n}^{k(n)}(h) = h = T(h)$$

and for every $h \in A(K)$ (see (6.2.19) and (6.1.42))

$$T_\lambda(t)(h^2) = \lim_{n \to \infty} S_{n,a_n}^{k(n)}(h^2)$$

$$= \lim_{n \to \infty} \left(\frac{n-1}{n}\frac{1}{1+a_n}\right)^{k(n)} h^2 + \left(1 - \left(\frac{n-1}{n}\frac{1}{1+a_n}\right)^{k(n)}\right)T(h^2)$$

$$= T(h^2) + \lim_{n \to \infty} \left(\frac{n-1}{n} \frac{1}{1+a_n} \right)^{k(n)} (h^2 - T(h^2))$$

$$= T(h^2) + \exp(-t\lambda)(h^2 - T(h^2)). \tag{2}$$

Consequently, we obtain $\lim_{t \to \infty} T_\lambda(t)(h) = T(h)$ for every $h \in H \cup A(K)^2$; this yields (6.2.20) by Remark 1 to Theorem 3.3.3.

Assertion (6.2.21) follows from the fact that, since $T(f) \in H$, then $T(f)$ takes its maximum and minimum value on $\partial_H^+ K$ (see Corollary 2.6.5) where $T(f)$ is equal to f by virtue of Proposition 3.3.1.

Thus, the proof is complete for Stancu-Schnabl operators and in particular for Bernstein-Schnabl operators. In the remaining part of the proof which is devoted to Lototsky-Schnabl operators we denote by A_1 the generator of the semigroup associated with Bernstein-Schnabl operators.

Now, suppose that λ is a strictly positive function in $B_1^+(\mathscr{C}(K))$ and consider the sequence $(L_{n,\lambda})_{n \in \mathbb{N}}$ of Lototsky-Schnabl operators associated with λ. Furthermore, let Z_λ be the operator defined by (6.2.22).

Since A_1 generates a positive semigroup, the Arendt-Dorroh Theorem 1.6.11 implies that the operator λA_1 is also the generator of a positive contraction semigroup. We also observe that $A_\infty(K)$ is a core for λA_1 because $A_\infty(K)$ is a core for A_1 and for every $u \in D(\lambda A_1) = D(A_1)$,

$$\|u\|_{\lambda A_1} = \|u\| + \|\lambda A_1(u)\| \le \|u\| + \|A_1(u)\| = \|u\|_{A_1},$$

where $\|\cdot\|_{A_1}$ denotes the graph norm of A_1. This allows us to conclude that $(I_{D(A_1)} - \lambda A_1)(A_\infty(K))$ is dense in $\mathscr{C}(K)$ and consequently the range of $R(I_{D(Z_\lambda)} - Z_\lambda)$ is dense in $\mathscr{C}(K)$ as well. Thus, we can again apply Trotter's theorem so that there exists a positive contraction semigroup $(T_\lambda(t))_{t \ge 0}$ on $\mathscr{C}(K)$ which satisfies (6.2.19), (6.2.24) and (6.2.25).

We have finally to show the validity of (6.2.20).

If λ is constant, the proof is similar to the case of Stancu-Schnabl operators by using (6.1.55) and (6.1.56). If λ is not constant, we set $m = \min\{\lambda(x)|x \in K\}$ and we consider the operator $B = mA_1$. The operator B generates a strongly continuous positive semigroup $(T_m(t))_{t \ge 0}$ and $\lim_{t \to \infty} T_m(t) = T$ strongly on $\mathscr{C}(K)$ because m is a positive constant.

We now observe that, if $u \in D(B) = D(A_1) = D(A_\lambda)$ and $A_1(u) \ge 0$, then $A_\lambda(u) = \lambda A_1(u) \ge mA_1(u) = B(u) \ge 0$. Furthermore, according to Theorem 1.6.1, (4), $A_1 T_m(t)u = T_m(t)A_1 u \ge 0$ and $A_\lambda T_\lambda(t)u = T_\lambda(t)A_\lambda u \ge 0$ for every $t \ge 0$ and consequently, for sufficiently large $\mu \in \mathbb{R}$,

$$A_1 R(\mu, B)u = \int_0^{+\infty} \exp(-\mu t) A_1 T_m(t)u \, dt \ge 0$$

and

$$A_\lambda R(\mu, A_\lambda)u = \int_0^{+\infty} \exp(-\mu t) A_\lambda T_\lambda(t)u \, dt \geq 0,$$

where $R(\mu, B) = (\mu I_{D(B)} - B)^{-1}$ and $R(\mu, A_\lambda) = (\mu I_{D(A_\lambda)} - A_\lambda)^{-1}$ (see Section 1.6, (1.6.20)).

On the other hand, we point out that, if a function $f \in D(A_1) = D(A_\lambda)$ is convex, then by Corollary 6.1.15, $f \leq L_{n,\lambda}(f) \leq T(f)$ as well as $f \leq B_n(f) \leq T(f)$ for every $n \geq 1$, and hence, by (6.2.19), (6.1.14) and (6.1.54),

$$f \leq T_\lambda(t)(f) \leq T(f) \tag{3}$$

and

$$f \leq T_1(t)(f) \leq T(f) \tag{4}$$

for every $t \geq 0$.

Thus, $A_1(f) \geq 0$ so that, by the above reasoning, $A_1(R(\mu, B)u) \geq 0$ and $A_\lambda(R(\mu, B)f) \geq B(R(\mu, B)f)$.

So, we obtain

$$R(\mu, A_\lambda)f - R(\mu, B)f = R(\mu, A_\lambda)(A_\lambda - B)R(\mu, B)f \geq 0.$$

Moreover,

$$R(\mu, B)^2 f \leq R(\mu, B)R(\mu, A_\lambda)f \leq R(\mu, A_\lambda)R(\mu, A_\lambda)f = R(\mu, A_\lambda)^2 f,$$

and, by induction on $n \geq 1$, we have

$$R(\mu, B)^n f \leq R(\mu, A_\lambda)^n f.$$

Then, for every $t > 0$, by using the exponential formula (1.6.21) we obtain

$$T_m(t)f = \lim_{n \to \infty} \left(\frac{n}{t} R\left(\frac{n}{t}, B \right) \right)^n f \leq \lim_{n \to \infty} \left(\frac{n}{t} R\left(\frac{n}{t}, A_\lambda \right) \right)^n f = T_\lambda(t)f$$

and hence

$$T_m(t)f \leq T_\lambda(t)f \leq T(f). \tag{5}$$

This yields $\lim_{t \to \infty} T_\lambda(t)f = T(f)$ for every convex function $f \in D(A_1)$ and, in particular, for every $f \in A(K)^2$; by Remark 1 to Theorem 3.3.3 we obtain (6.2.20) and hence (6.2.21).

Suppose now that K is a convex compact subset of $\mathbb{R}^p$ having non-empty interior and assume that the given projection $T: \mathscr{C}(K) \to \mathscr{C}(K)$ satisfies the assumptions

$$T(A_m(K)) \subset A_m(K) \quad \text{for every } m \geq 1.$$

Again, consider first the case $\lambda \in \mathbb{R}$, $\lambda \geq 1$. Then each finite dimensional subspace $A_m(K)$ is invariant under $P_{n,\lambda} = S_{n,a_n}$ for every $n \geq 1$, by virtue of the equality (2) of the proof of Theorem 6.2.1.

Accordingly, Schnabl's theorem 1.6.8 implies that Z_λ is closable and its closure is the generator of a Feller semigroup which satisfies (6.2.19).

Furthermore, on account of Theorem 6.2.5, we obtain $\mathscr{C}^2(K) \subset D(Z_\lambda)$ and

$$A_\lambda = Z_\lambda = \lambda W_T \quad \text{on } \mathscr{C}^2(K).$$

On the other hand, given $m \geq 1$, the subspace $A_m(K)$ is also invariant under the semigroup $(T_\lambda(t))_{t \geq 0}$ by (6.2.19) and the fact that it is closed.

Consequently $A_\infty(K)$ is invariant under $(T_\lambda(t))_{t \geq 0}$ as well and hence it is a core for A_λ on account of Theorem 1.6.1, (2).

A fortiori $\mathscr{C}^2(K)$ is a core for A_λ and, clearly, (6.2.25) holds.

Finally, formulas (6.2.20) and (6.2.21) can be shown as in the preceding case.

In the case when λ is a strictly positive function lying in $B_1^+(\mathscr{C}(K))$, the proof proceeds exactly as in the preceding case because there it was sufficient that the result holds for $\lambda = 1$ as we have already emphasized. $\qquad\qquad\square$

Remark. Under the assumptions of Theorem 6.2.6, given $t \geq 0$, and denoting by $[nt]$ the integer part of nt, we can write

$$T_\lambda(t) = \lim_{n \to \infty} P_{n,\lambda}^{[nt]} \quad \text{strongly on } \mathscr{C}(K) \tag{6.2.27}$$

because $\displaystyle\lim_{n \to \infty} \frac{[nt]}{n} = t$.

One of the most important applications of the previous result rests primarily on the fact that, if under the assumptions of Theorem 6.2.6 we consider the abstract Cauchy problem

$$\begin{cases} \dfrac{\partial u}{\partial t}(x, t) = \lambda A_1(u(\cdot, t))(x), & x \in K, t \geq 0, \\[2mm] u(x, 0) = u_0(x), & u_0 \in D(A_1), \end{cases} \tag{6.2.28}$$

then (6.2.28) admits a unique solution $u: K \times [0, +\infty[\to \mathbb{R}$ given by

$$u(x, t) = T_\lambda(t)(u_0)(x) = \lim_{n \to \infty} P_{n,\lambda}^{[nt]}(u_0)(x) \tag{6.2.29}$$

and the limit is uniform with respect to $x \in K$ (see Section 1.6).

We recall that here either $\lambda \in \mathbb{R}$, $\lambda \geq 1$, or λ is a strictly positive function satisfying $0 \leq \lambda \leq 1$.

Moreover $A_1: D(A_1) \to \mathscr{C}(K)$ is the generator of a semigroup which is represented by Bernstein-Schnabl operators via (6.2.19).

The operator A_1 coincides with L_T on $A_\infty(K)$ and with the elliptic second order differential operator W_T on $\mathscr{C}^2(K)$ provided $K \subset \mathbb{R}^p$ has non empty interior.

If it will be necessary, we shall use the notation $u_\lambda(x,t)$ as well to denote the unique solution of (6.2.28).

Note also that the 'boundary' conditions for the differential problem (6.2.28) are implicitly contained in the domain of A_1.

In fact, if $u_0 \in D(A_1)$, then on account of (6.1.27), (6.1.51) and (6.2.29) we have

$$u(x,t) = u_0(x) \quad \text{for every } x \in \partial_H^+ K \text{ and } t \geq 0 \tag{6.2.30}$$

or, equivalently,

$$A_1(u(\cdot,t)) = 0 \quad \text{on } \partial_H^+ K \text{ for every } t \geq 0. \tag{6.2.31}$$

Condition (6.2.31) is called *Wentcel's boundary condition* and will be discussed in more detail in Section 6.3 where, in some more concrete cases, we shall be able to describe $D(A_1)$ precisely.

There we shall see how problem (6.2.28) becomes an ordinary Cauchy problem associated with a partial differential equation of diffusion-type and we shall compare our results with the existing literature.

It is also clear that from (6.2.29) one can hope to obtain qualitative properties of the solution of (6.2.28) by studying the positive approximation process $(P_{n,\lambda})_{n \in \mathbb{N}}$.

Here we collect some of them that follow directly from the results we have previously obtained.

However we wish to stress that there is much work to be done in this field.

6.2.7 Theorem. *Under the assumptions of Theorem 6.2.6, let $u_0 \in D(A_1)$ and denote by $u(x,t)$ ($x \in K$, $t \geq 0$) the unique solution of the Cauchy problem (6.2.28).*

Then the following statements hold:

(1) $\lim_{t \to \infty} u(x,t) = T(u_0)(x)$ *uniformly in $x \in K$.*

(2) $\lim_{t \to \infty} u(x,t) = 0$ *uniformly in $x \in K$ if and only if $u_0 = 0$ on $\partial_H^+ K$.*

(3) *If $\lambda \in B_1^+(\mathscr{C}(K))$ is strictly positive and if u_0 is T-convex (or if, in particular, u_0 is convex), then $u_0(x) \leq u(x,s) \leq u(x,t) \leq T(u_0)(x)$ for every $0 \leq s \leq t$ and $x \in K$.*

(4) *Assume that K is endowed with a metric d whose corresponding modulus of continuity satisfies (6.1.82), and consider a number $\lambda \in]0,1]$. If $T(\text{Lip}_1 1) \subset \text{Lip}_1 1$ and $u_0 \in \text{Lip}_M \alpha$ for some $M > 0$ and $0 < \alpha \leq 1$, then $u(\cdot,t) \in \text{Lip}_M \alpha$ for every $t \geq 0$.*

Proof. Taking formulas (6.2.20) and (6.2.21) into account, properties (1) and (2) directly follow from (6.2.29).

To show (3), we recall that by (6.1.66) we have

$$u_0 \leq L_{n,\lambda}(u_0) \leq T(u_0) \quad \text{for every } n \geq 1, \tag{1}$$

and hence, by (6.1.54) we obtain

$$u_0 \leq L_{n,\lambda}^{[nt]}(u_0) \leq L_{n,\lambda}^{[ns]}(u_0) \leq T(u_0) \quad \text{for } 0 \leq s \leq t.$$

Now the result follows again from (6.2.29).

Finally, statement (4) is a direct consequence of Corollary 6.1.22. □

We conclude this section by showing some probabilistic aspects of the results we have obtained.

For the necessary background we refer the reader to the last part of Section 1.6.

We shall suppose that K is a convex compact subset of a metrizable locally convex space E. Furthermore, we shall assume that the hypotheses of Theorem 6.2.6 are satisfied.

We denote by λ either a number ≥ 1 or a strictly positive continuous function on K less than or equal to 1, and we shall denote by $(T_\lambda(t))_{t \geq 0}$ the corresponding Feller semigroup.

By virtue of Theorem 1.6.14, $(T_\lambda(t))_{t \geq 0}$ is the transition semigroup of a right-continuous normal Markov process $(\Omega, \mathcal{U}, (P^x)_{x \in K}, (Z_t)_{t \geq 0})$ with state space K, whose paths have left-hand limits on $[0, \zeta[$ almost surely, where $\zeta : \Omega \to \mathbb{R}$ is the lifetime of the process defined by

$$\zeta(\omega) := \inf\{t \geq 0 \,|\, Z_t(\omega) \in \partial K\} \tag{6.2.32}$$

for every $\omega \in \Omega$.

Let $(P_t)_{t \geq 0}$ be the normal transition function on K associated with the above Markov process.

Intuitively, we may think of a particle which moves in K after a random experiment $\omega \in \Omega$. Then for every $t \geq 0$ and $x \in K$ and for every Borel set B of K, $Z_t(\omega)$ expresses the position (in K) of the particle at time t and $P^x(t, B)$ is the probability that a particle starting at position x will be found in B at time t.

For given $x \in K$, $t \geq 0$ and $n \geq 1$ denote by $\mu_{n,x,t} \in \mathcal{M}^+(K)$ the probability Radon measure

$$\mu_{n,x,t}(f) := P_{n,\lambda}^{[nt]}(f)(x) \quad (f \in \mathscr{C}(K)) \tag{6.2.33}$$

and by $\tilde{\mu}_{n,x,t}$ the unique Borel measure on K which corresponds to $\mu_{n,x,t}$ via the Riesz representation theorem.

On account of (1.6.46) and (6.2.19), we then obtain that

$$P_t(x, \cdot) = \lim_{n \to \infty} \tilde{\mu}_{n,x,t} \quad \text{weakly.} \tag{6.2.34}$$

Moreover, from (6.2.20) it also follows that

$$\lim_{t \to \infty} P_t(x, \cdot) = \tilde{\mu}_x^T \quad \text{weakly,} \tag{6.2.35}$$

where $\tilde{\mu}_x^T$ denotes the Borel measure on K corresponding to the Radon measure μ_x^T defined by (6.1.7).

Thus, since the support of $\tilde{\mu}_x^T$ (or of μ_x^T) is contained in $\partial_H^+ K$ (see (6.1.8)), formula (6.2.35) tells us that, if $B \in \mathcal{B}(K)$ is a Borel set whose topological boundary is disjoint from $\partial_H^+ K$, then

$$\lim_{t \to \infty} P_t(x, B) = 0 \quad \text{for every } x \in K. \tag{6.2.36}$$

By the same reasons, given a Borel measure $\tilde{\upsilon} \in \mathcal{M}^+(K)$, we obtain

$$P_t(\tilde{\upsilon}) = \lim_{n \to \infty} P_{n,\lambda}^{[nt]}(\tilde{\upsilon}) \quad \text{weakly} \tag{6.2.37}$$

and

$$\lim_{t \to \infty} P_t(\tilde{\upsilon}) = T(\tilde{\upsilon}) \quad \text{weakly,} \tag{6.2.38}$$

where for every Borel set B of K we define

$$P_t(\tilde{\upsilon})(B) := \int P_t(x, B) \, d\tilde{\upsilon}(x), \tag{6.2.39}$$

$$P_{n,\lambda}^{[nt]}(\tilde{\upsilon})(B) := \int \tilde{\mu}_{n,x,t}(B) \, d\tilde{\upsilon}(x) \tag{6.2.40}$$

and

$$T(\tilde{\upsilon})(B) := \int \tilde{\mu}_x^T(B) \, d\tilde{\upsilon}(x). \tag{6.2.41}$$

Intuitively, $P_t(\tilde{\upsilon})$ gives the distribution for the position of the particle at time t provided the distribution of the initial position of the particle is given by $\tilde{\upsilon}$.

In Section 6.3, we shall furnish some further properties of the Markov process and of the random variables Z_t in the particular setting of the interval $[0, 1]$.

Notes and references

Voronovskaja-type formulas (6.2.8) and (6.2.15) generalize a result of Voronovskaja [1932] who established a similar one for Bernstein operators on the interval $[0, 1]$.

In fact, she showed that, if $f \in \mathscr{C}^2([0, 1])$, then

$$\lim_{n \to \infty} n(B_n(f)(x) - f(x)) = \frac{x(1 - x)}{2} f''(x) \tag{1}$$

uniformly with respect to $x \in [0, 1]$ and, as we shall see in Section 6.3, when $K = [0, 1]$ the differential operator W_T defined by (6.2.11) becomes $W_T(u)(x) = \frac{x(1 - x)}{2} u''(x)$.

However, note that the asymptotic formula (1) holds for every $x \in [0, 1]$ and every bounded function f on $[0,1]$ having a second derivative at x.

Theorem 6.2.1 is due to Campiti [1992c] for the case $\lambda \geq 1$ and to Altomare [1989], [1992b] when $\lambda \in B_1^+(\mathscr{C}(K))$ (see also Altomare and Romanelli [1992]).

It extends similar results obtained by Schnabl [1968], Felbecker [1972] and Nishishiraho [1978] in particular settings.

Theorem 6.2.5 is due to Altomare [1992a], [1992b] for the case $\lambda \in B_1^+(\mathscr{C}(K))$ while the case $\lambda \geq 1$ appears here for the first time.

For $\lambda = 1$, i.e., for Bernstein-Schnabl operators, Theorem 6.2.5 also appears in Blümlinger [1987] when K is a ball of $\mathbb{R}^p$ and T is the Dirichlet operator (3.3.16).

A Voronovskaja-type formula for the canonical simplex of $\mathbb{R}^p$ and for Bernstein operators first appeared in Stancu [1960] (see also [1959]).

Stancu also established a similar result for the operators introduced by him in [1968], [1969c] and [1973] for the unit interval and the multidimensional simplex, respectively.

For Bernstein operators on the product of finite dimensional simplices see Dahmen and Micchelli [1990].

As a matter of fact, many other positive approximation processes satisfy a Voronovskaja-type formula.

For instance, all those approximation processes $(P_n)_{n \in \mathbb{N}}$ of probabilistic-type (5.2.3) which are associated with a random scheme $Z(n, x) = \frac{1}{n} \sum_{k=1}^{n} Y(k, x)$ $(n \geq 1, x \in I)$, where for every $x \in I$ the random variables $Y(n, x)$, $n \geq 1$, are independent and identically distributed, satisfy the formula

$$\lim_{n \to \infty} n(P_n(f)(x) - f(x)) = \frac{\sigma^2(x)}{2} f''(x)$$

for every $f \in \mathscr{C}^2(I)$ and $x \in I$, where $\sigma^2(x) = \text{Var}(Y(n, x))$ (which is independent of $n \geq 1$).

This result is due to Cismasiu [1985].

Other results can be found in Becker, Butzer and Nessel [1976], Berens and Xu [1991a], Butzer and Nessel [1971], Cheney and Sharma [1964a], Derrienic [1981], Karlin and Ziegler [1970], Lupaş and Lupaş [1987], Mastroianni [1979], Mastroianni and Occorsio [1978b], Müller [1989], Sauer [1992a], Sikkema [1970a], Stancu [1983].

Theorem 6.2.6 was obtained by Altomare [1989], [1992a], [1992b], [1993] both for Bernstein-Schnabl and Lototsky-Schnabl operators, and by Campiti [1992c] for Stancu-Schnabl operators.

Particular cases of it include results by Altomare and Romanelli [1992] ($0 \leq \lambda \leq 1$, λ constant), Blümlinger [1987] (K a ball of $\mathbb{R}^p$, $\lambda = 1$), Dahmen and Micchelli [1990] (K product of finite dimensional simplices and $\lambda = 1$), Da Silva [1987] ($K = [0, 1] \times [0, 1]$ and $\lambda = 1$), Felbecker [1972] (K Bauer simplex, $\lambda \geq 1$), Nishishiraho [1978], [1987] and Schnabl [1969], [1972] (K Bauer simplex and $\lambda = 1$).

For the classical Bernstein operators on $[0, 1]$, Theorem 6.2.6 was in essence obtained by Karlin and Ziegler [1970] and Micchelli [1973b] (see also Da Silva [1985]).

There further properties of the semigroup can be found.

Finally, the qualitative properties of the semigroups or of the solutions of the corresponding Cauchy problems are taken from Altomare [1992b], [1993] (see also [1991]) and Rasa [1993].

The probabilistic considerations we developed in the last part of this section are quite recent. A deeper analysis of the Markov process described in the last part of this section does not seem to be devoid of interest.

In this respect the reader should also consult recent books of Taira [1988], [1992].

6.3 Miscellaneous examples and degenerate diffusion equations on convex compact subsets of $\mathbb{R}^p$

In the preceding sections we associated with a given positive projection T: $\mathscr{C}(K) \to \mathscr{C}(K)$ satisfying conditions (6.1.3) and (6.1.4) different mathematical objects, namely several positive approximation processes as well as some classes of Feller semigroups with their corresponding generators.

The interplay among these objects is expressed by formulas (6.2.7), (6.2.11), (6.2.19), (6.2.20), (6.2.24) and (6.2.26).

We used them fruitfully to approximate continuous functions and to derive qualitative properties of the solutions of some differential problems.

In this final section, we shall consider some cases of particular interest where a positive projection arises in a natural way. In all these cases we shall give the explicit expression of Bernstein-Schnabl, Stancu-Schnabl and Lototsky-Schnabl operators. Moreover, if K is contained in a finite dimensional space, then we shall describe the second order elliptic differential operator W_T defined by (6.2.11) and its closure A_1; in some cases we shall state some supplementary properties of the solution of the Cauchy problem (6.2.28) as well.

We begin by considering the class of Bauer simplices. This case is of particular interest because, among other things, it furnishes some examples of approximation processes also in an infinite dimensional setting.

6.3.1 Bauer simplices. We assume that K is a metrizable Bauer simplex and we consider the canonical positive projection $T: \mathscr{C}(K) \to \mathscr{C}(K)$ associated with K (see (1.5.18) and Corollary 1.5.9); thus, for every $f \in \mathscr{C}(K)$ and $x \in K$,

$$Tf(x) := \mu_x(f), \tag{6.3.1}$$

where $\mu_x \in \mathscr{M}_1^+(K)$ is the unique probability Radon measure on K satisfying

$$\mathrm{Supp}(\mu_x) \subset \overline{\partial_e K} = \partial_e K \tag{6.3.2}$$

and $r(\mu_x) = x$ (see Theorem 1.5.8, (ii)). Moreover, by Propositions 2.6.3 and 3.3.1 we have

$$\partial_e K = \partial_H^+ K = \{x \in K | \mu_x = \varepsilon_x\}. \tag{6.3.3}$$

We recall that T is the unique positive projection on $\mathscr{C}(K)$ such that $T(\mathscr{C}(K)) = A(K)$ (see Corollary 1.5.9) and consequently conditions (6.1.3) and (6.1.4) are obviously satisfied.

In the framework of Bauer simplices, it is interesting to point out the following sharper result concerning the different notions of convexity studied in Section 6.1 in connection with the monotonicity properties of Bernstein-Schnabl operators.

Actually, we show that the notion of axial convexity coincides with that of T-convexity and it is preserved by Bernstein-Schnabl operators.

In this respect we recall that, except for the case of the interval $[0,1]$, where a function $f \in \mathscr{C}([0,1])$ is convex if and only if all $B_n(f)$ are convex (see Corollary 6.3.8), in general the Bernstein polynomials do not preserve the convexity.

A simple counterexample is furnished by the function $u = |pr_1 - pr_2|$ defined on the two-dimensional simplex $K_2 = \{(x,y) \in \mathbb{R}^2 | x, y \geq 0, x + y \leq 1\}$ (see Sauer [1991, p. 468] for some details; for another counterexample see Chang and Davis [1984]).

6.3.2 Theorem. *Under the above assumptions, consider the sequence* $(B_n)_{n \in \mathbb{N}}$ *of Bernstein-Schnabl operators associated with* T. *Then, for a given* $f \in \mathscr{C}(K)$, *the following statements are equivalent*:

(i) f *is* T-*convex*.
(ii) f *is axially convex*.
(iii) $B_n(f)$ *is axially convex for every* $n \geq 1$.

Proof. First, we observe that, given $u, v \in K$ satisfying $u - v = \beta(p - q)$ for some $\beta \geq 0$ and $p, q \in \partial_e K$, then there exists $z \in K$ such that

$$u = (1 - \beta)z + \beta p \quad \text{and} \quad v = (1 - \beta)z + \beta q. \tag{1}$$

Indeed, $\mu_p = \varepsilon_p$ and $\mu_q = \varepsilon_q$ (see (6.3.3)) and hence, since $u + \beta q = v + \beta p$, for every $g \in \mathscr{C}(K)$ we obtain

$$\frac{T(g)(u)}{1 + \beta} + \frac{\beta g(q)}{1 + \beta} = \frac{T(g)(u)}{1 + \beta} + \frac{\beta}{1 + \beta} T(g)(q)$$

$$= T(g)\left(\frac{1}{1 + \beta}u + \frac{\beta}{1 + \beta}q\right) = T(g)\left(\frac{1}{1 + \beta}v + \frac{\beta}{1 + \beta}p\right)$$

$$= \frac{T(g)(v)}{1 + \beta} + \frac{\beta}{1 + \beta}g(p),$$

i.e., $\mu_u + \beta\varepsilon_q = \mu_v + \beta\varepsilon_p$.

Moreover, we can always write $\mu_u = \gamma\varepsilon_p + \delta\varepsilon_q + (1 - \gamma - \delta)v$, where $v \in \mathcal{M}_1^+(K)$ with $v^*(\{p,q\}) = 0$ and $\gamma, \delta \geq 0$ with $\gamma + \delta \leq 1$. Indeed, take $\gamma := \mu^*(\{q\})$, $\delta := \mu^*(\{q\})$ (see (1.2.31)) and $A := K\setminus\{p,q\}$. If $\gamma + \delta = 1$, denote by $v \in \mathcal{M}_1^+(K)$ an arbitrary Radon measure having its support contained in $K\setminus\{p,q\}$. If $\gamma + \delta < 1$, put $v(f) = \dfrac{1}{1 - \gamma - \delta}\displaystyle\int 1_A f \, d\mu_u$ for every $f \in \mathscr{C}(K)$ (see (1.2.29)).

Hence, if we denote by $r \in K$ the barycenter of v, then we can write $u = \gamma p + \delta q + (1 - \gamma - \delta)r$ and therefore $u - \beta p = (\gamma - \beta)p + \delta q + (1 - \gamma - \delta)r$. In particular, it follows that $0 \le \beta \le \gamma \le 1$.

Now, if $\beta = 1$, then $\gamma = 1$, $\delta = 0$, $u = p$ and $v = q$ and consequently (1) is satisfied by taking an arbitrary point $z \in K$. If $\beta < 1$, then (1) is satisfied by taking

$$z := \frac{1 - \gamma - \delta}{1 - \beta} r + \frac{\gamma - \beta}{1 - \beta} p + \frac{\delta}{1 - \beta} q.$$

At this point, we proceed to show that statements (i)–(iii) are equivalent.

(i) $\Rightarrow$ (ii): Suppose that f is T-convex. On account of (6.3.2), it is enough to prove that $f(\alpha u + (1 - \alpha)v) \le \alpha f(u) + (1 - \alpha)f(v)$ whenever $\alpha \in [0, 1]$ and $u, v \in K$ satisfy $u - v = \beta(p - q)$ for some $p, q \in \partial_e K$ and $\beta \ge 0$. In this case, indeed, if we consider $z \in K$ satisfying (1), we have (see (6.1.1))

$$f(\alpha u + (1 - \alpha)v) = f_{z,\beta}(\alpha p + (1 - \alpha)q) \le T(f_{z,\beta})(\alpha p + (1 - \alpha)q)$$

$$= \alpha T(f_{z,\beta})(p) + (1 - \alpha)T(f_{z,\beta})(q)$$

$$= \alpha f_{z,\beta}(p) + (1 - \alpha)f_{z,\beta}(q) = \alpha f(u) + (1 - \alpha)f(v).$$

(ii) $\Rightarrow$ (i): This follows by the Remark to Theorem 6.1.14.

(ii) $\Rightarrow$ (iii): Fix $n \in \mathbb{N}$ and let $u, v \in K$ and $p, q \in \partial_e K$ be such that $u - v = \beta(p - q)$ with $\beta \ge 0$. Moreover, consider $z \in K$ satisfying condition (1) and define the function $G: [0, 1] \to \mathbb{R}$ by

$$G(\alpha) := B_n(f)((1 - \alpha)u + \alpha v) \ (= B_n(f)((1 - \beta)z + \beta(1 - \alpha)p + \beta\alpha q))$$

for every $\alpha \in [0, 1]$.

In other words, if $\alpha \in [0, 1]$ and if we set $x := (1 - \beta)z + \beta(1 - \alpha)p + \beta\alpha q$, then $\mu_x = (1 - \beta)\mu_z + \beta(1 - \alpha)\varepsilon_p + \beta\alpha\varepsilon_q$ and

$$G(\alpha) = B_n(f)(x) = \int_K \cdots \int_K f\left(\frac{t_1 + \cdots + t_n}{n}\right) d\mu_x(t_1) \ldots d\mu_x(t_n)$$

$$= \sum_{k=0}^{n} \binom{n}{k} \beta^{n-k}(1 - \beta)^k \sum_{i=0}^{n-k} \binom{n - k}{i} \alpha^i(1 - \alpha)^{n-k-i} \int_K$$

$$\cdots \int_K f\left(\frac{iq + (n - k - i)p + t_1 + \cdots + t_k}{n}\right) d\mu_z(t_1) \ldots d\mu_z(t_k).$$

Finally, for every $\theta \in [0, 1]$ and $k = 0, \ldots, n$ consider the functions $\varphi_{\theta_k}: K \to \mathbb{R}$ and $\phi_k: [0, 1] \to \mathbb{R}$ defined by

$$\varphi_{\theta_k}(x) := f((1 - k/n)(\theta q + (1 - \theta)p) + kx/n) \quad (x \in K)$$

and

$$\phi_k(\theta) := B_k(\varphi_{\theta_k})(z) \quad (\theta \in [0, 1]).$$

Then we easily obtain

$$G(\alpha) = \sum_{k=0}^{n} \binom{n}{k} \beta^{n-k}(1 - \beta)^k \hat{B}_{n-k}(\phi_k)(\alpha), \tag{2}$$

where $\hat{B}_m$ ($m \geq 1$) denotes the m-th classical Bernstein operator on the interval $[0, 1]$ (see (5.2.36)) and $\hat{B}_0$ is the identity operator on $\mathscr{C}([0, 1])$. Since f is axially convex, it follows that the function $\theta \mapsto \varphi_{\theta_k}(y)$ ($\theta \in [0, 1]$) is convex for every $y \in K$ as well, and consequently ϕ_k is also convex. Now, the classical Bernstein operators preserve convex functions (see Corollary 6.3.8) and therefore by (2), the function G is convex. Hence we conclude that $B_n(f)((1 - \alpha)u + \alpha v) \leq (1 - \alpha)B_n(f)(u) + \alpha B_n(f)(v)$.

(iii) $\Rightarrow$ (ii): This is a consequence of Theorem 6.1.3, (1). □

Remark. As a consequence of Theorem 6.1.14, we also observe that each of statements (i), (ii) or (iii) implies the monotonicity of the sequence $(B_n(f))_{n \in \mathbb{N}}$, namely $B_n(f) \geq B_{n+1}(f)$ for every $n \geq 1$.

The canonical positive projection (6.3.1) also satisfies condition (6.2.17) and therefore Theorem 6.2.6 holds in this context.

So, there exists an operator $A_1: D(A_1) \to \mathscr{C}(K)$ which is the generator of a Feller semigroup and is uniquely determined on $A_\infty(K)$ by (6.2.7).

In analogy with the finite dimensional case, we shall call it as the *canonical abstract elliptic operator associated with the Bauer simplex K*.

A deeper analysis of this operator seems to be of interest.

However, we study in more details these questions only for some particular examples of Bauer simplices, where we can give an explicit expression of the projection T.

We begin with the case where K is the canonical simplex of $\mathbb{R}^p$.

6.3.3 Finite dimensional canonical simplices. Denote by K_p the canonical simplex in $\mathbb{R}^p$, i.e.,

$$K_p := \{(x_1, \ldots, x_p) \in \mathbb{R}^p | 0 \leq x_i, i = 1, \ldots, p \text{ and } x_1 + \cdots + x_p \leq 1\}.$$

In this case, the canonical projection $T_p\colon \mathscr{C}(K_p) \to \mathscr{C}(K_p)$ is defined by

$$T_p(f)(x_1,\ldots,x_p) := \sum_{\substack{0 \le h_1 + \cdots + h_p \le 1 \\ h_i \in \mathbb{N}_0}} \alpha_f(h_1,\ldots,h_p) x_1^{h_1} \ldots x_p^{h_p} \left(1 - \sum_{i=1}^{p} x_i\right)^{1 - \sum_{i=1}^{p} h_i}$$

$$= \left(1 - \sum_{i=1}^{p} x_i\right) f(0) + \sum_{i=1}^{p} x_i f(e_i) \qquad (6.3.4)$$

for every $f \in \mathscr{C}(K_p)$ and $(x_1,\ldots,x_p) \in K_p$, where

$$\alpha_f(h_1,\ldots,h_p) := f(\delta_{h_1 1},\ldots,\delta_{h_p 1}) \qquad (6.3.5)$$

and $e_i = (\delta_{ij})_{1 \le j \le p}$ for $i = 1, \ldots, p$ (δ_{ij} is the Kronecker symbol).

By (6.3.4) we can explicitly write the Bernstein-Schnabl, Stancu-Schnabl and Lototsky-Schnabl operators; as a matter of fact, we obtain the classical Bernstein operators on the p-dimensional simplex already described in Example 5.2.11 and a generalization of the Stancu operators considered in Example 5.2.7.

More precisely, for every $f \in \mathscr{C}(K_p)$ and $x = (x_1,\ldots,x_p) \in K_p$, we have

$$B_n(f)(x_1,\ldots,x_p)$$

$$= \sum_{\substack{h_1,\ldots,h_p \in \mathbb{N}_0 \\ h_1 + \cdots + h_p \le n}} f\left(\frac{h_1}{n},\ldots,\frac{h_p}{n}\right) \frac{n!}{h_1! \ldots h_p!(n - h_1 - \cdots - h_p)!} x_1^{h_1}$$

$$\ldots x_p^{h_p} \left(1 - \sum_{i=1}^{p} x_i\right)^{n - \sum_{i=1}^{p} h_i}. \qquad (6.3.6)$$

More generally, if $P = (p_{nj})_{n \ge 1, j \ge 1}$ is a lower triangular stochastic matrix, then (see (6.1.10))

$$B_{P,n}(f)(x_1,\ldots,x_p)$$

$$= \sum_{\substack{h_1,\ldots,h_p \in \mathbb{N}_0 \\ h_1 + \cdots + h_p \le n}} (\sum f(p_{ni_{1,1}} + \cdots + p_{ni_{1,h_1}},\ldots,p_{ni_{p,1}} + \cdots + p_{ni_{p,h_p}})) x_1^{h_1}$$

$$\ldots x_p^{h_p} \left(1 - \sum_{i=1}^{p} x_i\right)^{n - \sum_{i=1}^{p} h_i}, \qquad (6.3.7)$$

where the sum is extended over all pairwise disjoint subsets of different integers $\{i_{1,1},\ldots,i_{1,h_1}\}, \ldots, \{i_{p,1},\ldots,i_{p,h_p}\}$ of $\{1,\ldots,n\}$ with the convention that, if some $h_j = 0$, then $\{i_{j,1},\ldots,i_{j,h_j}\} = \varnothing$ and $p_{ni_{j,1}} + \cdots + p_{ni_{j,h_j}} = 0$.

Consequently, given a sequence $(a_n)_{n \in \mathbb{N}}$ of real numbers, from (6.1.24) we obtain

$$S_{n,a_n}(f)(x_1,\ldots,x_p)$$

$$= \frac{1}{p_n(a_n)} \sum_{k=1}^{n} \frac{n!}{k!} a_n^{n-k} \sum_{|v|_k=n} \frac{1}{v_1 \ldots v_k} \sum_{\substack{h_1,\ldots,h_p \in \mathbb{N}_0 \\ h_1+\cdots+h_p \le k}} x_1^{h_1} \ldots x_p^{h_p} \left(1 - \sum_{i=1}^{p} x_i\right)^{k - \sum_{i=1}^{p} h_i}$$

$$\times \sum f\left(\frac{v_{i_{1.1}} + \cdots + v_{i_{1.h_1}}}{n}, \ldots, \frac{v_{i_{p.1}} + \cdots + v_{i_{p.h_p}}}{n}\right), \tag{6.3.8}$$

where the sum is extended over all pairwise disjoint subsets $\{i_{1.1},\ldots,i_{1.h_1}\}, \ldots,$ $\{i_{p.1},\ldots,i_{p.h_p}\}$ of $\{1,\ldots,k\}$ with the convention that, if some $h_j = 0$, then $\{i_{j.1},\ldots,i_{j.h_j}\} = \varnothing$ and $v_{i_{j.1}} + \cdots + v_{i_{j.h_j}} = 0$.

As a matter of fact, by using the method of Felbecker [1972, pp. 61–67], it is possible to give another representation of Stancu-Schnabl operators (6.3.8), namely

$$S_{n,a_n}(f)(x_1,\ldots,x_p)$$

$$= \frac{1}{p_n(a_n)} \sum_{\substack{h_1,\ldots,h_p \in \mathbb{N}_0 \\ h_1+\cdots+h_p \le n}} f\left(\frac{h_1}{n},\ldots,\frac{h_p}{n}\right) \frac{n!}{h_1!\ldots h_p!(n - h_1 - \cdots - h_p)!}$$

$$\times \prod_{i_1=0}^{h_1-1} (x_1 + i_1 a_n) \prod_{i_2=0}^{h_2-1} (x_2 + i_2 a_n) \ldots \prod_{i_p=0}^{h_p-1} (x_p + i_p a_n) \prod_{i_0=0}^{n-h_1-\cdots-h_p}$$

$$\times (1 - x_1 - \cdots - x_p + i_0 a_n), \tag{6.3.9}$$

where if $h_i = 0$ for some $i = 1, \ldots, p$, the corresponding product must be taken to be equal to 1 by convention.

These last operators were first introduced by Stancu [1972b]. In the same paper the reader can find other interesting properties of them.

Finally, as regards the Lototsky-Schnabl operators $(L_{n,\lambda})_{n \ge 1}$ on the simplex K_p, by virtue of (6.1.52), for every $\lambda \in \mathscr{C}(K_p)$, $0 \le \lambda \le 1$, we have

$$L_{n,\lambda}(f)(x_1,\ldots,x_p)$$

$$= \sum_{k=0}^{n} \binom{n}{k} \lambda(x)^k (1 - \lambda(x))^{n-k} \sum_{\substack{h_1,\ldots,h_p \in \mathbb{N}_0 \\ h_1+\cdots+h_p \le k}} \frac{k!}{h_1!\ldots h_p!(k - h_1 - \cdots - h_p)!} x_1^{h_1}$$

$$\ldots x_p^{h_p} \left(1 - \sum_{i=1}^{p} x_i\right)^{k - \sum_{i=1}^{p} h_i} f\left(\frac{h_1}{n} + \left(1 - \frac{k}{n}\right) x_1, \ldots, \frac{h_p}{n} + \left(1 - \frac{k}{n}\right) x_p\right). \tag{6.3.10}$$

If we consider the ℓ^q-norm $\|\cdot\|_q$ on $\mathbb{R}^p$, $1 \le q \le +\infty$ and if $f \in \mathrm{Lip}_1 1$, then for $q < +\infty$ by using Hölder's inequality, we easily obtain (see (6.3.4))

$$|T_p(f)(x) - T_p(f)(y)|^q = \left| \sum_{i=1}^{p} (x_i - y_i)(f(e_i) - f(0)) \right|^q \le \left(\sum_{i=1}^{p} |x_i - y_i| \right)^q$$

$$\le p^{q-1} \|x - y\|_q^q,$$

for every $x, y \in K_p$ and hence

$$|T_p(f)(x) - T_p(f)(y)| \le p^{1-1/q} \|x - y\|_q.$$

If $q = +\infty$, we have

$$|T_p(f)(x) - T_p(f)(y)| \le p \|x - y\|_\infty.$$

Thus, in any case, we have shown that

$$T_p(f) \in \mathrm{Lip}_c 1 \quad \text{for every } f \in \mathrm{Lip}_1 1 \text{ with } c = p^{1-1/q}. \tag{6.3.11}$$

This allows us to apply Theorem 6.1.21 and to obtain the following results:
(1) If a function $\lambda \in \mathscr{C}(K_p)$, $0 \le \lambda \le 1$, belongs to $\mathrm{Lip}_N 1$ then

$$L_{n,\lambda}(f) \in \mathrm{Lip}_{cM+nN\|T(f)-f\|} 1 \quad \text{for every } f \in \mathrm{Lip}_M 1 \text{ and } n \in \mathbb{N}. \tag{6.3.12}$$

(2) If λ is constant, then

$$L_{n,\lambda}(f) \in \mathrm{Lip}_{cM} 1 \tag{6.3.13}$$

and in particular, for $\lambda = 1$, we have

$$B_n(f) \in \mathrm{Lip}_{cM} 1. \tag{6.3.14}$$

Note that, if we consider the ℓ^1-norm on $\mathbb{R}^p$, then we have $c = 1$ and therefore $T_p(\mathrm{Lip}_M 1) \subset \mathrm{Lip}_M 1$ for every $M > 0$ (see (6.3.11) and (6.1.84)). Accordingly, by the preceding property (2) we have $L_{n,\lambda}(\mathrm{Lip}_M 1) \subset \mathrm{Lip}_M 1$ for every $n \ge 1$ and $\lambda \in {]}0, 1]$.

If $q = 2$ and if we consider the function $e(x) := \|x\|^2$ ($x \in K_p$), then we have (see (6.3.4))

$$T_p(e)(x) - e(x) = \sum_{i=1}^{p} x_i(1 - x_i)$$

for every $x \in K_p$, and therefore $\|T(e) - e\| \leq \dfrac{p}{4}$. From Proposition 6.1.12, it then follows that, for every $f \in \mathrm{Lip}_M 1$ and $n \geq 1$,

$$\|B_n(f) - f\| \leq \frac{M}{2} \sqrt{\frac{p}{n}}, \tag{6.3.15}$$

$$\|S_{n, a_n}(f) - f\| \leq \frac{M}{2} \sqrt{\frac{p(1 + na_n)}{n(1 + a_n)}}, \tag{6.3.16}$$

$$\|L_{n, \lambda}(f) - f\| \leq \frac{M}{2} \sqrt{\frac{p}{n}} \|\lambda\|. \tag{6.3.17}$$

Estimates of the rate of convergence for an arbitrary function $f \in \mathscr{C}(K_p)$ can be obtained by (6.1.17), (6.1.44) and (6.1.58), by recalling that in this case $\Omega(f, \delta) \leq \omega(f, \delta)$ for every $\delta > 0$ (see (5.1.14)).

Finally, we point out that the projection T_p satisfies both conditions (6.2.17) and (6.2.18); moreover, we can write the explicit expression of the differential operator A_1 associated with T_p (see (6.2.11) and (6.2.26)). Indeed, by (6.3.4) and (6.2.12) we have

$$A_1(u)(x) = \frac{1}{2} \sum_{i=1}^{p} x_i(1 - x_i) \frac{\partial^2 u(x)}{\partial x_i^2} - \sum_{1 \leq i < j \leq p} x_i x_j \frac{\partial^2 u(x)}{\partial x_i \partial x_j} \tag{6.3.18}$$

for every $u \in \mathscr{C}^2(K_p)$ and $x = (x_1, \ldots, x_p) \in K_p$.

In the case of the interval $[0, 1]$ more information can be obtained about the differential operator A_1; therefore we consider this case separately.

6.3.4 Approximation processes and degenerate diffusion equations on the interval [0, 1].

If $p = 1$, the projection $T_1 : \mathscr{C}([0, 1]) \to \mathscr{C}([0, 1])$ defined by (6.3.4) becomes

$$T_1(f)(x) = (1 - x)f(0) + xf(1) \tag{6.3.19}$$

($f \in \mathscr{C}([0, 1])$, $x \in [0, 1]$).

In this case the Bernstein-Schnabl and the Stancu-Schnabl operators associated with T_1 are exactly the Bernstein and the Stancu operators studied in Subsections 5.2.5 and 5.2.7. Indeed, for every $f \in \mathscr{C}([0, 1])$, $x \in [0, 1]$ and $n \geq 1$, we clearly have

$$B_n(f)(x) = \sum_{k=0}^{n} f\left(\frac{k}{n}\right) \binom{n}{k} x^k (1 - x)^{n-k} \tag{6.3.20}$$

and (see (6.3.8))

$$S_{n,a_n}(f)(x) = \frac{1}{p_n(a_n)} \sum_{k=1}^{n} \frac{n!}{k!} a_n^{n-k} \sum_{|v|_k=n} \frac{1}{v_1 \ldots v_k}$$

$$\sum_{h=0}^{k} \left(\sum_{\{i_1,\ldots,i_h\} \in C(k,h)} f\left(\frac{v_{i_1} + \cdots + v_{i_h}}{n}\right) \right) x^h (1-x)^{k-h}, \qquad (6.3.21)$$

where $(a_n)_{n \in \mathbb{N}}$ is a sequence of positive real numbers and $C(k,h)$ denotes the set of all subsets of $\{1,\ldots,k\}$ having h different elements, provided $h \geq 1$, while, if $h = 0$, we set $C(k,h) = \varnothing$ and $v_{i_1} + \cdots + v_{i_h} = 0$.

Clearly, by virtue of (6.3.9), Stancu-Schnabl operators on $[0,1]$ coincide with Stancu operators defined in (5.2.52).

Moreover, for a given $\lambda \in \mathscr{C}([0,1])$, $0 \leq \lambda \leq 1$, the Lototsky-Schnabl operators $(L_{n,\lambda})_{n \in \mathbb{N}}$ associated with T_1 and λ are explicitly given by

$$L_{n,\lambda}(f)(x) = \sum_{k=0}^{n} \sum_{h=0}^{k} \binom{n}{k}\binom{k}{h} \lambda(x)^k (1-\lambda(x))^{n-k} x^h (1-x)^{k-h} f\left(\frac{h}{n} + \left(1-\frac{k}{n}\right)x\right).$$
$$(6.3.22)$$

Among the main properties of the above operators we observe that in this case we have $T_1(\text{Lip}_1 1) \subset \text{Lip}_1 1$ and consequently, by Theorem 6.1.21

$$L_{n,\lambda}(\text{Lip}_M 1) \subset \text{Lip}_{M+nN\|T(f)-f\|} 1 \qquad (6.3.23)$$

for every $n \geq 1$ and $\lambda \in \mathscr{C}([0,1])$, $0 \leq \lambda \leq 1$ satisfying $\lambda \in \text{Lip}_N 1$. If $\lambda \in \,]0,1]$ is constant, we obtain

$$L_{n,\lambda}(\text{Lip}_M 1) \subset \text{Lip}_M 1 \qquad (6.3.24)$$

and hence in particular

$$B_n(\text{Lip}_M 1) \subset \text{Lip}_M 1. \qquad (6.3.25)$$

Some quantitative estimates of the rate of convergence can be directly derived from (6.3.15)–(6.3.17).

Besides the probabilistic interpretation of Bernstein and Stancu operators already studied in Subsections 5.2.5 and 5.2.7, here we provide an analogous one for the operators $L_{n,\lambda}$.

For every $x \in [0,1]$ we consider an independent sequence $(Z(n,x))_{n \in \mathbb{N}}$ of discrete and identically distributed random variables defined on a fixed probability space $(\Omega, \mathscr{F}, P)$ and we assume that every $Z(n,x)$ takes the values 0, x and 1 with probability

$$P\{Z(n, x) = 0\} = \lambda(x)(1 - x), \qquad (6.3.26)$$

$$P\{Z(n, x) = x\} = 1 - \lambda(x), \qquad (6.3.27)$$

$$P\{Z(n, x) = 1\} = \lambda(x)x. \qquad (6.3.28)$$

Then, for every $n \geq 1$, $f \in \mathscr{C}([0, 1])$ and $x \in [0, 1]$,

$$L_{n, \lambda}(f)(x) = E\left(f\left(\frac{1}{n} \sum_{k=1}^{n} Z(k, x)\right)\right), \qquad (6.3.29)$$

where $E(\cdot)$ denotes, as usual, the expected value operator.
Moreover,

$$E(Z(n, x)) = x, \qquad (6.3.30)$$

and the variance $\sigma^2(Z(n, x))$ of $Z(n, x)$ is given by

$$\sigma^2(Z(n, x)) = x(1 - x)\lambda(x). \qquad (6.3.31)$$

Hence we can estimate the difference $|L_{n, \lambda}(f)(x) - f(x)|$ with the methods used in Section 5.2; for example, by (5.2.31) we obtain

$$|L_{n, \lambda}(f)(x) - f(x)| \leq M\omega_2\left(f, \sqrt{\frac{x(1 - x)\lambda(x)}{n}}\right) \qquad (6.3.32)$$

for every $f \in \mathscr{C}([0, 1])$ and $x \in [0, 1]$, M being a constant independent of f and x.
Furthermore, by applying Formula (5.2.21) we find the estimate

$$|L_{n, \lambda}(x) - f(x)| \leq (1 + \sqrt{x(1 - x)\lambda(x)})\omega(f, n^{-1/2}) \qquad (6.3.33)$$

which holds for every $f \in \mathscr{C}([0, 1])$ and $x \in [0, 1]$ and which generalizes (5.2.39).
If in addition f is differentiable and $f' \in \mathscr{C}([0, 1])$, then by (5.2.23) we also have

$$|L_{n, \lambda}(f)(x) - f(x)| \leq \sqrt{\frac{x(1 - x)\lambda(x)}{n}}(1 + \sqrt{x(1 - x)\lambda(x)})\omega(f', n^{-1/2}). \qquad (6.3.34)$$

In the case of the interval $[0, 1]$, we can say much more about the differential operator A_1 associated with T_1 and the corresponding Cauchy problem (6.2.28).

Clearly, by (6.3.18) we have

$$A_1(u)(x) = \frac{1}{2}x(1-x)u''(x) \tag{6.3.35}$$

for every $u \in \mathscr{C}^2([0,1])$ and $x \in [0,1]$.

More generally we have the following result.

6.3.5 Theorem. *Let $A_1: D(A_1) \to \mathscr{C}([0,1])$ be the generator of the semigroup $(T_1(t))_{t\geq 0}$ associated with the Bernstein operators (see (6.2.19)).*
Then

$$D(A_1) = \left\{ u \in \mathscr{C}([0,1]) \,|\, u \in \mathscr{C}^2(]0,1[), \; \lim_{x \to 0^+} \frac{x(1-x)}{2}u''(x) \right.$$

$$\left. = \lim_{x \to 1^-} \frac{x(1-x)}{2}u''(x) = 0 \right\}, \tag{6.3.36}$$

and, for every $u \in D(A_1)$ and $x \in [0,1]$,

$$A_1(u)(x) = \begin{cases} \dfrac{x(1-x)}{2}u''(x), & \text{if } 0 < x < 1, \\[2mm] 0, & \text{if } x = 0 \text{ or } x = 1. \end{cases} \tag{6.3.37}$$

Proof. By a result of Clément and Timmermans [1986, Theorem 2] the operator A_0 defined by the right-hand side of (6.3.37) generates a strongly continuous contraction semigroup $(T(t))_{t\geq 0}$ on $\mathscr{C}([0,1])$. Moreover, $\mathscr{C}^2([0,1]) \subset D(A_0)$ and for every $u \in \mathscr{C}^2([0,1])$ we have $A_0(u) = A_1(u)$ by (6.3.35). Now, the result follows from Theorem 6.2.6, (6), because $\mathscr{C}^2([0,1])$ is a core for the generator of the semigroup $(T_1(t))_{t\geq 0}$ associated with the Bernstein operators. $\qquad\Box$

As a consequence of Theorems 6.3.5 and 6.2.6, (5), for every $u \in D(A_\lambda) = D(A_1)$ we obtain

$$A_\lambda(u)(x) = \begin{cases} \lambda(x)\dfrac{x(1-x)}{2}u''(x), & \text{if } 0 < x < 1, \\[2mm] 0, & \text{if } x = 0 \text{ or } x = 1, \end{cases} \tag{6.3.38}$$

if $\lambda \in B_1^+(\mathscr{C}([0,1]))$ is strictly positive or, if $\lambda \in \mathbb{R}$, $\lambda \geq 1$,

$$A_\lambda(u)(x) = \begin{cases} \lambda\dfrac{x(1-x)}{2}u''(x), & \text{if } 0 < x < 1, \\[2mm] 0, & \text{if } x = 0 \text{ or } x = 1, \end{cases} \tag{6.3.39}$$

where we recall that the operator A_λ is the generator of the semigroup $(T_\lambda(t))_{t \geq 0}$ associated with the approximation process $(P_{n, \lambda})_{n \in \mathbb{N}}$ defined according to (6.2.2)–(6.2.4).

By using these results we can now give a representation of the solutions of a large class of Cauchy problems with Wentcel-type boundary conditions (see 6.2.31)).

6.3.6 Corollary. *Let* $\alpha \colon [0, 1] \to \mathbb{R}$ *be a positive continuous function which vanishes only in 0 and 1 and suppose that it is differentiable at 0 and 1 with* $\alpha'(0) \neq 0 \neq \alpha'(1)$.

Then the problem

$$
\begin{cases}
\dfrac{\partial u}{\partial t}(x, t) = \alpha(x)\dfrac{\partial^2 u}{\partial x^2}(x, t), \quad 0 < x < 1, t \geq 0, \\[2mm]
\displaystyle\lim_{x \to 0^+} \alpha(x)\dfrac{\partial^2 u}{\partial x^2}(x, t) = \lim_{x \to 1^-} \alpha(x)\dfrac{\partial^2 u}{\partial x^2}(x, t) = 0, \quad t \geq 0, \\[2mm]
u(x, 0) = u_0(x), \quad u_0 \in \mathscr{C}([0, 1]) \cap \mathscr{C}^2(\,]0, 1[), \\[2mm]
\displaystyle\lim_{x \to 0^+} \alpha(x)u_0''(x) = \lim_{x \to 1^-} \alpha(x)u_0''(x) = 0,
\end{cases}
\tag{6.3.40}
$$

has a unique solution given by

$$
u(x, t) = \lim_{n \to \infty} L_{n, \lambda}^{[cnt]}(u_0)(x), \quad x \in [0, 1], t \geq 0,
\tag{6.3.41}
$$

where

$$
c := \sup_{0 < x < 1} \frac{2\alpha(x)}{x(1 - x)}
\tag{6.3.42}
$$

and the strictly positive continuous function $\lambda \colon [0, 1] \to [0, 1]$ *is defined by*

$$
\lambda(x) :=
\begin{cases}
\dfrac{2}{c}\alpha'(0), & \text{if } x = 0, \\[3mm]
\dfrac{2}{c}\dfrac{\alpha(x)}{x(1 - x)}, & \text{if } 0 < x < 1, \\[3mm]
\dfrac{2}{c}\alpha'(1), & \text{if } x = 1.
\end{cases}
\tag{6.3.43}
$$

Proof. For every $x \in [0, 1]$ we have $\alpha(x) = c\dfrac{x(1 - x)}{2}\lambda(x)$ and therefore the result follows by Theorem 6.2.6 (see also the discussion concerning problem (6.2.28)) because the solution semigroup of (6.3.40) is $(T_\lambda(ct))_{t \geq 0}$.   ☐

Problem (6.3.40) has been investigated by several mathematicians (see the final notes). The main feature of our results rests primarily in the representation formula (6.3.41) by means of which one can derive qualitative properties of the corresponding solutions.

Some of them are listed in Theorem 6.2.7. In addition we show that, if $u_0 \in D(A_1)$ is convex, then $u(\cdot, t)$ is convex too for every $t \geq 0$.

This will be an immediate consequence of the next result that would be compared with Theorem 6.1.14.

6.3.7 Proposition. *Let* $\lambda \in \mathscr{C}(K)$, $0 \leq \lambda \leq 1$, *be a strictly positive function and consider the sequence* $(L_{n,\lambda})_{n \in \mathbb{N}}$ *of Lototsky-Schnabl operators and the corresponding semigroup* $(T_\lambda(t))_{t \geq 0}$ *with generator* A_λ. *Then, for every* $f \in \mathscr{C}([0,1])$, *the following statements are equivalent*:
(i) f *is convex.*
(ii) $L_{n+1, \lambda}(f) \leq L_{n, \lambda}(f)$ *for every* $n \geq 1$.
(iii) $f \leq L_{n, \lambda}(f)$ *for every* $n \geq 1$.
(iv) $f \leq T_\lambda(t)f$ *for every* $t \geq 0$.
Moreover, if $f \in D(A_\lambda) = D(A_1)$, *then statements* (i)–(iv) *are equivalent to the further one*:
(v) $T_\lambda(t)f$ *is convex for every* $t \geq 0$.

Proof. (i) $\Rightarrow$ (ii): This follows from Corollary 6.1.15.
(ii) $\Rightarrow$ (iii): For every $n, m \in \mathbb{N}$ we have $L_{n+m, \lambda}(f) \leq L_{n, \lambda}(f)$; so, letting $m \to \infty$, we obtain (iii) by virtue of Theorem 6.1.10, (1).
(iii) $\Rightarrow$ (iv): This is a consequence of (6.2.19).
(iv) $\Rightarrow$ (i): For every $t > 0$, we have $A_\lambda(\int_0^t T_\lambda(s)f\, ds) = T_\lambda(t)f - f \geq 0$ (see Theorem 1.6.1, (5)) and therefore, taking (6.3.38) into account, the function $\int_0^t T_\lambda(s)f\, ds$ is convex. Consequently, $f = \lim\limits_{t \to 0^+} \dfrac{1}{t}\int_0^t T_\lambda(s)f\, ds$ is convex too.

Suppose now that $f \in D(A_\lambda)$. If f is convex, then $A_\lambda(f) \geq 0$ and hence $A_\lambda(T_\lambda(t)f) = T_\lambda(t)(A_\lambda(f)) \geq 0$. Thus $T_\lambda(t)f$ is convex.

Conversely, if for every $t \geq 0$, $T_\lambda(t)f$ is convex, then $\int_0^t T_\lambda(s)f\, ds$ is convex. So $T_\lambda(t)f - f = A_\lambda(\int_0^t T_\lambda(s)f\, ds) \geq 0$ and the result follows.   ☐

A refined form of Proposition 6.3.7 can be established in the special case $\lambda = 1$, i.e., for Bernstein polynomials $(B_n)_{n \in \mathbb{N}}$ and the corresponding semigroup $(T_1(t))_{t \geq 0}$.

6.3.8 Corollary. *For a given $f \in \mathscr{C}([0,1])$ the following statements are equivalent*:

 (i) f *is convex.*
 (ii) $B_{n+1}(f) \leq B_n(f)$ *for every $n \geq 1$.*
 (iii) $f \leq B_n(f)$ *for every $n \geq 1$.*
 (iv) $f \leq T_1(t)f$ *for every $t \geq 0$.*
 (v) $B_n(f)$ *is convex for every $n \geq 1$.*
 (vi) $T_1(t)f$ *is convex for every $t \geq 0$.*

Proof. We have only to show the equivalences (i) $\Leftrightarrow$ (v) $\Leftrightarrow$ (vi).
(i) $\Rightarrow$ (v): For a given $n \geq 1$ and $x \in [0,1]$ we have

$B_n(f)'(x)$

$= -n(1-x)^{n-1}f(0)$

$\quad + \sum_{k=1}^{n-1} \binom{n}{k} f\left(\frac{k}{n}\right)(kx^{k-1}(1-x)^{n-k} - (n-k)x^k(1-x)^{n-k-1}) + nx^{n-1}f(1)$

$= -n(1-x)^{n-1}f(0) + nf\left(\frac{1}{n}\right)(1-x)^{n-1}$

$\quad + n\sum_{k=2}^{n-1} \binom{n-1}{k-1} f\left(\frac{k}{n}\right)x^{k-1}(1-x)^{n-k}$

$\quad - n\sum_{k=1}^{n-2} \binom{n-1}{k} f\left(\frac{k}{n}\right)x^k(1-x)^{n-k-1} - nf\left(\frac{n-1}{n}\right)x^{n-1} + nf(1)x^{n-1}$

$= n\sum_{h=0}^{n-1} \binom{n-1}{h}\left(f\left(\frac{h+1}{n}\right) - f\left(\frac{h}{n}\right)\right)x^h(1-x)^{n-1-h}$

$= n\sum_{h=0}^{n-1} \binom{n-1}{h}\Delta_{1/n}f\left(\frac{h}{n}\right)x^h(1-x)^{n-1-h}. \tag{1}$

By repeated differentiation we obtain

$$B_n(f)^{(m)}(x) = n(n-1)\dots(n-m+1)\sum_{h=0}^{n-m} \Delta_{1/n}^m f\left(\frac{h}{n}\right)\binom{n-m}{h}x^h(1-x)^{n-m-h}. \tag{2}$$

In particular, for $m = 2$, we obtain

$$B_n(f)^{(2)}(x) = n(n-1)\sum_{h=0}^{n-2} \Delta_{1/n}^2 f\left(\frac{h}{n}\right)\binom{n-2}{h}x^h(1-x)^{n-2-h}. \tag{3}$$

Therefore, if f is convex,

$$\Delta_{1/n}^2 f\left(\frac{h}{n}\right) = f\left(\frac{h}{n} + \frac{2}{n}\right) - 2f\left(\frac{h}{n} + \frac{1}{n}\right) + f\left(\frac{h}{n}\right) \geq 0 \quad (h = 0,\dots,n-2)$$

and hence, according to (3), $B_n(f)^{(2)}(x) \geq 0$, i.e., $B_n(f)$ is convex.

The implication (v) $\Rightarrow$ (vi) follows from the representation formula (6.2.19) and, finally, the implication (vi) $\Rightarrow$ (iv) can be proved as in the proof of the preceding theorem. $\square$

Remark. From formula (1) of the above proof it follows that, if f is increasing, then all $B_n(f)$ are increasing.

More generally, formula (2) shows, in fact, that Bernstein polynomials preserve the property of being convex of any order (a function $f \in \mathscr{C}([0,1])$ is said to be *convex of order n* if all its divided differences at $n+1$ points $x_0 < \cdots < x_n$ are positive (see Lupaş [1967a], Lorentz [1986b]).

Finally, from (6.3.13) and (6.3.14) with $c = 1$, it follows that Bernstein operators (in fact, Lototsky-Schnabl operators associated with a constant $\lambda \in]0,1]$) preserve Hölder continuous functions. This implies that, if the initial value u_0 in (6.3.40) belongs to some $\mathrm{Lip}_M \alpha$, then the solution function $u(\cdot,t)$ belongs to the same set for every $t \geq 0$.

A similar property also holds for Stancu operators as has been showed by Della Vecchia [1989] (see also Della Vecchia and Rasa [1993]).

In the setting of the interval $[0,1]$ the probabilistic aspects we developed in the last part of Section 6.2 grow rich of further details.

From now on we shall always refer to the diffusion problem (6.3.40). Thus we can write

$$\alpha(x) = c\frac{x(1-x)}{2}\lambda(x), \quad (x \in [0,1]), \tag{6.3.44}$$

where c and λ are given by (6.3.42) and (6.3.43), respectively. Moreover the corresponding semigroup is $(T_\lambda(ct))_{t\geq 0}$.

Actually, the differential equation in (6.3.40) is the so-called *backward equation* of a normal Markov process

$$(\Omega, \mathscr{U}, (P^x)_{x \in [0,1]}, (Z_t)_{t\geq 0}), \tag{6.3.45}$$

having $[0,1]$ as state space, with absorbing barriers at 0 and 1 and with mean instantaneous velocity 0 and variance instantaneous velocity $\alpha(x)$ at position $x \in [0,1]$ (see Taira [1988, Section I.4]).

According to the terminology introduced by Feller [1952, Section 11], Wentcel's boundary condition in (6.3.40) means that 0 and 1 are *exit boundary points*. From a probabilistic point of view, this means that the probability that a particle, located in the interior of $[0, 1]$, will reach 0 or 1 after a finite lapse of time, is strictly positive.

Furthermore, on account of (1.6.40), (1.6.43), (6.2.19), (6.1.41) and (6.1.54), for every $x \in [0, 1]$ and $t \geq 0$ we have

$$E_x(Z_t) := \int Z_t \, dP^x = \int y \, dP^x_{Z_t}(y) = T_\lambda(ct)(e_1)(x) = x, \qquad (6.3.46)$$

where, as usual, e_1 denotes the function $e_1(x) = x$ $(x \in [0, 1])$.

On the other hand, if $\alpha(x) = m\dfrac{x(1 - x)}{2}$ with $m > 0$ (and, thus, $\lambda = 1$ and $c = m$) then, by using formula (2) of the proof of Theorem 6.2.6, we obtain

$$E_x(Z_t^2) = \int y^2 \, dP^x_{Z_t}(y) = T_1(mt)(e_2)(x)$$

$$= T_1(e_2)(x) + \exp(-mt)(x^2 - T_1(e_2)(x))$$

$$= x + \exp(-mt)(x^2 - x) \qquad (6.3.47)$$

where $e_2(x) = x^2$ $(x \in [0, 1])$, so that

$$\mathrm{Var}_x(Z_t) = (1 - \exp(-mt))x(1 - x). \qquad (6.3.48)$$

For an arbitrary function α we can use the above identity and formula (5) of the proof of Theorem 6.2.6 to obtain

$$(1 - \exp(-mt))x(1 - x) \leq \mathrm{Var}_x(Z_t) \leq x(1 - x), \qquad (6.3.49)$$

where $m = \min\{\lambda(x)|0 \leq x \leq 1\}$.
In particular,

$$\lim_{t \to \infty} \mathrm{Var}_x(Z_t) = x(1 - x). \qquad (6.3.50)$$

We also point out that in this case $\mu_x^{T_1} = (1 - x)\varepsilon_0 + x\varepsilon_1$ and hence from (6.2.35) we obtain

$$\lim_{t \to \infty} P_t(x, \cdot) = (1 - x)\varepsilon_0 + x\varepsilon_1 \quad \text{weakly} \qquad (6.3.51)$$

for each $x \in [0, 1]$, i.e.,

$$\lim_{t \to \infty} Z_t = Z_\infty^x \quad \text{in distribution (with respect to } P^x), \qquad (6.3.52)$$

where Z_∞^x denotes a Bernoulli random variable on Ω with parameters 0 and x. (We recall that $(P_t)_{t \geq 0}$ denotes the normal transition function of the process (6.3.45)).

Another aspect we want to point out is the fact that our process (6.3.45) can be interpreted as a limit of random walks.
More precisely, for a given $n \geq 1$, set

$$\Sigma_n := \left\{ \frac{h}{n^k} \middle| h, k \in \mathbb{N}_0, h \leq n^k \right\} \qquad (6.3.53)$$

and for every $x \in \Sigma_n$ consider the Borel measure $\upsilon_{n, \lambda, x}$ on Σ_n defined by

$$\upsilon_{n, \lambda, x} := \sum_{k=0}^{n} \sum_{h=0}^{k} \binom{n}{k} \binom{k}{h} \lambda(x)^k (1 - \lambda(x))^{n-k} x^h (1 - x)^{k-h} \varepsilon_{h/n + (1 - k/n)x}. \quad (6.3.54)$$

Thus,

$$\text{Supp}(\upsilon_{n, \lambda, x}) = \begin{cases} \left\{ \frac{h}{n} \middle| 0 \leq h \leq n \right\}, & \text{if } \lambda(x) = 1, \\[2mm] \left\{ \frac{h}{n} + \frac{n-k}{n} x \middle| 0 \leq k \leq n, 0 \leq h \leq k \right\} & \\[2mm] \quad = \left\{ \frac{p}{n} + \frac{q}{n} x \middle| p, q \in \mathbb{N}_0, p + q \leq n \right\}, & \text{if } \lambda(x) < 1. \end{cases} \qquad (6.3.55)$$

In any case $\text{Supp}(\upsilon_{n, \lambda, x}) \subset \Sigma_n$.
Consider the random walk

$$\left\{ \Sigma_n, \frac{1}{n}, (\upsilon_{n, \lambda, x})_{x \in \Sigma_n} \right\}, \qquad (6.3.56)$$

having Σ_n as support, $\dfrac{1}{n}$ basic time-interval and $(\upsilon_{n, \lambda, x})_{x \in \Sigma_n}$ as one-step transition probability distributions. This means that, if a particle is at a position $x \in \Sigma_n$ at time $\dfrac{k}{n}$ $(k \in \mathbb{N}_0)$, then it remains at x during the interval $\left[\dfrac{k}{n}, \dfrac{k+1}{n} \right[$ and at time

$\dfrac{k+1}{n}$ it jumps so that the probability that it goes to any Borel subset B of $[0,1]$
is $\upsilon_{n,\lambda,x}(B)$ (see Doob [1953, p. 190 ff.] for more details).

The family $(\upsilon_{n,\lambda,x})_{x \in \Sigma_n}$ determines a linear operator $T_{n,\lambda} \colon \mathscr{B}_0(\Sigma_n) \to \mathscr{B}_0(\Sigma_n)$ defined by

$$T_{n,\lambda}(f)(x) := \int_{\Sigma_n} f \, d\upsilon_{n,\lambda,x} \tag{6.3.57}$$

for every $f \in \mathscr{B}_0(\Sigma_n)$ and $x \in \Sigma_n$ (see (1.6.41) for the definition of $\mathscr{B}_0(\Sigma_n)$).

The operator $T_{n,\lambda}$ is called the *one-step transition operator* of the random walk
and clearly

$$T_{n,\lambda}(f_{|\Sigma_n}) = L_{n,\lambda}(f)_{|\Sigma_n} \quad \text{for every } f \in \mathscr{C}([0,1]). \tag{6.3.58}$$

Note that, if a measure $\mu \in \mathscr{M}_1^+(\Sigma_n)$ gives the distribution of the initial position
of the particle, then the Borel probability measure $T_{n,\lambda}(\mu)$ on Σ_n defined by

$$T_{n,\lambda}(\mu)(B) := \int_{\Sigma_n} \upsilon_{n,\lambda,x}(B) \, d\mu(x), \quad (B \in \mathfrak{B}(\Sigma_n)), \tag{6.3.59}$$

gives the distribution for its position after the first jump and, for $m \geq 2$, $T_{n,\lambda}^m(\mu)$
is the distribution for its position after the m-th jump, where

$$T_{n,\lambda}^m(\mu) := T_{n,\lambda}(T_{n,\lambda}^{m-1}(\mu)). \tag{6.3.60}$$

We note that the sequence $(\Sigma_n)_{n \in \mathbb{N}}$ becomes dense in $[0,1]$, i.e., every open
subset of $[0,1]$ intersects Σ_n for sufficiently large n (see Trotter [1958, p. 892]).

Moreover, on account of (6.2.19) and (6.3.58), it is easy to show that

$$T_\lambda(t) = \lim_{n \to \infty} T_{n,\lambda}^{[nt]} \quad \text{for every } t \geq 0, \tag{6.3.61}$$

the type of convergence being that introduced by Trotter [1958, Section 2, Example (2)] (see also Pazy [1983, Section 3.6]).

Accordingly, by a result of Trotter [1958, Section 6, p. 907] we conclude that

> *The sequence of random walks (6.3.56) converges*
> *to the Markov process (6.3.45).* (6.3.62)

This means that, if a measure $\mu \in \mathscr{M}_1^+([0,1])$ gives the initial position of the
particle, then there exists a sequence $(\mu_n)_{n \in \mathbb{N}}$ in $\mathscr{M}_1^+([0,1])$ such that the support

of every μ_n is contained in Σ_n, $\lim_{n\to\infty} \mu_n = \mu$ weakly and for every $t \geq 0$

$$P_t(\mu) = \lim_{n\to\infty} T_{n,\lambda}^{[cnt]}(\mu_n) \quad \text{weakly,} \qquad (6.3.63)$$

where $P_t(\mu)$ is defined by (1.6.48).

We also point out that besides formula (6.2.34), we have at our disposal another representation formula for the transition function $(P_t)_{t\geq 0}$, or equivalently, of the semigroup $(T_1(mt))_{t\geq 0}$ (see (1.6.42) and (1.6.46)), provided $\alpha(x) = m\dfrac{x(1-x)}{2}$ ($x \in [0,1]$).

In fact, Karlin and Ziegler [1970] showed that, for every $f \in \mathscr{C}([0,1])$ and $x \in [0,1]$

$$T_1(mt)(f)(x) = T_1(f)(x) + \int_0^1 p(mt,x,y)(f(y) - T_1(f)(y))\,dy, \quad (6.3.64)$$

where

$$p(t,x,y) := \frac{1}{y(1-y)} \sum_{n=2}^{\infty} \exp\left(-\frac{n(n-1)}{2}t\right)(n-1)n(2n-1)Q_n(x)Q_n(y) \quad (6.3.65)$$

and

$$Q_n(x) := x(x-1)P_{n-2}^{(1,1)}(1-2x), \qquad (6.3.66)$$

$P_{n-2}^{(1,1)}(x)$ being the Jacobi polynomial of parameters $(1,1)$ normalized to be 1 at $x = 1$ (for more details see also Fichera [1992], Martini and Boer [1974], Da Silva [1985]).

Finally we briefly discuss the connection between the initial value problem (6.3.40) and a stochastic model from genetics that was proposed by Moran [1958] to study the fluctuations of gene frequency under the influence of mutation and selection (see also Feller [1951] and Kimura [1957]). Here we closely follow Karlin and McGregor [1962].

The formulation of the model is as follows.

Consider an infinite population of gametes which are either of type 0 or 1.

For a given $t \geq 0$, Z_t denotes the frequency of 0-gametes in the population at time t.

A change of state occurs when a single individual reproduces and is replaced by a new individual. The total population size remains constant.

It is assumed that the probability that the state changes during the time interval $(t, t + dt)$ is $a\,dt + o(dt)$, $(a > 0)$. Furthermore, it is assumed that the probability of two or more matings occurring in a time interval of length dt is $o(dt)$.

It is postulated that the matings occur according to the following procedures.

At the occurrence of mating an individual from the population is selected at random to be fertilized. Another individual is then selected to fertilize (self-fertilization is allowed). However, after the mating, the type of the new individual is the same as of the fertilizer.

Finally, it is assumed that a gamete can mutate type with probability 0.

The above mentioned process is of diffusion type and it can be described by the initial value problem (6.3.40) with

$$\alpha(x) = m\frac{x(1 - x)}{2}, \quad (x \in [0, 1], m > 0) \tag{6.3.67}$$

(for details, see Karlin and McGregor [1962]).

In this case the transition function $(P_t)_{t \geq 0}$ that is represented by (6.2.34) or (6.3.63) describes the distribution of the frequency of gametes of type 0 at time $t \geq 0$.

Formulas (6.3.46) and (6.3.48) give the expected value and the variance of this frequency at time $t > 0$, if at the initial time the frequency is x.

Formula (6.3.51) (or (6.3.52)) gives some information about the asymptotic behavior of these frequences.

In the next subsection we shall discuss the approximation process and the Feller semigroup associated with the Dirichlet operator (3.3.15) on balls of $\mathbb{R}^p$.

Although some results also hold for an arbitrary convex compact subset of $\mathbb{R}^p$, we restrict ourselves to consider only the case of balls.

6.3.9 Approximation processes and degenerate diffusion equations on balls of $\mathbb{R}^p$.
Let $\Omega = B(x_0, r)$ be the open ball of center x_0 and radius $r > 0$ in $\mathbb{R}^p$ and consider the Dirichlet operator $T: \mathscr{C}(\bar{\Omega}) \to \mathscr{C}(\bar{\Omega})$ (see (3.3.15)). Thus, for every $f \in \mathscr{C}(\bar{\Omega})$, $T(f)$ denotes the unique solution of the Dirichlet problem

$$\begin{cases} \Delta v = 0 \quad \text{on } \Omega, \quad v \in \mathscr{C}(\bar{\Omega}) \cap \mathscr{C}^2(\Omega), \\ v_{|\partial\Omega} = f_{|\partial\Omega}. \end{cases} \tag{6.3.68}$$

In Corollary 3.3.6, we have shown that T is a positive projection with range

$$H = H(\Omega) \, (= \{u \in \mathscr{C}(\bar{\Omega}) \cap \mathscr{C}^2(\Omega) | \Delta u = 0 \text{ on } \Omega\}). \tag{6.3.69}$$

Consequently, the Dirichlet operator satisfies condition (6.1.3) and (6.1.4). Moreover, by virtue of Proposition 2.6.2, we also have

$$\partial_H^+ \bar{\Omega} = \partial\Omega = \partial_e \bar{\Omega}. \tag{6.3.70}$$

By the Poisson formula for the solution of the Dirichlet problem for a ball, the Dirichlet operator $T: \mathscr{C}(\overline{\Omega}) \to \mathscr{C}(\overline{\Omega})$ is also given by

$$
Tf(x) = \begin{cases} \dfrac{r^2 - \|x - x_0\|^2}{r\sigma_p} \displaystyle\int_{\partial\Omega} \dfrac{f(z)}{\|z - x\|^p} d\sigma(z), & \text{if } \|x - x_0\| < r, \\[4mm] f(x), & \text{if } \|x - x_0\| = r, \end{cases}
\tag{6.3.71}
$$

for every $f \in \mathscr{C}(\overline{\Omega})$ and $x \in \overline{\Omega}$ (see (3.3.16)), where σ_p denotes the surface area of the unit sphere of $\mathbb{R}^p$ and σ is the surface area on $\partial\Omega$.

Consequently, we can explicitly describe the Bernstein-Schnabl, Stancu-Schnabl and Lototsky-Schnabl operators corresponding to the projection T (see (6.1.10), (6.1.23) and (6.1.52)).

Indeed, for every $f \in \mathscr{C}(\overline{\Omega})$, $x \in B(x_0, r)$ and $n \geq 1$, we have

$$
B_n(f)(x)
$$

$$
= \left(\frac{r^2 - \|x - x_0\|^2}{r\sigma_p}\right)^n \int_{\partial\Omega} \cdots \int_{\partial\Omega} \frac{f\left(\frac{1}{n}(x_1 + \cdots + x_n)\right)}{\|x_1 - x\|^p \dots \|x_n - x\|^p} d\sigma(x_1)\dots d\sigma(x_n),
\tag{6.3.72}
$$

$$
S_{n,a_n}(f)(x)
$$

$$
= \frac{1}{p_n(a_n)} \sum_{k=1}^{n} \frac{n!}{k!} a_n^{n-k} \left(\frac{r^2 - \|x - x_0\|^2}{r\sigma_p}\right)^k \sum_{|v|_k=n} \frac{1}{v_1 \dots v_k} \int_{\partial\Omega}
$$

$$
\cdots \int_{\partial\Omega} \frac{1}{\|x_1 - x\|^p \dots \|x_k - x\|^p} f\left(\frac{v_1 x_1 + \cdots + v_k x_k}{n}\right) d\sigma(x_1)\dots d\sigma(x_k),
\tag{6.3.73}
$$

where $(a_n)_{n \in \mathbb{N}}$ is a sequence of positive real numbers, and

$$
L_{n,\lambda}(f)(x) = \sum_{k=0}^{n} \binom{n}{k} \lambda(x)^k (1 - \lambda(x))^{n-k} \left(\frac{r^2 - \|x - x_0\|^2}{r\sigma_p}\right)^k \int_{\partial\Omega}
$$

$$
\cdots \int_{\partial\Omega} \frac{1}{\|x_1 - x\|^p \dots \|x_k - x\|^p} f\left(\frac{x_1}{n} + \cdots + \frac{x_k}{n}\right.
$$

$$
\left. + \left(1 - \frac{k}{n}\right)x\right) d\sigma(x_1)\dots d\sigma(x_k),
\tag{6.3.74}
$$

where $\lambda \in \mathscr{C}(\overline{\Omega})$, $0 \leq \lambda \leq 1$, and

$$
B_n(f)(x) = S_{n,a_n}(f)(x) = L_{n,\lambda}(f)(x) = f(x), \quad \text{if } x \in \partial\Omega.
\tag{6.3.75}
$$

To obtain some estimates of the rate of convergence of these approximation processes it suffices to use formulas (6.1.17), (6.1.44) and (6.1.58). However, in this case, we have $\Omega(f, \delta) \leq \omega(f, (r + \|x_0\|^2)\delta)$ for every $f \in \mathscr{C}(\bar{\Omega})$ and $\delta > 0$, by virtue of (5.1.14).

Moreover, denoting by pr_i the i-th projection on $\bar{\Omega}$, it is easy to show that for every $i, j = 1, \ldots, p$,

$$T(\mathrm{pr}_i \mathrm{pr}_j) = \begin{cases} \mathrm{pr}_i \mathrm{pr}_j, & \text{if } i \neq j, \\ \dfrac{1}{p}\left(r^2 - \sum_{h=1}^{p} (\mathrm{pr}_h - \mathrm{pr}_h(x_0))^2 \right) + \mathrm{pr}_i^2, & \text{if } i = j. \end{cases} \tag{6.3.76}$$

Accordingly, after a simple calculation, we obtain

$$Te(x) - e(x) = r^2 - \|x - x_0\|^2, \tag{6.3.77}$$

where $e(x) := \|x\|^2$, $(x \in \bar{\Omega})$, and hence Proposition 6.1.12 implies that

$$\|B_n(f) - f\| \leq Mr\,\frac{1}{\sqrt{n}}, \tag{6.3.78}$$

$$\|S_{n, a_n}(f) - f\| \leq Mr\,\sqrt{\frac{1 + na_n}{n(1 + a_n)}} \tag{6.3.79}$$

and

$$\|L_{n, \lambda}(f) - f\| \leq Mr\,\sqrt{\frac{\|\lambda\|}{n}} \tag{6.3.80}$$

provided $f \in \mathrm{Lip}_M 1$.

Dirichlet operator (6.3.71) satisfies condition (6.2.18) because the solution of Dirichlet problem (6.3.68) corresponding to a polynomial is a polynomial of the same degree (this result is due to Brelot and Choquet [1954, Theorem 6]; see also Armitage [1972, Theorems 2 and 4]). Therefore we can apply Theorem 6.2.6.

To describe the corresponding differential operator W_T (and hence the operator A_1) defined by (6.2.11), we shall use (6.3.76).

In fact, recalling (6.2.12) we have

$$a_{ij}(x) = \begin{cases} 0, & \text{if } i \neq j, \\ \dfrac{r^2 - \|x - x_0\|^2}{p}, & \text{if } i = j \end{cases} \tag{6.3.81}$$

for every $x \in \bar{\Omega}$, so that

$$A_1(u)(x) = W_T(u)(x) = \frac{r^2 - \|x - x_0\|^2}{2p} \Delta u \qquad (6.3.82)$$

for every $u \in \mathscr{C}^2(\bar{\Omega})$.

As in the case of the unit interval, it is possible to describe exactly the generator A_1 and its domain $D(A_1)$ by virtue of a result of Blümlinger [1987].

To this end we have to recall the notion of generalized Laplacian on Ω (see, e.g., Helms [1969, p. 65]).

Given $u \in \mathscr{C}(\Omega)$, the *generalized Laplacian of* u is the function $\tilde{\Delta}u$ defined on Ω by

$$\tilde{\Delta}u(x) := \lim_{\delta \to 0^+} 2p \frac{L(u; x, \delta) - u(x)}{\delta^2} \qquad (x \in \Omega), \qquad (6.3.83)$$

whenever the limit on the right-hand side exists.

Here we have set

$$L(u; x, \delta) := \frac{1}{\sigma_p \delta^{p-1}} \int_{\partial B(x, \delta)} u \, d\sigma. \qquad (6.3.84)$$

Since the Laplacian of a function can be obtained as limit of the averages (6.3.84) (see Helms [1969, p. 64]), we have

$$\tilde{\Delta}u = \Delta u \quad \text{on } \Omega \text{ provided } u \in \mathscr{C}^2(\Omega). \qquad (6.3.85)$$

After these preliminaries and by recalling Satz 6 of Blümlinger [1987], we obtain that

$$D(A_1) = \left\{ u \in \mathscr{C}(\bar{\Omega}) \mid \text{there exists } \tilde{\Delta}u \text{ on } \Omega, \tilde{\Delta}u \text{ is continuous on } \Omega \text{ and} \right.$$
$$\left. \lim_{x \to \bar{x}} (r^2 - \|x_0 - x\|^2)\tilde{\Delta}u(x) = 0 \text{ for every } \bar{x} \in \partial\Omega \right\} \qquad (6.3.86)$$

and

$$A_1(u)(x) = \frac{r^2 - \|x - x_0\|^2}{2p} \tilde{\Delta}u(x) \qquad (6.3.87)$$

for every $u \in D(A_1)$ and $x \in \bar{\Omega}$.

In this case, however, the solution of the Cauchy problem (6.2.28) associated with A_1 satisfies the properties stated in Theorem 6.2.7 except statement (4).

In fact, Hinkkanen [1988, Example (5.11) in Section 5] showed that, for $p = 2$,

$$T(\text{Lip}_1 1) \not\subset \text{Lip}_c 1 \quad \text{for every } c > 0, \tag{6.3.88}$$

while

$$T(\text{Lip}_M \alpha) \subset \text{Lip}_{C_\alpha} \alpha \quad \text{for every } 0 < \alpha < 1, \tag{6.3.89}$$

where C_α is a suitable constant which depends on α.

Similar results have also been proved for higher dimension (see also Rubel, Shields and Taylor [1975, p. 34] or Dankel [1979, p. 519]).

We also point out that in the present setting $\text{Supp}(\mu_x^T) = \partial\Omega$ provided $x \in \Omega$.

So, on account of (6.1.49), we see that for every $\lambda \in \mathscr{C}(\overline{\Omega})$, $0 \le \lambda \le 1$, $\lambda \ne 0$, the (T, λ)-convex functions are exactly the convex functions.

Furthermore, if $f \in \mathscr{C}(\overline{\Omega})$ is such a function, then the sequence $(L_{n,\lambda}(f))_{n \ge 1}$ is decreasing.

Another connection we want to show is that concerning subharmonic functions and T-convex functions.

In fact, if a function $u \in \mathscr{C}(\overline{\Omega})$ is subharmonic on Ω, then, for every $z \in \overline{\Omega}$ and $\alpha \in [0, 1]$ the function $u_{z,\alpha}$ defined by (6.1.1) is subharmonic too and hence $u_{z,\alpha} \le T(u_{z,\alpha})$ (see, e.g., Helms [1969, Theorem 4.11]).

Accordingly, u is T-convex (see (6.1.64)) so that we can apply Theorem 6.1.13 to obtain

$$u \le L_{n,\lambda}(u) \le B_n(u) \le T(u) \tag{6.3.90}$$

for every $n \ge 1$.

In this case, Theorem 6.1.17 gives, in fact, the maximum principle for subharmonic functions (see, e.g., Helms [1969, Corollary 4.3]).

We conclude this section by studying the tensor products of positive projections satisfying conditions (6.1.3), (6.1.4) and (6.2.18). This procedure allows us to construct in a natural way Bernstein-Schnabl, Stancu-Schnabl and Lototsky-Schnabl operators on other convex compact subsets of $\mathbb{R}^p$, such as, for example, the hypercube $[0, 1]^p$, the polydisk $\mathbb{D}^p$ in $\mathbb{R}^{2p}$, the cylinder in $\mathbb{R}^3$ and so on.

6.3.10 Positive approximation processes and Feller semigroups on a product space.

Consider a finite family $(K_i)_{1 \le i \le p}$ of convex compact subsets and suppose that every K_i is contained in some $\mathbb{R}^{p_i}$ and has non-empty interior. For every $i = 1, \ldots, p$ let $T_i \colon \mathscr{C}(K_i) \to \mathscr{C}(K_i)$ be a positive projection satisfying conditions (6.1.3), (6.1.4) and (6.2.18).

Put $K = \prod_{i=1}^{p} K_i$ and consider the tensor product

$$T := \bigotimes_{i=1}^{p} T_i \colon \mathscr{C}(K) \to \mathscr{C}(K) \tag{6.3.91}$$

of the family $(T_i)_{1 \leq i \leq p}$ (see the final part of Section 1.2). By virtue of Proposition 3.3.7, T is a positive projection on the product space K as well.

Moreover, by (3.3.19) the range H of T is given by

$$H = \overline{H_1 \otimes \cdots \otimes H_p} = \underset{i=1}{\overset{p}{\mathrm{M}}} H_i, \tag{6.3.92}$$

(see (2.6.24) and (2.6.25)) and therefore T satisfies conditions (6.1.3) and (6.1.4).

Furthermore

$$\partial_H^+ K = \prod_{i=1}^{p} \partial_{H_i}^+ K_i \tag{6.3.93}$$

(see Proposition 2.6.8).

For every $i = 1, \ldots, p$ and $n \geq 1$, let $B_{n,i}$ be the n-th Bernstein-Schnabl operator on K_i associated with the projection T_i; then, from the commutativity and associativity of the tensor product of measures (see Proposition 1.2.5), it follows that the n-th Bernstein-Schnabl operator on K associated with the projection T is the tensor product of the n-th Bernstein-Schnabl operators corresponding to each T_i, i.e.,

$$B_n = \bigotimes_{i=1}^{p} B_{n,i}. \tag{6.3.94}$$

Now, we want to investigate if Theorem 6.2.6 also holds for the projection T. To this end, for every $i = 1, \ldots, p$, denote by $(T_{1,i}(t))_{t \geq 0}$ the semigroup associated to the sequence $(B_{n,i})_{n \in \mathbb{N}}$ of Bernstein-Schnabl operators on K_i and let $A_{1,i} \colon D(A_{1,i}) \to \mathscr{C}(K_i)$ be its generator. Finally, consider the subspace

$$\bigotimes_{i=1}^{p} D(A_{1,i}) \tag{6.3.95}$$

generated by $\left\{ \bigotimes_{i=1}^{p} f_i \,\middle|\, f_i \in D(A_{1,i}), i = 1, \ldots, p \right\}$ (see (2.6.24)).

Then we have the following result.

6.3.11 Theorem. *The positive projection* $T: \mathscr{C}(K) \to \mathscr{C}(K)$ *defined by* (6.3.91) *satisfies condition* (6.2.18) *and the semigroup* $(T_1(t))_{t \geq 0}$ *associated with the sequence* $(B_n)_{n \in \mathbb{N}}$ *of Bernstein-Schnabl operators on the product space* K *is given by*

$$T_1(t) = \bigotimes_{i=1}^{p} T_{1,i}(t) \quad \text{for every } t \geq 0. \tag{6.3.96}$$

Moreover, if we denote by $A_1: D(A_1) \to \mathscr{C}(K)$ *the generator of the semigroup* $(T_1(t))_{t \geq 0}$, *then*

$$\bigotimes_{i=1}^{p} D(A_{1,i}) \quad \text{is a core for } A_1 \tag{6.3.97}$$

and A_1 *is the closure of the operator* $\sum_{i=1}^{p} \bigotimes_{j=1}^{p} A_{1,i,j}$ *defined on* $\bigotimes_{i=1}^{p} D(A_{1,i})$, *where*

$$A_{1,i,j} := \begin{cases} I_{\mathscr{C}(K_j)} & \text{if } j \neq i, \\ A_{1,i} & \text{if } j = i. \end{cases} \tag{6.3.98}$$

Proof. For every $i, j = 1, \ldots, p$ and $f_i \in A(K_i)$ and $g_j \in A(K_j)$, we have

$$T((f_i \circ \mathrm{pr}_i)(g_j \circ \mathrm{pr}_j)) = (T_i(f_i) \circ \mathrm{pr}_i)(T_j(g_j) \circ \mathrm{pr}_j) = (f_i \circ \mathrm{pr}_i)(g_j \circ \mathrm{pr}_j) \in A_2(K),$$

if $i \neq j$, whereas, if $i = j$,

$$T((f_i \circ \mathrm{pr}_i)(g_i \circ \mathrm{pr}_i)) = T_i(f_i g_i) \circ \mathrm{pr}_i \in A_2(K),$$

because every T_i satisfies condition (6.2.18).

Now every function $f \in A(K)$ is of the form $\alpha_0 + \sum_{i=1}^{p} \alpha_i(f_i \circ \mathrm{pr}_i)$ with $f_i \in A(K_i)$, $i = 1, \ldots, p$, and $\alpha_0, \alpha_1, \ldots, \alpha_p \in \mathbb{R}$, and therefore $T(A_2(K)) \subset A_2(K)$.

By induction it is easy to show that $T(A_m(K)) \subset A_m(K)$ for every $m \geq 1$ and hence T satisfies condition (6.2.18).

Now denote by A_0 the closure of the operator $\sum_{i=1}^{p} \bigotimes_{j=1}^{p} A_{1,i,j}$ defined on $\bigotimes_{i=1}^{p} D(A_{1,i})$ by (6.3.98). Then A_0 is the generator of the semigroup $(S(t))_{t \geq 0}$ with

$$S(t) = \bigotimes_{i=1}^{p} T_i(t) \text{ for every } t \geq 0.$$

Moreover $\bigotimes_{i=1}^{p} D(A_{1,i})$ is a core for A_0 (see, e.g., R. Nagel (Ed.) [1986, Section A-I-3.7, p. 23]).

Our result will follow if we show that $A_0 = A_1$ (and hence $S(t) = T_1(t)$ for every $t \geq 0$). To this end it is enough to show that A_0 and A_1 coincide on a core of A_1 contained in $D(A_0)$.

Indeed, from Theorem 6.2.6, (4), we already know that $A_\infty(K)$ is a core for A_1.

Now, for every $i = 1, \ldots, p$, consider $m_i \geq 1$ and $f_i \in A_{m_i}(K_i)$ and put $f :=$
$\bigotimes_{i=1}^{p} f_i \in \bigotimes_{i=1}^{p} D(A_{1,i})$. We have

$$\lim_{n \to \infty} n(B_n(f) - f)$$

$$= \lim_{n \to \infty} n \left(\bigotimes_{i=1}^{p} B_{n,i}(f_i) - \bigotimes_{i=1}^{p} f_i \right)$$

$$= \lim_{n \to \infty} n \left(\sum_{i=1}^{p} B_{n,1}(f_1) \otimes \cdots \otimes B_{n,i-1}(f_{i-1}) \otimes (B_{n,i}(f_i) - f_i) \otimes f_{i+1} \otimes \cdots \otimes f_p \right)$$

$$= \sum_{i=1}^{p} \lim_{n \to \infty} B_{n,1}(f_1) \otimes \cdots \otimes B_{n,i-1}(f_{i-1}) \otimes (n(B_{n,i}(f_i) - f_i)) \otimes f_{i+1} \otimes \cdots \otimes f_p$$

$$= \sum_{i=1}^{p} f_1 \otimes \cdots \otimes f_{i-1} \otimes A_{1,i}(f_i) \otimes f_{i+1} \otimes \cdots \otimes f_p = A_0(f).$$

On account of (6.2.24) this reasoning shows that $A_1(f) = A_0(f)$ for every $f \in \bigotimes_{i=1}^{p} A_{m_i}(K_i)$. Since

$$A_m(K) \subset \bigcup_{m_1 + \cdots + m_p \leq m} \bigotimes_{i=1}^{p} A_{m_i}(K_i) \subset \bigotimes_{i=1}^{p} D(A_{1,i}),$$

we conclude that $A_1 = A_0$ on every $A_m(K)$ and hence on the subspace $A_\infty(K)$ which is included in $\bigotimes_{i=1}^{p} D(A_{1,i}) \subset D(A_0)$. So the proof is complete. $\qquad \square$

Remark. The proof of the above theorem together with Theorem 6.2.5 shows, in fact, that the subspace $\bigotimes_{i=1}^{p} \mathscr{C}^2(K_i) \subset \bigotimes_{i=1}^{p} D(A_{1,i})$ is a core for A_1 and $A_1 = \sum_{i=1}^{p} \bigotimes_{j=1}^{p} A_{1,i,j}$ on $\bigotimes_{i=1}^{p} \mathscr{C}^2(K_i)$ as well.

We explicitly point out that, if every K_i is a finite dimensional simplex, then on every $\mathscr{C}(K_i)$ a natural projection satisfying (6.1.3), (6.1.4) and (6.2.18) is defined, namely the canonical projection (6.3.1).

Thus, according to the results of Section 6.1 and Theorem 6.3.11, on the product space we can construct several natural positive approximation processes and their corresponding Feller semigroups.

The simplest case we want to describe is that when $K_i = [0,1]$ for every $i = 1, \ldots, p$.

6.3.12 Approximation processes and diffusion equations on the hypercube of $\mathbb{R}^p$.

Let $K = [0,1]^p$ be the hypercube of $\mathbb{R}^p$ and for every $i = 1, \ldots, p$ let T_i: $\mathscr{C}([0,1]) \to \mathscr{C}([0,1])$ be the canonical projection defined by (6.3.19). In this case we shall denote by S_p the tensor product of the family $(T_i)_{1 \le i \le p}$. Thus, for every $f \in \mathscr{C}(K)$ and $x = (x_1, \ldots, x_p) \in [0,1]^p$,

$$S_p(f)(x_1, \ldots, x_p) = \sum_{h_1, \ldots, h_p = 0}^{1} \alpha_f(h_1, \ldots, h_p) x_1^{h_1}(1 - x_1)^{1 - h_1} \ldots x_p^{h_p}(1 - x_p)^{1 - h_p},$$

(6.3.99)

where $\alpha_f(h_1, \ldots, h_p) = f(\delta_{h_1 1}, \ldots, \delta_{h_p 1})$ (see Example 3.3.10).

According to Example 3.3.10, the range H of S_p is the subspace of $\mathscr{C}(K)$ generated by the set $\{1\} \cup \left\{ \prod_{i \in J} \mathrm{pr}_i \,\middle|\, J \subset \{1, \ldots, p\} \right\}$ (see (6.3.92)) and

$$\partial_H^+ [0,1]^p = \{(\delta_{h_1 1}, \ldots, \delta_{h_p 1}) | h_1, \ldots, h_p \in \{0, 1\}\}.$$

(6.3.100)

To describe the Bernstein-Schnabl operators on the hypercube $[0,1]^p$, we use (6.3.94) and then, for every $f \in \mathscr{C}([0,1]^p)$, $x = (x_1, \ldots, x_p) \in [0,1]^p$ and $n \ge 1$, we obtain

$$B_n(f)(x)$$

$$= \sum_{h_1, \ldots, h_p = 0}^{n} \binom{n}{h_1} \cdots \binom{n}{h_p} f\left(\frac{h_1}{n}, \ldots, \frac{h_p}{n}\right) x_1^{h_1}(1 - x_1)^{n - h_1} \ldots x_p^{h_p}(1 - x_p)^{n - h_p}.$$

(6.3.101)

More generally, if $P = (p_{nj})_{n \ge 1, j \ge 1}$ is a lower triangular stochastic matrix, then

$$B_{P,n}(f)(x_1, \ldots, x_p)$$

$$= \sum_{h_1, \ldots, h_p = 0}^{n} (\sum f(p_{ni_{1,1}} + \cdots + p_{ni_{1,h_1}}, \ldots, p_{ni_{p,1}}$$

$$+ \cdots + p_{ni_{p,h_p}})) x_1^{h_1}(1 - x_1)^{n - h_1} \ldots x_p^{h_p}(1 - x_p)^{n - h_p},$$

(6.3.102)

where the sum is extended over all subsets $\{i_{1,1}, \ldots, i_{1,h_1}\} \in C(n, h_1), \ldots,$ $\{i_{p,1}, \ldots, i_{p,h_p}\} \in C(n, h_p)$ (for the definition of the sets $C(n, h_1), \ldots, C(n, h_p)$ see (6.3.21)).

Accordingly, from (6.1.24) we obtain

$$S_{n,a_n}(f)(x) = \frac{1}{p_n(a_n)} \sum_{k=1}^{n} \frac{n!}{k!} a_n^{n-k} \sum_{|v|_k=n} \frac{1}{v_1 \cdots v_k} \sum_{h_1,\ldots,h_p=0}^{k} x_1^{h_1}(1-x_1)^{k-h_1}$$

$$\ldots x_p^{h_p}(1-x_p)^{k-h_p} \sum f\left(\frac{v_{i_{1.1}} + \cdots + v_{i_{1.h_1}}}{n}, \ldots, \frac{v_{i_{p.1}} + \cdots + v_{i_{p.h_p}}}{n}\right),$$

$$(6.3.103)$$

where $(a_n)_{n \in \mathbb{N}}$ is a sequence of positive real numbers and the last sum is extended over all subsets $\{i_{1,1}, \ldots, i_{1,h_1}\} \in C(k, h_1), \ldots, \{i_{p,1}, \ldots, i_{p,h_p}\} \in C(k, h_p)$.

Finally, taking (6.1.52) into account, we have

$$L_{n,\lambda}(f)(x) = \sum_{k=0}^{n} \binom{n}{k} \lambda(x)^k (1 - \lambda(x))^{n-k} \sum_{h_1,\ldots,h_p=0}^{k} \binom{k}{h_1} \cdots \binom{k}{h_p}$$

$$\times f\left(\frac{h_1}{n} + \left(1 - \frac{k}{n}\right)x_1, \ldots, \frac{h_p}{n} + \left(1 - \frac{k}{n}\right)x_p\right)$$

$$\times x_1^{h_1}(1-x_1)^{k-h_1} \ldots x_p^{h_p}(1-x_p)^{k-h_p}, \qquad (6.3.104)$$

where $\lambda \in \mathscr{C}([0,1]^p)$, $0 \leq \lambda \leq 1$.

In particular we reobtain Bernstein operators which have been introduced in 5.2.12 by probabilistic methods.

In our case we have

$$S_p(\mathrm{pr}_i\mathrm{pr}_j) = \begin{cases} \mathrm{pr}_i\mathrm{pr}_j, & \text{if } i \neq j, \\ \mathrm{pr}_i, & \text{if } i = j \end{cases} \qquad (6.3.105)$$

for every $i, j = 1, \ldots, p$, so that

$$S_p(e)(x) - e(x) = \sum_{i=1}^{p} x_i(1 - x_i) \qquad (6.3.106)$$

for every $x = (x_1, \ldots, x_p) \in [0,1]^p$ (where $e(x) := \|x\|^2$).

So estimates (6.3.15)–(6.3.17) as well as estimates (6.1.17), (6.1.44) and (6.1.58) hold for these approximation processes too. (Note that, again, $\Omega(f, \delta) \leq \omega(f, \delta)$ for $\delta > 0$).

Moreover the canonical projection on $[0,1]$ satisfies condition (6.2.18) and therefore we can apply Theorem 6.3.11. According to Remark to Theorem 6.3.11, for every $u \in \bigotimes_{i=1}^{p} \mathscr{C}^2([0,1])$ and $x = (x_i)_{1 \leq i \leq p} \in K$ we have

$$A_1(u)(x) = \left(\sum_{i=1}^{p} \bigotimes_{j=1}^{p} A_{1,i,j}(u) \right)(x) = \sum_{i=1}^{p} \frac{x_i(1-x_i)}{2} \frac{\partial^2 u(x)}{\partial x_i^2}. \qquad (6.3.107)$$

As a matter of fact, by using (6.3.105), (6.2.12) and (6.2.26), it is possible to show that (6.3.107) holds for $u \in \mathscr{C}^2([0,1]^p)$ as well.

A simple calculation shows that $S_p(\mathrm{Lip}_1 1) \subset \mathrm{Lip}_1 1$ provided we equip $[0,1]^p$ with the metric induced by the ℓ^1-norm.

So Corollary 6.1.22 and Theorem 6.2.7 apply to (Bernstein-) Lototsky-Schnabl operators (6.3.104) and to the solution of the Cauchy problem (6.2.28) corresponding to the operator A_1 given by (6.3.107), respectively.

Finally we notice that, according to a result of Rasa [1994], a function $f \in \mathscr{C}([0,1]^p)$ is S_p-convex if and only if it is convex with respect to each variable. Thus for such a function f we have

$$f \leq L_{n,\lambda}(f) \leq B_n(f) \leq S_p(f). \qquad (6.3.108)$$

For $p = 2$ some conditions ensuring the convexity of $B_n(f)$ can be found in a paper of Cavaretta and Sharma [1992].

By using similar methods one could discuss Bernstein- Stancu- and Lototsky-Schnabl operators and their corresponding Feller semigroups on the polydisk $\mathbb{D}^p$ by considering the Dirichlet operator (6.3.71) on the single factors $\mathbb{D}$ (see Altomare [1994, (4.3.6), (4.3.7)]).

As another possibility, one could study the cylinder $\mathbb{D} \times [0,1]$ by considering on $\mathbb{D}$ the Dirichlet operator and on $[0,1]$ the canonical projection (6.3.19) (see Altomare [1989c]).

Notes and references

The examples and the analysis we carried out in this section closely follow the paper of Altomare [1994].

Theorem 6.3.2 and its proof are due to Rasa [1993, Theorem 4.1]; in the finite dimensional case Theorem 6.3.2 is due independently to Dahmen [1991, Theorem 4.9 and (4.18)] and Sauer [1991, Theorem 3]. In the same paper Sauer also showed that, for finite dimensional simplices, the notion of polyhedral convexity is preserved by Bernstein polynomials. Conditions ensuring the convexity of the functions $B_n(f)$ in multidimensional settings have been established by Chang and Davis [1984], Chang and Feng [1984], Dahmen and Micchelli [1988], Cavaretta and Sharma [1992].

Theorem 6.3.5, Corollary 6.3.6 and Proposition 6.3.7 first appear in Altomare [1991c], [1992a].

As explained in Section 6.3, problem (6.3.40) arises from the study of one dimensional Markov processes of diffusion type.

The boundary condition in (6.3.40) which involves the second derivatives is usually called *Wentcel's boundary condition*. It was introduced by Wentcel [1959] to characterize multidimensional diffusion processes as well (see also Sato and Ueno [1965] and Taira [1988], [1991], [1992]).

As a matter of fact, a more general version of problem (6.3.40) has been considered recently by several mathematicians.

More precisely, given continuous real functions α and β on the open interval $]0, 1[$ with α strictly positive, consider the linear operator $A_1^*: D(A_1^*) \to \mathscr{C}([0, 1])$ defined by

$$A_1^*(u) := \alpha u'' + \beta u', \quad (u \in D(A_1^*)),$$

where

$$D(A_1^*) = \left\{ u \in \mathscr{C}([0, 1]) | u \in \mathscr{C}^2(]0, 1[) \text{ and} \right.$$

$$\left. \lim_{x \to 0^+} \alpha(x)u''(x) + \beta(x)u'(x) = \lim_{x \to 1^-} \alpha(x)u''(x) + \beta(x)u'(x) = 0 \right\}.$$

Improving preceding results by Martini [1973], [1975] and Martini and Boer [1974], Clément and Timmermans [1986] showed that the operator A_1^* generates a contraction C_0-semigroup on $\mathscr{C}([0, 1])$ if and only if the following two conditions hold:
(1) $W \in L^1(]0, \frac{1}{2}[)$ or $\int_0^{1/2} W(x) \int_0^x \alpha(s)^{-1} W(s)^{-1} \, ds \, dx = \infty$ or both,
(2) $W \in L^1(]\frac{1}{2}, 1[)$ or $\int_{1/2}^1 W(x) \int_x^1 \alpha(s)^{-1} W(s)^{-1} \, ds \, dx = \infty$ or both,
where $W(x) := \exp(\int_{1/2}^x - \beta(s)\alpha(s)^{-1} \, ds) \, (x \in [0, 1])$.

In particular, if $\beta = 0$, then A_1^* (i.e., our operator A_1 defined by (6.3.37)) always generates a contraction semigroup on $\mathscr{C}([0, 1])$.

A similar problem has been studied by Timmermans [1988] by considering the same differential operator A_1^* defined on the bigger domain $\{u \in \mathscr{C}([0, 1]) | u \in \mathscr{C}^2(]0, 1[)$ and $A_1^*(u) \in \mathscr{C}([0, 1])\}$.

Problem (6.3.40) has been recently studied by different methods also by Fichera [1992]. He considered the differential equation in (6.3.40) and proved existence and uniqueness theorems in various function spaces for a boundary value problem. Thereafter, by means of this boundary value problem, Fichera solved problem (6.3.40).

Several multidimensional versions of problem (6.3.40) have been considered in the literature. In fact some of them are also considered in this section (see in particular Subsection 6.3.9).

In particular many authors studied degenerate elliptic-parabolic equations whose coefficients behave like a power of the distance from the boundary of the domain (see, for instance, Lumer, Redheffer and Walter [1982], [1988], Sato and Ueno [1965], Taira [1988], [1991], [1992], Triebel [1978], Vespri [1986] and the references quoted therein).

In the late eighties a systematic study of nonlinear degenerate parabolic boundary value problems has been also carried out by J.A. Goldstein [1993], Goldstein and Lin [1987a], [1987b], [1989], Lin [1989] as well as by Dorroh and Goldstein [1991], [1993].

Corollary 6.3.8 includes the results obtained by several mathematicians by different methods.

The fact that Bernstein polynomials map continuous convex functions into continuous convex functions was first recognized by Popoviciu [1935].

The monotonicity property (ii) (and hence (iii)) of Corollary 6.3.8 was first studied by Temple [1954] and Aramă [1957] (see also Pólya and Schoenberg [1958]). Among other things, Aramă gave a representation of the differences $B_{n+1}(f) - B_n(f)$ and $B_n(f) - f$ in terms of divided differences.

Later on, Kosmák [1960], [1967] and Moldovan [1962] showed the converse implication (ii) $\Rightarrow$ (i) and, finally, the implication (iii) $\Rightarrow$ (i) was proved by Moldovan [1962].

The monotonicity of the sequence of derivatives of Bernstein polynomials was studied by Stancu [1967], [1979].

Several mathematicians took these results as a starting point for investigating similar properties for other positive approximation processes.

Without pretence of completeness we quote, for instance, the papers of Cheney and Sharma [1964b] and Horová [1968], [1982] as regards the monotonicity of the sequence of Szász-Mirakjan operators, Della Vecchia [1987], [1988] for the monotonicity of the sequence of derivatives of Szász-Mirakjan operators, Baskakov operators and other positive approximation processes, Adell, Badía and de la Cal [1993], Adell and de la Cal [1993a], Adell, de la Cal and San Miguel [1994], M.K. Khan [1991] and R.A. Khan [1980], [1991], who use probabilistic techniques based on martingale-type properties and a conditional version of Jensen's inequality, and Sauer [1992a] for Bernstein-Durrmeyer operators.

The monotonicity and the convexity preserving properties of Stancu operators have been investigated by Stancu [1968a] and Mastroianni and Occorsio [1978], respectively.

Finally we refer to Kocić and Stanković [1984], Kocić and Lacković [1986] and Ziegler [1968] (see also Karlin and Ziegler [1970]) for a general study concerning positive approximation processes which verify property (iii) of Corollary 6.3.8 or which preserve the convexity.

The nice property that the spaces of Hölder (Lipschitz, respectively) continuous functions on [0, 1] are invariant under Bernstein operators was first discovered by Lindvall [1982] (Hájek [1965], respectively) who used probabilistic methods. Afterwards Brow, Elliot and Paget [1987] provided a simpler proof.

As we pointed out in the final notes of Section 6.1 and in the Remark to Corollary 6.3.8, the corresponding results for Lototsky-Schnabl operators (associated with a constant $\lambda \in \,]0,1]$) and for Stancu operators are due to Rasa [1993] and Della Vecchia [1989] (see also Della Vecchia and Rasa [1993]), respectively.

In several recent papers these kinds of results have been generalized to other positive approximation processes. See, for instance, Anastassiou, Cottin and Gonska [1991], Della Vecchia [1989], Della Vecchia and Rasa [1993], Kratz and Stadtmüller [1988].

Adell and de la Cal [1993a], Adell, de la Cal and San Miguel [1994], Cismasiu [1987b] and Khan and Peters [1989] studied similar problems by using a probabilistic approach.

The probabilistic consideration we developed in Subsection 6.3.4 seems to be new except the discussion of the genetic model of Moran that we took from Karlin and McGregor [1962].

It seems to be interesting to investigate other models of fluctuations of gene frequences which are governed by a diffusion equation like (6.3.40) where α has the general form (6.3.44) rather than (6.3.67).

The fact that the Markov process (6.3.45) can be represented as 'a limit' of the random walks (6.3.56) generalizes a similar result obtained by Trotter [1958] in the case where α is given by (6.3.67).

Subsection 6.3.9 is largely based on the results of Blümlinger [1987] and Altomare [1989a] (see also Campiti [1991c]), although some results, such as estimates (6.3.78)–(6.3.80), seem to be new.

For Bernstein-Schnabl operators inequalities (6.3.90) were obtained by Blümlinger [1987, Lemma 1].

Finally, the tensor product of positive projections has been considered by Altomare [1991b] (see also [1989c]), to whom Theorem 6.3.11 is due.

Similar results together with the corresponding Theorem 6.2.6 (for $\lambda = 1$) have been obtained by Dahmen and Micchelli [1990] in the case of a product of finite dimensional simplices. In this setting they also obtain (6.1.72).

The case of the hypercube $[0,1]^p$ was also studied by Altomare [1989a] by a more direct method.

We also mention the paper of Stancu [1970], where a more general extension of Stancu operators are investigated.

Very recently Rasa [1994] studied the notion of T-convexity and the property of preserving spaces of Hölder continuous functions in the general framework of product spaces and for the tensor product of positive projections.

In particular he showed that projection (6.3.91) maps $\text{Lip}_1 1$ in $\text{Lip}_1 1$ provided the single factor projections satisfy a similar property.

Appendix A

Korovkin-type approximation theory on commutative Banach algebras

by Michael Pannenberg

By the well-known Gelfand representation theory, a commutative complex Banach algebra A with a non trivial structure space Δ_A of all its maximal regular ideals, respectively characters, may be continuously represented as a subalgebra of the algebra $\mathscr{C}_0(\Delta_A)$ of all continuous complex-valued functions on the locally compact Hausdorff space Δ_A.

Even if this representation is in general neither algebraically faithful nor topologically faithful (i.e., a homeomorphism onto its image), it often allows to generalize a theory valid for complex $\mathscr{C}_0(X)$-spaces (X locally compact Hausdorff space) to a more general theory valid for the class of all commutative Banach algebras with non trivial structure space.

This is the case for Korovkin-type approximation theory: The transfer from real $\mathscr{C}(X)$-spaces (X compact Hausdorff space) to complex $\mathscr{C}(X)$-spaces (= complex commutative unital C^*-algebras) by Briem [1979], Altomare [1981] and Altomare and Boccaccio [1982] and then to commutative unital Banach algebras by Altomare [1981], [1982a], [1984a], [1986] served as a starting point for the development of a theory of Korovkin-type approximation on different classes of commutative complex Banach algebras. As predecessors, we also quote the pioneering papers of Choda and Echigo [1963], Nakamoto and Nakamura [1965], Arveson [1970], Limaye and Shirali [1976], Limaye and Namboodiri [1979], [1986], Priestley [1976] and Takahasi [1979]. The main interest has been concentrated on unital algebras and on algebras possessing a continuous symmetric involution as well as a bounded approximate identity; however, it is possible to develop the theory for a general commutative complex Banach algebra with non-trivial structure space, as carried out in Pannenberg [1992b].

It is the purpose of the present appendix to indicate some of the possible generalizations of Korovkin-type theorems to complex commutative Banach algebras and to convince the reader that these generalizations are not generalizations just for their own sake but contribute both to our understanding of Korovkin-type approximation theory and commutative Banach algebra theory.

To this end, we refrain from giving a complete and detailed survey on the research carried out in this direction, just referring the reader to Sections 7.1–7.7 of the subject classification of Appendix D, and concentrate instead on (perhaps) the two most prominent examples of commutative Banach algebras different from $\mathscr{C}_0(X)$, namely the disk algebra $\mathscr{A}(\mathbb{D})$ and any group algebra $L^1(G)$ (G a locally compact abelian group).

These two examples, which are relatively well understood, allow to give a concrete picture of the possible generalizations and are very interesting in their own right, since in both cases the study of Korovkin-type approximation problems sheds some light on the structure of these algebras.

Since the Gelfand representation of both algebras is well-known in a very concrete form, the study of Korovkin-type approximation problems on both algebras even adds to our understanding of Korovkin-type approximation theory on $\mathscr{C}(\mathbb{D})$ (respectively, $\mathscr{C}_0(\hat{G})$) by answering the question if one can find a finite universal Korovkin subset for $\mathscr{C}(\mathbb{D})$ (respectively, $\mathscr{C}_0(\hat{G})$) consisting of functions analytic on the interior of $\mathbb{D}$ (respectively, having an absolutely convergent Fourier transform on G).

The same question may be asked for differentiable functions, and our next example deals with this case, thereby indicating that there are also interesting generalizations to commutative topological algebras which may not be endowed with a topologically equivalent norm.

We also consider an example that in some sense shares essential features with both main examples discussed in this appendix, namely the algebra of generalized analytic functions: There is a group in the background, and some sort of analyticity plays an important role.

But Korovkin-type approximation theory does not only lead to a deeper understanding of some specific though important examples of commutative Banach algebras: It even sheds some light on general results in Banach algebra theory like the Gleason-Kahane-Zelazko theorem or a problem of Bonsall and Duncan solved by the author in Pannenberg [1986].

This appendix is organized as follows:

In the first section, we give a brief survey on the main theorems on (universal) Korovkin-type approximation theory in commutative Banach algebras. The theory is developed for unital as well as for non-unital algebras, and we try to point out the differences as well as the common features of both settings. The non-unital setting is generally developed under the additional assumption of the existence of a symmetric involution and a bounded approximate identity, which makes life easier. However, most of the results also remain true without this additional assumption, and we prefer to give here a general treatment in the spirit of the first and fourth section of our paper [1992], whose results are slightly extended.

The second section studies commutative group algebras as well as commutative Beurling algebras. Besides describing the Korovkin theory of these algebras, we show precisely how to characterize structural properties of the underlying group (respectively, the underlying weight function) as well as structural properties of the associated group algebra (respectively, Beurling algebra) by a purely approximation-theoretic property, namely the existence of a finite universal Korovkin subset in the group (respectively, Beurling) algebra.

In the third section, we show that approximate normality of the algebra as well as metrizability, separability and finite dimensionality of its structure space are necessary conditions for the existence of a finite universal Korovkin subset. As a consequence, algebras like the polydisk algebras, which have analytic structure in their spectrum, do not possess any finite universal Korovkin subset. However, we give a method to calculate universal Korovkin closures of finite test sets in these algebras. The section closes with a discussion of algebras of infinitely differentiable functions, for which we give an explicit method of constructing a finite universal Korovkin subset.

In the fourth section, we concentrate on algebras of generalized analytic functions and their generalizations, namely uniform algebras generated by inner functions.

The fifth and final section contains two results indicating how Korovkin-type approximation theory may serve to shed some light on theorems in Banach algebra theory which apparently are very far from approximation theory: One is a characterization of those commutative unital Banach algebras for which each character is an extreme spectral state, the other deals with generalizations of the Gleason-Kahane-Zelazko theorem.

Nearly everything in this appendix is published somewhere in the mathematical literature, so that proofs of most of the statements are not given here. Some of the results, however, did not appear in the literature in the form we decided to display them here—in this case, we sketch a proof. The results of the fifth section are published for the first time in this appendix.

Finally, the author would like to thank George Maltese for introducing him to the subject some years ago, as well as Francesco Altomare and Michele Campiti for giving to him the possibility of writing this appendix.

A.1 Universal Korovkin-type approximation theory on commutative Banach algebras

The starting point of the development of a Korovkin-type approximation theory in any specific setting is the description of the class of operators we want to consider in our approximation procedures.

Let A be a commutative (complex) Banach algebra. We will denote by Δ_A the *spectrum* of A, which is its space of maximal regular ideals (or non-trivial characters) endowed with the Gelfand (= relative weak*) topology, and by ρ_A its spectral radius.

For the basic theory of Banach algebras the reader is referred for instance to Bonsall and Duncan [1973] and Rickart [1966].

For unital algebras, the natural generalization of unital positive linear operators on $\mathscr{C}(\Delta_A)$ are the unital spectral contractions:

If A and B are commutative unital Banach algebras, a linear operator L: $A \to B$ is called a *spectral contraction* if $\rho_B(Lx) \leq \rho_A(x)$ holds for each $x \in A$. Denoting by $\wedge$ the respective Gelfand transformations, one easily shows that a linear operator $L: A \to B$ is a spectral contraction if and only if there exists a contraction $\hat{L}: \hat{A} \to \hat{B}$ such that $(Lx)^\wedge = \hat{L}\hat{x}$ holds for each $x \in A$.

In case the algebra is no longer unital, the natural generalization of positive linear operators on $\mathscr{C}_0(\Delta_A)$ are the algebra-valued spectral states:

If A and B are commutative Banach algebras, a linear operator $L: A \to B$ is called a *B-valued spectral state* if

$$\mathrm{co}(\sigma_B(Lx)) \subset \mathrm{co}(\sigma_A(x)), \quad (x \in A),$$

where co denotes the convex hull and σ the spectrum in the respective algebra.

Let A_1 be the unitization of A; then Δ_{A_1} is the one-point compactification of Δ_A.

One has to be careful with this definition: For non-unital A, one has the relation $\sigma_A(x) = \sigma_{A_1}(x)$, while for unital A one has $\sigma_{A_1}(x) = \sigma_A(x) \cup \{0\}$ $(x \in A)$. Since adding the point 0 to the spectrum may drastically change its convex hull, we often distinguish for clarity's sake between the unital and the non-unital case. Both cases may be dealt with simultaneously by introducing the algebra A_e, which is A if A is unital but A_1 if A is non-unital: Then $\sigma_A(x) = \sigma_{A_e}(x)$ in any case.

Immediate examples of B-valued spectral states are given by algebra homomorphisms, which have to be unital for unital A and B. The set of all B-valued spectral states is invariant under the formation of convex combinations and pointwise limits. In the non-unital case, 0 is a B-valued spectral state, so that λL is a B-valued spectral state for each $\lambda \in [0, 1]$ and each B-valued spectral state L. This is perhaps the main difference between the unital and the non-unital case.

In case $B = \mathbb{C}$, we have $\mathrm{co}(\sigma_B(Lx)) = \{Lx\}$, so that the $\mathbb{C}$-valued spectral states in our sense are exactly the spectral states of Bonsall and Duncan [1971] (see also Pannenberg [1990a]). The weak* compact convex set of all $\mathbb{C}$-valued spectral states on A will be denoted by Ω_A, and q_A will denote the sublinear functional defined by

$$q_A(x) := \sup\{\mathscr{R}e\,\lambda \mid \lambda \in \sigma_A(x)\}$$

for each $x \in A$.

We recall that Ω_A contains exactly the unital spectral contractions if A is unital. In the non-unital case, Ω_A consists exactly of the restrictions to A of spectral states (i.e., unital spectral contractions) on A_1 (see Bonsall and Duncan [1971]). This is no longer true in the unital case: Since the unit u of A has spectrum $\{0, 1\}$ in A_1, spectral states on A_1 map the unit 1 of A_1 onto 1 but take an arbitrary real value between 0 and 1 in u. In any case, Ω_A is affinely homeomorphic to Ω_{A_e} via the restriction operator.

The classes of operators considered for unital (respectively, non-unital) A may be related as follows:

A.1.1 Proposition. *Let A, B be commutative unital Banach algebras and $L: A \to B$ be a continuous linear operator. Suppose that B is semisimple. Then the following assertions are equivalent*:
 (i) *L is a B-valued spectral state.*
 (ii) *L is a unital spectral contraction.*
 (iii) *$L'\Omega_B \subset \Omega_A$.*
 (iv) *$q_B(Lx) \leq q_A(x)$ for each $x \in A$.*
 (v) *$\rho_B(\exp(Lx)) \leq \|\exp(x)\|$ for each $x \in A$.*

Here $L': B' \to A'$ denotes the adjoint of L.

A proof of this proposition and its successors is given in our paper [1992].

So, in the unital case, B-valued spectral states are just unital spectral contractions. In the non-unital case, B-valued spectral states are (as in the complex-valued case) just restrictions to A of B_1-valued spectral states (i.e., unital spectral contractions) defined on A_1. This is the content of the following proposition:

A.1.2 Proposition. *Let A, B be non-unital commutative Banach algebras and $L: A \to B$ be a continuous linear operator. Suppose that B is semisimple. Then the following assertions are equivalent*:
 (i) *$L'\Omega_B \subset \Omega_A$.*
 (ii) *L is a B-valued spectral state.*
 (iii) *There is a unital spectral contraction $L_1: A_1 \to B_1$ such that $L_{1|A} = L$.*
 (iv) *$\rho_{B_1}(\exp(Lx)) \leq \|\exp(x)\|$ for each $x \in A$.*
 (v) *$q_B(Lx) \leq q_A(x)$ for each $x \in A$.*

Remark. If B is not semisimple, the above assertions remain equivalent if in condition (iii) L_1 is supposed to be a B_1-valued spectral state instead of a unital spectral contraction. An equivalent formulation of $L'\Omega_B \subset \Omega_A$ is given by $L'\Delta_B \subset \Omega_A$, and (iv) is equivalent to $\rho_{B_1}(\exp(Lx)) \leq \rho_{A_1}(\exp(x))$ for each $x \in A$.

The following result shows that in a certain sense algebra-valued spectral states are a substitute in the non-involutive situation for the positive linear operators, which in general are used for the approximation procedures on commutative Banach algebras with an approximate identity of norm at most one and a symmetric involution:

A.1.3 Proposition. *Let A, B be commutative Banach algebras with a symmetric involution and an approximate identity of norm at most one, and suppose that B is semisimple. Let L: A → B be a continuous linear operator.*

   If A and B are both non-unital, the following assertions are equivalent:
   (i) *L is a B-valued spectral state.*
   (ii) *L is a Schwarz operator, i.e., $(Lx)(Lx)^* \leq L(xx^*)$ for every $x \in A$.*
   (iii) *L is involutive, and $(Lu)^2 \leq L(u^2)$ for every self-adjoint $u \in A$.*
   (iv) *L is a positive spectral contraction.*

   If A and B are both unital, these conditions remain equivalent if additionally L is supposed to be unital in conditions (ii)–(iv).

   In any case, the B-valued spectral contractions on A are exactly the restrictions to A of B_e-valued unital positive spectral contractions on A_e mapping A into B.

In conditions (ii), (iii) and (iv), we use the order on B induced by the closed wedge B_+ generated by all elements of the form bb^* $(b \in B)$. Since B has a bounded approximate identity, B admits factorization $B^2 = B$ by Cohen's factorization theorem, so that a result of Render [1989] allows to conclude that for semisimple B we have $x \in B_+$, if and only if the Gelfand transform $\hat{x}$ is a positive function on the spectrum Δ_B.

Remark. Let us point out that the semi-simplicity assumed in the last two propositions will not be a restriction on B for the Korovkin-type approximation results. In fact, if in our context we replace B by its quotient modulo its radical and the operator by its composition with the quotient operator, then the assertions will not change, since the spectral radius remains the same. Furthermore, the existence of a bounded approximate identity has to be assumed in the last result only in order to guarantee the extendability of positive linear functionals on A to positive linear functionals on A_e; the result remains true under this weaker assumption.

We now turn to the universal Korovkin-type approximation theory, beginning with the unital case.

Let A be a commutative unital Banach algebra, and let H be an arbitrary subset of A. A notion of central importance is the universal Korovkin closure $K_u(H)$. This notion was first introduced in a somewhat different form by Altomare implicitly in [1982c] and explicitly in [1982a] (see also [1984a], [1986]); then it was drawn on by Pannenberg [1985–1992b]. Loosely spoken, it consists of all elements y of A which have the property that every approximation procedure working for the elements of H also works for y, where by an approximation procedure we understand a net of spectral contractions used to approximate the values of a unital algebra homomorphism. To be more precise, we set

$$K_u(H) = K_{u,A}(H) := \{ y \in A \,|\, \textit{For every commutative unital Banach algebra } B,$$

$$\textit{for every net } (L_i)_{i \in I}^{\leq} \textit{ of spectral contractions from}$$

$$A \textit{ into } B \textit{ and for every unital algebra homomorphism}$$

$$L\colon A \to B \textit{ such that } \lim_{i \in I}{}_{\leq} \rho_B(L_i x - Lx) = 0 \textit{ for every}$$

$$x \in H, \textit{ we also have } \lim_{i \in I}{}_{\leq} \rho_B(L_i y - Ly) = 0 \}.$$

In fact, one gets the same definition as the one we choose, if one considers only uniform algebras B, or only commutative C^*-algebras $B = \mathscr{C}(X)$ for a compact Hausdorff space X, or even as the only possibility $B = \mathbb{C}$; furthermore, for semi-simple B (or even only $B = \mathbb{C}$) it is sufficient to consider stationary nets $L_i \equiv L_0$ ($i \in I$) and replace convergence by equality (see Altomare [1982c], Pannenberg [1986a]). The question if we may restrict ourselves to the only possibility $B = A$ and $L =$ the identity operator on A is a lot more subtle; we will return to it later on.

We now turn to the non-unital case:

Let A be a commutative Banach algebra, and let H be a subset of A. Its universal closure with respect to algebra-valued spectral states is defined by

$$K_u^\sigma(H) = K_{u,A}^\sigma(H) := \{ y \in A \,|\, \textit{For every commutative Banach algebra } B,$$

$$\textit{for every net } (L_i)_{i \in I}^{\leq} \textit{ of spectral states from } A \textit{ into } B$$

$$\textit{and for every algebra homomorphism } L\colon A \to B$$

$$\textit{such that } \lim_{i \in I}{}_{\leq} \rho_B(L_i x - Lx) = 0 \textit{ for every } x \in H,$$

$$\textit{we also have } \lim_{i \in I}{}_{\leq} \rho_B(L_i y - Ly) = 0 \}.$$

Since in this definition B in general is non-unital, we are working in the non-unital context even if A has a unit. In particular, all spectra are calculated in A_1, and the spectral states on A arise as restrictions to A of spectral states on A_1.

The remarks on the definition of $K_u(H)$ established above apply in a similar form to the definition of $K_u^\sigma(H)$—this will be clear by the following proposition.

Setting $H_1 := H \cup \{1\} \subset A_1$ we denote by $K_{u,A_1}(H_1)$ its universal Korovkin closure with respect to the approximation of unital algebra homomorphisms by spectral contractions on the unital algebra A_1 as defined before. Then the relation of both universal Korovkin closures is given by

A.1.4 Proposition. *Let A be a commutative Banach algebra, and consider a subset H of A. Then $K^\sigma_{u,A}(H) = A \cap K_{u,A_1}(H_1)$, so that $K_{u,A_1}(H_1) = \{x + \lambda | x \in K^\sigma_{u,A}(H), \lambda \in \mathbb{C}\}$.*

The above proposition reduces everything to the unital case if one is willing to pass to A_1 and H_1 by adjoining a unit to both A and H.

Since the case of algebras with a symmetric involution and an approximate identity of norm at most one is often dealt with separately, we give an interpretation of $K_{u,A}(H)$, respectively of $K^\sigma_{u,A}(H)$, for this special case:

A.1.5 Theorem. *Let A be a commutative Banach algebra with a symmetric involution and an approximate identity of norm at most one. For each subset H of A we have*

$K^\sigma_{u,A}(H) = \{y \in A \,|\, For\ every\ commutative\ Banach\ algebra\ B\ with\ a\ symmetric$

involution and an approximate identity of norm at most one,

for every net $(L_i)^\le_{i \in I}$ of Schwarz operators from A into B and

for every involutive algebra homomorphism $L: A \to B$

such that $\lim_{\substack{\le \\ i \in I}} \rho_B(L_i x - Lx) = 0$ for every $x \in H$,

we also have $\lim_{\substack{\le \\ i \in I}} \rho_B(L_i y - Ly) = 0\}$.

In the defining property of the set on the right-hand side, Schwarz operators may be replaced by positive spectral contractions as well as by positive norm contractions.

If A is unital and H contains the unit of A, we have:

$K_{u,A}(H) = \{y \in A \,|\, For\ every\ commutative\ unital\ Banach\ algebra\ B\ with\ a$

symmetric involution, for every net $(L_i)^\le_{i \in I}$ of positive spectral

contractions from A into B and for every unital algebra

homomorphism $L: A \to B$ such that $\lim_{\substack{\le \\ i \in I}} \rho_B(L_i x - Lx) = 0$

for every $x \in H$, we also have $\lim_{\substack{\le \\ i \in I}} \rho_B(L_i y - Ly) = 0\}$.

Again, positive spectral contractions may be replaced by positive norm contractions.

This is an immediate consequence of Proposition A.1.3.

We now show how to compute the universal Korovkin closures:

A.1.6 Theorem. *Let A be a commutative Banach algebra and suppose H is a subset of A. Then*

$$K_{u,A}^{\sigma}(H) = \{y \in A \mid f(y) = m(y) \text{ for every } m \in \Delta_A \cup \{0\} \text{ and}$$

$$f \in \Omega_{A_1} \text{ satisfying } f_{|H} = m_{|H}\}.$$

If A is unital and H contains the unit of A, we have:

$$K_{u,A}(H) = \{y \in A \mid f(y) = m(y) \text{ for every } m \in \Delta_A \text{ and}$$

$$f \in \Omega_A \text{ satisfying } f_{|H} = m_{|H}\}.$$

This is a consequence of Altomare [1982a], [1982c], Pannenberg [1986a] and Proposition A.1.4.

As a corollary, we obtain the following result on the approximation of characters by nets of spectral states:

A.1.7 Corollary. *Let A be a commutative Banach algebra, H a subset of A and $y \in A$. Then $y \in K_{u,A}^{\sigma}(H)$ if and only if $\lim_{i \in I} \leq f_i(y) = m(y)$ for every net $(f_i)_{i \in I}^{\leq}$ of spectral states of A and every $m \in \Delta_A \cup \{0\}$ such that $\lim_{i \in I} \leq f_i(x) = m(x)$ holds for every $x \in H$.*

In the case of unital A, an analogous result holds for $K_u(H)$, where it is sufficient to consider $m \in \Delta_A$. If A possesses a symmetric involution and a bounded approximate identity of norm at most one, one may consider nets of positive linear functionals (since the latter coincide with spectral states in this context).

Since each spectral state on A is the restriction to A of a spectral state of A_1, the integral representation for spectral states on unital Banach algebras given by Maltese in [1979] shows that in the above corollary we may suppose f to be of the form

$$f(x) = \int \hat{x} \, d\mu, \quad (x \in A),$$

where μ is a probability measure on $\Delta_A \cup \{0\}$, respectively Δ_A. Defining the *Choquet boundary of H with respect to A* by the equation

$$\partial_A^\sigma H := \{m \in \Delta_A \cup \{0\} \mid f_{|A} = m \text{ for every } f \in \Omega_{A_1} \text{ such that } f_{|H} = m_{|H}\}$$

$$= \Big\{ m \in \Delta_A \cup \{0\} \mid \text{For every probability measure } \mu \text{ on } \Delta_A \cup \{0\}$$

$$\text{such that } m(x) = \int \hat{x} \, d\mu \text{ for all } x \in H, \text{ we already have}$$

$$m(x) = \int \hat{x} \, d\mu \text{ for all } x \in A \Big\},$$

in the non-unital situation (respectively

$$\partial_A H := \{m \in \Delta_A \mid f = m \text{ for every } f \in \Omega_A \text{ such that } f_{|H} = m_{|H}\}$$

$$= \Big\{ m \in \Delta_A \mid \text{For every probability measure } \mu \text{ on } \Delta_A \text{ such that}$$

$$m(x) = \int \hat{x} \, d\mu \text{ for all } x \in H, \text{ we already have}$$

$$m(x) = \int \hat{x} \, d\mu \text{ for all } x \in A \Big\},$$

in the unital situation with H containing the unit of A), we observe that

$$\partial_A^\sigma H = \partial_{A_1} H_1,$$

so that in general there is no loss in generality in assuming that A is unital and H contains the unit of A.

Now the *key observation* (established in our paper [1986a]) of the theory of finite universal Korovkin subsets is the fact that for test sets whose linear hull has finite dimension we may replace Ω_A by the convex hull $\mathrm{co}(\Delta_A)$ of the structure space Δ_A (respectively, Ω_{A_1} by $\mathrm{co}(\Delta_A \cup \{0\})$) in Theorem A.1.6 as well as in the definition of the Choquet boundaries given above. In other words, the probability measure μ occurring in the integral representation of spectral states may be chosen to have finite support. We refer the reader to our articles [1985] and [1986] for a discussion of this phenomenon.

If the linear hull of H has finite dimension (in particular, if H is finite), the following description of the Choquet boundary, which uses the *joint spectrum*

$$\sigma_A(x_1, \ldots, x_n) := \{0\} \cup \{(m(x_1), \ldots, m(x_n)) \mid m \in \Delta_A\} = \sigma_{A_1}(x_1, \ldots, x_n),$$

if A is non unital (respectively

$$\sigma_A(x_1,\ldots,x_n) := \{(m(x_1),\ldots,m(x_n))|m \in \Delta_A\}$$

if A is unital), is a direct consequence of the key observation and of great practical use, since it allows to visualize the Choquet boundary by geometric intuition in many cases (see Pannenberg [1986a] for a proof):

A.1.8 Theorem. *Let A be a commutative Banach algebra and suppose that H is a subset of A, whose linear hull has a finite linear basis $\{x_1,\ldots,x_n\}$ such that the Gelfand transforms $\{\hat{x}_1,\ldots,\hat{x}_n\}$ separate the points of $\Delta_A \cup \{0\}$. Then*

$$\partial_A^\sigma H = \{m \in \Delta_A \cup \{0\}|(m(x_1),\ldots,m(x_n)) \text{ is an extreme point of}$$

$$\mathrm{co}(\sigma_A(x_1,\ldots,x_n))\}.$$

If A is unital, H contains the unit of A and the linear hull of H has a finite linear basis $\{1,x_1,\ldots,x_n\}$ such that $\{1,\hat{x}_1,\ldots,\hat{x}_n\}$ separate the points of Δ_A, the Choquet boundary of H is given by

$$\partial_A H = \{m \in \Delta_A|(m(x_1),\ldots,m(x_n)) \text{ is an extreme point of } \mathrm{co}(\sigma_A(x_1,\ldots,x_n))\}.$$

Now theorem A.1.6 may be reformulated as follows:

A.1.9 Corollary. *Let A be a commutative Banach algebra and suppose H is a subset of A. Then $K_{u,A}^\sigma(H) = A$ if and only if $\partial_A^\sigma H = \Delta_A \cup \{0\}$.*
If A is unital and H contains the unit of A, we have $K_{u,A}(H) = A$ if and only if $\partial_A H = \Delta_A$.

In the sequel, subsets H satisfying $K_{u,A}^\sigma(H) = A$ (respectively $K_{u,A}(H) = A$) will be called *universal Korovkin subsets for A with respect to algebra-valued spectral states*, or, more briefly, K_u^σ-subsets in A (respectively, *universal Korovkin subsets in A or K_u-subsets in A*).

As a consequence of the last two theorems, we have that H is a K_u-subset in A whenever $\hat{H}$ is a K_u-subset in $\mathscr{C}(\Delta_A)$. Furthermore, we also get a geometric criterion for finite subsets H whose set of Gelfand transforms separates the points of $\Delta_A \cup \{0\}$ (respectively, Δ_A) (see Altomare [1984a], Pannenberg [1986a]).

A.1.10 Corollary. *Let $\{x_1,\ldots,x_n\}$ be a finite subset of a commutative Banach algebra A such that $\{\hat{x}_1,\ldots,\hat{x}_n\}$ separate the points of $\Delta_A \cup \{0\}$.*
Then $K_{u,A}^\sigma(H) = A$, if and only if $\sigma_A(x_1,\ldots,x_n)$ coincides with the extreme points of its convex hull.

If A is unital and $H = \{1, x_1, \ldots, x_n\}$ is a finite subset of A such that $\{1, \hat{x}_1, \ldots, \hat{x}_n\}$ separate the points of Δ_A, we may conclude that $K_{u, A}(H) = A$, if and only if $\sigma_A(x_1, \ldots, x_n)$ coincides with the extreme points of its convex hull.

For example, an easily drawn picture in the plane shows

$$\{x, x^2\} \text{ is a } K_u^\sigma\text{-subset in } \mathscr{C}_0(]0, 1]),$$

and the same picture shows that

$$\{1, x, x^2\} \text{ is a } K_u\text{-subset in } \mathscr{C}([0, 1]),$$

where obviously x^2 may be replaced by any strictly convex or strictly concave continuous real-valued function on the unit interval.

Using his geometric intuition, the reader may convince himself in the same way that

$$\{1, z, h \circ abs\} \text{ is a } K_u\text{-subset in } \mathscr{C}(\mathbb{D}) \quad \text{and} \quad \{1, z\} \text{ is a } K_u\text{-subset in } \mathscr{C}(\mathbb{T})$$

where $\mathbb{D}$ is the unit disk in the complex plane, $\mathbb{T}$ is its boundary, z is the complex coordinate function on $\mathbb{D}$, $abs(z) = |z|$ for all $z \in \mathbb{D}$ and h is a strictly convex continuous real-valued function on the unit interval—here the joint spectrum is obtained by rotating the graph of h around the y-axis.

This also shows that—contrary to the case of the unit interval—strictly concave functions h will not do the job.

In the same way, one sees that $\{\delta_0, \delta_1\}$ is a K_u-subset in the discrete group algebra $A = \ell^1(\mathbb{Z})$, where δ denotes the Dirac measure having the indicated one-point support: It is sufficient to observe that $\sigma_A(\delta_1) = \mathbb{T}$. We will return to group algebras later on.

Just as for $\mathscr{C}_0(X)$, the appearance of a Choquet boundary allows to consider upper and lower envelopes as well as H-affine elements in the context of commutative Banach algebras:

Let A be a commutative Banach algebra with non trivial structure space and H a subspace of A. Adapting an idea of Asimov and Ellis [1980] we represent A as a space of continuous real-valued functions on the weak* compact subset $\Sigma_A := \Delta_A \cup (-i\Delta_A) \cup \{0\}$ of the dual space A via

$$\gamma_x(\chi) := \mathscr{R}e\,\chi(x) \quad (x \in A, \chi \in \Sigma_A),$$

and define upper and lower envelopes by

$$(\gamma_x)^*(\chi) := \inf\{\mathscr{R}e\,\chi(x) | y \in H, \gamma_y \geq \gamma_x \text{ on } \Sigma_A\},$$

$$(\gamma_x)_*(\chi) := \sup\{\mathscr{R}e\,\chi(x) | y \in H, \gamma_y \leq \gamma_x \text{ on } \Sigma_A\},$$

for every $x \in A$ and every $\chi \in \Sigma_A$. If A is unital and H contains the unit of A, we consider $\Sigma_A := \Delta_A \cup (-i\Delta_A)$, which again is weak* compact, and leave the other definitions unchanged. Then again, $\Sigma_{A_1} = \Sigma_A$ for non-unital A, and for $x \in A$, we get the same representation when considering it as an element of A or A_1, so that also the envelopes coincide.

Note that by the definition of Σ_A, γ_x contains information on both $\mathscr{R}e\,\hat{x}$ and $\mathscr{I}m\,\hat{x}$, since $\mathscr{R}e\,\hat{x}\,(-im) = \mathscr{I}m\,\hat{x}(m)$ holds for $x \in A$ and $m \in \Delta_A$.

An element x of A will be called H-affine if its upper and lower envelopes coincide on Σ_A. Using this notion, we obtain

A.1.11 Proposition. *Let A be a commutative Banach algebra, and suppose that H is a subspace of A. Then, for every $x \in A$, the following assertions are equivalent:*
(i) *$x \in K_u^\sigma(H)$.*
(ii) *$(\gamma_x)^* = (\gamma_x)_*$.*
In other words, the universal Korovkin closure of H consists exactly of the H-affine elements of A. The same characterization is true in the context of a unital algebra.

The proof is clear from the unital case and the fact that the Choquet boundary of H depends only on the real and imaginary parts of the Gelfand transforms of the elements of H (compare Pannenberg [1985] and [1988]).

The next propositions of our survey of general results show how to imitate the construction of the classical universal Korovkin subset $\{x, x^2\}$ in $\mathscr{C}_0(]0, 1])$ (respectively, $\{1, x, x^2\}$ in $\mathscr{C}([0, 1])$) in the context of commutative Banach algebras with a symmetric involution; they use the notations $H^*H := \{x^*x | x \in H\}$ and $\hat{H} := \{\hat{x} | x \in H\}$. The existence of a bounded approximate identity with norm 1 is not assumed; however, if there is one, the universal Korovkin closure may be described as in Theorem A.1.5.

A.1.12 Theorem. *Let A be a commutative Banach algebra with a symmetric involution, and let M be a subset of A. Then*

$$K_u^\sigma(M \cup M^*M) = \{y \in A | m_1(y) = m_2(y) \text{ for all } m_1, m_2 \in \Delta_A \cup \{0\}$$

$$\text{such that } m_{1|M} = m_{2|M}\}.$$

In particular, $K_u^\sigma(M \cup M^*M)$ is a closed, *-closed subalgebra of A, and we have

$$K_u^\sigma(M \cup M^*M) = A \quad \Leftrightarrow \quad \hat{M} \text{ separates the points of } \Delta_A \cup \{0\}.$$

In the unital context, the theorem takes the following form.

A.1.13 Theorem. *Let A be a commutative unital Banach algebra with a symmetric involution, and let M be a subset of A. Then*

$$K_u(\{1_A\} \cup M \cup M^*M) = \{y \in A \mid m_1(y) = m_2(y) \text{ for all } m_1, m_2 \in \Delta_A$$

$$\text{such that } m_{1|M} = m_{2|M}\}.$$

In particular, $K_u(\{1_A\} \cup M \cup M^*M)$ is a closed, *-closed subalgebra of A, and we have

$$K_u(\{1_A\} \cup M \cup M^*M) = A \quad \Leftrightarrow \quad \hat{M} \text{ separates the points of } \Delta_A.$$

The proof of Theorem A.1.13 is contained in Altomare's papers [1982a], [1984]), and Theorem A.1.12 follows from the unital case via Proposition A.1.4.

Remark. It is of some interest for the applications that in both propositions the set M^*M may be replaced for a finite test set $M = \{x_1, \ldots, x_n\}$ by the single-ton $\left\{ \sum_{k=1}^{n} x_k^* x_k \right\}$—this is an immediate consequence of the proof of the above Propositions.

The last two propositions are the standard tools to construct universal Korovkin subsets in commutative Banach algebras with a symmetric involution: they will be used for group algebras and the Waelbroeck algebra of all infinitely differentiable functions on a compact manifold in the second and third section. That the symmetry of the involution is a vital assumption will be shown by the example of the disk algebra in the third section.

Let us now consider the case $A = \mathscr{C}(X)$, X compact Hausdorff, more closely: In this case, $M^*M = \{|h|^2 \mid h \in M\} := |M|^2$, and $K_u(\{1_A\} \cup M \cup |M|^2)$ coincides with the closed, *-closed subalgebra generated by M—this is the complex version of Theorem 4.4.6. If M also separates the points of X, then $\{1\} \cup M \cup |M|^2$ is a K_u-subset in $\mathscr{C}(X)$ (for more details see Altomare [1982a], [1984a], [1986]).

In particular, if each function of M has constant modulus on X, then even $\{1\} \cup M$ is a K_u-subset in $\mathscr{C}(X)$. This latter situation occurs, for instance, when one considers a compact abelian group G and a subset M of the character group Γ of G separating the points of G: By duality theory, these subsets M are exactly those subsets of Γ which generate Γ as a closed subgroup. M may be chosen to be finite, if and only if G is a subgroup of some finite dimensional torus, i.e., a finite product of some finite-dimensional torus and some finite cyclic groups (see Pannenberg [1990d] for a proof).

As a simple application of this last result, consider a net $(\mu_i)_{i \in I}^{\leq}$ of Radon measures on G satisfying $\|\mu_i\| \leq 1$ for every $i \in I$. Suppose that the associated net of Fourier transforms converges pointwise on M to the function constantly one, i.e., $\lim_{\substack{\le \\ i \in I}} \hat{\mu}_i(\gamma) = 1$ for each $\gamma \in M$. Then the associated convolution operators on $\mathscr{C}(G)$ (respectively, $L^p(G)$, $1 \le p < +\infty$) converge pointwise to the identity on $\mathscr{C}(G)$ (respectively, $L^p(G)$), i.e.,

$$\lim_{\substack{\le \\ i \in I}} f * \mu_i = f \quad (f \in \mathscr{C}(G) \text{ (respectively, } f \in L^p(G))).$$

For more details, see Altomare [1981], Bloom and Sussich [1980] and Pannenberg [1990d]. We finally point out that for locally compact G the universal Korovkin closure $K_u(\{1\} \cup M)$ associated to a subset M of the character group Γ in the algebra of all bounded uniformly continuous functions on G coincides with the subalgebra of all almost periodic functions in A (see Altomare [1982c]).

An immediate consequence of the last two results is the following general theorem:

A.1.14 Theorem. *Let A be a commutative Banach algebra with a symmetric involution. Then A possesses a finite universal Korovkin subset, if and only if A is finitely continuously generated.*

Here the last assertion means that there exists a finite subset $H = \{x_1, \ldots, x_n\}$ of A such that each $x \in A$ 'is' a continuous function of $x_1, \ldots, x_n$ in the sense $\hat{x} = f \circ (\hat{x}_1, \ldots, \hat{x}_n)$ for an appropriate continuous function f on the joint spectrum $\sigma_A(x_1, \ldots, x_n)$ of H (see Rickart's monograph [1966]).

The deduction of the above theorem from the preceding may be found in our paper [1985] for the unital case; the non-unital case follows again from Proposition A.1.4.

Applied to $\mathscr{C}(X)$, this shows (together with the Stone-Weierstrass theorem and the Menger-Nöbeling theorem, see Pannenberg [1990b]) that $\mathscr{C}(X)$ possesses a finite universal Korovkin subset if and only if X is metrizable and has finite covering dimension. In case X is a compact group, the functions forming a finite universal Korovkin subset may be chosen to reflect the group structure; see Pannenberg [1990d] for more details.

We finish our general survey by a consideration of Korovkin approximation versus universal Korovkin approximation:

Instead of considering the approximation of arbitrary algebra homomorphisms, one may restrict attention to the approximation of the identity operator as in Korovkin's original theorems (see Theorems 4.2.7 and 4.2.8).

We define the *Korovkin closure $K^\sigma(H)$ of a subset H* of a commutative Banach algebra A by

$$K^{\sigma}(H) = K_A^{\sigma}(H) = \left\{ y \in A \,|\, Whenever \,(L_i)_{i \in I}^{\leq} \text{ is a net of } A\text{-valued spectral} \right.$$

$$\text{states on } A, \text{ the condition } \lim_{\substack{\leq \\ i \in I}} \rho_A(L_i x - x) = 0 \text{ for}$$

$$\left. \text{every } x \in H \text{ already implies } \lim_{\substack{\leq \\ i \in I}} \rho_A(L_i y - y) = 0 \right\}.$$

Again A is considered as being non-unital, i.e., the spectral states occurring in the definition are defined by using the spectrum in A_1, in order to be able to compare $K^{\sigma}(H)$ with $K_u^{\sigma}(H)$.

For unital A, we consider the Korovkin closure $K(H) = K_A(H)$ defined in the same way as $K^{\sigma}(H)$ but using nets of unital spectral contractions instead of A-valued spectral states.

In general, the properties of $K^{\sigma}(H)$ are analogous to the (very few) facts known for the unital case (respectively, the case of algebras possessing a symmetric involution and a bounded approximate identity of norm one). We refer the reader to our papers [1985] and [1988] as well as the joint work with Romanelli [1991] and the references cited therein and focus our attention only on the case of algebras having a symmetric involution (but not necessarily an approximate identity).

A.1.15 Theorem. *Let A be a commutative Banach algebra with a symmetric involution, and let H be a subset of A. Then $K^{\sigma}(H) = K_u^{\sigma}(H)$.*

If A is unital and H contains the unit of A, we have $K(H) = K_u(H)$.

The *proof* follows the lines of the proof of Proposition 4.1 in our joint paper with Romanelli [1991]. Again, the disk algebra shows that the symmetry assumption is vital, even though it may be weakened for particular Beurling algebras, as the second section will show.

Remark. The importance of the symmetry assumption in the last results lies in the fact that it allows the construction of peak functions using the fact that $(x^*x)^\wedge \geq 0$, not in the extendability of positive linear functionals as in Proposition A.1.5—this explains why an approximate identity is not necessary and why the disk algebra, which contains no non-constant positive function, yields a counterexample showing that the symmetry assumption is necessary.

We finally consider a sufficient condition on a subset H to be a Korovkin subset, i.e., to satisfy the equation $K(H) = A$ (respectively, $K^{\sigma}(H) = A$). It uses the Shilov boundary of A_e, which will be denoted by Γ_{A_e}.

We recall that the *Shilov boundary* Γ_A of a commutative unital Banach algebra A is the smallest closed subset F of Δ_A such that $\rho_A(x) = \max_{m \in F} |m(x)|$ for every $x \in A$.

A.1.16 Theorem. *Let A be a commutative Banach algebra with non-void spectrum* Δ_A, *and let H be a subset of A. Then, the following statements are true:*

(1) *If A is non-unital,* $K^\sigma(H)$ *contains the set*

$$\{y \in A \mid f(y) = m(y) \text{ for all } m \in \Gamma_{A_1} \text{ and } f \in \Omega_{A_1} \text{ such that } f_{|H} = m_{|H}\}.$$

(2) *If A is unital and H contains the unit of A, K(H) contains the set*

$$\{y \in A \mid f(y) = m(y) \text{ for all } m \in \Gamma_A \text{ and } f \in \Omega_A \text{ such that } f_{|H} = m_{|H}\}.$$

(3) *If A is non-unital and* Γ_{A_1} *is contained in* $\partial_A^\sigma(H)$, *H is a Korovkin subset in A.*

(4) *If A is unital, H contains the unit of A and* Γ_A *is contained in* $\partial_A(H)$, *H is a Korovkin subset in A.*

Proof. Assertion (2) is proved by Altomare in [1984] (respectively, the author in [1985]). To prove assertion (1), we observe that $A \cap K(H_1) \subset K^\sigma(H)$ by Proposition A.1.2 and then use (2) to deduce (1). Now (3) and (4) easily follow from the definitions of the Choquet boundaries.   □

For example, let $A = \mathscr{A}(\mathbb{D}^n)$ be a polydisk algebra, and consider the test set $H = \{1, z_1, \ldots, z_n\}$ consisting of the unit of A and the n coordinate functions. Then the Choquet boundary of H is the polytorus, which coincides with the extreme points of the convex hull of $\sigma_A(z_1, \ldots, z_n)$ as well as with the Shilov boundary of A. This shows that H is a Korovkin system for the polydisk algebra. We will come back to this example in the last section.

It is an open problem if the sufficient conditions given here are also necessary (see Altomare's survey [1986] for a discussion of this problem).

Note that Theorem A.1.15 may be interpreted as giving a partial answer to this question: Indeed, it implies that inclusion may be replaced by equality in (1) and (2) whenever A has symmetric involution.

We further point out that again one may use the integral representation of spectral states in this context; as it is well-known, one may choose the probability measure μ even with support in Γ_{A_1} (respectively, Γ_A). In case H has a finite-dimensional linear hull, one may even assume that the support of μ is a finite subset of the Shilov boundary—this version of the key observation will be used in the third section.

A.2 Commutative group algebras

Korovkin's original theorems, which initiated the theory we are discussing, deal with spaces of continuous functions on the circle and the unit interval—it is for this reason that spaces like $\mathscr{C}(X)$ (X compact Hausdorff) and $\mathscr{C}_0(X)$ (X locally compact Hausdorff) seemed to be the natural candidates for generalizations of the original theorems. However, it quickly became clear by the work of Wolff [1973–1978] and others that the structure of a Banach lattice is the decisive ingredient for the validity of a number of theorems, and L^1-spaces offered themselves as a natural concrete candidate (different from $\mathscr{C}_0(X)$) for the study of Korovkin approximation, culminating in the study of Korovkin theorems in L^p-spaces by Donner [1981], [1982].

If one wishes to consider, however, uniform convergence of the Fourier transforms instead of L^1-convergence of the functions itself, the Banach algebra structure of $L^1(G)$, G a locally compact abelian (LCA) group, is the decisive ingredient for Korovkin approximation, and the general results of the first section may be applied to the particular Banach algebra $L^1(G)$.

As a consequence of the author's work on Korovkin approximation in $L^1(G)$, it turned out that the validity of a Korovkin-type theorem for $L^1(G)$ allows to characterize a class of LCA groups as well as a class of commutative group algebras—so the study of Korovkin approximation of $L^1(G)$ adds both to our knowledge of the structure of LCA groups as well as the structure of commutative group algebras.

In order to display the relevant properties of G (respectively, $L^1(G)$), we begin by recalling the relevant definitions:

The *covering dimension* of a topological space Γ is the least integer n such that every finite open covering of Γ has an open refinement of order not exceeding n, or is ∞ if there is no such integer (see Pears [1975]). The *torsion-free* rank of a discrete abelian group G is by definition the cardinal number of any maximal linearly independent subset of the $\mathbb{Z}$-module G (Hewitt and Ross [1979]); if $\mathbb{Q} \otimes_{\mathbb{Z}} G$ denotes the tensor product in the category of $\mathbb{Z}$-modules, it is an easy exercise in algebra to show that $\mathbb{Q} \otimes_{\mathbb{Z}} G$ is a vector space over the rationals, whose dimension equals the torsion-free rank of G.

For discrete groups, the central theorem connecting Korovkin approximation and the structure theory of abelian groups is the following result, for a proof of which we refer the reader to our papers [1989], [1990b] and [1991b]:

A.2.1 Theorem. *Let G be a discrete abelian group with associated group algebra $L^1(G)$ and character group $\Gamma = \hat{G}$. Then the following assertions are equivalent:*

 (i) *$L^1(G)$ possesses a finite universal Korovkin subset.*

 (ii) *G is countable and has finite torsion-free rank.*

(iii) *Γ is a metric space of finite covering dimension.*

(iv) *G is countable, and the vector space over the rationals $\mathbb{Q} \otimes_{\mathbb{Z}} G$ has finite dimension.*

 (v) *G is countable, and for some $n \in \mathbb{N}$ there exists an exact sequence*

$$0 \mapsto \mathbb{Z}^n \mapsto \mathbb{Q} \otimes_{\mathbb{Z}} G \mapsto (\mathbb{Q}/\mathbb{Z})^n \mapsto 0.$$

(vi) *For some $n \in \mathbb{N}$, there exists an exact sequence $0 \mapsto \mathbb{Z}^n \mapsto G \mapsto H \mapsto 0$, where H is a countable torsion group.*

(vii) *G is an inductive limit of a countable number of groups of the form $\mathbb{Z}^n \times \Phi_j$, where $n \in \mathbb{N}$ is fixed and each Φ_j is a finite abelian group.*

Loosely spoken, the existence of a finite universal Korovkin subset in $L^1(G)$ characterizes (among discrete abelian groups) the extensions of $\mathbb{Z}^n$ ($n \in \mathbb{N}_0$) by a countable torsion group.

To give a taste of the concrete construction of a finite universal Korovkin subset (or a finite Korovkin subset, which amounts to the same by Theorem A.1.15) we consider a finitely generated discrete abelian group G. If E is a finite set of generators for G containing zero, the associated set $H = \{1_g | g \in E\}$ of characteristic functions of the elements of E is a finite universal Korovkin system for $L^1(G)$ (see our paper [1988]), whose cardinality equals the number of generators of G. If a large number of these generators has finite order, one can even do better: By the well-known structure theorem, G is isomorphic to a finite direct sum of some $\mathbb{Z}^n$ ($n \in \mathbb{N}_0$) with some finite cyclic groups $H_1, \ldots, H_m$ ($m \in \mathbb{N}_0$), each of order a power of a prime. Now choose generators $x_1, \ldots, x_n$ of $\mathbb{Z}^n$ and $g_1, \ldots, g_m$ of $H_1, \ldots, H_m$. Then for appropriately chosen $\delta_1, \ldots, \delta_m > 0$, the set $H :=$
$$\left\{ 1_{x_1}, \ldots, 1_{x_n}, \sum_{k=1}^{m} \delta_k 1_{g_k} \right\}$$
yields via Theorem A.1.13 a finite universal Korovkin subset

$$\left\{ 1_0, 1_{x_1}, \ldots, 1_{x_n}, \sum_{k=1}^{m} \delta_k 1_{g_k}, \sum_{k=1}^{m} \sum_{j=1}^{m} \delta_k \delta_j 1_{g_k - g_j} \right\},$$

whose cardinality equals the torsion-free rank of G plus 3 and is independent of the number of finite cyclic factors of G (see our paper [1990b] for a proof).

An example of a three element universal Korovkin subset in the group algebra $L^1(G)$, where $G := \Omega_r$ is the locally compact abelian group of all r-adic numbers for some integer $r \geq 2$, is given by the author in [1990d].

Theorem A.2.1 even permits a new look at Korovkin's second theorem 4.2.8: The three-element universal Korovkin subset $\{1, \sin(t), \cos(t)\}$ for the space of all 2π-periodic continuous functions on the line (i.e., continuous functions on the one-dimensional torus) may be constructed in much more general situations if one is willing to add at most three function, whereas sine- and cosine functions alone only yield a universal Korovkin subset for a subgroup of some finite-dimensional torus. This is the content of our next theorem, which we proved in [1990d], where the result is stated for real-valued continuous functions on G—of course. Any notion of universal Korovkin subset will work in this context.

A.2.2 Theorem. *Let G be a metrizable compact abelian group of finite covering dimension n. Then there exist continuous group homomorphisms $t_1, \ldots, t_n$: $G \to \mathbb{R}/\mathbb{Z}$ and continuous real-valued functions $h_1, h_2 \in \mathscr{C}(G)$ such that the set*

$$H := \{1, \sin(2\pi t_1), \ldots, \sin(2\pi t_n), \cos(2\pi t_1), \ldots, \cos(2\pi t_n), h_1, h_2, h_1^2 + h_2^2\},$$

is a universal Korovkin subset in $\mathscr{C}(G)$.

Furthermore, there exists a universal Korovkin subset of the form

$$\{1, \sin(2\pi t_1), \ldots, \sin(2\pi t_n), \cos(2\pi t_1), \ldots, \cos(2\pi t_n)\},$$

if and only if G is isomorphic to a subgroup of an n-dimensional torus.

We now pass to the case of a general locally compact abelian group G, whose character group will again be denoted by $\hat{G} = \Gamma$.

The fact that $L^1(G)$ possesses a finite universal Korovkin subset is no longer equivalent to the assertion that G be countable and have finite torsion-free rank; the "right" property turns out to be the finite covering dimension of Γ, together with the second axiom of countability: For discrete G, both assertions are equivalent by a well-known theorem of Pontryagin, which is displayed in Hewitt's and Ross's monograph [1979]. In the general case, Γ has finite covering dimension, if and only if a certain discrete factor of G has finite torsion-free rank: This has been observed by the author in [1990b].

So far so good for the structure of G—concerning the structure of $L^1(G)$, the right notion here turns out to be the stable rank well-known from non-stable K-theory (see, e.g., Rieffel's paper [1983]) as well as separability. To recall the relevant definitions, let A be a Banach algebra and $n \in \mathbb{N}$ a positive integer. We set $e_1 := (1, 0, \ldots, 0)$ and

$$U_n(A_1, A) := \left\{ a = (a_1, \ldots, a_n) \in A_1^n \,|\, a \equiv e_1 \bmod A_n, \right.$$

$$\left. \text{and there exists } b \in A^n \text{ such that } \sum_{j=1}^{n} b_j a_j = 1 \right\}.$$

Using these (A_1, A)-unimodular vectors, two notions of stable rank are defined:

– Bass' *algebraic stable rank* is by definition the least integer $n \in \mathbb{N}$ such that for every $a = (a_1, \ldots, a_{n+1}) \in U_{n+1}(A_1, A)$ there exists $(x_1, \ldots, x_n) \in A^n$ such that $(a_1 + x_1 a_{n+1}, \ldots, a_n + x_n a_{n+1}) \in U_n(A_1, A)$.

– Rieffel's *topological stable rank* is by definition the least integer $n \in \mathbb{N}$ such that $U_n(A_1, A)$ is dense in the set $\{a \in A_1^n \,|\, a \equiv e_1 \bmod A^n\}$ (with respect to the product topology).

We refer the reader to the literature for some more background and only mention the fact that both definitions, though in general leading to different notions, yield the same positive integer $sr(A)$, if A is the group algebra of a locally compact abelian group G. Relying heavily on results of Corach, Larotonda and Suarez, the integer $sr(A)$ has been computed by the author [1990c]:

$$sr(L^1(G)) = \left[\frac{d}{2}\right] + 1,$$

where [ ] are the Gauss brackets and d equals the covering dimension of Γ. This makes the connection of the stable rank of $L^1(G)$ to the dimension of Γ and therefore to the existence of a finite universal Korovkin system obvious and allows to prove the following result:

A.2.3 Theorem. *Let G be a locally compact abelian group. Then the following assertions are equivalent*:
  (i) *$L^1(G)$ possesses a finite universal Korovkin subset.*
 (ii) *$L^1(G)$ possesses a finite Korovkin subset.*
(iii) *There exists a finite number of functions $f_1, \ldots, f_n \in L^1(G)$ whose Fourier transforms separate strongly the points of $\hat{G}$.*
 (iv) *Γ is separable and has finite covering dimension.*
  (v) *$L^1(G)$ is separable and has finite stable rank.*
 (vi) *$L^1(G)$ is finitely continuously generated.*

A proof of this and the following theorem is given by the author in [1990c]. The case of a compact group G, to which we now turn, has very special features: $L^1(G)$ has finite stable rank (namely, zero), so that separability alone is the decisive property.

A.2.4 Theorem. *Let G be a compact abelian group. Then the following assertions are equivalent*:
  (i) *$L^1(G)$ possesses a finite (universal) Korovkin subset.*
 (ii) *$L^1(G)$ possesses a countable (universal) Korovkin subset.*
(iii) *$L^1(G)$ is separable.*
 (iv) *G is metrizable.*
  (v) *$L^1(G)$ is finitely continuously generated.*
 (vi) *$L^1(G)$ is finitely generated for its norm topology.*
  In any of these cases, $L^1(G)$ is singly generated and possesses a two-element universal Korovkin subset.

We now pass from commutative group algebras to commutative Beurling algebras—a weight function enters the scene, and the study of universal

Korovkin approximation again yields some insight into its properties as well as the structural properties of the associated Beurling algebra.

To define a Beurling algebra, we consider a weight function ω on an LCA group G, i.e., a real-valued function on G having the properties

$$\omega(x) \geq 1, \quad (x \in G),$$

$$\omega(x + y) \leq \omega(x)\omega(y), \quad (x, y \in G),$$

ω is measurable and locally bounded.

By $L^1_\omega(G)$ we denote the associated *Beurling algebra* of all $f \in L^1(G)$ for which $f\omega \in L^1(G)$, too—see the books of Guichardet [1968], Reiter [1968] and Wang [1977] for more information on it. This is a semisimple commutative Banach algebra under convolution and the norm

$$\|f\|_\omega := \|f\omega\|_1 \quad (f \in L^1_\omega(G)).$$

It has a bounded approximate identity and contains the continuous functions with compact support $\mathscr{C}_c(G)$ as a dense subalgebra. In general, $L^1_\omega(G)$ is not *-closed in $L^1(G)$.

The property of a weight function that is characterized via Korovkin approximation in the associated Beurling algebra is best expressed via its *rate of growth* Ω, which is given by

$$\Omega(x) := \lim_{n \to \infty} \omega(nx)^{1/n} \quad (x \in G).$$

Ω is a continuous weight function, which is homogeneous (i.e., we have $\Omega(nx) = \Omega(x)^n$ for every $x \in G$ and $n \in \mathbb{N}_0$) and satisfies $1 \leq \Omega \leq \omega$; it coincides with ω if and only if ω is homogeneous. The property which turns out to be the right one for our purposes is $\Omega \equiv 1$, i.e., Ω is minimal.

The property of a commutative Beurling algebra, that has to be added to the properties already known from the group algebra case, is the approximate regularity of $L^1_\omega(G)$: Following Wilken [1966], we call a commutative Banach algebra A *approximately regular* if Δ_A is non-void and if for each $m \in \Delta_A$, each closed set $F \subset \Delta_A$ not containing m and each $\varepsilon > 0$ there exists some $x \in A$ with $m(x) = 1$ and $|\gamma(x)| < \varepsilon$ for every $\gamma \in F$.

Obviously, each regular A and each A endowed with a symmetric involution is approximately regular; the converse, however, does not hold.

It turns out that ω has minimal rate of growth ($\Omega = 1$), if and only if $L^1_\omega(G)$ is approximately regular; this fact proved by the author in [1992] shows that approximate regularity is the property we are looking for and allows to prove

A.2.5 Theorem. *Let G be a locally compact abelian group, and let ω be an upper semi-continuous weight function on G. Then the following assertions are equivalent:*

(i) *$L_\omega^1(G)$ possesses a finite universal Korovkin subset.*

(ii) *The character group $\Gamma = \hat{G}$ is a second countable space of finite covering dimension, and ω has minimal rate of growth.*

(iii) *$L_\omega^1(G)$ is separable, approximately regular and has finite stable rank.*

For the case of a discrete abelian group of finite torsion-free rank, the fact that ω has minimal rate of growth may easily be checked:

Remark. If G has finite torsion-free rank r and $\{x_1, \ldots, x_r\}$ is a maximal linearly independent subset of the $\mathbb{Z}$-module G, then $\Omega = 1$ is equivalent to

$$\log \omega(nx_k) = o(n), \quad (n \to \infty), \quad (1 \leq k \leq r).$$

Our last example for a characterization of a class of LCA groups via Korovkin theory is contained in the following theorem, which describes those LCA groups G, for which any Beurling algebra on G possesses a finite universal Korovkin subset:

A.2.6 Theorem. *Let G be a locally compact abelian group. Then the following assertions are equivalent:*

(i) *G is second countable and equals the union of its compact subgroups.*

(ii) *For each weight function ω on G, the associated Beurling algebra $L_\omega^1(G)$ possesses a finite universal Korovkin subset.*

Again a proof may be found in the author's paper [1992].

A.3 Finitely generated commutative Banach algebras and polydisk algebras

To motivate the discussion to follow, we take up the topic dealt with in Theorem A.2.2 and consider the following variant of this result:

A.3.1 Theorem. *Let G be a compact abelian group. Then $\mathscr{C}(G)$ possesses a finite universal Korovkin subset, if and only if G is metrizable and has finite covering dimension n.*

In this case, there exist characters $\gamma_1, \ldots, \gamma_n$ of G and a continuous positive definite function φ on G (which is an absolutely and uniformly convergent series of positive multiples of characters of G) such that the set $H := \{1, \gamma_1, \ldots, \gamma_n, \varphi, |\varphi|^2\}$ is a universal Korovkin subset in $\mathscr{C}(G)$.

Furthermore, there exists a universal Korovkin subset in $\mathscr{C}(G)$ of the form

$$H = \{1, \gamma_1, \ldots, \gamma_n\}$$

if and only if G is a subgroup of the n-dimensional torus, i.e., the finite product of some k-dimensional torus with some finite cyclic groups ($k \leq n$).

The proof of this theorem is contained in our paper [1990d]. To come to the point, we remark that the test set H consists entirely of functions having an absolutely convergent Fourier transform on the discrete dual Γ of G—in other words the given test set is in fact a subset of the *Wiener algebra* $W(G)$, which is the commutative unital Banach algebra (under pointwise operations) of all continuous functions on G having an absolutely convergent Fourier series on Γ, with the L^1-norm of Fourier transforms as norm. It is well-known that $\Delta_{W(G)} \cong G$ and, in fact, $W(G) \cong L^1(\hat{G})$. The point we want to stress is that Theorem A.3.1 holds true word by word if $\mathscr{C}(G)$ is replaced by $W(G)$—which means that there is a finite universal Korovkin subset in $W(G)$ which at the same time is a universal Korovkin subset in the larger algebra $\mathscr{C}(G)$.

This of course prompts two questions:
- Which additional properties (described, e.g., by membership in a subalgebra) may the test functions of a suitably chosen finite universal Korovkin subset in $\mathscr{C}(X)$ possess?
- Under which circumstances do the Gelfand transforms of the elements of a finite universal Korovkin subset in a commutative unital Banach algebra B form a universal Korovkin subset even in $\mathscr{C}(\Delta_B)$?

To formulate an answer to the second question, the following definition seems natural for our purposes: A commutative unital Banach algebra A is called a *superalgebra of B*, if B is algebraically isomorphic to a subalgebra C of A, such that the unit of B is mapped on the unit of A and such that the set $\hat{C}$, which is the image of C under the Gelfand transformation of A, separates the points of Δ_A. By

$$U_n(B) := \left\{ (b_1, \ldots, b_n) \in B^n \,|\, \text{There are } \beta_1, \ldots, \beta_n \in B \text{ such that } \sum_{j=1}^{n} \beta_j b_j = 1 \right\},$$

we will denote the set of all unimodular vectors in B^n ($n \in \mathbb{N}$). Then the answer to the second question is a particular case of the following result:

A.3.1 Proposition. *Let H be a finite universal Korovkin subset in B, consisting of n elements, and let A be a superalgebra of B. Suppose $U_n(B) = U_n(A) \cap B^n$.*
Then H is also a finite universal Korovkin subset in A.

A proof of this theorem is given in [1991b] by the author. Choosing $A = \mathscr{C}(\Delta_B)$, the condition on the unimodular vectors is satisfied for any n; consequently, every finite universal Korovkin subset in B induces via the Gelfand transformation a finite universal Korovkin subset in $\mathscr{C}(\Delta_B)$: This answers the second question completely. We remark that by passing to unitizations a variant of this result for the non-unital case may be obtained; the condition on the unimodular vectors may even take the form

$$U_n(B_1, B) = U_n(A_1, A) \cap \{b \in B_1^n \,|\, b \equiv e_1 \bmod B^n\}.$$

The answer to the first question depends on the following necessary condition on a commutative Banach algebra to possess a finite universal Korovkin subset:

A.3.2 Proposition. *Let A be a commutative Banach algebra with non-void spectrum Δ_A, and let us suppose that A possesses a finite universal Korovkin subset. Then the following statements hold:*
(1) *A is approximately normal, i.e., for any closed subset F and any compact subset E of Δ_A with empty intersection, there exists some $x \in A$ such that $|\hat{x} - 1| < \varepsilon$ on E and $|\hat{x}| < \varepsilon$ on F.*
(2) *Δ_A is a locally compact, separable metric space of finite local dimension.*

Proof. (1): By assumption, A_1 possesses a finite universal Korovkin subset, so the Choquet boundary of A_1 coincides with the structure space of A_1 by Proposition 8.1 of our paper [1986a]. But then A_1 and A are approximately regular by Lemma 2.2 of the author's paper [1992], and an inspection of the proof of Lemma 2.2 given by Wilken [1966] shows that A_1 and A are even approximately normal.
(2): Since A_1 possesses a finite universal Korovkin subset, it follows that the one-point compactification Δ_{A_1} of Δ_A is homeomorphic to a compact subset of some $\mathbb{C}^n$, which immediately implies the assertion (see our paper [1990c] for a discussion of the dimension theoretic aspects). $\qquad\square$

The necessary condition just established immediately implies the following result, which shows that a lot of finitely generated algebras do not possess any finite universal Korovkin subset and shows that membership in these algebras is not a property we may expect of the functions forming some finite universal Korovkin subset in $\mathscr{C}(\Delta_A)$:

A.3.3 Proposition. *Let A be a commutative unital Banach algebra. Suppose that there exists a closed ideal $I \subset A$ such that A/I has analytic structure in its spectrum. Then A does not possess any finite universal Korovkin subset.*

Here a commutative unital Banach algebra B is said to have *analytic structure in its spectrum* if and only if there exists a homeomorphism φ from the open unit disk D in the plane onto an open subset of Δ_B such that $\hat{x} \circ \varphi$ is analytic on D for every $x \in B$. A proof of this theorem is given by the author in [1991b]— alternatively, we note that A/I cannot be approximately normal due to the maximum principle for analytic functions, so A/I and hence A do not possess any finite universal Korovkin subset.

We now get lots of examples of finitely generated commutative unital Banach algebras with an isometric (but non-symmetric) involution that nevertheless do not possess any finite universal Korovkin subset:

The polydisk algebras $\mathscr{A}(\mathbb{D}^n)$ serve as counterexamples, as well as the algebras

$$\mathscr{A}^N(\mathbb{D}) := \{f \in \mathscr{A}(\mathbb{D}) | f^{(k)} \in \mathscr{A}(\mathbb{D}) \quad \text{for every } 0 \le k \le N\}$$

consisting of all functions on $\mathbb{D}$ whose derivatives of order at most N belong to $\mathscr{A}(\mathbb{D})$, which have been introduced by Novinger in [1991].

Another interesting example is Hoffman's "tomato can algebra" of all continuous functions on a solid cylinder, which are analytic on the disk at one end of the cylinder (see Hoffman [1966]): It obviously contains $\mathscr{A}(\mathbb{D})$ as a factor, so does not possess any finite universal Korovkin subset.

A class of related examples is given by the classical uniform algebras on planar sets. Let $K \subset \mathbb{C}$ be compact. We consider the following commutative unital Banach algebras (with pointwise operations and the uniform norm):

$$\mathscr{P}(K) := \{f \in \mathscr{C}(K) | f \text{ can be approximated uniformly on } K \text{ by polynomials}\},$$

$$\mathscr{R}(K) := \{f \in \mathscr{C}(K) | f \text{ can be approximated uniformly on } K \text{ by rational}$$

$$\text{functions with poles off } K\},$$

$$\mathscr{A}(K) = \{f \in \mathscr{C}(K) | f \text{ is holomorphic on the interior of } K\}.$$

Whenever one of these algebras is non-trivial in the sense that it differs from $\mathscr{C}(K)$, it does not contain any finite universal Korovkin subset:

A.3.4 Theorem. *Let $K \subset \mathbb{C}$ be compact, and let A be one of the algebras $\mathscr{A}(K)$, $\mathscr{R}(K)$, $\mathscr{P}(K)$. Then A possesses a finite universal Korovkin subset if and only if $A = \mathscr{C}(K)$.*

A *proof* of this theorem is given by the author in [1987], where the reader also finds conditions on K which characterize the validity of the equation $A = \mathscr{C}(K)$. The theorem remains true if A is allowed to be a T-invariant algebra. For compact subsets $K \subset \mathbb{C}^n$ ($n \geq 2$), it remains obviously true for $\mathscr{A}(K)$, but becomes false for $\mathscr{R}(K)$ and $\mathscr{P}(K)$, as we remarked in [1987].

In case we put additional requirements on the form of the universal Korovkin subset H, we may weaken the hypothesis on A:

A.3.5 Theorem. *Let A be a uniform algebra. Then the following assertions are equivalent*:
(i) *A possesses a two-element universal Korovkin subset $H = \{1, f\}$.*
(ii) *A is isometrically isomorphic to $\mathscr{C}(K)$, where K is a compact planar set which coincides with the extreme points of its convex hull.*

Here by a *uniform algebra* we mean a (commutative) unital Banach algebra whose norm is equivalent to the spectral radius.

We refer to our paper [1987] for a proof of this theorem; there the reader also finds an example of a uniform algebra, which is not a C^*-algebra but possesses a universal Korovkin subset of the form $H = \{1, f, g\}$.

Summing up, analyticity is not a property the test functions of a suitably chosen finite universal Korovkin subset in $\mathscr{C}(X)$ may have. At the end of this section, we will show that nevertheless infinite differentiability of the test functions may be achieved.

Before doing so, we will pursue the discussion of the finitely generated algebras investigated above a little further. Since these algebras do not contain any finite universal Korovkin subset, one knows in particular that no finite dimensional set H of polynomials is a universal Korovkin set in any of these algebras, though this might be the first example to try by Korovkin's second theorem.

This remark, however, does not answer the question what $K_u(H)$ looks like, even if we know that it differs from the whole of A. To approach this problem, we consider a compact subset X of some $\mathbb{C}^n$ and a Banach function algebra A on it, i.e., a unital point-separating subalgebra A of $\mathscr{C}(X)$, which contains the n coordinate functions $z_1, \ldots, z_n$ and is a Banach algebra under some norm dominating the uniform norm, which is supposed to be the spectral radius of A. Then via point evaluation, $\Gamma_A \subset X \subset \Delta_A$.

An important class of examples is given by the finitely generated commutative unital Banach algebras: If $x_1, \ldots, x_n$ are generators of such an algebra, a well-known theorem of Shilov implies that A is isomorphic to a Banach function algebra on $X := \sigma_A(x_1, \ldots, x_n)$.

For brevity's sake let us call a representation $x = \sum_{k=1}^{n} \lambda_k x_k$ of a point $x \in X$ as a non-trivial convex combination of pairwise different points $x_1, \ldots, x_n \in X \backslash \{x\}$

an *affine relation in X*. A function f defined on X is said to be *subject* to this affine relation if and only if $f(x) = \sum_{k=1}^{n} \lambda_k f(x_k)$ is true. Using this notational convenience, we have

A.3.6 Proposition. *Let X be a compact subset of $\mathbb{C}^n$, A a Banach function algebra on X and $H \subset A$ a finite subset containing $1, z_1, \ldots, z_n$. Then the following statements hold:*

(1) *$K(H)$ contains every function $f \in A$ which is subject to every affine relation in Γ_A to which every function $h \in H$ is subject.*

(2) *Suppose $X \cong \Delta_A$. Then $K_u(H)$ consists exactly of all functions $f \in A$ that are subject to every affine relation in X to which every function $h \in H$ is subject.*

Proof. The second assertion is proved in Section 8 of our survey paper [1988a]. The arguments given there also show that an affine relation in Γ_A to which every function $h \in H$ is subject corresponds exactly to a point evaluation supported by Γ_A and a discrete probability measure with finite support in Γ_A, whose restrictions to H coincide. Now, Theorem A.1.16 and our key observation finish the proof.   □

By means of the preceding proposition, the determination of $K(H)$ and $K_u(H)$ may be carried out using subsets of functional equations given by the set of all affine relations to which every function in H is subject.

As an example, let us consider a bounded region $G \subset \mathbb{C}^n$ and its compact closure K. The set of all polynomials in n complex variables of total degree at most m, regarded as functions on K, will be denoted by $H(m, n)$. Then it may be shown that

$$K_{u, \mathscr{A}(K)}(H(m, n)) = H(m, n), \tag{A.3.1}$$

$$K_{u, \mathscr{C}(K)}(H(m, n)) = H(m, n) + H(m, n)^*, \tag{A.3.2}$$

where $H(m, n)^*$ consists of all complex conjugates of the functions in $H(m, n)$.

The set of affine relations used to calculate $K_u(H(m, n))$ is given by the family of equations

$$p(z) = (N + 1)^{-1} \sum_{k=0}^{n} p(z + w\zeta^k) \quad (z \in G), \tag{A.3.3}$$

where $N \geq m$, $\zeta \in \mathbb{C}$ is a primitive root of unity of order $N + 1$ and w varies in a circled neighborhood V of the origin of $\mathbb{C}^n$ such that $z + V \subset G$. Every polynomial $p \in H(m, n)$ is subject to each of these affine relations, and they suffice to characterize the elements of $K_u(H(m, n))$: The polynomials $p \in H(m, n)$ are the only functions in $\mathscr{C}(K)$ which are subject to each of these affine relations. We refer to [1986b], where we give detailed proofs.

The functional equations (A.3.3) also allow to determine the universal Korovkin closure of the set $L := H(m_1, k_1) \otimes H(m_2, k_2) \otimes \cdots \otimes H(m_s, k_s) \subset \mathscr{A}(\mathbb{D}^n)$, where $n = k_1 + k_2 + \cdots + k_s$ and $m_1, m_2, \ldots, m_s \in \mathbb{N}$:

$$K_{u, \mathscr{A}(\mathbb{D}^n)}(L) = L, \tag{A.3.4}$$

$$K_{u, \mathscr{C}(\mathbb{D}^n)}(L) = L + L^*, \tag{A.3.5}$$

$$K_{\mathscr{A}(\mathbb{D}^n)}(L) = \mathscr{A}(\mathbb{D}^n), \tag{A.3.6}$$

$$K_{\mathscr{C}(\mathbb{D}^n)}(L) = L + L^*. \tag{A.3.7}$$

The first two equalities are proved by applying the above arguments to each of the factors $H(m_j, k_j)$. The last equality follows from the coincidence of $K(H)$ with $K_u(H)$ in $\mathscr{C}(X)$ for any H and any compact X. Finally, (A.3.6) follows from the fact that the Shilov boundary of the polydisk algebra is the polytorus, which coincides with the extreme points of its convex hull (which is the polydisk): Therefore, the Choquet boundary of L in $\mathscr{A}(\mathbb{D}^n)$ contains the Shilov boundary of $\mathscr{A}(\mathbb{D}^n)$, and L is a Korovkin subset in $\mathscr{A}(\mathbb{D}^n)$ by Theorem A.1.16. Alternatively, we note that by the extreme point argument there are no affine relations in the polytorus, so again L must be a Korovkin subset by Proposition A.3.6.

As our last example in this circle of ideas, we fix some complex numbers a, b, w, where $|w| < 1$, and consider the function f defined on the closed unit disk $\mathbb{D}$ by

$$f(z) := a(z - w)(1 - w^* z)^{-1} + b \quad (z \in \mathbb{D})$$

which belongs to the disk algebra $\mathscr{A}(\mathbb{D})$. Observing that f maps an open disk containing $\mathbb{D}$ biholomorphically onto another open disk somewhere in the plane and preserves circles, we conclude that the torus is the inverse image of the boundary of $f(\mathbb{D})$, which coincides with the extreme points of the convex hull of $\sigma(f)$. Consequently, $K(\{1, f\}) = \mathscr{A}(\mathbb{D})$. Since the operator $h \mapsto h \circ f^{-1}$ is an isometric isomorphism of $\mathscr{A}(\mathbb{D})$ onto $\mathscr{A}(f(D))$ transforming f into z, we conclude from the above that again $K_u(\{1, f\}) = \{\alpha + \beta f \,|\, \alpha, \beta \in \mathbb{C}\}$.

We now return to the beginning of this section, taking up the question if there is a finite universal Korovkin subset in $\mathscr{C}(M)$ consisting of infinitely (real) differentiable functions, where M is a compact Hausdorff manifold.

To see that the answer is always yes, we consider a class of topological algebras which is larger than the class of commutative unital Banach algebras, namely the (commutative) Waelbroeck algebras. All results on Korovkin approximation used in this paper for the more familiar class of commutative unital Banach algebras are also valid in the context of Waelbroeck algebras, as we found out in [1985], [1986a].

This observation is not merely a purely formal generalization, but in fact allows the consideration of other interesting algebras such as the algebra $\mathscr{C}^\infty(M)$ of all $\mathscr{C}^\infty$-functions on M, endowed with the topology of uniform convergence of the functions and all their derivatives, which is a complete commutative unital locally-m-convex Q-algebra, hence a complete Waelbroeck algebra.

A.3.7 Proposition. *Let M be a compact Hausdorff manifold. Then $\mathscr{C}^\infty(M)$ always possesses a finite universal Korovkin subset.*

To see this, use the fact that by Whitney's embedding theorem every compact Hausdorff manifold can be regularly embedded in $\mathbb{R}^n$ for some $n \in \mathbb{N}$ to find finitely many real-valued functions $f_1, \ldots, f_n \in \mathscr{C}^\infty(M)$ such that $\varphi = (f_1, \ldots, f_n): M \to \mathbb{R}^n$ is a regular embedding; in particular, the set $\{f_1, \ldots, f_n\}$ is a finite subset of $\mathscr{C}^\infty(M)$ separating the points of M. But the structure space of $\mathscr{C}^\infty(M)$ is homeomorphic to M via point evaluation, so that the set $H := \left\{1, f_1, \ldots, f_n, \sum_{k=1}^{n} f_k^2\right\}$ is a universal Korovkin subset for $\mathscr{C}^\infty(M)$.

One easily sees that the set $H = \{1, f_1 + if_1^2, \ldots, f_n + if_n^2\}$ would have served as well.

In both cases, H is (as in the Banach algebra case) automatically a universal Korovkin subset in $\mathscr{C}(M)$, so that our first question is answered.

To get more concrete examples, let us denote by $\mathscr{C}_{2\pi}^\infty(\mathbb{R}^d)$ the algebra of all $\mathscr{C}^\infty$-functions on $\mathbb{R}^d$ that are 2π-periodic in each coordinate direction, endowed with the topology of uniform convergence of the functions and all their derivatives. Then the obvious identification of $\mathscr{C}_{2\pi}^\infty(\mathbb{R}^d)$ with $\mathscr{C}^\infty(\mathbb{T}^d)$ (where $\mathbb{T}^d$ is the d-dimensional torus) is a topological isomorphism of complete Waelbroeck algebras. In this way we see that $\mathscr{C}_{2\pi}^\infty(\mathbb{R}^d)$ always possesses a finite universal Korovkin subset. Examples may be constructed using a regular embedding of $\mathbb{T}^d$ in some $\mathbb{R}^n$, e.g., the diffeomorphism from $\mathbb{T}^2$ onto the anchor ring in $\mathbb{R}^3$.

A.4 Generalized analytic functions and algebras generated by inner functions

We now consider a class of algebras which in some sense shares essential features with both classes of examples discussed in the previous chapters: There is a group in the background, and some sort of analyticity plays an important role.

Let G be a discrete abelian group, which contains a subsemigroup G_+ such that G_+ contains the identity of G and generates G. Let Γ be the compact dual group, and let λ be the normalized Haar measure on Γ. By $L^1(G_+)$ we will denote the unital subalgebra of $L^1(G)$ of those summable functions which vanish outside G_+: then clearly $L^1(G_+)$ is closed in $L^1(G)$, hence a commutative unital Banach algebra.

Via convolution, $L^1(G_+)$ determines a unital algebra of operators on $L^2(G)$, whose completion with respect to the operator norm yields a commutative unital Banach algebra $\mathscr{A}(G)$, *the algebra of all generalized analytic functions* discovered by Arens and Singer. The structure space of $\mathscr{A}(G)$ is, in the usual fashion, homeomorphic to the compact Hausdorff space Γ_+ of all characters of G_+, i.e., of all continuous complex-valued semigroup homomorphisms ξ of G_+, which are not identically zero and take their values in the closed unit disk $\mathbb{D}$. Via Fourier transformation, the norm of $\mathscr{A}(G)$ is just the uniform norm on Γ, and $\mathscr{A}(G)$ is a uniform algebra which may be identified with

$$\mathscr{A}(\Gamma) := \{ f \in \mathscr{C}(\Gamma) | \hat{f}(g) = 0 \text{ for every } g \in G_+ \}.$$

This explains the name generalized analytic function—just think of the case $G = \mathbb{Z}$, $G_+ = \mathbb{N}$, $\Gamma = H$ (the one dimensional torus), $\Gamma_+ = D$: The functions in $\mathscr{A}(\Gamma)$ are characterized by the fact that they have analytic extensions (via Gelfand transformation) to D, and we arrived at another possibility of defining the disk algebra.

This example suggests that Γ may be identified with the Shilov as well as the Choquet boundary of $\mathscr{A}(G) \cong \mathscr{A}(\Gamma)$, and indeed this is true (and proved, e.g., by Arens and Singer in [1956], respectively Suciu in [1975]).

Since $\mathscr{A}(\Gamma)$ is just the closure with respect to the spectral radius norm of the image of $L^1(G_+)$ under its Gelfand transformation, the identifications of the structure space, the Shilov boundary and the Choquet boundary remain true for $L^1(G_+)$ as well.

We now turn to Korovkin approximation on these algebras, showing that most of them do not possess any universal Korovkin subset:

A.4.1 Proposition. *Let G be a discrete abelian group with character group Γ, and let G_+ be a subsemigroup such that G_+ contains the identity of G and generates G.*

Then the following assertions are equivalent:
(i) *$\mathscr{A}(G)$ possesses a finite universal Korovkin subset.*
(ii) *$G_+ = G$ and G is a countable group with finite torsion-free rank.*
(iii) *$\mathscr{A}(\Gamma) = \mathscr{C}(\Gamma)$, and Γ is a metric space of finite covering dimension.*
(iv) *$\Gamma_+ = \Gamma$, and Γ is a metric space of finite covering dimension.*
(v) *$L^1(G_+)$ possesses a finite universal Korovkin subset.*

Proof. Suppose $\mathscr{A}(G)$ possesses a finite universal Korovkin subset; then its Shilov boundary coincides with its structure space, so that $\Gamma_+ = \Gamma$ by the above remarks. Since $\mathscr{A}(G)$ possesses a finite universal Korovkin subset, its structure space is a compact metric space of finite covering dimension, so that (i) implies (iv).

Since $L^1(G_+)$ has the same structure space and Shilov boundary as $\mathscr{A}(G)$, we conclude from (iv) that the Shilov boundary and structure space of $L^1(G_+)$ coincide. Observing that Γ_+ separates the points of G_+ (since Γ separates the points of G), we may now apply Theorem 6.4 of Comfort's paper [1961] to conclude that G_+ is in fact a union of groups. However, being a subset of the group G, G_+ contains exactly one idempotent element, so that G_+ must be a group and therefore coincides with G. Therefore, (iv) implies (ii).

But if $G_+ = G$, we necessarily have $\mathscr{A}(\Gamma) = \mathscr{C}(\Gamma)$ by the above definition of $\mathscr{A}(\Gamma)$; consequently, (ii) implies (iii).

Now (iii) obviously implies (i), since by the assumption on Γ, $\mathscr{C}(\Gamma)$ possesses a finite universal Korovkin subset. This shows that the assertions (i)–(iv) are indeed equivalent. Since obviously (ii) implies (v) and (v) implies (i), this finishes the proof. ☐

For some concrete examples of finitely generated discrete abelian groups, one may find easy examples of universal Korovkin subsets in $\mathscr{A}(G)$ in Altomare [1984].

We recall that a non-zero element of a uniform algebra will be called *inner*, if its associated multiplication operator is an isometry on A; this happens exactly if the element has unit modulus on the Shilov boundary of A—see, e.g., Suciu's monograph [1975] for a proof.

Using the relevant structure theory, we now characterize the existence of finite universal Korovkin subsets in uniform algebras generated by inner elements:

A.4.2 Proposition. *Let A be a uniform algebra generated by inner functions. Then the following assertions are equivalent*:
 (i) *A possesses a finite universal Korovkin subset.*
(ii) *$A \cong \mathscr{C}(\Delta_A)$, and Δ_A is a compact metric space of finite covering dimension.*

Proof. If (i) holds, the Shilov boundary and structure space of A coincide. A being generated by inner functions, its Shilov boundary coincides with those characters of A which have absolute value identically one on the inner elements of A (see Suciu [1975]). Taken together, these facts imply that every inner element of A is invertible, and that an element of A is inner exactly if it has absolute value identically one on the structure space of A, so that in particular its inverse and its complex conjugate coincide. For this reason, the algebra generated by the inner elements (which is A by our initial assumption) is closed with respect to the formation of complex conjugates and therefore coincides with $\mathscr{C}(\Delta_A)$ by the Stone-Weierstrass theorem. Since the remaining implication is obvious this finishes the proof. ☐

The last two propositions are closely related in the following way: Since $\mathscr{A}(G)$ is an example of a uniform algebra generated by inner functions, part of

Proposition A.4.1 is a consequence of Proposition A.4.2. Conversely, under some additional assumptions on the inner elements described by Suciu [1975] and Mochizuki [1964], uniform algebras generated by inner functions are already isomorphic to some $\mathscr{A}(G)$, so that in these cases, Proposition A.4.2 may be deduced from Proposition A.4.1. We exploit this last point of view in our final Proposition:

A.4.3 Proposition. *Let A be a uniform algebra on a compact Hausdorff space X. Suppose there exists a probability measure μ on X and a semigroup P of inner functions of A containing the unit of A, such that the elements of P are pairwise orthogonal in $L^2(\mu)$ and the closed subspace generated by them is the kernel of μ in A. Then the following assertions are equivalent:*
 (i) *A possesses a finite universal Korovkin subset.*
 (ii) *$A = \mathscr{C}(X)$, and X is a metric space of finite covering dimension.*
 (iii) *P is a countable group with finite torsion-free rank.*

Proof. By Theorem 1 of Mochizuki [1964], A is isometrically isomorphic to $\mathscr{A}(PP^{-1})$, so that the assertions follow immediately from Proposition A.4.1. $\square$

In Mochizuki [1964] it is shown that every $\mathscr{A}(G)$ is in fact isometrically isomorphic to an algebra A having the properties of the last result related to μ and P—we refer to this paper for further details.

A.5 Extreme spectral states and the Gleason-Kahane-Zelazko property

Let A be a commutative unital Banach algebra. If additionally A has a symmetric involution, it is well known that the structure space coincides with the set of extreme points of the weak-*-compact convex set of all unital positive linear functionals on A. In the general case, the set Ω_A of all spectral states is compact and convex as well, and a standard argument shows that it is the closed convex hull of Δ_A, even of Γ_A—therefore, the set $\partial_e \Omega_A$ of extreme points of Ω_A is contained in Γ_A by Milman's theorem. Since in many cases Ω_A is the natural analogue of the set P_A of all unital positive linear functionals (states) in the non-involutive situation, the following problem posed by Bonsall and Duncan in [1971] naturally occurs:

Characterize those A such that $\partial_e \Omega_A = \Delta_A$.

This problem has been solved in Pannenberg [1986a]. The solution is contained in the following theorem, where we denote the Choquet boundary of A by ∂_A:

A.5.1 Theorem. *Let A be a commutative unital Banach algebra. Then the following assertions are equivalent:*

(i) $\partial_e \Omega_A = \Delta_A$.

(ii) $\partial_A = \Delta_A$.

(iii) $\hat{A}$ *is a universal Korovkin subset in* $\mathscr{C}(\Delta_A)$.

(iv) A *is approximately regular.*

Proof. We refer the reader to Pannenberg [1986a] for the proof of the equivalence of the assertions (i)–(iii); there it is also shown that $\partial_e \Omega_A = \partial_A$. Now Lemma 2.2 of Pannenberg [1992b] yields the equivalence of (ii) and (iv), which proves the theorem and solves the problem posed by Bonsall and Duncan. $\square$

The last result shows that in some cases Korovkin-type results characterize properties of A which apparently have nothing to do with Korovkin theory. This is also true for the next theorem, which was motivated by recent work of Badea and Rasa [1993].

Again, let A be a commutative unital Banach algebra. If $E, F \subset A'$, $M \subset \Delta_A$ and $H \subset A$ are arbitrary subsets, we say that

$$(F, M, H) \text{ has the Gleason-Kahane-Zelazko property}$$

if every $f \in F$, which is such that for every $x \in H$ there exists some $m_x \in M$ with $f(x) = m_x(x)$, already belongs to M.

This definition is due to Badea and Rasa [1993]. We extend this definition and say that

$$(E, F, M, H) \text{ has the Gleason-Kahane-Zelazko property}$$

if every $g \in E$, which is such that for every $x \in A$ there exists some $f_x \in F$ with $g(x) = f_x(x)$ and for every $x \in H$ there exists some $m_x \in M$ with $g(x) = m_x(x)$, already belongs to M.

Stated in a different terminology, (E, F, M, H) has the Gleason-Kahane-Zelazko property if every $g \in E$ belonging locally to F on A and locally to M on H already belongs to M (globally).

Obviously, (F, M, H) has the Gleason-Kahane-Zelazko property, if and only if (F, F, M, H) has the Gleason-Kahane-Zelazko property.

A.5.2 Lemma. *Let A be a commutative Banach algebra and H be a subset of A. Then (Ω_A, Δ_A, H) has the Gleason-Kahane-Zelazko property, if and only if $(A', \Omega_A, \Delta_A, H)$ has the Gleason-Kahane-Zelazko property.*

Proof. Just observe that every $f \in A'$ belonging locally to Ω_A on A already belongs globally to Ω_A, since for any $x \in A$ we have $\text{co}(\sigma(x)) = \{h(x) | h \in \Omega_A\}$. $\square$

Following Badea and Rasa, we observe that the Gleason-Kahane-Zelazko theorem is equivalent to the assertion that (A', Δ_A, A^{-1}) has the Gleason-Kahane-Zelazko property.

We now show that subsets H such that (Ω_A, Δ_A, H) has the Gleason-Kahane-Zelazko property are necessarily universal Korovkin subsets in A—this generalizes a corresponding result for the set P_A of states on an algebra with a symmetric involution, which is proved in Badea and Rasa [1993].

A.5.3 Theorem. *Let A be a commutative unital Banach algebra, and let H be a subset of A containing the unit of A. Suppose (Ω_A, Δ_A, H) has the Gleason-Kahane-Zelazko property. Then H is a universal Korovkin subset in A.*

Proof. We adapt the argument given by Badea and Rasa and begin by showing that H separates the points of Δ_A, which in any case is necessary for H to be a universal Korovkin subset in A. To this end, consider two characters m_1, $m_2 \in \Delta_A$, which coincide on H, and the midpoint φ of the segment they define in A'. Being a convex combination of two characters, φ is in Ω_A, and by construction, φ coincides on H with a character, so in particular belongs locally to Δ_A on H. Since (Ω_A, Δ_A, H) has the Gleason-Kahane-Zelazko property, φ belongs to Δ_A (globally). But $\Delta_A = \partial_e \operatorname{co}(\Delta_A)$, as proved, e.g., in Pannenberg [1985], so that φ can only be the midpoint of a line joining two characters if these characters coincide. Consequently, $m_1 = m_2$, and we have proved that H separates the points of Δ_A.

The proof is finished with an application of Theorem A.1.6: Suppose $f \in \Omega_A$ and $m \in \Delta_A$ coincide on H. Then in particular f belongs locally to Δ_A on H, so by assumption f belongs to Δ_A globally. Now f and m are two characters of A coinciding on H, and since H separates the points of Δ_A, this shows that f coincides with m. Therefore, H is a universal Korovkin subset in A. □

A.5.4 Corollary. *Let A be a commutative unital Banach algebra, and let H be a subset of A containing the unit of A. Suppose H determines spectral selection, i.e., whenever $\varphi \in A'$ has the selection property*

$$\varphi(x) \in \operatorname{co}(\sigma(x)) \quad \text{for every } x \in A, \tag{A.5.1}$$

$$\varphi(x) \in \sigma(x) \qquad \text{for every } x \in H, \tag{A.5.2}$$

we may conclude that even

$$\varphi(x) \in \sigma(x) \quad \text{for every } x \in A. \tag{A.5.3}$$

Then H is a universal Korovkin subset in A.

The *proof* is an immediate consequence of the preceding theorem, Lemma A.5.2 and the following lemma:

A.5.5 Lemma. *Let A be a commutative unital Banach algebra, and let H be a subset of A containing the unit of A. Then H determines spectral selection, if and only if $(A', \Omega_A, \Delta_A, H)$ has the Gleason-Kahane-Zelazko property.*

Proof. The conditions above just mean that φ belongs locally to Ω_A on A (A.5.1), that φ belongs locally to Δ_A on H (A.5.2), and that φ belongs globally to Δ_A (A.5.3), where the last assertion is the Gleason-Kahane-Zelazko theorem. This proves the lemma.   □

The following theorem, proved in Badea and Rasa [1993], connects Korovkin-type approximation to the Gleason-Kahane-Zelazko theorem:

A.5.6 Theorem. *Let A be a commutative unital Banach algebra with a symmetric involution. Then the following assertions are equivalent:*
 (i) *A possesses a finite universal Korovkin subset.*
 (ii) *There exists a finite subset H of A containing the unit of A such that H determines spectral selection.*
 (iii) *There exists a finite subset H of A containing the unit of A such that (P_A, Δ_A, H) has the Gleason-Kahane-Zelazko property.*

Proof. Just use Theorem 4 of Badea and Rasa [1993] as well as Lemmas A.5.2 and A.5.5.   □

With the help of the following lemma, we will generalize the last theorem to certain non-involutive algebras:

A.5.7 Lemma. *Suppose A and B are commutative unital Banach algebras, such that B is contained algebraically as a unital subalgebra in A, which is full in A and dense with respect to the spectral radius of A. If H is a subset of B containing the common unit of A and B we may conclude that H determines spectral selection in B, if and only if H determines spectral selection in A.*

Proof. Consider the selection properties (A.5.1), (A.5.2) and (A.5.3) for A as well as for B: Since the spectrum of any element of B is the same, whether computed in A or in B, we see at once that the selection property (A.5.2) is the same in A and in B, and that the selection properties (A.5.1) and (A.5.3) are fulfilled in B whenever they are fulfilled in A. Conversely, if one of these latter properties is fulfilled in B, we know that φ is a spectral state (respectively, character) of B, which may be extended to a unique spectral state (respectively, character) of A by our assumption, so that (A.5.1) (respectively, (A.5.3)) is fulfilled in B.

Consequently, each of the selection properties is the same in A as well as in B—this obviously proves the lemma. □

We now generalize the result of Badea and Rasa to commutative Beurling algebras, which in general are not involutive.

A.5.8 Theorem. *Let G be a locally compact abelian group, and let ω be an upper semi-continuous weight function on G. Then the following assertions are equivalent:*
(i) *$L^1_\omega(G)$ possesses a finite universal Korovkin subset.*
(ii) *There exists a finite subset H in $L^1_\omega(G)$ which determines spectral selection.*

Proof. Proceeding as in the proof of the Theorem 3.1 of Pannenberg [1992b], we may find a weight v on G such that $B := L^1_v(G)$ has a symmetric involution and has all the properties which are required for B in Lemma A.5.7 with $A := L^1_\omega(G)$.

Now suppose (i) holds. Then B possesses a finite universal Korovkin subset as well by the main result of Pannenberg [1992b], so that by Theorem 4 of Badea and Rasa [1993], (ii) holds for B, as well as for A by the preceding Lemma.

Conversely, if (ii) holds, the same condition holds for B, and again by Badea's and Rasa's result, B possesses a finite universal Korovkin subset H. By Theorem 6.2 of Pannenberg [1986a], H is a universal Korovkin subset in A as well, which proves (i). □

Appendix B

Korovkin-type approximation theory on C^*-algebras

by Ferdinand Beckhoff

By a well-known theorem of Gelfand and Naimark the algebras $\mathscr{C}_0(X)$ and $\mathscr{C}(X)$, the main actors of the present book, are up to isomorphism the most general commutative C^*-algebras. Recall that a C^*-*algebra* is a Banach algebra A equipped with an involution * such that $\|x^*x\| = \|x\|^2$ for all $x \in A$. All these algebras may be faithfully represented as *-closed and norm closed subalgebras of $L(\mathscr{H})$, the C^*-algebra of all linear bounded operators on $\mathscr{H}$ for a suitable Hilbert space $\mathscr{H}$ (this is also due to Gelfand and Naimark). For more details on this result and on basic theory of C^*-algebras the reader is referred, for instance, to Bonsall and Duncan [1973] and to Sakai [1971]. In C^*-algebras we have an order structure given by $x \geq 0 :\Leftrightarrow x = y^*y$ for some $y \in A$. If A is represented in $L(\mathscr{H})$ we have $x \geq 0 :\Leftrightarrow \langle x\xi, \xi \rangle \geq 0$ for every $\xi \in \mathscr{H}$, which might be a more familiar notion of positivity.

Having an order we may talk about positive functionals and positive operators and so can transfer the definitions of Korovkin closures into the setting of arbitrary C^*-algebras. Even more general Banach-*-algebras may be taken into consideration, we only need a reasonable order structure.

As an example for such a definition: If A and B are C^*-algebras, H a subset of A and $T: A \to B$ a linear continuous operator, then $K^1_+(H, T)$ is defined to be the set of all $a \in A$ such that $\lim_{\substack{\leq \\ i \in I}} P_i a = Ta$ whenever $(P_i)^{\leq}_{i \in I}$ is a net of positive linear contractions from A into B satisfying $\lim_{\substack{\leq \\ i \in I}} P_i h = Th$ for all $h \in H$. Here the convergence means norm convergence which will always be the case unless stated otherwise.

In a similar way, one can define $K_+(H, T)$ by considering equicontinuous nets of positive linear operators from A into B.

As in the commutative theory T will be required to have special properties, say 'T is a *-homomorphism' or '$T = I_A$' where $A = B$.

If H is a subset of a C^*-algebra, then

$$K^+_u(H) = K^+_{u, A}(H) := \bigcap \{K^1_+(H, T) \mid T: A \to B \text{ a *-homomorphism, } B \text{ a } C^*\text{-algebra}\}$$

is called the *universal Korovkin closure of H*. If the intersection is taken over all surjective *-homomorphisms from A onto B we call it the *surjective Korovkin closure* and denote it by $K_u^{+,s}(H) = K_{u,A}^{+,s}(H)$.

Some results from the commutative theory, and from the Korovkin-type approximation theory on $\mathscr{C}_0(X)$, can be extended to non-commutative algebras. This report intends to give a taste of what these results look like. To keep this report to an appendix-like length it was impossible to have a closer look at the methods used to prove them.

This appendix is divided into two parts. In the first part pointwise convergence of positive linear functionals $\phi_i\colon A \to \mathbb{C}$ $(i \in I)$ to another linear functional ϕ is considered. Under what conditions may we conclude $\lim_{\substack{\leq \\ i \in I}} \phi_i(a) = \phi(a)$ for all $a \in A$ provided we know $\lim_{\substack{\leq \\ i \in I}} \phi_i(a) = \phi(a)$ for all a in a test set H? In the second part we ask analogous questions for the pointwise convergence of positive operators between C^*-algebras and even more general Banach algebras. In this context various topologies may be taken into account.

But before beginning this I would like to express my thanks to Professor F. Altomare and Professor M. Campiti for having invited me to write this Appendix which was an honor and a pleasure for me, and to Professor G. Maltese who introduced me to Korovkin-type approximation theory some years ago.

B.1 Approximation by positive linear functionals

One important special case in the above approximation situation is $B = \mathbb{C}$, i.e., approximation of positive linear functionals. In the commutative situation the first paper in this setting was written by Choda and Echigo [1963]. Probably the earliest paper about this subject where non commutative algebras have been admitted is from 1976 by B.V. Limaye and S. Shirali. Here pointwise convergence of positive linear functionals to multiplicative linear functionals is considered (see also Wille [1985] for this). Let us describe a result of the subsequent paper of B.V. Limaye and M.N.N. Namboodiri from 1979 that was stated in the more general context of *-normed algebras:

Let A be a *-normed algebra. Denote by P_A the convex set of all continuous positive linear functionals with norm ≤ 1 and by $\partial_e P_A$ the set of its extreme points. The theorem is the following:

B.1.1 Theorem. *Let B be a *-normed algebra containing an approximate identity, and let $\phi, \phi_1, \phi_2, \ldots$ be positive linear functionals on B such that $\lim_{n \to \infty} \|\phi_n\| = \|\phi\|$.*

Let H be a subset of B and assume that for all $b \in H$ we have

$$\lim_{n \to \infty} \phi_n(b) = \phi(b),$$

$$\lim_{n \to \infty} \phi_n(b^*b) = \phi(b^*b) = |\phi(b)|^2,$$

and

$$\lim_{n \to \infty} \phi_n(bb^*) = \phi(bb^*) = |\phi(b)|^2.$$

Then, $\lim_{n \to \infty} \phi_n(a) = \phi(a)$ for all elements a in the closed *-subalgebra A of B generated by H. If moreover $\phi_{|A} \in \partial_e P_A$, and ϕ is the only extension of $\phi_{|A}$ to an element of $\partial_e P_B$, then $\lim_{n \to \infty} \phi_n(b) = \phi(b)$ for all $b \in B$.

This generalizes the commutative situation, the extreme positive linear functionals play the role of the characters, i.e., the point evaluations in the case of $\mathscr{C}(X)$. Another result of this spirit can be found in Takahasi's paper, also published in 1979:

B.1.2 Theorem. *Let ϕ be an extreme state of a C^*-algebra A and let $x \in A_+$ peak for ϕ, i.e., the supports of x and ϕ in the enveloping von Neumann algebra add up to 1. Let $(\phi_i)_{i \in I}^{\leq}$ be a net of positive linear functionals of A and assume $\lim_{\substack{\leq \\ i \in I}} \phi_i(1) = 1$ and $\lim_{\substack{\leq \\ i \in I}} \phi_i(x) = 0$.*

Then, we have $\lim_{\substack{\leq \\ i \in I}} \phi_i = \phi$ pointwise on A.

If A is commutative, ϕ an extreme state (i.e., a point evaluation), and if $a \in A_+$ generates $\mathrm{Ker}(\phi)$ as a closed ideal, then a peaks for ϕ. So for commutative C^*-algebras the above theorem is, in fact, Theorem 2.5.4.

For other results of this kind in the setting of commutative Banach algebras we refer to Altomare [1982c] and Romanelli [1989].

In the case of the C^*-algebra $L(\mathscr{H})$, $\mathscr{H}$ a Hilbert space, ϕ a vector state, better results can be achieved (see Limaye-Namboodiri [1979], Altomare [1987], Dieckmann [1992]). One of the results of Altomare [1987] is

B.1.3 Theorem. *Let $T \in L(\mathscr{H}) \setminus \{0\}$ be a compact operator, λ a simple eigenvalue of T such that $\|T\| = |\lambda|$, x a corresponding unit eigenvector, and let $\phi := \langle \cdot x, x \rangle$ be the vector state corresponding to x, i.e., $\phi(S) = \langle Sx, x \rangle$ for every $S \in L(\mathscr{H})$. Put*

$$S := I_{\mathscr{H}} + \frac{1}{|\lambda|^2} T^*T - \frac{1}{\lambda}T - \left(\frac{1}{\lambda}T\right)^*.$$

Then, we have $K_+(\{I_{\mathscr{H}}, S\}, \phi) = L(\mathscr{H})$.

Here $T^* \in L(\mathscr{H})$ denotes the adjoint of T.

Dieckmann [1992] improved this result by weakening the assumption $\|T\| = |\lambda|$ to $\lambda \neq 0$.

Instead of $L(\mathscr{H})$ one can consider the C^*-algebra

$$\mathscr{A}(\mathscr{H}) := \{\lambda I_{\mathscr{H}} + T \mid \lambda \in \mathbb{C}, T \in L(\mathscr{H}) \text{ compact}\},$$

where $\dim(\mathscr{H}) = \infty$.

B.1.4 Theorem. *Let $T \in L(\mathscr{H})$ be a strictly positive (i.e., $\langle Tx, x \rangle > 0$ for every $x \in \mathscr{H} \backslash \{0\}$) compact operator and let $\phi \colon \mathscr{A}(\mathscr{H}) \to \mathbb{C}$ be the homomorphism defined by $\phi(\lambda I_{\mathscr{H}} + S) = \lambda, S \in L(\mathscr{H})$ compact. Then, $K_+(\{I_{\mathscr{H}}, T\}, \phi) = \mathscr{A}(\mathscr{H})$.*

This is a reformulation of a result also taken from Altomare [1987].

B.2 Approximation by positive linear operators

Now let us turn to the approximation of positive linear operators.

Already in 1970, W.B. Arveson considered the approximation of *-homomorphisms $T \colon \mathscr{C}(X) \to A$ by positive linear operators, where X is a compact Hausdorff space and A another C^*-algebra. The support K_T of T is defined to be the closed subset of X corresponding to the ideal $\text{Ker}(\phi)$ (so $T(f) = 0 \Leftrightarrow f_{|K_T} = 0$).

We recall that the Choquet boundary $\partial_H^+ X$ of a subspace H of $\mathscr{C}(X)$ is the set of all points $x \in X$ such that the evaluation functional ε_x at x is the only positive linear functional on $\mathscr{C}(X)$ extending $\varepsilon_{x|H}$ (see Section 2.6).

Arveson's result can now be stated as follows:

B.2.1 Theorem. *Let $1 \in H \subset \mathscr{C}(X)$ and let $T \colon \mathscr{C}(X) \to A$ be a *-homomorphism into a C^*-algebra A such that $K_T \subset \partial_H^+ X$. Then $K_+(H, T) = \mathscr{C}(X)$.*

The proof relies on the lattice structure of the self-adjoint part of $\mathscr{C}(X)$. Since the self-adjoint part of a C^*-algebra A is a lattice in the natural order if and only if A is commutative, one has to look for other methods to investigate the approximation of *-homomorphisms between general C^*-algebras.

One of the early results in this generality has been proved by W.M. Priestley in 1976. A *J*-algebra* in a C^*-algebra is a *-closed and norm closed subspace which is also closed with respect to the Jordan product $x \circ y := \frac{1}{2}(xy + yx)$.

B.2.2 Theorem. *Let A be a C^*-algebra with unit 1 and $(P_i)_{i \in I}^{\leq}$ be a net of positive linear operators from A into itself such that $P_i 1 \leq 1$ for every $i \in I$. Then, the subset*

$$J := \left\{ x \in A \,\middle|\, \lim_{i \in I} {}_{\leq} P_i a = a \text{ for } a \in \{x, x^* \circ x, x^2\} \right\},$$

is a J^*-algebra in A.

For positive linear operators between C^*-algebras the condition $P_i 1 \leq 1$ ($i \in I$) says that P_i is a contraction. So if $J^*(H)$ denotes the J^*-algebra generated by $H \subset A$, then we easily deduce from Priestley's result:

$$J^*(H) \subset K_+^1(H \cup H^* \circ H \cup H^2).$$

Here $H^* \circ H := \{h^* \circ h \,|\, h \in H\}$ and $H^2 := \{h^2 \,|\, h \in H\}$.

If A is embedded as a subalgebra of $L(\mathcal{H})$, where $\mathcal{H}$ is a Hilbert space, W.M. Priestley obtained similar results for convergence in the weak and strong operator topology restricted to A. He also considered the trace class which is not a C^*-algebra but a Banach-$*$-algebra.

The key idea of W.M. Priestley was the use of the Kadison-Schwarz inequality

$$P(x)^* \circ P(x) \leq P(x^* \circ x) \quad \text{for all } x \in A,$$

which holds for all positive linear contractions on a C^*-algebra.

A more restrictive class of operators is the class of *Schwarz operators* P: $A \to A$ which are linear and satisfy

$$P(x)^* P(x) \leq P(x^* x) \quad \text{for all } x \in A.$$

Let us denote the Korovkin closure of a subset H of A with respect to Schwarz operators (Schwarz operators P satisfying $P(1) \leq 1$, respectively) by $K_S(H)$ ($K_S^1(H)$, respectively). In 1977, A.G. Robertson proved:

B.2.3 Theorem. *Let A be a C^*-algebra with unit 1 and $(P_i)_{i \in I}^{\leq}$ be a net of Schwarz operators from A in itself such that $P_i 1 \leq 1$ for every $i \in I$. Then, the subset*

$$K := \left\{ x \in A \,|\, \lim_{i \in I} {}_{\leq} P_i a = a \text{ for } a \in \{x, x^* x, xx^*\} \right\},$$

is a C^*-subalgebra of A.

This immediately leads to the inclusion

$$C^*(H) \subset K_S^1(H \cup H^*H \cup HH^*),$$

where H is a subset of A and $C^*(H)$ denotes the C^*-algebra generated by H, the notions H^*H and HH^* being clear.

These results have been improved by B.V. Limaye and M.N.N. Namboodiri in 1982.

B.2.4 Theorem. *Let A and B be C^*-algebras with units 1_A and respectively 1_B. Let $(P_i)_{i \in I}^{\leq}$ be a net of positive linear operators from A into B such that $P_i 1_A \leq 1_B$ for every $i \in I$, and let $T: A \to B$ be a C^*-homomorphism. Then, the subset*

$$J := \left\{ x \in A \,\middle|\, \lim_{\substack{\leq \\ i \in I}} P_i a = Ta \text{ for } a \in \{x, x^* \circ x\} \right\},$$

is a J^*-algebra in A.

If all P_i $(i \in I)$ are Schwarz operators, then J is a C^*-algebra.

This theorem shows that the condition $\lim_{\substack{\leq \\ i \in I}} P_i(x^2) = x^2$ used to define J in Priestley's result is unnecessary. Again norm convergence may be replaced by strong or weak operator convergence. The next step is an examination of the set

$$D := \left\{ x \in A \,\middle|\, \lim_{\substack{\leq \\ i \in I}} P_i a = Ta \text{ for } a \in \{x, x^*x\} \right\},$$

for Schwarz operators P_i $(i \in I)$, which turns out to be an algebra (see Limaye and Namboodiri [1984] together with Robertson [1986]).

More general Banach-*-algebras are considered in Beckhoff [1987], [1991]. The inclusion

$$J^*(H) \subset K_+^1(H \cup (H^* \circ H)),$$

which easily follows from the result of Limaye and Namboodiri stated above, turns out to be an equality for H^*-algebras and dual C^*-algebras (Beckhoff [1991]).

For universal Korovkin closures one can say more. If A is a type I C^*-algebra and H is a subset of A, then $K_u^+(H)$ is always contained in the C^*-algebra generated by H. We have the same result for the surjective Korovkin closure if A is liminal. Since commutative C^*-algebras are liminal and homomorphic images of commutative C^*-algebras clearly are commutative, we see that $K_u^+(H)$

coincides with the definition given by Altomare [1986]. These results, which are taken from Beckhoff [1992], also can be extended to more general Banach-* algebras by using enveloping C^*-algebras.

The idea of unique extensions has been worked out by Limaye and Namboodiri ([1984] and [1986]). Since one must use the existence of limit points of the nets used for the approximation one is lead to consider the weak operator topology, since bounded closed sets are compact in this topology. One of the results in Limaye and Namboodiri [1984] is the following:

B.2.5 Theorem. *Let S be an irreducible set in $L(\mathcal{H})$ which contains the identity operator and suppose that there are $T \in \mathrm{span}(S^* \cup S)$ and a compact operator $K \in L(\mathcal{H})$ such that $\|T - K\| < \|T\|$. Then, we have $K_{cp}^{1,w}(S) = L(\mathcal{H})$.*

The expression $K_{cp}^{1,w}(S)$ as a special kind of Korovkin closure is nearly self-explaining: it is the set of all $T \in L(\mathcal{H})$ such that $\lim_{\substack{\leq \\ i \in I}} P_i(T) = T$ in the weak operator topology (hence the notation w) for every net $(P_i)_{i \in I}^{\leq}$ of completely positive operators (hence the notation cp) satisfying $P_i 1 \leq 1$ for every $i \in I$ (hence 1) and $\lim_{\substack{\leq \\ i \in I}} P_i U = U$ for all $U \in S$. Such results were also obtained by Popescu [1989].

Limaye and Namboodiri [1986] gave the following application of matrix approximation.

B.2.6 Theorem. *Let $x_1, \ldots, x_m$ be distinct points in a compact Hausdorff space X and let $P_1, \ldots, P_m$ be mutually orthogonal non-zero self-adjoint projections in the algebra $M_k (= L(\mathbb{C}^k))$ of complex $k \times k$-matrices. Define $T: \mathscr{C}(X) \to M_k$ by $Tf := \sum_{i=1}^{m} f(x_i)P_i$ for every $f \in \mathscr{C}(X)$. Let H be a subset of $\mathscr{C}(X)$ such that $1 \in H$, and each $x_i (i = 1, \ldots, m)$ may be separated from every other point by using a function in H. Then, we have $K_+^1(H \cup H^*H, T) = \mathscr{C}(X)$.*

The proof depends on the specific knowledge of the extreme points of the set of positive linear contractions from $\mathscr{C}(X)$ into M_k. This can alternatively be deduced from Arveson's result: $K_T = \{x_1, \ldots, x_m\}$ is obvious, and the assumption on the points $x_1, \ldots, x_m$ yields $\{x_1, \ldots, x_m\} \subset \partial_{H \cup H^*H} X$. Of course, Limaye and Namboodiri achieved this example as a corollary of a theorem which is more general than Arveson's result.

The unique extension method can also yield Korovkin approximation results about norm convergence and this leads us back to the paper of S. Takahasi already mentioned. Let A be a C^*-algebra, H be a subset of A and let $E \subset A'$ be the state space of A. The Choquet boundary $\partial_H E$ is defined to be the set of all states $f \in E$ such that f is the only positive linear functional on A extending $f_{|H}$. Let $P \subset E$ be the set of pure states (i.e., extreme points) and let $\bar{P}$ denote its w^*-closure. Then, Theorem 3.4 of Takahasi [1979] can be stated as follows:

B.2.7 Theorem. *Let A be a C*-algebra with unit and H be a subset of A such that* $\bar{P} \subset \partial_H E$. *Then,* $A = K_+(H) := K_+(H, I_A)$.

This has been generalized by Altomare [1984a], where we find the definition

$$U(H, F_1, F_2) := \{a \in A \,|\, f_1(a) = f_2(a) \text{ whenever } f_1 \in F_1, f_2 \in F_2 \text{ satisfy } f_{1|S} = f_{2|S}\},$$

where $H \subset A$, $F_1 \subset A'$, $F_2 \subset A'$. Then, the condition $\bar{P} \subset \partial_H E$ easily translates to $U(H, \bar{P}, E) = A$. So Takahasi's result and the commutative results from Altomare's paper [1984a] lead to conjecture that in general $K_+(H) = U(H, \bar{P}, E)$ (Altomare [1986]). Unfortunately, there is counterexample (Beckhoff [1988]), and so we must be content with the inclusion $U(H, \bar{P}, E) \subset K_+(H)$.

Alternative and considerably simpler proofs of Takahasi's result have been given by Fujii [1987].

There is a lot of interesting questions still open.

It would be interesting to characterize those subspaces of a non-commutative C*-algebra which occur as Korovkin closures in a reasonable way. Since the Korovkin closure of a Korovkin closed subspace yields nothing new, this would give us upper estimates of Korovkin closures. In type I C*-algebras the sub-C*-algebras are such sets (Beckhoff [1992]).

So the Korovkin closure of a subset H is always contained in the sub-C*-algebra generated by H. Under what circumstances (additional conditions on H or on the C*-algebra) is this estimate an equality? We have only partial answers to this question.

In the approximation by positive linear functionals in many situations one may restrict attention to constant approximating nets (i.e., to a single positive linear functional) by using w*-limits. Something like this has been done for operator approximation with respect to the weak operator topology. Is there something like this for the Korovkin approximation by positive linear operators with respect to pointwise norm-convergence?

Some special C*-algebras may be interesting in this context. What properties should a locally compact (non-abelian) group have, so that $C^*(G)$ or $L^1(G)$ possess finite Korovkin systems? How can one compute such Korovkin systems in terms of G? The minimal possible length of a Korovkin system of course is an isomorphism invariant. Is this expressible by other invariants? These questions have been treated by M. Pannenberg in the abelian case.

For a complete list of references on this subject, we refer to Sections 7.1–7.7, subsection b), of the final subject index of Appendix D.

Appendix C

A list of determining sets and Korovkin sets

In this appendix we collect some of the main examples of determining sets and Korovkin sets that appear in the book. The reader who is mainly interested in their use in approximation theory should find this list useful in some respects.

We have organized the appendix in several sections to make the search simple and quick. For the terminology employed in the titles of sections we refer to the appropriate definitions given throughout the book.

To every example we added a reference to which the reader is referred for more details or proofs.

For the reader's convenience we also recall that:

− Korovkin sets for finitely defined operators of order 1 are just those for the identity operator (see Remark to Corollary 4.1.3).

− Korovkin sets (determining sets, respectively) with respect to positive linear operators (positive Radon measures) are also Korovkin sets (determining sets, respectively) with respect to positive linear contractions. The converse holds whenever the underlying space is compact and the subset contains the function **1**.

− Korovkin sets are also determining sets (see Theorems 3.1.2, 3.1.3, 3.4.7, 4.1.2).

− Korovkin sets enjoy other additional properties, such as universal Korovkin-type properties with respect to positive linear operators (contractions), monotone operators and, in the compact case, the Korovkin-type property with respect to linear contractions (see Theorems 3.2.1, 3.2.2, 3.2.3).

− If $\mu \in \mathcal{M}_b^+(X)$, then every Korovkin set in $\mathcal{C}_0(X)$ is a Korovkin set in $L^p(X,\mu)$, $1 \leq p < +\infty$ (see Corollary 4.1.7).

C.1 Determining sets in $\mathcal{C}_0(X)$ (X locally compact Hausdorff space)

C.1.1 Determining sets in $\mathcal{C}_0(X)$ for discrete measures with respect to bounded positive Radon measures

1. $\left\{ \prod_{i=1}^{n} f_i \right\} \cup \left\{ \prod_{\substack{j=1 \\ j \neq i}}^{n} f_j \middle| i = 1,\ldots,n \right\}$, in $\mathcal{C}_0(X)$ for every $\mu \in C_+(x_1,\ldots,x_n)$, where $x_1, \ldots, x_n \in X$ are fixed, $f_1, \ldots, f_n \in \mathcal{C}_0(X)$ and every f_i is positive and vanishes only at x_i ($n \geq 2$). For example,

$$\left\{ \exp(-n\|x\|) \prod_{i=1}^{n} \|x - x_i\|^2 \right\} \cup \left\{ \exp(-(n-1)\|x\|) \prod_{\substack{j=1 \\ j \neq i}}^{n} \|x - x_j\|^2 \mid i = 1, \ldots, n \right\},$$

in $\mathscr{C}_0(\mathbb{R}^p)$ for every $\mu \in C_+(x_1, \ldots, x_n)$, where $x_1, \ldots, x_n \in \mathbb{R}^p$ are fixed ($n \geq 2$) (Example 1 to Corollary 2.2.7).

2. $\{f_0\} \cup \bigcup_{h=1}^{n} \{f_0 \cdot f_1 \ldots f_h \mid f_j \in M \cup M^2, \ j = 1, \ldots, h\}$, in $\mathscr{C}_0(X)$ for every $\mu \in$ $C_+(x_1, \ldots, x_n)$, where $x_1, \ldots, x_n \in X$ are fixed and $M \subset \mathscr{C}_0(X)$ and $f_0 \in \mathscr{C}_0(X)$ satisfy (2.4.7), (2.4.8), (2.4.9) and (2.4.12) (Theorem 2.4.1).

3. $\{f_0\} \cup f_0 M \cup \cdots \cup f_0 M^{2n}$, in $\mathscr{C}_0(X)$ for every $\mu \in C_+(x_1, \ldots, x_n)$, where $x_1, \ldots,$ $x_n \in X$ are fixed and $M \subset \mathscr{C}_0(X)$ and $f_0 \in \mathscr{C}_0(X)$ satisfy (2.4.7), (2.4.8) and (2.4.19). In particular,

$$\{f_0, f_0^2, \ldots, f_0^{2n+1}\}, \quad \text{in } \mathscr{C}_0(X) \text{ for every } \mu \in C_+(x_1, \ldots, x_n),$$

where f_0 is strictly positive and, whenever $n \geq 2$, $f_0(x_i) \neq f_0(x_j)$ for every $i, j = 1,$ $\ldots, n$, $i \neq j$, and $f_0(x) \neq f_0(x_i)$ for every $x \in X \backslash \{x_1, \ldots, x_n\}$ and $i = 1, \ldots, n$ (Proposition 2.4.4).

C.1.2 Determining sets in $\mathscr{C}_0(X)$ for discrete measures with respect to positive contractive Radon measures

1. $\bigcup_{h=1}^{n} \{f_1 \ldots f_h \mid f_j \in M \cup M^2, j = 1, \ldots, h\}$, in $\mathscr{C}_0(X)$ for every $\mu \in \tilde{C}_+^1(x_1, \ldots, x_n)$, where $x_1, \ldots, x_n \in X$ are fixed and M is a subset of $\mathscr{C}_0(X)$ satisfying (2.4.7), (2.4.8) and (2.4.12) (Theorem 2.4.2).

2. $M \cup \cdots \cup M^{2n}$, in $\mathscr{C}_0(X)$ for every $\mu \in \tilde{C}_+^1(x_1, \ldots, x_n)$, where $x_1, \ldots, x_n \in X$ are fixed and M is a subset of $\mathscr{C}_0(X)$ satisfying (2.4.7), (2.4.8) and (2.4.19) (Proposition 2.4.4).

C.1.3 Determining sets in $\mathscr{C}_0(X)$ for Dirac measures with respect to bounded positive Radon measures

1. $\{f_0\} \cup f_0 M \cup f_0 M^2$, in $\mathscr{C}_0(X)$ for ε_x, where $x \in X$, $f_0 \in \mathscr{C}_0(X)$ is a strictly positive function satisfying (2.5.21) and M is a subset of $\mathscr{C}_0(X)$ satisfying (2.5.20) and which separates x in X (see (2.4.6)).

Furthermore, if $M = \{h_n | n \in \mathbb{N}\}$ is finite or countable and $\sum_{n=1}^{\infty} h_n^2$ is uniformly convergent, the same result holds by replacing the subset M^2 with the function $\sum_{n=1}^{\infty} h_n^2$ (Theorem 2.5.6 and Proposition 2.5.7).

2. $\{\exp(-\|x\|^2), \|x\|^2 \exp(-\|x\|^2)\}$, in $\mathscr{C}_0(\mathbb{R}^p)$ for ε_0 (Example 1 to Theorem 2.5.4).

C.1.4 Determining sets in $\mathscr{C}_0(X)$ for Dirac measures with respect to positive contractive Radon measures

1. $M \cup M^2$, in $\mathscr{C}_0(X)$ for ε_x, where $x \in X$, and M is a subset of $\mathscr{C}_0(X)$ satisfying (2.5.20) and which separates x in X (see (2.4.6)).

Furthermore, if $M = \{h_n | n \in \mathbb{N}\}$ is finite or countable and $\sum_{n=1}^{\infty} h_n^2$ is uniformly convergent, the same result holds by replacing the subset M^2 with the function $\sum_{n=1}^{\infty} h_n^2$ (Theorem 2.5.6 and the subsequent Remark, Proposition 2.5.7).

C.2 Determining sets in $\mathscr{C}(X)$ (X compact Hausdorff space)

C.2.1 Determining sets in $\mathscr{C}(X)$ for discrete measures with respect to positive Radon measures

1. $\left\{ \prod_{i=1}^{n} d^2(\cdot, x_i) \right\} \cup \left\{ \prod_{\substack{j=1 \\ j \neq i}}^{n} d^2(\cdot, x_j) | i = 1, \ldots, n \right\}$, in $\mathscr{C}(X)$ for every $\mu \in C_+(x_1, \ldots, x_n)$, where (X, d) is a (compact) metric space and $x_1, \ldots, x_n \in X$ are fixed ($n \geq 2$) (Example 1 to Corollary 2.2.7).

2. $\{1\} \cup \bigcup_{h=1}^{n} \{f_1 \ldots f_h | f_j \in M \cup M^2, \ j = 1, \ldots, h\}$, in $\mathscr{C}(X)$ for every $\mu \in C_+(x_1, \ldots, x_n)$, where $x_1, \ldots, x_n \in X$ are fixed and M is a subset of $\mathscr{C}(X)$ satisfying (2.4.15) and (2.4.12) (Theorem 2.4.3).

3. $\{1\} \cup M \cup \cdots \cup M^{2n}$, in $\mathscr{C}(X)$ for every $\mu \in C_+(x_1, \ldots, x_n)$, where $x_1, \ldots, x_n \in X$ are fixed and M is a subset of $\mathscr{C}(X)$ satisfying (2.4.15) and (2.4.19). In particular,

$$\{1, f, f^2, \ldots, f^{2n}\}, \quad \text{in } \mathscr{C}(X) \text{ for every } \mu \in C_+(x_1, \ldots, x_n),$$

if $f \in \mathscr{C}(X)$ is such that $f(x) \neq f(x_i)$ for every $i = 1, \ldots, n$ and $x \in X \setminus \{x_1, \ldots, x_n\}$ and, if $n \geq 2$, $f(x_i) \neq f(x_j)$ for each $i, j = 1, \ldots, n$, $i \neq j$ (Proposition 2.4.4).

4. Let X be a compact real interval or the unit circle. Every Chebyshev subspace of order $n + 1$ is a determining subspace for every $\mu \in C_+(x_1, \ldots, x_p)$, where $x_1, \ldots, x_p \in X$ satisfy $\sum_{i=1}^{p} \omega(x_i) \leq n$ (Theorem 2.3.5 and the subsequent examples).

5. $\{1, x^\alpha, x^\beta\}$, in $\mathscr{C}([0, 1])$ for every $\mu = \alpha_1 \varepsilon_0 + \alpha_2 \varepsilon_1$, where $\alpha_1 \geq 0$, $\alpha_2 \geq 0$ and $0 < \alpha < \beta$ (Example 2 to Corollary 2.2.7).

6. $\{1, x^{2p}, x^{4p-1}, x^{4p}\}$, in $\mathscr{C}([-1, 1])$ for every $\mu = \alpha_1 \varepsilon_{-1} + \alpha_2 \varepsilon_0 + \alpha_3 \varepsilon_1$, where $\alpha_1, \alpha_2, \alpha_3 \geq 0$ and $p \in \mathbb{N}$ (Example 3 to Corollary 2.2.7).

C.2.2 Determining sets in $\mathscr{C}(X)$ for Dirac measures with respect to positive Radon measures

1. $\{1, d^2(\cdot, x)\}$, in $\mathscr{C}(X)$ for ε_x, where X is a compact subset of a metric space (E, d) and $x \in X$ (Example 3 to Theorem 2.5.4).

2. $\{1\} \cup M \cup M^2$, in $\mathscr{C}(X)$ for ε_x, where M is a subset of $\mathscr{C}(X)$ which separates x in X (see (2.4.6)) (Theorem 2.5.6 and the subsequent Remark, Proposition 2.5.7).

Furthermore, if $M = \{h_n | n \in \mathbb{N}\}$ is finite or countable and $\sum_{n=1}^{\infty} h_n^2$ is uniformly convergent, the same result holds by replacing the subset M^2 with the function $\sum_{n=1}^{\infty} h_n^2$.

3. $\{1\} \cup M^2$, in $\mathscr{C}(X)$ for ε_x, where M is a subset of $\mathscr{C}(X)$ such that $h(x) = 0$ for each $h \in M$ and, for every $y \in X$, $y \neq x$, there exists $h \in M$ satisfying $h(y) \neq 0$.

It is possible to replace M^2 with $\sum_{n=1}^{\infty} h_n^2$ if $M = \{h_n | n \in \mathbb{N}\}$ is finite or countable and $\sum_{n=1}^{\infty} h_n^2$ is uniformly convergent (Proposition 2.5.8).

4. $\{1, x^\alpha\}$, in $\mathscr{C}([0, 1])$ for both ε_0 and ε_1, where $\alpha > 0$ (Example 2 to Theorem 2.5.4).

C.3 Korovkin sets in $\mathscr{C}_0(X)$ (X locally compact Hausdorff space)

C.3.1 Korovkin sets in $\mathscr{C}_0(X)$ for finitely defined operators of order n with respect to positive linear operators

1. $\{f_0\} \cup \bigcup_{h=1}^{n} \{f_0 \cdot f_1 \ldots f_h | f_j \in M \cup M^2, \ j = 1, \ldots, h\}$, in $\mathscr{C}_0(X)$, where M is a subset of $\mathscr{C}_0(X)$ satisfying suitable separation properties and $f_0 \in \mathscr{L}(M)$ is strictly positive (Theorem 3.4.12).

2. $\{f_0\} \cup f_0 M \cup \cdots \cup f_0 M^{2n}$, in $\mathscr{C}_0(X)$, where M is a subset of $\mathscr{C}_0(X)$ satisfying suitable separation properties and $f_0 \in \mathscr{L}(M)$ is strictly positive. In particular,

$$\{f_0, f_0^2, \ldots, f_0^{2n+1}\}, \quad \text{in } \mathscr{C}_0(X),$$

if f_0 is injective and strictly positive (Corollary 3.4.13).

3. $\{\exp(-\|x\|^2)\|x\|^{2r}\mathrm{pr}_1^{h_{1,r}} \ldots \mathrm{pr}_p^{h_{p,r}} | 0 \leq r \leq n, \ h_{1,r}, \ldots, h_{p,r} \in \mathbb{N}_0, \ h_{1,r} + \cdots + h_{p,r} \leq n - r\}$, in $\mathscr{C}_0(\mathbb{R}^p)$, (Example 1 to Corollary 3.4.8).

4. $\{x, x^2, \ldots, x^{2n+1}\}$, in $\mathscr{C}_0(]0, 1])$ (Example 2 to Corollary 3.4.8).

5. $\{x^{-1}, x^{-2}, \ldots, x^{-2n-1}\}$, in $\mathscr{C}_0([1, +\infty[)$ (Example 2 to Corollary 3.4.8).

6. $\{\exp(-\alpha x), \exp(-2\alpha x), \ldots, \exp(-(2n+1)\alpha x)\}$, in $\mathscr{C}_0([0, +\infty[)$, where $\alpha > 0$ (Example 2 to Corollary 3.4.8).

C.3.2 Korovkin sets in $\mathscr{C}_0(X)$ for finitely defined operators of order n with respect to positive linear contractions

1. $\bigcup_{h=1}^{n} \{f_1 \ldots f_h | f_j \in M \cup M^2, j = 1, \ldots, h\}$, in $\mathscr{C}_0(X)$, where M is a subset of $\mathscr{C}_0(X)$ satisfying suitable separation properties (Theorem 3.4.12).

2. $M \cup M^2 \cup \cdots \cup M^{2n}$, in $\mathscr{C}_0(X)$, where M is a subset of $\mathscr{C}_0(X)$ satisfying suitable separation properties. In particular,

$$\{f_0, f_0^2, \ldots, f_0^{2n+1}\}, \quad \text{in } \mathscr{C}_0(X),$$

if f is injective and strictly positive (Corollary 3.4.13).

C.3.3 Korovkin sets in $\mathscr{C}_0(X)$ for the identity operator with respect to positive linear operators

1. $\{f_0\} \cup f_0 \cdot M \cup f_0 \cdot M^2$, in $\mathscr{C}_0(X)$, where $f_0 \in \mathscr{C}_0(X)$ is strictly positive and M is a subset of $\mathscr{C}_0(X)$ which separates the points of X.

If $M = \{h_n | n \in \mathbb{N}\}$ is finite or countable and the series $\sum_{n=1}^{\infty} h_n^2$ is uniformly convergent, one can replace $f_0 \cdot M^2$ simply with the function $f_0 \sum_{n=1}^{\infty} h_n^2$. In particular,

$$\{f_0, f_0^2, f_0^3\}, \quad \text{in } \mathscr{C}_0(X),$$

if $f_0 \in \mathscr{C}_0(X)$ is injective and strictly positive (Theorem 4.4.6 and the subsequent Example 4).

2. $\{\exp(-\|x\|^2),\ \exp(-2\|x\|^2),\ x_1\exp(-2\|x\|^2),\ \ldots,\ x_p\exp(-2\|x\|^2),\ (1 + \|x\|^2)\exp(-3\|x\|^2)\}$, in $\mathscr{C}_0(\mathbb{R}^p)$, (Example 1 to Theorem 4.4.6).

3. $\left\{ \dfrac{1}{1 + \|x\|^2}, \dfrac{1}{(1 + \|x\|^2)^2}, \dfrac{x_1}{(1 + \|x\|^2)^2}, \ldots, \dfrac{x_p}{(1 + \|x\|^2)^2} \right\}$, in $\mathscr{C}_0(\mathbb{R}^p)$, (Example 1 to Theorem 4.4.6).

4. $\{\exp(-x^2), \exp(-x^2) \cdot \varphi(x)^{\lambda_1}, \exp(-x^2) \cdot \varphi(x)^{\lambda_2}\}$, in $\mathscr{C}_0(\mathbb{R})$, where $\varphi: \mathbb{R} \to]0, 1[$ is a strictly monotone continuous function satisfying $\lim\limits_{x \to -\infty} \varphi(x) = 0$ and $\lim\limits_{x \to +\infty} \varphi(x) = 1$ and $0 < \lambda_1 < \lambda_2$ (Proposition 4.2.5).

5. $\{x^{\lambda_1}, x^{\lambda_2}, x^{\lambda_3}\}$, in $\mathscr{C}_0(]0, 1])$, where $0 < \lambda_1 < \lambda_2 < \lambda_3$ (Proposition 4.2.4).

6. $\{\exp(-x), \exp(-x) \cdot (1 + x)^{\lambda_1}, \exp(-x) \cdot (1 + x)^{\lambda_2}\}$, in $\mathscr{C}_0([0, +\infty[)$, where $0 < \lambda_1 < \lambda_2$ (Proposition 4.2.5).

7. $\{\exp(-\lambda_1 x), \exp(-\lambda_2 x), \exp(-\lambda_3 x)\}$, in $\mathscr{C}_0([0, +\infty[)$, where $0 < \lambda_1 < \lambda_2 < \lambda_3$ (Proposition 4.2.5).

8. $\{\exp(-x), \exp(-x) \cdot x^{\lambda_1}, \exp(-x) \cdot x^{\lambda_2}\}$, in $\mathscr{C}_0([1, +\infty[)$, where $0 < \lambda_1 < \lambda_2$ (Proposition 4.2.5).

9. $\{x^{-\lambda_1}, x^{-\lambda_2}, x^{-\lambda_3}\}$, in $\mathscr{C}_0([1, +\infty[)$, where $0 < \lambda_1 < \lambda_2 < \lambda_3$ (Proposition 4.2.5).

10. $\{\exp(-\lambda_1 n), \exp(-\lambda_2 n), \exp(-\lambda_3 n)\}$, in c_0, where $0 < \lambda_1 < \lambda_2 < \lambda_3$ (Proposition 4.2.5).

11. $\{n^{-\lambda_1}, n^{-\lambda_2}, n^{-\lambda_3}\}$, in c_0, where $0 < \lambda_1 < \lambda_2 < \lambda_3$ (Proposition 4.2.5).

C.3.4 Korovkin sets in $\mathscr{C}_0(X)$ for the identity operator with respect to positive linear contractions

1. $M \cup M^2$, in $\mathscr{C}_0(X)$, where M is a subset of $\mathscr{C}_0(X)$ which separates strongly the points of X (Theorem 4.4.5).

If $M = \{h_n | n \in \mathbb{N}\}$ is finite or countable and the series $\sum\limits_{n=1}^{\infty} h_n^2$ is uniformly convergent, one can replace M^2 simply with the function $\sum\limits_{n=1}^{\infty} h_n^2$. In particular,

$$\{f_0, f_0^2\}, \quad \text{in } \mathscr{C}_0(X),$$

if f_0 is injective and strictly positive (Example 4 to Theorem 4.4.6).

2. $\{\exp(-\|x\|^2), x_1 \exp(-\|x\|^2), \ldots, x_p \exp(-\|x\|^2), (1 + \|x\|^2)\exp(-2\|x\|^2)\}$, in $\mathscr{C}_0(\mathbb{R}^p)$, (Example 1 to Theorem 4.4.6).

3. $\left\{ \dfrac{1}{1 + \|x\|^2}, \dfrac{x_1}{1 + \|x\|^2}, \ldots, \dfrac{x_p}{1 + \|x\|^2} \right\}$, in $\mathscr{C}_0(\mathbb{R}^p)$, (Example 1 to Theorem 4.4.6).

4. $\{x^{\lambda_1}, x^{\lambda_2}\}$, in $\mathscr{C}_0(]0, 1])$, where $0 < \lambda_1 < \lambda_2$ (Proposition 4.2.4).

5. $\{\exp(-\lambda_1 x), \exp(-\lambda_2 x)\}$, in $\mathscr{C}_0([0, +\infty[)$, where $0 < \lambda_1 < \lambda_2$ (Proposition 4.2.6).

6. $\{x^{-\lambda_1}, x^{-\lambda_2}\}$, in $\mathscr{C}_0([1, +\infty[)$, where $0 < \lambda_1 < \lambda_2$ (Proposition 4.2.6).

7. $\{\exp(-\lambda_1 n), \exp(-\lambda_2 n)\}$, in c_0, where $0 < \lambda_1 < \lambda_2$ (Proposition 4.2.6).

8. $\{n^{-\lambda_1}, n^{-\lambda_2}\}$, in c_0, where $0 < \lambda_1 < \lambda_2$ (Proposition 4.2.6).

C.4 Korovkin sets in $\mathscr{C}(X)$ (X compact Hausdorff space)

C.4.1 Korovkin sets in $\mathscr{C}(X)$ for a positive projection with respect to positive linear operators

1. $H \cup \{u\}$, in $\mathscr{C}(X)$ for an arbitrary positive projection T, whose range H contains the constants and separates the points of X. Here, X must be supposed metrizable, and $u = \sum\limits_{n=1}^{\infty} h_n^2$, where $(h_n)_{n \in \mathbb{N}}$ is an arbitrary sequence in H separating the points of X and such that the series $\sum\limits_{n=1}^{\infty} h_n^2$ is uniformly convergent on X (Theorem 3.3.3). In particular, if K is a metrizable convex compact set,

$$H \cup A(K)^2, \quad \text{in } \mathscr{C}(K) \text{ for } T,$$

(Remark 1 to Theorem 3.3.3).

2. $A(K) \cup \{u\}$, in $\mathscr{C}(K)$ for the canonical projection on $A(K)$, where K is a Bauer simplex, $A(K)$ denotes the space of all continuous affine functions on X and $u \in \mathscr{C}(X)$ is strictly convex (Corollary 3.3.4).

3. $\left\{1, \text{pr}_1, \ldots, \text{pr}_p, \sum\limits_{i=1}^{p} \text{pr}_i^2\right\}$, in $\mathscr{C}(K_p)$ for the canonical projection defined by (3.3.10), where K_p is the canonical simplex in $\mathbb{R}^p$ (Example 3.3.5).

4. $\{1, x, x^2\}$, in $\mathscr{C}([0,1])$ for the canonical projection defined by (3.3.12), (Example 3.3.5).

5. $H(\Omega) \cup \{u\}$, in $\mathscr{C}(\bar{\Omega})$ for the Dirichlet operator defined by (3.3.15), where Ω is a regular bounded open subset of $\mathbb{R}^p$, $H(\Omega)$ denotes the space of all continuous function on $\bar{\Omega}$ harmonic in Ω and $u \in \mathscr{C}(\bar{\Omega}) \cap \mathscr{C}^2(\Omega)$ satisfies $\Delta u > 0$ on Ω (Corollary 3.3.6).

6. $\{1\} \cup \left\{\prod\limits_{i \in J} \text{pr}_i | J \subset \{1, \ldots, p\}\right\} \cup \left\{\sum\limits_{i=1}^{p} \text{pr}_i^2\right\}$, in $\mathscr{C}([0,1]^p)$ for the projection defined by (3.3.20), (Example 3.3.10).

C.4.2 Korovkin sets in $\mathscr{C}(X)$ for finitely defined operators of order n with respect to positive linear operators

1. $\{1\} \cup \bigcup\limits_{h=1}^{n} \{f_1 \ldots f_h | f_j \in M \cup M^2, j = 1, \ldots, h\}$, in $\mathscr{C}(X)$, where M is a subset of $\mathscr{C}(X)$ satisfying suitable separation properties (Theorem 3.4.12).

2. $\{1\} \cup M \cup M^2 \cup \cdots \cup M^{2n}$, in $\mathscr{C}(X)$, where M is a subset of $\mathscr{C}(X)$ satisfying suitable separation properties. In particular,

$$\{1, f, f^2, \ldots, f^{2n}\}, \text{ in } \mathscr{C}(X),$$

if $f \in \mathscr{C}(X)$ is injective (Corollary 3.4.13).

3. Every Chebyshev subspace of order $2n + 1$ or $2n + 2$ in $\mathscr{C}(X)$, where X is a compact real interval or the unit circle (Corollary 3.4.10 and Examples to Theorem 2.3.5).

4. $\{\|x\|^{2r} \text{pr}_1^{h_{1,r}} \dots \text{pr}_p^{h_{p,r}} | 0 \leq r \leq n, h_{1,r}, \dots, h_{p,r} \in \mathbb{N}_0, h_{1,r} + \dots + h_{p,r} \leq n - r\}$, in $\mathscr{C}(X)$, where X is a compact subset of $\mathbb{R}^p$ (Example 1 to Corollary 3.4.8).

C.4.3 Korovkin sets in $\mathscr{C}(X)$ for the identity operator with respect to positive linear operators

1. $\{1\} \cup M \cup M^2$, in $\mathscr{C}(X)$, where M is a subset of $\mathscr{C}(X)$ which separates the points of X (Theorem 4.4.6).

If $M = \{h_n | n \in \mathbb{N}\}$ is finite or countable and the series $\sum\limits_{n=1}^{\infty} h_n^2$ is uniformly convergent, one can replace M^2 simply with the function $\sum\limits_{n=1}^{\infty} h_n^2$. In particular,

$$\{1, f, f^2\}, \quad \text{in } \mathscr{C}(X),$$

if $f \in \mathscr{C}(X)$ is injective (Example 5 to Theorem 4.4.6).

2. $A(K) \cup \mathscr{U}$, in $\mathscr{C}(K)$, where K is convex and compact, $A(K)$ is the space of all continuous affine functions on K, $\mathscr{U}$ is a subset of continuous convex functions on K which separates convexly the points of K. In particular,

$$A(K) \cup \{u\}, \quad \text{in } \mathscr{C}(K),$$

if $u \in \mathscr{C}(K)$ is strictly convex (Theorem 4.3.7 and Corollary 4.3.8) and

$$A(K) \cup A(K)^2, \quad \text{in } \mathscr{C}(K), \text{ (Example 3 to Theorem 4.4.6)}.$$

3. $\left\{1, \text{pr}_1, \dots, \text{pr}_p, \sum\limits_{i=1}^{p} \text{pr}_i^2\right\}$, in $\mathscr{C}(X)$, where X is a compact subset of $\mathbb{R}^p$ (Example 2 to Theorem 4.4.6).

4. $\{1, \text{pr}_1, \dots, \text{pr}_p\}$, in $\mathscr{C}(X)$, where X is a compact subset of $\mathbb{R}^p$ ($p \geq 2$) such that $X = \partial_e C$ for some convex compact subset C of $\mathbb{R}^p$ (in particular, if X is a closed subset of the unit sphere of $\mathbb{R}^p$) (Corollary 4.5.2).

5. $\{1, \text{pr}_1, \dots, \text{pr}_p, u\}$, in $\mathscr{C}(K)$, where K is a convex compact subset of $\mathbb{R}^p$ and $u \in \mathscr{C}(K)$ is strictly convex (Corollary 4.3.9).

6. Every Chebyshev subspace of order ≥ 3 in $\mathscr{C}(X)$, where X is a compact real interval or the unit circle (see Examples to Theorem 2.3.5) (Theorem 4.5.7).

7. $\{1, \mathrm{pr}_1, \mathrm{pr}_2\}$, in $\mathscr{C}(\mathbb{T})$ (i.e., $\{1, \sin, \cos\}$ in $\mathscr{C}_{2\pi}$), where $\mathbb{T}$ is the unit circle (Theorem 4.2.8).

8. $\{1, x^{\lambda_1}, x^{\lambda_2}\}$ and, in particular, $\{1, x, x^2\}$, in $\mathscr{C}([0,1])$, where $0 < \lambda_1 < \lambda_2$ (Proposition 4.2.4).

9. $\{1, x, \exp(\lambda x)\}$, in $\mathscr{C}([a,b])$, where $a < b$, $\lambda \neq 0$ (Example to Corollary 4.3.9).

10. $\{1, \exp(x), \exp(2x)\}$ and $\{1, x, x^2\}$, in $\mathscr{C}([a,b])$, where $a < b$ (Example 6 to Theorem 4.4.6).

C.4.4 Korovkin sets in $\mathscr{C}(X)$ for the identity operator with respect to positive linear contractions

1. See Examples 1 and 2 of Section C.3.4.

2. $\{\mathrm{pr}_1, \ldots, \mathrm{pr}_p\}$, in $\mathscr{C}(X)$, where X is a compact subset of $\mathbb{R}^p$ ($p \geq 2$) such that $0 \notin X$ and $X \subset \partial_e \mathrm{co}(X \cup \{0\})$ (in particular, if X is a closed subset of the unit sphere of $\mathbb{R}^p$) (Corollary 4.5.4).

C.5 Korovkin sets in $L^p(X, \mu)$-spaces

C.5.1 Korovkin sets for the identity operator with respect to positive linear operators

1. $\{\exp(-x), \exp(-x) \cdot (1 + x)^{\lambda_1}, \exp(-x) \cdot (1 + x)^{\lambda_2}\}$, in $L^p([0, +\infty[)$, where $1 \leq p < +\infty$ and $0 < \lambda_1 < \lambda_2$ (Proposition 4.2.5).

2. $\{\exp(-\lambda_1 x), \exp(-\lambda_2 x), \exp(-\lambda_3 x)\}$, in $L^p([0, +\infty[)$, where $1 \leq p < +\infty$ and $0 < \lambda_1 < \lambda_2 < \lambda_3$ (Proposition 4.2.5).

3. $\{\exp(-x), \exp(-x) \cdot x^{\lambda_1}, \exp(-x) \cdot x^{\lambda_2}\}$, in $L^p([1, +\infty[)$, where $1 \leq p < +\infty$ (Proposition 4.2.5).

4. $\{x^{-\lambda_1}, x^{-\lambda_2}, x^{-\lambda_3}\}$, in $L^p([1, +\infty[)$, where $1 \leq p < +\infty$ and $\dfrac{1}{p} < \lambda_1 < \lambda_2 < \lambda_3$ (Proposition 4.2.5).

5. $\{\exp(-x^2), \exp(-x^2) \cdot \varphi(x)^{\lambda_1}, \exp(-x^2) \cdot \varphi(x)^{\lambda_2}\}$, in $L^p(\mathbb{R})$, where $1 \le p < +\infty$, $\varphi \colon \mathbb{R} \to]0,1[$ is a strictly monotone continuous function satisfying $\lim\limits_{x \to -\infty} \varphi(x) = 0$ and $\lim\limits_{x \to +\infty} \varphi(x) = 1$ and $0 < \lambda_1 < \lambda_2$ (Proposition 4.2.5).

6. $\{\exp(-\lambda_1 n), \exp(-\lambda_2 n), \exp(-\lambda_3 n)\}$, in ℓ^p, where $1 \le p < +\infty$ and $0 < \lambda_1 < \lambda_2 < \lambda_3$ (Proposition 4.2.5).

7. $\{n^{-\lambda_1}, n^{-\lambda_2}, n^{-\lambda_3}\}$, in ℓ^p, where $1 \le p < +\infty$ and $\dfrac{1}{p} < \lambda_1 < \lambda_2 < \lambda_3$ (Proposition 4.2.5).

Appendix D

A subject classification of Korovkin-type approximation theory with a subject index

Here we present a subject classification of the Korovkin-type approximation theory which reflects the main lines along which this theory has been developed.

Of course this classification corresponds with our personal opinions (which are sufficiently diffuse) about the development of the theory and we apologize if some other sections are not explicitly included.

In order to explain our classification scheme we need some preliminary remarks.

Let E and F be topological vector spaces and $\mathscr{L}_0$ and $\mathscr{L}$ classes of linear operators from E to F. Given a subspace H of E and an operator $T \in \mathscr{L}_0$, we define the Korovkin closure of H for T with respect to $\mathscr{L}$ to be the subspace of all elements f of E such that, if $(L_i)_{i \in I}^{\leq}$ is an equicontinuous net of operators of $\mathscr{L}$ such that $\lim_{\substack{\leq \\ i \in I}} L_i(h) = T(h)$ for every $h \in H$, then $\lim_{\substack{\leq \\ i \in I}} L_i(f) = T(f)$.

As it is well-known, the main problems of the theory are the characterization of the Korovkin closures and (or) the determination of sufficient (and necessary) conditions under which the Korovkin closures coincide with E. In this last case we say that H is a Korovkin subspace in E for T with respect to $\mathscr{L}$.

With these definitions, our subject classification should be sufficiently clear.

Furthermore, as explained in the Introduction, every article which appears in the bibliography of the book and which deals with the Korovkin-type approximation theory, has been classified with one or several numbers according with this subject classification.

We hope we have made a useful service to the scientific community by also presenting a subject index of the above mentioned articles in the last part of this appendix.

D.1 Subject classification (SC)

1. **Korovkin-type approximation theory in the space $\mathscr{C}(X)$ of all continuous real-valued functions defined on a compact Hausdorff space X.**

   1.1   Research surveys, monographs.
   1.2   KAT for the identity operator with respect to positive linear operators.
   1.3   KAT for the identity operator with respect to linear contractions.
   1.4   KAT for the identity operator with respect to positive linear contractions.
   1.5   KAT for the identity operator with respect to other classes of operators.
   1.6   KAT for a lattice homomorphism with respect to positive linear operators.
   1.7   KAT for a finitely defined operator with respect to positive linear operators.
   1.8   KAT for a positive operator with respect to positive linear operators.
   1.9   KAT for a positive linear functional with respect to positive linear functionals.
   1.10 Probabilistic aspects of the KAT.
   1.11 Korovkin closures, universal Korovkin closures.
   1.12 Other aspects of the KAT.

2. **Korovkin-type approximation theory in the space $\mathscr{C}_0(X)$ of all continuous real-valued functions defined on a locally compact Hausdorff space X, which vanish at infinity.**

   2.1   Research surveys, monographs.
   2.2   KAT for the identity operator with respect to positive linear operators.
   2.3   KAT for the identity operator with respect to positive linear contractions.
   2.4   KAT for the identity operator with respect to other classes of operators.
   2.5   KAT for a lattice homomorphism with respect to positive linear operators.
   2.6   KAT for a finitely defined operator with respect to positive linear operators.
   2.7   KAT for a positive linear operator with respect to positive linear operators.
   2.8   KAT for a positive linear functional with respect to positive linear functionals.
   2.9   Korovkin closures, universal Korovkin closures.
   2.10 Other aspects of the KAT.

3. **Korovkin-type approximation theory in $L^p(X, \mathscr{F}, \mu)$-Spaces, $(X, \mathscr{F}, \mu)$ being a measure space or in $L^p(X, \mu)$, where $\mu \in \mathscr{M}^+(X)$, and $1 \leq p \leq +\infty$.**

   3.1   Research surveys, monographs.
   3.2   KAT for the identity operator with respect to positive linear operators.
   3.3   KAT for the identity operator with respect to linear contractions.
   3.4   KAT for the identity operator with respect to positive linear contractions.
   3.5   KAT for a lattice homomorphism with respect to positive linear operators.
   3.6   Korovkin closures, universal Korovkin closures.
   3.7   Other aspects of the KAT.

4. **Korovkin-type approximation theory in spaces of continuous affine functions defined on a compact convex set.**

   4.1   Research surveys, monographs.
   4.2   KAT for the identity operator with respect to positive linear operators.
   4.3   KAT for the identity operator with respect to linear contractions.
   4.4   KAT for the identity operator with respect to positive linear contractions.
   4.5   KAT for a positive linear functional with respect to positive linear functionals.
   4.6   Korovkin closures, universal Korovkin closures.
   4.7   Other aspects of the KAT.

5. **Korovkin-type approximation theory in other function spaces such as:**
   a) **Function spaces on compact or locally compact Hausdorff spaces;**
   b) **Banach function spaces;**
   c) **Weighted function spaces;**
   d) **Adapted spaces;**
   e) **Spaces of differentiable functions;**
   f) **Sequence spaces;**
   g) **Spaces of measurable functions;**
   h) **Spaces of absolutely continuous functions;**
   i) **Spaces of Riemann integrable functions;**
   j) **None of above but in this section.**

   5.1   Research surveys, monographs.
   5.2   KAT for the identity operator with respect to positive linear operators.
   5.3   KAT for a positive operator with respect to positive linear operators.
   5.4   KAT for a positive linear functional with respect to positive linear functionals.
   5.5   Korovkin closures, universal Korovkin closures.
   5.6   Other aspects of the KAT.

6. **Korovkin-type approximation theory in Banach lattices (also in ordered normed spaces and in ordered topological vector spaces).**

   6.1 Research surveys, monographs.
   6.2 KAT for the identity operator with respect to positive linear operators.
   6.3 KAT for the identity operator with respect to positive linear contractions.
   6.4 KAT for a lattice homomorphism with respect to positive linear operators.
   6.5 KAT for a positive linear operator with respect to positive linear operators.
   6.6 KAT for a positive linear functional with respect to positive linear functionals.
   6.7 Korovkin closures, universal Korovkin closures.
   6.8 Other aspects of the KAT.

7. **Korovkin-type approximation theory in Banach algebras:**
   a) **General Banach (involutive) algebras (also topological (involutive) algebras);**
   b) **C^*-algebras;**
   c) **Function algebras on a compact or a locally compact Hausdorff space;**
   d) **Function algebras on a topological group;**
   e) **Banach algebras of vector valued functions.**

   7.1 Research surveys, monographs.
   7.2 KAT for the identity operator with respect to linear contractions.
   7.3 KAT for the identity operator with respect to positive linear operators.
   7.4 KAT for the identity operator with respect to other classes of operators.
   7.5 KAT for a (positive) linear functional with respect to (positive) linear functionals.
   7.6 Korovkin closures, universal Korovkin closures.
   7.7 Other aspects of the KAT.

8. **Korovkin-type approximation theory in Banach spaces and in locally convex spaces (also in spaces of vector valued functions).**

   8.1 Research surveys, monographs.
   8.2 KAT for the identity operator with respect to linear contractions.
   8.3 KAT for the identity operator with respect to other classes of operators.
   8.4 KAT for a linear functional with respect to linear functionals.
   8.5 Korovkin closures, universal Korovkin closures.
   8.6 Other aspects of the KAT.

9. Korovkin-type approximation theory in other settings.

9.1   Research surveys, monographs.
9.2   KAT for the identity operator.
9.3   Korovkin closures, universal Korovkin closures.
9.4   Other aspects of the KAT.

D.2 Subject index

Shashkin, Yu.A.: [1960], [1962], [1965a],
[1967], [1972]
Schäfer, E.: [1989]
Vinogradov, O.M.: [1981]
Volkov, V.I.: [1957], [1958],
[1960]
Yoshinaga, K. and Tamura, S.:
[1976]
Zifarelli, C.L.: [1988]

1.3 KAT for the identity operator with respect to linear contractions.

Berens, H. and Lorentz, G.G.: [1975]
Gusak, I. Ja. and Shashkin, Yu. A.:
[1979]
Kutateladze, S.S.: [1973a]
Lorentz, G.G.: [1972b]
Min'kova, R.M.: [1974]
Shashkin, Yu. A.: [1969]
Wulbert, D.E.: [1968]

1.4 KAT for the identity operator with respect to positive linear contractions.

Berens, H. and Lorentz, G.G.: [1975]
Campiti, M.: [1990]
Lorentz, G.G.: [1972b]
Zifarelli, C.L.: [1988]

1.5 KAT for the identity operator with respect to other classes of operators.

Abakumov, Ju. G. and Lampe, E.I.:
[1974]
Baskakov, V.A.: [1973b]
Franchetti, C.: [1970]
Korovkin, P.P.: [1962], [1965]
Labsker, L.G.: [1970], [1977], [1979a],
[1979b], [1980], [1989b], [1990]
Lupaş, A.: [1971]
Min'kova, R.M.: [1973], [1974]
Min'kova, R.M. and Shashkin, Yu. A.:
[1969]
Zybin, L.M.: [1966]

1.6 KAT for a lattice homomorphism with respect to positive linear operators.

Berens, H. and Lorentz, G.G.: [1973]
Fakhoury, H.: [1974]
Franchetti, C.: [1969]
Krasnosel'skiĭ, M.A. and Lifshits, E.A.:
[1968a]
Krasnosel'skiĭ, M.A., Lifshits, E.A. and
Sobolev, A.V.: [1989]
Kutateladze, S.S. and Rubinov, A.M.:
[1971a]
Nishishiraho, T.: [1987]
Rasa, I: [1985]

1.7 KAT for a finitely defined operator with respect to positive linear operators.

Altomare, F.: [1987c]
Cavaretta, A.S.Jr.: [1973]
Drozhzhin, I.A.: [1989]
Jiménes Poso, M.A.: [1975]
Labsker, L.G.: [1975], [1979a], [1979b],
[1980]
Micchelli, C.A.: [1973a]
Rasa, I: [1981]
Rusk, M.D.: [1975]
Shashkin, Yu. A.: [1965b]
Takahasi, Sin-Ei: [1990a], [1990b]
Zifarelli, C.L.: [1988]

1.8 KAT for a positive operator with respect to positive linear operators.

Altomare, F.: [1980b], [1989a]
Campiti, M.: [1988]
Felbecker, G.: [1972]
Ferguson, L.B.O. and Rusk, M.D.:
[1976]
Krasnosel'skiĭ, M.A. and Lifshits, E.A.:
[1968a]
Krasnosel'skiĭ, M.A., Lifshits, E.A. and
Sobolev, A.V.: [1989]
Micchelli, C.A.: [1975]
Nagel, J.: [1978]
Netuka, I.: [1985]
Nishishiraho, T.: [1976], [1987]

Scheffold, E.: [1973a]
Sendov, Bl.: [1965]
Takahasi, Sin-Ei: [1990a], [1990b],
    [1993a], [1993b]
Ustinov, G.M.: [1980]
Vasil'ev, R.K.: [1974b]
Vitale, R.A.: [1979]
Xie, Dun Li: [1985]
Yang, Li Hua: [1985]
Zybin, L.M.: [1966]

2. Korovkin-type approximation theory in the space $\mathscr{C}_0(X)$ of all continuous real-valued functions defined on a locally compact Hausdorff space X, which vanish at infinity.

2.1 Research surveys, monographs.

Altomare, F.: [1980a]
Donner, K.: [1982]
Sussich, J.F.: [1982]
Zifarelli, C.L.: [1988]

2.2 KAT for the identity operator with respect to positive linear operators.

Bauer, H. and Donner, K.: [1978]
Briem, E.: [1985]
Donner, K.: [1982]
Xiao, Chang Bai: [1988]
Zifarelli, C.L.: [1988]

2.3 KAT for the identity operator with respect to positive linear contractions.

Bauer, H. and Donner, K.: [1986]
Zifarelli, C.L.: [1988]

2.4 KAT for the identity operator with respect to other classes of operators.

Bauer, H. and Donner, K.: [1986]

2.5 KAT for a lattice homomorphism with respect to positive linear operators.

Fakhoury, H.: [1974]

2.6 KAT for a finitely defined operator with respect to positive linear operators.

Altomare, F.: [1987c]
Zifarelli, C.L.: [1988]

2.7 KAT for a positive linear operator with respect to positive linear operators.

Altomare, F.: [1980b], [1987c]
Campiti, M.: [1988]
Zifarelli, C.L.: [1988]

2.8 KAT for a positive linear functional with respect to positive linear functionals.

Altomare, F.: [1989b], [1991a]
Campiti, M.: [1988]
Zifarelli, C.L.: [1988]

2.9 Korovkin closures, universal Korovkin closures.

Altomare, F.: [1980b]
Bauer, H. and Donner, K.: [1978],
    [1986]
Donner, K.: [1982]
Fakhoury, H.: [1974]
Zifarelli, C.L.: [1988]

2.10 Other aspects of the KAT.

Altomare, F.: [1980b], [1991a]

3. Korovkin-type approximation theory in $L^p(X, \mathscr{F}, \mu)$-spaces, $(X, \mathscr{F}, \mu)$ being a measure space or in $L^p(X, \mu)$, where $\mu \in \mathscr{M}^+(X)$, and $1 \le p \le +\infty$.

Bibliography

Abakumov, Ju.G.
  1972   *The extension of systems of three functions to K-systems in the periodic case*
           (Russian), A collection of articles on the Constructive Theory of Functions
           and the Extremal Problems of Functional Analysis, 11–14, Kalinin. Gos.
           Univ., Kalinin, 1972. **MR** 51 # 13538. **SC** 1.2, 1.12.
Abakumov, Ju.G. and Lampe, E.I.
  1974   *A certain theorem of the type of P.P. Korovkin's theorem* (Russian), Applica-
           tion of Functional Analysis in Approximation Theory, no. **2**, 3–7, Kalinin.
           Gos. Univ., Kalinin, 1974. **MR** 51 # 13539. **SC** 1.5.
Abel, N.H.
  1881   *Solution de quelques problèmes à l'aide intégrales définies*, in: Oeuvres com-
           plétes de Niels Henrik Abel I (Nouv. Ed.; Ed. L. Sylow—S. Lie) Grøndahl &
           Søn, Christiania 1881, 11–27.
Adell, J.A., Badía, F.G. and de la Cal, J.
  1993   *Beta-type operators preserve shape properties*, to appear in Stochastic Process
           Appl. (1993).
Adell, J.A., Badía, F.G., de la Cal, J. and Plo, F.
  1992   *On the property of monotonic convergence for Beta operators*, preprint, 1992.
Adell, J.A. and de la Cal, J.
  1992   *Limiting properties of inverse beta and generalized Bleimann-Butzer-Hahn op-
           erators*, preprint, 1992.
  1993a   *Using stochastic processes for studying Bernstein-type operators*, Proceedings
           of the 2nd International Conference in Functional Analysis and Approxima-
           tion Theory, September 1992, Acquafredda di Maratea (Italy), 1992, Suppl.
           Rend. Circ. Mat. Palermo **33** (1993), 125–141.
  1993b   *On a Bernstein-type operator associated with the inverse Pòlya-Eggenberg dis-
           tribution*, Proceedings of the 2nd International Conference in Functional
           Analysis and Approximation Theory, September 1992, Acquafredda di
           Maratea (Italy), 1992, Suppl. Rend. Circ. Mat. Palermo **33** (1993), 143–154.
Adell, J.A., de la Cal, J. and San Miguel, M.
  1994   *Inverse Beta and generalized Bleimann-Butzer-Hahn operators*, J. Approx.
           Theory **76** (1994), no. 1, 54–64.
Akhadov, R.A.
  1981   *On convergence of a sequence of linear operators in the space of functions
           analytic in a disk* (Russian), Izv. Akad. Nauk Azerbaĭdžan SSR Ser. Fiz.—
           Tehn. Mat. Nauk (1981), no. 1, 67–71. **ZBL** 468. 47010. **SC** 5.2.a, 7.3.c.
Alfsen, E.M.
  1971   Compact Convex Sets and Boundary Integrals, *Ergebnisse der Mathematik
           und ihrer Grenzgebiete* **57**, *Berlin-Heidelberg-New York, Springer-Verlag,
           1971.* **MR** 56 # 3615.

Altomare, F.

1977   *Proiettori positivi, famiglie risolventi e problema di Dirichlet*, Ric. Mat. **26**
(1977), no. 1, 63–78. **MR** 58 # 30099.

1978   *Operatori di Lion sul prodotto di spazi compatti, semigruppi di operatori
positivi e problema di Dirichlet*, Ric. Mat. **28** (1978), no. 1, 33–58. **MR** 80d:
47048.

1979   *Théorèmes de convergence de type Korovkin relativament à une application
linéaire positive*, Boll. Un. Mat. Ital. (5) **16**-B (1979), 1013–1031. **MR** 81d:
41028. **SC** 5.2.b, 6.5, 6.7.

1980a   *Su alcuni aspetti della teoria dell'approssimazione di tipo Korovkin*, Quaderno
dell'Istituto di Analisi Matematica dell'Università di Bari, 1980. **SC** 1.1, 2.1, 3.1.

1980b   *Teoremi di approssimazione di tipo Korovkin in spazi di funzioni*, Rend. Mat.
(6) **13** (1980), no. 3, 409–429. **MR** 82h: 41029. **SC** 1.8, 2.7, 2.9, 2.10.

1981   *Quelques remarques sur les ensembles de Korovkin dans les espaces des fonctions continues complexes*, Séminaire d'Initiation à l'Analyse (G. Choquet-M.
Rogalski-J. Saint Raymond) **20**e Année 1980–81, no. 4, 12 pp.; Publ. Math.
Univ. Pierre et Marie Curie, Univ. Paris VI, Paris, 1981. **MR** 83j: 41026. **SC**
7.2.c, 7.2.d, 7.3.c.

1982a   *Frontières abstraites et convergence de familles filtrées de formes linéaires sur
les algèbres de Banach commutatives*, Séminaire d'Initiation à l'Analyse (G.
Choquet-M. Rogalski-J. Saint Raymond) **21**e Annèe 1981–82, no. 6, 19 pp.;
Publ. Math. Univ. Pierre et Marie Curie, Univ. Paris VI, Paris, 1982. **ZBL**
514. 43032. **SC** 7.5.a, 7.6.a, 7.6.c, 7.6.d, 7.7.a.

1982b   *Erratum "Teoremi di approssimazione di tipo Korovkin in spazi di funzioni"*,
Rend. Mat. (7) **2** (1982), no. 2, 421. **MR** 84a: 41026.

1982c   *On the Korovkin approximation theory in commutative Banach algebras*, Rend.
Mat. (7) **2** (1982), no. 4, 755–767. **MR** 84g: 41019. **SC** 7.4.a, 7.6.a.

1984a   *On the universal convergence sets*, Ann. Mat. Pura Appl. (4) **138** (1984), 223–
243. **MR** 86h: 41024. **SC** 6.7, 7.6.a, 7.6.b, 7.6.c, 7.6.d, 9.3.

1984b   *Korovkin closures in spaces of continuous affine functions*, Semesterbericht
Funktionalanalysis, Tübingen, Wintersem. 84/85. **SC** 4.6, 8.5.

1986   *Korovkin closures in Banach algebras*, Operator Theory: Advances and Applications, Vol. **17**, 35–42, Birkhäuser Verlag Basel, 1986. **MR** 88g: 46062. **SC**
7.1.a, 7.1.b, 7.1.c, 7.6.b, 7.6.c.

1987a   *Nets of positive operators in spaces of continuous affine functions*, Boll. Un.
Mat. Ital. (7) **1**-B (1987), 217–233. **MR** 88k: 46024. **SC** 4.6, 8.5.

1987b   *Korovkin-type theorems for positive functionals in spaces of continuous affine
functions*, Rend. Circ. Mat. Palermo (2) **36** (1987), 167–181. **MR** 90c: 46035.
SC 1.9, 4.5, 5.4.a, 7.5.b.

1987c   *Approximation of finitely defined operators in function spaces*, Note Mat. **7**
(1987), 211–229. **MR** 90i: 47029. **SC** 1.7, 2.6, 2.7, 5.3.a.

1989a   *Limit semigroups of Bernstein-Schnabl operators associated with positive projections*, Ann. Sc. Norm. Sup. Pisa, Cl. Sci. (4), (16) (1989), no. 2, 259–279. **MR**
91a: 47036. **SC** 1.2, 1.8.

1989b   *Positive linear forms and their determining subspaces*, Ann. Mat. Pura Appl.
154 (4) (1989), 243–258. **MR** 91d: 46003. **SC** 1.9, 2.8, 6.6.

1989c   *On a sequence of Bernstein-Schnabl operators defined on a cylinder*, in: Approximation Theory VI, C.K. Chui, L.L. Schumaker and J.D. Wards (eds.), pp. 1–4, Academic Press, Boston, 1989.

1991a   *Convergence subspaces associated with discrete measures*, Atti Sem. Mat. Fis. Univ. Modena **39** (1991), 473–486. **MR** 93c: 28012. **SC** 2.8, 2.10.

1991b   *Positive projections, approximation processes and degenerate diffusion equations*, in: Proceedings of the International Meeting in Mathematical Analysis and its Applications on the Occasion of the Seventieth Birthday of Professor G. Aquaro (Bari, Italy, November, 9–10, 1990), Conf. Sem. Mat. Univ. Bari **241** (1991), 43–68. **MR** 93i: 41015.

1991c   *Lototsky-Schnabl operators on the unit interval*, C. R. Acad. Sci. Paris **313**, Série I (1991), 371–375. **MR** 92j: 41035.

1992a   *Lototsky-Schnabl operators on the unit interval and degenerate diffusion equations*, in: Progress in Functional Analysis, K.D. Bierstedt, J. Bonet, J. Horvath and M. Maestre (Eds.), Proceedings of the International Functional Analysis Meeting on the Occasion of the Sixtieth Birthday of Professor M. Valdivia (Peñiscola, Spain, October, 21–27, 1990), North-Holland Mathematics Studies **170**, North-Holland Amsterdam 1992, 259–277. **MR** 93e: 41034.

1992b   *Lototsky-Schnabl operators on compact convex sets and their associated limit semigroups*, Mh. Math. **114** (1992), 1–13.

1994   *On some approximation processes and their associated parabolic problems*, to appear in Conf. Sem. Mat. Fis. Univ. Milano, 1994.

Altomare, F. and Boccaccio, C.

1982   *On Korovkin-type theorems in spaces of continuous complex-valued functions*, Boll. Un. Mat. Ital. (6) **1**-B (1982), no. 1, 75–86. **MR** 83j: 41028. **SC** 7.2.c, 7.3.c.

Altomare, F. and Campiti, M.

1989   *A bibliography on the Korovkin-type approximation theory (1952–1987)*, in "Functional Analysis and Approximation", Proc. Internat. Conf., Bagni di Lucca, May 16–20, 1988, edited by P.L. Papini, Pitagora Editrice, Bologna, 1989, 34–79. **MR** 90m: 41001.

Altomare, F. and Rasa, I.

1989   *Approximation by positive operators in the space $\mathscr{C}^{(p)}([a,b])$*, Anal. Numér. Théor. Approx. **18**, 1 (1989), 1–11. **MR** 91i: 41018. **SC** 5.2.e, 5.3.e, 5.4.e.

Altomare, F. and Romanelli, S.

1992   *On some classes of Lototsky-Schnabl operators*, Note Mat. **12** (1992), 1–13.

Amir, D. and Ziegler, Z.

1978   *Korovkin shadows and Korovkin systems in $\mathscr{C}(S)$-spaces*, J. Math. Anal. Appl. **62** (1978), no. 3, 640–675. **MR** 57 # 10321. **SC** 1.2, 1.9, 1.11, 1.12.

Anastassiou, G.A.

1985   *A discrete Korovkin theorem*, J. Approx. Theory **45** (1985), no. 4, 383–388. **MR** 87d: 41035. **SC** 5.2.f, 5.6.e.

1990   *On a discrete Korovkin theorem*, J. Approx. Theory **61** (1990), no. 3, 384–386. **MR** 91c: 41056. **SC** 5.2.f.

1991   *A discrete stochastic Korovkin theorem*, Internat. J. Math. Math. Sci. **14** (1991), no. 4, 679–682. **MR** 92g: 60048. **SC** 8.6, 9.2.

558   Bibliography

Anastassiou, G.A., Cottin, C. and Gonska, H.H.
  1991a   *Global smoothness of approximating functions*, Analysis **11** (1991), 43–57. **MR** 92h: 41047.
  1991b   *Global smoothness preservation by multivariate approximation operators*, in: Israel Mathematical Conference Proceedings, vol. IV (ed. by S. Baron and D. Leviatan), Ramat-Gan; Bar-Ilan University (1991), 31–44.
Andreev, V.I.
  1963   *Some problems relating to the convergence of positive linear operators* (Russian), Kalinin. Gos. Ped. Inst. Učen. Zap. **29** (1963), 3–14. **MR** 28 # 2392. **SC** 1.2, 7.3.c.
Andrica, D. and Badea, C.
  1986   *Jensen's and Jessen's inequality, convexity preserving and approximating polynomial operators and Korovkin's theorem*, Seminar on Math. Analysis, Babeş-Bolyai Univ., preprint no. 4 (1986), 7–16. **SC** 1.2.
Andrica, D. and Mustăţa, C.
  1989   *An abstract Korovkin type theorem and applications* (Romanian summary), Studia Univ. Babeş-Bolyai Math. **34** (1989), no. 2, 44–51. **MR** 91j: 41043. **SC** 1.2.
Anger, B. and Lembcke, J.
  1974   *Hahn-Banach type theorems for hypolinear functionals*, Math. Ann. **209** (1974), 127–151. **MR** 50 # 934.
Anger, B. and Portenier, C.
  1992   Radon Integrals, an Abstract Approach to Integration and Riesz Representation Through Function Cones, *Series Progress in Mathematics* **103**, Birkhäuser, Boston, 1992.
Anselone, P.M.
  1969   *Abstract Riemann integrals, monotone approximations and generalizations of Korovkin's theorem*, Iterationsverfahren, Numerische Mathematik, Approximations-theorie (Takung Numer. Method. Approximationstheorie, Oberwolfach, 1969), 107–114; Internat. Schr. Numer. Math., Vol. **15**, Birkhäuser, Basel, 1970. **MR** 51 # 3763. **SC** 5.3.a, 6.5, 6.6.
Aramă, O.
  1957   *Proprietăti privind monotonia şirului polinoamelor de interpolare ale lui S.N. Bernstein şi aplicarea lor la studiul approximării funcţiilor*, Acad. R. P. Rom. Fil. Cluj, Studii Cerc. Mat. **8** (1957), 195–210. **MR** 23 # A1986a.
Arató, M. and Rényi, A.
  1957   *Probabilistic proof of a theorem on the approximation of continuous functions by means of generalized Bernstein polynomials*, Acta Math. Acad. Sci. Hungar. **8** (1957), 91–98. **MR** 19, p. 411.
Arendt, W., Chernoff, P. and Kato, T.
  1982   *A generalization of dissipativity and positive semigroups*, J. Operator Theory **8** (1982), 167–180. **MR** 84e: 47047.
Arens, R. and Singer, I.M.
  1956   *Generalized analytic functions*, Trans. Amer. Math. Soc. **81** (1956), 379–393. **MR** 17, p. 1226.

Armitage, D.H.
1972   *A linear function from a space of polynomials onto a space of harmonic polyno-mials*, J. London Math. Soc. (2), **5** (1972), 529–538. **MR** 47 # 3701.

Arveson, W.B.
1970   *An approximation theorem for function algebras*, preprint, University of Tex-as at Austin, 1970. **SC** 1.2, 7.3.c.

Asimow, L. and Ellis, A.J.
1980   Convexity Theory and its Applications in Functional Analysis, *London Math. Soc. Mon.* **16**, *Academic Press, London 1980.* **MR** 82m: 46009.

Attalienti, A.
1994   *Uniqueness subspaces and Stone-Weierstrass theorems for adapted spaces*, to appear in Math. Japon., 1994. **SC** 5.3.d, 5.6.d.

Badea, C. *(see also Andrica, D.)*
1990   *On a Korovkin-type theorem for simultaneous approximation*, J. Approx. The-ory **62** (1990), 223–234. **MR** 92j: 41029. **SC** 5.2.e, 5.6.e.

Badea, C., Badea, I. and Cottin, C.
1988   *A Korovkin-type theorem for generalizations of Boolean sum operators and approximation by trigonometric pseudopolynomials*, Anal. Numér. Théor. Ap-prox. **17** (1988), no. 1, 7–17. **MR** 90f: 41019. **SC** 5.2.e.

Badea, C., Badea, I. and Gonska, H.H.
1986   *A test function theorem and approximation by pseudopolynomials*, Bull. Aus-tral. Math. Soc. **34** (1986), no. 1, 53–64. **MR** 87g: 41048. **SC** 5.2.j.

Badea, C. and Cottin, C.
1991   *Korovkin-type theorems for generalized Boolean sum operators*, Approxima-tion Theory (Proc. Conf. Kecskemét, Hungary, 1990). Colloquia Mathe-matica Soc. János Bolyai, 58, North-Holland Publ. Comp. Amsterdam-Oxford-New York, 1991, 51–68. **SC** 5.2.e.

Badea, C. and Rasa, I.
1993   *The Gleason-Kahane-Żelazko property and Korovkin systems in symmetric involutive Banach algebras*, Arch. Math. **61** (1993), 163–169. **SC** 7.6.a.

Badea, I. *(see Badea, C.)*

Badía, F.G. *(see Adell, J.A.)*

Bari, N.K.
1964   A Treatise on Trigonometric Series, *Vols. I and II, The Macmillan Co., New York, 1964.* **MR** 30 # 1347.

Baskakov, V.A.
1957   *An example of a sequence of linear positive operators in the space of continu-ous functions*, Dokl. Akad. Nauk. SSSR **113** (1957), 249–251. **MR** 20 # 1153.

1961   *On various convergence criteria for linear positive operators* (Russian), Usp. Mat. Nauk **16** (1961), no. 1 (97), 131–134. **MR** 23 # A3470. **SC** 1.2.

1972   *Generalization of certain theorems of P.P. Korovkin on positive operators for periodic functions* (Russian), A collection of articles on the Constructive The-ory of Functions and the Extremal Problems of Functional Analysis, 40–44, Kalinin. Gos. Univ., Kalinin, 1972. **MR** 51 # 8691. **SC** 1.2, 1.12.

1973a   *Generalization of certain theorems of P.P. Korovkin on positive operators*, Mat. Zametki **13** (1973), 785–794. **MR** 48 # 6776. **SC** 1.12.

1973b   *Operators that are nearly positive* (Russian), Application of Functional Analysis in Approximation Theory, no. **1**, 20–23, Kalinin. Gos. Univ., Kalinin, 1973. **MR** 50 # 2769. **SC** 1.5, 1.12.

1975   *A generalization of theorem of Korovkin on conditions for and the convergence of a sequence of positive operators*, in: Approximation Theory (Proc. Conf., Inst. Math., Adam Mickiewicz Univ., Poznán, 1972), 11–19, Reidel, Dordrecht, 1975. **MR** 56 # 9148. **SC** 1.2.

Bauer, H. *(see also Nöbeling, G.)*

1958   *Un problème de Dirichlet pour la frontière de Shilov d'un espace compact*, C. R. Acad. Sci. Paris **247** (1958), 843–846.

1959   *Frontière de Shilov et problème de Dirichlet*, Séminaire de Théorie du Potentiel **3** (1958–1959), no. 7, 23 p. (Institut H. Poincaré, Paris).

1961   *Shilovscher Rand und Dirichletsches Problem*, Ann. Inst. Fourier (Grenoble) **11** (1961), 89–136. **MR** 25 # 443.

1963   *Kennzeichnung kompakter Simplexe mit abgeschlossener Extremalpunktmenge*, Arch. Math. **14** (1963), 415–421. **MR** 29 # 1352.

1973   *Theorems of Korovkin type for adapted spaces* (French summary), Ann. Inst. Fourier (Grenoble) **23** (1973), fasc. 4, 245–260. **MR** 50 # 10643. **SC** 5.2.d, 5.5.d, 5.6.d.

1974   *Convergence of monotone operators*, Math. Z. **136** (1974), 315–330. **MR** 50 # 14184. **SC** 5.2.d, 5.5.d, 5.6.d.

1978a   *Approximation and abstract boundaries*, Amer. Math. Monthly **85** (1978), 632–647. **MR** 56 # 12723. **SC** 1.1, 1.2, 1.11.

1978b   *Funktionenkegel und Integralungleichungen*, Sitz. Ber. math. naturw. kl. Bayer. Akad. Wiss. München 1977 (1978), 53–61. **MR** 80f: 46023.

1979   *Korovkin approximation in function spaces*, in: Approximation Theory and Functional Analysis (Prolla, J.B. Ed.), Proc. Internat. Sympos. Approximation Theory (Univ. Estadual de Campinas, Campinas, 1977), 19–36; North-Holland Math. Studies **35**, North-Holland, Amsterdam, 1979. **MR** 81a: 41052. **SC** 1.1, 1.11, 5.2.a, 5.5.a.

1981   Probability Theory and Elements of Measure Theory, *Academic Press, London, New York, Toronto, Sydney, San Francisco, 1981.* **MR** 82k: 60001.

1986   *A class of means and related inequalities*, Manuscripta Math. **55** (1986), 199–211. **MR** 87j: 26027.

1987   *Mittelwerte und Funktionalgleichungen*, Sitz. Ber. math. naturw. kl. Bayer. Akad. Wiss. München 1986 (1987), 1–9. **MR** 89a: 39010.

1992   Maß- und Integrationstheorie, *2. Auflage, Walter de Gruyter, Berlin-New York, 1992.*

Bauer, H. and Donner, K.

1978   *Korovkin approximation in $\mathscr{C}_0(X)$*, Math. Ann. **236**, (1978), no. 3, 225–237. **MR** 58 # 12115. **SC** 2.2, 2.9.

1986   *Korovkin closures in $\mathscr{C}_0(X)$*, in: Aspects of Mathematics and its applications, 145–167; North-Holland Math. Library **34**, North Holland, Amsterdam-New York, 1986. **MR** 87k: 46050. **SC** 2.3, 2.4, 2.9.

Bauer, H., Leha, G. and Papadopoulou, S.
1979   *Determination of Korovkin closures*, Math. Z. **168** (1979), no. 3, 263–274. **MR**
        80k: 46019. **SC** 1.11.
Becker, M., Butzer, P.L. and Nessel, R.J.
1976   *Saturation for Favard operators in weighted function spaces*, Studia Math. **59**
        (1976), no. 2, 139–153. **MR** 55 # 10927.
Becker, M. and Nessel, R.J.
1975   *Iteration von Operatoren und Saturation in lokalkonvexen Räumen*, Forsch-
        ungsberichte des Landes Nordrhein-Westfalen, no. 2470, Westdeutscher
        Verlag Opladen, 1975, 27–49.
1981   *On global saturation for Kantorovich polynomials*, in: Approximation and
        Function Spaces, Proc. Conf. Gdánsk, 1979, North-Holland, Amsterdam,
        1981, 89–101. **MR** 83c: 41005.
Beckhoff, F.
1987   *Korovkin-Theorie in Algebren*, Schriftenr. Math. Inst. Univ. Münster **2** Ser.
        45 (1987). **MR** 89b: 46067. **SC** 7.3.a, 7.3.b, 7.6.b.
1988   *A counterexample in Korovkin theory*, Rend. Circ. Mat. Palermo (2) **37** (1988),
        no. 3, 469–473. **MR** 90k: 46154. **SC** 7.6.b.
1991   *Korovkin theory in normed algebras*, Studia Math. **100** (1991), 219–228. **MR**
        92j: 46098. **SC** 7.6.b.
1993   *Korovkin theory in Banach *-algebras*, Math. Slovaca, 43 (1993), no. 5, 631–
        642. **SC** 7.6.b.
Bell, H.T.
1973   *Order summability and almost convergence*, Proc. Amer. Math. Soc. **38** (1973),
        548–552. **MR** 46 # 9587.
Belleni-Morante, A.
1979   Applied Semigroups and Evolution Equations, *The Clarendan Press, Oxford
        University Press, New York, 1979.* **MR** 82f: 47001.
Bellman, R.
1961   A Brief Introduction to Theta Functions, *Holt, Rinehart and Winston, New
        York, 1961.* **MR** 23 # A2556.
Berens, H. *(see also Butzer, P.L.)*
Berens, H. and Lorentz, G.G.
1972   *Inverse theorems for Bernstein polynomials*, Indiana J. Math. **21** (8) (1972),
        693–708. **MR** 45 # 5638.
1973   *Theorems of Korovkin type for positive linear operators on Banach lattices*, in:
        Approximation Theory (Proc. Internat. Sympos., Univ. Texas, Austin, Tex.,
        1973), 1–30; Academic Press, New York, 1973. **MR** 49 # 5663. **SC** 1.2, 1.6,
        1.11, 3.2, 3.4, 3.6, 5.2.b, 5.5.b, 6.4.
1974a   *Sequences of contractions on L^1-spaces*, J. Functional Analysis **15** (1974),
        155–165. **MR** 50 # 865. **SC** 3.3, 3.6.
1974b   *Korovkin theorems for sequences of contractions on L^p-spaces*, in: Linear op-
        erators and approximation II (Proc. Conf., Oberwolfach, 1974), 367–375;
        Internat. Ser. Numer. Math., Vol. 25, Birkhäuser, Basel, 1974. **MR** 52 #
        3817. **SC** 3.3, 3.4, 6.3.

1975   *Geometric theory of Korovkin sets*, J. Approx. Theory **15** (1975), no. 3, 161–
        189. **MR** 52 # 11424. **SC** 1.1, 1.2, 1.3, 1.4, 1.11.
1976   *Convergence of positive operators*, J. Approx. Theory **17** (1976), no. 4, 307–
        314. **MR** 54 # 10947. **SC** 1.2, 1.11, 1.12.

Berens, H., Schmidt, H.J. and Xu, Y.
1992   *Bernstein-Durrmeyer polynomials on a simplex*, J. Approx. Theory **68** (1992),
        no. 3, 247–261. **MR** 93a: 41007.

Berens, H. and Xu, Y.
1991a   *On Bernstein-Durrmeyer polynomials with Jacobi weights*, in: Approximation
        Theory and Functional Analysis (C.K. Chui Ed.), Academic Press, Boston,
        1991, 25–46.
1991b   *On Bernstein-Durrmeyer polynomials with Jacobi weights: the cases $p = 1$ and
        $p = \infty$*, in: Approximation, Interpolation and Summation (S. Baron and D.
        Leviatan Eds.), Israel Mathematical Conference Proceedings, Vol. 4, 51–62,
        Weizmann Science Press of Israel, 1991.
1991c   *K-moduli, moduli of smoothness and Bernstein polynomials on a simplex*,
        Indag. Math. (N.S.) **2** (1991), no. 4, 411–421. **MR** 93d: 41007.

Bernau, S.J.
1974   *Theorems of Korovkin type for L^p-spaces*, Pacific J. Math. **53** (1974), 11–19.
        MR 52 # 14786. **SC** 3.3, 3.6.

Bernau, S.J., Huijsmans, C.B. and De Pagter, B.
1992   *Sum of lattice homomorphisms*, Proc. Amer. Math. Soc. **115** (1992), no. 1,
        151–156. **MR** 92h: 46049.

Bernstein, S.N.
1912   *Démonstration du théorème de Weierstrass fondée sur le calcul des probabilités*,
        Commun. Soc. Math. Kharkow (2) **13** (1912–13), 1–2.

Bézier, P.
1972   Numerical Control, *Mathematics and Applications, John Wiley and Sons,
        London, 1972.*

Bishop, E.
1959   *A minimal boundary for function algebra*, Pacific J. Math. **9** (1959), 629–642.
        MR 22 # 191.

Bishop, E. and de Leeuw, K.
1959   *The representation of linear functionals by measures on sets of extreme points*,
        Ann. Inst. Fourier (Grenoble) **9** (1959), 305–331.

Bleimann, G., Butzer, P.L. and Hahn, L.
1980   *A Bernstein-type operator approximating continuous functions on the semi-
        axis*, Indag. Math. **42** (1980), 255–262. **MR** 81m: 41023.

Bloom, W.R. and Sussich, J.F.
1980   *Positive linear operators and the approximation of continuous functions on
        locally compact abelian groups*, J. Austral. Math. Soc. (Series A) **30** (1980),
        180–186. **MR** 82e: 43004. **SC** 7.3.c, 7.3.d.

Blümlinger, M.
1987   *Approximation durch Bernsteinoperatoren auf kompakten konvexen Mengen*,
        Osterreich. Akad. Wiss. Math.-Natur. kl. Sitzungsber. II **196** (1987), no. 4–7,
        181–215. **MR** 89j: 41035.

Boccaccio, C. *(see Altomare, F.)*

Boer, H.R. *(see Martini, R.)*

Bohman, H.
1952   On approximation of continuous and analytic functions, Ark. Math. **2** (1952–54), 43–56. **MR** 14 # 254. **SC** 1.2, 1.12.

Bojanic, R.
1969   A note on the precision of interpolation by Hermite-Fejér polynomials, in: Proc. Conf. Constructive Theory of Functions, Budapest, 1969, 69–76.

Boltiansky, V.G., Ryshkov, S.S. and Shashkin, Yu. A.
1960   On k-regular imbeddings and their applications to the theory of approximation of functions, Uspehi Mat. Nauk. **15** (1960), no. 6. **MR** 23 # A2867.

Bonnesen, T. and Fenchel, W.
1934   Theorie der konvexen Körper, *Ergebnisse der Mathematik und ihrer Grenzgebiete* **3**, *no. 1, Springer-Verlag, Berlin, 1934.*

Bonsall, F.F. and Duncan, J.
1971   Numerical Ranges of Operators on Normed Spaces and of Elements of Normed Algebras, Vol. I, *London Math. Soc. Lect. Notes* **2**, *Cambridge University Press, Cambridge, 1971.* **MR** 44 # 5779.

1973   Complete Normed Algebras, *Ergebnisse der Mathematik und ihrer Grenzgebiete* **80**, *Springer-Verlag, Berlin-Heidelberg-New York, 1973.* **MR** 54 # 11013.

Borsuk, K.
1957   On the k-independent subsets of the euclidean spaces and of the Hilbert spaces, Bull. Acad. Pol. Sci. Cl. III **5** (1957), 351–356. **MR** 19, p. 567.

Bourbaki, N.
1965   Éléments de Mathématique, Livre III, Topologie Générale, Ch. 1–2, *Actualités Scientifiques et Industrielles, 1343 (1965), Hermann, Paris.*

1969   Éléments de Mathématique, Livre VI, Integration, Ch. 1–9, *Actualités Scientifiques et Industrielles, 1343 (1969), Hermann, Paris.* **MR** 36 # 2183.

Boyanov, B.D. and Veselinov, V.M.
1970   A note on the approximation of functions in an infinite interval by linear positive operators, Bull. Math. Soc. Sci. Math. R.S. Roumanie (N.S.) **14 (62)**, (1970), no. 1, 9–13 (1971). **MR** 48 # 2627. **SC** 5.2.a.

Bratteli, O. and Jørgensen, P.E.T.
1984   Positive Semigroups of Operators and Applications, *Special Issue of Acta Appl. Math.* **2** *(1984), Reidel, Dordrecht, 1984.*

Brelot, M. and Choquet, G.
1954   Polynômes harmoniques et polyharmoniques, Second Colloque sur les équations aux dérivées partielles, Bruxelles, 1954. **MR** 16, p. 1108.

Briem, E.
1979   Korovkin-type theorems for subspaces of $\mathscr{C}_{\mathbb{C}}(X)$, preprint series **11**, Köbenhavens Universität, Matematisk Institut, 1979. **SC** 7.2.c.

1985   Convergence of sequences of positive operators on L^p-spaces, in: Proceedings of the Nineteenth Nordic Congress of Mathematicians, Reykjavik, 1984, 126–131; Visindafel. Isl., XLIV, Icel. Math. Soc. Reykjavik, 1985. **MR** 87g: 41054. **SC** 2.2, 3.2.

Brodskii, M.L.
   1961   *On a necessary and sufficient criterion for a system of functions for which P.P.*
             Korovkin's theorem holds (Russian), in: Research on Current Problem in the con-
             structive Theory of Functions, Fizmatgiz, Moscow (1961), 318–323. **SC** 1.2.
Brosowski, B.
   1979a   *The completion of partially ordered vector spaces and Korovkin's theorem*, in:
             Approximation Theory and Functional Analysis (Proc. Internat. Sympos.
             Approximation Theory, Univ. Estadual de Campinas, Campinas, 1977), 63–
             69; North. Holland Math. Studies **35**, North-Holland, Amsterdam, 1979. **MR**
             81d: 06030. **SC** 6.8.
   1979b   *Die Anwendung eines verallgemeinerten Korovkin Satzes auf die Konvergenz*
             gewisser Differenzen-verfahren, in: Multivariate Approximation Theory (Proc.
             Conf. Math. Res. Inst., Oberwolfach, 1979), ISNM Vol. **51** (1979), 46–56.
             ZBL 421. 65059. **SC** 1.12.
   1980   *A Korovkin-type theorem for differentiable functions*, in: Approximation The-
             ory III (Proc. Conf. Univ. Texas, Austin, Tex., 1980), 255–260; Academic
             Press, New York, 1980. **MR** 82d: 41048. **SC** 5.6.d.
   1981   *An application of Korovkin's theorem to certain partial differential equations*,
             in: Functional Analysis, Holomorphy and Approximation Theory (Proc.
             Semin. Univ. Fed. Rio de Janeiro, Rio de Janeiro, 1978), 150–162; Lecture
             Notes in Math. **843** (1981), Springer, Berlin, 1981. **MR** 82f: 46005. **SC** 1.12.
   1983   *On an elementary proof of the Stone-Weierstrass theorem and some extensions*,
             Functional Analysis, Holomorphy and Approximation Theory Rio de
             Janeiro, 1979), 1–9; Lecture Notes in Pure and Appl. Math. **83**, Dekker, New
             York, 1983. **MR** 84f: 46043. **SC** 8.3, 8.6.
Brown, B.M., Elliot, D. and Paget, D.F.
   1987   *Lipschitz constants for the Bernstein polynomial of a Lipschitz continuous*
             function, J. Approx. Theory **31** (1987), 59–66. **MR** 88d: 41023.
Bunce, J.W.
   1980   *Approximating maps and a Stone-Weierstrass theorem for C*-algebras*, Proc.
             Amer. Math. Soc. **79** (1980), no. 4, 559–563. **MR** 81h: 46082. **SC** 7.4.b.
Burov, V.N.
   1983   *On a construction of linear positive operators with continuous values* (Russian),
             Operators and their applications, Interuniv. Collect. sci. Works, Leningrad
             1983 (1983), 23–27. **ZBL** 558. 47030. **SC** 1.2.
Butzer, P.L. *(see also Becker, M.; Bleimann, G.)*
   1953   *On two-dimensional Bernstein polynomials*, Canad. J. Math. **5** (1953), 107–113.
             MR 14, p. 641.
Butzer, P.L. and Berens, H.
   1967   Semi-Groups of Operators and Approximation, *Die Grundlehren der mathe-*
             matischen Wissenschaften 145, Springer-Verlag, Berlin-Heidelberg-New York,
             1967.
Butzer, P.L. and Hahn, L.
   1979   *General theorems on rates of convergence in distribution of random variables*,
             I, General limit theorems, J. Multivariate Analysis **8** (1978), 181–201. **MR**
             80a: 60026a.

Butzer, P.L. and Nessel, R.J.
1971   Fourier Analysis and Approximation, Vol. 1, *Academic Press, New York, 1971.* **MR** 58 # 23312.

Calvert, B.
1975   *Convergence sets in reflexive Banach spaces,* Proc. Amer. Math. Soc. **47** (1975), no. 2, 423–428. **MR** 50 # 8008. **SC** 8.6.

Campiti, M. *(see also Altomare, F.)*
1987a   *A Korovkin-type theorem in the space of Riemann integrable functions,* Collect. Math. **38** (1987), 199–228. **MR** 90g: 41030. **SC** 5.2.i.

1987b   *A Korovkin-type theorem for set-valued Hausdorff continuous functions,* Le Matematiche **42** (1987), Fasc. I–II, 29–35. **MR** 91b: 41017. **SC** 1.11, 1.12, 9.2.

1988   *Determining subspaces for continuous positive discrete linear forms,* Ric. Mat. **37** (1) (1988), 97–112. **MR** 90i: 46047. **SC** 1.8, 1.9, 2.7, 2.8.

1990   *Riemann sequential approximation of continuous functions* (Italian summary), Boll. Un. Mat. Ital. (7) **4-B** (1990), 143–154. **MR** 91e: 41019. **SC** 1.4, 1.12.

1991a   *Approximation of continuous set-valued functions in Fréchet spaces, I,* Anal. Numér. Théor. Approx. **20** (1991), no. 1–2, 15–23. **SC** 8.3, 8.6, 9.2, 9.4. **MR** 93g: 41016.

1991b   *Approximation of continuous set-valued functions in Fréchet spaces, II,* Anal. Numér. Théor. Approx. **20** (1991), no. 1–2, 25–38. **SC** 8.3, 8.6, 9.2, 9.4. **MR** 93g: 41016.

1991c   *A generalization of Stancu-Mühlbach operators,* Constr. Approx. **7** (1991), 1–18. **MR** 91m: 41043.

1992a   *Korovkin theorems for vector-valued continuous functions,* in: Approximation Theory, Spline Functions and Applications (Maratea, 1991), 293–302, Nato Adv. Sci. Inst. Ser. C: Math. Phys. Sci., 356, Kluwer Acad. Publ., Dordrecht, 1992. **MR** 93e: 41047. **SC** 8.1, 8.3.

1992b   *Convergence of nets of monotone operators between cones of set-valued functions,* Atti Accad. Sci. Torino **126** (1992), 3–4, 39–54. **SC** 8.3, 8.6, 9.4.

1992c   *Limit semigroups of Stancu-Mühlbach operators associated with positive projections,* Ann. Sc. Norm. Sup. Pisa, Cl. Sci. (4), (19) (1992), no. 2, 51–67. **MR** 93j: 41016.

1993   *Convexity-monotone operators in Korovkin theory,* Proceedings of the 2nd International Conference in Functional Analysis and Approximation Theory, September 1992, Acquafredda di Maratea (Italy), 1992, Suppl. Rend. Circ. Mat. Palermo **33** (1993), 229–238. **SC** 8.3.

Campiti, M. and Rasa, I.
1991   *Sets of parabolic functions,* Atti Sem. Mat. Fis. Univ. Modena **34** (1991), 513–526. **MR** 93a: 46041. **SC** 1.11.

Cao, Jia Ding
1989   *On Sikkema-Kantorovich polynomials of order k,* Approx. Theory Appl. **5** (1989), no. 2, 99–109. **MR** 90m: 41012.

Carathéodory, C.
1911   *Über den Variabilitätsbereich der Fourier'schen Konstanten von positiven harmonischen Funktionen,* Rend. Circ. Mat. Palermo **32** (1911), no. 2, 193–217.

van Casteren, J.
1985   Generators of Strongly Continuous Semigroups, *Pitman, Boston-London-Melbourne, 1985.*

Cavaretta, A.S. Jr.
1973   *A Korovkin theorem for finitely defined operators*, in: Approximation Theory (Proc. Internat. Sympos., Univ. Texas, Austin, Tex., 1973), 299–305; Academic Press, New York, 1973. **MR** 48 # 11872. **SC** 1.7, 1.9.

Cavaretta, A.S. Jr. and Sharma, A.
1992   *Variation diminishing properties and convexity for the tensor product of Bernstein operators*, in: Functional Analysis and Operator Theory, Proceedings of the Conference held in Memory of U.N. Singh, New Delhi, India, 2–6 August, 1990, B.S. Yadav and D. Singh (Eds.), Lecture Notes in Mathematics, **1511**, Springer-Verlag, Berlin, 1992, 18–32.

Censor, E.
1971   *Quantitative results for positive linear approximation operators*, J. Approx. Theory **4** (1971), 442–450. **MR** 44 # 4441.

Chakalov, V.L.
1977   *A note on a theorem of Korovkin* (Russian), Serdica **3** (1977), 242–248. **ZBL** 437. 47022.

Chang, G.-Z. and Davis, P.J.
1984   *The convexity of Bernstein polynomials over triangles*, J. Approx. Theory **40** (1984), no. 1, 11–28. **MR** 85c: 41001.

Chang, G.-Z. and Feng, Y.-Y.
1984   *An improved condition for the convexity of Bernstein-Bézier surfaces over triangles*, Computer Aided Geometric Design **1** (1984), 279–283.

Chang, G.-Z. and Zhang, J.-Z.
1990   *Converse theorems of convexity for Bernstein polynomials over triangles*, J. Approx. Theory **61** (1990), no. 3, 265–278. **MR** 91j: 41016.

Chao, M.T. and Strawderman, W.E.
1972   *Negative moments of positive random variables*, J. Amer. Statist. Assoc. **67** (1972), 429–431.

Chen, W. and Ditzian, Z.
1991   *Best polynomial and Durrmeyer approximation in $L_p(S)$*, Indag. Math., New Ser. **2** (1991), no. 4, 437–452. **MR** 93c: 41011.

Cheney, E.W.
1982   Introduction to Approximation Theory, *2nd Ed., Chelsea, New York, 1982.*

Cheney, E.W. and Sharma, A.
1964a   *On a generalization of Bernstein polynomials*, Riv. Mat. Univ. Parma (2) **5** (1964), 77–84. **MR** 33 # 6233.
1964b   *Bernstein power series*, Canadian J. of Math. **16**, 2 (1964), 241–264. **MR** 31 # 3770.

Chernoff, P. *(see Arendt, W.)*

Chihara, T.S.
1978   An Introduction to Orthogonal Polynomials, *Gordon and Breach Science Publishers, New York-London-Paris, 1978.* **MR** 58 # 1979.

Chlodovsky, I.
1937   *Sur le développement des fonctions définies dans un interval infini en séries de polynômes de M.S. Bernstein*, Compositio Math. **4** (1937), 380–393.

Choda, H. and Echigo, M.
1963   *On the theorems of Korovkin*, Proc. Japan Acad. **39** (1963), 107–108. **MR** 27 # 601. **SC** 7.5.a, 7.5.c.

Choquet, G. *(see also Brelot, M.)*
1953   *Theory of capacities*, Ann. Inst. Fourier **5** (1953), 131–295.
1956   *Existence et unicité des représentations intégrales*, Séminaire Bourbaki, Décembre 1956, 139, 15 p.
1969   Lecture on Analysis, Vol. I, II, III, *W. A. Benjamin Inc., New York-Amsterdam, 1969*. **MR** 40 # 3252–3253–3254.

Choquet, G. and Meyer, P.A.
1963   *Existence et unicité des représentations intégrales dans les convexes compacts quelconques*, Ann. Inst. Fourier (Grenoble) **13** (1963), 139–154. **MR** 26 # 6748.

Cimoca, Gh.
1974   *On the theorem of Korovkin* (Romanian, English summary), Anal. Numér. Théor. Approx. **3** (1974), no. 1, 53–59. **MR** 52 # 3818. **SC** 1.2.

Cismasiu, C.S.
1984   *About an infinitely divisible distribution*, in: Proceedings of the Colloquium on Approximation and Optimization, Cluj-Napoca, October 25–27 (1984), 53–58.
1985   *A probabilistic interpretation of Voronovskaja's theorem*, Bul. Univ. Brasov, seria C, **27** (1985), 7–12. **MR** 88d: 41033.
1987a   *A linear positive operator associated with the Pearson's-χ^2 distribution*, Studia Univ. Babes-Bolyai Mathematica **32** (1987), no. 4, 21–23. **MR** 90f: 41026.
1987b   *Some properties of the linear positive operators of probabilistic type*, Seminar on Numerical and Statistical Calculus (Cluj-Napoca, 1987), 129–133, preprint 87-9, Univ. Babes-Bolyai, Cluj-Napoca, 1987. **MR** 89j: 41036.

Ciupa, A.
1986   *Approximation of functions of two variables by means of an operator of Bernstein type* (Romanian. English Summary), Studia Univ. Babes-Bolyai Math. **31** (1986), no. 1, 51–57. **MR** 88f: 41037.

Clément, Ph. (et al.)
1987   One-Parameter Semigroups, *North-Holland Publ. Co., Amsterdam-New York, 1987*. **MR** 89b: 47058.

Clément, Ph. and Timmermanns, C.A.
1986   *On C_0-semigroups generated by differential operators satisfying Wentcel's boundary conditions*, Indag. Math. **89** (1986), 379–387. **MR** 88c: 47075.

Comfort, W.W.
1961   *The Silov boundary induced by a certain Banach algebra*, Trans. Amer. Math. Soc. **98** (1961), 501–517. **MR** 22 # 11282.

Comtet, L.
1970   Analyse Combinatoire, II, *Presses Universitaires de France, Paris, 1970*. **MR** 41 # 6697.

568   Bibliography

Cottin, C. *(see Anastassiou, G.A.; Badea, C.)*
Curtis, P.C.
　1959　*n-parameter families and best approximation*, Pacific J. Math. **9** (1959), 1013–
　　　　1027. **MR** 21 # 7385.
Dahmen, W.
　1991　*Convexity and Bernstein-Bézier polynomials*, in: Curves and Surfaces, P.J.
　　　　Laurent, A. Le Méhauté and L.L. Schumaker (Eds.) 107–134, Academic
　　　　Press, Boston, 1991.
Dahmen, W. and Micchelli, C.A.
　1988　*Convexity of multivariate Bernstein polynomials and box spline surfaces*,
　　　　Studia Sci. Math. Hungar. **23** (1988), 265–287. **MR** 90g: 41005.
　1990　*Convexity and Bernstein polynomials on k-simploids*, Acta Math. Appl. Sinica
　　　　6 (1990), no. 1, 50–66. **MR** 91f: 41006.
Dankel Jr., T.
　1979　*On moduli of continuity of analytic and harmonic functions*, Rocky Mountain
　　　　J. of Math., **9** (1979), no. 3, 519–526. **MR** 81b: 31003.
Da Silva, M.R.
　1985　*Nonnegative order iterates of Bernstein polynomials and their limiting semi-
　　　　groups*, Portugal Math. **42** (1983–84), no. 3 (1985), 225–248. **MR** 87j: 41063.
　1987　*Iteração dos polinómias de Bernstein em duas variáveis reais independentes e
　　　　identificação dos operadores lineares que comutam com eles*, Publ. Dept. Math.
　　　　University of Coimbra, Portugal, special volume in honour of Professor Luis
　　　　de Albuquerque, 1987.
Davies, E.B.
　1980　One-Parameter Semigroups, *Academic Press, London-New York-San Fran-
　　　　cisco, 1980*. **MR** 82j: 47060.
Davis, P.J. *(see Chang, G.-Z.)*
de la Cal, J. *(see also Adell, J.A.)*
　1994　*On Stancu-Mühlbach operators and some connected problems concerning prob-
　　　　ability distributions*, to appear in J. Approx. Theory.
de la Cal, J. and Luquin, F.
　1991a　*Approximating Szász and Gamma operators by Baskakov operators*, preprint,
　　　　1991.
　1991b　*Probabilistic methods in Approximation Theory: a general setting*, Atti Sem.
　　　　Mat. Fis. Univ. Modena **39** (1991), 211–221.
　1992　*A note on limiting properties of some Bernstein-type operators*, J. Approx.
　　　　Theory **68** (1992), no. 3, 322–329.
de Leeuw, K. *(see Bishop, E.)*
Della Vecchia, B.
　1987　*On the monotonicity of the derivatives of the sequences of Favard and
　　　　Baskakov operators*, Ric. Mat. **36** (1987), no. 2, 263–269. **MR** 90c: 41045.
　1988　*On monotonicity of some linear positive operators*, in: Numerical Methods and
　　　　Approximation Theory III (Niš, August 18–21, 1987), G.V. Milovanovič Ed.,
　　　　Univ. Niš, Niš, 1988, 165–178. **MR** 90c: 41014.
　1989　*On the preservation of Lipschitz constants for some linear operators*, Boll. Un.
　　　　Mat. Ital. B (7) **3** (1989), no. 1, 125–136. **MR** 90i: 41031.

1991   *On the approximation by the D.D. Stancu type operators,* Studia Univ. Babes-Bolyai, Math. (1991).

Della Vecchia, B. and Rasa, I.

1993   *Bernstein-type operators, convexity and Lipschitz classes,* preprint, 1993.

De Pagter, B. *(see Bernau, S.J.)*

Derriennic, M.M.

1981   *Sur l'approximation de fonctions intégrable su* [0,1] *par des polynômes de Bernstein modifies,* J. Approx. Theory **31** (1981), 325–343. **MR** 82m: 41004.

1985   *On multidimensional approximation by Bernstein type polynomials,* J. Approx. Theory **45** (1985), 155–166. **MR** 87d: 41033.

De Vore, R.A.

1972   The Approximation of Continuous Functions by Positive Linear Operators, *Lecture Notes in Math.* **293**, *Springer-Verlag, Heidelberg-Berlin-New York, 1972.* **MR** 54 # 8100.

Dickmeis, W., Mevissen, H., Nessel, R.J. and van Wickeren, E.

1988   *Sequential convergence and approximation in the space of Riemann integrable functions,* J. Approx. Theory **55** (1988), 65–85. **MR** 90f: 41017. **SC** 5.2.i.

Dieckmann, G.

1991   *A characterization of the Korovkin closure with respect to functionals,* Arch. Math. **57** (1991), 379–384. **MR** 92h: 46013. **SC** 4.5.

1992   *Korovkin-Approximation in Räumen von affinen stetigen Funktionen,* Dissertation Universität Münster, 1992. **SC** 4.1, 4.2, 4.3, 4.4, 4.5, 4.6, 7.1.b, 7.3.b, 7.5.b, 8.1, 8.2, 8.5.

Dinghas, A.

1951   *Über einige Identitäten von Bernsteinschem Typus,* Norske Vid. Selsk. Fohr. Trondheim **24** (1951), 96–97. **MR** 14 # 167.

Ditzian, Z. *(see also Chen, W.)*

1975   *Convergence of sequences of linear positive operators: remarks and applications,* J. Approx. Theory **14** (1975), no. 4, 296–301. **MR** 51 # 13540. **SC** 5.2.a.

1979   *A global inverse theorem for combination of Bernstein polynomials,* J. Approx. Theory **26** (1979), 277–292. **MR** 80m: 41003.

1985   *Rate of approximation by linear processes,* Acta Sci. Math. **48** (1985), 1–4, 103–128. **MR** 87g: 41049.

Ditzian, Z. and Ivanov, K.G.

1989   *Bernstein-type operators and their derivatives,* J. Approx. Theory **56** (1989), 72–90. **MR** 90c: 41042.

Ditzian, Z. and May, C.P.

1976 L_p-*saturation and inverse theorems for modified Bernstein polynomials,* Indiana Univ. Math. J. **25** (1976), 733–751. **MR** 54 # 10945.

Ditzian, Z. and Totik, V.

1987   Moduli of Smoothness, *Springer-Verlag, New York Inc., 1987.* **MR** 89h: 41002.

Donner, K. *(see also Bauer, H.)*

1975   *Korovkin theorems for positive linear operators,* Collection of articles dedicated to G.G. Lorentz on the occasion of his sixty-fifth birthday, IV; J. Approx. Theory **13** (1975), 443–450. **MR** 51 # 6237. **SC** 6.4, 6.7.

1979   *Korovkin closures for positive linear operators*, J. Approx. Theory **26** (1979), no. 1, 14–25. **MR** 80f: 41023. **SC** 6.4, 6.8.

1980   *Korovkin theorems for L^p-spaces*, in: Approximation Theory III (Proc. Conf. Univ. Texas, Austin, Tex., 1980), 355–360; Academic Press, New York, 1980. **MR** 82d: 41026. **SC** 3.2.

1981   *Korovkin theorems in L^p-spaces*, J. Functional Analysis **42** (1981), no. 1, 12–28. **MR** 82i: 41025. **SC** 3.2.

1982   Extension of Positive Operators and Korovkin Theorems, *Lecture Notes in Math.* **904***, Springer-Verlag, Berlin, 1982.* **MR** 83i: 46008. **SC** 2.1, 2.2, 2.9, 3.2, 3.6, 5.2.a, 5.3.a, 6.4, 6.7.

Doob, J.L.
1953   Stochastic Processes, *John Wiley and Sons, New York, 1953.* **MR** 15, p. 445.

Dorroh, J.R.
1966   *Contraction semi-groups in a function space*, Pacific. J. Math. **19** (1966), 35–38. **MR** 34 # 1860.

Dorroh, J.R. and Goldstein, G.R.
1993   *Quasilinear diffusion*, preprint, 1993.

Dorroh, J.R. and Rieder, G.R.
1991   *A singular quasilinear parabolic problem in one space dimension*, J. Diff. Equations **91** (1991), no. 1, 1–23. **MR** 92d: 35156.

Drozhzhin, I.A.
1984   *Supremal generators of finite order* (Russian), Primen. Funkts. Anal. Teor. Priblizh. 1984, 23–30 (1984). **ZBL** 607. 46020. **SC** 1.2.

1985   *Triviality of the germ of a linear positive functional* (Russian), Geometric questions in the theory of functions and sets, 42–46, Kalinin. Gos. Univ. Kalinin, 1985. **MR** 88b: 46047. **SC** 1.9.

1989   *Supremally generating cones of a space of continuous functions* (Russian), Mat. Zametki **46** (1989), no. 6, 46–52, 127; translation in Math. Notes **46** (1989), no. 5–6, 920–924 (1990). **MR** 91g: 46025. **SC** 1.7.

Duncan, J. *(see Bonsall, F.F.)*
Durrmeyer, J.L.
1967   *Une formule d'inversion de la transformeé de Laplace: Applications á la theorie des moments*, Thèse de 3e cycle, Faculté des Sciences de l'Université de Paris, 1967.

Dynkin, E.B.
1961   Theory of Markov Processes, *Prentice-Hall, Inc. Englewood Cliffs, N. J., Pergamon Press, Oxford, 1961.* **MR** 24 # A1747.

Džafarov, A.S.
1975   *Spherical analogues of P.P. Korovkin's theorems* (Russian, Azerbaijani summary), Azerbaĭdžan. Gos. Univ. Učen. Zap. Ser. Fiz.-Mat. Nauk (1975), no. **4**, 9–16. **MR** 56 # 16219. **SC** 1.2.

Dzjadyk, V.K.
1966   *Approximation of functions by positive linear operators and singular integrals* (Russian), Mat. Sb. (N.S.) **70 (112)** (1966), 508–517. **MR** 34 # 8053. **SC** 3.2.

Echigo, M. *(see Choda, H.)*

Edwards, D.A.
1966   *A class of Choquet boundaries that are Baire spaces*, Quart. J. Math. Oxford **17** (1966), 282–284. **MR** 33 # 7882.

Edwards, R.E.
1967   Fourier Series: A Modern Introduction, I, II, *Holt, Rinehart and Winston, New York-Montreal-London, 1967*. **MR** 35 # 7062.

Efendiev, R.O.
1984   *Conditions for convergence of linear operators to derivatives* (Russian), Akad. Nauk Azerbaĭdžan SSR Dokl. **40** (1984), no. 12, 3–6. **MR** 86j: 41020. **SC** 5.6.e.

Eisenberg, B.
1976   *Another look at the Korovkin theorems*, J. Approx. Theory **17** (1976), no. 4, 359–365. **MR** 54 # 5694. **SC** 1.2, 1.10.

Eisenberg, S.M.
1979   *Korovkin's theorems*, Bull. Malaysian Math. Soc. (2) **2** (1979), no. 1, 13–29. **MR** 80j: 41034. **SC** 1.1, 1.2, 5.1.a, 5.2.a.

Eisenberg, S. and Wood, B.
1970   *Approximating unbounded functions by linear operators generated by moment sequences*, Studia Math. **35** (1970), 299–304. **MR** 42 # 6468.

1976   *On the degree of approximation by extended Hermite-Fejér operators*, J. Approx. Theory **18** (1976), 169–173. **MR** 55 # 946.

Elliot, D. *(see Brown, B.M.)*

Ellis, A.J. *(see Asimow, L.)*

Engelking, R.
1978   Dimension Theory, *North-Holland Math. Library* **19**, *North-Holland Publishing Co., Amsterdam-Oxford-New York, 1978*. **MR** 58 # 2753b.

1989   General Topology, *Sigma Series in Pure Mathematics* **6**, *Heldermann Verlag, Berlin, 1989*. **MR** 91c: 54001.

Fakhoury, H.
1974   *Le théorème de Korovkin dans $\mathscr{C}(X)$ et $L^p(\mu)$*, Séminaire d'Initiation à l'Analyse (G. Choquet-M. Rogalski-J. Saint Raymond), **13** Année, 1973/74, no. 9, 20 pp.. **ZBL** 337. 46008. **SC** 1.6, 1.11, 2.5, 2.9, 3.2, 3.6, 4.6, 6.4, 6.7.

Favard, J.
1944   *Sur les multiplicateurs d'interpolation*, J. Math. Pures Appl. **23** (9) (1944), 219–247. **MR** 7, p. 436.

Fejér, L.
1900   *Sur les fonctions bornées et intégrables*, C. Rendus Hebdomadaries, Séances de l'Académie de Sciences (Paris) **131** (1900), 984–987.

1904   *Untersuchungen über Fouriersche Reihen*, Math. Ann. **58** (1904), 51–69.

1916a   *Über trigonometrische Polynome*, J. Reine Angew. Math. **146** (1916), 53–82.

1916b   *Über Interpolation*, Nachrichten der Akademie der Wissenschaften in Göttingen, 1916, 66–91.

1930   *Über Weierstrassche Approximation besonders durch Hermitesche Interpolation*, Math. Ann. **102** (1930), 707–725.

Felbecker, G.
  1972   *Approximation und Interpolation auf Räumen Radonscher Wahrscheinlichkeit-smaße*, Dissertation, Ruhr-Universität Bochum, 1972. **SC** 1.2, 1.8, 1.12.
  1973   *Über Verallgemeinerte Stancu-Mühlbach-Operatoren*, Z. Angew. Math. Mech. **53** (1973), 188–189. **MR** 49 # 7661.
Felbecker, G. and Schempp, W.
  1971   *A generalization of Bohman-Korovkin's theorem*, Math. Z. **122** (1971), 63–70. **MR** 45 # 789. **SC** 1.2.
Feller, W.
  1951   *Diffusion processes in genetics*, in: Proc. Second Berkeley Symposium on Mathematical Statistics and Probability (Berkeley, 1951), 227–246. **MR** 13, p. 671.
  1952   *The parabolic differential equation and the associated semigroups of transformations*, Annals of Math. (2) **55** (1952), 468–519. **MR** 13, p. 948.
  1957   An Introduction to Probability Theory and its Applications, Vol. I, 2nd edition, *John Wiley, New York, 1957*. **MR** 19, p. 466.
  1966   An Introduction to Probability Theory and its Applications, Vol. II, 3rd edition, *John Wiley, New York, 1966*. **MR** 35 # 1048.
Fenchel, W. *(see Bonnesen, T.)*
Feng, Y.-Y. *(see also Chang, G.-Z.)*
  1990   *Lipschitz constant for the Bernstein polynomials defined over a triangle* (Chinese summary), J. Math. Res. Exposition **10** (1990), no. 1, 105–108. **MR** 91d: 41001.
Ferguson, L.B.O. and Rusk, M.D.
  1976   *Korovkin sets for an operator on a space of continuous functions*, Pacific J. Math. **65** (1976), no. 2, 337–345. **MR** 54 # 8117. **SC** 1.8.
Fernández Muñiz, J.L.
  1982   *Qualitative theorems of Korovkin type for sequences of operators of class $\tilde{R}$* (Spanish), Cienc. Mat. (Havana) **3** (1982), no. 1, 57–69. **MR** 84m: 41032. **SC** 1.12, 5.6.a, 7.4.c, 7.7.c.
  1987   *On qualitative Korovkin theorems in $H^\infty(D)$ and with A-Distance*, preprint University of Havana, 1987. **SC** 7.4.c, 7.7.c.
  1988   *On qualitative Korovkin theorems with A-distance*, Approximation and Optimization (Havana, 1987), 136–139, Lecture Notes in Math., 1354, Springer, Berlin-New York, 1988. **MR** 90d: 41038. **SC** 5.6.j.
  1992a *Bohman-Korovkin theorems with R_F-convergence*, Cienc. Mat. **13** (1992), no. 2.
  1992b *Qualitative Korovkin-type theorems for R_F-convergence*, preprint 1992. **SC** 5.6.i.
  1992c *Bohman-Korovkin theorems with A-convergence*, preprint 1992. **SC** 5.6.j.
Fichera, G.
  1992   *On a degenerate evolution problem*, Pitman Research Notes in Math., 263, Longman, 1992.
Flösser, H.O.
  1978   *A Korovkin-type theorem in locally convex M-spaces*, Proc. Amer. Math. Soc. **72** (1978), no. 3, 456–460. **MR** 80f: 46005. **SC** 6.7.
  1979   *Korovkin-closures of finite sets*, Arch. Math. (Basel) **32** (1979), no. 6, 600–608. **MR** 81d: 41030. **SC** 5.5.f, 6.7, 6.8.

1980a   *Korovkin closures in AM-spaces*, in: Approximation Theory III (Proc. Conf. Univ. Texas, Austin, Tex., 1980), 409–417; Academic Press, New York, 1980. **MR** 82c: 41031. **SC** 6.1, 6.7.

1980b   *Korovkinsche Approximation stetiger Funktionen*, Yearbook: Surveys of Mathematics, 1980 (German), 93–119; Bibliographisches Inst. Mannheim, 1980. **MR** 84m: 41035. **SC** 1.1, 6.1.

1981   *Sequences of positive contractions on AM-spaces*, J. Approx. Theory **31** (1981), no. 2, 118–137. **MR** 82h: 41043. **SC** 6.3, 6.4, 6.7.

Flösser, H.O., Irmisch, R. and Roth, W.

1981   *Infimum-stable convex cones and approximation*, Proc. London Math. Soc. (3) **42** (1981), no. 1, 104–120. **MR** 82b: 46009. **SC** 6.7.

Flösser, H.O. and Roth, W.

1979   *Korovkinhüllen in Funktionenräumen*, Math. Z. **166** (1979), no. 2, 187–203. **MR** 80f: 46024. **SC** 5.3.a, 5.6.a.

Forrest, A.R.

1971   *Computational geometry*, Proc. Roy. Soc. London Ser. A **321** (1971), 187–195.

Franchetti, C.

1969   *Disuguaglianza e teoremi del tipo di Korovkin sugli operatori positivi in* $\mathscr{C}([0,1])$, Boll. Un. Mat. Ital. **6** (1969), 641–647. **MR** 41 # 5849. **SC** 1.6, 1.11.

1970   *Convergenza di operatori in sottospazi dello spazio* $\mathscr{C}(Q)$, Boll. Un. Mat. Ital. (4) **3** (1970), 668–676. **MR** 42 # 6490. **SC** 1.5, 1.11.

Freud, G.

1964   *Über positive lineare Approximationsfolgen von stetigen reellen Funktionen auf kompakten Mengen*, in: On Approximation Theory (Proc. Conf., Oberwolfach, 1963), 233–238; Birkhäuser, Basel, 1964. **MR** 31 # 6088. **SC** 1.2.

Fucheng, Q.

1985   *A Korovkin type theorem on weak convergence for (series of) positive linear operators* (Chinese), J. Math. Res. Expo. **5** (1985), no. 3, 55–58. **ZBL** 606. 41027. **SC** 3.2, 3.7.

Fujii, J.I.

1987a   *Korovkin's theorem and strict positivity*, Math. Japon. **32** (1987), 17–20. **MR** 88c: 46067. **SC** 7.3.b, 7.7.b.

1987b   *A simple proof to non-commutative Korovkin's theorem*, Math. Japon. **32** (1987), no. 4, 531–532. **MR** 88k: 46066. **SC** 7.3.b.

Gadzhiev, A.D.

1974a   *Linear k-positive operators in a space of regular functions, and theorems of P.P. Korovkin type* (Russian, Azerbaijani and English summary), Izv. Akad. Nauk Azerbaĭdžan SSR Ser. Fiz.—Tehn. Mat. Nauk (1974), no. 5, 49–53. **MR** 55 # 3264. **SC** 7.7.c.

1974b   *A problem on the convergence of a sequence of positive linear operators on unbounded sets, and theorems that are analogous to P.P. Korovkin's theorem* (Russian), Dokl. Akad. Nauk SSSR **218** (1974) no. 5, 1001–1004; translated in Soviet Math. Dokl. **15** (1974), no. 5, 1433–1436 (1975). **MR** 51 # 3764. **SC** 5.2.c, 5.6.c.

1976   *Theorems of the type of P.P. Korovkin's theorems* (Russian), Presented at
the International Conference on the Theory of Approximation of Functions
(Kaluga, 1975), Mat. Zametki **20** (1976), no. 5, 781–786; translated in Math.
Notes **20** (1976), no. 5–6, 995–998, (1977). **MR** 58 # 12118. **SC** 5.6.c, 7.3.c.

1980   *Positive linear operators in weighted spaces of functions of several variables*
(Russian), Izv. Akad. Nauk Azerbaǐdžan SSR Ser. Fiz.—Tehn. Mat. Nauk **1**
(1980), no. 4, 32–37. **MR** 82g: 41022. **SC** 5.2.c.

Garkavi, A.L.

1967   *The theory of best approximation in normed linear spaces* (Russian), Mate-
matičeskiǐ Analiz (1967), 75–132; translated in Progress in Mathematics, Vol.
8: Mathematical Analysis, Plenum Press, N.Y., 1970, 83–150. **MR** 43 # 7843.
SC 1.1.

1972   *A certain criterion for a K-system of continuous functions* (Russian), Izv. Vysš.
Učebn. Zaved. Matematika **119** (1972), no. 4, 55–59. **MR** 46 # 5903. **SC**
1.2.

Gale, D.

1956   *Neighboring vertices on a convex polyhedron*, in: Linear Inequalities and Related Systems, H.W. Kuhn and A.W. Tucker Eds., Annals of Mathematics
Studies **38**, 255–263, Princeton University Press, Princeton, N. J., 1956. **MR**
19, p. 57.

Goldstein, G.R. *(see also Dorroh, J.R.)*

1993   *Non linear singular diffusion with non linear boundary conditions*, to appear in
Math. Methods in the Applied Sciences, 1993.

Goldstein, J.A.

1975   *Some applications of the law of large numbers*, Bol. Soc. Bras. Mat. **6** (1975),
25–38. **MR** 56 # 16739.

1985   Semigroups of Operators and Applications, *The Clarendan Press, Oxford
University Press, New York, 1985.* **MR** 87c: 47056.

1993   *Nonlinear semigroups and applications to degenerate parabolic partial differential equations*, to appear in the Proceedings of the 4th Mathematics Work-
shop organized by the Korea Institute of Technology, 1993.

Goldstein, J.A. and Lin, C.-Y.

1987a   *Singular nonlinear parabolic boundary value problems in one space dimension*,
J. Diff. Equations **68** (1987), 429–443. **MR** 89e: 35079.

1987b   *Degenerate nonlinear parabolic boundary problems*, Nonlinear Analysis and
Applications (Arlington, Tex. 1986), 189–196, Lecture Notes in Pure and
Appl. Math., **109**, Dekker, New York, 1987). **MR** 88k: 35106.

1989   *Highly degenerate parabolic boundary value problems*, Differential Integral
Equations **2** (1989), no. 2, 216–227. **MR** 90b: 35131.

Gonska, H.H. *(see also Anastassiou, G.A.; Badea, C.)*

1975   *Konvergenzsätze vom Bohman-Korovkin-Typ für positive lineare Operatoren*,
Dissertation, Universität Bochum, 1975. **SC** 3.5, 5.3.b, 6.4.

Gonska, H.H. and Meier-Gonska, J.

1983   *A bibliography on approximation of functions by Bernstein-type operators
(1955–1982)*, in: Approximation Theory, IV (College Station, Tex., 1983),
739–785, Academic Press, New York, 1983. **MR** 85h: 41001.

1984   *Quantitative theorems on approximation by Bernstein-Stancu operators*, Cal-
        colo **21** (1984), no. 4, 317–335 (1985). **MR** 87g: 41055.
1986   *A bibliography on approximation of functions by Bernstein-type operators
        (supplement 1986)*, in: Approximation Theory, V (College Station, Tex.,
        1986), 621–654, Academic Press, Boston, Mass., 1986. **MR** 88k: 41001.

Gordon, W.J. and Riesenfeld, R.F.
1974   *Bernstein-Bézier methods for the computer-aided design of free-form curves
        and surfaces*, J. Assoc. Comut. Mach. **21** (1974), 293–310. **MR** 50 # 6108.

Goullet de Rugy, A.
1972   *La structure idéale des M-espaces*, J. Math. Pures Appl. **51** (1972), 331–373.
        MR 52 # 6373.

Grossman, M.W.
1965   *A Choquet boundary for the product of two compact spaces*, Proc. Amer.
        Math. Soc. **16** (1965), 967–971. **MR** 31 # 5189.
1970   *Limits and colimits in certain categories of spaces of continuous functions*,
        Dissertationes Mathematicae **79** (1970), 1–40.
1974   *Note on a generalized Bohman-Korovkin theorem*, J. Math. Anal. Appl. **45**
        (1974), 43–46. **MR** 49 # 947. **SC** 1.2.
1976a   *Korovkin theorems for adapted spaces with respect to a positive operator*,
        Math. Ann. **220** (1976), no. 3, 253–262. **MR** 53 # 1124. **SC** 5.3.d.
1976b   *Lototsky-Schnabl functions on compact convex sets*, J. Math. Anal. Appl. **55**
        (1976), no. 2, 525–530. **MR** 56 # 6237.

Grünbaum, B.
1967   *Convex Polytopes, Interscience, New York, 1967*. **MR** 37 # 2085.
1970   *Polytopes, groups and complexes*, Bull. Amer. Math. Soc. **76** (1970), no. 6,
        1131–1201. **MR** 42 # 959.

Guendouz, F.
1990   *Opérateurs linéaires positifs, leur convergence et leur ordre d'approximation*,
        Thesis, University of Annaba, Algeria, 1990. **SC** 1.1, 1.2, 5.2.b.

Guichardet, A.
1968   Analyse Harmonique Commutative, *Monographies Universitaires de Mathé-
        matiques* **26**, *Dunod, Paris, 1968*. **MR** 39 # 1901.

Guo, Zhu-Rui and Zhou, Ding-Xuan
1994   *Approximation theorems for modified Szasz operators*, to appear in Acta Sci.
        Hungar., 1994.

Gusak, I.Ja. and Shashkin, Yu.A.
1979   *Korovkin systems in certain finite-dimensional spaces*, (Russian), Trudy Inst.
        Mat. i. Meh. Ural. Naučn Centr. Akad. Nauk SSSR, no. **31**, Obobščen.
        Funkcii. i Vektor. Mery (1979), 68–74, 92. **MR** 81i: 41019. **SC** 1.3, 1.11, 3.3,
        3.6.

Gustin, W.
1947   *Sets of finite planar order*, Duke Math. J. **14** (1947), no. 1, 51–66. **MR** 8,
        p. 524.

Habib, A. and Umar, S.
1980   *On generalized Bernstein polynomials*, Indian J. Pure Appl. Math. **11** (1980),
        1177–1189. **MR** 82d: 41028.

Hahn, L. *(see also Bleimann, G.; Butzer, P.L.)*
1982   *A note on stochastic methods in connection with approximation theorems for positive linear operators*, Pacific J. of Math. **101** (1982), no. 2, 307–319. **MR** 84d: 41041.

Hájek, O.
1965   *Uniform polynomial approximation*, Amer. Math. Monthly **72** (1965), 681.

Hardy, G.H. and Rogosinski, W.W.
1944   Fourier Series, *Cambridge Univ. Press, 1944.*

Hasumi, M.
1975   *A note on Korovkin sets*, Bull. Fac. Sci. Ibaraki Univ., Math. Ser. A, no. 7 (1975), 39–41. **MR** 52 # 1125. **SC** 1.1, 7.1.c.

Haupt, O. and Künneth, H.
1967   Geometrische Ordnungen, *Die Grundlehren der mathematischen Wissenschaften in Einzeldarstellungen* **133**, *Springer-Verlag, Berlin-Heidelberg-New York, 1967.* **MR** 37 # 3491.

Heilmann, M.
1988 L_p-*saturation of some modified Bernstein operators*, J. Approx. Theory **54** (1988), 260–281. **MR** 89h: 41049.

Helms, L.L.
1969   Introduction to Potential Theory, *Wiley-Interscience, New York, 1969.* **MR** 41 # 5638.

Hewitt, E. and Ross, K.
1979   Abstract Harmonic Analysis, I, *Die Grundlehren der mathematischen Wissenschaften in Einzeldarstellungen* **115**, *Springer-Verlag (2nd Edition 1979), Berlin-Heidelberg-New York, 1979.* **MR** 54 # 446.

Hildebrandt, T.H. and Schoenberg, I.J.
1933   *On linear functional operations and the moment problem*, Ann. of Math. (2), **34** (1933), 317–328.

Hille, E. and Phillips, R.S.
1957   Functional Analysis and Semigroups, *Amer. Math. Soc. Coll. Publ.* **31**, *Providence, R.I., 1957.* **MR** 19, p. 664.

Hinkkanen, A.
1988   *Modulus of continuity of harmonic functions*, J. Analyse Math. **51** (1988), 1–29. **MR** 89m: 31002.

Hoffmann, K.
1966   *Lectures on sup norm algebras*, in: Proc. Summer School on topological algebra theory, Bruges, 1966, Les Presses Universitaires de Bruxelles, Bruxelles, 1966.

Homagk, F.
1979   *Zur intuitionistischen Kennzeichnung reeller Funktionen auf substantiellen Intervallen*, Math.—Phys. Semesterber. **26** (1979), no. 1, 95–113. **MR** 81f: 03069. **SC** 1.12.

Horová, I.
1968   *Linear positive operators of convex functions*, Mathematica **10** (33), no. 2 (1968), 275–283. **MR** 40 # 604.

1982   *A note on the sequence formed by the first order derivatives of the Szász-Mirakyan operators*, Mathematica, **24** (47), no. 1–2 (1982), 49–52. **MR** 84i: 41034.

Horváth, J.

1966   Topological Vector Spaces and Distributions, *Addison-Wesley Publ. Comp., Reading, Massachusetts, 1966.* **MR** 34 # 4863.

Hsiao, Shêng Yen

1980   *Approximation of unbounded functions and applications to representations of semigroups*, J. Approx. Theory **28** (1980), no. 3, 238–259. **MR** 81f: 41027. **SC** 8.3, 8.6.

Hsiao, Shêng Yen and Yeh, Cheh-Chih

1989   *Rates for approximation of unbounded functions by positive linear operators*, J. Approx. Theory **57** (1989), 278–292. **MR** 90f: 41028. **SC** 5.2.a.

Hsu, L.C.

1961   *Approximation of non-bounded continuous functions by certain sequences of linear positive operators or polynomials*, Studia Math. **21** (1961/62), 37–43. **MR** 25 # 2357. **SC** 5.2.a.

1964   *On a kind of Fejér-Hermite interpolation polynomials*, Acta Math. Acad. Sci. Hungar. **15** (1964), 325–328. **MR** 29 # 2026.

Hsu, L.C. and Wang, H.J.

1964   *General increasing multiplier methods and approximation of unbounded continuous functions by certain concrete polynomial operators*, Dokl. Akad. Nauk. SSSR **156** (1964), 264–267. **MR** 29 # 2588.

Huijsmans, C.B. *(see Bernau, S.J.)*

Hurewicz, W. and Wallman, H.

1948   Dimension Theory, *Princeton Math. Series **9**, Princeton, Princeton University Press, 1948.*

Irmisch, R. *(see Flösser, H.O.)*

Ivanov, K.G. *(see Ditzian, Z.)*

Jackson, D.

1930   The Theory of Approximation, *(Amer. Math. Soc. Colloq. Publ. 11)* Amer. Math. Soc., New York *(1930)*, 1958.

Jain, P.C. and Pai, D.V.

1971   *Approximation theorems of the Korovkin-type for complex-valued functions*, Indian J. Math. **13** (1971), 123–129. **MR** 49 # 966. **SC** 7.4.c, 7.5.c.

Jakimovski, A. and Leviatan, D.

1969   *Generalized Szász operators for the approximation in the infinite interval*, Mathematica (Cluj) **11** (34) (1939), no. 1, 97–103. **MR** 41 # 7348.

James, R.L.

1973   *The extension and convergence of positive operators*, J. Approx. Theory **7** (1973), 186–197. **MR** 50 # 2771. **SC** 3.2, 5.2.a, 6.5.

1974   *Korovkin sets in locally convex function spaces*, J. Approx. Theory **12** (1974), 205–209. **MR** 50 # 7915. **SC** 5.2.a, 5.4.a.

Jensen, J.L.W.V.

1902   *Sur une identité d'Abel et sur d'autres formules analogues*, Acta Math. **26** (1902), 307–318.

Jiménes Poso, M.A.

1974   *Sur les opérateurs linéaires positifs et la méthode des fonctions tests*, C. R. Acad. Sci. Paris Sér. A **278** (1974), 149–152. **MR** 48 # 11861. **SC** 5.2.a.

1975   *Sopre la convergencia de operatores lineales y el metodo de funciones des prueba*, Rev. Centro Univ. Las Villas **3** (1975), 67–72. **SC** 1.7, 1.9.

1980   *Déformation de la convexité et thèorémes du type Korovkin*, C. R. Acad. Sci. Paris, Ser. A **290** (1980), 213–215. **MR** 81a: 41054.

1981   *Convergence of sequences of linear functionals*, Z. Angew. Math. Mech. **61** (1981), no. 10, 495–500. **MR** 83b: 41024. **SC** 1.9.

Jiménez Poso, M.A. and López Nuñez, E.F.

1987   *Korovkin type theorems for certain classes of analytic functions*, Cienc. Mat. (Havana) **8** (1987), no. 2, 25–29. **MR** 89c: 30098. **SC** 7.2.c.

1990   *Qualitative theorems of Korovkin type theorems for the space of Riemann integrable functions* (Spanish. English summary), Cienc. Mat. (Havana) **11** (1990), no. 2, 123–135. **MR** 92k: 41031. **SC** 5.6.i.

Jørgensen, P.E.T. *(see Bratteli, O.)*

Jurkat, W.B. and Peyerimhoff, A.

1971a   *Fourier effectiveness and order summability*, J. Approx. Theory **4** (1971), 231–244. **MR** 43 # 7855.

1971b   *Inclusion theorems and order summability*, J. Approx. Theory **4** (1971), 245–262. **MR** 43 # 7856.

Kakutani, S.

1941   *Concrete representation of abstract* (M)-*spaces*, Ann. of Math. (2) **42** (1941), 994–1024. **MR** 3,p. 205.

Kantorovich, L.V.

1930   *Sur certains développements suivant les polynômes de la forme de S. Bernstein*, I, II, C. R. Acad. URSS (1930), 563–568, 595–600.

Karlin, S. and McGregor, J.

1962   *On a genetics model of Moran*, Proc. Cambridge Phil. Soc. **58** (1962), 299–311. **MR** 25 # 1591.

Karlin, S. and Studden, W.

1966   Tchebycheff Systems: with Applications in Analysis and Statistic, *Pure and Appl. Math.* **15**, *Interscience, New York, 1966.* **MR** 34 # 4757.

Karlin, S. and Ziegler, Z.

1970   *Iteration of positive approximation operators*, J. Approx. Theory **3** (1970), 310–339. **MR** 43 # 3715.

Kato, T. *(see Arendt, W.)*

Keimel, K. and Roth, W.

1988   *A Korovkin type approximation theorem for set-valued functions*, Proc. Amer. Math. Soc. **104** (1988), 819–824. **MR** 90g: 41033. **SC** 1.11, 1.12, 9.2.

1992   Ordered Cones and Approximation, *Lecture Notes in Math.* **1517**, *Springer-Verlag, Berlin, 1992.* **MR** 93i: 46017. **SC** 9.1, 9.2, 9.4.

Keldish, M.V.

1941a   *Sur la résolubilité et la stabilité du problème de Dirichlet* (en Russe), Uspehi. Mat. Nauk. **8** (1941), 171–231. **MR** 3, p. 123.

1941b   *On the Dirichlet problem* (Russian), Dokl. Akad. Nauk. SSSR **32** (1941), 308–309.

Kelisky, R.P. and Rivlin, T.J.
  1967   *Iterates of Bernstein polynomials*, Pacific J. Math. **21** (1967), no. 3, 511–520.
          MR 35 # 3328.

Kendall, D.G.
  1962   *Simplexes and vector lattices*, J. London Math. Soc. **37** (1962), 365–371. **MR**
          25 # 2423.

Khan, M.K.
  1989   *On the Bernstein-type operator of Bleimann, Butzer, and Hahn*, J. Natur. Sci.
          Math. **29** (1989), no. 2, 133–148. **MR** 93b: 41010.

  1991   *Approximation properties of Beta operators*, in: Progress in Approximation
          Theory, P. Nevai and A. Pinkus Eds., 483–495, Academic Press, Boston MA,
          1991. **MR** 92m: 41047.

Khan, M.K. and Peters, M.A.
  1989   *Lipschitz constants for some approximation operators of a Lipschitz continu-
          ous function*, J. Approx. Theory **59** (1989), 307–315. **MR** 90k: 41031.

Khan, R.A.
  1980   *Some probabilistic methods in the theory of approximation operators*, Acta
          Math. Acad. Sci. Hungar. **39** (1980), 193–203. **MR** 81m: 41024.

  1988   *A note on a Bernstein-type operator of Bleimann, Butzer, and Hahn*, J. Approx.
          Theory **53** (1988), 295–303. **MR** 89j: 41033.

  1991   *On some properties of a Bernstein-type operator of Bleimann, Butzer, and
          Hahn*, in: Progress in Approximation Theory, P. Nevai and A. Pinkus Eds.,
          497–504, Academic Press, Boston MA, 1991. **MR** 92h: 41015.

Kimura, M.
  1957   *Some problems of stochastic processes in genetics*, Ann. Math. Statist. **28**
          (1957), 882–901.

King, J.P.
  1966   *The Lototsky transform and Bernstein polynomials*, Canadian J. Math. **18**
          (1966), 89–91. **MR** 32 # 6113.

  1975   *Probability and positive linear operators*, Rev. Roumaine Math. Pures Appl.
          20 (1975), no. 3, 325–327. **MR** 51 # 8693. **SC** 1.2, 1.10.

  1980a   *Probabilistic interpretation of some positive linear operators*, Rev. Roumaine
          Math. Pures Appl. **25** (1980), no. 1, 77–82. **MR** 81f: 41026. **SC** 1.2, 1.10.

  1980b   *Probabilistic analysis of Korovkin's theorem*, J. Indian Math. Soc. (N.S.) **44**
          (1980), no. 1–4, 51–58 (1982). **MR** 85g: 41029. **SC** 1.2, 1.10.

Kitto, W. and Wulbert, D.E.
  1976   *Korovkin approximations in L^p-spaces*, Pacific J. Math. **63** (1976), no. 1, 153–
          167. **MR** 54 # 5708. **SC** 3.2, 5.2.f.

Klimov, V.S., Krasnosel'skiĭ, M.A. and Lifshits, E.A.
  1965   *The convergence of positive functionals and operators* (Russian), Dokl. Akad.
          Nauk SSSR **162** (1965), 258–261; translated in Soviet Math. Dokl. **6** (1965),
          162, no. 2, 644–647. **MR** 33 # 4679. **SC** 6.2, 6.6.

  1966   *Points of smoothness of a cone and convergence of positive functionals
          and operators* (Russian), Trudy Moskov Mat. Obšč. **15** (1966), 55–69;
          translated in Trans. Moscow Math. Soc. 1966, 61. **MR** 34 # 8149. **SC** 6.2,
          6.6, 8.3.

Knoop, H.B. and Pottinger, P.

1976   *Ein Satz vom Korovkin-Typ für $\mathscr{C}^k$-Räume*, Math. Z. **148** (1976), no. 1, 23–32. **MR** 54 # 3259. **SC** 5.2.e, 5.6.e.

1987   *On simultaneous approximation by certain linear positive operators*, Arch. Math. (Basel) **48** (1987), 511–520. **MR** 88i: 41036. **SC** 5.2.e, 5.6.e.

Kocić, L.M. and Lacković, I.B.

1986   *Convexity criterion involving linear operators*, Facta Universitatis (Niš), Ser. Math. Inform. **1** (1986), 13–22.

Kocić, L.M. and Stanković, M.S.

1984   *Dominated approximation of convex functions by positive linear operators*, in: Constructive Theory of Functions, Proc. Internat. Conf., Sofia, (1984), 463–467.

Kolomeicev, V.I.

1970   *On the question of the convergence of a sequence of linear operators* (Russian), Interuniv. Sci. Conf. on the Problem "Application of Functional Analysis in Approximation Theory" (Proc. Conf. Kalinin) (Russian), 72–75. Kalinin. Gos. Ped. Inst. Kalinin, 1970. **MR** 46 # 5905. **SC** 1.2.

Komornik, V.

1979a   *Korovkin type theorems in general topological spaces*, preprint Eötvos Loránd University, Budapest, 1979, no. 9. **SC** 5.5.a.

1979b   *Korovkin closures in finite discrete spaces*, preprint Eötvos Loránd University, Budapest, 1979, no. 10. **SC** 1.11, 1.12.

1980   *A Korovkin type theorem for locally compact spaces*, Studia Sci. Math. Hungar. **15** (1980), no. 1–3, 309–311. **MR** 84c: 41014. **SC** 5.5.a.

1981   *On the Korovkin closure*, Acta Sci. Math. (Szeged) **43** (1981), no. 1–2, 41–43. **MR** 82g: 41026. **SC** 5.5.a.

Korneichuk, N.P.

1991   Exact Constants in Approximation Theory, *Encyclopaedia of Mathematics and its Applications, Vol. 38, Cambridge University Press, Cambridge, 1991.* **MR** 92m: 41002.

Korovkin, P.P.

1953   *On convergence of linear positive operators in the space of continuous functions* (Russian), Dokl. Akad. Nauk SSSR (N.S.) **90** (1953), 961–964. **MR** 15 # 236. **SC** 1.2.

1958a   *The conditions for the uniqueness of the problem of moments and the convergence of sequences of linear operators* (Russian), Kalinin. Gos. Ped. Inst. Učen Zap. **26** (1958), 95–102. **SC** 1.2.

1958b   *An asymptotic property of positive methods of summation of Fourier series and best approximation of functions of class Z_2 by linear positive polynomial operators* (Russian), Uspehi Mat. Nauk. **13**, no. 6 (84), 1958, 99–103. **MR** 21 # 253.

1960   Linear Operators and Approximation Theory, *translated from the Russian ed. (1959), Russian Monographs and Texts on Advanced Mathematics and Physics, Vol. III, Gordon and Breach Publishers, Inc., New York, Hindustan Publishing Corp. (India), Delhi, 1960.* **MR** 27 # 561. **SC** 1.2, 1.9, 1.12.

1962   *Convergent sequences of linear operators* (Russian), Usp. Mat. Nauk **17** (1962), no. 4 (106), 147–152. **MR** 25 # 4356. **SC** 1.5, 1.12.

1965   *On conditions of convergence of a sequence of operators* (Russian), in: Studies Contemporary Problems Constructive Theory of Functions (Proc. Second All-Union Conf., Baku, 1962), 95–97; Izdat. Akad. Nauk Azerdaĭdžan SSR, Baku, 1965. **MR** 33 # 3023. **SC** 1.2, 1.5.

Kosmák, L.

1960   *A note on Bernstein polynomials of convex functions*, Mathematica (Cluj), **2** (25) (1960), no. 2, 281–282.

1967   *Les polynomes de Bernstein des fonctions convexes*, Mathematica (Cluj), **9** (32) (1967), no. 1, 71–75. **MR** 35 # 4357.

Krasnosel'skiĭ, M.A. *(see also Klimov, V.S.)*

Krasnosel'skiĭ, M.A. and Lifshits, E.A.

1968a   *A principle of convergence of sequences of positive linear operators* (Russian), Studia Math. **31** (1968), 455–468. **MR** 38 # 6372. **SC** 1.2, 1.6, 1.8, 5.2.b, 6.5, 8.3.

1968b   *The convergence of sequences of positive operators in linear topological spaces*, Usp. Mat. Nauk **23** (1968), no. 2, 213–214. **SC** 1.2, 6.2.

Krasnosel'skiĭ, M.A., Lifshits, E.A. and Sobolev, A.V.

1989   Positive Linear Systems, *Sigma Series in Applied Mathematics* **5**, *Heldermann Verlag, Berlin, 1989*. **MR** 91f: 47051. **SC** 1.2, 1.6, 1.8, 6.5, 8.3.

Kratz, W. and Stadtmüller, U.

1988   *On the uniform modulus of continuity of certain discrete approximation operators*, J. Approx. Theory **54** (1988), 326–337. **MR** 90a: 41019.

Kubo, F. and Takashima, Y.

1985   *Probabilistic proof of Korovkin's theorem*, Math. Japon. **30** (1985), no. 1, 109–110. **MR** 87c: 41019. **SC** 1.10.

Kudryavcev, G.I.

1970   *The convergence of sequences of linear operators to derivatives* (Russian), Proc. Central Regional Union of Math. Departments **1**, Kalinin Gos. Ped. Inst. Kalinin (1970), 122–136. **MR** 45 # 790. **SC** 5.2.e, 5.6.e.

1972   *Certain questions on the convergence of sequences of linear operators* (Russian), A collection of articles on the Constructive Theory of Functions and the Extremal Problems of Functional Analysis, 77–86, Kalinin. Gos. Univ., Kalinin, 1972. **MR** 51 # 13541. **SC** 1.12, 5.6.e.

1977   *Convergent sequences of linear operators* (Russian), Mathematical Analysis and the Theory of Functions, no. 8, 102–105; Moskov. Oblast. Ped. Inst. im. Krupskoĭ, Moscow, 1977, 169 pp. 1.00 r.. **MR** 81c: 32001.

1979   *Convergence of the derivatives of linear convex and smooth operators* (Russian), in "Application of Functional Anal. in Approx. Theory" (V.N. Nikol'skii, Ed.), 61–65. Kalinin: Kalinin Gos. Univ. 1979. **MR** 82c: 41023. **SC** 5.6.e.

Kudryavcev, G.I. and Kudryavceva, A.M.

1979   *Conditions for the convergence of sequences of functionals in the class S_m to the functional $\Phi_r(f) = \sum_{i=1}^{r} \lambda_i f(x_i)$* (Russian), in "Application of Functional

Anal. in Approx. Theory" (V.N. Nikol'skii, Ed.), 66–72. Kalinin: Kalinin Gos. Univ. 1979. **SC** 1.9, 1.12.

Kudryavceva, A.M. *(see Kudryavcev, G.I.)*

Künneth, H. *(see Haupt, O.)*

Kurtz, L.C.

1975   *Unique Hahn-Banach extensions and Korovkin's theorem*, Proc. Amer. Math. Soc. **47** (1975), 413–416. **MR** 52 # 8887. **SC** 8.4, 8.6.

1976   *On uniform convergence and positive operators*, Indian J. Math. **18** (1976), 35–40. **ZBL** 396. 41019. **SC** 6.2, 8.3.

Kurtz, T.G.

1970   *A general theorem on the convergence of operator semigroups*, Trans. Amer. Math. Soc. **148** (1970), 23–32. **MR** 41 # 867.

Kutateladze, S.S.

1973a   *Supremal generators and convergence of nonexpansive operators* (Russian), Mat. Zametki **13** (1973), no. 1, 55–65; translated in Math. Notes **13** (1973), 33–38. **MR** 47 # 7506. **SC** 1.3, 8.2.

1973b   *Certain theorems on the convergence of operators* (Russian), Dokl. Akad. Nauk SSSR **208** (1973), 771–774; translated in Soviet Math. Dokl. **14** (1973), no. 1, 166–169. **MR** 48 # 945. **SC** 1.12, 6.5, 8.3.

Kutateladze, S.S. and Rubinov, A.M.

1971a   *Supremal generators* (Russian), Dokl. Akad. Nauk SSSR **199** (1971), 776–777; translated in Soviet Math. Dokl. **12** (1971), no. 4, 1179–1181. **MR** 44 # 5753. **SC** 1.2, 1.6, 6.5.

1971b   *Supremal generators, and convergence of sequences of operators* (Russian), Collection of articles dedicated to L.V. Kantorovič on the occasion of his sixtieth birthday, Optimizatcija Vyp. **3** (20) (1971), 120–153. **MR** 51 # 1336. **SC** 1.2, 1.9, 1.12, 5.2.g, 6.2, 6.6, 6.8.

1972   *Minkowski duality and its applications* (Russian), Usp. Mat. Nauk **27** (1972), no. 3 (165), 127–176; translated in Russian Math. Surveys **27** (1972), no. 3, 137–191. **MR** 52 # 14922. **SC** 1.2, 6.5.

Labsker, L.G.

1970   *Certain sufficient conditions for the approximation of continuous functions by operators of the class* S_m^* (Russian. Azerbaijani summary), Akad. Nauk Azerbaĭdžan. SSR Dokl. **26** (1970), no. 9, 3–7. **MR** 44 # 1974. **SC** 1.5.

1971   *The weak convergence of sequences of linear positive functionals* (Russian), Dokl. Acad. Nauk SSSR **197** (1971), no. 6, 1264–1267; translated in Soviet Math. Dokl. **12** (1971), 669–673. **MR** 43 # 6697. **SC** 1.9, 6.6, 8.4.

1972   *The strong convergence of sequences of positive linear operators in Banach spaces* (Russian), Dokl. Akad. Nauk SSSR **206** (1972), no. 3, 525–528, translated in Soviet. Math. Dokl. **13** (1972), no. 5, 1229–1233. **MR** 47 # 4053. **SC** 6.8, 8.6.

1975   *Some necessary conditions for the convergence of sequences of linear positive operators to operators of the set* Φ_k^0 (Russian), Application of Functional Analysis in Approximation Theory, no. 6, 59–69, Kalinin. Gos. Univ., Kalinin, 1975. **MR** 58 # 29656. **SC** 1.7, 1.12.

1977   *Differing sets and the convergence of sequences of linear functionals and operators of class S_m* (Russian), Application of Functional Analysis in Approximation Theory, no. 7, 79–91, 155, Kalinin. Gos. Univ., Kalinin, 1977. **MR** 58 # 23277. **SC** 1.5, 1.9, 1.12.

1979a   *Korovkin sets in the space of continuous functions for operators of class S_m^0* (Russian), Mat. Zametki **25** (1979), no. 4, 521–536, 636; translated in Math. Notes **25** (1979), no. 4, 270–278. **MR** 81h: 41026. **SC** 1.5, 1.7, 1.12.

1979b   *On the question of criteria for Korovkin (S_m^0, ϕ_k^0)-sets* (Russian), Application of Functional Analysis in Approximation Theory **161**, 73–83, Kalinin. Gos. Univ., Kalinin 1979. **MR** 82f: 41028. **SC** 1.5, 1.7.

1980   *Some necessary and sufficient criteria for Korovkin sets for operators of class S_m^0* (Russian), Sibirsk. Mat. Z. **21** (1980), no. 2, 128–138, 238; translated in Siberian Math. J. **21** (1980), no. 2, 242–249. **MR** 81i: 41030. **SC** 1.5, 1.7, 1.12.

1982   *Korovkin sets in a Banach space for sets of linear functionals* (Russian), Mat. Zametki **31** (1982), no. 1, 93–112, 159; translated in Math. Notes **31** (1982), no. 1–2, 47–56. **MR** 83f: 46016. **SC** 1.9, 8.4.

1985   *On the question on weak convergence of sequences of linear functionals* (Russian), Anal. Numér. Théor. Approx. **14** (1985), 33–57. **MR** 88a: 41016. **SC** 1.2, 1.9, 6.6, 8.4.

1989a   *On a test set in the Banach space for a set of linear operators*, preprint, Moscow University, 1989. **SC** 6.8, 8.6.

1989b   *The test sets in Banach spaces* (Russian), Moscow Institute for High Qualification for Specialist of Chemical Industry, Moscow 1989. **SC** 1.1, 1.5, 1.9.

1990   *Test sets for approximation operators and functionals of class S_m* (Russian), Moscow Institute for High Qualification for Specialist of Chemical Industry, Moscow, 1990. **SC** 1.1, 1.5, 1.9.

Lacković, I.B. *(see Kocić, L.M.)*

Lampe, E.I. *(see Abakumov, Ju.G.)*

Leha, G. *(see also Bauer, H.)*

1976   *Korovkin-Sätze für Funktionenräume*, Dissertation, Universität Erlangen-Nürnberg, 1976. **SC** 4.2, 4.6, 5.2.a, 5.5.a.

1977   *Relative Korovkin-Sätze und Ränder*, Math. Ann. **229** (1977), no. 1, 87–95. **MR** 58 # 12151a. **SC** 4.6, 5.5.a.

Leha, G. and Papadopoulou, S.

1978   *Nachtrag zu "Relative Korovkin-Sätze und Ränder" (Math. Ann. 229 (1977), no. 1, 87–95) by Leha*, Math. Ann. **233** (1978), no. 3, 273–274. **MR** 58 # 12151b. **SC** 4.6, 5.5.a.

Lehnhoff, H.G.

1979   *Lokale Approximationsmasse und Nikolskii-konstanten für positive lineare Operatoren*, Dissertation, Universität Dortmund, 1979.

1981   *Local Nikolskii constants for a special class of Baskakov operators*, J. Approx. Theory **33** (1981), 236–247. **MR** 84e: 41025a.

Lembcke, J. *(see Anger, B.)*

Lenze, B.

1990   *Bernstein-Baskakov-Kantorovich operators and Lipschitz-type maximal functions*, Seminarberichte Fernuniversität Hagen, 1990, no. 39.

Leonte, A. and Vîrtopeanu, I.
1983   *On a result of L.G. Labsker concerning Korovkin sets* (Romanian), An. Univ. Craiova Mat. Fiz.-Chim. **11** (1983), 31–34. **MR** 87h: 41032. **SC** 8.2.

Leviatan, D. *(see Jakimovski, A.)*

Lifshits, E.A. *(see Klimov, V.S.; Krasnosel'skiĭ, M.A.)*

Limaye, B.V. and Namboodiri, M.N.N.
1979   *Approximation by positive functionals*, J. Indian Math. Soc. (N.S.) **43** (1979), no. 1–4, 195–202 (1980). **MR** 84e: 46059. **SC** 7.5.a, 7.5.b, 7.5.c.

1982   *Korovkin-type approximation on C*-algebras*, J. Approx. Theory **34** (1982), no. 3, 237–246. **MR** 83g: 46052. **SC** 7.3.c, 7.7.c.

1984a   *Weak Korovkin approximation by completely positive linear maps on β(H)*, J. Approx. Theory **42** (1984), no. 3, 201–211. **MR** 86m: 47066. **SC** 7.4.b, 7.7.b.

1984b   *A generalized non commutative Korovkin theorem and *-closedness of certain sets of convergence*, Ill. J. Math. **28** (1984), 267–280. **MR** 85k: 46066. **SC** 7.7.b.

1986   *Weak approximation by positive maps on C*-algebras*, Math. Slovaca **36** (1986), no. 1, 91–99. **MR** 87f: 46102. **SC** 7.3.b.

Limaye, B.V. and Shiraly, S.
1976   *Korovkin's theorem for positive functionals on * normed algebras*, J. Indian Math. Soc. (N.S.) **40** (1976), no. 1–4, 163–172 (1977). **MR** 58 # 12396. **SC** 7.5.a, 7.5.b.

Lin, C.-Y. *(see also Goldstein, J.A.)*
1989   *Degenerate non linear parabolic boundary value problems*, Nonlinear Anal. TMA **13** (1989), no. 11, 1303–1315. **MR** 91d: 35121.

Lindvall, T.
1982   *Bernstein polynomials and the law of large numbers*, Math. Scientist **7** (1982), 127–139. **MR** 84d: 41009.

Liukkonen, J. and Mosak, R.
1977   *Harmonic analysis and centers of Beurling algebras*, Comment. Math. Helvetici **52** (1977), 297–315. **MR** 57 # 10359.

López Nuñez, E.F. *(see Jiménez Poso, M.A.)*

Lorentz, G.G. *(see also Berens, H.)*
1937   *Zur Theorie der Polynome von S. Bernstein*, Math. Sbornik (N.S.) **2** (1937), 543–556.

1948   *A contribution to the theory of divergent sequences*, Acta Math. **80** (1948), 167–190. **MR** 10, p. 367.

1964   *Inequalities and saturation classes of Bernstein polynomials*, in: On Approximation Theory, P. Butzer et al. Editors), Birkhäuser Basel, 1964, 200–207. **MR** 31 # 6083.

1972a   *Positive and monotone approximation*, in: Linear Operators and Approximation (Proc. Conf., Oberwolfach, 1971), 284–291; Internat. Ser. Numer. Math., Vol. **20**, Birkhäuser, Basel, 1972. **MR** 51 # 8695. **SC** 1.2, 1.12.

1972b   *Korovkin sets (Sets of convergence)*, Regional Conference at the University of California, Riverside, June 15–19, 1972, Center for Numerical Analysis, no. 58, The University of Texas at Austin, 1972. **SC** 1.2, 1.3, 1.4, 3.2, 5.2.b, 6.2.

1985   *Korovkin theorems for contractions and positive contractions*, in: Approximation Theory and Applications (Proc. Internat. Conf. 75th Birthday G.G.

Lorentz, St. John's/Newfoundland, 1984), Research Notes in Mathematics, no. 133, Pitman Publ., Boston, 1985, 104–114. **MR** 88e: 41074. **SC** 3.3, 3.4, 5.2.b, 6.2.

1986a   Approximation of Functions, 2nd Ed., *Chelsea Publ. Comp., New York, NY, 1986.* **MR** 88j: 41001. **SC** 1.1, 1.2.

1986b   Bernstein Polynomials, 2nd Ed., *Chelsea Publ. Comp., New York, NY, 1986.* **MR** 88a: 41006.

Lorentz, G.G. and Schumaker, L.L.

1972   *Saturation of positive operators,* J. Approx. Theory **5** (1972), 413–424. **MR** 49 # 9495.

Lubinsky, D.S.

1993   *Weierstrass' theorem in the twentieth century: a selection,* International Conference on Abstract Analysis, Berg-en-Dal, Gruger Park, April 6–17, 1993, preprint, 1993.

Lumer, G., Redheffer, R. and Walter, W.

1982   *Comportement des solutions d'inequations différentielles dégénérées du second ordre et applications aux diffusion,* C. R. Acad. Sci. Paris **294**, Série I, 1982, 617–620. **MR** 83f: 34021.

1988   *Estimates for solutions of degenerate second-order differential equations and inequalities with applications to diffusion,* Nonlinear Anal. TMA **12** (1988), no. 10, 1105–1121. **MR** 89m: 34018.

Lupaş, A.

1967a   *Some properties of the linear positive operators* (I), Mathematica (Cluj), **9** (32) (1967), 77–83. **MR** 35 # 7052.

1967b   *Some properties of the linear positive operators* (II), Mathematica (Cluj), **9** (32) (1967), 295–298. **MR** 37 # 6647.

1971   *On the approximation by linear operators of the class S_m* (Romanian summary), An. Şti. Univ. "Al. I. Cuza" Iaşi Secţ. I a Mat. (N.S.) **17** (1971), 133–137. **MR** 46 # 5911. **SC** 1.5, 1.11.

1974a   *Some properties of the linear positive operators* (III), Anal. Numér. Théor. Approx. **3** (1974), 47–61. **MR** 52 # 1104.

1974b   *Teoreme de medie pentru transformari liniare şi pozitive* (Romanian) *[Mean-value theorems for linear and positive transformations],* Anal. Numér. Théor. Approx. **3** (1974), 121–140. **MR** 52 # 11426.

1976   *A generalization of Hadamard's inequalities for convex functions,* Univ. Beograd Publ. Elektrotehn. Fak. Ser. Mat. Fiz. no. 544–no. 576 (1976), 115–121. **MR** 56 # 3212. **SC** 1.2.

1988   *Jensen-type inequalities in approximation theory,* Graz. Mat. Ser. Metodica (Bucarest) **9** (1988), no. 1, 41–48. **SC** 1.2.

Lupaş, A. and Lupaş, L.

1987   *Polynomials of binomial type and approximation operators,* Studia Univ. Babes-Bolyai, Mathematica **32** (1987), no. 4, 61–69. **MR** 90a: 41020.

Lupaş, A. and Müller, M.W.

1967   *Approximationseigenschaften der Gammaoperatoren,* Math. Z. **98** (1967), 208–226. **MR** 35 # 7053.

586   Bibliography

Lupaş, L. *(see Lupaş, A.)*
Luquin, F. *(see de la Cal, J.)*
Maier, V.
  1978   *The L_1-saturation class of the Kantorovich operators*, J. Approx. Theory **22**
            (1978), 223–232. **MR** 57 # 10324.
Maier, V., Müller, M.W. and Swetits, J.
  1981   *The local L_1-saturation class of the method of integrated Meyer-König and
            Zeller operators*, J. Approx. Theory **32** (1981), 27–31. **MR** 83g: 41026.
Mairhuber, J.C.
  1956   *On Haar's theorem concerning Chebyshev approximation problems having
            unique solutions*, Proc. Amer. Math. Soc. **7** (4) (1956), 609–615. **MR** 18, p. 125.
Maligranda, L.
  1987   *Korovkin theorem in symmetric spaces*, Ann. Soc. Math. Pol. Series I: Com-
            ment. Math. **27** (1987), 135–140. **MR** 89a: 41031. **SC** 5.2.b.
Maltese, G.
  1979   *Integral representation theorems via Banach algebras*, L'Enseignement Math.
            25 (1979), 273–284. **MR** 82j: 46067.
Mamedov, R.G.
  1962   *On the order of the approximation of differentiable functions by linear positive
            operators* (Russian), Doklady SSSR **146** (1962), 1013–1016.
Marlewski, A.
  1980   *Asymptotic form of Bernstein-Kantorovich approximation*, Fasciculi Math. **12**
            (1980), 99–102. **MR** 82h: 41007.
  1982   *Die Grenze der Folge der Potenzen von verallgemeinerten Bernsteinoperatoren
            auf dem Simplex*, Funct. Approx. Comment. Math. **12** (1982), 63–67. **MR** 86k:
            41028.
  1984   *Powers of Bernstein-type operators*, in: Proceedings of the International Con-
            ference on Constructive Theory of Functions (Varna, May 27–June 2, 1984)
            Publishing House of the Bulgarian Academy of Sciences (1984), 583–587.
Marsden, M.J. and Riemenschneider, S.D.
  1974   *Korovkin theorems for integral operators with kernels of finite oscillation*,
            Canad. J. Math. **26** (1974), no. 6, 1390–1404. **MR** 50 # 7902. **SC** 3.2,
            3.7.
Martini, R.
  1973   *A relation between semi-groups and sequences of approximation operators*,
            Indag. Math. **35** (1973), 456–465. **MR** 49 # 3394.
  1975   *Differential operators degenerating at the boundary as infinitesimal generators
            of semi-groups*, Dissertation, T.H. Delft, 1975.
Martini, R. and Boer, H.R.
  1974   *On the construction of semi-groups of operators*, Indag. Math. **36** (1974), 392–
            405. **MR** 52 # 6496.
Mastroianni, G.
  1979   *Su un operatore lineare e positivo*, Rend. Acc. Sc. Fis. Mat., Napoli, Serie IV,
            Vol. 46 (1979), 161–176 (1980). **MR** 81i: 47028.
  1980a   *Una generalizzazione dell'operatore di Mirakyan*, Rend. Acc. Sc. Fis. Mat.,
            Napoli, Serie IV, Vol. 48 (1980/81), 237–252. **MR** 85b: 41035.

1980b   *Su una classe di operatori lineari e positivi*, Rend. Acc. Sc. Fis. Mat., Napoli, Serie IV, Vol. 48 (1980/81), 217–235 (1982). **MR** 85b: 41034.

Mastroianni, G. and Occorsio, M.R.

1978a   *Sulle derivate dei polinomi di Stancu*, Rend. Accad. Sci. Fis. Mat. Napoli **45** (1978), 273–281. **MR** 82a: 41019.

1978b   *Una generalizzazione dell'operatore di Stancu*, Rend. Accad. Sci. Fis. Mat. Napoli **45** (1978), 495–511 (1978). **MR** 81e: 41034.

Matsuda, M.

1978   *Some properties of general denting points*, Math. Japon. **23**, (1978/1979), no. 6, 581–585. **MR** 81g: 46008. **SC** 5.2.a, 5.4.a.

May, C.P. *(see Ditzian, Z.)*

Mazhar, S.M. and Siddiqi, A.H.

1967   *On F_A-summability and A_B-summability of a trigonometric sequence*, Indian J. Math. **9** (1967), 461–466.

McGregor, J. *(see Karlin, S.)*

Meier-Gonska, J. *(see Gonska, H.H.)*

Meinardus, G.

1967   *Approximation of Functions: Theory and Numerical Methods*, *Springer-Verlag, Berlin, 1967*. **MR** 36 # 571.

Meir, A.

1976   *A Korovkin type theorem on weak convergence*, Proc. Amer. Math. Soc. **59** (1976), no. 1, 72–74. **MR** 53 # 13943. **SC** 3.4, 3.7.

Mevissen, H., *(see also Dickmeis, W.)*

Mevissen, H., Nessel, R.J. and van Wickeren, E.

1988   *A quantitative Bohman-Korovkin theorem and its sharpness to Riemann integrable functions*, Zeitschrift für Analysis und ihre Anwendungen **7** (1988), no. 4, 367–376. **MR** 90f: 41016. **SC** 5.2.i.

1989   *On the Riemann convergence of positive linear operators*, Rocky Mountain J. Math. **19** (1989), no. 1, 271–280. **MR** 91h: 41027. **SC** 5.2.i.

Meyer, P.A. *(see Choquet, G.)*

Meyer-König, W. and Zeller, K.

1960   *Bernsteinsche Potenzreihen*, Studia Math. **19** (1960), 89–94. **MR** 22 # 2823.

Micchelli, C.A. *(see also Dahmen, W.)*

1973a   *Chebyshev subspaces and convergence of positive linear operators*, Proc. Amer. Math. Soc. **40** (1973), no. 2, 448–452. **MR** 48 # 6787. **SC** 1.7, 1.9.

1973b   *The saturation class and iterates of the Bernstein polynomials*, J. Approx. Theory **8** (1973), 1–18. **MR** 49 # 9496.

1975   *Convergence of positive linear operators on $\mathscr{C}(X)$*, Collection of articles dedicated to G.G. Lorentz on the occasion of his sixty-fifth birthday, III; J. Approx. Theory **13** (1975), 305–315. **MR** 52 # 3819. **SC** 1.8, 1.9.

Min'kova, R.M.

1970   *The convergence of the derivatives of linear operators* (Russian), C. R. Acad. Bulgare Sci. **23** (1970), 627–629. **MR** 43 # 7830. **SC** 1.12.

1972   *The convergence of nonexpansive operators in the space of continuous abstract functions* (Russian), Sibirsk. Mat. Ž. **13** (1972), 790–804; translated in Siberian Math. J. **13** (1972), 546–556 (1973). **MR** 47 # 4059. **SC** 8.2, 8.6.

1973   *Korovkin systems for operators of the class S_m* (Russian), Mat. Zametki **13** (1973), 147–158; translated in Math. Notes **13** (1973), no. 1–2, 87–93. **MR** 48 # 767. **SC** 1.5.

1974   *Approximation of continuous functions by certain classes of linear operators* (Russian), Application of Functional Analysis in Approximation Theory, no. **2**, 85–100, Kalinin. Gos. Univ., Kalinin, 1974. **MR** 51 # 6240. **SC** 1.3, 1.5, 1.11.

1979   *Weak Korovkin spaces*, Mat. Zametki **25** (1979), no. 3, 435–443, 447; translated in Math. Notes **25** (1979), no. 3–4, 47–56. **MR** 82g: 46011. **SC** 1.12, 3.7, 8.6.

Min'kova, R.M. and Shashkin, Yu.A.

1969   *Convergence of linear operators of class S_m* (Russian), Mat. Zametki, **6** (1969), no. 5, 591–598; translated in Math. Notes **6** (1969), 816–820. **MR** 41 # 4083. **SC** 1.5, 1.12.

Mirakjan, G.M.

1941   *Approximation of continuous functions with the aid of polynomials* (Russian), Dokl. Akad. Nauk SSSR **31** (1941), 201–205.

Mitjagin, B.S. and Semenov, E.M.

1977   *Lack of interpolation of linear operators in spaces of smooth functions*, Math. USSR. Izv. **11** (1977), 1229–1266. **MR** 58 # 2234.

Mochizuki, N.

1964   *A characterization of the algebra of generalized analytic functions*, Tôhoku Math. J. **16** (1964), 313–319. **MR** 30 # 461.

Moldovan, E.

1962   *Observation sur la suite des polinômes de S.N. Bernstein d'une fonction continue*, Mathematica (Cluj), **4** (27) (1962), no. 2, 289–292.

Mond, B. *(see Shisha, O.)*

Moran, P.A.P.

1958   *Random processes in genetics*, Proc. Cambridge Phil. Soc. **54** (1958), 60–71. **MR** 23 # B1034.

Morozov, E.N.

1958   *Convergence of a sequence of positive linear operators in the space of continuous 2π-periodic functions of two variables* (Russian), Kalinin. Gos. Ped. Inst. Učen. Zap. **26** (1958), 129–142. **MR** 24 # A957. **SC** 1.2.

Mosak, R. *(see Liukkonen, J.)*

Mühlbach, G.

1969   *Über das Approximationverhalten gewisser positiver linearer Operatoren*, Dissertation, Universität Hannover, 1969.

1970   *Verallgemeinerung der Bernstein- und der Lagrange Polynome*, Rev. Roumaine Math. Pures Appl. **15** (1970), 1235–1252. **MR** 42 # 8134.

Müller, M.W. *(see also Lupaş, A.; Maier, V.)*

1967   *Die Folge der Gammaoperatoren*, Dissertation, Universität Stuttgart, 1967.

1968   *Punktweise und gleichmässige Approximation durch Gammaoperatoren*, Math. Z. **103** (1968), 227–238. **MR** 37 # 4470.

1972   *Sätze vom Bohman-Korovkin-Typ für Banachsche Funktionenräume*, in: Linear Operators and Approximation (Proc. Conf., Oberwolfach, 1971), 292–299; Internat. Ser. Numer. Math., Vol. **20**, Birkhäuser, Basel, 1972. **MR** 55 # 10937. **SC** 5.2.b.

1977   *Approximation unbeschränkter Funktionen bezüglich einer Korovkin-Metrik*,
          in: The Theory of the Approximation of Functions, (Proc. Internat.
          Conf. Collect. Artic., Kaluga, 1975) (1977), 269–272. **MR** 80j: 41035. **SC**
          1.12.

1978 L_p-*approximation by the method of integral Meyer-König and Zeller opera-*
          tors, Studia Math. **63** (1978), 81–88.

1989   *Approximation by Cheney-Sharma-Kantorovich polynomials in the L_p-metric*,
          Rocky Mountain J. of Math. **19** (1989), no. 1, 281–291. **MR** 90k: 41034.

Müller, M.W. and Walk, H.

1972   *Konvergenz—und Güteaussagen für die Approximation durch Folgen linearer*
          positiver Operatoren, in: Constructive Theory of Functions (Proc. Internat.
          Conf., Varna, May 19–25, 1970), 221–233; Izdat. Bolgar. Akad. Nauk, Sofia,
          1972. **MR** 51 # 3761. **SC** 1.12, 5.6.a.

Muni, G.

1979   *Algebre e spazi di Riesz-Stone*, Ric. Mat. **28** (1) (1979), 183–201. **MR** 81e:
          46006.

Mustăţa, C. *(see Andrica, D.)*

Nagel, J.

1978   *Sätze Korovkinschen Typs für die Approximation linearer positiver Opera-*
          toren, Dissertation, Universität Essen, 1978. **ZBL** 435. 41015. **SC** 1.8, 1.9,
          5.2.a, 5.4.a.

1980   *Asymptotic properties of powers of Bernstein operators*, J. Approx. Theory **29**
          (1980), 323–335. **MR** 82f: 41014.

1982   *Asymptotic properties of powers of Kantorovich operators*, J. Approx. Theory
          36 (1982), no. 3, 268–275. **MR** 84g: 41018.

Nagel, R. (Ed.)

1986   One-Parameter Semigroups of Positive Operators, *Lecture Notes in Math.*
          1184*, Springer-Verlag, Berlin, 1986.* **MR** 88i: 47022.

Nakamoto, R. and Nakamura, M.

1965   *On theorems of Korovkin, II*, Proc. Japan Acad. **41** (1965), 433–435. **MR** 32
          # 4449. **SC** 7.3.a, 7.3.c.

Nakamura, M. *(see Nakamoto, R.)*

Namboodiri, M.N.N. *(see Limaye, B.V.)*

Németh, A.B.

1969   *Korovkin's theorem for nonlinear 3-parameter families*, Mathematica (Cluj) **11**
          (34) (1969), no. 1, 135–136. **MR** 41 # 4085. **SC** 1.2.

Nessel, R.J. *(see Becker, M.; Butzer, P.L.; Dickmeis, W.; Mevissen, H.)*

Netuka, I.

1985   *Extensions of operators and the Dirichlet problem in potential theory*, Suppl.
          Rend. Circ. Mat. Palermo (2) **10** (1985), 143–163 (1986). **MR** 89f: 31003. **SC**
          1.8, 1.12.

Nishishiraho, T.

1974   *A generalization of the Bernstein polynomials and limit of its iterations*, Sci.
          Rep. Kanazawa Univ. **19** (1974), no. 1, 1–7. **MR** 50 # 10654.

1976   *Saturation of positive linear operators*, Tôhoku Math. J. (2) **28** (1976), no. 2,
          239–243. **MR** 54 # 3244. **SC** 1.8.

1977   *The degree of convergence of positive linear operators*, Tôhoku Math. J. (2) **29** (1977), no. 1, 81–89. **MR** 55 # 10924. **SC** 1.2.

1978   *Saturation of bounded linear operators*, Tôhoku Math. J. **30** (1978), 69–81. **MR** 58 # 12510.

1981   *Quantitative theorems on linear approximation processes of convolution operators in Banach spaces*, Tôhoku Math. J. **33** (1981), 109–126. **MR** 82m: 41034.

1982a   *Saturation of multiplier operator in Banach spaces*, Tôhoku Math. J. **34** (1982), 23–42. **MR** 84f: 41028.

1982b   *Quantitative theorems on approximation processes of positive linear operators*, in: Multivariate Approximation Theory II (Proc. Conf. Math. Res. Inst. Oberwolfach 1982; ed. by W. Schempp and K. Zeller), ISNM 61, 297–311. Birkhäuser Verlag, Basel-Boston-Stuttgart, 1982. **MR** 84k: 41024.

1983   *Convergence of positive linear approximation processes*, Tôhoku Math. J. (2) **35** (1983), no. 3, 441–458. **MR** 85i: 41024. **SC** 1.2, 1.12.

1987   *The convergence and saturation of iterations of positive linear operators*, Math. Z. **194** (1987), 397–404. **MR** 88m: 47065. **SC** 1.6, 1.8.

1988a   *The order of approximation by positive linear operators*, Tôhoku Math. J. **40** (1988), no. 4, 617–632. **MR** 90c: 41033.

1988b   *Convergence of positive linear functionals*, Ryukyu Math. J. **1** (1988), 73–94. **MR** 93e: 46029. **SC** 1.9, 4.5, 5.4.a.

1989   *Saturation of iterations for approximation processes on Banach spaces*, Ryukyu Math. J. **2** (1989), 49–81. **MR** 92m: 41057.

1990   *Saturation of iterations of multiplier operators in Banach spaces*, Ryukyu Math. J. **3** (1990), 45–84. **MR** 92m: 41058.

1991   *Convergence of quasi-positive linear operators*, Atti Sem. Mat. Fis. Univ. Modena **29** (1991), 367–374. **SC** 8.3.

Nöbeling, G. and Bauer, H.

1955   *Allgemeine Approximationskriterien mit Anwendungen*, Jber. Deutsch. Math. Verein **58** (1955), 54–72.

Novinger, W.P.

1971   *Holomorphic functions with infinitely differentiable boundary values*, Illinois J. Math. **15** (1971), 80–90.

Occorsio, M.R. *(see Mastroianni, G.)*

Paget, D.F. *(see Brown, B.M.)*

Pai, D.V. *(see Jain, P.C.)*

Pannenberg, M.

1985   *Korovkinapproximation in Waelbroeckalgebren*, Schriftenr. Math. Inst. Univ. Münster, (2) **37** (1985). **MR** 87f: 46091. **SC** 7.1.a, 7.2.a, 7.3.a, 7.4.a, 7.6.a.

1986a   *Korovkin-Approximation in Waelbroeck-algebras*, Math. Ann. **274** (1986), 423–437. **MR** 87i: 46118. **SC** 7.2.a, 7.3.a, 7.4.a, 7.6.a.

1986b   *A discrete integral representation for polynomials of fixed maximal degree and their universal Korovkin closure*, J. Reine Angew. Math. **370** (1986), 74–82. **MR** 87j: 41083. **SC** 7.6.c, 7.7.c.

1987   *A characterization of $\mathscr{C}(X)$ among algebras on planar sets by the existence of a finite universal Korovkin system*, Math. Ann. **277** (1987), 785–792. **MR** 88j: 46048. **SC** 7.6.c, 7.7.c.

1988 *Topics in qualitative Korovkin approximation*, Conf. Sem. Mat. Univ. Bari **226** (1988), 31 p. **MR** 90c: 46064. **SC** 1.1, 6.1, 7.1.a, 7.1.b, 8.1.

1990a *Some remarks on spectral states on LMC-algebras*, Rend. Circ. Mat. Palermo (2) **39** (1990), no. 1, 79–98. **MR** 91f: 46070. **SC** 7.6.a.

1990b *A characterization of a class of abelian groups via Korovkin theory*, Math. Z. **204** (1990), 451–464. **MR** 92j: 22007. **SC** 7.2.d, 7.7.d.

1990c *Stable rank of Fourier algebras and an application to Korovkin theory*, Math. Scand. **67** (1990), 299–319. **MR** 92b: 46077. **SC** 7.2.d, 7.7.d.

1990d *A new look at Korovkin's second theorem*, Exposition. Math. **8** (1990), no. 1, 91–96. **MR** 91c: 41062. **SC** 1.2, 7.3.d.

1991 *Finite universal Korovkin systems in tensor product of commutative Banach algebras*, Rend. Circ. Mat. Palermo (2) **40** (1991), 215–240. **MR** 93a: 46110. **SC** 7.6.a, 7.6.e, 7.7.e.

1992a *When does a commutative Banach algebra possesses a finite universal Korovkin system?*, Atti Sem. Mat. Fis. Univ. Modena **40** (1992), no. 1, 89–99. **MR** 93h: 46067. **SC** 7.6.a, 7.6.d.

1992b *Approximate regularity of commutative Beurling algebras and Korovkin approximation*, Mh. Math. **113** (1992), no. 4, 275–310. **MR** 93f: 46085. **SC** 7.4.d, 7.6.a, 7.6.d.

Pannenberg, M. and Romanelli, S.

1992 *Korovkin approximation of operators in commutative involutive Banach algebras*, Atti Sem. Mat. Fis. Univ. Modena **40** (1992), 101–113. **MR** 93h: 46068. **SC** 7.6.a, 7.7.a.

Papadopoulou, S. *(see also Bauer, H.; Leha, G.)*

1979 *Über den stationären Korovkin-Abschluss eines Funktionenraumes*, Habilitationsschrift Universität Erlangen-Nürnberg, 1979. **SC** 1.12, 4.7, 5.6.a.

Pazy, A.

1983 Semigroups of Linear Operators and Applications to Partial Differential Equations, *Springer-Verlag, Berlin, 1983.* **MR** 85g: 47061.

Pears, A.R.

1975 Dimension Theory of General Spaces, *Cambridge University Press, Cambridge, 1975.* **MR** 52 # 15405.

Peetre, J.

1963 A Theory of Interpolation of Normed Spaces, *Lecture Notes, Brasilia, 1963.*

Peters, M.A. *(see Khan, M.K.)*

Pethe, S.

1983 *On linear positive operators*, J. London Math. Soc. **27** (1983), no. 2, 55–62. **MR** 84d: 41044.

Peyerimhoff, A. *(see Jurkat, W.B.)*

Phelps, R.R.

1966 Lectures on Choquet's Theorem, *Van Nostrand Math. Studies 7, Princeton, N.J., Van Nostrand, 1966.* **MR** 33 # 1690.

Phillips, R.S. *(see Hille, E.)*

Plo, F. *(see Adell, J.A.)*

Pólya, G. and Schoenberg, I.J.
1958   *Remarks on de la Valleé Poussin means and convex conformal maps of the circle*, Pacific J. Math. **8** (1958), 295–334. **MR** 20 # 7181.

Popescu, G.
1988   *Korovkin type theorems in locally convex vector lattices*, in: Proceedings of the National Conference on Geometry and Topology (Romanian) (Tirgovişte, 1986), 231–234, Univ. Bucureşt, Bucharest, 1988. **MR** 90b: 46049. **SC** 6.7, 6.8.
1989   *Korovkin theorems in Banach algebras*, Complex Analysis and Applications '87 (Varna, 1987), 413–416, Bulgar. Acad. Sci. Sofia, 1989. **MR** 92h: 46106. **SC** 7.4.b, 7.7.b.

Popov, V.A. *(see Sendov, B.)*

Popoviciu, T.
1935   *Sur l'approximation des fonctions convexes d'ordre superieur*, Mathematica (Cluj) **10** (1935), 49–54.
1951   *Asupra demonstratiei teoremei lui Weierstrass cu ajutorul polinoamelor de interpolare [On the proof of Weieratrass theorem with the help of interpolation polynomials]*, Lucrările Sesiunii generale ştiinţifice (2–12 Iunie 1950), 1664–1667, Editura Academiei R. P. Române, Bucuresti, 1951.

Portenier, C. *(see Anger, B.)*

Pottinger, P. *(see also Knoop, H.B.)*
1976   *Zur linearen Approximation in Raum $\mathscr{C}^k(I)$*, Habilitationsschrift, Universität Duisburg, 1976. **SC** 5.2.e, 5.6.e.

Precup, P.
1980   *Asupra unei teoreme de tip Popoviciu-Korovkin* (Romanian) *[On a Popoviciu-Korovkin theorem]*, in: Seminar Itinerant de Ecuaţii Funcţionale, Aproximare and Convexitate (Timişoara, November 7–8, 1980), 148–153; Univ. of Timişoara Press, 1980. **SC** 8.3.

Priestley, W.M.
1976   *A noncommutative Korovkin theorem*, J. Approx. Theory **16** (1976), no. 3, 251–260. **MR** 53 # 1128. **SC** 7.4.b, 7.7.b.

Prolla, J.B.
1988   *A generalized Bernstein approximation theorem*, Math. Proc. Cambridge Phil. Soc. **104** (1988), 317–330. **MR** 90g: 41055. **SC** 1.2.

Rasa, I. *(see also Altomare, F.; Badea, C.; Campiti, M.; Della Vecchia, B.)*
1980   *On some results of C.A. Micchelli*, Anal. Numér. Théor. Approx. **9** (1980), no. 1, 125–127. **MR** 82g: 46025. **SC** 1.2, 1.11.
1981   *Determining sets for finitely defined operators*, Anal. Numér. Théor. Approx. **10** (1981), no. 7, 89–93. **MR** 84i: 41033. **SC** 1.7.
1984   *A Korovkin type theorem for the Lion operators*, Itinerant Seminar on Functional Equations, Approximation and Convexity, 157–158, Universitatea "Babeş-Bolyai", Facultatea de Matematica, Cluj-Napoca, 1984. **MR** 86f: 00008. **SC** 1.8.
1985   *On some Korovkin subspaces*, Anal. Numér. Théor. Approx. **14** (1985), no. 2, 127–130. **MR** 87h: 41034. **SC** 1.2, 1.6, 1.8.
1986a   *Sets on which concave functions are affine and Korovkin closures*, Anal. Numér. Théor. Approx. **15** (1986), no. 2, 163–165. **MR** 89a: 41045. **SC** 1.11.

1986b   *The Choquet boundary and the Korovkin closure of some function spaces*, preprint Babeş-Bolyai Univ., Fac. Math., Res. Semin. **7** (1986), 231–238. **ZBL** 619. 46006. **SC** 1.2, 1.11.

1987a   *Uniqueness closures and Korovkin closures of some function spaces*, Studia Univ. Babeş-Bolyai Math. **32** (1987), no. 2, 32–33. **MR** 89g: 46050. **SC** 1.11.

1987b   *On some Korovkin closures in* $\mathscr{C}(X,\mathbb{C})$, Itinerant Seminar on Functional Equations, Approximation and Convexity, Cluj-Napoca (1987), 273–274, preprint, 87–6, Univ. Babeş-Bolyai, Cluj-Napoca, 1987. **MR** 90d: 41058. **SC** 7.6.c.

1987c   *On some test systems in approximation by linear operators*, Itinerant Seminar on Functional Equations, Approximation and Convexity, Cluj-Napoca (1987), 49–54, preprint 87–6 Univ. Babeş-Bolyai, Cluj-Napoca (1987). **MR** 90i: 41033. **SC** 1.11, 1.12.

1988a   *Generalized Bernstein operators and convex functions*, Studia Univ. "Babes-Bolyai", Math. **33** (1988), no. 2, 36–39. **MR** 90f: 41005.

1988b   *On the monotonicity of sequences of Bernstein-Schnabl operators*, Anal. Numér. Théor. Approx. **17** (1988), no. 2, 185–187.

1991   *Korovkin Approximation and parabolic functions*, Conf. Sem. Mat. Univ. Bari **236** (1991), 25 p. **MR** 93c: 41038. **SC** 1.1, 1.11.

1993   *Altomare projections and Lotosky-Schnabl operators*, Proceedings of the 2nd International Conference in Functional Analysis and Approximation Theory, September 1992, Acquafredda di Maratea (Italy), 1992, Suppl. Rend. Circ. Mat. Palermo **33** (1993), 439–451.

1994   *On some properties of Altomare projections*, to appear in Conf. Sem. Math. Univ. Bari, 1994.

Redheffer, R. *(see Lumer, G.)*

Reiter, H.

1968   Classical Harmonic Analysis and Locally Compact Groups, *Oxford University Press, Oxford, 1968.*

Render, H.

1989   *An order theoretic approach to involutive algebras*, Studia Math. **92** (1989), 177–186. **MR** 90c: 46068.

Rényi, A. *(see also Arató, M.)*

1959   *Summation methods and probability theory*, Magyar Tud. Akad. Mat. Kutató Int. Közl. **4** (1959), 389–399. **MR** 24 # A561.

Rickart, C.E.

1966   General Theory of Banach Algebras, *Univ. Series in Higher Mathematics, New York, 1966.*

Rieder, G.R. *(see Dorroh, J.R.)*

Rieffel, M.A.

1983   *Dimension and stable rank in the K-theory of C*-algebras*, Proc. London Math. Soc. (3) **46** (1983), 301–333. **MR** 84g: 46085.

Riemenschneider, S.D. *(see also Marsden, M.J.)*

1975   *Korovkin theorems for a class of integral operators*, J. Approx. Theory **13** (1975), 316–326. **MR** 52 # 1106. **SC** 5.6.a, 5.6.b.

1978   *The L_p-saturation of the Kantorovich-Bernstein polynomials*, J. Approx. Theory **23** (1978), 158–162. **MR** 80g: 41018.

Riesenfeld, R.F. *(see Gordon, W.J.)*

Riordan, J.
1963   Introduction to the Combinatorial Analysis, *M, IL, 1963.*

Rivlin, T.J. *(see Kelisky, R.P.)*

Robertson, A.G.
1977   *A Korovkin theorem for Schwarz maps on C*-algebras*, Math. Z. **156** (1977),
no. 2, 205–207. **MR** 58 # 23625. **SC** 7.4.b, 7.7.b.

1986   *Asymmetric invariant sets for completely positive maps on C*-algebras*, Bull.
Austral. Math. Soc. **33** (1986), no. 3, 471–473. **MR** 87f: 46104. **SC** 7.4.b.

Rogalski, M.
1968   *Opérateurs de Lion, projecteurs boréliens et simplexes analytiques*, J. Funct.
Anal. **2** (1968), no. 4, 458–488. **MR** 38 # 5057.

Rogosinski, W.W. *(see Hardy, G.H.)*

Romanelli, S. *(see also Altomare, F.; Pannenberg, M.)*
1989   *Determining subspaces for discrete-type linear forms on a Banach algebras*,
Rend. Circ. Mat. Palermo (2) **38** (1989), no. 3, 455–476 (1990). **MR** 91e: 46073.
SC 7.5.a.

1990   *Universal Korovkin closures with respect to linear operators on commutative
Banach algebras*, Math. Japon. **37** (1992), no. 3, 427–443. **MR** 93c: 46092.
SC 7.6.a.

Ross, K. *(see Hewitt, E.)*

Roth, W. *(see also Flösser, H.O.; Keimel, K.)*
1989   *Families of convex subsets and Korovkin-type theorems in locally convex
spaces*, Rev. Roumaine Math. Pures Appl. **34** (1989), no. 4, 329–346. **MR** 90h:
46017. **SC** 8.6.

Rubel, L.A., Shields, A.L. and Taylor, B.A.
1975   *Mergelyan sets and the modulus of continuity of analytic functions*, J. Approx.
Theory **15** (1975), 23–40. **MR** 54 # 521.

Rubinov, A.M. *(see also Kutateladze, S.S.)*
1971   *A certain theorem of V.S. Klimov, M.A. Krasnosel'skiĭ and E.A. Lifshits* (Russian), Collection of articles dedicated to L.V. Kantorovič on the occasion of
his sixtieth birthday, Optimizacija Vyp. **3** (20) (1971), 154–158. **MR** 49 #
1063. **SC** 6.8.

Rublev, V.S.
1969a   *A special class of finite-dimensional subspaces of a Banach space*, Dokl. Akad.
Nauk SSSR **184** (1969), no. 5, 1038–1040. **MR** 39 # 4648. **SC** 5.6.f.

1969b   *The saturated and completely saturated subspaces of certain coordinate spaces*
(Russian), Dokl. Akad. Nauk SSSR **188** (1969), no. 2, 290–293. **MR** 40 #
4742. **SC** 5.6.f.

Rusk, M.D. *(see also Ferguson, L.B.O.)*
1975   *Korovkin type theorems for finitely defined operators*, Dissertation, University of California, Riverside, 1975. **SC** 1.7, 1.8.

1977   *Determining sets and Korovkin sets on the circle*, J. Approx. Theory **20** (1977),
no. 3, 278–283. **MR** 56 # 6240. **SC** 1.8, 1.9, 1.12.

Ryshkov, S.S. *(see also Boltiansky, V.G.)*
1959   *On k-regular imbeddings* (Russian), Doklady Akad. Nauk. SSSR **127** (1959), no. 2. **MR** 22 # 1872.

Sakai, S.
1971   *C*-Algebras and W*-Algebras, Ergebnisse der Mathematik und ihrer Grenzgebiete* **60**, *Springer-Verlag, Berlin-Heidelberg-New York, 1971*. **MR** 56 # 1082.

San Miguel, M. *(see Adell, J.A.)*

Sato, R.
1991   *A counterexample to a discrete Korovkin theorem*, J. Approx. Theory **64** (1991), no. 2, 235–237. **MR** 92a: 41013. **SC** 5.2.f.

Sato, R. and Ueno, T.
1965   *Multi-dimensional diffusion and the Markov process on the boundary*, J. Math. Kyoto Univ. **4** (1965), 529–605.

Sauer, T.
1991   *Multivariate Bernstein polynomials and convexity*, Computer Aided Geometric Design **8** (1991), 465–478. **MR** 93a: 41012.

1992a   *Ein Bernstein-Durrmeyer-Operator auf dem Simplex*, Dissertation, Universität Erlangen-Nürnberg, 1992.

1992b   *On the maximum principle of Bernstein polynomials on a simplex*, J. Approx. Theory **71** (1992), no. 1, 121–122. **MR** 93j: 41008.

1993   *The genuine Bernstein-Durrmeyer operators on a simplex*, submitted to J. of Fourier Analysis and Applications, 1993.

Schaefer, H.H.
1974   Banach Lattices and Positive Operators, *Die Grundlehren der mathematischen Wissenschaften* **215**, *Springer-Verlag, Berlin-Heidelberg-New York, 1974*. **MR** 54 # 11023.

Schäfer, E.
1989   *Korovkin's theorems: A unifying version*, Funct. Approx. Comment. Math. **18** (1989), 43–49. **MR** 91e: 41031. **SC** 1.1, 1.2.

Scheffold, E.
1973a   *Über die punktweise Konvergenz von Operatoren in 𝒞(X)* (English summary), Rev. Acad. Ci. Zaragoza (2) **28** (1973), 5–12. **MR** 48 # 948. **SC** 1.12, 7.6.c, 7.7.c.

1973b   *Ein allgemeiner Korovkin-Satz für lokalkonvexe Vektorverbände*, Math. Z. **132** (1973), 209–214. **MR** 48 # 9365. **SC** 6.4, 6.7.

1977   *Über Konvergenz linearer Operatoren*, Mathematica (Cluj) **20** (**43**) (1977), no. 2, 193–198. **MR** 80d: 47064. **SC** 6.2, 7.2.c, 8.2.

1978   *Fixräume regulärer Operatoren und ein Korovkin-Satz* (English summary), J. Functional Analysis **30** (1978), no. 2, 147–161. **MR** 80b: 47049. **SC** 3.3, 6.8, 8.2.

1981   *Über die Konvergenz positiver Operatoren mit Werten in atomaren Banachverbänden mit ordnungsstetiger Norm*, J. Approx. Theory **31** (1981), no. 1, 90–96. **MR** 82m: 47029. **SC** 6.5, 6.8.

Schempp, W. *(see also Felbecker, G.)*
1971   *Zur Lototsky-Transformation über kompakten Räumen von Wahrscheinlichkeitsmassen*, Manuscripta Math. **5** (1971), 199–211. **MR** 50 # 5271.

1972   *A note on Korovkin test families*, Arch. Math. (Basel) **23** (1972), 521–524. **MR** 47 # 7400. **SC** 5.2.a.

Schmidt, H.J. *(see Berens, H.)*

Schnabl, R.

1968   *Eine Verallgemeinerung der Bernsteinpolynome*, Math. Ann. **179** (1968), 74–82. **MR** 38 # 4881.

1969a   *Zum Saturationsproblem der verallgemeinerten Bernsteinoperatoren*, in: Abstract Spaces and Approximation (Proc. Conf. Oberwolfach, July 18–27, 1968, P.L. Butzer and B.Sz.-Nagy Eds.), Birkhäuser Basel, 1969, 281–289. **MR** 42 # 6493.

1969b   *Zur Approximation durch Bernstein Polynome auf gewissen Räumen von Wahrscheinlichkeitsmassen*, Math. Ann. **180** (1969), 326–330. **MR** 39 # 4564.

1972   *Über gleichmäßige Approximation durch positive lineare Operatoren*, in: Constructive Theory of Functions (Proc. Internat. Conf., Varna, 1970), 287–296; Izdat. Bolgar. Akad. Nauk, Sofia, 1972. **MR** 51 # 6241. **SC** 7.5.c.

Schoenberg, I.J. *(see Hildebrandt, T.H.; Pólya, G.)*

Schumaker, L.L. *(see Lorentz, G.G.)*

Schurer, F.

1962   *Linear positive operators in approximation theory*, Math. Inst. Techn. Univ. Delft report, 1962.

1965   *On linear positive operators in approximation theory*, Dissertation, University of Delft, 1965.

1966   *On the construction of linear positive operators in approximation theory*, Mathematica **8** (31), 2 (1966), 365–371. **MR** 35 # 3333.

Schwartz, J. *(see Dunford, N.)*

Seidman, T.I.

1970   *Approximation of operator semigroups*, J. Funct. Anal. **5** (1970), 160–166. **MR** 40 # 7866.

Semenov, E.M. *(see Mitjagin, B.S.)*

Sendov, B.

1965   *The convergence of a sequence of positive linear operators* (Bulgarian, English summary), Annuaire Univ. Sofia Fac. Math. **60** (1965/66), 279–296 (1967). **MR** 37 # 666. **SC** 1.12.

Sendov, B. and Popov, V.A.

1969   *The convergence of the derivatives of positive linear operators* (Russian), C.R. Acad. Bulgare Sci. **22** (1969), 507–509. **MR** 40 # 4644. **SC** 5.6.e.

1988   The Averaged Moduli of Smoothness, *Pure and Applied Mathematics, John Wiley & Sons, 1988.* **MR** 91b: 41013.

Sharma, A. *(see Cavaretta, A.S.Jr.; Cheney, E.W.)*

Shashkin, Yu.A. *(see also Boltiansky, V.G.; Gusak, I.Ja.; Min'kova, R.M.)*

1960   *Convergence of positive linear operators in the space of continuous functions*, Dokl. Akad. Nauk SSSR (N.S.) **131** (1960), 525–527; translated in Soviet Math. Dokl. **1**, 303–305. **MR** 22 # 8351. **SC** 1.2, 1.12.

1962   *Korovkin systems in spaces of continuous functions* (Russian), Izv. Akad. Nauk SSSR Ser. Mat. **26** (1962), 495–512; translated in Amer. Math. Soc. Tansl. Ser. **2**, 54 (1966), 125–144. **MR** 26 # 5418. **SC** 1.2, 1.12.

1965a   *Topological properties of sets connected with approximation theory*, Izv. Akad. Nauk SSSR, Ser. Matem. **29** (5) (1965), 1085–1094. **MR** 34 # 3549. **SC** 1.2, 1.12.

1965b   *Finitely defined linear operators in spaces of continuous functions* (Russian), Usp. Mat. Nauk **20** (1965), no. 6 (126), 175–180. **MR** 33 # 1715. **SC** 1.7, 1.12.

1967   *The Mil'man-Choquet boundary and the theory of approximation*, Functional Anal. i Priložen. **1** (1967), no. 2, 95–96. **MR** 35 # 3343. **SC** 1.2, 5.2.a.

1969   *The convergence of contractive operators* (Russian), Mathematica (Cluj) **11** (**34**), 2 (1969), 355–360. **MR** 42 # 871. **SC** 1.3.

1972   *The convergence of linear operators* (Russian), in: Constructive Theory of Functions (Proc. Internat. Conf., Varna, 1970), 119–125; Izdat. Bolgar. Akad. Nauk, Sofia, 1972. **MR** 51 # 3765. **SC** 1.2, 1.11, 1.12.

1973   *Abstract harmonic functions and the problem of the convergence of operators* (Russian), Mathematica (Cluj), **15** (**38**) (1973), 143–148. **MR** 50 # 13583. **SC** 1.11.

1974   *Topological properties of the set of extreme points with applications to the problem of moments* (Russian), in: Some Applications of Measure Theory (Proc. Conf. Inst. Mat. Mech. Sverdlovsk), 1974, Vol. 13, Izdat. Akad. Nauk. SSSR, Sverdlovsk, 1974, 96–119.

Shaw, Sen-Yen *(see Hsiao, Shêng Yen)*

Shields, A.L. *(see Rubel, L.A.)*

Shilov, G.

1951   *On rings of functions with uniform convergence*, Ukrain. Mat. Zurnal **3** (1951), 404–411. **MR** 14, p. 884.

Shiraly, S. *(see Limaye, B.V.)*

Shisha, O. and Mond, B.

1968   *The degree of convergence of linear positive operators*, Proc. Nat. Acad. Sci. U.S.A. **60** (1968), 1196–1200. **MR** 37 # 5582.

Siddiqi, A.H. *(see Mazhar, S.M.)*

Sikkema, P.C.

1961   *Der Wert einiger Konstanten in der Theorie der Approximation mit Bernstein-Polynomen*, Numer. Math. **3** (1961), 107–116. **MR** 23 # A459.

1966   *Über Potenzen von Verallgemeinerten Bernstein Operatoren*, Mathematica (Cluj) 8–31, (1966), no. 1, 173–180. **MR** 35 # 632.

1970a   *On some linear positive operators*, Indag. Math. **32** (1970), 327–337. **MR** 42 # 4931.

1970b   *On some research in linear positive operators in approximation theory*, Nieuw Archief voor Wiskunde **18** (1970), no. 3, 36–60. **MR** 41 # 5851.

Singer, I.M. *(see Arens, R.)*

Sobolev, A.V. *(see Krasnosel'skii, M.A.)*

Stadler, S.

1974   *Über 1-positive lineare Operatoren*, in: Linear Operators and Approximation, II (Proc. Conf., Oberwolfach Math. Res. Inst., Oberwolfach, 1974) 391–403; Internat. Ser. Numer. Math., Vol. **25**, Birkhäuser, Basel, 1974. **MR** 52 # 6265. **SC** 5.2.h, 5.6.h.

Stadtmüller, U. *(see Kratz, W.)*
Stancu, D.D.
1959   *On the approximation by polynomials of Bernstein type of functions of two variables* (Romanian), Comunicările Acad. R. P. Rom. **9** (1959), 773–777.
1960   *On some polynomials of Bernstein type*, Acad. R. P. Romîne Fil. Iaşi. Stud. Cerc. Şti. Mat. **11** (1960), 221–233. **MR** 24 # A374a.
1967   *On the monotonicity of the sequence formed by the first order derivative of the Bernstein polynomials*, Math. Z. **98** (1967), 46–51. **MR** 35 # 3018.
1968a   *Approximation of functions by a new class of linear polynomial operators*, Rev. Roum. Math. Pures et Appl. **13** (1968), no. 8, 1173–1194. **MR** 38 # 6278.
1968b   *On a new positive linear polynomial operator*, Proc. Japan Acad. **44** (1968), 221–224. **MR** 38 # 461.
1969a   *Asupra unei generalizări a polinoamelor lui Bernstein [On a generalization of the Bernstein polynomials]*, Studia Univ. Babes-Bolyai Math. (Cluj) **14** (1969), 31–45. **MR** 43 # 775.
1969b   *Use of probabilistic methods in the theory of uniform approximation of continuous functions*, Rev. Roum. Math. Pures et Appl. **14** (1969), no. 5, 673–691. **MR** 40 # 606.
1970   *Approximation of functions of two and several variables by a class of polynomials of Bernstein type* (Roumanian), Stud. Cerc. Mat. **22** (1970), 335–345. **MR** 47 # 2242.
1972a   *Approximation of functions by means of some new classes of positive linear operators*, in: Numerische Methoden der Approximationstheorie, Bd. 1, Proc. Conf. Math. Res. Inst. Oberwolfach, 1971; L. Collatz and G. Meinardus Eds., Birkhäuser, Basel, 1972, 187–203. **MR** 52 # 1107.
1972b   *A new class of uniform approximating polynomial operators in two and several variables*, Proceedings of the Conference on the Constructive Theory of Functions (Approximation Theory, Budapest, 1969), 443–455, Akadémiai Kiadó, Budapest, 1972. **MR** 52 # 14785.
1972c   *On the approximation of functions of two variables by means of a class of linear operators*, in: Constructive theory of functions (Proc. Internat. Conf., Varna, 1970) (Russian), 327–336. Izdat. Bolgar. Akad. Nauk, Sofia, 1972. **MR** 52 # 3823.
1979   *Application of divided differences to the study of monotonicity of the derivatives of the sequence of the Bernstein polynomials*, Calcolo **16** (1979), no. 4, 431–445 (1980). **MR** 82b: 41008.
1983   *Approximation of functions by means of a new generalized Bernstein operator*, Calcolo **20** (1983), no. 2, 211–229 (1984). **MR** 85j: 41015.
Stark, E.L.
1984   *A bibliography of the Bernstein power series of operators of Meyer-König and Zeller and their generalizations*, Anniversary volume on approximation theory and functional analysis (Oberwolfach, 1983), 303–314, Internat. Schriftenreihe Numer. Math., **65**, Birkhäuser, Basel-Boston, Mass. 1984. **MR** 87g: 41053.

Stanković, M.S. *(see Kocić, L.M.)*

Stone, M.H.
  1937   *Applications of the theory of boolean rings to general topology,* Trans. Amer.
          Math. Soc. **41** (1937), 375–481.
  1948   *The generalized Weierstrass approximation theorem,* Math. Mag. **21** (1948),
          167–184, 237–254. **MR** 10, p. 255.
  1962   *A generalized Weierstrass theorem,* Studies in Analysis (R.C. Burck ed.),
          Prentice Hall (1962), 30–87.

Strawderman, W.E. *(see Chao, M.T.)*

Studden, W. *(see Karlin, S.)*

Suciu, I.
  1975   Function Algebras, *Editura Academiei Republicii Socialiște Romania,*
          Bucuresti, Noordhoff International Publishing, Leyden, 1975. **MR** 51 #
          6428.

Sussich, J.F. *(see also Bloom, W.R.)*
  1982   *Korovkin's theorem for locally compact abelian groups,* Dissertation, Murdoch
          University, Perth, 1982. **SC** 2.1, 3.1, 3.2, 6.1, 7.2.d, 7.3.d, 7.7.d.

Swetits, J. *(see also Maier, V.)*

Swetits, J. and Wood, B.
  1973a   *On a class of positive linear operators,* Canad. Math. Bull., **16** (1973), 557–559.
          MR 50 # 10630.
  1973b   *Generalized Bernstein power series,* Rev. Roum. Math. Pures et Appl. **13**
          (1973), no. 3, 461–471. **MR** 47 # 7278.

Szász, O.
  1950   *Generalization of S. Bernstein's polynomials to the infinite interval,* J. Res. Nat.
          Bur. Standards **45** (1950), 239–245. **MR** 13, p. 648.

Sz-Nagy, B.
  1950   *Méthodes de sommation des séries de Fourier,* I, Ac. Sz. **12** (1950), 204–210.
          MR 11, p. 656.

Taylor, B.A. *(see Rubel, L.A.)*

Taira, K.
  1988   Diffusion Processes and Partial Differential Equations, *Academic Press,*
          Boston-San Diego-London-Tokyo, 1988. **MR** 90m: 60089.
  1991   Elliptic Boundary Value Problems, Analytic Semigroups and Markov Pro-
          cesses, *Lecture Notes in Math.* **1499**, *Springer-Verlag, Berlin-Heidelberg-New*
          York-Tokyo, 1991.
  1992   On the Existence of Feller Semigroups with Boundary Conditions, *Memoirs*
          of Amer. Math. Soc. **99** *(1992), no. 475, American Mathematical Society,*
          Providence, Rhode Island. **MR** 93f: 47050.

Takahasi, Sin-Ei
  1979   *Korovkin's theorems for C*-algebras,* J. Approx. Theory **27** (1979), no. 3,
          197–202. **MR** 81f: 46073. **SC** 7.3.b, 7.5.b.
  1990a   *Korovkin type theorem on $\mathscr{C}([0,1])$,* Approximation, Optimization and Com-
          puting: Theory and Applications, A.G. Law and C.L. Wang (eds.) Elsevier
          Science Publ. B. V. North.-Holland (1990), 189–192. **MR** 93f: 41031. **SC** 1.7,
          1.12.

1990b   *Bohman—Korovkin—Wulbert operators on* $\mathscr{C}([0,1])$ *for* $\{1, x, x^2, x^3, x^4\}$, Nihonkai Math. J. **1** (1990), no. 2, 155–159. **MR** 92e: 41019. **SC** 1.7, 1.12.

1993a   *Bohman—Korovkin—Wulbert operators on normed spaces*, J. Approx. Theory **72** (1993), 174–184. **SC** 1.12, 7.7.c.

1993b   *BKW-operators on function spaces*, Proceedings of the 2nd International Conference in Functional Analysis and Approximation Theory, September 1992, Acquafredda di Maratea (Italy), 1992, Suppl. Rend. Circ. Mat. Palermo **33** (1993), 479–487. **SC** 1.12, 5.6.f, 8.6.

Takashima, Y. *(see Kubo, F.)*

Tamura, S. *(see Yoshinaga, K.)*

Temple, W.B.

1954   *Stieltjes integral representation of convex functions*, Duke Math. J. **21** (1954), 527–531. **MR** 16, p. 22.

Tihomirov, N.B.

1973   *The probabilistic analogues of a theorem of P.P. Korovkin* (Russian), Application of Functional Analysis in Approximation Theory, no. **1**, 132–146, Kalinin. Gos. Univ., Kalinin, 1973. **MR** 49 # 9911. **SC** 1.10.

Timmermans, C.A. *(see also Clément, Ph.)*

1988   *On C_0-semigroup in a space of bounded continuous functions in the case of entrance or natural boundary points*, in: Approximation and Optimization, Proceeding of the International Seminar held at the University of Havana, Havana, January, 12–16, 1987, Edited by J. A. Gómez Fernández et al., Lecture Notes in Mathematics **1354**, 209–216, Springer-Verlag, Berlin-New York, 1988, 209–216. **MR** 90f: 47058.

Totik, V. *(see also Ditzian, Z.)*

1983a   *Approximation by Meyer-König and Zeller type operators*, Math. Z. **182** (1983), 425–446. **MR** 84g: 41013.

1983b L_p $(p > 1)$-*approximation by Kantorovich polynomials*, Analysis **3** (1983), 79–100. **MR** 86c: 41013.

1983c   *Problems and solutions concerning Kantorovich operators*, J. Approx. Theory **37** (1983), no. 1, 51–68. **MR** 84c: 41003.

1983d   *Uniform approximation by Baskakov and Meyer-König and Zeller operators*, Periodica Math. **14** (1983), 209–228. **MR** 85c: 41038.

1983e   *Uniform approximation by Szász-Mirakjan-type operators*, Acta Math. Acad. Sci. Hungar. **41** (1983), 291–307. **MR** 84j: 41033.

1983f   *Approximation in L^1 by Kantorovich polynomials*, Acta Sci. Math. **46** (1983), 211–222. **MR** 85k: 41028.

1984a   *Uniform approximation by Bernstein-type operators*, Indag. Math. **46** (1984), 87–93. **MR** 86c: 41014.

1984b   *Uniform approximation by positive operators on infinite intervals*, Anal. Math. **10** (1984), 163–182. **MR** 86e: 41043.

1991   *Approximation by Bernstein polynomials*, preprint, 1991.

Triebel, H.

1978   Interpolation Theory, Function Spaces, Differential Operators, *North-Holland Publishing Co., Amsterdam-New York, 1978.* **MR** 80i: 46032b.

Trotter, H.F.
1958 *Approximation of semi-groups of operators,* Pacific J. Math. **8** (1958), 887–919. **MR** 21 # 2190.

Ueno, T. *(see Sato, R.)*

Umar, S. *(see Habib, A.)*

Ustinov, G.M.
1975 *Certain properties of subspaces of a space of affine functions* (Russian), Mat. Zametki **17** (1975) no. 2, 307–318; translated in Math. Notes **17** (1975), no. 1–2, 177–182. **MR** 52 # 14968. **SC** 3.2, 3.7.

1979 *Convergence of positive operators* (Russian), Mat. Zametki **26** (1979), no. 2, 201–216, 316; translated in Math. Notes **26** (1979), no. 1–2, 593–599 (1980). **MR** 81j: 41039. **SC** 4.2, 4.7, 5.3.a, 5.6.a.

1980 *Finite-dimensional Korovkin spaces in spaces of affine functions* (Russian), Mat. Zametki **28** (1980), no. 6, 883–897, 961; translated in Math. Notes **28** (1980), no. 5–6, 896–904, 1981. **MR** 83b: 46031. **SC** 1.8, 1.12, 4.2, 4.7, 6.2, 6.8.

Ustinov, G.M., Vasil'čenko, A.A.
1977 *Convergence of operators in the space of differentiable functions* (Russian), Ural. Gos. Univ. Mat. Zap. **10**, no. 2, Issled. Sovremen. Mat. Anal. 8–14, 213 (1977). **MR** 57 # 17371. **SC** 5.2.e, 5.6.e.

Vasil'čenko, A. A. *(see also Ustinov, G.M.)*
1978 *Korovkin systems in some function spaces* (Russian), Studies in functional analysis, 6–14, Ural. Gos. Univ. Sverdlovsk, 1978. **MR** 82c: 46031. **SC** 5.2.e, 5.2.g, 5.2.h.

Vasil'ev, R.K.
1970a *Converging sequences of linear operators in partially ordered spaces* (Russian), Mat. Zametki **8** (1970), no. 4, 475–486; translated in Math. Notes **8** (1970), 736–741. **MR** 43 # 2543. **SC** 5.2.g, 5.6.g, 6.5, 6.8.

1970b *The convergence of positive linear operators and of singular integrals in certain partially ordered spaces* (Russian), Izv. Vysš Učebn. Zaved. Matematika **97** (1970), no. 6, 35–45. **MR** 44 # 828. **SC** 5.2.g, 6.2.

1972 *Conditions for the convergence of isotone operators in partially ordered sets with convergence classes* (Russian), Mat. Zametki **12** (1972), no. 3, 337–348; translated in Math. Notes **12** (1972), no. 3, 632–637 (1973). **MR** 51 # 11185. **SC** 6.2, 6.8.

1974a *Certain questions on the convergence of isotone and positive additive operators* (Russian), Dokl. Akad. Nauk SSSR **217** (1974), 260–263; translated in Soviet Math. Dokl. **15** (1974), 1031–1035 (1975). **MR** 50 # 2994. **SC** 3.2, 5.2.b, 5.6.b, 6.2, 6.8.

1974b *Normed lattices without countable Korovkin systems* (Russian), Izv. Vysš Učebn. Zaved. Matematika **140** (1974), no. 1, 21–30; translated in Soviet Math. (Iz. VUZ) **18** (1974), 16–22. **MR** 50 # 7998. **SC** 1.12, 3.7, 6.2, 6.8.

1975 *Erratum "Normed lattices without countable Korovkin systems"* (Russian), Izv. Vysš Učebn. Zaved. Matematika **152** (1975), no. 1, 130. **MR** 50 # 7998.

1976 *Sur les conditions de convergence des applications isotones dans les espaces euclidiens,* Ann. de l'Univ. J.B. Bokassa **1** (1976), 171–176. **SC** 5.6.j, 6.8.

1977   *Convergence of isotonic operators to the identity operator* (Russian), in: The Theory of the Approximation of Functions (Proc. Internat. Conf., Kaluga, 1975), 70–77, "Nauka", Moscow, 1977. **MR** 80c: 41009. **SC** 6.1.

1979   *Equivalence of convergence conditions for isotone and positive additive operators* (Russian), Trudy Moscov Mat. Obšč **39** (1979), 213–234, 237. **MR** 80h: 41011. **SC** 6.1, 6.2, 6.8.

Veselinov, V.M. *(see Boyanov, B.D.)*

Vespri, V.

1986   *Analytic semigroups and continuous interpolation spaces for a class of degenerate elliptic equations*, preprint, Dipartimento di Matematica, Seconda Università degli Studi di Roma, 1986.

Vinogradov, O.M.

1981   *Some classes of Korovkin systems* (Russian), Application of Functional Analysis in Approximation Theory, **148**, 25–36, Kalinin. Gos. Univ., Kalinin, 1981. **MR** 84a: 41028. **SC** 1.2.

Vîrtopeanu, I. *(see Leonte, A.)*

Vitale, R.A.

1979   *Approximation of convex set-valued functions*, J. Approx. Theory **26** (1979), no. 4, 301–316. **MR** 81a: 41063. **SC** 1.11, 1.12, 9.2.

Volkov, V.I.

1957   *On the convergence of sequences of linear positive operators in the space of continuous functions of two variables* (Russian), Dokl. Akad. Nauk SSSR (N.S.), **115** (1957), 17–19. **MR** 20 # 1205. **SC** 1.2.

1958   *Conditions for convergence of a sequence of positive linear operators in the space of continuous functions of two variables* (Russian), Kalinin. Gos. Ped. Inst. Učen. Zap. **26** (1958), 27–40. **MR** 24 # A955, **MR** 24 # A956. **SC** 1.2.

1960   *On conditions for convergence of sequences of linear positive operators in the space of continuous functions assigned on closed surfaces* (Russian), Usp. Mat. Nauk **15**, (1960), no. 1 (91), 181–184. **MR** 23 # A2037. **SC** 1.2.

Voronovskaja, E.V.

1932   *The asymptotic properties of the approximation of functions with Bernstein polynomials*, Dokl. Akad. Nauk. SSSR, A (1932), 79–85.

Walk, H. *(see also Müller, M.W.)*

1969   *Approximation durch Folgen linearer positiver Operatoren*, Arch. Math. (Basel) **20** (1969), 398–404. **MR** 42 # 2229. **SC** 5.2.g.

1970   *Approximation unbeschränkter Funktionen durch lineare positive Operatoren*, Habilitationsschrift, Stuttgart, 1970. **SC** 5.2.a, 5.4.a.

1975   *Über die Approximation unbeschränkter Funktionen durch lineare positive Operatoren*, J. Reine Angew. Math. **276** (1975), 83–94. **MR** 53 # 13944. **SC** 5.2.a, 5.4.a.

1980   *Probabilistic methods in the approximation by linear positive operators*, Nederl. Akad. Wetensch. Indag. Math. **42** (1980), no. 4, 445–455. **MR** 82c: 41024. **SC** 1.9, 1.10.

Wallman, H. *(see Hurewicz, W.)*

Walter, W. *(see Lumer, G.)*

Wang, H.C.
1977   Homogeneous Banach Algebras, *Lecture Notes Pure Appl. Math.* **29**, *Marcel Dekker, New York, 1977.* **MR** 56 # 16258.

Wang, H.J. *(see Hsu, L.C.)*

Watanabe, H.
1975   *Some remarks on theorems of Korovkin type for adapted spaces,* Natur. Sci. Rep. Ochanomizu Univ. **26** (1975), no. 1, 1–6. **MR** 52 # 1128. **SC** 5.2.d.

1979a   *Convergence of monotone operators,* Natur. Sci. Rep. Ochanomizu Univ. **28** (1977), no. **1**, 7–19. **MR** 58 # 12124. **SC** 5.3.a, 6.5, 6.8.

1979b   *Theorems of Korovkin type in an ordered vector space with a locally convex topology,* Natur. Sci. Rep. Ochanomizu Univ. **30** (1979), no. 2, 37–46. **MR** 81h: 41028. **SC** 5.3.a, 6.5, 6.8.

1988   *Korovkin-type theorems for a countably sublinear functional,* Exposition. Math. **6** (1988), 185–191. **MR** 89e: 41032. **SC** 5.5.b.

Weba, M.
1986   *Korovkin systems of stochastic processes,* Math. Z. **192** (1986), 73–80. **MR** 88a: 60100. **SC** 6.8, 8.6.

Weierstrass, K.
1885   *Über die analytische Darstellbarkeit sogenannter willkürlicher Funktionen einer reellen Veränderlichen,* Sitzungsberichte der Akademie zu Berlin, 1885, 633–639, 789–805.

Wentcel, A.D.
1959   *On boundary conditions for multidimensional diffusion processes,* Theory Prob. Appl. **4** (1959), 164–177.

Westermann, K.G.
1980   *Approximation periodischer Banachraumwertiger Abbildungen,* Dissertation, Abteilung für Mathematik der Ruhr-Universität Bochum, VII. 116 S. (1980). **ZBL** 469. 41029. **SC** 8.1, 8.6.

van Wickeren, E. *(see Dickmeis, W.; Mevissen, H.)*

Wilken, D.R.
1966   *Approximate normality and function algebras on the interval and circle,* In: Function Algebras (ed. F.T. Birtel), Scott, Foreman and Co., Chicago, 1966, 98–111. **MR** 33 # 4712.

Wille, R.
1985   *Korovkin approximation of homomorphisms in involution algebras,* Arch. Math. (Basel) **45** (1985), no. 6, 549–554. **MR** 87f: 46098. **SC** 7.5.a.

Wolff, M.
1969   *Über das Spektrum von Verbandshomomorphismen,* Math. Ann. **182** (1969), 161–169.

1973a   *Darstellung von Banachverbänden und Sätze vom Korovkin-Typ,* Math. Ann. **200** (1973), 47–67. **MR** 48 # 6884. **SC** 6.2, 6.7.

1973b   *Über Korovkin-Sätze in lokalkonvexen Vektorverbänden,* Math. Ann. **204** (1973), 49–56. **MR** 51 # 1238. **SC** 6.2, 6.7.

1973c   *A general theorem of Korovkin type for vector lattices,* in: Approximation Theory (Proc. Internat. Sympos., Univ. Texas, Austin, Tex., 1973), 517–521; Academic Press, New York, 1973. **MR** 51 # 6357. **SC** 6.7, 6.8.

1975   *Über die Korovkinhülle von Teilmengen in lokalkonvexen Vektoverbänden*, Math. Ann. **213** (1975), 97–108. **MR** 52 # 8874. **SC** 6.2, 6.7.

1976   *Über die Charakterisierung von 𝒞(X) durch einen optimalen Satz vom Korovkintyp*, Jber. Deutsch. Math.-Verein **78** (1976/77), no. 2, 78–80. **MR** 55 # 948. **SC** 6.8.

1977   *On the theory of approximation by positive operators in vector lattices*, in: Functional Analysis: Surveys and Recent Results (Proc. Conf., Paderborn, 1976), 73–87; North-Holland Math. Studies, Vol. 27; Notas de Mat., no. 63, North-Holland, Amsterdam, 1977. **MR** 57 # 6979. **SC** 6.1, 6.7.

1978   *On the universal Korovkin closure of subsets in vector lattices*, J. Approx. Theory **22** (1978), no. 3, 243–253. **MR** 57 # 10323. **SC** 6.7.

Wolik, N.

1985   *Approximation durch verallgemeinerte Bernstein-Kantorovich-Operatoren*, Universität Dortmund, 1985.

Wood, B. *(see Eisenberg, S.; Swetits, J.)*

Wulbert, D.E. *(see also Kitto, W.)*

1968   *Convergence of operators and Korovkin's theorem*, J. Approx. Theory **1** (1968), 381–390. **MR** 38 # 3679. **SC** 1.3, 3.3.

1975   *Contractive Korovkin approximations*, J. Functional Analysis **19** (1975), 205–215. **MR** 52 # 6385. **SC** 3.3, 8.2.

Wood, B.

1985   *Order of approximation by linear combinations of positive linear operators*, J. Approx. Theory **45** (1985), 375–382. **MR** 87d: 41036.

Xiao, Chang Bai

1988   *The Korovkin approximation in 𝒞₀(X)* (Chinese. English summary), Nanjing Daxue Xuebao Ziran Kexue Ban **24** (1988), no. 2, 178–180. **MR** 90i: 41046. **SC** 2.2.

Xie, Dun Li

1985   *Some results on approximation by positive linear operators* (Chinese, English summary), J. Math. Res. Exposition **5** (1985), no. 2, 49–54. **MR** 87h: 41037. **SC** 1.12, 5.6.a.

Xu, Y. *(see Berens, H.)*

Yang, Li Hua

1985   *Korovkin-type theorems on simultaneous approximation* (Chinese, English summary), Hunan Shifan Daxue Ziran Kexue Ban **8** (1985), no. 4, 5–8. **MR** 87j: 41055. **SC** 1.12.

Yeh, Cheh-Chih *(see Hsiao, Shêng Yen)*

Yoshinaga, K. and Tamura, S.

1976   *On a Korovkin theorem of uniform convergence*, Bull. Kyushu Inst. Tech. Math. Natur. Sci. (1976), **23**, 1–9. **MR** 54 # 13407. **SC** 1.2, 1.9, 5.2.a.

Zaric'ka, Z.V.

1967   *Approximation of function by linear positive operators in the Lᵖ metric* (Ukrainian, Russian and English summary), Dopovīdī Akad. Nauk Ukraïn RSR Ser. A, no. 1 (1967), 14–17. **MR** 35 # 2053. **SC** 3.2.

Zeller, K. *(see Meyer-König, W.)*

Zhang, J.-Z. *(see Chang, G.-Z.)*

Zhou, Ding-Xuan *(see also Guo, Zhu-Rui)*
  1990a   *Inverse theorems for some multidimensional operators,* Approx. Theory & Appl. **6** (1990), 25–40. **MR** 92c: 41017.
  1990b   *Uniform approximation by some Durrmeyer operators,* Approx. Theory & Appl. **6** (1990), 87–100. **MR** 91k: 41043.
  1992a   *Inverse theorems for multidimensional Bernstein-Durrmeyer operators in L_p,* J. Approx. Theory **70** (1992), no. 1, 68–93. **MR** 93e: 41029.
  1992b   *Global approximation theorems for Bernstein-Durrmeyer operators,* preprint, 1992.

Ziegler, Z. *(see also Amir, D.; Karlin, S.)*
  1968   *Linear approximation and generalized convexity,* J. Approx. Theory **1** (1968), 420–443. **MR** 39 # 1872.

Zifarelli, C.L.
  1988   *Teoremi di convergenza di tipo Korovkin per operatori positivi,* Thesis, University of Bari, 1987/88. **SC** 1.1, 1.2, 1.4, 1.7, 1.8, 1.9, 1.11, 2.1, 2.2, 2.3, 2.6, 2.7, 2.8, 2.9.

Zybin, L.M.
  1966   *On conditions for the convergence of a sequence of linear operators of class S_m* (Russian), Novgorod. Golovn. Gos. Ped. Inst. Učen. Zap. **7** (1966), 37–43. **MR** 37 # 4473. **SC** 1.5, 1.12.

Zygmund, A.
  1959   Trigonometric Series, I, II, *Cambridge Univ. Press, 1959.* **MR** 21 # 6498.

Symbol index

1. General symbols

$\mathbb{N}$	natural numbers 10
$\mathbb{N}_0$	$\mathbb{N} \cup \{0\}$ 10
$\mathbb{Z}, \mathbb{Z}_+$	integers, positive integers 10
$\mathbb{Q}, \mathbb{Q}_+$	rational numbers, positive rational numbers 10
$\mathbb{R}, \mathbb{R}_+$	real numbers, positive real numbers 10
$\mathbb{C}$	complex numbers 10
$\mathbb{K}$	real or complex numbers 10
$\tilde{\mathbb{R}}$	extended real numbers 10
$\bar{z}, \mathscr{R}\epsilon\, z, \mathscr{I}m\, z, \|z\|$	conjugate, real part, imaginary part, modulus of z 10
$[a,b], [a,b[,]a,b],]a,b[$	real intervals 10
$\varnothing$	empty set 10
$\mathbb{T}, \mathbb{D}$	unit circle, unit disk 11
$\mathbb{S}_n$	unit sphere of $\mathbb{R}^{n+1}$ 11
ω	point at infinity 18
$\mathscr{T}_\omega$	18 [(1.1.3)]
X_ω	(Alexandrov) one-point-compactification 18
$(\Omega, \mathscr{F}, P)$	probability space 37
$(\Omega, \mathscr{U}, (P^x)_{x \in X}, (Z_t)_{t \geq 0})$	Markov process 70 [(1.6.31)]
$\mathfrak{B}(X)$	Borel sets 25
$\mathfrak{B}_0(X)$	Baire sets 25
$\mathfrak{B}^*(X)$	μ-measurable sets 30
δ_{ij}	Kronecker symbol 10
$\partial_e K$	extreme points 51
$\delta_{x_0} := \{\lambda x_0 \| \lambda > 0\}$	ray with direction x_0 52 [(1.4.16)]
K_p	canonical simplex of $\mathbb{R}^p$ 173 [(3.3.9)]
$\mathfrak{R}(\mathscr{U})$	233
μ-a.e., μ almost everywhere	29
o, O	Landau symbols 13
$r(\mu)$	resultant 55 [(1.5.1)]
$\sigma(E', E)$	weak*-topology 46
σ_p	surface area of a ball of radius 1 125
σ	surface area 125
$\|v\|_k = n$	385 [(6.1.21)]
ω_0	growth bound 62 [(1.6.7)]

2. Operations and relations

3. Mappings, functions and measures

4. Classes of functions and operators

5. Positive approximation processes

Subject index

We refer to every item by indicating the page number(s) of the main places where it appears.

Items concerning notions and results which are quoted in the final Notes and References of each section or in minor remarks are not listed.